Advanced Seminar on Generalized Inverses and Applications

Generalized Inverses and Applications

Edited by M. Zuhair Nashed

Proceedings of an Advanced Seminar
Sponsored by the Mathematics Research Center
The University of Wisconsin–Madison
October 8 - 10, 1973

Academic Press
New York • San Francisco • London 1976
A Subsidiary of Harcourt Brace Jovanovich, Publishers

ACADEMIC PRESS, INC.
111 Fifth Avenue, New York, New York 10003

United Kingdom Edition published by
ACADEMIC PRESS, INC. (LONDON) LTD.
24/28 Oval Road, London NW1

Library of Congress Cataloging in Publication Data

Advanced Seminar on Generalized Inverses and Applications, Madison, Wis., 1973.
Generalized inverses and applications.

(Publication of the Mathematics Research Center, University of Wisconsin–Madison ; no. 32)
Bibliography: p.
Includes index.
1. Matrix inversion–Congresses. 2. Pseudoinverses–Congresses. I. Nashed, M. Zuhair. II. Wisconsin. University–Madison. Mathematics Research Center. III. Title. IV. Series: Wisconsin. University–Madison. Mathematics Research Center. Publication ; no. 32
QA3.U45 no. 32 [QA188] 510'.8s [512.9'43]
ISBN 0–12–514250–1 76-4938

PRINTED IN THE UNITED STATES OF AMERICA

GENERALIZED INVERSES AND APPLICATIONS

Publication No. 32
of the Mathematics Research Center
The University of Wisconsin—Madison

ACADEMIC PRESS RAPID MANUSCRIPT REPRODUCTION

Contents

Foreword ix
Preface xi

I. Theory of Generalized Inverses

A Unified Operator Theory of Generalized Inverses 1
M. Zuhair Nashed
G. F. Votruba
Georgia Institute of Technology, Atlanta, Georgia and
University of Montana, Missoula, Montana

Algebraic Aspects of the Generalized Inverse of a Rectangular Matrix 111
R. E. Kalman
University of Florida, Gainesville, Florida

Some Topics in Generalized Inverses of Matrices 125
Adi Ben-Israel
Thomas N. E. Greville
Technion-Israel Institute of Technology and Northwestern University,
Evanston, Illinois, and University of Wisconsin, and National Center for
Health Statistics, Rockville, Maryland

II. Generalized Inverses in Analysis

The Fredholm Pseudoinverse – An Analytic Episode in the
History of Generalized Inverses 149
L. B. Rall
Mathematics Research Center, University of Wisconsin-Madison

A Role of the Pseudo-Inverse in Analysis 175
Magnus R. Hestenes
The University of California at Los Angeles and
IBM Research, Yorktown Heights, New York

Aspects of Generalized Inverses in Analysis and Regularization 193
M. Zuhair Nashed
Georgia Institute of Technology, Atlanta, Georgia

III. Computational Methods and Approximation Theory

Methods for Computing the Moore-Penrose Generalized Inverse, and Related Matters 245
B. Noble
Mathematics Research Center, University of Wisconsin–Madison

Differentiation of Pseudoinverses, Separable Nonlinear Least Squares Problems and Other Tales 303
G. H. Golub
V. Pereyra
Stanford University, Stanford, California and Universidad Central de Venezuela, Caracas, Venezuela

Perturbations and Approximations for Generalized Inverses and Linear Operator Equations 325
M. Zuhair Nashed
Georgia Institute of Technology, Atlanta, Georgia

IV. Applications

The Operator Pseudoinverse in Control and Systems Identification 397
Frederick J. Beutler
William L. Root
University of Michigan, Ann Arbor, Michigan

Applications of Generalized Inverses to Programming, Games and Networks 495
Adi Ben-Israel
Technion–Israel Institute of Technology and Northwestern University, Evanston, Illinois

Statistical Applications of the Pseudo Inverse 525
Arthur Albert
Boston University, Boston, Massachusetts

Estimation and Aggregation in Econometrics: An Application of the Theory of Generalized Inverses 549
John S. Chipman
University of Minnesota, Minneapolis, Minnesota

Foreword

This volume contains the complete texts of thirteen addresses and an invited contribution to the Advanced Seminar on Generalized Inverses and Applications, held in Madison on October 8-10, 1973, under the sponsorship of the Mathematics Research Center, The University of Wisconsin–Madison. The conference and preparation of the proceedings were supported by the U.S. Army under Contracts DA-31-124-ARO-D-462 and DAAG29-75-C-0024.

The Program Committee, chaired by the Editor, consisted of Professors Adi Ben-Israel, M. R. Hestenes, L. B. Rall, J. Barkley Rosser, and Mrs. Gladys G. Moran, Secretary.

The conference was opened with words of welcome by Professor R. C. Buck, then Acting Director of the MRC.

There were five sessions covering the following areas: I. Basic theory of generalized inverses; II. Applications to analysis and operator equations; III. Numerical analysis and approximation methods; IV. Applications to statistics and econometrics; V. Applications to optimization, system theory, and operations research. The sessions were presided over by the following chairmen:

Professor Randall E. Cline, *University of Tennessee*
Professor Robert H. Moore, *University of Wisconsin–Milwaukee*
Professor Carl de Boor, *Mathematics Research Center*
Professor Patrick L. Odell, *University of Texas at Dallas*
Professor Stephen M. Robinson, *Mathematics Research Center*

There was also an informal session devoted to open problems and discussions. Professor Kalman was unable to attend the meeting. The text of Professor Kahan's fine lecture was, unfortunately, not received. The conference was attended by 196 registrants.

Mrs. Gladys Moran served as Program Secretary and conducted the efficient preparation of the conference with intelligence and dedication. Mrs. Doris Whitmore is to be particularly commended for her painstaking efforts in the typing of the manuscript and preparation of the volume for publication. To them both I wish to extend my appreciation for their valuable assistance.

This volume reflects the main thrust of research in generalized inverses and applications, and establishes strong contacts with the current frontiers. I wish to express my appreciation to the speakers at the conference for their important contributions. With their help it was possible to provide a broad coverage of the area.

FOREWORD

Generalized inverses have been an active area of research at the MRC since the early 1960s. In the tradition of previous proceedings of MRC seminars, it is hoped that the goal of high-level exposition is met in this volume, and that the papers will provide useful surveys in the field.

This is the heaviest book that I have read cover-to-cover in a long time. (I passed Math 7, which has since become 16.007, long before the 3-lb calculus book became fashionable on the college scene.) All in all, this has been a pleasurable and educational experience. On a more personal note, I would like to thank Professor J. Barkley Rosser, for inviting me to chair this program, and the staff of the MRC for providing the best support it is possible to get.

M. Zuhair Nashed

Preface

Le juge: Accusé, vous tâcherez d'être bref.
L'accusé: Je tâcherai d'être clair.
–G. Courteline.

The theory of generalized inverses has its genetic roots essentially in the context of so-called "ill-posed" linear problems. These include problems in which one either specifies too much information, or too little. These problems cannot be solved in the sense of a solution of a nonsingular problem. However, there is a sense in which there is still a solution (and in fact a unique "solution") if one adopts, for example, the notion of "least-squares solution" (of minimal norm).

It is well known that if A is a (square) nonsingular matrix, then there exists a unique matrix B, which is called the inverse of A, such that $AB = BA = I$, where I is the identity matrix. If A is a singular or a rectangular matrix, no such matrix B exists. Now if A^{-1} exists, then the system of linear equations $Ax = b$ has a unique solution $x = A^{-1}b$. On the other hand, in many cases, solutions of a system of linear equations exist even when the inverse of the matrix defining these equations does not. Also, in the case when the equations are inconsistent, one is often interested in least-squares solutions, i.e., vectors that minimize the sum of the squares of the residuals. These problems, along with many others in numerical linear algebra, optimization and control, statistics, and other areas of analysis and applied mathematics, are readily handled via the concept of a *generalized inverse* (or pseudoinverse) of a matrix or of a linear operator. Roughly speaking, a generalized inverse must possess some of the following properties to be useful: (i) it must reduce to A^{-1} *if* A is nonsingular; (ii) it must always exist (or in the case of linear operators, it must exist for a larger class than the class of invertible operators); (iii) it must have some of the properties of the inverse (or appropriate modifications thereof), such as $(A^{-1})^{-1} = A$, $(A^{-1})^* = (A^*)^{-1}$, etc; (iv) when used in place of the inverse (when the latter fails to exist), it should provide sensible answers to important questions such as consistency of the equations, or least-squares solutions.

E. H. Moore in 1920 generalized the notion of "inverse" of a matrix to include all matrices, rectangular as well as singular. Moore's definition of the generalized inverse of an $m \times n$ matrix A, is equivalent to the existence of an $n \times m$ matrix G such that $AG = P_A$ and $GA = P_G$, where P_X is the orthogonal projector onto the space spanned by the columns of the matrix X. Unaware of Moore's work, R. Penrose showed in 1955 that there exists a unique matrix B satisfying the four relations:

$$ABA = A, \quad BAB = B, \quad (AB)^* = AB \quad \text{and} \quad (BA)^* = BA,$$

where * denotes "conjugate transpose". These conditions are equivalent to Moore's conditions. The unique matrix B that satisfies these relations is now known as the Moore-Penrose inverse, and is denoted by $A^\dagger$. Penrose also showed in 1956 that this generalized inverse possesses the following *least-squares* property: The vector $A^\dagger y$ minimizes $\|Ax-y\|^2 = (Ax-y)^* (Ax-y)$, and is of smallest (Euclidean) norm among all vectors that minimize this quantity. It is this property of the Moore-Penrose inverse, together with its natural extension to generalized inverses of linear operators, that is relevant to some of the applications. Thus when the equation $Ax = y$ does not have a solution in the traditional sense, the vector $x = A^\dagger y + z$ where z is any solution of the homogeneous system $Au = 0$, gives a *best approximate* solution in the sense of least squares, i.e., Ax is as "close" to y as it can be made. This situation arises often in optimization, control pattern recognition, least-squares fit, and other statistical applications. For example, prediction theory (with quadratic error norm), exact solutions correspond to perfect prediction and occur only in trivial problems; best approximate solutions are more important in this case.

During the past two decades, many authors have proposed and investigated various types of generalized inverses of matrices. In particular, generalized inverses that satisfy some of the four properties of the Moore-Penrose inverse, as well as several modifications thereof, have been considered. Generalized inverses of singular linear operators, as well as notions of generalized inversion of elements in various algebraic and topological structures, have also emerged and have been studied extensively. At the present time, the theory is elegant, the applications are diverse, and some outstanding computational aspects still await better methods. The literature on the theory and applications of generalized inverses encompasses over 1700 papers. The purpose of these proceedings is to bring together in a unified framework many of the results on generalized inverses, and to discuss a wide range of their applications.

In many of the papers on generalized inverses of linear transformations from one vector space into another that have appeared in the last two decades, the topic has been treated *de novo*, leading to a diversity of definitions of generalized inverses. This situation is further complicated when the spaces are endowed with a topology. Depending on the defining conditions that are used, a variety of generalized inverses can be constructed to suit different purposes. The first paper in this volume presents a unified approach to generalized inverses of linear operators, with particular emphasis on algebraic, topological, projectional, and extremal-proximinal properties. This approach provides a structural classification within the hierarchy of various generalized inverses.

The second paper develops *ab initio* the theory of generalized inverses of matrices over arbitrary fields.

The third paper deals with several topics in generalized inverses of matrices, which include several types of matrix generalized inverses and their representations, extremal properties, partitioned matrices, and spectral inverses.

The next three papers deal with various aspects of generalized inverses in linear and nonlinear analysis and optimization, with particular reference to integral equations of the first and second kinds, partial differential equations, linear operator equations and regularization of ill-posed problems.

Next there are three papers devoted to numerical analysis and approximation theory aspects of generalized inverses of matrices and linear operators. These include numerical methods for computing the Moore-Penrose generalized inverse; differentiation of pseudo-inverses and computational aspects of linear least squares problems and of nonlinear least squares problems whose variables separate; perturbation theory for generalized inverses; and approximate, iterative, and projectional methods for generalized inverses of linear operators and least-squares solutions.

The final four papers address several important applications of generalized inverses to several fields: control and systems identification; games, networks, and mathematical programming; statistics; and econometrics, with particular reference to the theory of estimation and aggregation.

An emphasis has been placed in this volume on the operator theory of generalized inverses, uses in analysis, and other applications that are not covered in existing books. Some new results, which have not appeared elsewhere, are also included in several papers in this volume.

The literature on generalized inverses has become so extensive that it would be impossible to reflect the various directions of its development, proliferation, and applications in a few papers. To compensate partially for omissions, an extensive bibliography has been prepared and classified. Also, detailed annotations have been written to provide an overview of the field and to guide the reader through the literature. The Bibliography and the Annotations are at the end of this volume.

Some of the history of generalized inverses of matrices is too well known to be repeated here. Nevertheless, a few historical comments, some of which are not widely known, are in order.

Attempts at defining or using explicitly an "inverse" of a singular matrix have been made from time to time during the past 60 years. Several historical facts stand out: Moore was the first to offer a systematic study of generalized inverses, but his work received practically no attention in the period 1920–1950. Penrose, unaware of Moore's work, made fundamental contributions to the field. These contributions mark the rebirth and the beginning of the renaissance in generalized inverses. Various important contributions in the 1950s were made by Bjerhammar, Greville, C. R. Rao, and others.

Penrose did not pursue much the subject of generalized inverses after his fundamental contributions[1] [1002], [1003]. However, there were several results that he had at the time and never published.[2] One result concerns generalized matrices over rings more general than the complex numbers. Another result, which does not seem to be well known, concerns a decision procedure for †-identities, i.e., expressions like $(aa^\dagger + bb^\dagger)^\dagger (aa^\dagger - bb^\dagger) - aa^\dagger = (b^\dagger aa^\dagger)(b^\dagger - aa^\dagger b)^\dagger$ and other expressions which are considerably less obvious. He obtained a general result that concerned †-identities involving just two Hermitian idempotents e, f (and the identity 1 if desired). Penrose also developed an algorithm for deciding whenever a † (and *) -expression in e, f, and 1 is an identity or not. However the solution to the general problem seems much more difficult. Penrose conjectures[2] that there is no decision procedure in the general case.

A generalized inverse enthusiast can easily find many precursors to, or explicit occurrences

[1]Numbers in [ ] refer to the Bibliography at the end of this volume. The numbers in ( ) refer to date of the publication.

[2]R. Penrose, Private communications, 24 August 1973 and 3 October 1973.

of generalized inverses of matrices and linear operators in contexts of algebra, analysis, statistics, and applications in other fields. Many of these have been cited, for example, in recent books [125], [190], [1054], [1113] (see also the Annotations at the end of this volume). We mention a few more. These occurrences were given in specific contexts and offered no systematic study.

To handle the multicollinearity in the problem of specification of error in econometrics, Koopmans (1937) introduced a concept of a "partial inverse" which is equivalent to the Moore-Penrose inverse. As pointed out by Chipman, Koopmans' approach was to diagonalize the matrix, replace the nonzero diagonal elements by their reciprocals, and apply the diagonal transformation in reverse. This approach was also adopted by C. R. Rao (1966). It also appears in the construction of a certain contravariant symmetric tensor in a physics paper by Bergmann *et al.* [1730], where in fact all the generalized inverses are constructed, and a particular one is singled out. An "inverse" of a singular matrix appears also in a classical paper of Yates and Hale (1939), in lecture notes by R. C. Bose (1959), and others. Bjerhammar [156] mentions that the specific solution of Moore has been presented earlier by the German geodesist F. R. Helmert (1866–1916).

H. Chernoff [246; Appendix A] extended the concept of the inverse to nonnegative definite symmetric matrices and demonstrated the relevance of this extension to certain statistical problems in design of experiments. Let A and B be nonnegative definite symmetric matrices. He defined the inverse of A relative to B by $A_B^{-1} = \lim_{\lambda \to 0^+} (A + \lambda B)^{-1}$. This seems to be a precursor to what is now called ridge analysis.

Episodes in the history of generalized inverses in the context of analysis (differential and integral equations) may be traced to the work of Fredholm (1903), Hurwitz (1912), Hilbert (1912), and others. It is in the setting of integral operators, rather than in the algebraic case, that the "general reciprocal" was first introduced.

There is also support for the following proposition [1145]: "Gauss did not formulate the notion of a generalized inverse, but he did provide the essential ingredients to produce one."

One can cite several implicit appearances of generalized inverses in functional analysis settings, with definite roots in problems of analysis. However, the earliest work explicitly devoted to the study of generalized inverses of linear operators in Hilbert space seems to be that of Tseng [1328], [1329], [1330], [1331] (all of the results of Tseng are given without proofs). Tseng's contributions were not independent of those E. H. Moore. I am grateful to Professor W. T. Reid for pointing out[3] that Tseng was a student of E. H. Moore and R. W. Barnard at Chicago and later spend some time at the Institute for Advanced Study and at Yale. In his Ph.D. dissertation at Chicago in 1933, Tseng used the Moore generalized inverse, and an appendix to his dissertation is concerned with its pertinent properties. Also, Tseng's 1949 *Comptes Rendus* paper was published under his original name, Yuan-Yung Tseng.

After E. H. Moore's death, there appeared only two volumes on his *General Analysis.* They were compiled and edited by R. W. Barnard, who possessed to a great degree the character of E. H. Moore in deferring publication until a manuscript was polished and repolished. Professor Reid recalls from conversations with Professor Barnard of long ago, that the initial plan was to publish four volumes (on Moore's *General Analysis*), and that the results of Tseng's thesis were to be presented in the fourth one. (This may account for the fact that Tseng never published the proofs of his results.) However, this program went very slowly and with the altered conditions of World War II the project came to an end.

M. Zuhair Nashed

[3]W. T. Reid, Private communication, 9 February 1973.

A Unified Operator Theory of Generalized Inverses

M. Zuhair Nashed
G.F. Votruba

Dedicated to Professor Lamberto Cesari on the occasion of his 65th birthday.

ABSTRACT

The main purpose of this paper is to develop a unified approach to generalized inverses of linear operators, with particular emphasis on algebraic, topological, extremal, and proximinal properties.

CONTENTS

Introduction

The main part of the operator theory of generalized inverses is developed in Sections 1-3, where we also provide complete proofs of the results announced in [1], [2]. Sections 5, 6 and 7 also contain some new results. A glance at the Table of Contents will give the reader a view of the topics treated in this paper.

Our approach includes as special cases various definitions of generalized inverses of matrices and linear operators in Hilbert spaces. In addition it provides results for generalized inverses in the case of normed spaces, and develops the first treatment of generalized inverses of arbitrary linear operators between two topological vector spaces. The main point of departure between our approach and the various functional-analytic and/or geometric approaches to generalized inverses in analysis is delineated in the introduction to Section 2. For a treatment and a survey of generalized inverses in functional analysis up to 1970, see Nashed [1] and the references cited therein. For other aspects of generalized inverses of linear operators see also the papers by Beutler and Root, and Nashed in this volume, and the related cited references. See also the annotated bibliography in this volume.

1. Algebraic Theory of Generalized Inverses

Although this paper is primarily concerned with operator-theoretic aspects of generalized inverses in analysis, we consider first the algebraic setting. Some of the notions of generalized inverses are, after all, purely algebraic. Many of the concepts and proofs become clearer if the unneccessary additional structures are dispensed with initially.

Let $\mathcal{V}$ and $\mathcal{W}$ be (algebraic) vector spaces over the same field, and let L be a linear transformation from $\mathcal{V}$ into $\mathcal{W}$. We assume that L is singular (i.e. non-invertible); otherwise the theory of generalized inverses has little content.

We use the following notation:

$\Lambda(\mathcal{V}, \mathcal{W})$:= the vector space of all linear transformations from $\mathcal{V}$ into $\mathcal{W}$.

$\mathcal{R}(L)$:= the range of L.

$\mathcal{N}(L)$:= the null space of L.

$L|\mathcal{M}$:= the restriction of L to a subspace $\mathcal{M}$.

$\mathcal{V}_1 + \mathcal{V}_2$:= the algebraic sum of two subspaces $\mathcal{V}_1$ and $\mathcal{V}_2$. That is, $\mathcal{V}_1 + \mathcal{V}_2 = \{y + z : y \in \mathcal{V}_1, z \in \mathcal{V}_2\}$.

$\mathcal{V}_1 \dot{+} \mathcal{V}_2$:= the algebraic direct sum of $\mathcal{V}_1$ and $\mathcal{V}_2$. That is, $\mathcal{V} = \mathcal{V}_1 \dot{+} \mathcal{V}_2 \iff$ each element $v \in \mathcal{V}$ can be uniquely written as the sum of an element $v_1 \in \mathcal{V}_1$ and an element $v_2 \in \mathcal{V}_2$.

$$\mathcal{V} = \mathcal{V}_1 \dot{+} \mathcal{V}_2 \iff \mathcal{V} = \mathcal{V}_1 + \mathcal{V}_2 \text{ and } \mathcal{V}_1 \cap \mathcal{V}_2 = \{0\}.$$

If $\mathcal{V} = \mathcal{V}_1 \dot{+} \mathcal{V}_2$, we say that $\mathcal{V}_2$ is an (algebraic) complementary subspace of $\mathcal{V}_1$ in $\mathcal{V}$, and that $\mathcal{V}$ is decomposed into two complementary subspaces $\mathcal{V}_1$ and $\mathcal{V}_2$.

A map $P : \mathcal{V} \to \mathcal{V}$ is said to be idempotent if $P^2 = P$. A linear idempotent map is called an (algebraic) projector.

The theory of generalized inverses is intimately connected with properties of algebraic complements and projectors. We shall use the following known and easy-to-establish properties (Halmos [1], Raikov [1]):

1. Every algebraic subspace of a vector space has an algebraic complement.

2. For every projector P the subspaces $\mathcal{R}(P)$ and $\mathcal{N}(P)$ are complementary; i.e., P induces an algebraic direct sum decomposition:

$$\mathcal{V} = \mathcal{R}(P) \dot{+} \mathcal{N}(P) .$$

3. Conversely, each direct sum decomposition of $\mathcal{V}$ determines two complementary projectors P and I - P. For suppose $\mathcal{V} = \mathcal{V}_1 \dot{+} \mathcal{V}_2$. Then each $v \in \mathcal{V}$ can be uniquely decomposed as $v = v_1 + v_2$ with $v_1 \in \mathcal{V}_1$ and $v_2 \in \mathcal{V}_2$. The mapping P defined by $Pv = v_1$ is a projector with $\mathcal{R}(P) = \mathcal{V}_1$. We say that P is a projector onto $\mathcal{V}_1$ along $\mathcal{V}_2$. Note that P is a projector onto $\mathcal{R}(P)$ along $\mathcal{R}(I - P)$.

Thus for any linear map $P: \mathcal{V} \to \mathcal{V}$, the following conditions are equivalent:

(i) $P^2 = P$

(ii) $Px = x$ for all $x \in \mathcal{R}(P)$

(iii) $\mathcal{R}(I - P) = \mathcal{N}(P)$ or $\mathcal{R}(P) = \mathcal{N}(I - P)$,

where I is the identity map

(iv) $\mathcal{V} = \mathcal{R}(P) \dot{+} \mathcal{R}(I - P)$ or $\mathcal{R}(P) \cap \mathcal{R}(I - P) = \{0\}$.

If $\mathcal{V}$ is of finite dimension, then these conditions are equivalent to:

(v) rank P + rank $(I - P) = \dim \mathcal{V}$.

A (not necessarily linear) map $M: \mathcal{W} \to \mathcal{V}$ is said to be an inner inverse of the linear transformation $L: \mathcal{V} \to \mathcal{W}$ if $LML = L$. Clearly M is not unique unless L has an inverse.

Proposition 1.1. Let $L: \mathcal{V} \to \mathcal{W}$ be a linear transformation.

(a) If M is an inner inverse of L, then LM and ML are idempotent, $\mathcal{R}(ML) \subset \mathcal{R}(M)$, $\mathcal{R}(LM) = \mathcal{R}(L)$, and $\mathcal{N}(M) \subset \mathcal{N}(LM)$.

(b) If M is an inner inverse of L, then $\mathcal{R}(ML)$ contains one and only one element v from each coset $\mathcal{C}$ of $\mathcal{N}(L)$ in $\mathcal{V}$.

PROOF. (a) $(LM)^2 = (LML)M = LM$ and $(ML)^2 = M(LML) = ML$. Also $\mathcal{R}(L) = \mathcal{R}(LML) \subset \mathcal{R}(LM) \subset \mathcal{R}(L)$.

(b) Suppose $LML = L$ and let $\mathcal{C}$ be an arbitrary coset of $\mathcal{N}(L)$. Suppose v_1 and v_2 are both contained in $\mathcal{R}(ML)$ and in $\mathcal{C}$. Then $v_1 = MLx_1$, $v_2 = MLx_2$ for some $x_1, x_2 \in \mathcal{V}$ and $Lv_1 = Lv_2$. Thus $LMLx_1 = LMLx_2$, $Lx_1 = Lx_2$ and $v_1 = v_2$.

Now we show that $\mathcal{R}(ML)$ contains an element v for which $Lv = L(\mathcal{C})$. Let $w \in L(\mathcal{C})$ and let $v := Mw$. Then $v = MLx$ for some $x \in \mathcal{V}$. Thus $v \in \mathcal{R}(ML)$. Also $Lv = LMLx = Lx = w \in L(\mathcal{C})$. Hence $v \in \mathcal{C}$. Therefore $v \in \mathcal{C} \cap \mathcal{R}(ML)$.

Conversely suppose there is a rule which selects a unique element v from each coset $\mathcal{C}$. Let $w := Lv$ and define $M_1: \mathcal{W} \to \mathcal{V}$ by the rule: $M_1 w = v$ for $w \in \mathcal{R}(L)$. Let $\mathcal{S}$ denote some complement of $\mathcal{R}(L)$ in $\mathcal{W}$ and let Q denote the (algebraic) projector of $\mathcal{W}$ onto $\mathcal{R}(L)$ along $\mathcal{S}$. Let $M := M_1 Q$. Then $LM = LM_1 = I|\mathcal{R}(L)$ on $\mathcal{R}(L)$; hence $LML = L$ on $\mathcal{V}$. (Here and throughout the identity transformation is denoted by I). ■

Clearly the transformation M constructed above need not be linear, unless L is invertible (in which case M is unique). For any inner inverse M of L, and any transformation $M_0: \mathcal{W} \to \mathcal{V}$ such that $M_0 w = 0$ for $w \in \mathcal{R}(L)$, the transformation $M + M_0$ is also an inner inverse.

With the aid of the above construction, a simple example shows the construction of a nonlinear inner inverse M (Sheffield [1]). Let $\mathcal{V} = \mathcal{D}(L) = \mathbb{R}^2$, and

$$L = \begin{bmatrix} 0 & 1 \\ 0 & 1 \end{bmatrix} .$$

Then $\mathcal{R}(L) = \{(a, a): a \in \mathbb{R}\}$, $\eta(L) = \{(a, 0): a \in \mathbb{R}\}$, and each coset of $\eta(L)$ is a line $x_2 = k$ in $\mathbb{R}^2$. Let $\mathcal{G}$ denote the graph of any continuous monotone function. Then $\mathcal{G} \cap \mathcal{C}$ is a singleton for each coset $\mathcal{C}$. Then for $w \in \mathcal{C} \cap \mathcal{R}(L)$, define M_1 by the rule $M_1 w = \mathcal{G} \cap \mathcal{C}$. Clearly $LM_1L = L$, but M_1 is nonlinear. This M_1 is defined on $\mathcal{R}(L)$. To construct M which is defined on all of $\mathbb{R}^2$, we proceed as in the above construction.

(Linear) Inner Inverses

We have seen that linearity of M is independent of the condition $LML = L$. From here on we shall impose linearity as a preferential condition.

Definition 1.2. Let L be a linear transformation from $\mathcal{V}$ into $\mathcal{W}$. A linear map M from $\mathcal{W}$ into $\mathcal{V}$ such that

(1.1) $$LML = L$$

is called an inner inverse of L. The set of all inner inverses of L is denoted by $\mathcal{J}(L)$. Clearly $\mathcal{J}(L)$ is an affine manifold in $\Lambda(\mathcal{V}, \mathcal{W})$.

An inner inverse is also called a partial inverse, a 1-inverse, etc.

Proposition 1.3. Every linear transformation has a (linear) inner inverse.

PROOF. Let $\mathcal{M}$ be an (algebraic) complementary subspace of $\eta(L)$ in $\mathcal{V}$, and consider $\tilde{L} = L|\mathcal{M}$. Then $\tilde{L}$ is one to one from $\mathcal{M}$ onto $\mathcal{R}(L)$; hence $\tilde{L}$ has a (linear) inverse $\tilde{M}$ from $\mathcal{R}(L)$ into $\mathcal{V}$, with $\mathcal{R}(\tilde{M}) = \mathcal{M}$. Let $\mathcal{S}$ be an (algebraic) complementary subspace of $\mathcal{R}(L)$ in $\mathcal{W}$, and let M be an arbitrary linear extension of $\tilde{M}$ to $\mathcal{W} = \mathcal{R}(L) \dot{+} \mathcal{S}$. Any $v \in \mathcal{V}$ can be written as $v = v_0 + v_1$, where $v_0 \in \eta(L)$ and $v_1 \in \mathcal{M}$. Hence $LMLv = LMLv_1 = Lv_1 = Lv$.

Another way to introduce an inner inverse is by means of the linear equation

(1.2) $$Lx = y \ .$$

We seek a linear transformation $M: \mathcal{Y} \to \mathcal{V}$ such that $x = My$ is a solution of (1.2) whenever a solution exists, i.e., when $y \in \mathcal{R}(L)$. Such a transformation always exists and is an inner inverse. Conversely, every inner inverse has this property.

The relevance of inner inverses to the general theory of linear equations is indicated in the following proposition, whose proof is immediate.

Proposition 1.4. Let $L \in \Lambda(\mathcal{V}, \mathcal{Y})$. Then:

(a) For any inner inverse M, the transformation $I - ML$ is an algebraic projector whose range is $\mathcal{N}(L)$, so that $(I - ML)z$, where z is an arbitrary element in $\mathcal{V}$, is a solution of the homogeneous equation $Lx = 0$.

(b) For a given $y \in \mathcal{Y}$, equation (1.2) has a solution if and only if $LMy = y$, where M is any inner inverse.

(c) For a given $y \in \mathcal{R}(L)$, the general solution of (1.2) is given by $My + (I - ML)z$, where $z \in \mathcal{V}$.

(d) For any $y \in \mathcal{Y}$, My is an "approximate" solution of (1.2) in the sense that it is a solution of the "projectional" equation

$$Lx = Qy$$

where Q is the algebraic projector of $\mathcal{Y}$ onto $\mathcal{R}(L)$ along $\mathcal{N}(LM)$.

The following propositions follow easily from Definition 1.2 and the elementary properties of projectors and algebraic complements cited earlier.

Proposition 1.5. If M is an inner inverse of L, then LM and ML are idempotent, $\mathcal{R}(L) \cap \mathcal{N}(M) = \{0\}$, $\mathcal{R}(ML) \subset \mathcal{R}(M)$, $\mathcal{R}(LM) = \mathcal{R}(L)$, $\mathcal{N}(M) \subset \mathcal{N}(LM)$, $\mathcal{N}(ML) = \mathcal{N}(L)$, and $\mathcal{V}$ and $\mathcal{W}$ have the following algebraic direct sum decompositions:

$$\mathcal{V} = \mathcal{N}(ML) \dotplus \mathcal{R}(ML) = \mathcal{N}(L) \dotplus \mathcal{R}(ML) \ ,$$

and

$$\mathcal{W} = \mathcal{R}(LM) \dotplus \mathcal{N}(LM) = \mathcal{R}(L) \dotplus \mathcal{N}(LM) \ .$$

Proposition 1.6. Let $L \in \Lambda(\mathcal{V}, \mathcal{W})$. Then the following statements are equivalent for any linear transformation $M: \mathcal{W} \to \mathcal{V}$:

(a) $M \in \mathcal{J}(L)$;

(b) LM is idempotent and $\mathcal{R}(LM) = \mathcal{R}(L)$;

(c) LM is idempotent and $\mathcal{W} = \mathcal{R}(L) \dotplus \mathcal{N}(LM)$;

(d) LM is idempotent and $\mathcal{V} = \mathcal{R}(M) + \mathcal{N}(L)$;

(e) ML is idempotent and $\mathcal{N}(ML) = \mathcal{N}(L)$;

(f) ML is idempotent and $\mathcal{V} = \mathcal{N}(L) \dotplus \mathcal{R}(ML)$;

(g) ML is idempotent and $\mathcal{R}(L) \cap \mathcal{N}(M) = \{0\}$;

(h) My gives a solution of a consistent equation $Lx=y$.

Propositions 1.5 and 1.6 contain many of the properties and characterizations of inner inverses (Rao and Mitra [1], Boullion and Odell [1], Deutsch [1] and others). We emphasize that $\mathcal{N}(L) \cap \mathcal{R}(M)$ is not necessarily the zero vector; thus in general the sum in Proposition 1.6(d) is not a direct sum. Also the inclusions $\mathcal{R}(ML) \subset \mathcal{R}(M)$ and $\mathcal{N}(M) \subset \mathcal{N}(LM)$ are, in general, proper inclusions, as illustrated by the following example: Let $L: \mathbb{R}^3 \to \mathbb{R}^3$ be defined by $L(x, y, z) = (x, 0, 0)$. Define $M(x, y, z) = (x, y, 0)$. Then $LML(x, y, z) = LM(x, 0, 0) = L(x, 0, 0) = (x, 0, 0) = L(x, y, z)$, so that M is an inner inverse of L. We have:

$$\mathcal{N}(M) = \{(0, 0, c): c \in \mathbb{R}\} ;$$

$$\mathcal{N}(L) = \mathcal{N}(LM) = \{(0, b, c): b, c \in \mathbb{R}\} ;$$

$$\mathcal{R}(M) = \{(a, b, 0): a, b \in \mathbb{R}\} ;$$

and

$$\mathcal{R}(L) = \mathcal{R}(LM) = \{(a, 0, 0): a \in \mathbb{R}\} .$$

From Proposition 1.5 we see that an inner inverse determines a particular complementary subspace to $\mathcal{N}(L)$ in $\mathcal{V}$, and a particular complementary subspace to $\mathcal{R}(L)$ in $\mathcal{W}$. Conversely, we next show that selecting different complementary subspaces gives rise to different partial inverses and we relate these partial inverses to each other; the latter result seems to be new. In order to do this, we first characterize all of the linear indempotents whose range is a given subspace S of a vector space V. See also Sobczyk [1].

Let $\Lambda(\mathcal{V})$ denote the set of all (algebraic) linear transformations from $\mathcal{V}$ into $\mathcal{V}$, let

$$\mathcal{P}_{\mathcal{S}} = \{P \in \Lambda(\mathcal{V}): P^2 = P \text{ and } \mathcal{R}(P) = \mathcal{S}\}$$

and

$$\mathcal{A}_{\mathcal{S}} = \{A \in \Lambda(\mathcal{V}): \mathcal{R}(A) \subset \mathcal{S} \subset \mathcal{N}(A)\} .$$

<u>Proposition 1.7</u>. The set $\mathcal{P}_{\mathcal{S}}$ is an affine manifold, and the subspace parallel to $\mathcal{P}_{\mathcal{S}}$ is $\mathcal{A}_{\mathcal{S}}$.

PROOF. For P_1 and P_2 in $\mathcal{P}_{\mathcal{S}}$ and α an arbitrary scalar, let $P_3 = \alpha P_1 + (1-\alpha)P_2$. Clearly P_3 is idempotent since $P_1P_2 = P_2$ and $P_2P_1 = P_1$. We next show that $\mathcal{R}(P_3) = \mathcal{S}$. Take $v \in \mathcal{S}$. Then $P_1v = P_2v = v$, so $P_3v = v$ and $\mathcal{S} \subset \mathcal{R}(P_3)$. But $P_1P_3 = \alpha P_1 + (1-\alpha)P_2 = P_3$ so $\mathcal{R}(P_3) \subset \mathcal{R}(P_1) = \mathcal{S}$. This proves that $\mathcal{P}_{\mathcal{S}}$ is an affine manifold.

Now select $P_0 \in \mathcal{P}_{\mathcal{S}}$ and let P be any other element of $\mathcal{P}_{\mathcal{S}}$. Then $A = P - P_0$ belongs to the subspace parallel to $\mathcal{P}_{\mathcal{S}}$. We show that $A \in \mathcal{A}_{\mathcal{S}}$. Take $v \in \mathcal{R}(P - P_0)$, i.e. $v = (P - P_0)u$ for some $u \in \mathcal{V}$. Then $v = Pv$ since $PP_0 = P_0$ so $v \in \mathcal{R}(P) = \mathcal{S}$. Thus $\mathcal{R}(A) \subset \mathcal{S}$. Next, take $v \in \mathcal{S}$. Then $v \in \mathcal{N}(P - P_0)$ since $Pv = P_0 v = v$; i.e. $\mathcal{S} \subset \mathcal{N}(P - P_0)$. We have thus shown that the subspace parallel to $\mathcal{P}_{\mathcal{S}}$ is contained in $\mathcal{A}_{\mathcal{S}}$.

To complete the proof we need to show that for any $A \in \mathcal{A}_{\mathcal{S}}$, $A + P_0 \in \mathcal{P}_{\mathcal{S}}$. We note that $(A + P_0)^2 = A^2 + AP_0 + P_0 A + P_0^2 = A + P_0$ since $A^2 = AP_0 = 0$, $P_0 A = A$ and $P_0^2 = P_0$. Thus $A + P_0$ is idempotent. If $v \in \mathcal{R}(A + P_0)$, then $(A + P_0)v = v = P_0 v$ so $\mathcal{R}(A + P_0) \subset \mathcal{R}(P_0) = \mathcal{S}$. If $v \in \mathcal{S}$, then $(A + P_0)v = P_0 v = v$, so $\mathcal{S} \subset \mathcal{R}(A + P_0)$. ■

Now let M be an inner inverse of L. We have seen that M gives rise to direct sum decompositions $\mathcal{V} = \mathcal{R}(P) \dot{+} \mathcal{N}(P)$ and $\mathcal{W} = \mathcal{R}(Q) \dot{+} \mathcal{N}(Q)$ where $P = I - ML$ and $Q = LM$. Here $\mathcal{R}(P) = \mathcal{N}(L)$ and $\mathcal{R}(Q) = \mathcal{R}(L)$. Suppose we take a different linear idempotent Q' whose range is also $\mathcal{R}(L)$. Then by Proposition 1.7, $B = Q' - Q \in \mathcal{A}_{\mathcal{R}(L)}$. We seek a different inner inverse M' such that $LM' = Q'$; to do this we determine C so that $M' = M + C$ satisfies $LM' = Q' = Q + B$. Thus we want $LC = B$. But $B = QB = LMB$ so we can take $C = MB$ and $M' = M(I+B) = M(I-Q+Q')$.

Similarly let us take a different linear idempotent P' with $\mathcal{R}(P') = \mathcal{N}(L)$ and determine M' so $I - M'L = P'$. We have $A = P' - P \in \mathcal{A}_{\mathcal{N}(L)}$. Writing $M' = M + C$ it follows that C must satisfy $(M+C)L = I - P' = I - P - A$; i.e. $CL = -A$. But $AP = 0$, so $CL = -A(I-P) = -AML$. Thus we can take $M' = (I-A)M = (I+P-P')M$. We have proved the following theorem:

THEOREM 1.8. _Let_ M _be an inner inverse of_ L _and write_ $P = I - ML$ _and_ $Q = LM$. _Let_ P' _and_ Q' _be linear idempotents with_ $\mathcal{R}(P') = \mathcal{N}(L)$ _and_ $\mathcal{R}(Q') = \mathcal{R}(L)$. _Then_

$$M' = (I+P-P')M(I-Q+Q')$$

(1.3) $$= (2I - ML - P')M(I - LM + Q')$$

is an inner inverse of L which satisfies $I - M'L = P'$ and $LM' = Q'$.

On the other hand, if M and M' are two inner inverses of L, then we have $(M - M')L \in \mathcal{A}_{\mathcal{N}(L)}$ and $L(M - M') \in \mathcal{A}_{\mathcal{R}(L)}$.

Corollary 1.9. If M is an inner inverse of L, then $M - MLM$ is invariant under change of projectors, i.e. if $M' = (I - A)M(I+B)$, where $A = P' - P$ and $B = Q' - Q$, then $M - MLM = M' - M'LM'$.

PROOF. $LA = 0$ and $BL = 0$ so $M'LM' = (I - A)MLM(I+B)$. Thus $M' - M'LM' = (I - A)[M - MLM](I+B)$. But $AP = 0$ so $(I - A)(M - MLM) = (I - A)PM = PM = M - MLM$. Also $QB = B$, so $(M - MLM)(I+B) = M(I - Q)(I+B) = M(I - Q) = M - MLM$, and the result follows. ■

Outer Inverses

We next consider the relation $MLM = M$. A map M from $\mathcal{W}$ into $\mathcal{V}$ which satisfies this relation is called an outer inverse. Note at the outset that this defining equation is quadratic in M in contrast with the defining equation for an inner inverse, which is linear in M. Also the zero transformation is always an outer inverse of every transformation. This dramatizes the fact that an outer inverse may have a range that is too "small" and a null space that is too "big" in general. An outer inverse need not be linear. The following proposition is immediate.

Proposition 1.10. Let $L \in \Lambda(\mathcal{V}, \mathcal{W})$. If M is an outer inverse of L, then LM and ML are idempotent, $\mathcal{R}(ML) = \mathcal{R}(M)$ and $\mathcal{R}(LM) \subset \mathcal{R}(L)$. If M is linear, then also $\mathcal{N}(M) = \mathcal{N}(LM)$, $\mathcal{N}(LM) \subset \mathcal{N}(M)$, and $\mathcal{V}$ and $\mathcal{W}$ have the following direct sum decompositions:

$$\mathcal{V} = \mathfrak{R}(ML) \dotplus \mathcal{N}(ML) = \mathfrak{R}(M) \dotplus \mathcal{N}(ML)$$

and

$$\mathcal{W} = \mathfrak{R}(LM) \dotplus \mathcal{N}(LM) = \mathfrak{R}(LM) \dotplus \mathcal{N}(M) \quad .$$

<u>Proposition 1.11.</u> Every nonzero linear transformation has a nonzero linear outer inverse.

PROOF. In the proof of Proposition 1.3, extend $\tilde{M}$ to M so that M takes $\mathfrak{g}$ into θ .

Again we impose linearity as a preferential condition in what follows. More formally we have

<u>Definition 1.12.</u> Let L be a linear transformation from $\mathcal{V}$ into $\mathcal{W}$. A <u>linear</u> map M from $\mathcal{W}$ into $\mathcal{V}$ such that

$$(1.4) \qquad MLM = M$$

is called an <u>outer inverse</u> of L . The set of all outer inverses of L is denoted by $\mathfrak{O}(L)$. (An outer inverse is also called a semi-inverse, a 2-inverse, a subinverse, among other names). We believe that the terminology: inner and outer inverses is suggestive. It seems ironic that one way to hope for the adoption of a uniform terminology is to introduce terms which are not among the numerous names already adopted !

Clearly $M \in \mathfrak{O}(L) \iff L \in \mathfrak{I}(M)$. Thus results analogous to Proposition 1.6 and Theorem 1.8 follow immediately.

<u>Proposition 1.13.</u> Let $L \in \Lambda(\mathcal{V},\mathcal{W})$. Then the following statements are equivalent for any linear transformation $M: \mathcal{W} \to \mathcal{V}$:

(a) $M \in \mathfrak{O}(L)$;

(b) ML is idempotent and $\mathfrak{R}(ML) = \mathfrak{R}(M)$;

(c) ML is idempotent and $\mathcal{V} = \mathfrak{R}(M) \dotplus \mathcal{N}(ML)$;

(d) ML is idempotent and $\mathcal{W} = \mathcal{R}(L) + \mathcal{N}(M)$;

(e) LM is idempotent and $\mathcal{N}(LM) = \mathcal{N}(M)$;

(f) LM is idempotent and $\mathcal{W} = \mathcal{N}(M) \dot{+} \mathcal{R}(LM)$;

(g) LM is idempotent and $\mathcal{R}(M) \cap \mathcal{N}(L) = \{0\}$.

We note that if $M_1, M_2 \in \mathcal{J}(L)$, then $M_1LM_2 \in \mathcal{O}(L) \cap \mathcal{J}(L)$.

Algebraic Generalized Inverses

Since (1.4) is the symmetric analog of (1.1), it is quite natural to consider the subclass of inner inverses which are also outer inverses. It is a fair proposition in today's competition to want L to do to M what M does to L .

Definition 1.14. Let $L: \mathcal{V} \to \mathcal{W}$ be linear. If M is a (linear) inner inverse of L and L is a (linear) inner inverse of M, then L and M are called algebraic generalized inverses (A. G. I.) of each other. Equivalently if

$$M \in \mathcal{J}(L) \cap \mathcal{O}(L) ,$$

then M is an algebraic generalized inverse of L .

Proposition 1.15. Every linear map $L: \mathcal{V} \to \mathcal{W}$ has an algebraic generalized inverse. In particular, if M_1 and M_2 are any two inner inverses of L, then M_1LM_2 is an (algebraic) generalized inverse.

PROOF. Extend $\tilde{M}$ to M so that M takes $\mathcal{S}$ to the zero element in the proof of Proposition 1.3. Also $L(M_1LM_2)L = (LM_1L)M_2L = LM_2L = L$, and $(M_1LM_2)L(M_1LM_2) = M_1(LM_2L)M_1LM_2 = M_1(LM_1L)M_2 = M_1LM_2$. ∎

Other terminology used for A.G.I.: (1, 2)-inverse, reflexive partial inverse, reflexive semi-inverse.

Proposition 1.16. If $L \in \Lambda(\mathcal{V},\mathcal{W})$ and $M \in \Lambda(\mathcal{W},\mathcal{V})$ are algebraic generalized inverses of each other, then $\mathcal{R}(L) = \mathcal{R}(LM)$, $\mathcal{R}(ML) = \mathcal{R}(M)$, $\mathcal{N}(LM) = \mathcal{N}(M)$, $\mathcal{N}(ML) = \mathcal{N}(L)$, and we have the following direct sum decompositions:

$$\mathcal{V} = \mathcal{R}(M) \dotplus \mathcal{N}(L), \quad \mathcal{W} = \mathcal{R}(L) \dotplus \mathcal{N}(M) \ .$$

L determines a one-to-one mapping of $\mathcal{R}(M)$ onto $\mathcal{R}(L)$ and M determines a one-to-one mapping of $\mathcal{R}(L)$ onto $\mathcal{R}(M)$.

Hence, LM is the (algebraic) projector of $\mathcal{W}$ onto $\mathcal{R}(L)$ along $\mathcal{N}(M)$ and ML is the projector of $\mathcal{V}$ onto $\mathcal{R}(M)$ along $\mathcal{N}(L)$. Thus L is surjective (onto) iff M is a right inverse for L. Also L is injective iff M is a left inverse for L.

Let P be the (algebraic) projector corresponding to the direct sum decomposition $\mathcal{V} = \mathcal{N}(L) \dotplus \mathcal{M}$, and let Q be the projector corresponding to the decomposition $\mathcal{W} = \mathcal{R}(L) \dotplus \mathcal{S}$. In other words, P and Q are linear and idempotent transformations with $\mathcal{R}(P) = \mathcal{N}(L)$, $\mathcal{N}(P) = \mathcal{M}$, $\mathcal{R}(Q) = \mathcal{R}(L)$ and $\mathcal{N}(Q) = \mathcal{S}$. Also, let i and j denote the injections $i: \mathcal{M} \to \mathcal{V}$ and $j: \mathcal{R}(T) \to \mathcal{W}$. Then there exists a <u>unique</u>

algebraic generalized inverse of L (with respect to the choice of P and Q), denoted by $L^{\#}_{P,Q}$. The construction is illustrated by the following commutative diagram where $\tilde{L} = L|\mathcal{M}$:

$$
\begin{array}{ccccc}
\mathcal{V} & \xrightarrow{\;L\;} & \mathcal{W} & \xrightarrow{\;L^{\#}_{P,Q}\;} & \mathcal{V} \\
i \uparrow \downarrow I-P & & j \uparrow \downarrow Q & & i \uparrow \downarrow I-P \\
\mathcal{M} & \xrightarrow{\;\tilde{L}\;} & \mathcal{R}(L) & \xrightarrow{\;\tilde{L}^{-1}\;} & \mathcal{M}
\end{array}
$$

The generalized inverse $L^{\#}_{P,Q}$ is given by

$$L^{\#}_{P,Q} = i\,\tilde{L}^{-1}Q\ .$$

(We write $L^{\#}$ rather than $L^{\dagger}$ for algebraic generalized inverses, reserving the latter for use when topological structures are invoked.) It is easy to verify that $L^{\#}_{P,Q}$ satisfies the equations:

$$LX = jQ, \quad XL = i(I - P), \quad XLX = X, \quad \text{and} \quad LXL = L\ .$$

The uniqueness of the generalized inverse X satisfying these equations is also immediate.

The injections i and j have now outlived their usefulness. From this point on, we consider P as taking $\mathcal{V}$ into $\mathcal{V}$ and Q as taking $\mathcal{W}$ into $\mathcal{W}$, and we will write the equations without the injections i and j. Thus $L^{\#}_{P,Q}$ satisfies the equations

$$
(1.5) \qquad \begin{cases}
LL^{\#}_{P,Q}L = L \\
L^{\#}_{P,Q}L\,L^{\#}_{P,Q} = L^{\#}_{P,Q} \\
L^{\#}_{P,Q}L = I - P \\
L\,L^{\#}_{P,Q} = Q
\end{cases}
.
$$

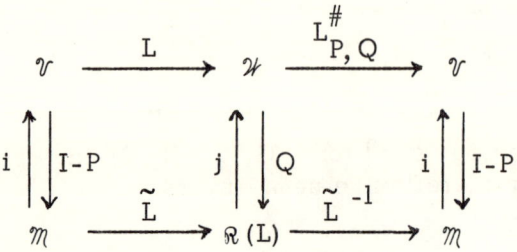

Note that the first equation in (1.5) is redundant since it follows from $LL^{\#}_{P,Q} = Q$. The equations $LX = Q$ and $XL = I - P$ are always solvable. Moreover if we impose the additional condition $XLX = X$, then the equations are uniquely solvable and the solution is the (unique) generalized inverse with respect to P, Q.

We note that the map

$$(P, Q) \to L^{\#}_{P,Q}$$

is a bijection. Also the map

$$(\mathcal{M}, \mathcal{S}) \to (P, Q)$$

from the pairs of algebraic complements for $\mathcal{N}(L)$ and $\mathcal{R}(L)$, respectively, to the pairs of projectors P, Q, defined above, is a bijection. We shall denote $L^{\#}_{P,Q}$ also by $L^{\#}_{\mathcal{M},\mathcal{S}}$ whenever it is desirable to refer in the generalized inverse directly to the <u>complements</u> rather than to the <u>projectors</u>.

<u>Proposition 1.17</u>. Let $L \in \Lambda(\mathcal{V}, \mathcal{U})$. Then the following statements are equivalent for $M \in \Lambda(\mathcal{U}, \mathcal{V})$:

(a) M is an algebraic generalized inverse of L;

(b) LM is idempotent and $\mathcal{V} = \mathcal{N}(L) \dot{+} \mathcal{R}(M)$;

(c) ML is idempotent and $\mathcal{U} = \mathcal{R}(L) \dot{+} \mathcal{N}(M)$;

(d) LM is idempotent, $\mathcal{R}(LM) = \mathcal{R}(L)$, and $\mathcal{N}(LM) = \mathcal{N}(M)$;

(e) ML is idempotent, $\mathcal{R}(ML) = \mathcal{R}(M)$, and $\mathcal{N}(ML) = \mathcal{N}(L)$.

(f) $M = M_{\mathcal{M},\mathcal{S}} = \tilde{L}^{-1}Q$, where $\tilde{L} := L|\mathcal{M}$, $\mathcal{V} = \mathcal{N}(L) \dot{+} \mathcal{M}$, $\mathcal{U} = \mathcal{R}(L) \dot{+} \mathcal{S}$ for some $\mathcal{M}$ and $\mathcal{S}$, and Q is the (algebraic) projector of $\mathcal{U}$ onto $\mathcal{R}(L)$ along $\mathcal{S}$.

(g) $M = M_{\mathcal{M},\mathcal{S}} = (I - P)\, L^{(-1)}Q$, where $L^{(-1)}$ is a selection of an inverse, and P is the projector of $\mathcal{V}$ on $\mathcal{N}(L)$ along $\mathcal{M}$.

(h) The same as (g), where $L^{(-1)}$ is replaced by any inner inverse L^- : $LL^-L = L$.

(i) M is the map $M_{\mathcal{M},\mathcal{S}} : \mathcal{W} \to \mathcal{V}$ defined by

$$M_{\mathcal{M},\mathcal{S}}\, y = (L|\mathcal{M})^{-1}\, y_1$$

where $y = y_1 + y_2$, $y_1 \in \mathcal{R}(L)$, $y_2 \in \mathcal{S}$.

(j) M is the linear map on $\mathcal{W}$ into $\mathcal{V}$ satisfying: $MLx = x$ for each $x \in \mathcal{M}$, $My = 0$ for $y \in \mathcal{S}$, and $M(y_1 + y_2) = My_1$ for $y = y_1 + y_2$, $y_1 \in \mathcal{R}(L)$, $y_2 \in \mathcal{S}$.

The following proposition also follows from the construction of $L^{\#}_{P,Q}$ (see also the diagram).

Proposition 1.18. Let $L \in \Lambda(\mathcal{V},\mathcal{W})$ and let $\mathcal{M}, \mathcal{S}, P$, and Q be as before. Then

$$\mathcal{R}(L^{\#}_{P,Q}) = \mathcal{M} \quad\text{and}\quad \mathcal{N}(L^{\#}_{P,Q}) = \mathcal{S}\ .$$

Also, $L^{\#}_{P,Q}L$ is the projector of $\mathcal{V}$ onto $\mathcal{M}$ along $\mathcal{N}(L)$ and $LL^{\#}_{P,Q}$ is the projector of $\mathcal{W}$ onto $\mathcal{R}(L)$ along $\mathcal{S}$.

Remark. We have noted (Corollary 1.9), that if M is an inner inverse of L, then $M - MLM$ is invariant under change of projectors. Thus we may use the transformation $M - MLM$ to measure the "amount" that an inner inverse misses being a generalized inverse. It is also important to note that if M is a generalized inverse and we change projectors, then we are sure that the M' given by (1.3) is guaranteed to be a generalized inverse and not just an inner inverse.

Proposition 1.19. Let M_1, M_2 be inner inverses of L. Denote the corresponding induced projectors by

$$Q_1 = LM_1, \; Q_2 = LM_2, \; I-P_1 = M_1L, \text{ and } I-P_2 = M_2L .$$

Then M_1LM_2 is a generalized inverse of L relative to the projector P_1 and Q_2. In particular if P, P' are two projectors of $\mathcal{V}$ onto $\mathcal{N}(L)$ and Q, Q' are two projectors of $\mathcal{W}$ onto $\mathcal{R}(L)$, then

$$L^{\#}_{P,Q} = L^{\#}_{P,Q'} \, L \, L^{\#}_{P',Q} .$$

The approach which we have taken to generalized inverses leads naturally to the following questions:

(a) Given two linear idempotent maps P and Q on $\mathcal{V}$ and $\mathcal{W}$ respectively with $\mathcal{R}(P) = \mathcal{N}(L)$ and $\mathcal{R}(Q) = \mathcal{R}(L)$, we obtain a unique generalized inverse $L^{\#}_{P,Q}$. If we change P and Q to some new (algebraic) projectors $P' = P + A$ and $Q' = Q + B$, where $\mathcal{R}(P') = \mathcal{R}(P) = \mathcal{N}(L)$ and $\mathcal{R}(Q') = \mathcal{R}(Q) = \mathcal{R}(L)$, we obtain a new generalized inverse $L^{\#}_{P',Q'}$. How is $L^{\#}_{P',Q'}$ related to $L^{\#}_{P,Q}$?

(b) Given two algebraic generalized inverses $L^{\#}$ and $L^{¶}$, they induce four linear idempotent maps $Q := LL^{\#}$, $P := I - L^{\#}L$, $Q' := LL^{¶}$, and $P' := I - L^{¶}L$. How are the pairs $\{P, Q\}$ and $\{P', Q'\}$ related?

These questions are answered in the following theorem which plays a central role in unifying various approaches to generalized inverses in the algebraic context, as well as in the topological considerations of Section 2.

Theorem 1.20. Let $L \in \Lambda(\mathcal{V}, \mathcal{W})$.

(a) Given linear idempotent maps P and Q on $\mathcal{V}$ and $\mathcal{W}$ respectively, with $\mathcal{R}(P) = \mathcal{N}(L)$ and $\mathcal{R}(Q) = \mathcal{R}(L)$, there is a unique solution $X = L^{\#}_{P,Q}$ of the system:

$$LX = Q \ ,$$

$$XL = I - P \ ,$$

and

$$XLX = X \ .$$

This solution is an algebraic generalized inverse of L. Any other A. G. I. is given by

$$\begin{aligned} L^{\#}_{P',Q'} &= (2I - ML - P')\, L^{\#}_{P,Q}(I - LM + Q') \\ &= (I + P - P')\, L^{\#}_{P,Q}(I - Q + Q') \\ &= (I - P')\, L^{\#}_{P,Q}\, Q' \end{aligned}$$

for some linear idempotent maps P' and Q' on $\mathcal{V}$ and $\mathcal{W}$ respectively, with $\mathcal{R}(P') = \mathcal{N}(L)$ and $\mathcal{R}(Q') = \mathcal{R}(Q)$.

(b) Using the notation of question (b) above, the following relations hold:

$$P' = P + A \ \text{ and } \ Q' = Q + B \ ,$$

where $A \in \mathcal{A}_{\mathcal{N}(L)}$, i.e., $\mathcal{R}(A) \subset \mathcal{N}(L) \subset \mathcal{N}(A)$, and $B \in \mathcal{A}_{\mathcal{R}(L)}$, i.e., $\mathcal{R}(B) \subset \mathcal{R}(L) \subset \mathcal{N}(B)$. Equivalently,

$$P' = P + (L^{\#} - L^{\P})L \ \text{ and } \ Q' = Q + L(L^{\P} - L^{\#}) \ .$$

PROOF. The theorem follows easily from Theorem 1. 8 invoking the uniqueness of the algebraic generalized inverse $L^{\#}_{P,Q}$ relative to specified projectors P, Q. ■

We note as a by-product of Theorem 1. 20 that any inner inverse serves as a building block for the construction of all generalized inverses. For if M is an inner inverse of L, then MLM is a generalized inverse. Then all generalized inverses (with respect to different linear idempotents) can be obtained by Theorem 1.20.

We emphasize again that each pair $\mathcal{M}, \mathcal{S}$ of complements to $\mathcal{N}(L)$ in $\mathcal{V}$ and $\mathcal{R}(L)$ in $\mathcal{W}$, respectively, determines a unique generalized inverse of L. Conversely, each generalized inverse M of L induces a complement $\mathcal{M}$ to $\mathcal{N}(L)$ in $\mathcal{V}$ and a complement $\mathcal{S}$ to $\mathcal{R}(L)$ in $\mathcal{W}$. Thus there is a one-to-one correspondence between the following sets for any given $L \in \Lambda(\mathcal{V}, \mathcal{W})$:

$$\{(\mathcal{M}, \mathcal{S}): \mathcal{V} = \mathcal{N}(L) \dotplus \mathcal{M}, \quad \mathcal{W} = \mathcal{R}(L) \dotplus \mathcal{S}\}$$

and

$$\{M: LML = L, \ MLM = M, \ L \in \Lambda(\mathcal{V}, \mathcal{W}), \ M \in \Lambda(\mathcal{W}, \mathcal{V})\}.$$

In particular if the domain of M is restricted to $\mathcal{R}(L)$, then we have the following characterization of $\{M: LML = L, L \in \Lambda(\mathcal{V}, \mathcal{W}), M \in \Lambda(\mathcal{W}, \mathcal{V})\}$: there exists a one-to-one correspondence between the complements of the null space $\mathcal{N}(L)$ and the inner inverses of L.

Finally we note that the results of this section hold for any linear transformation on any vector space (not necessarily finite dimensional) over an arbitrary field. The treatment is purely algebraic; no topological concepts or analysis are used. Inner products, orthogonal projectors, or adjoints did not enter into consideration. In contrast, algebraic complements and projectors were essential to the development.

The approach subsumes various generalized inverses introduced earlier, e.g., the Moore-Penrose inverse, oblique generalized inverses, weighted generalized inverses, generalized inverses with specified range and null space, etc. Each of these generalized inverses represents a particular way of choosing the projectors P, Q, or equivalently the complements $\mathcal{M}, \mathcal{S}$. We have provided structural characterizations for the sets of inner inverses and generalized inverses, as well as a formula for the transformation of these inverses under changes of projectors. Crucial to these structural characterizations and formulas was the characterization of all linear idempotents whose range is a given subspace of a vector space.

We shall consider specific cases in Section 4.

2. Generalized Inverses in Topological Vector Spaces

Algebraic aspects of generalized inverses do not suffice for applications of generalized inverses (G. I.) in analysis which are intimately connected, among other things, with the topological structure of the space and continuity properties of the operator. For example we are interested in properties of a linear operator which assure the existence of a continuous G. I. in a normed or a topological space. Best approximation properties of generalized inverses in Hilbert spaces suggest also that we take a closer look at projectional, extremal and proximinal properties of generalized inverses. This brings in the geometry of the spaces and concepts of orthogonality in normed spaces.

In this section we develop a unifying approach to the operator theory of generalized inverses when some topology is endowed on the domain and/or the range space of the operator. Our approach has a great deal of simplicity, even though it requires a higher level of abstraction.

The point of departure between various functional-analytic and/or geometric approaches to G. I. of linear operators in Hilbert or Banach spaces and the approach we take here is the following: In our approach we superimpose the analysis on the algebraic structures and consider G. I. in topological spaces starting from algebraic inner inverses; the existence of the latter is guaranteed by Theorem 1.20. In contrast, the approaches that have been advanced in the literature treat G. I. of operators *ab initio*, ignoring the fact that algebraic generalized inverses always exist regardless of the topology. This complicates questions of existence of G. I. and one does not get a clear picture of how the difficulties creep in with the analysis and topology.

It may occur to the novice that the various forms and properties of G. I. of a matrix carry over to operators in topological spaces, and in particular Hilbert and Banach spaces, in a direct fashion using standard tools of abstraction from functional analysis. This turns out not to be always the case. There are several sources of difficulties which do not arise in the algebraic context, or for operators defined on a finite dimensional space. We encounter these difficulties in questions of existence of G. I. in topological spaces, continuous projectors, best approximation properties, etc. We face questions, pathologies and distinctions which have no counterpart in the finite dimensional context. It is only through experience and involvement that one can grasp the scope of these difficulties and differences. Nevertheless, it may be appropriate to allude to some of their sources.

(i) While every (algebraic) subspace has an algebraic complement, not every closed subspace of a topological vector space has a (topological) complement.

(ii) The algebra of operators in functional analysis raises possible complexities. For example, combinations of densely defined closed linear operators need not give operators with this property, and may in fact be defined only on the zero element.

(iii) Domains of definitions, both of the operator and the generalized inverse, must be carefully treated. In many cases of interest in analysis, either the operator or its generalized inverse will not be everywhere defined.

(iv) The range of a linear operator in a topological space is not necessarily a closed subspace, and the generalized inverse, if it exists, is not necessarily bounded. This complicates questions of existence of G. I., projectional, extremal and operational properties of G. I.

(v) When using G. I. of linear operators in functional analysis, the objectives are not necessarily the same as in the algebraic context. New questions have to be addressed.

(vi) The approximation theory of G.I. of linear operators has many subtle points involving several modes of convergence, analytic and computational tractability, and techniques which are not extensions of those used in the matrix case.

Throughout this section, $\mathcal{V}, \mathcal{W}$ denote (algebraic) vector spaces, not necessarily equipped with any topology, and $\mathcal{X}, \mathcal{Y}$ denote topological vector spaces (TVS). The space of all continuous linear operators on $\mathcal{X}$ into $\mathcal{Y}$ is denoted by $\mathcal{L}(\mathcal{X}, \mathcal{Y})$. $\mathcal{L}(\mathcal{X}, \mathcal{X})$ is abbreviated $\mathcal{L}(\mathcal{X})$. To avoid possible confusion, we emphasize that $\Lambda(\mathcal{V}, \mathcal{Y})$, for example, denotes the vector space of all (not necessarily continuous) linear transformations on $\mathcal{V}$ into $\mathcal{Y}$.

As to be expected, the theory of generalized inverses in TVS is closely connected with projectors. However these are no longer merely algebraic projectors. A <u>projector</u> P on a TVS $\mathcal{X}$ is a <u>continuous</u> linear idempotent operator, i.e., $P \in \mathcal{L}(\mathcal{X})$ and $P^2 = P$.

Suppose $\mathcal{X} = \mathcal{M} \dot{+} \mathcal{N}$ (algebraic direct sum) and consider the associated linear idempotent map P with $\mathcal{R}(P) = \mathcal{M}$ and $\mathcal{N}(P) = \mathcal{R}(I - P) = \mathcal{N}$. If further $P \in \mathcal{L}(\mathcal{X})$, so that P is a projector on the TVS $\mathcal{X}$, we say that $\mathcal{X}$ is the <u>topological direct sum</u> of $\mathcal{M}$ and $\mathcal{N}$ and write

$$\mathcal{X} = \mathcal{M} \oplus \mathcal{N} = \mathcal{R}(P) \oplus \mathcal{N}(P) .$$

A subspace $\mathcal{M}$ of $\mathcal{X}$ is said to have a <u>topological complement</u> if and only if there exists a subspace $\mathcal{N}$ such that $\mathcal{X} = \mathcal{M} \oplus \mathcal{N}$; in this case we say that $\mathcal{M}$ and $\mathcal{N}$ are <u>topological complements.</u>

If $\mathcal{X}$ is Hausdorff and if $\mathcal{X} = \mathcal{M} \oplus \mathcal{N}$, then $\mathcal{M}$ and $\mathcal{N}$ are closed. In a Banach space $\mathcal{B}$, if $\mathcal{B} = \mathcal{M} \dot{+} \mathcal{N}$ and if $\mathcal{M}$ and $\mathcal{N}$ are closed, then $\mathcal{B} = \mathcal{M} \oplus \mathcal{N}$. However this is not true if $\mathcal{B}$ is not complete. It can happen, even in a Hilbert space, that $\mathcal{M}$ and $\mathcal{N}$ are closed, $\mathcal{M} \cap \mathcal{N} = \{0\}$ but $\mathcal{M} + \mathcal{N}$ is <u>not</u> closed. Thus in the inner product space $\mathcal{M} + \mathcal{N} = \mathcal{X}$ we can have $\mathcal{X} = \mathcal{M} \dot{+} \mathcal{N}$ with $\mathcal{M}, \mathcal{N}$ closed, but $\mathcal{X}$ is not the topological direct sum of $\mathcal{M}$ and $\mathcal{N}$. However the orthogonal direct sum of two closed subspaces (i.e. if $\mathcal{M} \subset \mathcal{N}^{\perp}$ and $\mathcal{X} = \mathcal{M} \dotplus \mathcal{N}$) is closed, and $\mathcal{X}$ is the topological direct sum of $\mathcal{M}$ and $\mathcal{N}$.

We recall that if $\mathcal{M}$ is any closed subspace of a Hilbert space, then $\mathcal{M}$ has a topological complement. $\mathcal{M}^{\perp}$ is one such complement. However, a closed subspace of a TVS need not necessarily have a topological complement; even the "nice" spaces $\mathcal{L}_p$ and ℓ_p, $p \neq 2$, have closed subspaces which have no topological complements (i.e. no algebraic complement can be closed). This was shown first by Murray [1]. However, the construction given by Sobczyk [1] is easier. For a recent survey on complemented subspaces, see Kadets and Mityagin [1].

A projector $P \in \mathcal{L}(\mathcal{X})$ induces a decomposition of $\mathcal{X}$ into two topological complements $P\mathcal{X}$ and $(I - P)\mathcal{X}$. On the other hand, a subspace $\mathcal{M}$ of $\mathcal{X}$ has a topological complement if and only if there exists a projector $P \in \mathcal{L}(\mathcal{X})$ with $\mathcal{R}(P) = \mathcal{M}$. See Bourbaki [1; Chapter I, pp. 14-19].

Unless otherwise explicitly stated, projectors in this section will mean continuous projectors and will be denoted by P, Q, P', Q', etc. When possible confusion may arise with algebraic projectors, which were also denoted by the same symbols, the word algebraic will be stressed.

We consider generalized inverses of a linear operator L for the following cases:

(i) L maps a vector space into a topological vector space;

(ii) L maps a topological vector space into a vector space;

(iii) L is a mapping between two topological vector spaces.

We assume that our topological spaces are Hausdorff.

2.1. Right-Topological Inner Inverses (R-T. I. I)

We first replace in the setting of Section 1 the (algebraic) vector space $\mathcal{Y}$ by a topological vector space $\mathcal{Y}$.

Consider $L \in \Lambda(\mathcal{V}, \mathcal{Y})$ with the property that $\overline{\mathcal{R}(L)}$, the closure in $\mathcal{Y}$ of $\mathcal{R}(L)$, has a topological complement in $\mathcal{Y}$, and write

$$\mathcal{Y} = \overline{\mathcal{R}(L)} \oplus \mathcal{S} \quad . \tag{2.1}$$

Let Q be the (continuous) projector onto $\overline{\mathcal{R}(L)}$ along $\mathcal{S}$. Let

$$\mathcal{D}_Q := \mathcal{R}(L) \dot{+} \mathcal{S} = \mathcal{R}(L) \dot{+} \mathcal{N}(Q) \quad . \tag{2.2}$$

Now $\mathcal{D}_Q$, as a TVS, is dense in $\mathcal{Y}$. However, we shall also consider $\mathcal{D}_Q$ as a vector space on its own, i.e., we strip it for the moment from its topology. We consider L as being in $\Lambda(\mathcal{V}, \mathcal{D}_Q)$. Let $\tilde{Q} := Q|\mathcal{D}_Q$ and consider $\tilde{Q}$ as an algebraic projector of $\mathcal{D}_Q$ onto $\mathcal{R}(L)$. Let $M_{\tilde{Q}} \in \Lambda(\mathcal{D}_Q, \mathcal{V})$ be any (algebraic) inner inverse of L which satisfies $LM_{\tilde{Q}} = \tilde{Q}$; i.e., $LM_{\tilde{Q}} = Q$ on $\mathcal{D}_Q$. This can be done by the algebraic theory of inner inverses (section 1). We call such an $M_{\tilde{Q}}$ a <u>right-topological inner inverse</u> (abbreviated as R-T.I.I.) of L. (The hyphen is used to emphasize that "right" modifies "topological", rather than "inner", reflecting the fact that the topology is on $\mathcal{Y}$.) We use the notation $L_r^- = L_{r,Q}^-$ for a R-T.I.I.

Let M be an (algebraic) inner inverse of L such that LM is an <u>algebraic</u> projector of $\mathcal{D}_Q$ onto $\mathcal{R}(L)$. Then it follows from Theorem 1.8 (see equation (1.3)) that

$$M_{\tilde{Q}} = M(I - LM + \tilde{Q}) \tag{2.3}$$

is an inner inverse which satisfies $LM_{\tilde{Q}} = \tilde{Q}$. Hence $M(I - LM + \tilde{Q})$ is a R-T.I.I. of L.

We thus have:

<u>Proposition 2.1</u>. Let $L \in \Lambda(\mathcal{V}, \mathcal{Y})$, where $\mathcal{Y}$ is a topological

vector space. Suppose $\overline{\mathfrak{R}(L)}$ has a topological supplement $\mathfrak{Z}$ in $\mathcal{Y}$ and let Q be the (continuous) projector of $\mathcal{Y}$ onto $\overline{\mathfrak{R}(L)}$ along $\mathfrak{Z}$. Then L has a right-topological inner inverse; i.e., there exists a linear operator $L_r^- = L_{r,Q}^-$ such that:

$$
(2.4) \qquad \begin{cases} L_r^- : \mathfrak{R}(L) \oplus \eta(Q) \to \mathcal{V} \\ LL_r^- L = L \quad \text{on } \mathcal{V} \\ LL_r^- = Q \quad \text{on } \mathfrak{R}(L) \oplus \eta(Q) . \end{cases}
$$

In particular any operator given by (2.3) is a R-T. I. I. of L.

We note that $L_{r,Q}^-$ is only densely defined and that its domain depends on the choice of the projector Q whose range is $\overline{\mathfrak{R}(L)}$, unless L is an operator whose range is closed in $\mathcal{Y}$. Given two (continuous) projectors on $\mathcal{Y}$ with range $\overline{\mathfrak{R}(L)}$, it is not necessarily true that $\mathfrak{R}(L) \dot{+} \eta(Q) = \mathfrak{R}(L) \dot{+} \eta(Q')$, i.e., the subspace $\mathfrak{R}(L) \dot{+} \eta(Q)$ might not be invariant with respect to Q'.

Theorem 2.2. *Let $L \in \Lambda(\mathcal{V}, \mathcal{Y})$ have the property that $\overline{\mathfrak{R}(L)}$ has a topological complement, and let Q and Q' be two (continuous) projectors with range $\overline{\mathfrak{R}(L)}$. Let $M = L_{r,Q}^-$ and $M' = L_{r,Q'}^-$ be two corresponding R-T. I. I.'s.*

(a) $\mathcal{D}(M) = \mathcal{D}(M')$ *if and only if* $\mathfrak{R}(Q' - Q) \subset \mathfrak{R}(L)$.

(b) *If* $\mathfrak{R}(L)$ *is closed then every R-T. I. I. is defined on all of* $\mathcal{Y}$.

(c) *If* $\mathfrak{R}(L)$ *is not closed and if* $\mathcal{Y}$ *is Hausdorff and locally convex, then for any* $y_0 \notin \overline{\mathfrak{R}(L)}$ *we can select* Q *and* Q' *so that* $y_0 \in \mathcal{D}(M)$ *and* $y_0 \notin \mathcal{D}(M')$.

PROOF. (a) Write $y = Qy + (I - Q)y = y_1 + y_2$, and also write $y = Q'y + (I - Q')y = y_3 + y_4$. Then $y_3 = y_1 + (Q' - Q)y$ and $y_4 = (I - Q')y_2$. If $\mathcal{D}(M) = \mathcal{D}(M')$ then $y_1 \in \mathfrak{R}(L)$ and

$y_3 \in \mathcal{R}(L)$ so $(Q' - Q)y = y_3 - y_1 \in \mathcal{R}(L)$ for all $y \in \mathcal{Y}$ since $\mathcal{R}(L)$ is a subspace. Conversely if for all $y \in \mathcal{Y}$ we have $(Q' - Q)y \in \mathcal{R}(L)$, then $y_1 \in \mathcal{R}(L)$ if and only if $y_3 \in \mathcal{R}(L)$, i.e., $y \in \mathcal{D}(M)$ if and only if $y \in \mathcal{D}(M')$.

(b) It is always true that $\mathcal{R}(Q - Q') \subset \overline{\mathcal{R}(L)}$, so if L has closed range, then by part (a), every R-T. I. I. has domain all of $\mathcal{Y}$.

(c) If $\mathcal{Y}$ is a Hausdorff locally convex topological vector space and $y_0 \notin \overline{\mathcal{R}(L)}$, it follows from the Hahn-Banach theorem that there exists a continuous linear functional f on $\mathcal{Y}$ such that $f(y_0) = 1$ and $f(z) = 0$ if $z \in \overline{\mathcal{R}(L)}$. Let Q be any (continuous) projector with range $\overline{\mathcal{R}(L)}$ and take $z_0 \in \overline{\mathcal{R}(L)}$. Then by the characterization of Proposition 1.7, $Q'y = Qy - f(y)[Qy_0 - z_0]$ defines a projector with range $\overline{\mathcal{R}(L)}$, and $Q'y_0 = z_0$. If we take $z_0 \in \overline{\mathcal{R}(L)} - \mathcal{R}(L)$ then $y_0 \notin \mathcal{D}(M')$. On the other hand we could also take z_0 to be anything we like in $\mathcal{R}(L)$ which would put $y_0 \in \mathcal{D}(M')$. ■

We note that if $\mathcal{V}$ and $\mathcal{Y}$ are finite dimensional spaces, if $\mathcal{Y}$ is endowed with the standard inner product, and if Q is chosen to be the orthogonal projector, then L has a particular R-T.I.I. which is characterized by

$$LL_r^- L = L$$

$$(LL_r^-)^* = LL_r^- .$$

That is, L_r^- is a (1, 3)-inverse in the terminology of Ben-Israel and Greville [1] and others; i.e., it satisfies the "first" and "third" equations of the Moore-Penrose inverse. (1, 3)-inverses can be characterized by the concept of least-squares solutions: For any $y \in \mathcal{Y}$, $\|Lx - y\|$ is minimized when $x = L_r^- y$; conversely, if $M \in \Lambda(\mathcal{Y}, \mathcal{V})$ has the property that for all $y \in \mathcal{Y}$, $\|Lx - y\|$ is minimum when $x = My$, then M is a (1, 3)-inverse. See, e.g., Ben-Israel and Greville [1], Rao and Mitra [1].

More generally we have:

Proposition 2.3. Let $\mathcal{V}$ be a vector space and let $\mathcal{K}$ be a Hilbert space, both over the real or the complex numbers. Let $L: \mathcal{V} \to \mathcal{K}$ be a linear operator (not necessarily bounded or closed). Then L has a R-T. I. I. In particular R-T. I. I.'s $L^{\dagger}_{r,Q}$ corresponding to the choice of the orthogonal projector Q of $\mathcal{K}$ onto $\overline{\mathcal{R}(L)}$ exist and are linear operators $M: \mathcal{R}(L) \oplus \mathcal{R}(L)^{\perp} \to \mathcal{V}$ such that

$$LML = L \quad \text{and} \quad LM = Q|\mathcal{R}(L) \oplus \mathcal{R}(L)^{\perp} .$$

Furthermore, such $L^{\dagger}_{r,Q}$ are characterized by the extremal property cited above, which holds for all $y \in \mathcal{R}(T) \oplus \mathcal{R}(T)^{\perp}$.

PROOF. This follows from Proposition 2.1 using the fact that every closed subspace in a Hilbert space has a complement (in particular an orthogonal complement). (The extremal property is a special case of Theorem 3.3 below. More general results and related extremal properties are given in Sections 3 and 5).

2.2. Left-Topological Inner Inverses (L-T. I. I.)

We consider now a linear operator L whose domain $\mathcal{D}(L)$ lies in a topological vector space $\mathcal{X}$ and whose range is considered to be in an (algebraic) vector space $\mathcal{Y}$. $\mathcal{D}(L)$ is not necessarily dense in $\mathcal{X}$.

The case of a L-T. I. I. is more complicated. It is not sufficient to assume $\overline{\eta(L)}$ has a topological complement. Rather we need a "decomposability condition" on the domain of L.

For example, let f be a discontinuous linear functional (so $\eta(f)$ is dense in $\mathcal{X}$) and let $f(x_0) = 1$. Then $f^- : \mathbb{R} \to \mathcal{X}$ defined by $f^-(t) = tx_0$ is a R-T. I. I. but not a L-T.I.I. since f^-f is zero on a dense subset of $\mathcal{X}$ and yet $f^-f(x_0) = x_0$, so that f^-f is not a continuous projector.

We introduce a notion of decomposiability which is weaker than a related notion used in Hilbert spaces by Tseng [2], Hestenes [1], Arghiriade [1] and others.

Definition 2.4. Let $\mathcal{D}(L) \subset \mathcal{X}$. We say L is domain decomposable with respect to the projector $P \in \mathcal{L}(\mathcal{X})$ if $\mathcal{N}(L) \subset \mathcal{R}(P)$, if $Px \in \mathcal{N}(L)$ for all $x \in \mathcal{D}(L)$, and if $\mathcal{D}(L) \cap \mathcal{N}(P)$ is dense in $\mathcal{N}(P)$. In this case we call

$$(2.5) \qquad C_P := \mathcal{D}(L) \cap \mathcal{N}(P)$$

the carrier of L with respect to P.

Hestenes [1], Arghiriade [1], Arghiriade and Dragomir [1], and Arghiriade and Boros [1] say that a linear operator T on a Hilbert (or inner product) space $\mathcal{H}$ has a decomposable domain if

$$(2.6) \qquad \mathcal{D}(T) = \mathcal{N}(T) \oplus (\mathcal{D}(T) \cap \mathcal{N}(T)^{\perp}) ,$$

and call $\mathcal{D}(T) \cap \mathcal{N}(T)^{\perp}$ the carrier of T. Note that if an operator T has a decomposable domain in this sense, then T is domain decomposable in the sense of Definition 2.4 with respect to the orthogonal projector $P \in \mathcal{L}(\mathcal{H})$ on $\overline{\mathcal{N}(T)}$, with $\mathcal{N}(P) = \mathcal{N}(T)^{\perp}$.

Condition (2.6) was also used by Tseng [2], [3], [4]. See also Nashed [1].

We note that if T is a closed densely defined linear operator with $\overline{\mathcal{D}(T)} = \mathcal{H}$ or if T is a bounded linear operator in Hilbert space $\mathcal{H}$, then $\mathcal{N}(T)$ is closed and T is domain decomposable in our sense with respect to the orthogonal projector P of $\mathcal{H}$ onto $\mathcal{N}(T)$, and hence also in the sense of Hestenes and others. However, T is also domain decomposable in our sense with respect to other (not necessarily orthogonal) projectors of $\mathcal{H}$ onto $\mathcal{N}(T)$.

The definition of domain decomposability in the sense of Hestenes and others is restricted to orthogonal projectors and complements in Hilbert space, whereas our definition allows any projector and makes sense in topological vector spaces.

We point out here that L may be decomposable with respect to one projector but not with respect to another projector whose range contains $\mathcal{N}(L)$. In particular if $\mathcal{D}(L)$ is dense in a Hilbert space $\mathcal{H}$, it may be the case that L is not decomposable with respect to the orthogonal (self-adjoint) projector but that L is decomposable with respect to some other (non-orthogonal) projector. See Example 2.11.

Definition 2.5. If $\mathcal{D}(L) \subset \mathcal{X}$ and L, considered as being in $\Lambda(\mathcal{D}(L), \mathcal{Y})$, has an inner inverse M with the property that ML has an extension to a projector $I - P \in \mathcal{L}(\mathcal{X})$ with $\mathcal{R}(I - P) = \overline{\mathcal{R}(ML)}$ (the closure being taken in $\mathcal{X}$), we say that M is a left-topological inner inverse of L (abbreviated as L-T. I. I.). We will use the notation $M = L^-_{\ell, P}$ to denote a L-T. I. I.

Theorem 2.6. *If $M = L^-_{\ell, P}$, then L is decomposable with respect to P, $\mathcal{D}(L) = \mathcal{N}(L) \oplus \mathcal{C}_P(L)$, and L can be extended to a densely defined operator $\hat{L}$ with a closed null space which also has M as a L-T. I. I.*

PROOF. If $z \in \mathcal{N}(L)$ then $MLz = (I - P)z = 0$ so $z \in \mathcal{R}(P)$ and thus $\mathcal{N}(L) \subset \mathcal{R}(P)$. If $x \in \mathcal{D}(L)$, then $Px = x - MLx \in \mathcal{N}(L)$. Also $\mathcal{N}(P) = \overline{\mathcal{R}(ML)}$ (closure in $\mathcal{X}$), so $\mathcal{D}(L) \cap \mathcal{N}(P)$ is dense in $\mathcal{N}(P)$. Thus the existence of a L-T. I. I. implies that L is decomposable with respect to P.

Take $x \in \mathcal{D}(L)$ and write $x = (I - ML)x + MLx$. This shows that $x \in \mathcal{N}(L) \oplus \mathcal{C}_P(L)$. Hence $\mathcal{D}(L) = \mathcal{N}(L) \oplus \mathcal{C}_P(L)$.

We extend L to $\hat{L}$ defined on $\mathcal{D}(\hat{L}) := \mathcal{R}(P) \oplus \mathcal{C}_P(L)$ as follows: $\hat{L}(Px + y) = Ly$ where $y = MLy$. Then $\hat{L}M\hat{L}(Px+y) = \hat{L}MLy = \hat{L}y = \hat{L}(Px + y)$, so M is an inner inverse of $\hat{L}$. In addition $M\hat{L}$ agrees with $I - P$ on $\mathcal{D}(\hat{L})$ and $\mathcal{R}(M\hat{L}) = \mathcal{R}(ML) = \mathcal{C}_P(L) = \mathcal{C}_P(\hat{L})$ so M is a L-T. I. I. of $\hat{L}$ and $\mathcal{N}(\hat{L})$ is closed. Now $\mathcal{X} = \mathcal{R}(P) \oplus \mathcal{N}(P) = \mathcal{R}(P) \oplus \overline{\mathcal{C}_P(L)}$ so $\mathcal{D}(\hat{L}) = \mathcal{R}(P) \oplus \mathcal{C}_P(L)$ is dense in $\mathcal{X}$. ■

Theorem 2.7. If $\mathcal{D}(L) \subset \mathcal{X}$ and L is decomposable with respect to P, then L has a L-T. I. I.

PROOF. Let M' be an algebraic inner inverse of $L \in \Lambda(\mathcal{D}(L), \mathcal{Y})$, and write $P' = I - M'L$. Let $\tilde{P}$ be the restriction of P to $\mathcal{D}(L)$. We use Theorem 1.8 to change projectors: $M = (2I - M'L - \tilde{P})M'$ satisfies $ML = I - \tilde{P}$. Also $\mathcal{R}(ML) = \mathcal{N}(\tilde{P}) = \mathcal{N}(P) \cap \mathcal{D}(L) = C_P(L)$ which is dense in $\mathcal{N}(P) = \mathcal{R}(I-P)$ so M is a L-T. I. I. of L. ∎

We note that if in Definition 2.5, $\mathcal{X}$ and $\mathcal{Y}$ are finite dimensional spaces, if $\mathcal{X}$ endowed with the standard inner product, and if P is chosen to be the orthogonal projector on $\mathcal{N}(L)$, then L has a particular L-T.I.I. which is characterized by

$$LL_\ell^- L = L$$

and

$$(L_\ell^- L)^* = L_\ell^- L \ .$$

That is, L_ℓ^- is a (1, 4)-inverse in the sense that it satisfies the "first" and "fourth" equations of the Moore-Penrose inverse. For any $y \in \mathcal{R}(L)$, $L_\ell^- y$ is the (unique) minimum norm solution of the equation $Lx = y$, and this property characterizes L_ℓ^- (see e.g. Rao and Mitra [1], Ben-Israel and Greville [1]). More generally we have:

Proposition 2.8. Let $\mathcal{H}$ be a Hilbert space and $\mathcal{Y}$ be a vector space, both over the real or complex numbers. Let $L: \mathcal{H} \to \mathcal{Y}$ be a densely defined closed linear operator or a bounded linear operator on $\mathcal{H}$. Then L has a L. T. I. I. In particular L. T. I. I.'s corresponding to the choice of the orthogonal projector of $\mathcal{H}$ onto $\mathcal{N}(L)$ exist and are equivalently characterized as linear operators $M: \mathcal{Y} \to \mathcal{H}$ such that $LML = L$ and $(ML)^* = ML$ on $\mathcal{D}(L)$.

More general results in normed spaces and related extremal properties are given in Sections 3 and 5.

2.3. Generalized Inverses in Topological Vector Spaces

We finally get to the case where T is a linear operator whose domain lies in a topological vector space $\mathcal{X}$ and whose range lies in a topological vector space $\mathcal{Y}$.

Definition 2.9. Let $T \in \Lambda(\mathcal{X},\mathcal{Y})$, where $\mathcal{X},\mathcal{Y}$ are topological vector spaces. If U is both a L-T. I. I. and a R-T. I. I. of T, and UTU = U, we say that U is a topological generalized inverse (abbreviated T. G. I.) of T and we write $U = T^{\dagger} = T^{\dagger}_{P,Q}$.

In view of Theorem 2.6, we might as well consider that either T is defined on all of $\mathcal{X}$ or that $\mathcal{D}(T)$ is dense in $\mathcal{X}$, since the extension $\hat{T}$ given in the proof of the Theorem 2.6 has the same range as T.

Theorem 2.10. Let $T \in \Lambda(\mathcal{X},\mathcal{Y})$, where $\mathcal{X},\mathcal{Y}$ are topological vector spaces. If T is decomposable with respect to a (continuous) projector P and if there exists a (continuous) projector Q onto $\mathcal{R}(T)$, then T has a (unique) T. G. I. $T^{\dagger} = T^{\dagger}_{P,Q}$ (with respect to the choice of P and Q) which satisfies:

$$(2.6)\quad \begin{cases} \mathcal{D}(T^{\dagger}) = \mathcal{R}(T) \oplus \mathcal{N}(Q) \\ \mathcal{R}(T^{\dagger}) = \mathcal{C}_P(T) \\ \mathcal{N}(T^{\dagger}) = \mathcal{N}(Q) \\ T^{\dagger}Tx = x - Px \quad \text{for all } x \in \mathcal{D}(T) \\ TT^{\dagger}y = Qy \quad \text{for all } y \in \mathcal{D}(T^{\dagger}) . \end{cases}$$

PROOF. We could have used algebraic generalized inverses instead of inner inverses in all of the work in subsections 2.1 and 2.2. In Theorem 2.7 we changed the projector P in the space $\mathcal{X}$ so that $T^{\dagger}T = I - P$ on $\mathcal{D}(T)$. Similarly we can change the projector in $\mathcal{Y}$ to get in addition $TT^{\dagger} = Q$ on $\mathcal{D}(T^{\dagger})$. ■

The construction of $T^{\dagger} = T^{\dagger}_{P,Q}$ is illustrated by the following commutative diagram, where an over ~ denotes restriction of an operator to an indicated domain, while i, j denote injections.

$$\begin{array}{ccccc}
\mathcal{X} = \mathcal{R}(P) \oplus \mathcal{N}(P) & & \mathcal{Y} = \overline{\mathcal{R}(T)} \oplus \mathcal{N}(Q) & & \mathcal{X} = \mathcal{R}(P) \oplus \mathcal{N}(P) \\
\cup & & \cup & & \cup \\
\mathcal{D}(T) = \mathcal{N}(T) \oplus \mathcal{C}_P(T) & \xrightarrow{\;T\;} & \mathcal{R}(T) \oplus \mathcal{N}(Q) & \xrightarrow{\;T^{\dagger}\;} & \mathcal{N}(T) \oplus \mathcal{C}_P(T) \\
i \uparrow \downarrow \widetilde{I-P} & & j \uparrow \downarrow \tilde{Q} & & i \uparrow \downarrow \widetilde{I-P} \\
\mathcal{C}_P(T) = \mathcal{D}(T) \cap \mathcal{N}(P) & \xrightarrow{\;\tilde{T}\;} & \mathcal{R}(T) & \xrightarrow{\;\tilde{T}^{-1}\;} & \mathcal{C}_P(T) = \mathcal{R}(T^{\dagger}T)
\end{array}$$

It follows from Theorem 2.10 that Definition 2.9 reduces to Tseng's definition of a maximal generalized inverse for the case where $\mathcal{X}$ and $\mathcal{Y}$ are Hilbert spaces, T is densely defined, and P and Q are taken to be self adjoint projectors. See Tseng [2]. (See also Section 5, and Nashed [1]). If in addition T is bounded, then our definition is equivalent to the usual definition. See Nashed [1] for a discussion of various definitions of generalized inverses in Hilbert space. See also Section 5.

Recall however that T may not be decomposable with respect to an orthogonal projector but it may be decomposable with respect to some other projector. Thus, in the Hilbert space case, one might as well select Q to be self-adjoint, but there is still some value in allowing P to not be self-adjoint.

The following example shows that for the Hilbert space case, Definition 2.9 can give a generalized inverse in cases where Tseng's definition fails because it requires P to be self-adjoint.

Example 2.11. Let I be an incomplete inner product space with the separable Hilbert space H as its completion, and let $\{e_n\}_{n=1}^{\infty}$ be a complete orthonormal sequence in H with $e_1 \in H - I$ and $e_n \in I$ for $n \geq 2$. (Olmstead's paper [1] shows how to construct such a sequence). Approximate e_1 by $g_1 \in I$. Then $g_1 = \sum_{k=1}^{\infty} \alpha_k e_k$ where α_1 is close to one and α_n is small for $n \geq 2$. Now take an operator T with domain I and null space the closed span of $\{e_n\}_{n=2}^{\infty}$. Then $\mathcal{N}(T)^{\perp} = \{te_1\}$, and since $e_1 \notin \mathcal{D}(T)$ we have $\mathcal{D}(T) \cap \mathcal{N}(T)^{\perp} = \mathcal{C}(T) = \{\theta\}$. However T is decomposable with respect to the non-orthogonal projector P defined by

$$Px = x - \frac{1}{\alpha_1}(e_1, x)g_1 .$$

We next consider the case where $\mathcal{X}$ and $\mathcal{Y}$ are topological vector spaces, and $T \in \Lambda(\mathcal{X}, \mathcal{Y})$. The next theorem gives conditions that will yield a continuous generalized inverse. Recall that $T \in \mathcal{L}(\mathcal{X}, \mathcal{Y})$ is a <u>topological homomorphism</u> if the image of every open set in $\mathcal{X}$ is open in $T(\mathcal{X})$ in the induced topology.

<u>Theorem 2.12</u>. If T is a <u>topological homomorphism whose null space has a topological complement in $\mathcal{X}$ and whose range has a topological complement in $\mathcal{Y}$, then T has a generalized inverse (defined on all of $\mathcal{Y}$) which is itself a topological homomorphism.</u>

PROOF. Since $\mathcal{X}$ and $\mathcal{Y}$ are automatically assumed to be Hausdorff, we have that $\mathcal{R}(T)$ and $\mathcal{N}(T)$ are closed. Also T is decomposable with respect to any (continuous) projector $P: \mathcal{X} \to \mathcal{N}(T)$. Thus Theorem 2.10 gives $T^{\dagger}$ satisfying $\mathcal{D}(T^{\dagger}) = \mathcal{Y}$, $\mathcal{R}(T^{\dagger}) = \mathcal{N}(P)$, $\mathcal{N}(T^{\dagger}) = \mathcal{N}(Q)$, $T^{\dagger}T = I - P$, and $TT^{\dagger} = Q$.

Since the natural map $\mathcal{X}/\mathcal{N}(T) \to \mathcal{N}(P)$ is an isomorphism for the topological vector space structure, we identify $\mathcal{X}/\mathcal{N}(T)$ with $\mathcal{N}(P)$. Since T is a topological homomorphism, the induced map $\tilde{T}$ of $\mathcal{N}(P)$ onto $\mathcal{R}(T)$ is a topological isomorphism. Hence $\tilde{T}^{-1}$ is a topological isomorphism of $\mathcal{R}(T)$ onto $\mathcal{N}(P)$. Identifying $\mathcal{R}(T)$ with $\mathcal{Y}/\mathcal{N}(Q)$ shows that $T^{\dagger}$ is a topological homomorphism. ∎

Theorem 2.12 was given by Votruba [1] and for the case of Fredholm operators by Pietsch [1].

Corollary 2.13. Suppose $T^\dagger$ exists. Then T is surjective (onto) if and only if $T^\dagger$ is a right inverse for T. Also T is injective (i.e., $\mathcal{N}(T) = \{0\}$) if and only if $T^\dagger$ is a left inverse for T. (Compare with Bourbaki [1, Chapter I, Propositions 13 and 14, pp. 17-18].) Hence if T is a topological isomorphism, then $T^\dagger = T^{-1}$.

PROOF. $TT^\dagger$ is a projector onto the range of T; thus T is surjective if and only if $TT^\dagger = I$. Similarly $I - T^\dagger T$ is a projector onto $\mathcal{N}(T)$; thus T is injective if and only if $T^\dagger T = I$. If T is a topological isomorphism, then T is both surjective and injective; thus $T^\dagger = T^{-1}$.

Corollary 2.14. If $\mathcal{X}$ and $\mathcal{Y}$ are complete metrizable topological vector spaces, then any $T \in \mathcal{L}(\mathcal{X}, \mathcal{Y})$ such that $\mathcal{N}(T)$ is complemented in $\mathcal{X}$ and $\mathcal{R}(T)$ is complemented in $\mathcal{Y}$ has a generalized inverse which is a topological homomorphism.

Corollary 2.15. If $\mathcal{X}$ and $\mathcal{Y}$ are Hilbert spaces, then any $T \in \mathcal{L}(\mathcal{X}, \mathcal{Y})$ such that $\mathcal{R}(T)$ is closed has a generalized inverse which is a topological homomorphism.

We consider other special cases in Section 5, where we also develop the basic theory of generalized inverses of bounded or densely defined closed linear operators on Banach or Hilbert spaces, with some reference to the framework of our approach.

We conclude this section with two important remarks.

Remark 2.16. All the generalized inverses that we have considered in this section are of maximal domain. With reference to the notation and construction of Section 2.3, it is evident that for each subspace $\mathcal{L} \subset \mathcal{N}(Q)$ there is a (unique) G.I. $T^{\ell}_{P,Q}$ with

$$\mathcal{D}(T^{\ell}_{P,Q}) = \mathcal{R}(L) \oplus \mathcal{L}$$

and

$$\mathcal{N}(T^{\ell}_{P,Q}) = \mathcal{L} ;$$

the other properties of $T^{\ell}_{P,Q}$ being the same as those of $T^{\dagger}_{P,Q}$. Note that $T^{\dagger}_{P,Q}$ is densely defined and has a closed null space by (2.6). It is clear that an operator T may have, under the assumptions of Theorem 2.10, infinitely many generalized inverses $T^{\ell}_{P,Q}$ with dense domain. Also, T may have infinitely many generalized inverses with closed null space, each obtained by choosing $\mathfrak{L}$ as a closed subspace of $\mathcal{N}(Q)$. However $T^{\dagger}_{P,Q}$ is the unique densely defined G.I. with a closed null space.

Remark 2.17. From Theorems 2.6 and 2.7, a linear operator $L: \mathfrak{X} \to \mathfrak{Y}$ has a L-T. I. I. iff $\mathfrak{D}(L)$ is decomposable (in the sense of Definition 2.4). However, if $\mathfrak{D}(L)$ is not decomposable, we can still associate with L a related L-T. I. I. as follows.

Let L_0 be the restriction of L defined by

$$\mathfrak{D}(L_0) = \mathcal{N}(L) \dotplus \mathfrak{C}_P(L) \quad \text{and} \quad \mathcal{N}(L_0) = \mathcal{N}(L) \quad ,$$

where P is any projector in $\mathfrak{L}(X)$ such that $\mathcal{N}(L) \subset \mathfrak{R}(P)$. Then L_0 is domain decomposable and so it has a L-T. I. I. One might then identify a L-T. I. I. of L with that of L_0. In this manner L-T. I. I. 's would always exist. Similarly, we can define a T. G. I. under the remaining assumptions of Definition 2.9.

However, this does not seem to be a very fruitful route since, among other things, $\mathfrak{R}(L)$ is not always in the domain of the T. G. I. or L-T. I. I. defined in the above manner.

3. Extremal and Proximinal Properties of Generalized Inverses. Right-Orthogonal, Left-Orthogonal, Orthogonal, and Metric Generalized Inverses.

In this section we develop proximinal properties of generalized inverses in normed spaces within the general setting of Section 2, and demonstrate the relation of inner and generalized inverses to extremal, minimal, and best approximate solutions. The results in normed spaces are new and include as special cases previously known results in the context of Hilbert spaces. This section is a sequel to Section 2; the notations are same.

Let N be a normed linear space. We write $\mathrm{sp}\, A$ for the span of a set A (i.e., $\mathrm{sp}\, A$ is the linear hull of A) and $d(x, A)$ for the distance from x to the set A (i.e., $d(x, A) = \inf\{\|x-z\| : z \in A\}$. The following definition of orthogonality is used (see James [1]):

$$x \perp y \text{ means } d(x, \mathrm{sp}\{y\}) = \|x\|$$

and

$$B \perp A \text{ means } d(x, A) = \|x\| \text{ for all } x \in B .$$

That is $B \perp A$ if and only if θ is a nearest point from A to each $x \in B$.

Note that this orthogonality is not a symmetric relation. For example in R^2 with the norm $\|(\xi, \eta)\| = |\xi| + |\eta|$ we have that $(0, 1)$ is orthogonal to $(1, 1)$ but $(1, 1)$ is not orthogonal to $(0, 1)$. As an aside, it is a curious fact that inner product spaces of dimension at least three can be characterized by the "symmetry of orthogonality," but in dimension two there are norms which do not come from inner products but which give rise to a symmetric orthogonality relation. See Schäffer [1].

Now let M be a subspace of N which has a topological complement, and consider the affine manifold $\mathcal{P}_M = \{P \in \mathfrak{L}(N) \mid P = P^2 \text{ and } \mathfrak{R}(P) = M\}$ in the normed linear space $\mathfrak{L}(N)$. Note that $d(I, \mathcal{P}_M) \geq 1$ since for any projector P we have $\|I - P\| \geq 1$.

Proposition 3.1. Let $P_0 \in \mathcal{P}_M$. The following three statements are equivalent:

(a) P_0 is a nearest point in $\mathcal{P}_M$ to I, and $d(I, \mathcal{P}_0) = 1$.

(b) $\mathfrak{N}(P_0) \perp \mathfrak{R}(P_0)$.

(c) For each $x \in N$ we have that $P_0 x$ is a nearest point in M to x.

PROOF. If (a) holds, then $\|I - P_0\| = 1$. Let $x \in \mathcal{N}(P_0)$ and $z \in \mathcal{R}(P_0)$. Then $\|x\| = \|(I - P_0)(x - z)\| \leq \|x - z\|$; i.e., if $x \in \mathcal{N}(P_0)$ then $d(x, \mathcal{R}(P_0)) = \|x\|$. This gives (b).

If (b) holds, let $x \in N$ and $z \in \mathcal{R}(P_0)$. Then $\|x - z\| = \|(I - P_0)x - P_0(z - x)\| \geq d((I - P_0)x, \mathcal{R}(P_0)) = \|x - P_0 x\|$ since $(I - P_0)x \in \mathcal{N}(P_0)$. This gives (c).

Assume (c) holds. Now any $x \in N$ can be written as $x = y + z$ where $y = P_0(x) \in \mathcal{R}(P_0)$ and $z = (I - P_0)x \in \mathcal{N}(P_0)$. We have $\|x\| = \|z - (-y)\| \geq \|z - P_0 z\| = \|z\| = \|(I - P_0)x\|$ so $\|I - P_0\| = 1$ and (a) holds. ■

If there exists a $P_0 \in \mathcal{P}_M$ such that any one (hence all three) of the statements given in Proposition 3.1 holds, we say that M is an <u>orthogonally complemented subspace</u> of N. We emphasize that if $M = \mathcal{R}(P_0)$ is orthogonally complemented by $\mathcal{N}(P_0)$, it is not necessarily true that $\mathcal{N}(P_0)$ is orthogonally complemented by $\mathcal{R}(P_0)$. (If $\|I - P_0\| = 1$, all we can say is that $\|P_0\| \leq 2$). Also, orthogonal complements are not necessarily unique. Hilbert spaces are an exceptional case; if M is a closed subspace of a Hilbert space, then $M^{\perp}$ is the unique orthogonal complement of M in the above sense, and also $M^{\perp}$ has M as an orthogonal complement. In this case the corresponding projector will be self-adjoint. Conversely if P is a self-adjoint projector in a Hilbert space, then P is a metric (i.e. closest-point) projector, $\mathcal{R}(P) \perp \mathcal{N}(P)$, and $\perp$ is reflexive.

We remark that in a normed linear space every one-dimensional subspace is an orthogonal complement for some hyperplane, for let $S = sp\{e\}$, $\|e\| = 1$. Take (by the Hahn-Banach theorem) $f \in N^*$ with $\|f\| = 1$ and $f(e) = 1$. Define P by $Px = x - f(x)e$. Then $\mathcal{R}(P) = \mathcal{N}(f)$ and $\|I - P\| = 1$.

In general, in normed linear spaces, orthogonally complemented subspaces are rare. The characterization problem of minimal projectors is discussed in Morris and Cheney [1].

If $T \in \Lambda(V, N_2)$ where N_2 is a normed linear space, then $v_0 \in V$ is called an extremal solution or a virtual solution of the equation $Tv = y$ if $v = v_0$ minimizes $\|Tv - y\|$. If $V = N_1$ is also a normed linear space, then an extremal solution of minimal norm is called a best approximate solution or a least extremal solution. The equation $Tx = y$ need not have an extremal solution for every y, and the existence of an extremal solution does not imply the existence of a best approximate solution.

Definition 3.2. Let N_2 be a normed linear space, and let $T \in \Lambda(\mathcal{V}, N_2)$. If $U = T^-_{r,Q}$ is a R-T. I. I. of T where $\|I - Q\| = 1$, we call U a right-orthogonal inner inverse (abbreviated as R-O. I. I.) of T, and denote it by $T^\dagger_{r,Q}$. On the other hand, if $\mathcal{D}(T) = \mathcal{V}$ is contained in a normed linear space N_1, $\mathcal{R}(T)$ is in an (algebraic) vector space $\mathcal{W}$, and $U = T^\dagger_{\ell,P}$ is a L-T. I. I. of T, where $\|I - P\| = 1$, we call U a left-orthogonal inner inverse (L-O. I. I.). If U is both a L-O. I. I. and a R-O. I. I. we call U an orthogonal inner inverse (O. I. I.) of T.

Note that by Proposition 3.1, the existence of a R-O. I. I. implies that $\mathcal{R}(T)$ is an orthogonally complemented subspace of N_2. Also, the existence of a L-O. I. I. implies that $\mathcal{C}_P(T) \perp \mathcal{N}(T)$ since $\mathcal{N}(P) \perp \mathcal{R}(P)$ by Proposition 3.1 and since $\mathcal{N}(T) \subset \mathcal{R}(P)$ and $\mathcal{C}_P(T) \subset \mathcal{N}(P)$ by Theorem 2.6 and Definition 2.4. Thus, in general, the existence of an O.P.I. is a rare phenomenon in normed spaces without inner products.

If $N_2 = H_2$ is a Hilbert space, there is a unique projector Q with range $\mathcal{R}(T)$ satisfying $\|I - Q\| = 1$. Thus for $T \in \Lambda(\mathcal{V}, H_2)$ a R-O. I. I. always exists. If $N_1 = H_1$ is also a Hilbert space, then T has an O. P. I. if and only if T is domain-decomposable with respect to a self-adjoint projector P in the sense of Definition 2.4.

The importance of right-orthogonal, left-orthogonal, and orthogonal inner inverses lies in the fact that they are connected with the existence of extremal solutions, minimal solutions, and best approximate solutions.

Theorem 3.3. If U is a right-orthogonal inner inverse of $T \in \Lambda(\mathcal{V}, N_2)$, then for $y \in \mathfrak{D}(U)$ the equation $Tv = y$ has $v = Uy$ as an extremal solution. If, in addition, nearest points from $\overline{\mathfrak{R}(T)}$ are unique, (in particular if N_2 is strictly convex), then the existence of an extremal solution to $Tv = y$ implies that $y \in \mathfrak{D}(U) = \mathfrak{R}(T) \oplus \mathcal{N}(Q)$.

PROOF. By Proposition 3.1, Qy is a nearest point in $\overline{\mathfrak{R}(T)}$ to y. If also $Qy \in \mathfrak{R}(T)$, then $y \in \mathfrak{D}(U) = \mathfrak{R}(T) \oplus \mathcal{N}(Q)$. Thus $Qy = TUy$ so $v = Uy$ minimizes $\|Tv - y\|$.

It is both well-known and easy to show that in a strictly convex space ($\|x\| = \|y\| = 1$ and $x \neq y \Longrightarrow \|x+y\| < 2$) nearest points from a subspace (if they exist) must be unique.

If Qy is the unique nearest point to y from $\overline{\mathfrak{R}(T)}$ and if $Qy \notin \mathfrak{R}(T)$, then clearly $\|Tv - y\| > \|y - Qy\| = d(y, \overline{\mathfrak{R}(T)}) = d(y, \mathfrak{R}(T))$ for all $v \in V$, so $Tv = y$ has no extremal solution. ∎

Theorem 3.4. If U is a left-orthogonal inner inverse of T and the equation $Tx = y$ has a solution, then Uy is a (not-necessarily unique) solution of minimal norm.

PROOF. By Proposition 1.4(c), the general solution of $Tx=y$ is $Uy + z$ where $z \in \mathcal{N}(T)$. To get a minimal solution, we get a nearest point in $\mathfrak{R}(P)$ to Uy. By Proposition 3.1, PUy is such a point; i.e., $\|(I - P)Uy\| \leq \|Uy + z\|$ for all $z \in \mathfrak{R}(P)$. But $\mathcal{N}(T) \subset \mathfrak{R}(P)$ and $PUy \in \mathcal{N}(T)$, so PUy is a nearest point to Uy from $\mathcal{N}(T)$. Thus $(I - P)Uy$ is a minimal solution of $Tx = y$. But $(I - P)Uy = UTUy = UQy = Uy$ since $y \in \mathfrak{R}(T)$. ∎

Theorem 3.5. If U is an orthogonal inner inverse of T, and if in N_2 nearest points from $\overline{\mathfrak{R}(T)}$ are unique, then for all $y \in \mathfrak{D}(U)$ we have that $x = UTUy = T^{\dagger}_{P,Q}y$ is a (not-necessarily unique) best approximate solution of $Tx = y$.

PROOF. From Theorem 3.3 we know that Uy is one extremal solution. Let $x = x_0$ be another. Since nearest points from $\overline{\mathfrak{R}(T)}$ are unique, we have that $Tx_0 = TUy = Qy$. In other words, the set of extremal solutions of $Tx = y$ is exactly the set of actual solutions of $Tx = Qy$. By Theorem 3.4 we have that $UQy = UTUy = T^{+}_{P,Q}y$ is a solution of $Tx = Qy$ of minimal norm; i.e., an extremal solution of $Tx = y$ of minimal norm. ∎

Corollary 3.6. Let T have $\mathcal{D}(T)$ in a Hilbert space H_1 and $\mathcal{R}(T)$ in a Hilbert space H_2. Then $Tx = y$ has an extremal solution if and only if the orthogonal projection of y onto $\overline{\mathcal{R}(T)}$ lies in $\mathcal{R}(T)$. In this case $Tx = y$ has a (unique) best approximate solution if and only if the orthogonal projection of any (every) extremal solution onto $\overline{\eta(T)}$ lies in $\eta(T)$.

If T has an orthogonal inner inverse, then $Tx = y$ has a (unique) best approximate solution whenever it has an extremal solution.

PROOF. In a Hilbert space, every closed subspace is orthogonally complemented, and a self-adjoint projector P has $\|P\| = \|I - P\| = 1$ and is the only projector with a given range which has $\|I - P\| = 1$. Also a Hilbert space is strictly convex, so nearest points are unique (see, e.g., E. W. Cheney, Introduction to Approximation Theory, McGraw-Hill, New York, 1966.)

Let $Q = Q^* = Q^2$ have $\mathcal{R}(Q) = \overline{\mathcal{R}(T)}$ and let U be a R-O. I. I. of T. By Theorem 3.3 $Tx = y$ has an extremal solution if and only if $Qy \in R(T)$. Let x_0 and x_1 be any two extremal solutions. Then, since nearest points are unique, $T(x_0 - x_1) = 0$. Thus the set of all extremal solutions is $x_0 + \eta(T)$. Let $P = P^* = P^2$ have $\mathcal{R}(P) = \overline{\eta(T)}$. Then Px_0 is the unique nearest point to x_0 from $\overline{\eta(T)}$. Thus $\|x_0 - Px_0\| \leq \|x_0 - z\|$ for all $z \in \eta(T)$, and strict inequality holds unless $Px_0 \in \eta(T)$.

If T has an O.I.I.; i.e., if T is decomposable with respect to a self-adjoint projector P, then Theorem 3.5 gives a best approximate solution, and strict convexity gives uniqueness.

This corollary basically contains results given in Erdelyi and Ben-Israel [1, §2].

For matrices, Ben-Israel and Greville [1] and Rao and Mitra [1] show that $\|Ax - b\|$ is minimized by $x = Ub$ if U satisfies $AUA = A$ and AU is self-adjoint. This is a special case of Theorem 3.3. They also show that if $Ax = b$ has a solution, then the unique solution of smallest norm is $x = Ub$ where $AUA = A$ and UA is self-adjoint. This follows as a special case of Theorem 3.4 plus the use of strict convexity which gives uniqueness. Finally the well-known result for Hilbert spaces is Corollary 3.6.

Some other special cases are also discussed in Sections 4 and 5.

The Metric Generalized Inverse

In view of Theorem 3.5, which gives the existence of best approximate solutions, it is useful to consider another kind of generalized inverse.

Definition 3.7. Let $T \in \Lambda(N_1, N_2)$, and consider a $y \in N_2$ such that $Tx = y$ has a best approximate solution in N_1. We define $T^{\partial}(y) = \{x \in N_1 | x$ is a best approximate solution to $Tx = y\}$ and call the set-valued mapping $y \to T^{\partial}(y)$ the metric generalized inverse. Here $\mathcal{D}(T^{\partial}) = \{y \in N_2 | Tx = y$ has a best approximate solution in $N_1\}$. A (in general nonlinear) function $T^{\sigma}: \mathcal{D}(T^{\partial}) \to N_1$ such that $T^{\sigma}(y) \in T^{\partial}(y)$ is called a selection for the metric generalized inverse.

From Theorem 3.3 we see that if T has a right-orthogonal inner inverse and if nearest points from $\mathcal{R}(T)$ are unique, then $\mathcal{D}(T^{\partial}) \subset \mathcal{R}(T) \oplus \eta(Q)$. If in addition T has an orthogonal partial inverse U, then $\mathcal{D}(T^{\partial}) = \mathcal{D}(U) = \mathcal{R}(T) \oplus \eta(Q)$ and $T^{\dagger}_{P,Q} = UTU$ is a linear selection for T^{∂}.

Remarks and Special Cases of the Metric Generalized Inverse

(i) Holmes [1] defines the generalized inverse of $T \in \mathcal{L}(N_1, N_2)$ to be T^{∂} in the case where $T^{\partial}(y)$ is always a singleton and gives some conditions which imply that T^{∂} is densely defined and/or continuous. He uses this approach to obtain the (linear) generalized inverse in the Hilbert space case.

(ii) Newman and Odell [1] have studied T^{∂} in the finite-dimensional case where the norms are strictly convex. In this case, there is always a unique best approximate solution of $Lx = y$ given by $B = (I - F)ME$ where M is any partial inverse of L, E is the nearest point map onto $\mathcal{R}(A)$, and F is the nearest point map onto $\eta(A)$. They call B the p-q generalized inverse of the (linear) mapping A.

Some of the results of Erdelsky [1] (see also Ben-Israel and Greville for an exposition of these results) are also subsumed by our theorems.

(iii) Erdelyi and Ben-Israel [1] considered T^{∂} for an arbitrary linear mapping with domain in a Hilbert space $\mathcal{H}_1$ and range in a Hilbert space $\mathcal{H}_2$. In this case $T^{\partial}(y)$ is always a singleton and T^{∂} is a linear map. They used the name g-inverse for the linear operator T^{∂} such that the minimal extremal solution x_0 of $Tx = y$ exists if and only if $y \in \mathcal{D}(T^{\partial})$; in this case $x_0 = T^{\partial}y$. It is easy to show that the maximal domain of T^{∂} is $\mathcal{D}(T^{\partial}) = T[\mathcal{C}(T)] \oplus^{\perp} \mathcal{R}(T)^{\perp}$ where $\mathcal{C}(T) = \mathcal{D}(T) \cap \eta(T)^{\perp}$ and that it has the following properties:

$$\mathcal{R}(T^{\partial}) = \mathcal{C}(T), \quad \eta(T^{\partial}) = \mathcal{R}(T)^{\perp},$$

$$TT^{\partial} = P_{\overline{\mathcal{R}(T)}} \,|\, \mathcal{D}(T^{\partial})$$

$$T^{\partial}T = P_{\overline{\mathcal{R}(T^{\partial})}} \,|\, \eta(T) \oplus^{\perp} \mathcal{C}(T) .$$

Note that for any linear operator $T\colon \mathcal{D}(T) \subset \mathcal{H}_1 \to \mathcal{H}_2$, a direct sum decomposition can be obtained via a linear manifold $\mathcal{A}(T)$, produced for example using a Hamel basis, such that

$$\mathcal{D}(T) = \mathcal{A}(T) \dotplus [(\mathcal{D}(T) \cap \overline{\eta(T)}) \oplus^{\perp} \mathcal{C}(T)] .$$

Then

$$\mathcal{R}(T) = T[\mathcal{A}(T)] + T[\mathcal{D}(T) \cap \overline{\eta(T)}] + T[\mathcal{C}(T)] \quad ;$$

thus in general

$$T[\mathcal{C}(T)] \subsetneq \mathcal{R}(T) \quad \text{and} \quad \mathcal{R}(T) \not\subset \mathcal{D}(T^{\partial}) \quad .$$

Also T^{∂} is not defined on $T[\mathcal{A}(T)] \cup T[\mathcal{D}(T) \cap \overline{\eta(T)}]$ so $\mathcal{D}(T^{\partial})$ is not dense in $\mathcal{V}_2$ unless

$$T[\mathcal{A}(T)] = \{0\} \Longrightarrow \mathcal{A}(T) = \{0\}$$

and

$$T[\mathcal{D}(T) \cap \overline{\eta(T)}] = \{0\} \Longrightarrow \mathcal{D}(T) \cap \overline{\eta(T)} = \eta(T) \; .$$

The facts that $\mathcal{D}(T^{\partial}) \not\supset \mathcal{R}(T)$ and that $\mathcal{D}(T^{\partial})$ is not necessarily dense are the weak points of T^{∂}; its strong point is that T^{∂} is a metric generalized inverse.

If $\mathcal{D}(T)$ is decomposable with respect to a self-adjoint projector P, and if Q is self-adjoint, then $T^{\partial} = T^{\dagger}_{P,Q}$. See Corollary 3.6. Also $(T^{\partial})^{\partial} = \hat{T}$ where $\hat{T}$ is the extension given by Theorem 2.6.

If $\mathcal{D}(T)$ is not decomposable with respect to self-adjoint P, then our (topological) generalized inverse (in the special case where P, Q are self-adjoint) of a restriction of T is an extension of T^{∂} . Let $\tilde{T} = T|\mathcal{D}_0$, where $\mathcal{D}_0 = \eta(T) \oplus^{\perp} \mathcal{C}(T)$. Then $\tilde{T}$ has an (orthogonal) generalized inverse $\tilde{T}^{\dagger}$ in the sense of Definition 2.9 with domain $T(\mathcal{C}) \oplus^{\perp} T(\mathcal{C})^{\perp}$, and the g-inverse turns out to be $T^{\partial} = \tilde{T}^{\dagger}|\mathcal{D}(T^{\partial})$ since $T(\mathcal{C}) \subset \mathcal{R}(T)$.

(iv) The problem of obtaining selections with nice properties for the metric generalized inverse merits study. Of course, there is no reason to restrict attention here to metric generalized inverse of linear operators.

We list here some references which may be useful. See also Holmes [1] and Singer [1], [2]. The survey paper of Vlasov contains an extensive bibliography on metric projections and related questions.

L. P. Vlasov, Approximative properties of sets in normed linear spaces, Russian Math. Survey, 28 (1973), 1-66.

E. Michael, Selected selection theorems, Amer. Math. Monthly, 63(1956), 233-238.

D. E. Wulbert, Continuity of metric projections, Trans. Amer. Math. Soc., 134 (1968), 335-341.

V. Blatter, P. D. Morris, and D. E. Wulbert, Continuity of the set-valued metric projection, Math. Ann. 178(1968), 12-24.

A. J. Lazar, P. D. Morris, and D. E. Wulbert, Continuous selections for metric projections, J. Functional Analysis, 3(1969), 193-216.

4. Specific Types of Algebraic Generalized Inverses

We continue to use the terminology and notation of Section 1. Let $L \in \Lambda(\mathcal{V}, \mathcal{W})$. Algebraic direct sum decompositions:

$$\mathcal{V} = \mathcal{N}(L) \dot{+} \mathcal{M} \quad \text{and} \quad \mathcal{W} = \mathcal{R}(L) \dot{+} \mathcal{S} \tag{4.1}$$

always exist, and the correspondence $(\mathcal{M}, \mathcal{S}) \longleftrightarrow (P, Q)$ is a bijection. Each pair of complements $\mathcal{M}, \mathcal{S}$ as in (4.1) determines a unique algebraic generalized inverse $L^{\#}$. We shall write $L^{\#}_{P,Q}$ (or $L^{\#}_{\mathcal{M},\mathcal{S}}$) when it is necessary to stress the dependence of $L^{\#}$ on the projectors P and Q (or equivalently on $\mathcal{M}$ and $\mathcal{S}$). There should be no confusion between complements and projectors since the formers are always designated by script letters.

In this section we focus on specific types of generalized inverses either in the algebraic context of Section 1, i.e. the spaces will not necessarily be endowed with topologies, or in finite dimensional spaces. In the latter case, we employ the standard notions of inner product, orthogonality, etc., and we designate linear maps in this case by A, B, etc. However, the results and proofs apply without major modification to bounded linear operators with closed range on Hilbert spaces. Section 5 is devoted to specific types of generalized inverses in infinite dimensional spaces. Our aim in both sections is to examine in the framework of our approach a diversity of generalized inverses that have been used in the literature.

4.1. The A.G.I. of L relative to K. For any $L \in \Lambda(\mathcal{V}, \mathcal{W})$, there exists always a (non-unique) map $K \in \Lambda(\mathcal{W}, \mathcal{V})$ such that

(4.2) $$\mathcal{V} = \mathcal{N}(L) \dotplus \mathcal{R}(K) \quad \text{and} \quad \mathcal{W} = \mathcal{R}(L) \dotplus \mathcal{N}(K) \ .$$

For any given $\mathcal{M}$ and $\mathcal{S}$ as in (4.1) there is a (non-unique) map $K \in \Lambda(\mathcal{W}, \mathcal{V})$ such that

(4.3) $$\mathcal{R}(K) = \mathcal{M} \quad \text{and} \quad \mathcal{N}(K) = \mathcal{S} \ .$$

The A.G.I. $L^{\#}_{\mathcal{R}(K), \mathcal{N}(K)}$ relative to the pair of complements (4.3) is called the A.G.I. of L relative to K, and is denoted by $L^{\#K}$. Note at the outset that every A.G.I. M satisfies the property $L^{\#M} = M$ since M induces the decompositions (see Proposition 1.16)

(4.4) $$\mathcal{V} = \mathcal{N}(L) \dotplus \mathcal{R}(M) \quad \text{and} \quad \mathcal{W} = \mathcal{R}(L) \dotplus \mathcal{N}(M) \ .$$

From (4.4) it follows immediately that

$$(L^{\#}_{P,Q})^{\#}_{I-Q,\, I-P} = L \quad \text{and} \quad (L^{\#}_{\mathcal{M}, \mathcal{S}})^{\#}_{\mathcal{R}(L), \mathcal{N}(L)} = L \ .$$

4.2. The Moore-Penrose Inverse. Penrose defined the generalized inverse $A^{\dagger}$ of a rectangular (m by n) matrix A with complex entries as the unique solution of the following equations

$$(4.4) \qquad \begin{cases} \text{(i)} & AXA = A \\ \text{(ii)} & XAX = X \\ \text{(iii)} & (AX)^* = AX \\ \text{(iv)} & (XA)^* = XA \end{cases}$$

where A^* denotes the adjoint (conjugate transpose) of A. Penrose [1] established the existence and uniqueness of a solution to the system (4.4) and in a subsequent note [2] that $A^\dagger$ has the best approximation property.

We now show by the results of Section 1 that the system (4.4) has a unique solution. We consider the A. G. I. $A^\#$ corresponding to the choice of orthogonal projectors on $\mathfrak{N}(A)$ and $\mathfrak{R}(A)$. Since every orthogonal projector is Hermitian, $A^\#$ will satisfy the system (4.4). Now the system (4.4) has a unique solution; for if X is a solution, then XA and AX are Hermitian projectors and hence orthogonal. Thus X and $A^\#$ must be equal since they give rise to the same projectors. Thus the system (4.4) is equivalent to the following system:

$$(4.5) \qquad XAX = X, \quad XA = P_{\mathfrak{R}(A)}, \quad AX = P_{\mathfrak{R}(A^*)} ,$$

where $P_{\mathfrak{M}}$ denotes the orthogonal projector on the subspace $\mathfrak{M}$.

Moore [2] also characterized $A^\dagger$ as the unique solution of the system

$$AX = P_{\mathfrak{R}(A)} \text{ and } XA = P_{\mathfrak{R}(X)} .$$

Clearly this system is equivalent to (4.4).

In view of the well-known decompositions:

$$\mathbb{C}^n = \mathfrak{N}(A) \oplus \mathfrak{R}(A^*)$$

$$\mathbb{C}^m = \mathfrak{R}(A) \oplus \mathfrak{N}(A^*) ,$$

we may consider the Moore-Penrose inverse as the A.G.I. of A relative to A^* (as defined in 4.1).

4.3. A.G.I. with Specified Range and Null Space. There exists a unique A.G.I. with a specified range and null space. We recall (Proposition 1.16) that $\mathcal{R}(L^\dagger)$ and $\mathcal{N}(L^\dagger)$ are complements to $\mathcal{N}(L)$ and $\mathcal{R}(L)$ respectively. $L^\dagger$ is then the unique A.G.I. induced by the decompositions:

$$\mathcal{V} = \mathcal{N}(L) \dotplus \mathcal{R}(L^\dagger) \text{ and } \mathcal{W} = \mathcal{R}(L) \dotplus \mathcal{N}(L^\dagger) \ .$$

4.4. A.G.I. Corresponding to Oblique Projectors. We refer to a nonorthogonal projector in an inner product space as an oblique projector. Various authors have developed a theory of A.G.I. based on oblique projectors in much the same way that the Moore-Penrose theory is based on orthogonal projectors. See Langenhop [1], Milne [1], Nashed [1] and Robinson [1]. This theory is contained in the framework of Section 1. The A.G.I. $A^{\#}_{P,Q}$ when P and Q are oblique projectors will be called an oblique generalized inverse.

4.5. Weighted A.G.I. Let W and V be (Hermitian) positive definite matrices of order m and n respectively. The following system of equations:

$$(4.6) \qquad \begin{cases} AXA = A \ , \\ XAX = X \\ (XA)^* = V^{-1}XAV \quad \text{and} \\ (AX)^* = WAXW^{-1} \end{cases}$$

has a unique solution X which is called a weighted generalized inverse of A (Chipman [1]). We establish the existence and uniqueness of the Chipman weighted G.I. by showing that it is the A.G.I. corresponding to orthogonal projectors defined relative to "ellipsoidal" inner products.

To this end, we let $U = V^{-1}$ and rewrite (4.6) in the equivalent form:

(4.7) $AXA = A,\ XAX = X,\ (WAX)^* = WAX,$ and $UXA = (UXA)^*$.

We consider A as a linear transformation on $\mathbb{C}^n$ into $\mathbb{C}^m$. We define new inner products on $\mathbb{C}^n$ and $\mathbb{C}^m$ respectively by:

(4.8) $$\langle x, x' \rangle_U = \langle Ux, x' \rangle ,$$

and

(4.9) $$\langle y, y' \rangle_W = \langle Wy, y' \rangle$$

where $\langle \cdot , \cdot \rangle$ is the standard inner product on Euclidean spaces. Then $AX: \mathbb{C}^m \to \mathbb{C}^m$ is Hermitian relative to the new inner product on $\mathbb{C}^m$ if and only if

$$\langle AXy, y' \rangle_W = \langle y, AXy' \rangle_W \quad \text{for all} \quad y, y' \in \mathbb{C}^n ,$$

or equivalently, in terms of the standard inner product,

$$\langle WAXy, y' \rangle = \langle Wy, AXy' \rangle = \langle y, WAXy' \rangle ,$$

i.e. if and only if $(WAX)^* = WAX$. Similarly for $UXA = (UXA)^*$.

Thus the (Chipman) weighted G.I. is characterized by the following equations:

(4.10) $AXA = A,\ XAX = X,\ AX = P^W_{\mathcal{R}(A)},$ and $XA = I - P^U_{\mathcal{N}(A)}$,

where $P^W_{\mathcal{R}(A)}$ denotes the orthogonal projector in the weighted inner product (4.9) onto $\mathcal{R}(A)$ and $P^U_{\mathcal{N}(A)}$ is the orthogonal projector in the weighted inner product (4.8) onto $\mathcal{N}(A)$.

Since every inner product $[\cdot\,,\,\cdot]$ in $\mathbb{C}^n$ can be represented in terms of the standard inner product $\langle\cdot\,,\,\cdot\rangle$ as $[x, y] = \langle Ux, y\rangle$ for some positive definite matrix, it follows that the set of weighted generalized inverses as V and W vary over (Hermitian) positive definite matrices is the same as the set of A. G. I. with respect to orthogonal projectors as the inner products vary over the set of all inner products on $\mathbb{C}^n$ and $\mathbb{C}^m$.

Ward, Boullion and Lewis [1] have shown that if $A^{\#}$ is a given A. G. I., then there exist positive definite matrices U and W such that $A^{\#}$ is the weighted G. I. determined by U and W.

In view of the characterization (4.10) we obtain a simple proof of the following known result.

Theorem 4.1. Let A be an $m \times n$ matrix, $b \in \mathbb{C}^m$, and let W and U be positive definite matrices of orders m and n respectively.

(a) Let X be an $m \times n$ matrix which satisfies

$$AXA = A, \quad (WAX)^* = WAX . \tag{4.11}$$

Then $x_0 = Xb$ minimizes $\|Ax - b\|_W$, where $\|y\|_W = \langle Wy, y\rangle^{\frac{1}{2}}$. Conversely if X is an $n \times m$ matrix such that Xb has this minimizing property for all b, then X satisfies (4.11).

(b) For any $b \in \mathcal{R}(A)$ the unique solution for which $\|x\|_U$ is smallest is given by $x_0 = Xb$, where X satisfies

$$AXA = A, \quad (UXA)^* = UXA . \tag{4.12}$$

Conversely, if a matrix X has the property that for each $b \in \mathcal{R}(A)$, $x_0 = Xb$ is the solution of $Ax = b$ for which $\|x\|_U$ is smallest, then X satisfies (4.12).

(c) If X satisfies (4.7), or equivalently (4.10), then

$x_0 = Xb$ minimizes $\|Ax - b\|_W$ and $\|Xb\|_U < \|z\|_U$ for all $z \neq Xb$ such that $\|Az - b\|_W = \min \|Ax - b\|_W$. Conversely, if $x_0 = Yb$ has this property for all b, then Y satisfies (4.7).

This theorem could be restated verbatim for bounded linear operators with closed range in Hilbert spaces. The cases of bounded or closed linear operators with nonclosed range require technical modifications which should be evident in view of similar results developed in Sections 3 (see also Section 5) and the characterization analogous to (4.10).

4.6. The Group Inverse of a Linear Transformation. Let L be a linear transformation of a vector space $\mathcal{V}$ into itself. The group inverse L^g of L is the unique solution, if there is a solution, of the following equations:

$$LXL = L, \quad XLX = X, \quad \text{and} \quad LX = XL \ . \tag{4.13}$$

It follows from Proposition 1.16 that the group inverse, if it exists, is an A.G.I. such that $\mathcal{R}(X) = \mathcal{R}(L)$ and $\mathcal{N}(X) = \mathcal{N}(L)$. Furthermore, L has a group inverse if and only if $\mathcal{V} = \mathcal{N}(L) \dotplus \mathcal{R}(L)$. A detailed exposition of the group inverse and spectral generalized inverses is given in Ben-Israel and Greville [1; Ch. 4], where further references are cited.

5. Generalized Inverses of Topological Homomorphisms in Topological Vector Spaces. Generalized Inverses in Banach and Hilbert Spaces

The topics of this title are subsumed under the unified approach to generalized inverses which was developed in Sections 2 and 3. There is merit, however, in stating the results of this section independently of this approach, as well as in examining these results in the framework of a unified approach. On the one hand, generalized inverses of bounded or densely defined closed linear operators and of topological homomorphisms in topological vector spaces arise frequently in applications, and therefore are of interest to many readers who may not necessarily be excited by the more encompassing general approach. On the other hand, it is always instructive to examine some concrete situations that a more abstract approach is supposed to unify. As a by-product, this section will provide a synopsis of properties of generalized inverses in the specific contexts cited above.

5.1. Generalized Inverses of Topological Homomorphisms in Topological Vector Spaces

Theorem 5.1. Let X and Y be topological vector spaces over the real or complex numbers, and let T be a linear map of X into Y. The following three statements are equivalent:

1. T is a topological homomorphism (i.e., the image of every open set in X is open in T(X) in the induced topology), the null space $\mathcal{N}(T)$ has a topological complement in X, and the range $\mathcal{R}(T)$ has a topological complement in Y.

2. There exists a linear topological homomorphism $T^{\dagger}$ of Y into X such that:

   (a) $TT^{\dagger}$ is a projector (continuous and idempotent) taking Y onto $\mathcal{R}(T)$,

   (b) $I-T^{\dagger}T$ is a projector taking X onto $\mathcal{N}(T)$ where I is the identity operator,

   (c) $T^{\dagger}TT^{\dagger} = T^{\dagger}$.

3. $\mathcal{N}(T)$ has a topological complement $\mathcal{M}$ in X, and there exists a linear topological homomorphism S of Y onto $\mathcal{M}$ whose null space $\mathcal{N}(S)$ has $\mathcal{R}(T)$ as a topological complement in Y.

Proof that $1 \Longrightarrow 2$. By assumption $X = \mathcal{N}(T) \oplus \mathcal{M}$ for some $\mathcal{M}$. Thus there is a projector $P:X \to X$ such that $P(X) = \mathcal{N}(T)$ and $(I-P)(X) = \mathcal{M}$. Moreover the natural map $X/\mathcal{N}(T) \to \mathcal{M}$ is an isomorphism for the topological vector space structure. We identify $X/\mathcal{N}(T)$ with $\mathcal{M}$. Similarly Y is the topological direct sum of $\mathcal{R}(T)$ and $\mathcal{S}$, where $\mathcal{S}$ is the given topological supplement of $\mathcal{R}(T)$ in Y. There is a projector $Q: Y \to Y$ such that $Q(Y) = \mathcal{R}(T)$ and $(I-Q)(Y) = \mathcal{S}$.

Also by assumption T is a topological homomorphism of X onto $\mathcal{R}(T)$. This means that the image under T of an open set in X is open in $\mathcal{R}(T)$ in the topology induced by Y. Thus, T induces a topological isomorphism of $X/\mathcal{N}(T)$ (identified with $\mathcal{M}$) onto $\mathcal{R}(T)$. We call this isomorphism $\tilde{T}$. Note that $\tilde{T}$ is the restriction of T to $\mathcal{M}$, since $(I-P)$ is a continuous linear extension of the identity map on $\mathcal{M}$. Let $j: \mathcal{R}(T) \to Y$ be the injection map. We can "factor" the map T as follows: $T = j\tilde{T}(I-P)$.

We now define $T^{\dagger}: Y \to X$ by the equation

$$T^{\dagger} = i\tilde{T}^{-1}Q \tag{5.1}$$

where i is the injection $\mathcal{M} \to X$. Since Q is a projector of Y onto $\mathcal{R}(T)$, the null space of $T^{\dagger}$ is $\mathcal{S}$ where $Y = \mathcal{R}(T) \oplus \mathcal{S}$. Moreover, $\tilde{T}^{-1}$ is $T^{\dagger}$ restricted to $\mathcal{R}(T)$.

Thus we have the following commutative diagram:

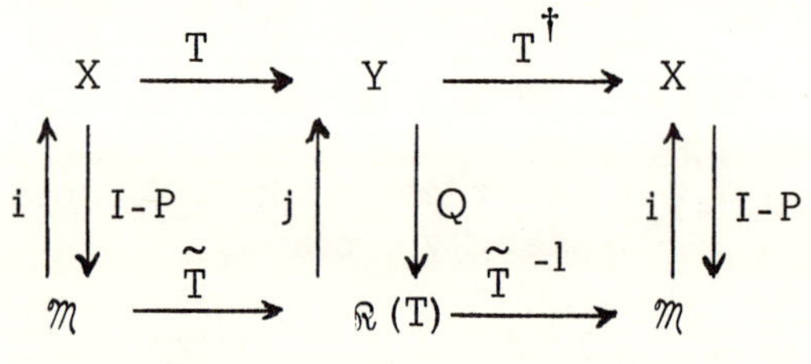

Now $T^{\dagger}$ is a linear topological homomorphism of Y onto $\mathcal{M}$ since the map of $Y/\mathcal{S}$ onto $\mathcal{M}$ induced by $T^{\dagger}$ is a topological isomorphism (we identify $Y/\mathcal{S}$ and $\mathcal{R}(T)$ and note that $\tilde{T}^{-1}$ is a topological isomorphism of $\mathcal{R}(T)$ onto $\mathcal{M}$). Next note that $TT^{\dagger} = [j\tilde{T}(I-P)][i\tilde{T}^{-1}Q] = j\tilde{T}[(I-P)i]\tilde{T}^{-1}Q = j\tilde{T}\tilde{T}^{-1}Q = jQ$; i.e., $TT^{\dagger}$ is a projector onto $\mathcal{R}(T)$. Thus the restriction of $TT^{\dagger}$ is a projector onto $\mathcal{R}(T)$. Thus the restriction of $TT^{\dagger}$ to $\mathcal{R}(T)$ is the identity on $\mathcal{R}(T)$, and $TT^{\dagger}T = T$.

It remains to prove statements (b) and (c). Now $T^{\dagger}T = [i\tilde{T}^{-1}Q][j\tilde{T}(I-P)] = i\tilde{T}^{-1}[Qj]\tilde{T}(I-P) = i\tilde{T}^{-1}\tilde{T}(I-P) = i(I-P)$; i.e., $T^{\dagger}T$ is a projector onto $\mathcal{M}$. Thus $I-T^{\dagger}T$ is a projector onto $\mathcal{N}(T)$, and the restriction of $T^{\dagger}T$ to $\mathcal{M}$ is the identity on $\mathcal{M}$. The statement $T^{\dagger}TT^{\dagger} = T^{\dagger}$ now follows from the fact that the range of $T^{\dagger}$ is $\mathcal{M}$.

The construction of $T^{\dagger}$ can be summarized as follows:

$$(5.2)\quad \begin{cases} T^{\dagger}y = i(I-P)x \text{ for any } x \text{ such that } Tx = jQy\ , \\ P \text{ is a projector onto the null space of } T\ , \\ Q \text{ is a projector onto the range of } T\ . \end{cases}$$

Proof that $2 \Longrightarrow 3$. If $I-T^{\dagger}T$ is a projector of X onto $\mathcal{N}(T)$ then $\mathcal{N}(T)$ has a topological complement $\mathcal{M}$; specifically $\mathcal{M} = T^{\dagger}T(X)$. Next we show that the range of the homomorphism $T^{\dagger}$ given by the hypothesis is $\mathcal{M}$. Clearly $\mathcal{M} \subset T^{\dagger}(Y)$ since $\mathcal{M} = T^{\dagger}T(X)$. To show that $T^{\dagger}(Y) \subset \mathcal{M}$ consider any $x \in T^{\dagger}(Y)$. We have the unique decomposition $x = x_1 + x_2$ where $Tx_1 = 0$ and where $x_2 \in \mathcal{M}$. Since $x \in T^{\dagger}(Y)$ we have $x_1 + x_2 = T^{\dagger}y$ for some $y \in Y$. Thus $T^{\dagger}T(x_1 + x_2) = T^{\dagger}TT^{\dagger}y = T^{\dagger}y$; i.e., $x_1 + x_2 = T^{\dagger}T(x_1 + x_2) = x_2$. Hence $x \in \mathcal{M}$ and thus $T^{\dagger}(Y) = \mathcal{M}$.

Finally, we must show that $\mathcal{N}(T^{\dagger})$ and $\mathcal{R}(T)$ are topological complements in Y. By hypothesis $TT^{\dagger}$ is a projector of Y onto $\mathcal{R}(T)$; i.e., $\mathcal{R}(T)$ has a topological complement $\mathcal{S} = (I-TT^{\dagger})(Y)$. We show that $\mathcal{S} = \mathcal{N}(T^{\dagger})$. Clearly $\mathcal{N}(T^{\dagger}) \subset \mathcal{S}$ for if $T^{\dagger}y = 0$ then $TT^{\dagger}y = 0$ and hence $y \in \mathcal{S}$. To show that $\mathcal{S} \subset \mathcal{N}(T^{\dagger})$, let $y \in \mathcal{S}$. Now $(I-TT^{\dagger})$ is a projector of Y onto $\mathcal{S}$, thus $(I-TT^{\dagger})y = y$ if $y \in \mathcal{S}$. Hence $T^{\dagger}(I-TT^{\dagger})y = T^{\dagger}y$. But $T^{\dagger}(I-TT^{\dagger})y = T^{\dagger}y - T^{\dagger}TT^{\dagger}y = T^{\dagger}y - T^{\dagger}y = 0$; i.e., $T^{\dagger}y = 0$ and $y \in \mathcal{N}(T^{\dagger})$. Thus $\mathcal{N}(T^{\dagger}) = \mathcal{S}$.

Proof that $3 \Longrightarrow 1$. The statement that S is a topological homomorphism of Y onto $\mathcal{M}$ is equivalent to the statement that the map of $Y/\mathcal{N}(S)$ onto $\mathcal{M}$ induced by S is a topological isomorphism. To show that T is a topological homomorphism, we will show that the map of $X/\mathcal{N}(T)$ onto $\mathcal{R}(T)$ induced by T is a topological isomorphism. Now since

$Y = \mathcal{R}(T) \oplus \mathcal{N}(S)$ we have that $Y/\mathcal{N}(S)$ and $\mathcal{R}(T)$ are topologically isomorphic. Thus $\mathcal{R}(T)$ and $\mathcal{M}$ are topologically isomorphic. Thus $\mathcal{R}(T)$ and $\mathcal{M}$ are topologically isomorphic. But since $X = \mathcal{N}(T) \oplus \mathcal{M}$ we have also that $X/\mathcal{N}(T)$ and $\mathcal{M}$ are topologically isomorphic. Thus we can conclude that T is a topological homomorphism of X onto $\mathcal{R}(T)$. ■

Theorem 5.2. Let T be a linear topological homomorphism such that $\mathcal{N}(T)$ has a topological supplement in X and $\mathcal{R}(T)$ has a topological supplement in Y . Let P and Q be given projectors onto $\mathcal{N}(T)$ and $\mathcal{R}(T)$ respectively. Then the linear topological homomorphism $T^\dagger$, or $T^\dagger_{PQ}$, of Y into X satisfying

$$(5.3) \qquad \begin{cases} TT^\dagger = jQ \\ T^\dagger T = i(I-P) \\ T^\dagger TT^\dagger = T^\dagger \end{cases}$$

is unique for given P and Q .

PROOF. The existence of such a $T^\dagger$ was given in the proof that $1 \Longrightarrow 2$. To show uniqueness, assume that $U: Y \to X$ satisfies

$$TU = jQ$$
$$UT = i(K-P)$$
$$UTU = U \quad .$$

Then since $Y = T(X) \oplus (I-Q)(Y)$, any $y \in Y$ can be written as $y = Tx + y_1$, for some $x \in X$ where $Qy_1 = 0$. Thus $Uy = UTx + Uy_1 = i(I-P)x + UTUy_1 = i(I-P)x + UjQy_1 = i(I-P)x$. But by (5.2) we have that $T^\dagger y = i(I-P)x$ since $jQy = jQTx + jQy_1 = Tx$. Thus $Uy = T^\dagger y$ for all $y \in Y$. ■

Remarks. 1. The existence of a topological complement of a subspace M is equivalent to the existence of a projector whose range is M . However a given subspace does not have a unique topological complement. Thus in the above theorem, $T^\dagger = T^\dagger_{PQ}$ depends on the particular choise of projectors P and Q. If X and Y are Hilbert spaces, then P and Q can always be taken to be orthogonal projectors (i.e., P and Q can be taken to be self-adjoint).

2. The injections i and j have now outlived their usefulness. From this point on, we consider P as taking X into X and Q as taking Y into Y, and we will write equations (5.3) without the injections i and j. We replace (5.3) by

$$(5.4) \qquad \begin{cases} TT^\dagger T = T \\ T^\dagger TT^\dagger = T^\dagger \\ T^\dagger T = I - P \\ TT^\dagger = Q \end{cases} .$$

Note that these equations are redundant since $TT^\dagger T = T$ follows from $TT^\dagger = Q$.

Definition 5.3. If T satisfies the hypothesis of Theorem 5.2, we say that the map $T^\dagger : Y \to X$ satisfying equation (5.4) is the generalized inverse of T relative to P and Q. We will write $T^\dagger_{P,Q}$ if we want to call attention to the projectors P and Q.

We note that the correspondence

$$(P, Q) \longleftrightarrow T^\dagger_{P,Q}$$

is a bijection.

We recall that a linear operator $T : \mathcal{D}(T) \subset X \to Y$, where X and Y are topological spaces, is said to be a closed operator if its graph

$$\mathcal{G}(T) := \{(x, Tx) : x \in \mathcal{D}(T)\}$$

is closed in $X \times Y$. Equivalently, in normed spaces, T is closed if $x_n \in \mathcal{D}(T)$, $x_n \to x$, and $Tx_n \to y$ imply $x \in \mathcal{D}(T)$

and $Tx = y$. It follows immediately that the null space of a closed linear operator is a closed subspace in X, and that the generalized inverse of T exists if $\mathcal{N}(T)$ and $\overline{\mathcal{R}(T)}$ have topological complements.

Following Atkinson [1], we say that an operator $T \in \mathcal{L}(X, Y)$ is relatively regular if there exists an operator $S \in \mathcal{L}(Y, X)$ such that $TST = T$ and $STS = S$. Thus T is relatively regular if and only if T has a bounded algebraic inverse. Theorem 5.1 implies the following corollary.

Corollary 5.4. Let $T \in \mathcal{L}(X, Y)$. Then

(a) T is relatively regular if and only if $\mathcal{N}(T)$ has a topological complement in X and $\mathcal{R}(T)$ has a topological complement in Y.

(b) If S is a bounded generalized inverse of T, then $\mathcal{R}(S)$ is a topological complement of $\mathcal{N}(T)$ and $\mathcal{N}(S)$ is a topological complement of $\mathcal{R}(T)$. Furthermore, ST is the (continuous) projector of X onto $\mathcal{R}(S)$ along $\mathcal{N}(T)$, and TS is the (continuous) projector of Y onto $\mathcal{R}(T)$ along $\mathcal{N}(S)$.

We observe that Theorems 5.1 and 5.2 do not cover the case of a linear operator with a nonclosed range. However, this case is covered by Theorem 2.10 for a wide class of linear operators, which are not necessarily topological homomorphisms. The statements and proof of Theorem 5.1 can be modified to cover the case when $\overline{\mathcal{R}(T)}$ has a topological complement in Y and $\mathcal{N}(T)$ has a topological complement in X.

Theorem 5.5. Let $T \in \mathcal{L}(X, Y)$, where X, Y are topological vector spaces. Suppose that $\mathcal{N}(T)$ has a topological complement in X and $\overline{\mathcal{R}(T)}$ has a topological complement in Y, say

$$X = \mathcal{N}(T) \oplus \mathcal{M} \quad \text{and} \quad Y = \overline{\mathcal{R}(T)} \oplus \mathcal{S} .$$

Let P denote the projector of X onto $\mathcal{N}(T)$ along $\mathcal{M}$ and let Q denote the projector of Y onto $\overline{\mathcal{R}(T)}$ along $\mathcal{S}$. Then

(a) T has a (unique) generalized inverse $T^\dagger = T^\dagger_{P,Q}$ (relative to the choice of P and Q) which satisfies the following properties:

$$(5.5)\quad \begin{cases} \mathcal{D}(T^\dagger) = \mathcal{R}(T) \oplus \mathcal{N}(Q) = \mathcal{R}(T) \oplus \mathcal{S} \ , \\ TT^\dagger T = T \ \text{ on } \ \mathcal{D}(T) \ , \\ T^\dagger T T^\dagger = T^\dagger \ \text{ on } \ \mathcal{D}(T^\dagger) \ , \\ T^\dagger T = (I - P) \,|\, \mathcal{D}(T) \ , \\ \text{and} \\ TT^\dagger = Q \,|\, \mathcal{D}(T^\dagger) \ . \end{cases}$$

(b) $T^\dagger = T^\dagger_{P,Q}$ is also characterized as the linear extension of $(T|\mathcal{M})^{-1}$ to $\mathcal{D}(T^\dagger) := \mathcal{R}(T) \oplus \mathcal{S}$ such that $\mathcal{N}(T^\dagger) = \mathcal{S}$.

PROOF. Define $T^\dagger$ by

$T^\dagger Tx = x$ if $x \in \mathcal{M}$

$T^\dagger y = 0$ if $y \in \mathcal{S}$

$T^\dagger y = T^\dagger y_1$ if $y \in \mathcal{R}(T) \oplus \mathcal{S}$, where $y_1 \in \mathcal{R}(T)$ and $y_2 \in \mathcal{S}$.

Note that the domain of $T^\dagger$ is independent of the choice of the complement $\mathcal{M}$, and that $T^\dagger$ may be written as $T^\dagger y = (T|\mathcal{M})^{-1}Qy$ for all $y \in \mathcal{D}(T^\dagger)$. ∎

5.2. Generalized Inverses in Banach Spaces

Nothing new can be said by considering Theorem 5.5 in the context of Banach spaces. The existence of topological complements to $\mathcal{N}(T)$ and $\overline{\mathcal{R}(T)}$ still has to be assumed even if we have that T is a bounded or a densely defined closed linear operator (in other words, the existence of a generalized inverse in the sense of (5.5) (or the existence of a topological generalized inverse in the sense of Definition 2.9) is not guaranteed). In a Banach space, a closed subspace need not have a topological complement; the fact that the closed subspace happens to be the null space of a closed (or bounded) linear operator, or the closure of the range of such an operator, does not contribute to the existence of topological complements. This is true even if $\mathcal{R}(T)$ itself is a closed subspace. However, in this latter case, the existence of the generalized inverse $T^\dagger$ implies that $T^\dagger$ is bounded.

Theorem 5.6. (a) Let X and Y be Banach spaces and let T be either a bounded linear operator on X or a densely defined closed linear operator with $\mathcal{D}(T) = X$, and let $\mathcal{R}(T)$ be in Y. Suppose that $\mathcal{N}(T)$ has a topological complement in X and $\overline{\mathcal{R}(T)}$ has a topological complement in Y, say

$$X = \mathcal{N}(T) \oplus \mathcal{M} \quad \text{and} \quad Y = \overline{\mathcal{R}(T)} \oplus \mathcal{S} .$$

Let P denote the projector of X onto $\mathcal{N}(T)$ and let Q denote the projector of Y onto $\overline{\mathcal{R}(T)}$. Then T has a (unique) topological generalized inverse $T^\dagger = T^\dagger_{P,Q}$ (relative to the choice of P and Q, or equivalently of the complements $\mathcal{M}$ and $\mathcal{S}$), which satisfies (5.5).

(b) Under the assumptions of part (a), $T^\dagger$ is bounded if and only if $\mathcal{R}(T)$ is closed in Y.

PROOF. We only need to prove (b). From the closed graph theorem it follows that $(T|\mathcal{M})^{-1}$ is continuous if and only if the range of T is closed. Thus $T^\dagger$ is continuous if and only if $\mathcal{R}(T)$ is closed, in which case $\mathcal{D}(T^\dagger) = \mathcal{R}(T) \oplus \mathcal{S} = Y$, and $TT^\dagger = Q$ on all of Y.

Thus if $\mathcal{R}(T)$ is not closed, then $T^\dagger$ will be an unbounded linear operator defined on a (proper) dense subspace of Y. ∎

Suppose now that new projectors P', Q' are chosen, corresponding to new complements $\mathcal{M}'$, $\mathcal{S}'$ of $\mathcal{N}(T)$ and $\mathcal{R}(T)$, respectively. Let $T^{\dagger}_{P',Q'}$ be the generalized inverse of T, relative to P', Q'. Then the restrictions of $T^{\dagger}_{P',Q'}$ and $T^{\dagger}_{P,Q}$ to $\mathcal{R}(T)$ are related as follows

$$T^{\dagger}_{P',Q'} = (2I - T^{\dagger}_{P,Q}T - P')T^{\dagger}_{P,Q}(I - TT^{\dagger}_{P,Q} + Q')$$

$$= (I - P')T^{\dagger}_{P,Q} Q' .$$

For an application of this relation and for some connections between generalized inverses and what might be called regularizing abstract spline functions, see Nashed [2], where other applications to regularization problems are also given.

5.3. Generalized Inverses in Hilbert Spaces

We now consider the important cases when X and Y are Hilbert spaces, and $T: X \to Y$ is either a bounded linear operator on X or a densely defined closed linear operator with $\overline{\mathcal{D}(T)} = X$. In either case $\mathcal{N}(T)$ is closed. Thus $\mathcal{N}(T)$ and $\mathcal{R}(T)$ have topological complements since every closed subspace of a Hilbert space has a topological complement (in fact this property characterizes Hilbert spaces). Thus the assumptions of Theorem 5.6 are satisfied. In particular we may choose $\mathcal{M} = \mathcal{N}(T)^{\perp}$ and $\mathcal{S} = \mathcal{R}(T)^{\perp}$. This choice implies that the corresponding projectors P and Q are self-adjoint (equivalently orthogonal). Conversely the choice of self-adjoint projectors P and Q implies that the induced topological complements $\mathcal{M}$ and $\mathcal{S}$ are orthogonal to $\mathcal{N}(T)$ and $\mathcal{R}(T)$ respectively. The generalized inverse corresponding to the decompositions:

$$(5.6) \quad \begin{cases} X = \overline{\mathcal{D}(T)} = \mathcal{N}(T) \oplus \mathcal{N}(T)^{\perp} = \mathcal{N}(T) \oplus \overline{\mathcal{R}(T^*)} \\ Y = \overline{\mathcal{R}(T)} \oplus \mathcal{R}(T)^{\perp} = \overline{\mathcal{R}(T)} \oplus N(T^*) \end{cases}$$

is the operator version of the Moore-Penrose inverse and will be called the orthogonal generalized inverse, denoted by $T^{\dagger}$.

From Theorem 5.6 it also follows that $T^{\dagger}$ is bounded if and only if $\mathcal{R}(T)$ is closed.

It follows immediately that the generalized inverse of a closed densely defined linear operator is a closed operator. $T^{\dagger}$ may be also defined by specifying its graph. To this end, we denote by $G(T)$ the graph of T, i.e. $G(T)$ is the subspace $\{(x,Tx) : x \in \mathcal{D}(T)\}$ of $X \times Y$. Then $G(T^{*}) = \{(u,v) \in Y \times X : (-v, u) \in G(T)^{\perp}\}$ where the orthogonal complement is, of course, in $X \times Y$. Let

$$G_X = G(T) \cap [X \times \{0\}], \quad G_Y = G(-T^{*}) \cap [\{0\} \times Y] .$$

Clearly G_X and G_Y are closed subspaces of $X \times Y$ which are isomorphic with $\mathcal{N}(T)$ and $\mathcal{N}(T^{*})$, respectively. Let

$$G_0 = G(T) \cap G_X^{\perp} \text{ and } G_0^{*} = G(T^{*}) \cap G_Y^{\perp} .$$

The closed subspace $G_Y \dotplus G_0$ is the (inverse) graph of the linear operator $T^{\dagger}$.

Theorem 5.7. Each of the following set of conditions characterizes the orthogonal generalized inverse $T^{\dagger}$ of a densely defined closed linear operator with domain $\mathcal{D}(T)$ in the Hilbert space $X = \overline{\mathcal{D}(T)}$ and range in the Hilbert space Y. Throughout $\mathcal{D}(T^{\dagger}) := \mathcal{R}(T) + \mathcal{R}(T)^{\perp}$, and $\mathcal{C}(T) = \mathcal{D}(T) \cap \mathcal{N}(T)^{\perp}$

(a) $T^{\dagger}$ is the unique linear extension of $(T|\mathcal{C}(T))^{-1}$ to $\mathcal{R}(T) + \mathcal{R}(T)^{\perp}$ so that $\mathcal{N}(T^{\dagger}) = \mathcal{N}(T^{*}) = \mathcal{R}(T)^{\perp}$.

(b)
$$\begin{cases} T^{\dagger}TT^{\dagger} = T^{\dagger} \text{ on } \mathcal{D}(T^{\dagger}) , \\ TT^{\dagger} = P_{\overline{\mathcal{R}(T)}} | \mathcal{D}(T^{\dagger}) , \\ T^{\dagger}T = P_{\mathcal{N}(T)^{\perp}} | \mathcal{D}(T) , \end{cases}$$

where $P_{\mathcal{M}}$ denotes the orthogonal projector onto the closed subspace $\mathcal{M}$.

(c) $$\begin{cases} TT^{\dagger}T = T \text{ on } \mathcal{D}(T) \\ T^{\dagger}TT^{\dagger} = T^{\dagger} \text{ on } \mathcal{D}(T^{\dagger}) \\ T^{\dagger}T \text{ and } TT^{\dagger} \text{ are symmetric operators} \end{cases}$$

(d) For $y \in \mathcal{D}(T^{\dagger})$, $T^{\dagger}y$ is the unique solution of minimal norm of the equation $Tx = P_{\overline{\mathcal{R}(T)}}\, y$.

(e) For $y \in \mathcal{D}(T^{\dagger})$, $T^{\dagger}y$ is the unique extremal (least-squares) solution of minimal norm of the equation $Tx = y$.

(f) For $y \in \mathcal{D}(T^{\dagger})$, $T^{\dagger}y$ is the unique solution of minimal norm of the equation $T^{*}(Tx - y) = 0$.

(g) The same as (d), (e), or (f) with "of minimal norm" replaced by "in $\mathcal{N}(T)^{\perp}$."

(h) $T^{\dagger}$ is the linear operator whose (inverse) graph is the closed subspace $G_Y + G_0$, where G_Y and G_0 are as defined above.

We note that $T^{\dagger}T$ and $TT^{\dagger}$ are densely defined and symmetric but not closed. In fact,

$T^{\dagger}T$ is closed $\Longleftrightarrow$ T is bounded

$TT^{\dagger}$ is closed $\Longleftrightarrow$ $T^{\dagger}$ is bounded (equivalently $\mathcal{R}(T)$ is closed).

Thus the generalized inverse of a bounded linear operator with closed range in Hilbert space can be characterized, as in the case of matrices, as the unique solution for the four well-known Moore-Penrose equations:

(1) $TT^{\dagger}T = T$

(2) $T^{\dagger}TT^{\dagger} = T^{\dagger}$

(3) $(TT^{\dagger})^{*} = TT^{\dagger}$

(4) $(T^{\dagger}T)^{*} = T^{\dagger}T$.

It is common to call an inner inverse satisfying (1) and (3) a $\{1,3\}$-inverse and to denote, say, the set of all $\{1,3\}$-inverses of T by $T\{1,3\}$.

5.4. Other Definitions of Generalized Inverses of Arbitrary Linear Operators in Hilbert Spaces

A. Let X and Y be two Hilbert spaces. Tseng [1, pp. 431-432] defined a generalized inverse of a densely defined linear operator $T: \mathcal{D}(T) \to Y$, where $\overline{\mathcal{D}(T)} = X$, as a linear operator T^g with $\overline{\mathcal{D}(T^g)} = Y$ such that

$$\mathcal{R}(T) \subset \mathcal{D}(T^g), \quad \mathcal{R}(T^g) \subset \mathcal{D}(T) ,$$

$$TT^g = P_{\overline{\mathcal{R}(T)}} \text{ on } \mathcal{D}(T^g) \text{ and } T^gT = P_{\overline{\mathcal{R}(T^g)}} \text{ on } \mathcal{D}(T),$$

where the projectors here are orthogonal. Note that this definition is symmetric in T and T^g, so T is a generalized inverse of T^g .

Let $\mathcal{L}$ be a subspace of a Hilbert space $\mathcal{H}$, and $\mathcal{S}$ a subspace of $\mathcal{L}$, neither $\mathcal{L}$ nor $\mathcal{S}$ is assumed to be closed. Tseng called $\mathcal{L}$ *decomposable with respect to* $\mathcal{S}$ if for each $x \in \mathcal{L}$, the *orthogonal* projection of x onto $\overline{\mathcal{S}}$ is in $\mathcal{S}$. It is easy to show that this is equivalent to the *direct orthogonal sum* decomposition:

$$\mathcal{L} = \mathcal{S} \oplus^{\perp} (\mathcal{L} \cap \mathcal{S}^{\perp}) .$$

Theorem 5.8. An operator $T: X \to Y$ $(\overline{\mathcal{D}(T)} = X)$ has a (Tseng) generalized inverse if and only if $\mathcal{D}(T)$ is *decomposable with respect to* $\mathcal{N}(T)$. In this case T has a unique *maximal generalized inverse* T^g_m for which $\mathcal{D}(T^g_m)$ is maximal,

$$\mathcal{D}(T^g_m) = \mathcal{R}(T) \oplus \mathcal{R}(T)^{\perp}, \quad \text{and} \quad \mathcal{N}(T^g_m) = \mathcal{R}(T)^{\perp} .$$

Any other generalized inverse T^g is a restriction of T^g_m. Furthermore T^g_m is the only generalized inverse possessing a closed null space.

This theorem was given by Tseng [1], [2]. Arghiriade [1] adopted the definition of Tseng except that he freed the definition of the generalized inverse from the restrictions relative to the domain of $\mathcal{D}(T)$ and $\mathcal{D}(T^g)$, and showed that Theorem 5.8 is also valid for an operator for which $\mathcal{D}(T)$ is not necessarily dense in X; in this case the second part of Tseng's theorem, as given above, has to be modified as follows: Among generalized inverses with dense domain, T^g_m is the only generalized inverse with a closed null space.

We note that if an operator T is decomposable with respect to $\mathcal{N}(T)$ in the sense of Tseng, then T is also domain-decomposable in our sense (Definition 2. 4) but not conversely. It is easy to see that Theorem 5.8 and its modified version when $\mathcal{D}(T)$ is not necessarily dense, are special cases of the theory of left-topological and right-topological inner inverses and their variants for not necessarily densely defined generalized inverses (see Remark 2.17). The maximal generalized inverse in Tseng's definition is a special case of topological generalized inverse in our sense.

B. An elegant development of some properties of generalized inverses of closed operators appears as a part of a classic paper of Hestenes [1]. A closed linear operator T on a dense domain $\mathcal{D}(T)$ in a Hilbert space establishes a one-to-one correspondence between its carrier $\mathcal{C}$ and its range. The linear extension of $(T|\mathcal{C})^{-1}$ to $\mathcal{R}(T) + \mathcal{R}(T)^{\perp}$ so that its null space is $\mathcal{R}(T)^{\perp}$ is called by Hestenes the <u>reciprocal</u> of T (which is of course also the generalized inverse; see Theorem 5.7). We have $\mathcal{C}(T^{\dagger}) = \mathcal{R}(T)$ and $\mathcal{R}(T^{\dagger}) = \mathcal{C}(T)$.

A definition of the generalized inverse which may appear to be suitable for "arbitrary" linear operators in Hilbert spaces is implicit in one section of this paper by Hestenes

[1]. For $T: X \to Y$, "define" an extension $\hat{T}$ with

$$(5.7) \qquad \begin{cases} \mathcal{D}(\hat{T}) = \mathcal{D}(T) + \overline{\mathcal{N}(T)} \\ \mathcal{N}(\hat{T}) = \overline{\mathcal{N}(T)} \ . \end{cases}$$

Since $\mathcal{N}(\hat{T})$ is closed, $\mathcal{D}(\hat{T})$ is decomposable:

$$\mathcal{D}(\hat{T}) = \mathcal{N}(\hat{T}) \oplus C(\hat{T}) = \mathcal{N}(\hat{T}) \oplus C(T) \ .$$

Hestenes then identified the generalized inverse of T with that of $\hat{T}$. (Of course $\hat{T}$ coincides with T if $\mathcal{N}(T)$ is closed.) However it should be pointed out that $\hat{T}$ is not well defined by (5.7) if

$$\mathcal{D}(T) \cap \overline{\mathcal{N}(T)} \neq \mathcal{N}(T) \quad \text{and} \quad \mathcal{N}(\hat{T}) \neq \mathcal{D}(\hat{T}),$$

which is the only case that calls for the use of $\hat{T}$, since otherwise $\mathcal{D}(T)$ is decomposable with respect to $\mathcal{N}(T)$ or $\hat{T}$ is identically zero. To show this, we note that by (5.7) there exist x_0 and x_1 such that $x_0 \in \mathcal{D}(T) \cap \overline{\mathcal{N}(T)}$, $x_0 \notin \mathcal{N}(T)$ and $x_1 \in \mathcal{D}(T)$, $x_1 \notin \overline{\mathcal{N}(T)}$. Then

$$\hat{T}(x_0 + x_1) = \hat{T}x_1 \quad \text{since} \quad x_0 \in \mathcal{N}(\hat{T}) \ ,$$

while on the other hand since $x_0, x_1 \in \mathcal{D}(T)$

$$\hat{T}(x_0 + x_1) = T(x_0 + x_1) \neq Tx_1 \quad \text{since} \quad x_0 \notin \mathcal{N}(T) \ .$$

C. Another function analytic approach to generalized inverses for arbitrary linear operators between Hilbert spaces was investigated by Erdelyi [1], where the generalized inverse of $T: H_1 \to H_2$ is defined by specifying its graph. $T^\dagger: \mathcal{D}(T^\dagger) \subset \mathcal{H}_2 \to \mathcal{H}_1$, where $\mathcal{D}(T^\dagger) = T[C(T)] \oplus T[C(T)]^\perp$, is the linear operator determined by the graph:

$$G^{-1}(T^{\dagger}) = \{(x, \ Tx + z): \ x \in \mathcal{C}(T), \ z \in (T[\mathcal{C}(T)])^{\perp}\} .$$

It is easy to show, however, that Erdelyi's definition of $T^{\dagger}$ is the same as the (Tseng) G.I. of a suitable restriction of T. Let T_0 be the restriction of T defined by

$$\mathcal{D}(T_0) = \mathcal{N}(T) \oplus^{\perp} \mathcal{C}(T), \ \mathcal{N}(T_0) = \mathcal{N}(T) .$$

The (Erdelyi) G.I. of T is the same as the (Tseng) G.I. of T_0, the latter G.I. exists since T_0 has a decomposable domain. Note that $T_0^{\dagger}$ is densely defined.

Since domain decomposability (see (2.6)) is a necessary and sufficient condition for the existence of the generalized inverse in the sense of Tseng, the essence of Erdelyi's approach is to provide (when $T^{\dagger}$ does not exist in the sense of Tseng) an operator which reduces to $T^{\dagger}$ when the latter exists and which still has some properties of $T^{\dagger}$, except for the serious drawback that $\mathcal{R}(T) \not\subset \mathcal{D}(T_0^{\dagger})$. Note also that this is a special case of Remark 2.17 where P is the orthogonal projector of $\mathcal{H}_1$ onto $\mathcal{N}(T)$. The following properties of $T_0^{\dagger}$ can be easily deduced:

$$\mathcal{D}(T_0^{\dagger}) = T[\mathcal{C}(T)] \oplus^{\perp} (T[\mathcal{C}(T)])^{\perp}$$

$$\mathcal{R}(T_0^{\dagger}) = \mathcal{C}(T), \quad \overline{\mathcal{R}(T_0^{\dagger})} = \mathcal{N}(T)^{\perp}$$

$$TT_0^{\dagger}y = P_{\overline{\mathcal{R}(T)}}\, y \quad \text{for all} \quad y \in \mathcal{D}(T_0^{\dagger})$$

$$T_0^{\dagger}Tx = P_{\overline{\mathcal{R}(T_0^{\dagger})}}\, x \quad \text{for all} \quad x \in \mathcal{D}(T_0) .$$

Hence, $TT_0^{\dagger}T = T$ restricted to $\mathcal{D}(T_0)$ and $T_0^{\dagger}TT_0^{\dagger} = T_0^{\dagger}$ restricted to $\mathcal{D}(T_0^{\dagger})$.

Also, $x_0 = T_0^{\dagger}y$ is the minimal extremal solution of the equation $Tx = y$ for $y \in T[\mathcal{C}(T)] \oplus \mathcal{R}(T)^{\perp} \subset \mathcal{D}(T_0^{\dagger})$. Thus the restriction of $T_0^{\dagger}$ to $T[\mathcal{C}(T)] \oplus \mathcal{R}(T)^{\perp}$ gives the metric generalized inverse in the Hilbert space case (see Section 3).

Other properties of $T_0^\dagger$ include:

$$\mathcal{D}((T_0^\dagger)_0^\dagger) = \overline{\mathcal{N}(T)} \oplus^\perp \mathcal{C}(T) \supset \mathcal{D}(T_0)$$

$$\mathcal{R}((T_0^\dagger)_0^\dagger) = T[\mathcal{C}(T)]$$

$$\mathcal{N}((T_0^\dagger)_0^\dagger) = \overline{\mathcal{N}(T)}$$

$$T \subset (T_0^\dagger)_0^\dagger \quad \text{if } \mathcal{D}(T) \text{ is decomposable}$$

$$T = (T_0^\dagger)_0^\dagger \iff \mathcal{N}(T) \text{ is closed.}$$

D. We remark that our approach to generalized inverses in topological vector spaces when specialized to the case of (not necessarily complete) inner product spaces contains the analog of the Moore-Penrose inverse. Let $\mathcal{E}_1$ and $\mathcal{E}_2$ be inner product spaces and let $T: \mathcal{E}_1 \to \mathcal{E}_2$. T is said to have an (orthogonal) generalized inverse S with domain $\mathcal{D}(S) \subset \mathcal{E}_2$ if

$$\begin{cases} TST = T \text{ on } \mathcal{D}(T), \quad STS = S \text{ on } \mathcal{D}(S) \\ \langle STx, u\rangle = \langle x, STu\rangle \quad \text{for } x, u \in \mathcal{D}(ST) \\ \langle TSy, v\rangle = \langle y, TSv\rangle \quad \text{for } y, v \in \mathcal{D}(TS) \end{cases}.$$

The operator T has a generalized inverse S if and only if

$$\mathcal{D}(T) = \mathcal{N}(T) \oplus (\mathcal{D}(T) \cap \mathcal{N}(T)^\perp) .$$

Then T has a unique maximal generalized inverse $T^\dagger$ and all other generalized inverses are restrictions of $T^\dagger$.

E. Other approaches to generalized inverses in Hilbert spaces, which have been proposed in the literature, lead to special cases of the results presented in Sections 2 and 3 (cf. Nashed [1]).

5.5. Generalized Inverses in Mixed Spaces

Another advantage of the approach to generalized inverses of arbitrary linear operators as developed in Sections 2 and 3 is that it provides immediately results on generalized inverses in mixed spaces. For example we may consider a linear operator from a vector space to a Hilbert space, or from a normed space into a Hilbert space, or from a Banach space to a topological vector space, etc. Algebraic, topological, projectional, and extremal properties of generalized inverses in these contexts are immediate byproducts of the approach.

6. Generalized Inverses of Adjoints and Operational Properties

Many operational properties and identities have been developed for generalized inverses of matrices (cf. Albert [1], Ben-Israel and Greville [1], Rao and Mitra [1]). Some of these properties carry over almost verbatim to the context of operators. In fact, operational properties of G.I.'s of bounded linear operators with closed range are almost identical with those for matrices (cf. Nashed [1]). In contrast, many properties of G.I. with nonclosed range as well as operators in the general context of topological vector spaces must be carefully treated. These operators do not necessarily enjoy all the properties of G.I. of matrices, although many of the properties and identities are retained in a weak version. The proofs also become more sophisticated and are not merely translations of the matrix arguments into the language of operator theory.

In this section we examine properties of the generalized inverse of the adjoints.

Proposition 6.1. Under the assumptions of Theorem 2.10, $(T^{\dagger}_{PQ})^{\dagger}_{I-Q,\,I-P}$ exists and is equal to T.

PROOF. We have

$$\mathcal{D}(T) = \mathcal{N}(T) \oplus \mathcal{C}_P(T)$$

$$\mathcal{Y} = \overline{\mathcal{R}(T)} \oplus \mathcal{S} = \overline{\mathcal{R}(T)} \oplus \mathcal{N}(Q)$$

$$\mathcal{N}(T^\dagger) = \mathcal{N}(Q) .$$

Since $\mathcal{N}(T^\dagger)$ is closed, it follows that $T^\dagger$ is domain decomposable with respect to Q. The desired result then follows from Theorem 2.10 using the indicated decompositions. ∎

We now consider generalized inverses of adjoints in the framework of Theorem 5.1 in order to avoid excessive technicalities.

Theorem 6.2. Let X and Y be Hausdorff and locally convex, and let $T: X \to Y$ be linear and continuous. Let X' and Y' be the duals of X and Y supplied with the weak topologies $\sigma(X', X)$ and $\sigma(Y', Y)$, and let T' be the transpose of T.

(a) If $T^\dagger_{PQ}$ exists, then $(T')^\dagger_{I-Q', I-P'}$ exists and $(T')^\dagger = (T^\dagger)'$.

(b) If X and Y are Fréchet spaces (complete, metrizable, and locally convex) and if $(T')^\dagger: X' \to Y'$ exists, then $T^\dagger$ exists.

PROOF. Suppose $T^\dagger$ exists on all of Y. Then $\mathcal{R}(T)$ is closed since Y is Hausdorff. Thus $\mathcal{R}(T)$ is closed for the topology $\sigma(Y, Y')$. (See Taylor [1], 3.81-C].) From this it follows that T' is a linear topological homomorphism of Y' onto $T'(Y')$ for the weak topologies $\sigma(Y', Y)$ and $\sigma(X', X)$. (See Bourbaki [1], Chapter IV, p. 101].)

We must show that $\mathcal{R}(T')$ has a topological supplement in X' and that $\mathcal{N}(T')$ has a topological supplement in Y'. Now since T is continuous for the initial topologies on X

and Y, T is continuous for the topologies $\sigma(X, X')$ and $\sigma(Y, Y')$, and T' is continuous for the weak topologies on Y' and X'. (See Bourbaki [1], Chapter IV, p. 103].) Also we have that $\mathcal{N}(T') = \mathcal{R}(T)^{\circ} = \{y' \in Y' | y'(Tx) = 0$ for all $x \in X\}$. Similarly we have that Q' is continuous for $\sigma(Y', Y)$, $\mathcal{N}(Q') = \mathcal{R}(Q)^{\circ} = \mathcal{R}(T)^{\circ}$, and $\mathcal{N}(I-Q') = \mathcal{R}(I-Q)^{\circ} = \mathcal{S}^{\circ}$. Thus I-Q' is a continuous idempotent whose range is $\mathcal{R}(T)^{\circ} = \mathcal{N}(T')$.

Similarly, since $(T')' = T$, we have that $\mathcal{N}(T) = \mathcal{R}(T')^{\circ}$, hence $\mathcal{N}(T)^{\circ} = \mathcal{R}(T')^{\circ\circ}$. But this is just the closure of $\mathcal{R}(T')$ for the topology $\sigma(X', X)$ by [Bourbaki, Chapter IV, p. 52]. Since T is a homomorphism, it is also a homomorphism for the topologies $\sigma(X, X')$ and $\sigma(Y, Y')$ (Bourbaki [1], Chapter IV, p. 104), hence $\mathcal{R}(T')$ is closed in X' for $\sigma(X', X)$. Thus $\mathcal{R}(T') = \mathcal{N}(T)^{\circ}$. Also P' is continuous for $\sigma(X', X)$, $\mathcal{N}(P') = \mathcal{R}(P)^{\circ} = \mathcal{N}(T)^{\circ}$, and $\mathcal{N}(I-P') = \mathcal{R}(I-P)^{\circ} = \mathcal{M}^{\circ}$. Thus I-P' is a continuous idempotent whose range is $\mathcal{N}(T)^{\circ} = \mathcal{R}(T')$.

To show that $(T')^{\dagger} = (T^{\dagger})'$, we have simply to note that equations (5.4) are satisfied with T replaced by T', $T^{\dagger}$ replaced by $(T^{\dagger})'$, Q replaced by I-P', and I-P replaced by Q'.

To prove part (b), assume X and Y are Fréchet spaces. We think of the projectors $I-Q': Y' \to \mathcal{N}(T')$ and $I-P': X' \to \mathcal{R}(T')$ as being given. A repetition of the argument above with T' replacing T, etc., shows that $(P')' = P$ is an idempotent which is continuous for the topology $\sigma(X, X')$ and whose range is $\mathcal{N}(T)$. Since the initial topology of a Fréchet space is identical with its Mackey topology (the finest topology on X consistent with the duality between X and X') by Proposition 6 of (Bourbaki [1], Chapter IV, p. 71), it follows that every linear map which is continuous for the topology $\sigma(X, X')$ is also continuous for the initial topology by (Bourbaki [1], Chapter IV, p. 104). Thus $\mathcal{N}(T)$ has a topological complement if X has its initial topology. Similarly $\mathcal{R}(T)$ has a topological complement. If we can show that $(T')' = T$ is continuous for the initial topologies, then we have that $T^{\dagger}$ exists for the initial topologies. But T is continuous since it is continuous for the topologies $\sigma(X, X')$

and $\sigma(Y, Y')$ and since the initial topologies on X and Y are the Mackey topologies.

Corollary 6.3. If X and Y are Hilbert spaces, and if $T^\dagger$ exists, then $(T^*)^\dagger$ exists and is equal to $(T^\dagger)^*$. The generalized inverses here are considered to be the orthogonal generalized inverses.

PROOF. We obtain this result as a corollary, although a direct proof is simple. For $x_0 \in X$, let $Ex_0 \in X'$ be the functional defined by $Ex_0(x) = (x, x_0)$. By the Riesz representation theorem E gives a one-to-one correspondence between X and X'. From the Cauchy-Schwarz inequality it follows that $\|Ex_0\| \leq \|x_0\|$. On the other hand $Ex_0(x_0) = \|x_0\|^2$ gives $\|x_0\| \leq \|Ex_0\|$. Thus E is norm preserving. Let F be the corresponding map between Y and Y'. From the definitions of T^* and T', it follows that $T^* = E^{-1}T'F$.

Thus by identifying X' and X, etc., and applying Theorem 6.2 we see that $(T^*)^\dagger$ exists for the weak topologies, hence $(T^{**})^\dagger$ exists for the weak topologies, hence $(T^*)^\dagger$ exists for the original topologies. The formula $(T^*)^\dagger = (T^\dagger)^*$ is clear. Clearly this is also valid in reflexive spaces.

We now construct an example, which is essentially due to Pietsch [1] of a bounded linear operator T on a non-reflexive Banach space with the property $(T')^\dagger$ exists but $T^\dagger$ does not.

Let m be the Banach space of bounded complex sequences with norm $\|\{x_i\}\| = \sup|x_i|$ and c the subspace of null sequences. It is well known that c is not complemented in m (cf. Phillips [1]). On the other hand, letting i be the inclusion map $i: c \to m$, then the null space of the dual map $i^*: m^* \to c^*$ is complemented in m^* (cf. Day [1], p. 30-31).

Let $K: m \to m \times m$ be defined by

$$(x_1, x_2, x_3, \ldots) \to ((x_1, x_3, x_5, \ldots), (x_2, x_4, x_6, \ldots)).$$

Define $S = LK^{-1}j$, where $j = c \times m \to m \times m$ is the natural imbedding and $Lx = (0, x)$ for $x \in m$. Thus $S \in \mathfrak{L}(c \times m)$

and $\mathcal{N}(S) = \{0\}$. But $\mathcal{R}(S) = (0)\times K^{-1}(c \times m)$ so $\mathcal{R}(S)$ is not complemented.

Now $S^* = j^*(K^{-1})^* L^*$. Then $\mathcal{R}(S^*) = c^* \times m^*$ and $\mathcal{N}(S^*) = c^* \times K^*(c^\perp \times (0))$. Thus $\mathcal{N}(S^*)$ is complemented.

7. Examples and Generalized Inverses for Special Classes of Operators

7.1. The Generalized Inverse of a Projector

Let X be a TVS. Let P be a projector on X. Then since

$$P^3 = P ,$$

$$P^2 = P \text{ is a projector onto } P(X) ,$$

$$I - P^2 \text{ is a projector onto } \mathcal{N}(P) ,$$

it follows from the uniqueness of the generalized inverse that

$$P^\dagger_{I-P,\,P} = P .$$

In particular if P is an orthogonal projector in Hilbert space, then P is its own Moore-Penrose inverse. Thus generalized inversion does to a projector what inversion does to the identity map.

7.2. The Generalized Inverse of a Continuous Linear Functional

Let f be a continuous nonzero linear functional on the topological space $\mathfrak{X}$. Then $\mathcal{V} = \{x \in X \mid f(x) = 0\}$ is a closed hyperplane in X, (i.e., $\mathcal{V}$ is a closed subspace of codimention one), f is a topological homomorphism, and $X = \mathcal{V} \oplus sp\{x_0\}$ where $f(x_0) = 1$. Thus if x is written in the form $s = h + \lambda x_0$ where $h \in \mathcal{V}$ and λ is a scalar, we see that $f(x) = \lambda$. The choice of x_0 determines a projector of X onto $\mathcal{N}$, and $f^{\dagger}$ relative to this projector is the map from the scalar field into X taking λ onto λx_0. Note that since f is onto, the projector Q onto $\mathcal{R}(f)$ does not play a role in this example.

7.3. Dependence of Generalized Inverses on the Given Projectors: An Example

A specific example will illustrate the way in which the generalized inverse is dependent upon the given projectors. Let $\mathcal{D}$ be the space of complex valued functions of n real variables which are infinitely differentiable and which are of compact support. A fundamental system of neighborhoods of $\{0\}$ is defined as follows: Let $\Omega = \{\emptyset, \Omega_1, \Omega_2, \ldots\}$ be an infinite sequence of open sets such that $\overline{\Omega}_{\nu-1} \subset \Omega_\nu$ and such that every compact set K of R^n is contained in some Ω_ν. Let $\{\varepsilon_\nu\}$ be a decreasing sequence of positive numbers converging to 0, and let $\{m_\nu\}$ be an increasing sequence of integers tending to $+\infty$. A fundamental neighborhood of $\{0\}$ is then taken to be $V(\{m_\nu\}, \{\varepsilon_\nu\}, \{\Omega_\nu\}) = \{\varphi \in \mathcal{D} \,|\, x \notin \Omega_\nu \Longrightarrow |D^P\varphi(x)| \leq \varepsilon_\nu$ if $|P| \leq m_\nu$ for all $\nu\}$ where $P = (P_1, \ldots, P_n)$ is an n-tuple of integers, $|P| = P_1 + P_2 + \ldots + P_n$, and $D^P = \dfrac{\partial^{|P|}}{\partial x_1^{P_1} \partial x_2^{P_2} \cdots \partial x_n^{P_n}}$. Letting $\{m_\nu\}$, $\{\varepsilon_\nu\}$, and $\{\Omega_\nu\}$ vary we obtain a fundamental system of neighborhoods for a locally convex topology with a nondenumerable base. $\mathcal{D}$ is an $\mathcal{LF}$ space. A distribution in the sense of Schwartz is a continuous linear functional on the space $\mathcal{D}$. (See Schwartz [1], Chapter III].)

In particular we take $n = 1$ and consider the Dirac measure as a distribution: $\delta \cdot \varphi = \varphi(0)$ for $\varphi \in \mathcal{D}$. If we select a function ψ in $\mathcal{D}$ that has the value 1 at $x = 0$, we can write any φ in $\mathcal{D}$ as the sum of a function in $\mathcal{D}$ that has the value zero at $x = 0$ and a function which is some multiple of ψ. Two such possible choices of ψ are indicated in Fig. 7.1. The generalized inverse of δ corresponding to the first choice takes a scalar λ onto $\lambda\psi_1$;

for the second choice each scalar λ is mapped into $\lambda\psi_2$.

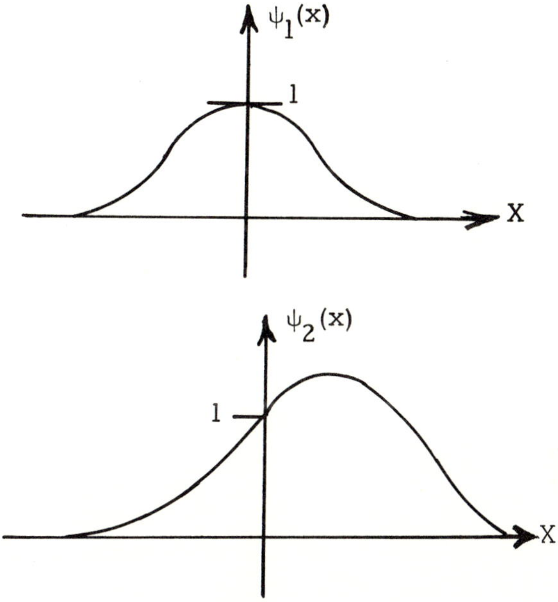

Figure 7.1

7.4. The Generalized Inverse of the Differentiation Operator for Distributions

Let $\mathcal{D}$ be the space defined above. $\mathcal{D}'$, the space of distributions, is the strong dual of $\mathcal{D}$. Moreover $\mathcal{D}$ is also the strong dual of $\mathcal{D}'$ (see Schwartz [1], Theorem XIV, Chapter III). For a distribution T, $\partial T/\partial x_1$ is defined by the statement

$$\frac{\partial T}{\partial x_1} \cdot \varphi = (-1)T \cdot \frac{\partial \varphi}{\partial x_1} .$$

Theorem IV of Chapter II of Schwartz [1] states that every distribution has an infinite number of primitives with respect to x_1, and any two such primitives differ by a distribution which is independent of x_1 . A distribution S is said to be independent of x_1 if $S \cdot \phi(x_1 + h, x_2, \ldots, x_n) = S \cdot \phi(x_1, x_2, \ldots, x_n)$ for all h .

We show that the linear operator $\partial/\partial x_1 : \mathcal{D}' \to \mathcal{D}'$ has a generalized inverse; this generalized inverse is a single valued antiderivative operator $\partial/\partial x_1^{\dagger} : \mathcal{D}' \to \mathcal{D}'$.

By the above remark, the operator $\partial/\partial x_1$ is surjective. Also in Theorem XVIII of Chapter II of Schwartz [1] it is shown that $\partial/\partial x_1$ is a topological homomorphism by obtaining a continuous right inverse A_1 . But this also yields the fact that the null space of $\partial/\partial x_1$ has a topological supplement and that $I - A_1 \frac{\partial}{\partial x_1}$ is a projector onto the null space of $\partial/\partial x_1$. (See Bourbaki [1], Chapter I, the proof of Proposition 13].) Thus $\partial/\partial x_1^{\dagger}$ exists, and since A_1 satisfies (5.4) we have $A_1 = \frac{\partial^{\dagger}}{\partial x_1}$.

Specifically, A_1 is defined as follows: Select a function φ_0 in $\mathcal{D}$ for R^1 such that $\int_{-\infty}^{\infty} \varphi_0(x_1)dx_1 = 1$. Then if φ is a function in $\mathcal{D}$ for R^n, let $\lambda_1(x_2, \ldots, x_n) = \int_{-\infty}^{\infty} \varphi(t, x_2, \ldots, x_n)dt$. We see that φ can be written as $\lambda_1(x_2, \ldots, x_n)\varphi_0(x_1) + \tau(x_1, \ldots, x_n)$ where $\int_{-\infty}^{\infty} \tau(t, x_2, \ldots, x_n)dt = 0$. Thus there is a unique $\psi \in \mathcal{D}$ for R^n such that $\partial\psi/\partial x_1 = \tau$; since $\psi(x_1, \ldots, x_n) = \int_{-\infty}^{x_1} \tau(t, x_2, \ldots, x_n)dt$ is the only antiderivative of τ with respect to x_1 which has compact support.

$A_1 : \mathcal{D}' \to \mathcal{D}'$ is defined by the statement

$$A_1(S) \cdot \varphi = -S \cdot \psi .$$

A_1 of course depends on the choice of φ_0 . In particular if $n = 1$ and δ is the Dirac measure, then $A_1(\delta)$ is the distribution defined by $A_1(\delta) \cdot \varphi = -\delta \cdot \psi = \int_{-\infty}^{0} \tau(t)dt$.

7.5. Generalized Inverses of Normal Operators of Finite Descent

Ascent and Descent of a Linear Operator:

Consider a linear operator T with $\mathcal{D}(T) = X$ and $\mathcal{R}(T) \subset X$, where X is an (algebraic) vector space. We define $T^0 = I$. Let $\mathcal{N}(T^k)$ and $\mathcal{R}(T^k)$ be the null space and range respectively of T^k where $k = 0, 1, 2, \ldots$. Then $\{0\} \subset \mathcal{N}(T) \subset \mathcal{N}(T^2) \subset \ldots \subset \mathcal{N}(T^k) \subset \ldots$ and $X \supset \mathcal{R}(T) \supset \mathcal{R}(T^2) \supset \ldots \supset \mathcal{R}(T\;) \supset \ldots$. T is said to have *finite descent* if there is a nonnegative integer β such that $\mathcal{R}(T^{\beta}) = \mathcal{R}(T^{\beta+k})$ for $k = 1, 2, 3, \ldots$. If T has finite descent we call the smallest such integer the *descent* of T; if T does not have finite descent, we say that the descent of T is ∞. Similarly T is said to have *finite ascent* if there is a nonnegative integer β such that $\mathcal{N}(T^{\beta}) = \mathcal{N}(T^{\beta+k})$ for $k = 1, 2, 3, \ldots$. If T has finite ascent, we call the smallest such integer the *ascent* of T; if T does not have finite ascent we say that the ascent of T is ∞. If T has finite ascent α and finite descent δ, then $\alpha = \delta$ and $X = \mathcal{R}(T^{\delta}) \dot{+} \mathcal{N}(T^{\delta})$ (algebraic direct sum). (See Taylor [1], Section 5.41).

Examples. 1. Let $X = \ell^2 = \{\{x_n\} \mid \sum |x_n|^2 < \infty\}$, and define $T: X \to X$ by $(x_1, x_2, \ldots) \to (x_2, x_3, \ldots)$. Then $\alpha = \infty$ and $\delta = 0$.

2. Let X be a normed linear space, and let K be a compact operator taking X into X. Then if $\lambda \neq 0$ the ascent and descent of $\lambda - K$ are finite (and equal).

Theorem 7.1. Let X be a Hilbert space, and let T be a continuous normal operator of finite descent taking X into X. (Normal means that $TT^* = T^*T$.) Then

(a) The orthogonal generalized inverse $T^{\dagger}$ exists on all of X, or equivalently the range of T is a closed subspace,

(b) $TT^{\dagger} = T^{\dagger}T$,

(c) $(T^n)^{\dagger} = (T^{\dagger})^n$.

PROOF. It is known that the ascent of a normal operator is either 0 or 1 (Taylor [1], Theorem 6.2F). Also since the descent is finite we have $\alpha = \delta$. There are two possible cases:

Case 1: $\alpha = \delta = 0$. Here $\mathcal{R}(T) = X$ and $\mathcal{N}(T) = \{0\}$, and the proof of the existence of $T^\dagger$ is trivial.

Case 2: $\alpha = \delta = 1$. In this case we know that $X = \mathcal{R}(T) \dot{+} \mathcal{N}(T)$ (algebraic direct sum). But $\mathcal{N}(T)$ is closed since T is continuous and X is Hausdorff. Thus $X = \mathcal{N}(T) \oplus \mathcal{N}(T)^\perp$ (topological direct sum), and hence if we show that $\mathcal{R}(T) = \mathcal{N}(T)^\perp$ we have that $\mathcal{R}(T)$ is closed, and the existence of $T^\dagger$ follows. To show that $\mathcal{R}(T) = \mathcal{N}(T)^\perp$, note first that $\mathcal{N}(T) = \mathcal{N}(T^*)$ since for a normal operator $\|Tx\| = \|T^*x\|$ for all $x \in X$. Thus $\mathcal{R}(T) \subset \mathcal{N}(T)^\perp$, for if $x \in \mathcal{R}(T)$ (i.e., $x = Ty$ for some y), then $\langle Ty, z\rangle = \langle y, T^*z\rangle = 0$ for all $z \in \mathcal{N}(T) = \mathcal{N}(T^*)$. On the other hand, $\mathcal{N}(T)^\perp \subset \mathcal{R}(T)$, since if $x \in \mathcal{N}(T)^\perp$ we can write $x = Ty + z$ where $z \in \mathcal{N}(T)$ (using the decomposition $X = \mathcal{R}(T) \oplus \mathcal{N}(T)$). Hence if $\langle x, w\rangle = 0$ for all $w \in \mathcal{N}(T)$ then $\langle Ty, w\rangle = -\langle z, w\rangle = 0$ since $\langle Ty, w\rangle = \langle y, T^*w\rangle = 0$ for $w \in \mathcal{N}(T) = \mathcal{N}(T^*)$. If we set $w = z$ we obtain $\|z\| = 0$; hence $x = Ty + z = Ty$, i.e., $x \in \mathcal{R}(T)$. Thus $\mathcal{N}(T)^\perp = \mathcal{R}(T)$ and thus $\mathcal{R}(T)$ is closed. Hence $T^\dagger$ exists.

We next show that T commutes with $T^\dagger$. An examination of equations (5.4) shows that T commutes with $T^\dagger$ if and only if $I - P = Q$. For the orthogonal generalized inverse, P and Q are self-adjoint. We simply need to observe that if P is the orthogonal projection onto $\mathcal{N}(T)$, then $I - P$ is the orthogonal projection onto $\mathcal{N}(T)^\perp = \mathcal{R}(T)$, and the result follows.

To show that $(T^n)^\dagger = (T^\dagger)^n$, we note that equations (5.4) are satisfied with T replaced by T^n and $T^\dagger$ replaced by $(T^\dagger)^n$ since T and $T^\dagger$ commute. We need only note in addition that $(I - P)$ is the orthogonal projection onto $\mathcal{R}(T^n)$ since $\mathcal{R}(T^n) = \mathcal{R}(T)$. The result follows from the uniqueness of the orthogonal generalized inverse.

7.6. Generalized Inverse of $\lambda I - K$

Theorem 7.2. Let X be a Banach space, and let K be a

compact operator taking X into X (i.e., if $\{x_n\}$ is a bounded sequence, then some subsequence of $\{Kx_n\}$ converges). Then if λ is a nonzero scalar, $(\lambda I-K)^{\dagger}$ exists relative to suitable projectors P and Q .

Remark. It is well known that, except for an at most countable set of complex numbers, λ, with 0 as the only possible accumulation point, the operator $\lambda I-K$ has a continuous inverse. In this case $(\lambda I-K)^{-1} = (\lambda I-K)^{\dagger}$. Thus we need only focus attention on the case where λ is a nonzero eigenvalue. The reader can consult Taylor [1] for a discussion of compact operators.

PROOF. Assume that $\lambda \neq 0$ is an eigenvalue. It is well known that the ascent α and descent δ of $\lambda I-K$ are finite (and equal). Let $T = \lambda I-K$. Then X is the algebraic direct sum of $\mathcal{N}(T^{\delta})$ and $\mathcal{R}(T^{\delta})$ where $\mathcal{N}(T^{\delta}) = \{x \mid T^{\delta}x = 0\}$ is a finite dimensional space. Moreover T is completely reduced by this decomposition: T restricted to $\mathcal{R}(T^{\delta})$ is an automorphism, and $T(\mathcal{N}(T^{\delta})) \subset \mathcal{N}(T^{\delta-1})$ where the inclusion is proper. We show that $T^{\dagger}$ exists, with $\mathcal{D}(T^{\dagger}) = X$.

We show that T is a continuous map, $\mathcal{N}(T)$ has a topological supplement, and $\mathcal{R}(T)$ has a topological supplement. The continuity of T is obvious. Next, we make use of the fact that if M is a closed subspace of X such that the dimension of X/M is finite, then every algebraic supplement of M is a topological supplement.

It is well known that the range of each T^k is closed; thus X is the topological direct sum of $\mathcal{R}(T^{\delta})$ and $\mathcal{N}(T^{\delta})$ since $\mathcal{N}(T^{\delta})$ is finite dimensional. To show that $\mathcal{R}(T)$ has a topological supplement, we write $\mathcal{N}(T^{\delta}) = \mathcal{M}_1 + \mathcal{M}_2$ where $\mathcal{M}_1 = \mathcal{R}(T) \cap \mathcal{N}(T^{\delta})$ and where $\mathcal{M}_2$ is a supplement of $\mathcal{M}_1$ in $\mathcal{N}(T^{\delta})$. Then $X = \mathcal{R}(T^{\delta}) + \mathcal{M}_1 + \mathcal{M}_2 = \mathcal{R}(T) + \mathcal{M}_2$ since $\mathcal{R}(T) \supset \mathcal{R}(T^{\delta})$. Furthermore X is the topological direct sum of $\mathcal{R}(T)$ and $\mathcal{M}_2$ since $\mathcal{R}(T)$ is closed and $\mathcal{M}_2$ is finite dimensional. Similarly, since $\mathcal{N}(T) \subset \mathcal{N}(T^{\delta})$, we can write $\mathcal{N}(T^{\delta}) = \mathcal{M}_3 + \mathcal{N}(T)$ for some finite dimensional space $\mathcal{M}_3$. Thus $X = \mathcal{R}(T^{\delta}) \oplus \mathcal{M}_3 \oplus \mathcal{N}(T)$, and $\mathcal{N}(T)$ has a topological supplement. Hence $T^{\dagger}$ exists.

7.7. A Canonical Representation of $(\lambda I - K)^{\dagger}$, $\lambda \neq 0$:

Let T_1 be the restriction of $\lambda I - K$ to $\mathcal{R}(T^{\delta})$. It is known that T_1 is an automorphism of $\mathcal{R}(T^{\delta})$ and that T_1^{-1} is continuous. Let T_2 be the restriction of $\lambda I - K$ to the finite dimensional subspace $\mathcal{N}(T^{\delta})$. Then it follows from the uniqueness of the generalized inverse that $T^{\dagger}x = T_1^{-1}x_1 + T_2^{\dagger}x_2$ where $x = x_1 + x_2$ with $x_1 \in \mathcal{R}(T^{\delta})$ and $x_2 \in \mathcal{N}(T^{\delta})$. We shall describe $T_2^{\dagger}$ explicitly.

We first take a canonical basis for the finite dimensional space $\mathcal{N}(T^{\delta})$ (see Riesz and Nagy [1], Chapter IV, Section 80). Write $\mathcal{N}(T^{\delta}) = \mathcal{N}(T^{\delta-1}) + \mathcal{D}_{\delta-1}$ and let $e_{11}, \ldots, e_{1n_1}$ be a basis for $\mathcal{D}_{\delta-1}$. Write $\mathcal{N}(T^{\delta-1}) = \mathcal{D}_{\delta-2} + T\mathcal{D}_{\delta-1} + \mathcal{N}(T^{\delta-2})$ and let $e_{21}, \ldots, e_{2n_2}$ be a basis for $\mathcal{D}_{\delta-2}$. We continue in this way: $\mathcal{N}(T^{\delta-2}) = \mathcal{D}_{\delta-3} + T\mathcal{D}_{\delta-2} + T^2\mathcal{D}_{\delta-1} + \mathcal{N}(T^{\delta-3})$ where $\mathcal{D}_{\delta-3}$ has a basis $e_{31}, \ldots, e_{3n_3}$ etc. The vectors $T^k e_{ji}$ where $i = 1, 2, \ldots, n_j$ span $T^k\mathcal{D}_j$ (where $k \leq j$) and are linearly independent. We obtain the following decomposition of $\mathcal{N}(T^{\delta})$.

$$\begin{aligned}
\mathcal{N}(T^{\delta}) = {} & \mathcal{D}_{\delta-1} + T\mathcal{D}_{\delta-1} + T^2\mathcal{D}_{\delta-1} + \ldots + T^{\delta-1}\mathcal{D}_{\delta-1} \\
& + \mathcal{D}_{\delta-2} + T\mathcal{D}_{\delta-2} + \ldots + T^{\delta-2}\mathcal{D}_{\delta-2} \\
& \vdots \\
& + \mathcal{D}_1 + T\mathcal{D}_1 \\
& + \mathcal{D}_0 .
\end{aligned}$$

Let $\mathcal{L}_{ji}$ be the space spanned by $e_{ji}, Te_{ji}, \ldots, T^j e_{ji}$. Then $T(\mathcal{L}_{ji}) \subset \mathcal{L}_{ji}$ since $T^{j+1}e_{ji} = 0$. Moreover

$\eta(T^{\delta}) = \sum_{j,i} \mathfrak{L}_{ji}$. Thus the matrix of T_2 with respect to the above basis consists of blocks down the main diagonal and zeros elsewhere, the blocks describing the behavior of T on the $\mathfrak{L}_{ji}$.

Let us order the $\mathfrak{L}_{ji}$ first with respect to j then with respect to i . The matrix of T_2 thus has the form

$$\mathrm{diag}(A_{\delta-1}, A_{\delta-2}, \ldots, A_0)$$

Here A_j describes T on $\mathfrak{L}_{j1} + \ldots + \mathfrak{L}_{jn_j}$, and thus itself consists of n_j diagonal blocks. Each block is a $(j+1) \times (j+1)$ square matrix with ones below the main diagonal and zeros everywhere else, where the images of the base vectors under T_2 appear as the columns and not as the rows. The matrix of T_2 must contain at least one such block of size $\delta \times \delta$ (otherwise δ would not be the descent of T); it need not, however, contain any blocks of any other size. However if the descent of T is 1, then this block consists simply of a single zero.

If we now consider $\eta(T^{\delta})$ as having the canonical inner product given by our canonical basis, we can compute the orthogonal generalized inverse of T_2 . First we note that since the matrix of T_2 consists of diagonal blocks, we have only to compute the generalized inverse of each block. This follows easily from the uniqueness of the generalized inverse; one simply obtains the matrix

$$\mathrm{diag}(A^{\dagger}_{\delta-1}, A^{\dagger}_{\delta-2}, \ldots, A^{\dagger}_0) \ .$$

Since each A_j itself consists of diagonal blocks, we have only to find the generalized inverse of matrices consisting of ones below the main diagonal and zeros elsewhere. It is easily seen that the generalized inverse of a matrix of this form consists of ones above the main diagonal and zeros elsewhere.

<u>Example.</u> Consider the space $\ell^1 = \{\{x_n\} \mid \sum |x_n| < \infty\}$. It is well known that a continuous linear operator A on ℓ^1 into itself can be represented by a matrix (a_{ij}) where $\|A\| = \sup\limits_j (\sum\limits_i |a_{ij}|)$. Conversely if a matrix (a_{ij}) has the property that $\sup\limits_j(\sum\limits_i |a_{ij}|) < \infty$, then (a_{ij}) determines a continuous linear operator on ℓ^1 into itself (Taylor [1], p. 220). Moreover such a matrix represents a compact operator if and only if $\sum\limits_{i=n}^{\infty} |a_{ij}| \to 0$ as $n \to \infty$, uniformly in j (Taylor, p. 278).

Thus consider the operator K defined by the matrix

$$\begin{bmatrix} 1 & & & & & \bigcirc \\ -1 & 1 & & & & \\ & & 1/2 & & & \\ & & & 1/3 & & \\ & & & & 1/4 & \\ & \bigcirc & & & & \ddots \end{bmatrix}$$

Clearly this defines a compact operator K. We first find the eigenvalues. The condition $(\lambda I - K)x = 0$ gives $(\lambda-1)x_1 = 0$, $-x_1 + (\lambda-1)x_2 = 0$, $(\lambda - 1/2)x_2 = 0$, $(\lambda - 1/3)x_3 = 0, \ldots$. Thus the eigenvalues are $1, 1/2, 1/3, \ldots$. We select $\lambda = 1$ and compute $(I - K)^\dagger$. We see that $\eta(T) = \{x \mid x_n = 0 \text{ if } n \neq 2\}$ and $\eta(T^2) = \{x \mid x_n = 0 \text{ if } n \neq 1, 2\} = \eta(T^3)$. We can thus write $\ell^1 = \eta(T^2) \oplus \Re(T^2)$ where $\Re(T^2) = \{x \mid x_1 = x_2 = 0\}$. Clearly $(I - K)^\dagger$ can be represented as

$$\begin{bmatrix} 0 & 1 & & & & \\ 0 & 0 & & & & \\ & & 2 & & & \\ & & & 3/2 & & \\ & & & & 4/3 & \\ & & & & & \ddots \end{bmatrix} .$$

Thus the transformation $(x_1, x_2, x_3, x_4, x_5, \ldots) \to (x_2, 0, 2x_3, 3x_4/2, 4x_5/3, \ldots)$ is a generalized inverse of $I - K$.

8. Generalized Inverses in Various Algebraic Structures

Concepts of generalized inverses make sense in a variety of mathematical structures. In this section we provide a brief exposition of concepts and results in several structures. We make no attempt to generalize or unify known results in the direction of our unified approach, although such extensions are possible in most cases. Our contribution in this section is at most editorial. However, we hope that the reader will note the common thread in concepts of generalized inverses in various structures and the canonical forms of the results and techniques, once the necessary transposition of terminology and concepts is made.

Penrose [1] considered generalized inverses of matrices over the complex field, and over more general rings (the results for rings were not published). Rado [1] extended some results on generalized inverses to matrices over any division ring with an involutory anti-automorphism. Generalized inverse concepts and related results in rings and/or semigroups were considered by von Neumann [1], Drazin [1], Munn and Penrose [1], Kaplansky [1], Foulis [1], Hartwig [1] and others. Set-theoretic and/or categorial settings were considered by Rabson [1], Davis and Robinson [1], Hansen and Robinson [1] and others.

8.1. Generalized Inverses of Morphisms

Let $\mathcal{C}$ be a category. Let X and Y be objects of $\mathcal{C}$. A morphism ϕ from X to Y, written $\phi: X \to Y$, is said to be:

- regular (von Neumann regular or quasi-invertible) if there exists a morphism $\psi: Y \to X$ such that $\phi\psi\phi = \phi$;

- an epimorphism if ϕ is right cancellative with respect to composition of morphisms, i.e.,

$\psi: Y \to Z, \quad \psi': Y \to Z, \quad \text{and} \quad \psi\phi = \psi'\phi \Longrightarrow \psi = \psi'$;

- a monomorphism if ϕ is left cancellative with respect to composition of morphisms, i.e.,

  $\psi: Z \to X, \quad \psi': Z \to X, \quad \text{and} \quad \phi\psi = \phi\psi' \Longrightarrow \psi = \psi'$;

- a retraction if there exists a morphism $\psi: Y \to X$ such that $\phi\psi = 1_Y$, i.e., a retraction is right invertible with respect to composition;

- a coretraction (or section) if there exists a morphism $\psi: Y \to X$ such that $\psi\phi = 1_X$, i.e., a coretraction is left invertible;

- an isomorphism if ϕ is both a retraction and coretraction.

A morphism $\phi: X \to Y$ is said to have a factorization if there exists an object Z, an epimorphism $\phi_e: X \to Z$ and a monorphism $\phi_m: Z \to Y$ such that $\phi = \phi_m\phi_e$. The category $\mathcal{C}$ is said to be a category with factorization if every morphism of $\mathcal{C}$ has a factorization. Most categories which are of interest in algebra are categories with factorization. Objects and morphisms in an abelian category, e.g., module and module-homomorphisms, are of particular interest. For such morphisms, the following conditions are equivalent (cf. MacLane [1, p. 264, Proposition 5.1]):

(i) ϕ is a regular morphism.

(ii) ϕ_e is a retraction, and ϕ_m is a coretraction;

(iii) ker ϕ has a left inverse, and coker ϕ has a right inverse.

If $\phi: X \to Y$ is regular (quasi-invertible), then any morphism $\psi: Y \to X$ such that $\phi\psi\phi = \phi$ is called an inner inverse (also called inside inverse, semi-inverse, partial inverse, 1-inverse, pseudoinverse, among other names). A morphism η is said to be an outer inverse to ϕ if $\eta\phi\eta = \eta$. Not every inner inverse is an outer inverse. However, if ψ' is an inner inverse of ϕ, then $\psi = \psi'\phi\psi'$ is both an inner inverse and outer inverse. That is, if ϕ is regular, then there exists a ψ such that $\phi\psi\phi = \phi$ and $\psi\phi\psi = \psi$. Such a ψ need not be unique. We call any such a ψ a quasi-inverse (or a generalized inverse).

A retraction is an epimorphism, and a right inverse is a quasi-inverse. Also, for an epimorphism, a quasi-inverse is a right inverse, so a regular epimorphism is a retraction, and conversely. Similarly, a regular monomorphism is a coretraction, and conversely.

Thus it is easy to see that, in a category $\mathcal{C}$ with factorization, every morphism of $\mathcal{C}$ is regular if and only if every epimorphism of $\mathcal{C}$ is a retraction and every monomorphism of $\mathcal{C}$ is a coretraction. In other words, in a category with factorization, every morphism is regular if and only if every epimorphism and every monomorphism is regular.

Theorem 8.1. (See Davis and Robinson [1]). Let $\mathcal{C}$ be a category. Let $\phi: X \to Y$ be a morphism of $\mathcal{C}$ with factorization $\phi = \phi_m\phi_e$, $\phi_e: X \to Z$, $\phi_m: Z \to Y$. Suppose there exist objects K, Q and morphisms $i: K \to X$, $\pi: Y \to Q$ such that $\phi_e i: K \to Z$ and $\pi\phi_m: Z \to Q$ are isomorphisms. Then ϕ is regular and there is a unique morphism such that

$$\phi\psi\phi = \phi ,$$

$$\psi\phi\psi = \psi ,$$

$$\psi\phi i = i, \quad \text{and}$$

$$\pi\phi\psi = \pi .$$

Useful characterizations of inner and generalized inverses can be obtained in concrete categories (see, for example, Maclane and Birkhoff [1, p. 64]). For X, an object of a concrete category $\mathcal{C}$, let $U(X)$ denote the underlying set of X. A morphism $i: M \to X$ is said to be an insertion of M in X provided $U(M) \subset U(X)$ and $i : U(M) \to U(X)$ is the natural injection $m \mapsto m$. A morphism $\pi: X \to X/F$ is said to be a congruence of X provided $U(X/F)$ is the quotient set $U(X)/F$ of $U(X)$ modulo an equivalence relation F and $\pi: U(X) \to U(X)/F$ is the natural projection $x \mapsto [x]$, where $[x]$ is the equivalence class that contains x. The category $\mathcal{C}$ is said to be balanced provided every morphism that is both a monomorphism and an epimorphism is also an isomorphism.

If $\phi: X \to Y$ is a morphism of $\mathcal{C}$, then it is well known that the function $\phi: U(X) \to U(Y)$ with equivalence kernel E_ϕ and image $\phi U(X)$ factors into a product of the natural projection ϕ_e of $U(X)$ onto the quotient set $U(X)/E_\phi$, a bijection ϕ_i of $U(X)/E_\phi$ onto $\phi U(X)$, and an injection ϕ_m of $\phi U(X)$ into $U(Y)$. (See, for example, MacLane and Birkhoff [1, p. 23]). We say that the morphism $\phi: X \to Y$ has a concrete factorization provided that there exist objects X/E_ϕ and $\operatorname{Im}\phi$ of $\mathcal{C}$ with underlying sets $U(X/E_\phi) = U(X)/E_\phi$ and $U(\operatorname{Im}\phi) = \phi U(X)$, respectively, such that $\phi_e: X \to X/E_\phi$, $\phi_i : X/E_\phi \to \operatorname{Im}\phi$, and $\phi_m: \operatorname{Im}\phi \to Y$ are morphisms of $\mathcal{C}$. Since ϕ_e is surjective and ϕ_m is injective, it follows that ϕ_e is an epimorphism and ϕ_m is a monomorphism. Furthermore, if $\mathcal{C}$ is balanced, then ϕ_i is an isomorphism. A concrete category in which every morphism $\phi: X \to Y$ has a concrete factorization

$$\phi_m \phi_i \phi_e: X \to X/E_\phi \to \operatorname{Im}\phi \to Y ,$$

is said to be a category with concrete factorization.

Theorem 8.2. (Davis and Robinson [1]). Let $\mathcal{C}$ be a balanced concrete category with concrete factorization. A morphism $\phi: X \to Y$ is regular if and only if there is an insertion $M \to X$ and a congruence $Y \to Y/F$ such that M represents X/E_ϕ and $\operatorname{Im}\phi$ represent Y/F.

Theorem 8.2. Let $\mathcal{C}$ be a balanced concrete category with concrete factorization. A morphism $\phi: X \to Y$ is regular if and only if there is an insertion $M \to X$ and a congruence $Y \to Y/F$ such that M represents X/E_ϕ and $\mathrm{Im}\,\phi$ represent Y/F.

Corollary 8.3. Let $\mathcal{C}$ be a balanced concrete category with concrete factorization. Let $\phi: X \to Y$ be a morphism of $\mathcal{C}$ and suppose $M \to X$ is an insertion of X and $Y \to Y/F$ is a congruence of Y such that M represents X/E_ϕ and $\mathrm{Im}\phi$ represents Y/F. Then there is one and only one morphism $\psi: Y \to X$ such that

$$\phi\psi\phi = \phi, \quad \psi\phi\psi = \psi, \quad U(\mathrm{Im}\,\psi) = U(M), \quad U(Y/E_\psi) = U(Y/F).$$

We now specialize to some categories and morphisms of interest.

Regularity of R-homomorphisms. Let R be an associative ring with identity. Let $\mathcal{C}$ be the category whose objects are (left) R-modules and whose morphisms are the R-homomorphisms between these R-modules. If M, N are objects of this category and if $\phi: M \to N$ is a morphism of this category, then it is well known that ϕ has the factorizations $\phi = \phi_2\phi_1$, $\phi_1: M \to M/\mathrm{Ker}\,\phi$, $\phi_2: M/\mathrm{Ker}\,\phi \to N$ and $\phi = \phi_2'\phi_1'$, $\phi_1': M \to \mathrm{Im}\phi$, $\phi_2': \mathrm{Im}\phi \to N$. The morphism ϕ_1 is the natural projection of M onto $M/\mathrm{Ker}\,\phi$ and ϕ_2' is the natural embedding of $\mathrm{Im}\,\phi$ onto N. Also there is a canonical isomorphism $\theta: M/\mathrm{Ker}\,\phi \to \mathrm{Im}\,\phi$ such that $\phi_1' = \theta\phi_1$ and $\phi_2 = \phi_2'\theta$. Therefore, ϕ is regular if and only if ϕ_1 is a retraction and ϕ_2' is a coretraction. However it is well known that a surjective R-homomorphism $\psi: K \to L$ of R-modules K, L is a retraction iff $\mathrm{Ker}\,\psi$ is a direct summand of K and an injective R-homomorphism $\psi: K \to L$ of R-modules is a coretraction iff $\mathrm{Im}\,\psi$ is a direct summand of L. Thus we have the following result.

Theorem 8.4. Let R be an associative ring with identiy. Let $\phi: M \to N$ be an R-homomorphism of R-modules M and N. Then ϕ is regular if and only if Ker ϕ is a direct summand of M and Im ϕ is a direct summand of M.

Abelian groups. Since every Abelian group can be considered to be a $\mathbb{Z}$-module (the integral group ring) we have the following.

Corollary 8.5. If $\phi: G \to H$ is a homomorphism of Abelian groups then ϕ is regular if and only if Ker ϕ is a direct summand of G and Im ϕ is a direct summand of H.

For homomorphisms of non-Abelian groups, Davis and Robinson [1] show that a homomorphism $\phi: G \to H$ of groups G, H is regular if and only if Ker ϕ is a normal semi-direct summand of G and Im ϕ is a semi-direct summand of H.

Set mappings. We consider now the concrete category whose objects are nonempty sets and whose morphisms are set mappings. This category is balanced and has concrete factorization. Also, insertions are determined by the natural injection of subsets into their supersets and congruences are determined by the natural projections of sets onto quotient sets given by an equivalence relation. Therefore we have

Theorem 8.6. Let $\phi: X \to Y$ be a mapping of sets X, Y. Then ϕ is regular if and only if there exist a subset M of X and an equivalence relation F on Y such that M is a set of representatives for X/E_ϕ and Im ϕ is a set of representatives for Y/F.

Let $\mathcal{C}$ be the concrete category of nonempty sets and nonempty mappings. The following statements are equivalent:

(i) Every morphism $\phi: X \to Y$ of $\mathcal{C}$ is regular.

(ii) The Axiom of Choice.

Regularity of the Composition of Two Morphisms. If α and β are regular morphisms, it does not necessarily follow of course that $\beta\alpha$ is regular. The following proposition asserts the regularity of $\beta\alpha$ under additional conditions. The following facts, which are easy to show, will be needed in the proof:

(i) If $\alpha\beta$ is regular and β is an epimorphism, then α is regular.

(ii) If $\alpha\beta$ is regular and α is a monomorphism, then β is regular.

Proposition 8.7. (see Noll [1]). Let $\mathcal{C}$ be an abelian category. Let α, β be regular morphisms, $\alpha: A \to E$, $\beta: E \to B$, and consider the "cross" diagram:

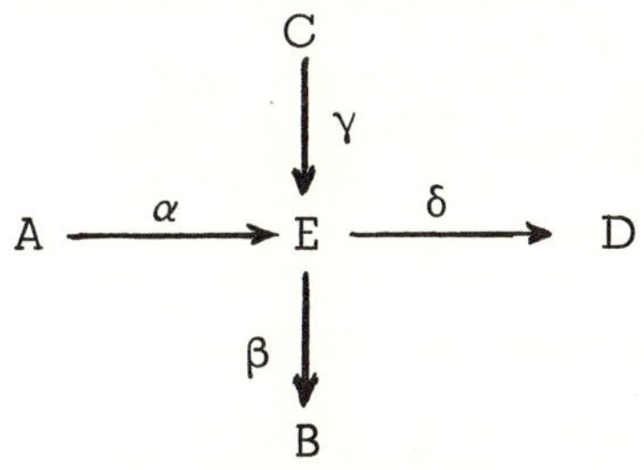

Assume that the row and column are exact, and that $\delta\gamma$ is regular. Then $\beta\alpha$ is also regular.

PROOF. Consider the standard factorizations $\gamma = \gamma_m\gamma_e$ and $\delta = \delta_m\delta_e$. Since $\delta\gamma = (\delta_m\delta_e\gamma_m)\gamma_e$ is regular it follows from (i) that $\delta_m\delta_e\gamma_m = \delta_m(\delta_e\gamma_m)$ is also regular. But then from (ii) $\delta_e\gamma_m$ is also regular. Note that there is not loss of generality if we assume that δ is an epimorphism and γ

a monomorphism since $\operatorname{Im}\gamma = \operatorname{Im}\gamma_m$ and $\operatorname{Ker}\delta = \operatorname{Ker}\delta_e$. In view of the exactness of the row and column of the diagram, we may actually assume that

$$\gamma = \operatorname{Ker}\beta, \qquad \delta = \operatorname{coker}\alpha \ .$$

Now let $\bar{\alpha}$ be an inner inverse of α, so that

$$(1_E - \alpha\bar{\alpha})\alpha = \alpha - \alpha\bar{\alpha}\alpha = 0 \ .$$

It follows that $1_E - \alpha\bar{\alpha}$ annihilates α and hence must factor through $\operatorname{coker}\alpha = \delta$. Thus, the exactness of the column of the diagram is expressed by

(8.1) $$\delta\alpha = 0, \quad 1_E - \alpha\bar{\alpha} = \bar{\delta}\delta$$

where $\bar{\delta}: D \to E$. Similarly, the exactness of the column of the diagram is expressed by

$$\beta\gamma = 0, \quad 1_E - \bar{\beta}\beta = \gamma\bar{\gamma} \ ,$$

where $\bar{\beta}$ is an inner inverse of β and $\bar{\gamma} = E \to C$. Let $\bar{\phi}$ be an inner inverse of $\delta\gamma$, so that

(8.2) $$\delta\gamma\bar{\phi}\delta\gamma = \delta\gamma \ ,$$

and put

$$\bar{\psi} := \bar{\alpha}(1_E - \gamma\bar{\phi}\delta)\bar{\beta} \ .$$

By (8.1) and (8.2) we obtain

$$\beta\alpha\bar{\psi}\beta\alpha = \beta\alpha\bar{\alpha}(1_E - \gamma\bar{\phi}\delta)\bar{\beta}\beta\alpha = \beta(1_E - \bar{\delta}\delta)(1_E - \gamma\bar{\phi}\delta)(1_E - \gamma\bar{\gamma})\alpha$$

$$= \beta(1_E + \bar{\delta}\delta\gamma\bar{\gamma} - \bar{\delta}\delta\gamma\bar{\phi}\delta\gamma\bar{\gamma})\alpha \ .$$

It follows from (8.2) that the last two terms cancel and hence $\bar{\psi}$ is an inner inverse of $\beta\alpha$.

The following result was given by Noll [1] and used in an investigation of annihilators of differential operators.

Theorem 8.8. Consider the "staircase" diagram in which dots denote unnamed objects. Assume that the diagram is commutative, that all rows and columns are exact, that the morphisms indicated by boldface arrows are regular, and that either ξ is an epimorphism or η a monomorphism. Under these conditions if λ is a monomorphism then μ is regular, and if μ is an epimorphism then λ is regular.

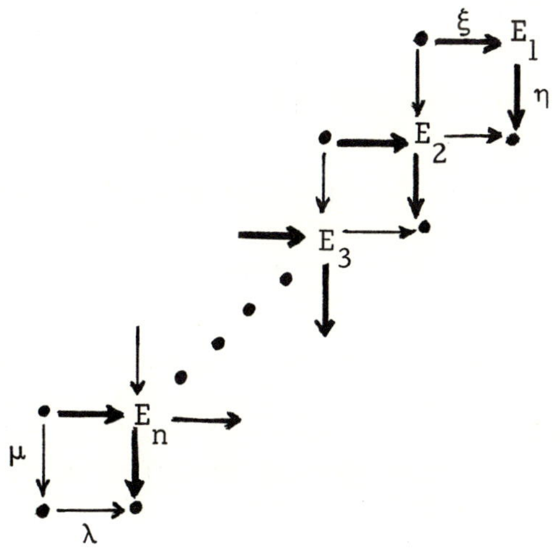

8.2. Algebraic Generalized Inverses for Nonlinear Mappings

Let X and Y be nonempty sets and f a mapping from X to Y. For each $y \in f[X]$ let

$$f^{-1}(y) := \quad \{x \in X: f(x) = y\}$$

and

$$X/f := \{f^{-1}(y): y \in f[X]\} \ .$$

We define the mapping

$$\varphi: X/f \rightarrow f[X]$$

by

$$\varphi(K) = f(x) \quad \text{for} \quad \mathcal{C} \in X/f, \quad x \in \mathcal{C} \ .$$

Obviously the value of $\varphi(\mathcal{C})$ is independent of the particular x in $\mathcal{C}$ which we choose to evaluate it, and φ is one-to-one and onto. Thus we have a mapping:

$$\varphi^{-1}: f[X] \rightarrow X/f \ .$$

Let ψ be a mapping from Y to $f[X]$ with the property

$$\psi(f(x)) = f(x) \quad \text{for any} \quad x \in X \ .$$

Let $\mathcal{M}$ be a subset of X with the property that, for each $K \in X/f$, $\mathcal{M} \cap \mathcal{C}$ consists of exactly one point of X (compare with Proposition (1.1)b. Let $\pi: X \rightarrow X/f$ be defined by $\pi(x) = \mathcal{C}$ where $\mathcal{C}$ is that element of X/f which contains x . Define $\pi^*: X/f \rightarrow X$ by $\pi^*(\mathcal{C}) = \mathcal{M} \cap \mathcal{C}$. Finally let

$$f^{\dagger} := \pi^* \varphi^{-1} \psi \ .$$

Then it is easy to verify that

$$ff^{\dagger}f = f \ ,$$

$$f^{\dagger}ff^{\dagger} = f^{\dagger} \ ,$$

$$ff^{\dagger} = \psi^{*}\psi \ ,$$

and

$$f^{\dagger}f = \pi^{*}\pi \ .$$

The construction of $f^{\dagger}$ and the maps defined above are illustrated in the following figure:

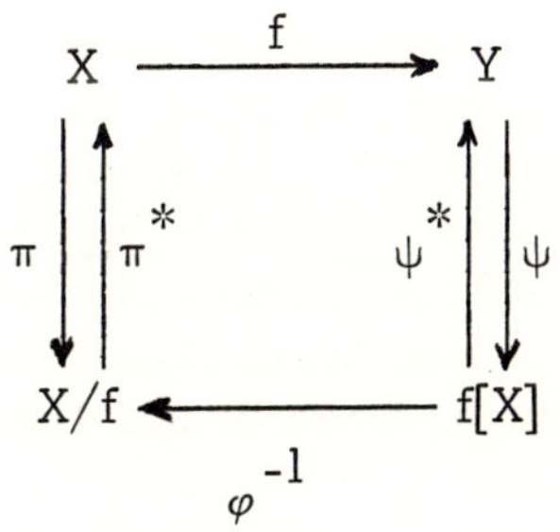

This construction stands merely as a straightforward technical exercise of little substance until the sets are enriched with sufficient structure to permit the selection of useful ψ and π^{*}, as we have seen for example in the case of topological homomorphisms on topological vector spaces, where ψ becomes the projector of Y onto $f[X]$, $\mathcal{M}$ a topological complement of $\mathcal{N}(f)$, ψ^{*} and π^{*} injections, and π the projector of X onto $\mathcal{C}$.

8.3. Generalized Inverses in *-Regular Rings

Von Neumann [1] introduced the concept of a regular ring, i.e. a ring R in which for every $a \in R$ there exists an element $x \in R$ such that $axa = a$. Von Neumann [2], [3] also discussed the relation between regular rings and continuous geometry.

The existence of the Moore-Penrose inverse, inner, outer, and other inverses for elements of a $*$-regular ring R is discussed by Hartwig [1]. In this lengthy paper, Hartwig also generalizes many of the matrix results in the literature of generalized inverses to $*$-regular rings and provides an excellent exposition of several techniques and applications. (The thrust of Hartwig's paper is in the direction of generalized inverses in the ring of 2×2 matrices over a ring R, with particular attention being given to computational formulae, bordered matrices, Schur complements, block-rank formulae and EP elements, etc., which we do not consider here).

An element $a \in R$ is regular iff there exist $a^- \in R$ such that $aa^-a = a$. The element a^- is called an inner or 1-inverse of a. A nonzero element $a \in R$ is called antiregular iff there exists nonzero $\hat{a} \in R$ such that $\hat{a}a\hat{a} = \hat{a}$. The element $\hat{a}$ is called an outer or 2-inverse of a. R is called anti-regular iff every nonzero element is anti-regular. It is well known (Rao and Mitra [1], p. 26), that if a is a regular element then the general solution to $axa = a$ is

$$(8.3) \qquad a^- + h - a^-ahaa^- \quad \text{or} \quad a^- + h - haa^- + k - a^-ak ,$$

where a^- is any particular 1-inverse and h, k are arbitrary. In an anti-regular ring, the general solution to $xax = x$ is $x = f(eaf)^{\wedge}e$, where a, f are arbitrary idempotents and $(\cdot)^{\wedge}$ denotes any 2-inverse.

If a is a nonzero regular element, then a^-aa^- is a nonzero 2-inverse of a; thus every regular ring is anti-regular. The converse is not necessarily true. Let $R = \mathcal{B}(\mathbb{R}^\infty)$ be the ring of all bounded real valued sequences with termwise addition and multiplication. If x_k is a nonzero term in (x_n), then $(0, \ldots, 0, x_k^{-1}, 0, \ldots, 0, \ldots)$ is a 2-inverse of (x_n), while the sequence $(1/n)$ has no bounded 1-inverse.

Every 1-2 inverse of a can be expressed in terms of any particular 1-inverse (Hartwig [1]):

$$a^+ = (a^- + h - a^-ahaa^-)a(a^- + h - a^-ahaa^-), \quad h \text{ arbitrary.}$$

This is not possible in terms of a particular 2-inverse.

Let $R_{n\times n}$ denote the ring of all $n \times n$ matrices over a ring R. It has been shown by von Neumann [1] and by Brown and McCoy [1] that if R is regular (with unity) then $R_{n\times n}$ is also regular (with unity). Hartwig generalized the proof of Brown and McCoy to obtain the following characterizations:

R is regular (with unity) $\Longleftrightarrow$ $R_{n\times n}$ is regular (with unity)

$\Longleftrightarrow$ each $A \in R_{m\times n}$ has a 1-inverse

R is anti-regular (with unity) $\Longleftrightarrow$ $R_{n\times n}$ is anti-regular (with unity)

$\Longleftrightarrow$ each $A \in R_{m\times n}$ has a 2-inverse.

In order to define 1-3 and 1-4 inverses and the analog of a Moore-Penrose inverse we introduce a "star" operation in R which plays a role similar in many ways to the "conjugate transpose" operation in the case of matrices. An involution $*$ of a ring R is an involutory anti-automorphism; i.e.,

$$(a^*)^* = a, \quad (a+b)^* = a^* + b^*, \quad (ab)^* = b^* a^*, \quad a^* = 0 \Longleftrightarrow a = 0 .$$

If $a^* = a$ we call the element Hermitian. An Hermitian idempotent element is called a projector.

A $*$-regular ring is a regular ring with involution $*$ such that $a^* a = 0 \Longleftrightarrow a = 0$. If moreover

$$\sum_{i=0}^{n} a_i^* a_i = 0 \Longrightarrow a_i = 0, \qquad i = 1, \ldots, n, \quad n = 1, 2, \ldots ,$$

then we call R $\sum$-* regular and the involution strong. A central easy result in a $*$-regular ring is the star cancellation law:

(8.4) $$a^* ab = a^* ac \Longrightarrow ab = ac .$$

It turns out that the strong involution partially replaces, through the star cancellation law (8.4), the concept of rank in "rank $A = 0 \iff A \neq 0$".

$\sum$-* regular rings always have zero characteristic. A classic example of such a ring is $C_{n\times n}$, the ring of all $n \times n$ complex matrices.

Let R be a *-regular ring. An element $x \in R$ is called a <u>3-inverse</u> of a iff $(ax)^* = ax$, and $y \in R$ a <u>4-inverse</u> iff $(ya)^* = ya$.

Combinations of the four types of inverses above are defined in the obvious fashion. We shall denote 1-2, 1-3, 1-4 and 1-2-3-4 inverses by $(\cdot)^+$, $(\cdot)^\sim$, $(\cdot)^V$ and $(\cdot)^\dagger$ respectively.

Penrose [1], in his proof of the existence of the generalized inverse used the star cancellation (8.4) and the three obvious equivalent forms of this, namely the "starred" version of (8.4) $baa^* = caa^* \iff ba = ca$ and two similar results with a and a^* interchanged. From these it follows easily that

(i) x is a 1-3 inverse of $a \iff a^*ax = a^*$

(ii) x is a 1-4 inverse of $a \iff aa^*x^* = a$

(iii) x is a 2-3 inverse of $a \iff xx^*a^* = x$

(iv) x is a 2-4 inverse of $a \iff axx^* = x^*$

(v) $a^* = a \implies ax = 0 \iff a^2x = 0$.

The proof of existence of a 1-2-3-4 inverse (or Moore-Penrose inverse) $a^\dagger$ for the elements of a *-regular ring goes back to Kaplansky [1], who proved it at about the same time that Penrose [1] proved the existence for matrices. Kaplansky based his results on the work of Rickart [1] and von Neumann [1], where Moore's original definition is transparent.

To assist in transposition of notions and techniques in matrix theory into ring theory and conversely, we display the following correspondence:

Right coset $aR = \{ax: x \in R\} \longleftrightarrow \mathcal{R}(A) = CS(A) = \{Ax: x \in \mathbb{R}^n\}$,

range or column space

Left coset $Ra = \{xa: x \in R\} \longleftrightarrow \mathcal{R}(A^*) = RS(A) = \{y^* A: y \in \mathbb{R}^m\}$,

row space of A

Right annihilator $a^o = \{x \in R: ax = 0\} \longleftrightarrow \eta(A) = \{x \mid Ax = 0\}$,

null space of A

Left annihilator $^o a = \{x \in R: xa = 0\} \longleftrightarrow \eta(A^*) = \{x \mid A^* x = 0\}$,

null space of A^* .

As in the case of matrices, the proof of existence of $a^\dagger$ in a $*$-regular ring could be based on the existence of 1-3 and 1-4 inverses. First one verifies, using the cancellation law, that

(8.5)
$$\begin{cases} \tilde{a} = (a^* a)^- a^* & \text{is a 1-3 inverse of } a \\ a^\vee = a^*(aa^*)^- & \text{is a 1-4 inverse of } a \end{cases}$$

and then uses these facts to verify that

(8.6)
$$a^\dagger = a^\vee a \tilde{a} = a^*(aa^*)^- a(a^* a)^- a^*$$

is a Moore-Penrose inverse and that $aa^\dagger = e$, $a^\dagger a = f$. The uniqueness follows from the uniqueness of the left and right projectors e and f which correspond to the unique projectors onto the range and row space of A respectively in the case of a matrix. The existence of a 1-3 inverse in a $*$-regular ring is intimately connected with the existence of a unique Hermitian projector and a related direct sum decomposition of R . More precisely:

Let R be a $*$-regular ring and $a \in R$. The following are equivalent:

(i) There exists a 1-3 inverse $\tilde{a}$ of a.

(ii) There exists a unique projector $e(= a\tilde{a})$ such that $aR = eR$, $e = e^2 = e^*$.

(iii) $R = aR \oplus (a^*)^{\circ}$.

Similarly, the following statements are equivalent:

(i) There exists a 1-4 inverse $\check{a}$ of a.

(ii) There exists a unique projector $f(= \check{a}a)$ such that $Ra = Rf$, $f = f^2 = f^*$.

(iii) $R = Ra \oplus {}^{\circ}(a^*)$.

The notion of a group inverse (a commuting generalized inverse), see Section 4.6, makes sense in any regular R. The following statements are equivalent in R:

(i) $axa = a$, $xax = x$, $ax = xa$ have a solution in R;

(ii) $a^2u = a$, $va^2 = a$ have solutions in R;

(iii) $aR = a^2R$, $a^{\circ} = (a^2)^{\circ}$;

(iv) $Ra = Ra^2$, ${}^{\circ}a = {}^{\circ}(a^2)$;

(v) $R = aR \oplus a^{\circ}$;

(vi) $R = Ra \oplus {}^{\circ}a$.

If any of these statements holds, then the solution to (i) is unique and $x = vau$ is called the group inverse.

Many useful results can be translated from matrix theory to the $*$-regular rings, and in most cases the necessary translations are evident.

8.4. The Drazin Inverse in a Semigroup and in $\mathcal{L}(X)$

The concepts of inner inverse, outer inverse, and algebraic generalized inverse as defined in Section 1 make sense in any semigroup. Let S be an (algebraic) semigroup, i.e. a set with an associative composition $(x,y) \to xy$. An element $a \in S$ is called regular if $axa = a$ for some $x \in S$, i.e. if a has an inner inverse in S. An element $b \in S$ is an outer inverse of $a \in S$ if $bab = b$, and b is called a (generalized) inverse of a if b is both an inner inverse and an outer inverse of a. An inverse semigroup is one in which each element has a unique (generalized) inverse. Inverse semigroups have been thoroughly investigated by algebraists; see the monograph by Clifford and Preston [1].

An element $a \in S$ is said to have a Drazin inverse if there exists $x \in S$ such that

(i) $a^m = a^{m+1}x$ for some non-negative integer m,

(ii) $x = x^2 a$,

and

(iii) $ax = xa$.

Note that the concept of a group inverse (a commuting generalized inverse), see Section 4.6, also makes sense in a semigroup and that the group inverse is just the case $m=1$ of the Drazin inverse. If a has a Drazin inverse, then the smallest possible non-negative integer m in (i) above is called the index $i(a)$ of a.

Note also that a Drazin inverse which is also a (generalized) inverse must be a group inverse, and conversely.

Drazin [1] established the following results:

Theorem 8.9. Let S be a semigroup. An element $a \in S$ has at most one Drazin inverse. If the Drazin inverse exists, it commutes with every $y \in S$ which commutes with a. Also a has a Drazin inverse if and only if there exist $x, y \in S$ and non-negative integers p and q such that

$$a^p = a^{p+1}x \quad \text{and} \quad a^q = ya^{q+1} .$$

Theorem 8.10. Assume $a \in S$ has a Drazin inverse $a^{\#}$. Then

(a) $a^{\#}$ has a Drazin inverse, $i(a^{\#}) \leq 1$, and $(a^{\#})^{\#} = a^2a^{\#}$.

(b) $(a^{\#})^{\#} = a \iff i(a) \leq 1$.

(c) $[(a^{\#})^{\#}]^{\#} = a^{\#}$.

(d) a^n has a Drazin inverse and $(a^n)^{\#} = (a^{\#})^n$.

We note that since the space $\mathcal{L}(X)$ of all continuous linear operators on a Banach space X into X is a multiplicative semigroup, we can talk about a Drazin inverse of a bounded linear operator. It is not difficult to show that $T \in \mathcal{L}(X)$ has a Drazin inverse and has index d if and only if the ascent and descent of T are equal to d.

Recently, Caradus [1] considered the question of the existence of a Drazin inverse for a bounded or a closed linear operator on a Banach space X and proved the following:

Theorem 8.11. $T \in \mathcal{L}(X)$ has a Drazin inverse D and has index d if and only if the resolvent $R_\lambda = (\lambda I - T)^{-1}$ has a pole of order d at $\lambda = 0$. In this case, R_λ has the Laurent expansion

$$R_\lambda = \left[\frac{T^{d-1}}{\lambda^d} + \frac{T^{d-2}}{\lambda^{d-2}} + \ldots + \frac{I}{\lambda} \right] (I - TD) - D - \lambda D^2 - \lambda^2 D^3 - \ldots$$

in the region $0 < |\lambda| < \frac{1}{\rho(D)}$, where $\rho(D)$ is the spectral radius of D.

For a closed linear operator T, a slight modification in the definition of a Drazin inverse is necessary to avoid problems about domains. An operator $D \in \mathfrak{L}(X)$ is called a Drazin inverse of T if $\mathfrak{R}(D) \subset \mathfrak{D}(T)$ and

(i) $TDx = DTx$ for all $x \in \mathfrak{D}(T)$

(ii) $D = D(TD)$

(iii) $T^m(I - TD) = 0$.

Theorem 8.11 holds for any closed linear operator with nonempty resolvent set on a Banach space.

In the case of a finite dimensional space, the ascent and descent of an operator (or of its associated matrix) must, of course, both be finite and hence equal. Thus the Drazin inverse of a finite (square) matrix always exists.

In another context, we mention that Dade, Taussky and Zassenhauss [1] have shown that an ideal class in an order of a quadratic field has as generalized inverse an element which satisfies the Drazin's axioms, but that this is not any longer true in general for number fields of higher degree. Since these ideal classes correspond to classes of matrices (under unimodular similarity) this can also be formulated as a matrix property.

8.5. Generalized Inverses of Matrices with Entries Taken from an Arbitrary Field

The theory of algebraic generalized inverses as presented in Section 1 specializes to matrices over an arbitrary field. The existence of a Moore-Penrose inverse for a matrix over an arbitrary field requires, however, special consideration. In particular, while algebraic generalized inverses of a matrix over an arbitrary field always exist; the existence of a Moore-Penrose depends on the chosen field.

To consider the analog of the equations $(AX)^* = AX$ and $(XA)^* = XA$ over an arbitrary field, we assume that the field $\mathfrak{F}$ has an involutory (anti-) automorphism $\lambda \to \bar{\lambda}$, and denote by A^* the matrix $\overline{A'} = \bar{A}'$, where A' is the transpose of A and $\bar{A}$ is obtained from A by replacing each entry a_{ij} by $\bar{a}_{ij}$. In general $A^*A = 0$ does not imply $A = 0$. For any matrix A over the complex field, A, A^*A and AA^* have the same rank. However, this is not necessarily true for a matrix over a field $\mathfrak{F}$. This equality of ranks is precisely the condition that is needed to assert the existence of $A^\dagger$ over $\mathfrak{F}$:

$$A^\dagger \text{ exists if and only if } r(A) = r(A^*A) = r(AA^*).$$

(The proof of uniqueness is independent of the field.) For generalized inverses of matrices over arbitrary fields, see Kalman [1] and Pearl [1].

REFERENCES

Albert, A. [1]: Regression and the Moore-Penrose Pseudoinverse, Academic Press, New York, 1972.

Arghiriade, E. [1]: Sur l' inverse généralisée d'un opérateur linéaire dans les espaces de Hilbert, Atti. Accad. Naz. Lincei Rend. Cl. Sci. Fis. Mat. Natur., Ser. VIII, 45 (1968), 471-477.

Arghiriade, E. and E. Boros [1]: L'inverse généralisée d'un opérateur linéarire dans un espace à produit intérieur, Atti. Accad. Naz. Lincei Rend. Cl. Sci. Fis. Mat. Natur., Ser. VIII, 46(1969), 646-649.

Arghiriade, E. and A. Dragomir [1]: Remarques sur quelques théorèmes relatives à l'inverse généralisée d'un opérateur lineaire dans les espaces de Hilbert, Atti. Accad. Naz. Lincei Rend. Cl. Sci. Fis. Mat. Natur., Ser. VIII, 46 (1969), 333-338.

Atkinson, F. W. [1]: The normal solvability of linear equations in normed spaces, Mat. Sbornik N.S., 28 (1951), 3-14.

Ben-Israel, A. and T.N.E. Greville [1]: Generalized Inverses: Theory and Applications, Wiley-Interscience, New York, 1974.

Boullion, T. L. and P. L. Odell [1]: Generalized Inverse Matrices, Wiley-Interscience, New York, 1971.

Bourbaki, N. [1]: Espaces Vectoriels Topologiques. Elements de Mathematique. Livre V, Hermann and Cie, Paris, tome I (1953), tome II (1955).

Brown, B. and N. H. McCoy [1]: The maximal regular ideal of a ring, Proc. Amer. Math. Soc., 1 (1950), 165-171.

Caradus, S. R. [1]: The Drazin inverse for operators in Banach spaces, Compositio Math., 47(1975), 409-412.

Chipman, J. S. [1]: On least-squares with insufficient observations, J. Amer. Statist. Assoc., 54 (1964), 1078-1111.

Clifford, A. H. and G. B. Preston [1]: The Algebraic Theory of Semigroups, I, Mathematical Surveys, Vol. 7, Amer. Math. Soc., Providence, R. I., 1961.

Dade, E. C., O. Taussky and H. Zassenhaus [1]: On the theory of orders, in particular on the semigroup of ideal classes and genera of an order in an algebraic number field, Math. Ann. 148 (1962), 31-64.

Davis, D. L. and D. W. Robinson [1]: Generalized inverses of morphisms, Linear Algebra and Appl., 5(1972), 329-338.

Day, M. M. [1]: Normed Linear Spaces, 3rd edition, Springer-Verlag, Berlin, 1973.

Deutsch, E. [1]: Semi-inverses, reflexive semi-inverses, and pseudoinverses of an arbitrary linear transformation, Linear Algebra and Appl., 4(1971), 313-322.

Drazin, M. P. [1]: Pseudo inverses in associative rings and semigroups, Amer. Math. Monthly, 65(1958), 506-514.

Erdelsky, P. J. [1]: Projections in a Normed Linear Space and a Generalization of the Pseudo-Inverse, Doctoral Dissertation, California Institute of Technology, Pasadena, 1969.

Erdelyi, I. [1]: A generalized inverse for arbitrary operators between Hilbert spaces, Proc. Cambridge Philos. Soc., 71(1972), 43-50.

Erdelyi, I. and A. Ben-Israel [1]: Extremal solutions of linear equations and generalized inversion between Hilbert spaces, J. Math. Anal. Appl., 39(1972), 298-313.

Foulis, D. J. [1]: Relative inverses in Baer *-semigroups, Michigan Math. J., 10(1963), 65-84.

Halmos, P. R. [1]: Finite Dimensional Vector Spaces, 2nd edition, Van Nostrand, Princeton, 1958.

Hansen, G. W. and D. W. Robinson [1]: On the existence of generalized inverses, Linear Algebra and Appl., 8(1974), 95-104.

Hartwig, R. E. [1]: Block generalized inverses, Arch. Rational Mech. Anal., to appear.

Hestenes, M. R. [1]: Relative self-adjoint operators in Hilbert space, Pacific J. Math., 11(1961), 225-245.

Holmes, R. B. [1]: A Course on Optimization and Best Approximation, Lecture Notes in Mathematics, Vol. 257, Springer-Verlag, Berlin-New York, 1972.

James, R. C. [1]: Orthogonality and linear functionals in normed linear spaces, Trans. Amer. Math. Soc., 61 (1947), 265-292.

Kalman, R. E. [1]: Algebraic aspects of the generalized inverse of a rectangular matrix, this Volume.

Kadets, M. I. and B. S. Mityagin [1]: Complemented subspaces in Banach spaces, Russian Math. Surveys, 28 (1973), 77-95.

Kaplansky, I. [1]: Any ortho-complemented complete modular lattice is a continuous geometry, Ann. of Math., 61(1955), 524-541.

Langenhop, C. E. [1]: On generalized inverse of matrices, SIAM J. Appl. Math., 15(1967), 1239-1246.

MacLane, S. [1]: Homology, Springer-Verlag, Berlin, 1963.

MacLane, S. and G. Birkhoff [1]: Algebra, MacMillan, New York, 1968.

Milne, R. D. [1]: An oblique matrix pseudoinverse, SIAM J. Appl. Math., 16(1968), 931-944.

Moore, E. H. [1]: On the reciprocal of the general algebraic matrix (abstract), Bull. Amer. Math. Soc., 26(1920), 394-395.

Moore, E. H. [2]: General Analysis, Memoirs of the American Philosophical Society, I (esp., pp. 147-209), Philadelphia, 1935.

Morris, P. D. and E. W. Cheney [1]: On the existence and characterization of minimal projections, Center for Numerical Analysis Technical Rept. CNA-37, University of Texas, Austin, Texas, 1972; to appear in Crelle's Journal.

Munn, W. D. [1]: Pseudoinverses in semigroups, Proc. Cambridge Philos. Soc., 57(1961), 247-250.

Munn, W. D. and R. Penrose [1]: A note on inverse semigroups, Proc. Cambridge Philos. Soc., 51(1955), 396-399.

Murray, F. J. [1]: On complementary manifolds and projections in spaces L_p and ℓ_p, Trans. Amer. Math. Soc., 41(1937), 138-172.

Nashed, M. Z. [1]: Generalized inverses, normal solvability, and iteration for singular operator equations in Nonlinear Functional Analysis and Applications, L. B. Rall, editor, pp. 311-359, Academic Press, New York, 1971.

Nashed, M. Z. [2]: Approximate regularized solutions to improperly posed linear integral and operator equations, in Constructive and Computational Methods for Differential and Integral Equations (Symposium Proceedings, Research Center for Applied Science, Indiana University, Bloomington, Indiana, February 17-20, 1974), D. Colton and R. P. Gilbert, editors, Lecture Notes in Mathematics, Vol. 430, pp. 289-322, Springer-Verlag, Berlin-New York.

Nashed, M. Z. and G. F. Votruba [1]: A unified approach to generalized inverses of linear operators: I. Algebraic, topological and projectional properties, Bull. Amer. Math. Soc., 80(1974), 825-830.

Nashed, M. Z. and G. F. Votruba [2]: A unified approach to generalized inverses of linear operators: II. Extremal and proximinal properties, Bull. Amer. Math. Soc., 80(1974), 831-835.

von Neumann, J. [1]: On regular rings, Proc. Nat. Acad. Sci. USA, 22(1936), 707-713.

von Neumann, J. [2]: Functional Operators, Vol. II. The Geometry of Orthogonal Spaces, Ann. of Math. Studies No. 29, Princeton Univ. Press, Princeton, N. J., 1950.

von Neumann, J. [3]: Continuous Geometry, Princeton Univ. Press, Princeton, N. J., 1960.

Newman, T. G. and P. L. Odell [1]: On the concept of a p-q generalized inverse of a matrix, SIAM J. Appl. Math., 17(1969), 520-525.

Noll, W. [1]: Quasi-reversibility in a staircase diagram, Proc. Amer. Math. Soc., 23(1969), 1-4.

Olmsted, J. M. H. [1]: Completeness and Parseval's equation, Amer. Math. Monthly, 65(1958), 343-345.

Pearl, M. H. [1]: Generalized inverses of matrices with entries taken from an arbitrary field, Linear Algebra and Appl., 1(1968), 571-587.

Penrose, R. [1]: A generalized inverse for matrices, Proc. Cambridge Philos. Soc., 51(1955), 406-413.

Penrose, R. [2]: On best approximate solutions of linear matrix equations, Proc. Cambridge Philos. Soc., 52 (1956), 17-19.

Phillips, R. S. [1]: On linear transformation, Trans. Amer. Math. Soc., 48(1940), 516-541.

Pietsch, A. [1]: Zur Theorie der σ-Transformationen in lokalkonvexen Vektörraumen, Math. Nach., 21(1960), 347-369.

Rabson, G. [1]: The generalized inverse in set theory and matrix theory, Notices of the Amer. Math. Soc., 1969.

Rado, R. [1]: Note on generalized inverses of matrices, Proc. Cambridge Philos. Soc., 52(1956), 600-601.

Raikov, D. A. [1]: Vector Spaces, translated from the first Russian Edition by L. F. Boron with the assistance of H. R. Stevens and G. F. Votruba, Noordhoff, Groningen, 1965.

Rao, C. R. and S. K. Mitra [1]: Generalized Inverse of Matrices and its Applications, Wiley and Sons, New York, 1971.

Rao, C. R. and S. K. Mitra [2]: Theory and applications of constrained inverse of matrices, SIAM J. Appl. Math., 24(1973), 473-488.

Rickart, C. E. [1]: Banach algebras with an adjoint operation, Ann. of Math., 47(1946), 528-550.

Robinson, D. W. [1]: On the generalized inverse of an arbitrary linear transformation, Amer. Math. Monthly, 69(1962), 412-416.

Schäffer, J. J. [1]: Another characterization of Hilbert space, Studia Math., 25(1965), 271-276.

Schwartz, L. [1]: Théorie des Distributions, tome I, Hermann, Paris, 1950.

Sheffield, R. D. [1]: On pseudo-inverses of linear transformations in Banach spaces, Oak Ridge Nat. Lab. Rept. #2133; Ph.D. Thesis, University of Tennessee, Knoxville, Tennessee, 1956.

Sobczyk, A. [1]: Projections in Minkowski and Banach spaces, Duke Math. J., 8(1941), 78-106.

Taylor, A. E. [1]: Introduction to Functional Analysis, John Wiley and Sons, New York, 1958.

Tseng, Yu. Ya [1]: Sur les solutions des équations operatrices fonctionnelles entre les espaces unitaries. Solutions extremales. Solutions virtuelles., C. R. Acad. Sci. Paris, 228(1949), 640-641.

Tseng, Yu. Ya [2]: Generalized inverses of unbounded operators between two unitary spaces, Dokl. Akad. Nauk SSSR (N. S.), 67(1949), 431-434 (reviewed in Math. Rev. 11(1950, p. 115).

Tseng, Yu. Ya [3]: Properties and classification of generalized inverses of closed operators, Dokl. Akad. Nauk. SSSR (N. S.), 67(1949), 607-610 (reviewed in Math. Rev. 11(1950, p. 115).

Tseng, Yu. Ya [4]: Virtual solutions and general inversions, Uspehi, Mat. Nauk. (N. S.), 11(1956), 213-215 (reviewed in Math. Rev., 18(1957), p. 749).

Votruba, G. F. [1]: Generalized Inverses and Singular Equations in Functional Analysis, Doctoral Dissertation, University of Michigan, Ann Arbor, Michigan, 1963.

Ward, J. F., T. L. Boullion and T. O. Lewis [1]: A note on the oblique matrix pseudoinverse, SIAM J. Appl. Math., 20(1972), 514-518.

Professor M. Zuhair Nashed
School of Mathematics
Georgia Institute of Technology
Atlanta, Georgia 30332

Professor G. F. Votruba
Mathematics Department
University of Montana
Missoula, Montana 59801

Algebraic Aspects of the Generalized Inverse of a Rectangular Matrix

R.E. Kalman

1. Introduction

When PENROSE [1955] rediscovered the notion of the generalized inverse of a rectangular matrix, his point of view and proofs were purely algebraic. It soon became clear that PENROSE's axioms were equivalent to the earlier definition of MOORE [1920], which was expressed in a rather different language. The technique used by subsequent authors (see, for example, GREVILLE [1959], DESOER and WHALEN [1963], and BEN-ISRAEL and CHARNES [1963]) gradually shifted from the algebraic toward what might be called "Hilbert-space geometric". Since HILBERT himself, with his great "Grundzüge einer allgemeinen Theorie ...", may be regarded as one of the inventors of the generalized inverse, this development was surely not unnatural. On the other hand, the generalized inverse is an algebraic notion and it is of some interest to ask what the structure of a Hilbert space (= vector space with a positive definite bilinear form) has to do with it. Very little, as it turns out. The important thing is the bilinear form. Assuming positivity yields some additional results (especially as regards optimization) but has little to do with the development of the formulas. In other words, positivity can be imposed on the theory ex post facto after the generalized inverse has been defined over an arbitrary field.

The purpose of this modest note is to develop the theory of the generalized inverse ab initio in a strictly linear-algebraic fashion. Most of the steps are either implicit or explicit in the literature and our accomplishment is primarily editorial. Still, it is possible to give more natural formulas and even to obtain some new results.

The generalized inverse is not an invariant notion. It is not a basis-free property of a linear transformation but a gadget constructed from a rectangular array of numbers. Neither a geometric nor an abstract-algebraic formalism is really appropriate; the most natural approach, very similar to PENROSE's, is to write all formulas in terms of matrices obtained directly from the given rectangular matrix A. We shall give a self-contained account of the theory in this style. We shall develop all the main facts of the theory and ask only very little from the reader by way of mathematical culture.

2. Definitions and uniqueness theorem

Let k be any fixed field (commutative but not necessarily of characteristic 0).

Let A be a fixed $m \times n$ matrix (with elements) over k and let X be an $n \times m$ matrix also over k. Denote the transpose by the prime. Consider the following relations, due mainly to PENROSE [1955]:

(I) $AXA = A,$

(II) $XAX = X,$

(III) $AX = X'A',$

(IV) $XA = A'X'.$

If (I) and (II) are satisfied, we call $X = A^{+}$ a weak generalized inverse of A; if (I) through (IV) are satisfied, we call $X = A^{\#}$ the generalized inverse of A.

To justify our terminology, we note first the

(1) UNIQUENESS THEOREM. A has at most one generalized inverse.

Proof. Let X, Y be two matrices satisfying (I) - (IV). Then

$$
\begin{aligned}
X &= XAX && \text{by (II),}\\
&= A'X'X && \text{by (IV),}\\
&= A'Y'A'X'X && \text{by transposing (I),}\\
&= A'Y'XAX && \text{by (IV),}\\
&= A'Y'X && \text{by (II),}\\
&= YAX && \text{by (IV),}\\
&= YAYAX && \text{by (II),}\\
&= YY'A'X'A' && \text{by (III),}\\
&= YY'A' && \text{by transposing (I),}\\
&= YAY && \text{by (III),}\\
&= Y.
\end{aligned}
$$

□

3. Existence theorems

A very large number of existence proofs and algorithmic constructions are available for the generalized and weak generalized inverses. The following procedure uses only the notion of rank and therefore has some claim to be the most elementary. It is certainly not the most efficient construction if computational considerations are important.

Let B be any $m \times r$ submatrix of A containing exactly $r = \text{rank } A$ linearly independent columns of A. (These columns need not be adjacent, of course.) Let K be any $m \times r$ matrix over k such that $\det K'B \neq 0$. Such a matrix K always exists; by the definition of "rank B" it is even possible to choose K so that it consists of 0's and 1's only. Then the formula

$$B^+ = (K'B)^{-1}K'$$

defines a weak generalized inverse of B.

Similarly, let C be any $r \times n$ submatrix of A containing exactly r linearly independent rows of A. Again, there is an $r \times n$ matrix L over k such that $\det CL' \neq 0$ and $C^+ = L'(CL')^{-1}$ is a weak generalized inverse of C.

Note that BB^+ is a left identity for A. In fact, $BB^+B = B$ (for instance, by (I)); this implies $BB^+A = A$ since the columns of A are linear combinations of the columns of B, by definition of B. Similarly, C^+C is a right identity for A.

Let $D = B^+AC^+$. Then $BDC = A$, by the results of the preceding paragraph.

Recall the elementary inequality $\operatorname{rank} UV \leq \min\{\operatorname{rank} U, \operatorname{rank} V\}$ which is valid for any two matrices U, V whose product is defined. It follows that

$$\operatorname{rank} A \geq \operatorname{rank} B^+AC^+ = \operatorname{rank} D \geq \operatorname{rank} BDC = \operatorname{rank} A,$$

so $\operatorname{rank} A = \operatorname{rank} D$. Since D is $r \times r$, $\det D \neq 0$. From $D = (K'B)^{-1}K'AL'(CL')^{-1}$, it follows that $\det K'AL' \neq 0$ also. From $A = BDC$, it is natural and possible to define the weak generalized inverse of A as

$$A^+ := C^+D^{-1}B^+ = L'(K'AL')^{-1}K'.$$

Property (II) is immediately verified. Property (I) is less obvious:

$$\begin{aligned} AA^+A &= (BDC)L'(K'AL')^{-1}K'(BDC), \\ &= BD(CL'(K'AL')^{-1}K'B)DC, \\ &= BDD^{-1}DC, \\ &= BDC, \\ &= A. \end{aligned}$$

If A^+ satisfies (I) and (II) it follows further that

$$\text{rank } A \geq \text{rank } A^+AA^+ = \text{rank } A^+ \geq \text{rank } AA^+A = \text{rank } A \ .$$

The preceding formulas prove the

(2) EXISTENCE THEOREM FOR WEAK GENERALIZED INVERSE. An $m \times n$ matrix A over a field k has at least one $n \times m$ weak generalized inverse A^+ which can be expressed (locally) by a rational formula in the elements of A. Moreover, rank A^+ = rank A for every weak generalized inverse.

The fact that A^+ is (in general) not unique is abundantly clear from the freedom involved in the choice of K and L. Consequently the construction of a (unique) generalized inverse must depend on a forced choice of K and L.

Setting $K = B$ gives

$$\begin{aligned} AA^+ &= AL'(B'AL')^{-1}B', \\ &= (BDC)L'(B'AL')^{-1}B', \\ &= BDD^{-1}(B'B)^{-1}B', \\ &= B(B'B)^{-1}B' = \text{symmetric}. \end{aligned}$$

Similarly, $L = C$ gives

$$A^+A = C'(CC')^{-1}C = \text{symmetric}.$$

In other words, the conditions $\det B'B \neq 0$ and $\det CC' \neq 0$ are sufficient for the existence of a generalized inverse which can be explicitly written as

$$A^\# := C'(B'AC')^{-1}B'.$$

On the other hand, suppose that $A^\#$ exists. Then

$$\begin{aligned} \text{rank } A &\geq \text{rank } A'A \ , \\ &\geq \text{rank } A'AA^{\#} \ , \\ &\geq \text{rank } A'(A^{\#})'A' \qquad \text{by (III)} \ ; \end{aligned}$$

from the PENROSE axioms and the Uniqueness Theorem $(A^{\#})' = (A')^{\#}$, so that

$$\text{rank } A \geq \text{rank } A'A \geq \text{rank } A' = \text{rank } A \ .$$

Thus (III) implies that rank $A'A = \text{rank } A$. Similarly, (IV) implies that rank $AA' = \text{rank } A$.

Conversely, suppose rank $A'A = r$. Since $A'A$ is symmetric, at least one of the $r \times r$ principal minors of $A'A$ must be nonzero (see MACDUFFEE [1933, p. 12, Theorem 8.5]). The columns corresponding to a principal minor define an $m \times r$ submatrix B of A, as above, and it follows that the value of the minor is exactly $\det B'B$. Thus it is clear that there is a B with $\det B'B \neq 0$ iff rank $A'A = \text{rank } A$.

This completes the proof of the

(3) EXISTENCE THEOREM FOR GENERALIZED INVERSES. An $m \times n$ matrix A over a field k has a generalized inverse $A^{\#}$ if and only if (i) rank $A'A = \text{rank } A$ and (ii) rank $AA' = \text{rank } A$. The first condition (with $K = B$) is necessary and sufficient for (III) and the second (with $L = C$) for (IV). If $A^{\#}$ exists, it can be expressed by the formula $C'(B'AC')^{-1}B'$; in other words, $A^{\#}$ is (locally) a rational function in the elements of A.

Insofar as generalized inverses have not been studied over an arbitrary field, this result appears to be new.

(4) REMARK. The uniqueness theorem for the generalized inverse shows that the above representation is independent of the choice of B, C; in particular, the expressions $AA^{\#} = B(B'B)^{-1}B'$ and $A^{\#}A = C'(CC')^{-1}C$ are also independent of these choices.

(5) EXAMPLE. The simplest case where the generalized inverse does not exist is

$$A = (1 \quad 1),$$

regarded over the 2-element field $\underline{\underline{Z}}/2\underline{\underline{Z}}$ of integers modulo 2. Clearly $\det AA' = 0$. A has exactly two weak generalized inverses, namely

$$A_1^+ = \begin{pmatrix} 1 \\ 0 \end{pmatrix} \quad \text{and} \quad A_2^+ \begin{pmatrix} 0 \\ 1 \end{pmatrix}.$$

We have $AA_i^+ = (1) =$ symmetric for $i = 1, 2$; in both cases $A_i^+ A \neq$ symmetric.

(6) COROLLARY. _If_ $k =$ _real field_, $A^{\#}$ _always exists._

Proof. By the well-known Binet-Cauchy formula (see GANTMAKHER [1959, p. 9]), $\det B'B$ is the sum of squares of $r \times r$ minors of B. At least one of these is nonzero since $r = \operatorname{rank} A$. Hence over the real field $\det B'B > 0$. The same argument shows also that $\det CC' > 0$. □

(7) REMARK. In the literature, the existence of the generalized inverse is often proved via arguments such as the diagonalization of symmetric matrices via an orthogonal transformation. (In fact, if $A = A'$ and $A = U\Lambda U'$, $U =$ orthogonal and $\Lambda =$ diagonal, then clearly $A^{\#} = U\Lambda^{\#}U'$ satisfies the axioms, while $\Lambda^{\#}$ is defined by $\lambda_{ii}^{\#} = \lambda_{ii}^{-1}$ if $\lambda_{ii} \neq 0$ and 0 otherwise.) Our proof shows that the use of this "advanced" theorem, while providing insight, is an avoidable mathematical detour; only the most basic property of a real field is actually needed.

(8) REMARK. If the field k admits an involutory automorphism $\lambda \to \bar{\lambda}$, all the preceding remains valid upon replacing

the operation $A \to A'$ by $A \to A^*$:= "conjugate" transpose. Then Corollary (6) depends on the single question: does $\lambda_1 \bar{\lambda}_1 + \ldots + \lambda_n \bar{\lambda}_n = 0$ imply $\lambda_1 = \ldots = \lambda_n = 0$? Over the complex field, this is true and we see why the Hermitian rather than ordinary transpose is needed to obtain the existence of the generalized inverse over the complex field.

(9) REMARK. If A is reduced to its classical canonical form exhibiting the rank (exactly rank A 1's on the main diagonal, 0's elsewhere), we see that the existence condition is trivially satisfied. Hence the existence of a generalized inverse is not an intrinsic property of A but depends on the chosen field and basis.

(10) REMARK. It has been frequently noted (see, for example, LANGENHOP [1967]) that a weak generalized inverse can be converted into a generalized inverse by a redefinition of the inner product in the source and target space of A. By writing $K'B = B'T_1'T_1B$ and $CL' = CT_2T_2'C'$ one obtains a generalized inverse with respect to the quadratic forms $T_1'T_1$ and T_2T_2' instead of I. Even this procedure, however, cannot yield a definition so that the generalized inverse exists in any specifically given field. For example, it is known (see SERRE [1973, p. 37]) that over the p-adic field $\mathbb{Q}_p$ every quadratic form in five or more variables admits a nontrivial isotropic vector, that is, the sum of 5 or more nonzero squares can vanish.

4. Orthogonality and positivity

In the space of $m \times n$ matrices (over a fixed field k), the usual scalar product (symmetric bilinear form) is defined as

$$\langle U, V \rangle := \sum_{i,j} u_{ij} v_{ij},$$

$$= \operatorname{tr} U'V = \operatorname{tr} VU' = \operatorname{tr} V'U = \operatorname{tr} UV' \text{ (tr means "trace")} .$$

From the scalar product, we get the usual Euclidean "norm" defined as $\|A\|^2 := \langle A, A \rangle$ (not necessarily positive).

(11) ORTHOGONALITY THEOREM. Let A be a fixed $m \times n$ matrix over k and let X be any fixed $n \times m$ matrix. Then the relation

$$\langle X, Y \rangle = \operatorname{tr} X'Y = 0$$

holds for all Y satisfying $AYA = 0$ (i) if and (ii) only if X is the generalized inverse of A.

Proof. (i) Suppose $X = A^{\#}$. Then

$$
\begin{aligned}
\operatorname{tr} Y'X &= \operatorname{tr} Y'XAX && \text{by (II),}\\
&= \operatorname{tr} Y'A'X'X && \text{by (IV),}\\
&= \operatorname{tr} Y'A'X'XAX && \text{by (II),}\\
&= \operatorname{tr} Y'A'X'XX'A' && \text{by (III),}\\
&= \operatorname{tr} A'Y'A'X'XX' && \text{by definition of trace,}\\
&= 0 && \text{by condition on } Y.
\end{aligned}
$$

(ii) Let Z_{pq} be a matrix (of suitable size) all of whose elements are 0 except $z_{pq} = 1$. Let $Y = (I-XA)Z_{pq}$. Then $\operatorname{tr} X'Y = \operatorname{tr} X'(I-XA)Z_{pq} = 0$. Since this must hold for all p, q, it follows that $X' = X'XA$. Multiplying on the left by A and transposing shows that X satisfies (IV).

Similarly, let $Y = Z_{pq}(I - AX)$. Then $\operatorname{tr} YX' = \operatorname{tr} Z_{pq}(I - AX)X' = 0$ implies $X' = AXX'$; multiplying on the right by A' and transposing proves (III).

Finally, consider $Y = (XA - I)'Z'_{pq}$. Because it was already shown that X satisfies (IV), Y is a solution of $AYA = 0$. So $\operatorname{tr} Y'X = 0$ implies $XAX = X$, proving (II). □

A more familiar form of this result is given by the following

(12) COROLLARY. Assume that char $k \neq 2$. Then the relation

$$\|X + Y\|^2 = \|X\|^2 + \|Y\|^2$$

holds for all Y satisfying $AYA = 0$ if and only if $X = A^{\#}$.

Proof. Obvious, since $\|X + Y\|^2 - \|X\|^2 - \|Y\|^2 = 2\langle X, Y\rangle$. □

(13) COROLLARY. If k = real field, $A^{\#}$ is that solution of $AXA = A$ which minimizes $\|X\|$.

Proof. Over the real field, $\| \ \|$ defines a (positive definite) norm. Hence by (12) it follows that $\|A^{\#}\| < \|X\|$ unless $A^{\#} = X$. □

A proof of this result may be found in KALMAN [1961, p. 124, Corollary (A. 8)] †; it utilizes the facts developed below together with some elementary properties of the tensor product.

Remark (8) holds for Corollary (13) exactly as it did for Corollary (6).

PENROSE [1956] initiated the study of linear equation systems in the most general case, that is, without assuming at first whether a solution exists or not. See also HEARON [1968]. These results are obtained most directly by first establishing some

(14) ALGEBRAIC IDENTITIES FOR EUCLIDEAN "NORM". Let A be a fixed $m \times n$ matrix such that $A^{\#}$ exists and let b be a fixed m-vector over k. Write $x^{\#} = A^{\#}b$ and $z = A(x - x^{\#})$. Then:

† Unfortunately, this material was left out of the published version (see KALMAN [1963, p. 379]) due to unauthorized interference with the author's manuscript.

(i) $\|Ax - b\|^2 = \|Ax^{\#} - b\|^2 + \|z\|^2,$

(ii) $\|x\|^2 = \|x^{\#} + A^{\#}z\|^2 + \|(I - AA^{\#})x\|^2.$

Proof. The first relation is verified by direct computation, using only (I) and (III). The second is verified similarly, using only (II) and (IV). □

(15) SOLUTIONS OF LINEAR EQUATIONS. With the hypotheses of (14), consider the equation $Ax = b$. Then:

(i) If there exists a solution, $x^{\#}$ is a solution.

(ii) If $x^{o} \neq x^{\#}$ is a solution, then

$$\|x^{o}\|^2 = \|x^{\#}\|^2 + \|(I - A^{\#}A)x^{o}\|^2.$$

(iii) $\|Ax^{o} - b\| = \|Ax^{\#} - b\|$ if and only if

$$x^{o} = x^{\#} + A^{\#}Ay + (I - A^{\#}A)x^{o} ,$$

where y is an n-vector with $\|Ay\|^2 = 0$ (either $Ay = 0$ or y belongs to an isotropy subspace of $A'A$).

Proof. (i) If x^{o} is a solution, using only (I) gives

$$b = Ax^{o} = AA^{\#}Ax^{o} = AA^{\#}b = Ax^{\#}.$$

(ii) If x^{o} is a solution, so is $x^{\#}$ by (i). Hence $z = A(x^{o} - x^{\#}) = 0$. The claim then follows by (14ii).

(iii) Suppose x^{o} is as given. By (14i), $\|Ax^{o} - b\|^2 = \|Ax^{\#} - b\|^2 + \|AA^{\#}Ay\|^2$. So if $\|Ay\|^2 = 0$ both x^{o} and $x^{\#}$ are approximate solutions of the same error norm. Conversely, if $\|Ax^{o} - b\| = \|Ax^{\#} - b\|$ then $y = x^{o} - x^{\#}$ satisfies $\|Ay\|^2 = 0$. Now x^{o} has a (unique, orthogonal) direct-sum decomposition

$$x^o = x^1 + x^2 = A^{\#}Ax^o + (I - A^{\#}A)x^o.$$

Consequently, $x^2 = (I - A^{\#}A)x^o$ and x^1 is given by

$$x^1 = A^{\#}Ax^o = A^{\#}Ax^{\#} + A^{\#}Ay = x^{\#} + A^{\#}Ay. \qquad \square$$

(16) COROLLARY. If k = real field, $x^{\#}$ is the unique approximate solution of $Ax = b$ in the least-squares sense (minimizing $\|Ax - b\|$); if $Ax = b$ has an exact solution, $x^{\#}$ is the unique solution of minimum norm.

Proof. Over the real field, there are no (nonzero) isotropic vectors. So, by (15iii), any least-squares or exact solution $x^o = x^{\#} + (I - A^{\#}A)x^o$. By (14ii) (with $z = 0$), $\|x^o\|^2 = \|x^{\#}\|^2 + \|x^2\|^2$. Hence over the real field x^o has minimum norm iff $x^2 = 0$, that is, iff $x^o = x^1 = x^{\#}$.

5. Conclusions

The most striking aspect of the generalized inverse, in the axiomatic form given to it by PENROSE [1955], is its purely algebraic character. From the analysis of the proofs in the style given above, it is clear that the ideas needed are nothing deeper than repeated application of the axioms, together with very elementary facts from linear algebra such as associativity of matrix multiplication and properties of "rank" and "transpose". (The only heuristic proof principle we have been able to find is "try to make the formulas longer because they will then simplify.") The base field plays a role only through the possibility that a sum of squares may vanish. The presentation of the generalized inverse through orthogonal projections, minimizing properties, etc. appears arificial to us; these notions can be used to derive some extra results, but their study can be postponed until after the basic algebraic theory has been developed.

It may be a question of some interest to try to apply mechanical theorem-proving techniques to the search for furthere results via algebraic manipulation. Since all known facts (in the finite-dimensional case) are accessible to elementary computational proofs, it does not seem unfair to pose such a task for automated creative mathematics.

6. Acknowledgements

The point of view of this paper was first developed more than a decade ago when interest in generalized inverses was at its height. The author wishes to thank the organizers of the Advanced Seminar for the opportunity to publish this material.

REFERENCES

A. BEN-ISRAEL and A. CHARNES

[1963] "Contributions to the theory of generalized inverses", SIAM J., 11: 667-699.

C. A. DESOER and B. H. WHALEN

[1963] "A note on pseudoinverses", SIAM J., 11: 442-447.

F. R. GANTMAKHER

[1959] The theory of matrices, vol. 1, Chelsea.

T. N. E. GREVILLE

[1960] "The pseudoinverse of a rectangular singular matrix and its application to the solution of systems of linear equations", SIAM Review, 1: 38-43.

J. Z. HEARON

[1968] "Generalized inverses and solutions of linear systems", J. Res. National Bureau of Standards, 72B: 303-308.

R. E. KALMAN

[1961] "New methods and results in linear prediction and filtering theory", RIAS Technical Report 61-1.

[1963] "New methods in Wiener filtering theory", in Proc. 1st Symp. on Engr. Appl. of Random Function Theory and Probability, Wiley, pp. 270-388.

C. E. LANGENHOP

[1967] "On generalized inverses of matrices", SIAM J. Appl. Math., 15: 1239-1246.

C. C. MACDUFFEE

[1933] The theory of matrices, Springer (Ergebnisse der Mathematik, N^O 2).

E. H. MOORE

[1920] Abstract in Bull. Am. Math. Soc., 26: 394-395.

R. PENROSE

[1955] "A generalized inverse for matrices", Proc. Cambridge Philosophical Society, 51: 406-413.

[1956] "On the best approximate solutions of linear matrix equations", Proc. Cambridge Phil. Soc., 52: 17-19.

J. P. SERRE

[1973] A course in arithmetic, Springer Graduate Texts in Mathematics, N^O 7.

Center for Mathematical System Theory
University of Florida
Gainesville, FL 32611

This research was supported in part by the US Air Force under Grant 72-2268 with the Center for Mathematical System Theory, University of Florida.

Some Topics in Generalized Inverses of Matrices

Adi Ben-Israel
Thomas N.E. Greville

1. Introduction

By a generalized inverse of a given matrix A we shall mean a matrix X associated in some way with A that (i) exists for a class of matrices larger than the class of nonsingular matrices, (ii) has some of the properties of the usual inverse, and (iii) reduces to the usual inverse when A is nonsingular.

The various generalized inverses that can so be defined, differ widely in their properties, applicability and computation. This variety is manifest in hundreds of papers and in several books which appeared since 1955, when Penrose [22] revived the interest in generalized inverses.

This paper is a brief outline of some aspects of generalized inverses of matrices, using the notation and terminology of [2], where further information, and the proofs omitted here, can be found.

2. Notation and terminology

For a given matrix $A \in C^{m \times n}$ consider the matrix equations

(1) $$AXA = A$$

(2) $$XAX = X$$

(3) $$(AX)^* = AX$$

(4) $$(XA)^* = XA ,$$

known as the Penrose equations, [22].

If $m = n$ and k is a positive integer, consider also

(1^k) $$A^k XA = A^k$$

(5) $$AX = XA .$$

Let $A\{i, j, \ldots, \ell\}$ denote the set of matrices $X \in C^{n \times m}$ which satisfy equations (i), (j), ..., (ℓ) from among the above equations. A matrix $X \in A\{i, j, \ldots, \ell\}$ is called an $\{i, j, \ldots, \ell\}$-inverse of A, and also denoted by $A^{(i, j, \ldots, \ell)}$.

In particular, for any $A \in C^{m \times n}$, the set $A\{1, 2, 3, 4\}$ consists of a single element, the Moore-Penrose inverse of A, denoted by $A^{\dagger}$, [22].

Also, for any $A \in C^{n \times n}$, the Drazin pseudoinverse $A^{(d)}$ is the unique $\{1^k, 2, 5\}$-inverse, which exists if and only if $k \geq$ index A, where

$$\text{index } A = \min\{k: k = 1, 2, \ldots; \text{ rank } A^k = \text{rank } A^{k+1}\} ,$$

see, e.g., [7] and ([2], Chapter 4).

The group inverse $A^{\#}$ of $A \in C^{n \times n}$ is its unique $\{1, 2, 5\}$-inverse, which exists if and only if index $A = 1$, see, e.g., [10] and [25].

3. Linear equations and {1}-inverses.

The existence of {1}-inverses is settled in the following theorem, which also describes their computation by Gaussian elimination.

3.1. Theorem. Let $A \in C_r^{m \times n}$ (= the $m \times n$ complex matrices of rank r), and let $E \in C_m^{m \times m}$ and $P \in C_n^{n \times n}$ be such that

$$EAP = \begin{bmatrix} I_r & K \\ 0 & 0 \end{bmatrix} .$$

Then, for any $L \in C^{(n-r)\times(m-r)}$, the $n\times m$ matrix

$$X = P \begin{bmatrix} I_r & 0 \\ 0 & L \end{bmatrix} E$$

is a {1}-inverse of A. ∎

Since rank $X = r +$ rank L, this proves the existence of {1}-inverses having specified ranks between r and $\min\{m, n\}$, see, e.g. [11] and ([2] p. 17 and §7.2).

The principal application of {1}-inverses is to the solution of linear equations, where they are used in much the same way as ordinary inverses in the nonsingular case:

3.2. Theorem (Penrose [22]). Let $A \in C^{m\times n}$, $B \in C^{p\times q}$, $D \in C^{m\times q}$. Then the matrix equation

$$AXB = D$$

is consistent if and only if for some $A^{(1)}$, $B^{(1)}$

$$AA^{(1)}DB^{(1)}B = D$$

in which case the general solution is

$$X = A^{(1)}DB^{(1)} + Y - A^{(1)}AYBB^{(1)}, \quad Y \in C^{n\times m} . \tag{6}$$ ∎

3.3. Corollary. Let $A \in C^{m\times n}$, $b \in C^m$. Then the equation

$$Ax = b$$

is consistent if and only if for some $A^{(1)}$

$$AA^{(1)}b = b$$

in which case the general solution is

$$x = A^{(1)}b + (I - A^{(1)}A)y, \quad y \in C^n . \qquad \blacksquare$$

Indeed, {1}-inverses can be characterized as solvers of linear equations:

3.4. Theorem (Bose). Let $A \in C^{m\times n}$, $X \in C^{n\times m}$. Then $x \in A\{1\}$ if and only if for all b such that $Ax = b$ is consistent, $x = Xb$ is a solution. ([2], p. 40). $\blacksquare$

Another characterization of $A\{1\}$, in terms of a particular element $A^{(1)}$, is given by (6) as

$$A\{1\} = \{A^{(1)} + Y - A^{(1)}AYAA^{(1)}: \ Y \in C^{n\times m}\} . \tag{7}$$

The number of parameters used in (7) equals nm, the number of elements in the arbitrary $Y \in C^{n\times m}$. In general, this number far exceeds the actual number of degrees of freedom available, $nm - r^2$, where $r = \text{rank } A$. An efficient characterization (i.e., with no more parameters than needed) of $A\{1\}$, in terms of a particular element $A^{(1)}$ and matrices $F \in C_{n-r}^{n\times(n-r)}$, $K^* \in C_{m-r}^{m\times(m-r)}$ and $B \in C_r^{n\times r}$ whose columns are bases for $N(A)$, $N(A^*)$ and $R(A^{(1)}A)$, respectively, is

$$A\{1\} = \{A^{(1)} + FY + BZK: \ Y \in C^{(n-r)\times m}, \ Z \in C^{r\times(m-r)}\} ,$$

see ([2], p. 77) where efficient characterizations of other classes of generalized inverses are also given.

4. Projection and minimal properties of generalized inverses

Let L, M be complementary subspaces of C^n, and let the projector on L along M be denoted by $P_{L,M}$, or just P_L in the case that $M = L^{\perp}$.

If $A \in C_r^{m \times n}$ and $X \in A\{1\}$, then both AX and XA are idempotents with rank r, and therefore

(8) $$AX = P_{R(A), S}, \quad XA = P_{T, N(A)}$$

where S is some subspace of C^m complementary to $R(A)$, and T is some subspace of C^n complementary to $N(A)$. Since Hermitian idempotents are orthogonal projectors it follows from (8) that

(9) $$X \in A\{1, 3\} \Longleftrightarrow AX = P_{R(A)}$$

(10) $$X \in A\{1, 4\} \Longleftrightarrow XA = P_{N(A)^{\perp}} = P_{R(A^*)} \ .$$

Let $\| \ \|$ denote the Euclidean norm. A vector x is called a <u>least-squares solution</u> of the equation

$$Ax = b$$

if it minimizes $\|b - Ax\|$, i.e., if it solves the (always consistent) equation

$$Ax = P_{R(A)} b \ .$$

Thus, (9) shows the $\{1, 3\}$-inverses to be the generalized inverses most appropriate for least-squares solutions of linear equations:

4.1. <u>Theorem</u>. Let $A \in C^{m \times n}$, $b \in C^m$. Then $\|Ax - b\|$ is smallest when $x = A^{(1,3)}b$, where $A^{(1,3)} \in A\{1, 3\}$.

Conversely, if $X \in C^{n\times m}$ has the property that, for all b, $\|Ax - b\|$ is smallest when $x = Xb$, then $X \in A\{1, 3\}$. ∎

Similarly, (10) can be restated as follows:

4.2. Theorem. Let $A \in C^{m\times n}$, $b \in C^m$. If $Ax = b$ has a solution, the unique solution for which $\|x\|$ is smallest is given by $x = A^{(1,4)}b$ where $A^{(1,4)} \in A\{1, 4\}$. Conversely, if $X \in C^{n\times m}$ is such that, whenever $Ax = b$ has a solution, $x = Xb$ is the solution of minimum norm, then $X \in A\{1, 4\}$. ∎

Now, for any $A^{(1,3)} \in A\{1, 3\}$ and $A^{(1,4)} \in A\{1, 4\}$, the matrix $A^{(1,4)}AA^{(1,3)}$ satisfies the Penrose equations (1)-(4), and therefore, by the uniqueness of the Moore-Penrose inverse

$$A^{\dagger} = A^{(1,4)}AA^{(1,3)}$$

which, combined with Theorems 4.1 and 4.2, gives:

4.3. Corollary (Penrose [23]). Let $A \in C^{m\times n}$, $b \in C^m$. Then, among the least-squares solutions of $Ax = b$, $A^{\dagger}b$ is the one of minimum norm. Conversely, if $X \in C^{n\times m}$ has the property that, for all b, Xb is the minimum-norm least-squares solution of $Ax = b$, then $X = A^{\dagger}$. ∎

Consider again equations (8). If we choose arbitrary subspaces S and T complementary to $R(A)$ and $N(A)$, respectively, does there exist a $\{1\}$-inverse X satisfying (8)? The following theorem (parts of which have appeared in the works of Robinson [26], Langenhop [17] and Milne [19]; see [2] Theorem 2.10 and Ex. 3.44) answers the question in the affirmative.

4.4. Theorem. Let $A \in C_r^{m\times n}$, and let S and T be arbitrary subspaces complementary to $R(A)$ and $N(A)$ respectively. Then:

(a) The general solution of (8) is

$$X = P_{T, N(A)} A^{(1)} P_{R(A), S} + (I_n - A^{(1)} A) Y (I_m - AA^{(1)})$$

where $A^{(1)}$ is a fixed (but arbitrary) element of $A\{1\}$ and Y is an arbitrary element of $C^{n \times m}$.

(b) The matrix

$$A^{(1,2)}_{T,S} = P_{T, N(A)} A^{(1)} P_{R(A), S} \tag{11}$$

is the unique $\{1, 2\}$-inverse of A having range T and null space S.

(c) For any $b \in C^m$, the vector

$$x = A^{(1,2)}_{T,S} b$$

minimizes $x^* Ux$ among all vectors minimizing $(Ax - b)^* W(Ax - b)$, where U and W are positive definite matrices satisfying

$$T = U^{-1} N(A)^{\perp}, \qquad S = W^{-1} R(A)^{\perp} . \tag{12}$$

Conversely, if for two given positive definite matrices $U \in C^{n \times n}$ and $W \in C^{m \times m}$, the matrix $X \in C^{n \times m}$ has the property that for all b, $x = Xb$ is the vector minimizing $x^* Ux$ among all minimizers of $(Ax - b)^* W(Ax - b)$, then $X = A^{(1,2)}_{T,S}$ where the subspaces T and S are given by (12). ■

If $T = N(A)^{\perp}$ and $S = R(A)^{\perp}$, then (11) reduces to $A^{\dagger}$ and Theorem 4.4(c) is in agreement with Corollary 4.3.

The above minimal properties of generalized inverses extend to certain norms other than the Euclidean norm. A norm β on C^n is called an essentially strictly convex (abbreviated e. s. c.) norm if

$$\left.\begin{array}{l} x \neq y \in C^n \\ \beta(x) = \beta(y) \\ 0 < \lambda < 1 \end{array}\right\} \Longrightarrow \beta(\lambda x + (1-\lambda)y) < \beta(x) .$$

If β is an e. s. c. norm on C^n, then for any subspace $L \subset C^n$ and any point $x \in C^n$ there is a unique point $y \in L$ closest to x (in the norm β):

$$\beta(y - x) = \inf \{\beta(\ell - x): \quad \ell \in L\} .$$

The β-metric projector on L, denoted by $P_{L,\beta}$, is the mapping $P_{L,\beta}: C^n \to L$ assigning to each point in C^n its (unique) closest point in L, see ([15], section 32). We note that $P_{L,\beta}$ is a continuous homogeneous mapping, but in general nonadditive, hence nonlinear.

Following Boullion and Odell ([4], pp. 43-44) we define generalized inverses associated with pairs of e. s. c. norms as follows: Let α and β be e. s. c. norms on C^m and C^n, respectively. For any $A \in C^{m\times n}$ we define the generalized inverse associated with α and β (called the α-β generalized inverse), denoted by $A^{(-1)}_{\alpha,\beta}$, as

$$(13) \qquad A^{(-1)}_{\alpha,\beta} = (I - P_{N(A),\beta}) A^{(1)} P_{R(A),\alpha}$$

where $A^{(1)}$ is any $\{1\}$-inverse of A. Properties of $A^{(-1)}_{\alpha,\beta}$ are summarized in:

4.5. Theorem (Erdelsky [9], Newman and Odell [20]).

(a) $A^{(-1)}_{\alpha,\beta}: C^m \to C^n$ is a homogeneous transformation.

(b) $A^{(-1)}_{\alpha,\beta}$ is additive (hence linear) if $P_{R(A),\alpha}$ and $P_{N(A),\beta}$ are additive.

(c) $$N(A^{(-1)}_{\alpha,\beta}) = P^{-1}_{R(A),\alpha}(0)$$

(d) $$R(A^{(-1)}_{\alpha,\beta}) = P^{-1}_{N(A),\beta}(0)$$

where, as in the case of linear transformations, we denote

$$N(A^{(-1)}_{\alpha,\beta}) = \{y \in C^m: \; A^{(-1)}_{\alpha,\beta}(y) = 0\}$$

$$R(A^{(-1)}_{\alpha,\beta}) = \{A^{(-1)}_{\alpha,\beta}(y): \; y \in C^m\} .$$

(e) $$AA^{(-1)}_{\alpha,\beta} = P_{R(A),\alpha}$$

(f) $$A^{(-1)}_{\alpha,\beta}A = I - P_{N(A),\beta}$$

(g) $$AA^{(-1)}_{\alpha,\beta}A = A$$

(h) $$A^{(-1)}_{\alpha,\beta}AA^{(-1)}_{\alpha,\beta} = A^{(-1)}_{\alpha,\beta}$$

(i) For any $b \in C^m$, a vector $x \in C^n$ minimizes $\alpha(Ax - b)$ if and only if

$$Ax = AA^{(-1)}_{\alpha,\beta}(b) .$$

Among all such minimizers, the unique vector of minimum β-norm is $x = A^{(-1)}_{\alpha,\beta}(b)$. ∎

If both norms α and β are Euclidean, then

$$P_{R(A),\alpha} = P_{R(A)}, \quad P_{N(A),\beta} = P_{N(A)}$$

and $A^{(-1)}_{\alpha,\beta}$, defined by (13), reduces to $A^\dagger$.

Another generalized inverse with particularly useful minimal properties is the Bott-Duffin inverse [3], see also ([2], §§ 2.9, 2.12 and 3.5).

5. Full-rank factorizations and partitioned matrices

A full-rank factorization of a matrix $A \in C_r^{m\times n}$, $r > 0$, is its representation as the product of two matrices of full rank

$$A = FG, \quad F \in C_r^{m\times r}, \quad G \in C_r^{r\times n}. \tag{14}$$

Such factorizations are useful since in general it is easier to study and compute the generalized inverses of full-rank matrices, e.g.,

$$F^\dagger = (F^*F)^{-1}F^*, \quad G^\dagger = G^*(GG^*)^{-1}. \tag{15}$$

5.1. Theorem. Let (14) be a full-rank factorization of $A \in C_r^{m\times n}$, $r > 0$.

Then:

(a) $\quad G^{(i)}F^{(1)} \in A\{i\} \qquad (i = 1, 2, 4)$

(b) $\quad G^{(1)}F^{(j)} \in A\{j\} \qquad (j = 1, 2, 3)$

(c)
$$A^\dagger = G^\dagger F^{(1,3)} = G^{(1,4)} F^\dagger$$
$$= G^\dagger F^\dagger$$
$$= G^*(GG^*)^{-1} (F^*F)^{-1} F^*$$
$$= G^*(F^*AG^*)^{-1} F^* \;. \qquad \blacksquare$$

Here $G^{(i)}$ denotes an arbitrary $\{i\}$-inverse of G, etc.

A full-rank factorization (14) can be computed as follows. Since $A \in C_r^{m\times n}$, $r > 0$, A has at least one nonsingular $r \times r$ submatrix A_{11}, which by a rearrangement of rows and columns can be brought to the top left corner of A, say

$$\text{(16)} \qquad PAQ = \begin{bmatrix} A_{11} & A_{12} \\ A_{21} & A_{22} \end{bmatrix}$$

where P and Q are permutation matrices. Thus

$$\text{(17)} \qquad A = P^T \begin{bmatrix} A_{11} & A_{12} \\ A_{21} & A_{22} \end{bmatrix} Q^T$$

$$= P^T \begin{bmatrix} I_r \\ S \end{bmatrix} A_{11}[I_r \quad T]Q^T$$

where

$$T = A_{11}^{-1} A_{12}, \quad S = A_{21} A_{11}^{-1} \;.$$

A full-rank factorization of A is, by (17),

$$F = P^T \begin{bmatrix} I_r \\ S \end{bmatrix} A_{11}, \quad G = [I_r \quad T] Q^T$$

The partition (17) can also be used directly to compute generalized inverses.

5.2. <u>Theorem</u>. Let $A \in C_r^{m\times n}$, $r > 0$, be partitioned as in (17). Then

(a) A $\{1, 2\}$-inverse of A is

$$A^{(1,2)} = Q \begin{bmatrix} A_{11}^{-1} & 0 \\ 0 & 0 \end{bmatrix} P \quad \text{(C. R. Rao [24])}$$

(b) A $\{1, 2, 3\}$-inverse of A is

$$A^{(1,2,3)} = Q \begin{bmatrix} A_{11}^{-1} \\ 0 \end{bmatrix} (I_r + S^*S)^{-1} [I_r \quad S^*]P$$

(Meyer and Painter [18])

(c) A $\{1, 2, 4\}$-inverse of A is

$$A^{(1,2,4)} = Q \begin{bmatrix} I_r \\ T^* \end{bmatrix} (I_r + TT^*)^{-1} [A_{11}^{-1} \quad 0]P$$

(d) The Moore-Penrose inverse of A is

$$A^{\dagger} = Q \begin{bmatrix} I_r \\ T^* \end{bmatrix} (I_r + TT^*)^{-1} A_{11}^{-1} (I_r + S^*S)^{-1} [I_r \quad S^*]P$$

(Noble [21]). ∎

For further results on generalized inverses of partitioned matrices see, e.g., [5] and ([2], Chapter 5).

6. Spectral generalized inverses

Let A be a square matrix, and let λ be an eigenvalue of A. A vector x such that

$$(A - \lambda I)^p x = 0, \quad (A - \lambda I)^{p-1} x \neq 0 ,$$

where p is a positive integer, is called a principal vector of A of grade p associated with the eigenvalue λ, abbreviated λ-vector of A of grade p. A is called diagonable if all its principal vectors have grade 1, i.e., are eigenvectors.

Spectral generalized inverses are generalized inverses having certain spectral properties, usually stated in terms of the principal vectors. In particular, X is called an S-inverse of A (or A and X S-inverses of each other) if they share the property that, for every $\lambda \in C$ and every vector x, x is a λ-vector of A of grade p if and only if it is a $\lambda^\dagger$-vector of X of grade p. Here

$$\lambda^\dagger = \begin{cases} \frac{1}{\lambda}, & \lambda \neq 0 \\ 0, & \lambda = 0 . \end{cases}$$

Also, X is an S'-inverse of A (or A and X are S'-inverses of each other) if, for all $\lambda \neq 0$ a vector x is a λ^{-1}-vector of X of grade p if and only if it is a λ-vector of A of grade p, and x is a 0-vector of X if and only if it is a 0-vector of A (without regard to grade). The best known spectral generalized inverses of a square matrix A are its Drazin pseudoinverse $A^{(d)}$ (the unique $\{1^k, 2, 5\}$-inverse, $k \geq$ index A) and its group inverse $A^{\#}$ (the unique $\{1, 2, 5\}$-inverse) when it exists, i.e., when index $A = 1$.

6.1. Theorem. Let $A \in C^{n \times n}$ have index 1. Then $A^{\#}$ is the unique S-inverse of A in $A\{1\} \cup A\{2\}$. If A is diagonable, $A^{\#}$ is the only S-inverse of A. ([2], Theorem 4.5). ∎

When it exists, the group inverse $A^{\#}$ can be computed by using a full-rank factorization of A.

6.2. Theorem (Cline [6]). Let a square matrix A have a full-rank factorization $A = FG$. Then A has a group inverse if and only if GF is nonsingular, in which case

$$A^{\#} = F(GF)^{-2}G \quad . \qquad \blacksquare$$

For other computations of $A^{\#}$ see [10] and [25].

The spectral properties of the Drazin pseudoinverse are the same as those of the group inverse with regard to nonzero eigenvalues, but weaker for 0-vectors.

6.3. Theorem. For every square matrix A, A and $A^{(d)}$ are S'-inverses of each other. ([2], Theorem 4.9). ■

The Drazin pseudoinverse of a square matrix A with index k is given by

$$A^{(d)} = A^k(q(A))^{k+1}$$

where the polynomial q is obtained from the minimum polynomial of A

$$m(\lambda) = c\,\lambda^k(1 - \lambda q(\lambda)), \qquad c \neq 0 \quad ,$$

see ([2], Theorem 4.7).

The following theorem is a variation of a well known result by Wedderburn [27].

6.4. Theorem. A square matrix A has a unique decomposition

$$A = B + N$$

such that B has index 1, N is nilpotent, and

$$BN = NB = 0 \ .$$

Moreover

$$B = (A^{(d)})^{\#} \ . \qquad \text{([2], Theorem 4.10)} \ . \qquad \blacksquare$$

This theorem shows that the Drazin pseudoinverse satisfies

$$(A^{(d)})^{(d)} = A \iff \text{index } A = 1$$

in which case it coincides with the group inverse.

From Theorem 6.4 we also conclude the following

6.5. <u>Theorem</u>. Let $A \in C^{n\times n}$. Then A and X are S'-inverses of each other if

(18) $$X^{(d)} = (A^{(d)})^{\#} \ .$$

Moreover, if $X \in A\{1\} \cup A\{2\}$ it is an S'-inverse of A only if (18) holds. ([2], Theorem 4.11). ■

7. A spectral theory for rectangular matrices

Generalized inverses extend to singular and rectangular matrices some of the privileges hitherto enjoyed only by nonsingular matrices. It is therefore not surprising that generalized inverses play a role, although a minor one, in the spectral theory for rectangular matrices, see, e.g., [22], [16], [13], [14] and ([2], Chapter 6).

The following theorem gives a variation of the singular value decomposition, as given by Eckart and Young [8], see also ([2], p. 242) and [12].

7.1. <u>Theorem</u>. Let $A \in C_r^{m\times n}$, $r > 0$, let

$$\alpha_1 \geq \alpha_2 \geq \dots \geq \alpha_r > 0$$

be the nonzero singular values of A and let $\{d_1, d_2, \dots, d_r\}$ be any complex scalars satisfying

(19) $$|d_i| = \alpha_i, \qquad i = 1, \dots, r \; .$$

Then there exist unitary matrices $U \in C^{m \times m}$ and $V \in C^{n \times n}$ such that the matrix

(20) $$D = U^* AV = \begin{bmatrix} d_1 & & & \vdots & \\ & \ddots & & \vdots & 0 \\ & & d_r & \vdots & \\ \cdots & \cdots & \cdots & \cdots & \cdots \\ & 0 & & \vdots & 0 \end{bmatrix}$$

is diagonal. ∎

Here, the first r columns of U are an orthonormal set $\{u_1, u_2, \dots, u_r\}$ of eigenvectors of AA^* corresponding to its r nonzero eigenvalues

$$AA^* u_i = \alpha_i^2 u_i, \qquad i = 1, \dots, r \; ,$$

and the first r columns of V

(21) $$v_i = \frac{1}{\bar{d}_i} A^* u_i, \qquad i = 1, \dots, r \; ,$$

are the corresponding eigenvectors of A^*A. The matrices

(22) $$E_i = u_i v_i^* \qquad i = 1, \dots, r$$

are then partial isometries of rank 1, and their sum

(23) $$E = \sum_{i=1}^{r} E_i$$

is a partial isometry mapping $R(A^*)$ onto $R(A)$,

$$EE^* = P_{R(A)}, \quad E^*E = P_{R(A^*)} .$$

In terms of these partial isometries, Theorem 7.1 can be restated as:

7.2. <u>Theorem</u> (Penrose [22]). Let A, $\{\alpha_1, \ldots, \alpha_r\}$ and $\{d_1, \ldots, d_r\}$ be as in Theorem 7.1. Then the partial isometries $\{E_1, \ldots, E_r\}$ and E satisfy

(24) $$E_i E_j^* = 0, \quad E_i^* E_j = 0, \qquad 1 \le i \ne j \le r .$$

(25) $$E_i E^* A = A E^* E_i$$

(26) $$A = \sum_{i=1}^{r} d_i E_i .$$ ■

If A is a normal matrix, and the scalars $\{d_1, \ldots, d_r\}$ in Theorem 7.1 are chosen as the nonzero eigenvalues of A, then (21) gives

$$v_i = u_i, \qquad i = 1, \ldots, r ,$$

and the partial isometries (22) and (23) are orthogonal projectors. Theorem 7.2 reduces then to the spectral theorem for normal matrices.

In the general rectangular case, Theorem 7.2 is the basis of a spectral theory, illustrated by the following

7.3. Theorem. Let $A \in C_r^{m \times n}$, $r > 0$, be represented by (26).

Let $\{\hat{d}_j : i = 1, \ldots, q\}$ be the set of distinct $\{d_i : i = 1, \ldots, r\}$, and for each $j = 1, \ldots, q$ let Γ_j be a contour (i.e., a closed rectifiable Jordan curve positively oriented in the customary way) surrounding $\hat{d}_j$ but no other $\hat{d}_k$. Then:

(a) For each $j = 1, \ldots, q$, the partial isometry

$$\hat{E}_j = \sum_{\substack{i \\ \{d_i = \hat{d}_j\}}} E_i$$

satisfies

$$\hat{E}_j^* = \frac{1}{2\pi i} \int_{\Gamma_j} (zE - A)^\dagger dz, \qquad j = 1, \ldots, q \ .$$

(b) If the function $f: C \to C$ is analytic in a domain containing the set surrounded by

$$\Gamma = \bigcup_{j=1}^{q} \Gamma_j$$

then

$$\sum_{j=1}^{r} f(d_j) E_j^* = \frac{1}{2\pi i} \int_{\Gamma} f(z)(zE - A)^\dagger dz \tag{27}$$

([2], Lemma 6.1) . ∎

In particular,

$$A^\dagger = \sum_{i=1}^{r} \frac{1}{d_i} E_i^* \qquad \text{(by Theorem 7.2)}$$

$$= \frac{1}{2\pi i} \int_{\Gamma} \frac{1}{z} (zE - A)^\dagger dz, \qquad \text{by (27)} \ .$$

8. Example: The orthogonal projector on the intersection of subspaces

The power of generalized inverses is demonstrated by the following theorem, giving a closed form expression for the orthogonal projector $P_{L\cap M}$ on the intersection of two given subspaces L, M in C^n.

8.1. Theorem (Anderson and Duffin [1]). Let P_L and P_M be the orthogonal projectors on L and M, respectively. Then

$$P_{L\cap M} = 2P_L(P_L + P_M)^\dagger P_M$$
$$= 2P_M(P_L + P_M)^\dagger P_L .$$

Proof. Since $M \subset L + M$ it follows that

$$P_{L+M} P_M = P_M = P_M P_{L+M}$$

which can be rewritten as

$$(28) \quad (P_L + P_M)(P_L + P_M)^\dagger P_M = P_M = P_M(P_L + P_M)^\dagger (P_L + P_M)$$

since

$$(29) \quad P_{L+M} = (P_L + P_M)(P_L + P_M)^\dagger$$
$$= (P_L + P_M)^\dagger(P_L + P_M) .$$

Subtracting $P_M(P_L + P_M)^\dagger P_M$ from the first and last expressions in (28) gives

$$P_L(P_L + P_M)^\dagger P_M = P_M(P_L + P_M)^\dagger P_L .$$

Now let

$$H = 2P_L(P_L + P_M)^{\dagger}P_M = 2P_M(P_L + P_M)^{\dagger}P_L \ .$$

Evidently, $R(H) \subset L \cap M$ and therefore

$$H = P_{L\cap M}H = P_{L\cap M}(P_L(P_L + P_M)^{\dagger}P_M + P_M(P_L + P_M)^{\dagger}P_L)$$

$$= P_{L\cap M}(P_L + P_M)^{\dagger}(P_L + P_M)$$

$$= P_{L\cap M}P_{L+M}, \quad \text{by (29)}$$

$$= P_{L\cap M}, \quad \text{since } L\cap M \subset L + M \ . \qquad \blacksquare$$

References

[1] Anderson, W. N. Jr.; Duffin, R. J., Series and parallel addition of matrices, J. Math. Anal. Appl. 26 (1969), 576-594.

[2] Ben-Israel, A.; Greville, T. N. E., Generalized Inverses: Theory and Applications, John Wiley & Sons, Inc., New York, 1974, xi + 395 pp.

[3] Bott, R.; Duffin, R. J., On the algebra of networks, Trans. Amer. Math. Soc. 74 (1953), 99-109.

[4] Boullion, T. L.; Odell, P. L., Generalized Inverse Matrices, John Wiley & Sons, Inc., New York, 1971, x + 103 pp.

[5] Cline, R. E., Representations for the generalized Inverse of a partitioned matrix, J. Soc. Indust. Appl. Math. 12 (1964), 588-600.

[6] Cline, R. E., Inverses of rank invariant powers of a matrix, SIAM J. Numer. Anal. **5** (1968), 182-197.

[7] Drazin, M. P., Pseudo inverses in associative rings and semigroups, Amer. Math. Monthly **65** (1958), 506-514.

[8] Eckart, C.; Young, G., A principal axis transformation for non-Hermitian matrices, Bull. Amer. Math. Soc. **45** (1939), 118-121.

[9] Erdelsky, P. J., Projections in a Normed Linear Space and a Generalization of the Pseudo-Inverse, doctoral dissertation in mathematics, California Institute of Technology, Pasadena, 1969.

[10] Erdelyi, I., On the matrix equation $Ax = \lambda Bx$, J. Math. Anal. Appl. **17** (1967), 119-132.

[11] Fisher, A. G., On construction and properties of the generalized inverse, SIAM J. Appl. Math. **15** (1967), 269-272.

[12] Golub, G. H.; Kahan, W., Calculating the singular values and pseudo-inverse of a matrix, SIAM J. Numer. Anal. **2** (1965), 205-224.

[13] Hestenes, M. R., Inversion of matrices by biorthogonalization and related results, J. Soc. Indust. Appl. Math. **6** (1958), 51-90.

[14] Hestenes, M. R., A ternary algebra with applications to matrices and linear transformations, Arch. Rational Mech. Anal. **11** (1962), 138-194.

[15] Holmes, R. B., A Course on Optimization and Best Approximation, Springer-Verlag, Berlin, 1972, viii+233 pp.

[16] Lanczos, C., Linear systems in self-adjoint form, Amer. Math. Monthly **65** (1958), 665-679.

[17] Langenhop, C. E., On generalized inverses of matrices, SIAM J. Appl. Math. **15** (1967), 1239-1246.

[18] Meyer, C. D.; Painter, R. J., Note on a least squares inverse for a matrix, J. Assoc. Comput. Mach. **17** (1970), 110-112.

[19] Milne, R. D., An oblique matrix pseudoinverse, SIAM J. Appl. Math. **16** (1968), 931-944.

[20] Newman, T. G.; Odell, P. L., On the concept of a p-q generalized inverse of a matrix, SIAM J. Appl. Math. **17** (1969), 520-525.

[21] Noble, B., A method for computing the generalized inverse of a matrix, SIAM J. Numer. Anal. **3** (1966), 582-584.

[22] Penrose, R., A generalized inverse for matrices, Proc. Cambridge Philos. Soc. **51** (1955), 406-413.

[23] Penrose, R., On best approximate solution of linear matrix equations, Proc. Cambridge Philos. Soc. **52** (1956), 17-19.

[24] Rao, C. R., Linear Statistical Inference and its Applications, John Wiley & Sons, Inc., New York, 1965, xviii + 522 pp.

[25] Robert, P., On the group-inverse of a linear transformation, J. Math. Anal. Appl. **22** (1968), 658-669.

[26] Robinson, D. W., On the generalized inverse of an arbitrary linear transformation, Amer. Math. Monthly **69** (1962), 412-416.

[27] Wedderburn, J. H. M., Lectures on Matrices, Amer. Math. Soc. Colloq. Publ. Vol. XVII, American Mathematical Society, Providence, R. I., 1934.

SOME TOPICS IN GENERALIZED INVERSES OF MATRICES

Adi-Ben-Israel, Department of Applied Mathematics, Technion-Israel Institute of Technology and Department of Industrial Engineering and Management Sciences, Northwestern University.

Thomas N. E. Greville, Mathematics Research Center, University of Wisconsin and National Center for Health Statistics.

Research supported by the National Science Foundation Grant GP 38867 and by the United States Army under Contract No. DA-31-124-ARO-D-462.

The Fredholm Pseudoinverse— An Analytic Episode in the History of Generalized Inverses

L.B. Rall

In 1903, Ivar Fredholm formulated a pseudoinverse for a linear integral operator which is not invertible in the ordinary sense. Closely related operators include the Hurwitz pseudoresolvent (1912) and an operator obtained by analytic continuation of the Fredholm resolvent. Properties of these pseudoinverses will be discussed from the standpoint of the theory of generalized inverses developed in recent years. In particular, comparisons of these operators with the Moore-Penrose generalized inverse and the Drazin pseudoinverse will be made.

1. Historical background.

In his famous paper, Fredholm [4] considered the problem of finding a continuous solution $f(x)$ of the "functional equation",

$$f(x) - \lambda \int_0^1 K(x,t)\, f(t)dt = g(x), \qquad 0 \leqq x \leqq 1 , \tag{1.1}$$

in which $g(x)$ and the kernel $K(x,t)$ are given continuous functions, and λ is a given numerical parameter. The basic result obtained by Fredholm was that except for certain isolated values of λ, equation (1.1) has a unique solution which can be expressed in the form

$$f(x) = g(x) + \lambda \int_0^1 \Gamma(x,t;\lambda)\, g(t)dt, \qquad 0 \leqq x \leqq 1 , \tag{1.2}$$

in which the resolvent kernel $\Gamma(x, t; \lambda)$ of $K(x, t)$ is continuous. For the exceptional values of λ, now generally called eigenvalues of $K(x, t)$, Fredholm showed that equation (1.1) has no solutions unless $g(x)$ satisfies certain conditions, in which case an infinite family of solutions exists. In the latter case, Fredholm was able to construct a solution of equation (1.1) in the form of an integral transformation of the type (1.2), which he termed the pseudoinverse of the integral transformation of $f(x)$ defined by the left side of (1.1).

The success of Fredholm with this problem led to a rapid development of the theory of integral equations and formed the basis of a number of concepts of functional analysis. Equation (1.1) is now called a linear integral equation of Fredholm type and second kind in recognition of his contribution. The idea of the pseudoinverse of an integral operator, however, was not pursued as intensively as some of the other concepts presented by Fredholm, although Hurwitz [9] added to the theory in 1912. The idea of the "general reciprocal" of a matrix, announced by E. H. Moore to his students sometime between 1903 and 1906 [1, 6, 11] also languished in obscurity for a number of years, but has since been found to be fundamental to the theory of generalized inverses. Because of the current interest in this subject, it is worthwhile to see how the early idea presented by Fredholm fits into the general picture. As Fredholm made extensive use of the theory of determinants in his treatment of equation (1.1), it might also be possible to find earlier anticipations of the notion of a pseudoinverse of an algebraic operator (see Hellinger and Toeplitz [8], pp. 1356, 1373-1374).

2. Pseudoinverses and the solution of equations.

The basic mathematical problem of solving an equation may be formulated abstractly as follows: Given an operator A which maps a set F into a set G, and an element g of G, find an element f of F such that

$$Af = g \quad . \tag{2.1}$$

If A is one-to-one and onto G, then there exists the _inverse_ operator A^{-1} to A which gives the unique solution of (2.1) as $f = A^{-1}g$. Otherwise, there exists at least one value of g for which equation (2.1) is _inconsistent_ (has no solution), or is _consistent_, but has more than one solution. In order to solve equation (2.1) if it is consistent, one needs an operator X from F into G such that

$$A(Xg) = g \quad . \tag{2.2}$$

It is evident that any operator X which satisfies the relationship

$$AXA = A \tag{1}$$

will be suitable for this purpose. Following the notation of Ben-Israel and Greville [1], any operator X satisfying (1) will be called a _{1}-pseudoinverse_ of A. It may be noted that (1) is satisfied automatically by $X = A^{-1}$ if A is invertible.

In addition to solving equation (2.1) if consistent, a {1}-pseudoinverse X of A may be used to define a _pseudosolution_ $u = Xg$ of an inconsistent equation, assuming, of course, that X is defined on all of G. Pseudosolutions of inconsistent equations are important in many applications, but will not be dealt with further in the present work.

The Fredholm integral equation (1.1) is the special case of (2.1) obtained by taking $F = G = C[0,1]$, the linear space of continuous functions on the interval $[0,1]$, and

$$A = I - \lambda K \quad , \tag{2.3}$$

where I denotes the identity operator, and K is the linear integral operator with the continuous kernel $K(x,t)$. The various kernels, functions, and parameters considered may take on complex values in general. The value $\lambda = 0$ leads to a trivial problem, and will be excluded throughout. Most of the results obtained in this classical setting may be extended to integral operators in more general function spaces without much difficulty, or specialized to square matrices.

By analogy with (1.2), if λ is an eigenvalue of $K(x,t)$, then it is natural to seek $\{1\}$-pseudoinverses of the operator (2.3) in the form

$$X = I + \lambda H(\lambda) \tag{2.4}$$

where the pseudoresolvent operator $H(\lambda)$ has the continuous kernel $H(x,t;\lambda)$. In this case, (1) becomes

$$H = K - \lambda K^2 + \lambda KH + \lambda HK - \lambda^2 KHK \ . \tag{2.5}$$

Recalling that the composition LM of linear integral operators with kernels $L(x,t)$ and $M(x,t)$, respectively, has the kernel

$$(LM)(x,t) = \int_0^1 L(x,s)\, M(s,t)ds, \qquad 0 \leqq x,\ t \leqq 1 \ , \tag{2.6}$$

the relationship (2.5) may be written as an identity for continuous functions of two variables. In actual practice, it is usually easier to verify (1) directly than to use (2.5) or the equivalent expression in terms of kernels.

3. Other properties of pseudoinverses.

In the development of the theory of pseudoinverses, it has been found useful to require these operators to satisfy more than one relationship characteristic of the inverse of an invertible operator. These more restricted classes of pseudoinverses have various interesting and useful properties which are dealt with extensively elsewhere. For the present purpose, it will be adequate to list the properties of some of these pseudoinverses for comparison with the types of operators arising from integral equations.

A pseudoinverse X of A may satisfy the relationship

$$XAX = X \ , \tag{2}$$

which may be interpreted as stating that A is a {1}-pseudoinverse of X .

In the case of linear integral operators, one may define the conjugate transpose K^* of K to be the operator with the kernel

$$K^*(x, t) = \overline{K(t, x)}, \qquad 0 \leqq x, \ t \leqq 1 \ , \tag{3.1}$$

with the bar denoting complex conjugation. One also has $I^* = I$ and $(A + B)^* = A^* + B^*$ for linear operators A, B of the form (2.3) in $C[0,1]$. An operator is said to be symmetric if it is equal to its conjugate transpose. The relationships

$$(AX)^* = AX \ , \tag{3}$$

$$(XA)^* = XA \ , \tag{4}$$

require AX and XA to be symmetric operators. The {1, 2, 3, 4}-pseudoinverse X of A is unique, and is the famous Moore-Penrose generalized inverse of A . This operator has been studied intensively (see [1] for historical details and a bibliography). It follows that A is also the Moore-Penrose generalized inverse of X, a close analogy to the case of an invertible operator and its inverse.

In the study of integral equations, one also encounters a pseudoinverse which is defined in a somewhat different fashion. For F = G, one may consider a commuting pseudoinverse X such that

$$AX = XA \ , \tag{5}$$

and require that

$$A^{k+1}X = A^k \tag{1^k}$$

for some positive integer k . The $\{2, 5, 1^k\}$-pseudoinverse of A is unique, and is called the Drazin pseudoinverse of A [2]. If (1^k) holds for k = 1, then the Drazin pseudoinverse is a $\{1\}$-pseudoinverse of A, and may be used to solve consistent equations of the form (2.1). In addition, if A and its Drazin pseudoinverse are symmetric, then (3) and (4) follow from (5), and the Drazin pseudoinverse and Moore-Penrose generalized inverses of A coincide.

On the other hand, suppose that the smallest value of k for which (1^k) holds is greater than one. Then,

$$AXA^k = A^k \quad , \tag{3.2}$$

and the Drazin pseudoinverse of A may be used to solve consistent equations of the form

$$Au = q \tag{3.3}$$

on the subset A^kF of F defined by

$$A^kF = \{u: \ u = A^kf, \quad f \in F\} \ . \tag{3.4}$$

4. The Fredholm resolvent.

Attention will now be returned to the integral equation (1.1). To avoid clutter, all the following notation will be based on the assumption that all quantities considered are real, the transition to the complex case being simple. First, a description of the construction of the resolvent kernel $\Gamma(x, t;\lambda)$ in (1.2) will be given, as the pseudoresolvent kernels to be described later are related to $\Gamma(x, t;\lambda)$ in one way or another.

If λ is not an eigenvalue of $K(x, t)$, then the integral transformation $I + \lambda\Gamma(\lambda)$ appearing in (1.2) is the inverse of $I - \lambda K$, and one has

$$(I - \lambda K)(I + \lambda\Gamma(\lambda)) = (I + \lambda\Gamma(\lambda))(I - \lambda K) = I \ , \tag{4.1}$$

from which follow the resolvent equations

(4.2) $$\Gamma(\lambda) = K + \lambda K\Gamma(\lambda) = K + \lambda\Gamma(\lambda)K ,$$

which incidentally show that K and $\Gamma(\lambda)$ commute. By analogy with linear systems of equations, one may look for a representation of $\Gamma(\lambda)$ as

(4.3) $$\Gamma(\lambda) = \frac{N(\lambda)}{\Delta(\lambda)} = \frac{K_1 + \lambda K_2 + \lambda^2 K_3 + \ldots}{1 + c_1\lambda + c_2\lambda^2 + \ldots} ,$$

where the operators $K_1, K_2, \ldots$ have continuous kernels $K_1(x, t), K_2(x, t), \ldots,$ and $c_1, c_2, \ldots$ are constants. Then,

(4.4) $$N(\lambda) = \Delta(\lambda)K + \lambda KN(\lambda) = \Delta(\lambda)K + \lambda N(\lambda)K ,$$

and formal manipulations with the infinite series in (4.3) give the relationships

(4.5) $$K_1 = K ,$$ $$K_{n+1} = c_n K + KK_n = c_n K + K_n K, \qquad n = 1, 2, \ldots .$$

This procedure does not determine the constants $c_1, c_2, \ldots$. For example, if one takes $c_1 = c_2 = \ldots = 0$, then this construction yields the <u>Neumann series</u>

(4.6) $$\Gamma(\lambda) = K + \lambda K^2 + \lambda^2 K^3 + \ldots$$

for the resolvent operator $\Gamma(\lambda)$. It is known that this series may converge only for $|\lambda|$ sufficiently small [10]. It is possible, however, to choose $c_1, c_2, \ldots$ in such a way that both the numerator and denominator of (4.3) are entire functions of λ, the series converging for all finite λ. This choice is

(4.7) $$c_n = -\frac{1}{n} \operatorname{tr} K_n = -\frac{1}{n}\int_0^1 K_n(x, x)dx, \qquad n = 1, 2, \ldots .$$

(For a given linear integral operator L with kernel $L(x, t)$, the number

$$\text{tr } L = \int_0^1 L(x, x)dx \tag{4.8}$$

is called the <u>trace</u> of L.)

Fredholm was able to establish the convergence of the series for $\Delta(\lambda)$ and $N(\lambda)$ by expressing them in terms of determinants. For

$$K\begin{pmatrix} x_1, x_2, \ldots, x_n \\ t_1, t_2, \ldots, t_n \end{pmatrix} = \begin{vmatrix} K(x_1, t_1) & K(x_1, t_2) & \ldots & K(x_1, t_n) \\ K(x_2, t_1) & K(x_2, t_2) & \ldots & K(x_2, t_n) \\ \vdots & \vdots & & \vdots \\ K(x_n, t_1) & K(x_n, t_2) & \ldots & K(x_n, t_n) \end{vmatrix}, \tag{4.9}$$

it can be shown by mathematical induction that

$$K_n(x, t) = \frac{(-1)^{n-1}}{(n-1)!} \int_0^1 \ldots \int_0^1 K\begin{pmatrix} x, s_1, \ldots, s_{n-1} \\ t, s_1, \ldots, s_{n-1} \end{pmatrix} ds_1 \ldots ds_{n-1} \tag{4.10}$$

and

$$c_n = \frac{(-1)^n}{n!} \int_0^1 \ldots \int_0^1 K\begin{pmatrix} s_1, \ldots, s_n \\ s_1, \ldots, s_n \end{pmatrix} ds_1 \ldots ds_n \tag{4.11}$$

satisfy (4.5) and (4.7) for $n = 1, 2, \ldots$. As $K(x, t)$ is continuous on $0 \leqq x, t \leqq 1$, a positive constant μ exists such that

$$|K(x, t)| \leqq \mu, \quad 0 \leqq x, t \leqq 1, \tag{4.12}$$

and Hadamard's inequality for determinants [10] asserts that

$$\left| K\begin{pmatrix} x_1, \ldots, x_n \\ t_1, \ldots, t_n \end{pmatrix} \right| \leqq \mu^n n^{\frac{n}{2}} , \tag{4.13}$$

from which follows the convergence of the series for $\Delta(\lambda)$ and $N(\lambda)$ for all finite λ .

The algorithm (4.5) and (4.7) thus provides a means for constructing the resolvent operator $\Gamma(\lambda)$ of K if $\Delta(\lambda) \neq 0$. In spite of its simplicity, this procedure is suspect from a numerical standpoint even for square matrices, in which case $\Delta(\lambda)$ and $N(\lambda)$ are of polynomial form [3]. It follows by continuity that (4.4) holds for all λ, and as $\Delta(\lambda)$ and $N(\lambda)$ are entire functions of λ, it may be differentiated any number of times with respect to λ to obtain the identities

$$\begin{cases} (I - \lambda K)\, N^{(m)}(\lambda) = \Delta^{(m)}(\lambda)K + mKN^{(m-1)}(\lambda) \\ \qquad = \Delta^{(m)}(\lambda)K + mN^{(m-1)}(\lambda)K \\ \qquad = N^{(m)}(\lambda)\,(I - \lambda K), \quad m = 1, 2, \ldots \ . \end{cases} \tag{4.14}$$

The important trace relationships

$$-\mathrm{tr}\ N^{(m)}(\lambda) = \Delta^{(m+1)}(\lambda), \qquad m = 0, 1, 2, \ldots , \tag{4.15}$$

where $N^{(0)}(\lambda) = N(\lambda)$, also follow directly from the series definitions of $\Delta(\lambda)$ and $N(\lambda)$.

It follows that if $\Delta(\lambda) \neq 0$, then the Fredholm integral operator $I - \lambda K$ has the inverse operator $I + \lambda\Gamma(\lambda)$, with $\Gamma(\lambda)$ given by (4.3), (4.5), and (4.7). On the other hand, suppose that $\Delta(\lambda) = 0$. Excluding the trivial case that $K(x, t) \equiv 0$, as $\Delta(\lambda)$ and $N(\lambda)$ are entire functions of λ which do not vanish identically, a positive integer p exists such that

$$\Delta(\lambda) = \Delta'(\lambda) = \ldots = \Delta^{(p-1)}(\lambda) = 0, \qquad \Delta^{(p)}(\lambda) \neq 0 \ . \tag{4.16}$$

By (4.15), a non-negative integer $q < p$ exists such that

$$N(\lambda) = \ldots = N^{(q-1)}(\lambda) = 0, \quad N^{(q)}(\lambda) \neq 0 , \tag{4.17}$$

which means that the continuous function $\frac{\partial^q}{\partial \lambda^q} N(x, t; \lambda)$ does not vanish identically, where $N(x, t; \lambda)$ is of course the kernel of $N(\lambda)$, and thus a value $t = t_1$ exists such that the continuous function

$$N_1^{(q)}(x) = \frac{\partial^q}{\partial \lambda^q} N(x, t_1; \lambda), \qquad 0 \leqq x \leqq 1 , \tag{4.18}$$

is not identically zero. However, from (4.14),

$$(I - \lambda K) N^{(q)}(\lambda) = 0 , \tag{4.19}$$

which means that $\varphi(x) = N_1^{(q)}(x)$ is a nontrivial continuous solution of the homogeneous integral equation

$$(I - \lambda K)\varphi = 0 , \tag{4.20}$$

and the operator $I - \lambda K$ cannot have an inverse in this case. This establishes the well-known fact that the condition $\Delta(\lambda) \neq 0$ is necessary and sufficient for the existence of the inverse of $I - \lambda K$. Consequently, the case $\Delta(\lambda) = 0$ will be the one of interest in what follows. A nontrivial function $\varphi(x)$ satisfying (4.20) will be called a <u>right eigenfunction</u> of K; similarly, if $\psi(t)$ does not vanish identically, and

$$\psi(t) - \lambda \int_0^1 \psi(x) K(x, t) dx = 0, \qquad 0 \leqq t \leqq 1 , \tag{4.21}$$

or

$$\psi(I - \lambda K) = 0 , \tag{4.22}$$

then $\psi(x)$ is said to be a <u>left eigenfunction</u> of K.

5. The Fredholm pseudoinverse.

In order to simplify the following discussion somewhat, it will be helpful to introduce some notation. An integral operator R is said to be of finite rank n if its kernel $R(x, t)$ is of the form

$$(5.1)\qquad R(x,t) = \sum_{i=1}^{n} u_i(x)\, v_i(t), \qquad 0 \leqq x, t \leqq 1 ,$$

with $u_1(x), \ldots, u_n(x)$ linearly independent, and $v_1(t), \ldots, v_n(t)$ also linearly independent. It is convenient to use the dyadic notation of Friedman [5] for an operator of this type, and write

$$(5.2)\qquad R = \sum_{i=1}^{n} u_i \rangle \langle v_i .$$

The notation

$$(5.3)\qquad \langle u, v \rangle = \int_0^1 u(s)\, v(s) ds$$

will be used for the inner product of continuous functions $u(x)$, $v(x)$.

Fredholm constructed a pseudoresolvent kernel $H(x, t; \lambda)$ for $K(x, t)$, λ an eigenvalue, in terms of what he called minors of the determinant $\Delta(\lambda)$, once again apparently inspired by similar results for linear systems of equations. In order to define these minors, one starts from the functions

$$(5.4)\quad K_{n,p}\begin{pmatrix} x_1, \ldots, x_p \\ t_1, \ldots, t_p \end{pmatrix} =$$

$$\frac{(-1)^{n-1}}{(n-1)!}\int_0^1 \cdots \int_0^1 K\begin{pmatrix} x_1, \ldots, x_p; s_1, \ldots, s_{n-1} \\ t_1, \ldots, t_p; s_1, \ldots, s_{n-1} \end{pmatrix} ds_1 \ldots ds_{n-1}$$

of the 2p variables $x_1, \ldots, x_p; t_1, \ldots, t_p$, where the determinant under the integral signs is given by (4.9). For $n = 1$, (5.4) is taken to mean

$$K_{1,p}\begin{pmatrix} x_1, \ldots, x_p \\ t_1, \ldots, t_p \end{pmatrix} = K\begin{pmatrix} x_1, \ldots, x_p \\ t_1, \ldots, t_p \end{pmatrix}, \tag{5.5}$$

while for $p = 0$,

$$K_{n,0} = \frac{(-1)^{n-1}}{(n-1)!}\int_0^1 \cdots \int_0^1 K\begin{pmatrix} s_1, \ldots, s_{n-1} \\ s_1, \ldots, s_{n-1} \end{pmatrix} ds_1 \ldots ds_{n-1} = c_{n-1} \tag{5.6}$$

by (4.11), with

$$K_{1,0} = c_0 = 1 \tag{5.7}$$

by definition. Note also that

$$K_{n,1}\begin{pmatrix} x \\ t \end{pmatrix} = \tag{5.8}$$

$$\frac{(-1)^{n-1}}{(n-1)!}\int_0^1 \cdots \int_0^1 K\begin{pmatrix} x, s_1, \ldots, s_{n-1} \\ t, s_1, \ldots, s_{n-1} \end{pmatrix} ds_1 \ldots ds_{n-1} = K_n(x,t),$$

$n = 1, 2, \ldots$, by (4.10). The <u>pth minor</u> of $\Delta(\lambda)$ is

$$D\begin{pmatrix} x_1, \ldots, x_p \\ t_1, \ldots, t_p \end{pmatrix}\lambda = \sum_{n=1}^{\infty} \lambda^{p+n-1} K_{n,p}\begin{pmatrix} x_1, \ldots, x_p \\ t_1, \ldots, t_p \end{pmatrix}, \tag{5.9}$$

an entire function of λ which is continuous in the 2p variables $x_1, \ldots, x_p; t_1, \ldots, t_p$. For $p = 0$ and $p = 1$,

$$D(\lambda) = \Delta(\lambda), \quad D\begin{pmatrix} x \\ t \end{pmatrix}\lambda = \lambda N(\lambda), \tag{5.10}$$

respectively.

By formal manipulation with determinants, one can verify for $p \geqq 1$ that the pth minors satisfy relationships similar to (4.4) and (4.14). In particular, one has

$$(5.11)\quad D\begin{pmatrix} x_1, \ldots, x_p \\ t_1, \ldots, t_p \end{pmatrix}\lambda = $$

$$\sum_{j=1}^{p} (-1)^{i+j} \lambda K(x_i, t_j)\, D\begin{pmatrix} x_1, \ldots, x_{i-1}, x_{i+1}, \ldots, x_p \\ t_1, \ldots, t_{j-1}, t_{j+1}, \ldots, t_p \end{pmatrix}\lambda +$$

$$+ \lambda \int_0^1 K(x_i, s)\, D\begin{pmatrix} x_1, \ldots, x_{i-1}, s, x_{i+1}, \ldots, x_p \\ t_1, \ldots, t_{i-1}, t_i, t_{i+1}, \ldots, t_p \end{pmatrix}\lambda\, ds ,$$

$i = 1, 2, \ldots, p$, and

$$(5.12)\quad D\begin{pmatrix} x_1, \ldots, x_p \\ t_1, \ldots, t_p \end{pmatrix}\lambda = $$

$$\sum_{i=1}^{p} (-1)^{i+j} \lambda K(x_i, t_j)\, D\begin{pmatrix} x_1, \ldots, x_{i-1}, x_{i+1}, \ldots, x_p \\ t_1, \ldots, t_{j-1}, t_{j+1}, \ldots, t_p \end{pmatrix}\lambda +$$

$$+ \lambda \int_0^1 D\begin{pmatrix} x_1, \ldots, x_{j-1}, x_j, x_{j+1}, \ldots, x_p \\ t_1, \ldots, t_{j-1}, s, t_{j+1}, \ldots, t_p \end{pmatrix}\lambda\, K(s, t_j)ds ,$$

$j = 1, 2, \ldots, p$ [10].

The pth minors of $\Delta(\lambda)$ also satisfy trace relationships of a generalized kind. If $L(x_1, \ldots, x_p; t_1, \ldots, t_p)$ is a continuous function of 2p variables, then its <u>trace</u> tr L is defined to be the continuous function

$$(5.13)\qquad \mathrm{tr}\, L = \int_0^1 L(x_1, \ldots, x_{p-1}, s; t_1, \ldots, t_{p-1}, s)ds$$

of 2p - 2 variables. This operation may be iterated p times to obtain

$$(5.14)\qquad (\mathrm{tr})^p L = \int_0^1 L(s_1, \ldots, s_p; s_1, \ldots, s_p) ds_1 \ldots ds_p ,$$

a number. For $K_{n,p}$ as defined by (5.4), one has

$$(5.15)\qquad -\mathrm{tr}\, K_{n,p} = n\, K_{n+1, p-1} .$$

Repeated use of this relationship gives

$$(5.16)\qquad (-\mathrm{tr})^p D\begin{pmatrix} x_1, \ldots, x_p & \\ t_1, \ldots, t_p & \lambda \end{pmatrix} = \lambda^p \Delta^{(p)}(\lambda) ,$$

which expresses the pth derivative of $\Delta(\lambda)$ in terms of the complete trace of the pth minor.

Now suppose that λ is an eigenvalue of K for which (4.16) holds. Equation (5.16) implies that there is a positive integer $q \leqq p$ for which the qth minor of $\Delta(\lambda)$ does not vanish identically, but the minors of lower order do. A Fredholm pseudoresolvent kernel $H(x, t;\lambda)$ can be obtained by choosing a fixed set of values of $x_1, \ldots, x_q; t_1, \ldots, t_q$ for which

$$(5.17)\qquad \Delta_q(\lambda) = D\begin{pmatrix} x_1, \ldots, x_q & \\ t_1, \ldots, t_q & \lambda \end{pmatrix} \neq 0 ,$$

putting

$$(5.18)\qquad N_q(x, t;\lambda) = D\begin{pmatrix} x, x_1, \ldots, x_q & \\ t, t_1, \ldots, t_q & \lambda \end{pmatrix}$$

for the same 2q values of the indicated parameters, and finally taking

$$(5.19)\qquad \lambda H(x, t;\lambda) = \frac{N_q(x, t;\lambda)}{\Delta_q(\lambda)}, \qquad 0 \leqq x, t, \leqq 1 .$$

It will now be shown that this process yields a {1}-pseudoinverse of $(I - \lambda K)$. Set

$$(5.20)\quad \varphi_i(x) = \frac{1}{\Delta_q(\lambda)} D\begin{pmatrix} x_1, \ldots, x_{i-1}, x, x_{i+1}, \ldots, x_q & \\ t_1, \ldots, t_{i-1}, t_i, t_{i+1}, \ldots, t_q & \lambda \end{pmatrix},$$

$$= \frac{(-1)^i}{\Delta_q(\lambda)} D\begin{pmatrix} x, x_1, \ldots, x_{i-1}, x_{i+1}, \ldots, x_q & \\ t_1, \ldots, t_{i-1}, t_i, t_{i+1}, \ldots, t_q & \lambda \end{pmatrix},$$

$i = 1, 2, \ldots, q$. These functions are linearly independent, as

$$(5.21)\qquad \varphi_i(x_j) = \delta_{ij} = \begin{cases} 0, & i \neq j , \\ 1, & i = j . \end{cases}$$

Similarly, for

$$(5.22)\quad \psi_i(t) = \frac{1}{\Delta_q(\lambda)} D\begin{pmatrix} x_1, \ldots, x_{i-1}, x_i, x_{i+1}, \ldots, x_q & \\ t_1, \ldots, t_{i-1}, t, t_{i+1}, \ldots, t_q & \lambda \end{pmatrix}$$

$$= \frac{(-1)^i}{\Delta_q(\lambda)} D\begin{pmatrix} x_1, \ldots, x_{i-1}, x_i, x_{i+1}, \ldots, x_q & \\ t, t_1, \ldots, t_{i-1}, t_{i+1}, \ldots, t_q & \lambda \end{pmatrix},$$

$i = 1, 2, \ldots, q$, one has

$$(5.23)\qquad \psi_i(t_j) = \delta_{ij} .$$

Furthermore, as the minor of $\Delta(\lambda)$ of order $q-1$ vanishes identically, it follows from (5.11) and (5.12) that

$$(5.24) \qquad \psi_i(I-\lambda K) = (I-\lambda K)\varphi_i = 0, \quad i = 1, 2, \ldots, q \ .$$

Hence, Fredholm's construction furnishes sets of q linearly independent left and right eigenfunctions of K. For

$$(5.25) \qquad \Theta_i(t) = K(x_i, t), \quad \Xi_j(x) = K(x, t_j) \ ,$$

(5.11) may be used to show that

$$(5.26) \qquad (I-\lambda K)(I+\lambda H(\lambda)) = I + \lambda \sum_{j=1}^{q} \Xi_j \rangle\langle \psi_j \ ,$$

from which

$$(5.27) \qquad (I-\lambda K)(I+\lambda H(\lambda))(I-\lambda K) = I - \lambda K$$

by (5.24). Thus, the Fredholm pseudoinverse $X = I + \lambda H(\lambda)$ is a $\{1\}$-pseudoinverse of $A = I - \lambda K$. Similarly,

$$(5.28) \qquad (I+\lambda H(\lambda))(I-\lambda K) = I + \sum_{i=1}^{q} \varphi_i \rangle\langle \Theta_i \ ,$$

which also gives (5.27). However, the Fredholm pseudoinverse does not appear to satisfy conditions (3), (4), or (5) in general. From (5.26) and (5.28),

$$(5.29) \qquad (I+\lambda H(\lambda))(I-\lambda K)(I+\lambda H(\lambda)) =$$

$$= I + \lambda H(\lambda) + \lambda \sum_{j=1}^{q} (I+\lambda H(\lambda))\Xi_j \rangle\langle \psi_j$$

$$= I + \lambda H(\lambda) + \lambda \sum_{i=1}^{q} \varphi_i \rangle\langle \Theta_i(I+\lambda H(\lambda)) \ ,$$

so that (2) will hold for the Fredholm pseudoinverse if $\Xi_1(x), \ldots, \Xi_q(x)$ are right eigenfunctions of $H(\lambda)$, and $\Theta_1(t), \ldots, \Theta_q(t)$ are left eigenfunctions. If one sets $x = x_i$ in the identity

$$H(x,t;\lambda) = K(x,t) + \sum_{j=1}^{q} \Xi_j(x)\,\psi_j(t) + \lambda \int_0^1 K(x,s)\,H(s,t;\lambda)ds \,, \tag{5.30}$$

and uses the fact that $H(x_i, t;\lambda) \equiv 0$ by (5.18) and (5.19), then one sees that

$$\Theta_i(I - \lambda H) = -\sum_{j=1}^{q} \Xi_j(x_i)\,\psi_j \,, \tag{5.31}$$

and hence (2) will be satisfied if

$$K(x_i, t_j) = 0, \qquad i, j = 1, 2, \ldots, n \,. \tag{5.32}$$

This choice of parameters cannot be made in general, for example, $K(x,t)$ could be positive for $0 \leqq x, t \leqq 1$.

Equation (5.26) shows that equation (1.1) will be consistent only for functions $f(x)$ such that

$$\langle \psi_j, f \rangle = 0, \qquad j = 1, 2, \ldots, q \,. \tag{5.33}$$

6. The Hurwitz pseudoresolvent.

The pseudoinverse operator of $I - \lambda K$ formulated by Hurwitz [9] in 1912 is also of interest. Given complete sets of left and right eigenvectors of K, it can be assumed without loss of generality that they are orthonormal, that is,

(6.1) $$\langle \psi_i, \psi_j \rangle = \langle \varphi_i, \varphi_j \rangle = \delta_{ij}, \quad i, j = 1, 2, \ldots, q .$$

Under this assumption, Hurwitz showed that an inverse of the operator $I - \lambda L$ exists, where

(6.2) $$L = K - \sum_{i=1}^{q} \psi_i \rangle \langle \varphi_i ,$$

and then established that the resolvent operator $H(\lambda)$ for L is a $\{1\}$-pseudoresolvent for K.

In fact,

(6.3) $$\lambda H(\lambda) = \lambda K + \lambda^2 H(\lambda) K - \sum_{i=1}^{q} \varphi_i \rangle \langle \varphi_i ,$$

and

(6.4) $$\lambda H(\lambda) = \lambda K + \lambda^2 K H(\lambda) - \sum_{i=1}^{q} \psi_i \rangle \langle \psi_i .$$

These follow from operating on φ_j by

(6.5) $$\lambda H(\lambda)(I - \lambda K) = \lambda K - \sum_{i=1}^{q} (I + \lambda H(\lambda))\psi_i \rangle \langle \varphi_i$$

to obtain (6.3), the other resolvent equation for $H(\lambda)$ and L giving (6.4). Thus,

(6.6) $$\begin{aligned}(I - \lambda K)(I + \lambda H(\lambda)) &= \\ &= I - \lambda K + \lambda H(\lambda) - \lambda^2 K H(\lambda) \\ &= I - \sum_{i=1}^{q} \psi_i \rangle \langle \psi_i ,\end{aligned}$$

and

$$(6.7)\qquad (I-\lambda K)(I+\lambda H(\lambda))(I-\lambda K) = I-\lambda K\ ,$$

and condition (1) is satisfied.

Similarly,

$$(6.8)\qquad (I+\lambda H(\lambda))(I-\lambda K) = I-\sum_{i=1}^{q} \varphi_i\rangle\langle\varphi_i\ .$$

Consequently, the Hurwitz pseudoinverse satisfies (3) and (4). If K is symmetric, then one may take $\varphi_i = \psi_i$, $i = 1, 2, \ldots, q$, and the Hurwitz pseudoinverse also satisfies (5). There is no hope, however, that this pseudoinverse can satisfy condition (2), as $I+\lambda H(\lambda)$ is the inverse of the operator $B = I-\lambda L$. If (2) is satisfied by $X = B^{-1}$, then $A=B$, contradicting the assumption that A is not invertible.

The Hurwitz pseudoresolvent $H(\lambda)$ does reduce to the resolvent $\Gamma(\lambda)$ in a natural way if λ is not an eigenvalue, for then the sets of left and right eigenfunctions of K are empty, $K = L$, and $H(x,t;\lambda) = \Gamma(x,t;\lambda)$.

Hurwitz was also able to show that <u>any</u> pseudoresolvent $H(\lambda)$ of K must satisfy the relationships

$$(6.9)\qquad \lambda H(\lambda) = \lambda K + \lambda^2 KH(\lambda) - \sum_{i=1}^{q} \Psi_i\rangle\langle\psi_i\ ,$$

and

$$(6.10)\qquad \lambda H(\lambda) = \lambda K + \lambda^2 H(\lambda)K - \sum_{i=1}^{q} \varphi_i\rangle\langle\Phi_i\ ,$$

where

$$(6.11)\qquad \langle\varphi_i,\Phi_j\rangle = \langle\Psi_i,\psi_j\rangle = \delta_{ij},\quad i,j = 1, 2, \ldots, q\ ,$$

and was able to calculate the required functions φ_i, Φ_i, ψ_i, Ψ_i for the Fredholm pseudoresolvent (5.19).

In order for the pseudoinverse $I+\lambda H(\lambda)$ to satisfy condition (2), it is thus necessary and sufficient that

$$(6.12) \qquad \Phi_i(I+\lambda H(\lambda)) = (I + \lambda H(\lambda))\Psi_i = 0, \quad i = 1, 2, \ldots, q \ .$$

If K is symmetric and $H(\lambda)$ is the resolvent operator for

$$(6.13) \qquad K - \sum_{i=1}^{q} \varphi_i \rangle\langle \varphi_i \ ,$$

then it is easy to verify that

$$(6.14) \qquad X(\lambda) = (I - \lambda K)(I+\lambda H(\lambda))^2$$

is the Moore-Penrose generalized inverse of $A = I - \lambda K$.

7. An analytic pseudoinverse.

A simple derivation of a pseudoinverse of $I - \lambda K$ may be made on the basis of the expression (4.3) of the Fredholm resolvent $\Gamma(\lambda)$. Again, suppose that (4.16) holds, and look for a solution of the equation

$$(7.1) \qquad (I - \lambda K)f = g$$

in the form

$$(7.2) \qquad f = g + \lambda \lim_{\kappa \to \lambda} \Gamma(\kappa)g \ .$$

As $\Delta(\kappa)$ is an entire function of κ, it follows from (4.16) that $\Delta(\kappa)$, $\Delta'(\kappa), \ldots, \Delta^{(p-1)}(\kappa)$ do not vanish in some deleted neighborhood of λ . Thus, one is justified in using L'Hospital's rule to evaluate the limit in (7.2), which exists for functions g such that

$$(7.3) \qquad N(\lambda)g = N'(\lambda)g = \ldots = N^{(p-1)}(\lambda)g = 0 \ ,$$

and is given by

(7.4) $$\lim_{\kappa \to \lambda} \Gamma(\kappa)g = \frac{N^{(p)}(\lambda)g}{\Delta^{(p)}(\lambda)} .$$

This suggests the use of the operator

(7.5) $$H(\lambda) = \frac{N^{(p)}(\lambda)}{\Delta^{(p)}(\lambda)}$$

as a possible pseudoresolvent of K.

It follows at once from (4.14) that

(7.6) $$(I - \lambda K)(I + \lambda H(\lambda)) = (I + \lambda H(\lambda))(I - \lambda K) ,$$

so that condition (5) holds. If K is symmetric, then $H(\lambda)$ is also symmetric, and (3) and (4) are satisfied in addition.

In order to investigate other properties of this pseudo-inverse, it will be helpful to define the operators V_j, $j = 0, 1, \ldots, p-1$, by

(7.7) $$V_j = \frac{1}{j!} K^{p-j} N^{(j)}(\lambda) .$$

It follows from (4.14) that

(7.8) $$V_j(I - \lambda K) = (I - \lambda K)V_j = V_{j-1}, \qquad j = 1, \ldots, p-1 ,$$

and from (4.4) that

(7.9) $$V_0(I - \lambda K) = (I - \lambda K)V_0 = 0 ,$$

using the fact that all powers of K commute with $N(\lambda)$ and its derivatives. As

(7.10) $$(I - \lambda K)^p V_{p-1} = V_{p-1}(I - \lambda K)^p = 0 ,$$

and, from (4.14),

$$(7.11) \qquad (I - \lambda K)\left(I + \frac{\lambda N^{(p)}(\lambda)}{\Delta_p(\lambda)}\right) = I + \frac{\lambda p!\, V_{p-1}}{\Delta_p(\lambda)} ,$$

one has

$$(7.12) \qquad (I - \lambda K)\left(I + \frac{\lambda N^{(p)}(\lambda)}{\Delta_p(\lambda)}\right)(I - \lambda K)^p = (I - \lambda K)^p ,$$

and thus (1^k) will be satisfied for some positive integer $k \leq p$. Equation (7.12) will also be satisfied by any operator of the form

$$(7.13) \qquad X(\lambda) = I + \lambda \sum_{i=0}^{p-1} \alpha_i V_i + \frac{\lambda N^{(p)}(\lambda)}{\Delta_p(\lambda)} ,$$

as

$$(7.14) \qquad (I - \lambda K)X(\lambda) = I + \lambda \sum_{i=0}^{p-2} \alpha_{i+1} V_i + \frac{\lambda p!\, V_{p-1}}{\Delta_p(\lambda)} .$$

If the constants $\alpha_0, \alpha_1, \ldots, \alpha_{p-1}$ can be selected in such a way as to satisfy condition (2), then $X(\lambda)$ will be the Drazin pseudoinverse of $I - \lambda K$. If K is symmetric, then $k = 1$ and this construction will give the Moore-Penrose generalized inverse of $I - \lambda K$. Direct computation with (7.13) and (7.14) gives

$$(7.15) \qquad \alpha_0 = \frac{2\Delta'(\lambda) - \Delta''(\lambda)}{2\Delta'(\lambda)^2}$$

for $p = 1$, and

$$(7.16)\qquad \begin{cases} \alpha_0 = \dfrac{2}{\Delta''(\lambda)}\left(1 - \dfrac{\lambda\Delta'''(\lambda)}{3\Delta''(\lambda)}\right)^2 - \dfrac{\lambda^2\Delta^{iv}(\lambda)}{6\Delta''(\lambda)^2} , \\ \alpha_1 = \dfrac{2}{\Delta''(\lambda)}\left(1 - \dfrac{\lambda\Delta'''(\lambda)}{3\Delta''(\lambda)}\right)^2 , \end{cases}$$

for $p = 2$, and so on.

These tedious computations can be avoided by using results of Greville [7] for the construction of the Drazin pseudoinverse of square matrices. Suppose that k is the smallest positive integer for which

$$(7.17)\qquad (I - \lambda K)^{k+1}\left(I + \frac{\lambda N^{(p)}(\lambda)}{\Delta_p(\lambda)}\right) = (I - \lambda K)^k ,$$

where $k \leqq p$ by (7.12), and define

$$(7.18)\qquad X(\lambda) = (I - \lambda K)^k\left(I + \frac{\lambda N^{(p)}(\lambda)}{\Delta_p(\lambda)}\right)^{k+1} .$$

By successive applications of (7.17), one verifies that

$$(7.19)\qquad (I - \lambda K)\, X(\lambda)^2 = X(\lambda)(I - \lambda K)\, X(\lambda) = X(\lambda) ,$$

and

$$(7.20)\qquad X(\lambda)(I - \lambda K)^{k+1} = (I - \lambda K)\, X(\lambda)(I - \lambda K)^k = (I - \lambda K)^k ,$$

the required conditions.

For the operator $X(\lambda)$ defined by (7.18),

$$(7.21)\qquad V_j\, X(\lambda) = X(\lambda) V_j = 0, \qquad j = p - k, \ldots, p - 1 ,$$

so that the range of $X(\lambda)$ does not contain the generalized eigenfunctions of K obtained from $V_{p-k+1}(x, t), \ldots, V_{p-1}(x, t)$.

REFERENCES

1. Ben-Israel, A. and Greville, T. N. E., Generalized inverses: Theory and applications. John Wiley & Sons, New York, 1974.

2. Drazin, M. P., Pseudo-inverses in associative rings and semigroups. Amer. Math. Monthly 65(1958), 506-514.

3. Forsythe, G. E. and Straus, L. W., The Souriau-Frame characteristic equation algorithm on a digital computer. J. Math. Phys. (M. I. T.) 34(1955), 152-156.

4. Fredholm, I., Sur une classe d'équations fonctionnelles. Acta Math. 27 (1903), 365-390.

5. Friedman, B., Principles and techniques of applied mathematics. John Wiley & Sons, New York, 1956.

6. Greville, T. N. E., Private communication.

7. Greville, T. N. E., The Souriau-Frame algorithm and the Drazin pseudoinverse. Linear Algebra and Appl. 6(1973), 205-208.

8. Hellinger, E. and Toeplitz, O., Integralgleichungen und Gleichungen mit undendlichvielen Unbekannten, Chelsea, New York, 1953. Reprint of Encyklopädie der mathematischen Wissenschaften IIC 13, pp. 1335-1616, B. Teubner, Leipzig, 1927.

9. Hurwitz, W. A., On the pseudo-resolvent to the kernel of an integral equation. Trans. Amer. Math. Soc. 13 (1912), 405-418.

10. Lovitt, W. V., Linear integral equations, Dover, New York, 1950. Reprint of first edition, McGraw-Hill, New York, 1924.

11. Reid, W. T., Generalized inverses of differential and integral operators. Theory and Applications of Generalized Inverses of Matrices, ed. by T. L. Boullion and P. L. Odell, pp. 1-25, Texas Technological College Mathematical Series No. 4, Lubbock, 1968.

Additional references, not cited in text:

12. Goursat, É., A course in mathematical analysis, Volume III, Part Two, Integral equations, Calculus of variations. Translated by H. G. Bergman, Dover, New York, 1964. Translation of Cours d'analyse mathématique, Fifth edition, Gauthier-Villars, Paris, 1956. The Fredholm theory was included in the third edition, published in 1923.

13. Kowalewski, G., Einführung in die Determanententheorie einschliesslich der Fredholmschen Determinanten. Chelsea, New York, 1948. Reprint of third edition, Walter de Gruyter, Berlin, 1942. First edition published by Viet and Co., Leipzig, 1909.

Sponsored by the U. S. Army under
Contract No. DAAG29-75-C-0024
and the Science Research Council
of Great Britain.

Mathematics Research Center
University of Wisconsin
Madison, Wisconsin 53706

A Role of the Pseudoinverse in Analysis

Magnus R. Hestenes

1. Introduction

An operator A from one Hilbert space to another has associated with it three operators namely, its adjoint A^*, its pseudoinverse A^{-1} and its *-reciprocal A^{-*}, the pseudoinverse of its adjoint. The purpose of this paper is to give a brief discussion of the role these operators play in linear analysis.

We begin with the concept of relative self adjointness of an operator A. It is shown that every operator A is self adjoint in the sense that there is a second operator U such that $A = UA^*U$, $U = UU^*U$. By considering the operator $A - \lambda U$ one can develop a spectral theory of A relative to U. The naturalness of this spectral theory is illustrated by simple examples. These developments leads us to the concept of a ternary algebra which appears to be of singificance in the study of linear transformations. In this algebra the *-reciprocal A^{-*} plays an important role.

We turn next to Hilbert's spaces generated by operators. It is shown that the Sobolev Spaces belong to this category and that the *-reciprocals leads us to the theory of "negative norms". An equivalence relation leads us to the concept of ellipticity for differential operators.

2. Operators

Let A be a closed dense linear operator from a Hilbert

space $\mathcal{H}$ to a second Hilbert space $\mathcal{H}^*$. Such a linear transformation will be called an operator from $\mathcal{H}$ to $\mathcal{H}^*$. Its domain $\mathcal{D}(A)$ is a linear manifold which is dense in $\mathcal{H}$. Its null space $\mathcal{N}(A)$, that is, the class of all vectors in $\mathcal{D}(A)$ such that $Ax = 0$, is a closed linear subspace of $\mathcal{H}$. Its orthogonal complement $\mathcal{N}(A)^{\perp}$ intersects $\mathcal{D}(A)$ in a linear manifold $\mathcal{C}(A) = \mathcal{D}(A) \cap \mathcal{N}(A)^{\perp}$ which we shall call the carrier of A. We have the direct sum $\mathcal{D}(A) = \mathcal{C}(A) + \mathcal{N}(A)$. The range $\mathcal{R}(A)$ of A is a linear manifold in $\mathcal{H}^*$. It need not be closed. The quantity

$$\|A\| = \sup \frac{|Ax|}{|x|} \quad \text{for all } x \neq 0 \text{ in } \mathcal{D}(A)$$

will be called the norm of A. Here $|x|$ is the length of x and $|Ax|$ is the length of Ax. We admit the value $\|A\| = \infty$. In this case A is said to be unbounded. If $\|A\| < \infty$, then A is said to be bounded. Since A is closed, A is bounded if and only if $\mathcal{D}(A) = \mathcal{H}$, that is, if and only if $\mathcal{D}(A)$ is closed. Observe that $\mathcal{D}(A)$ is closed if and only if $\mathcal{C}(A)$ is closed.

As is well known the operator A has associated with it a unique operator A^* from $\mathcal{H}^*$ to $\mathcal{H}$ called its adjoint. It is basically defined by the relation

(1) $\langle Ax, y \rangle = \langle x, A^*y \rangle$ for all x in $\mathcal{D}(A)$ and all y in $\mathcal{D}(A^*)$,

where $\langle \, , \rangle$ denotes the inner product in the appropriate space. The relation (1) signifies that the domain $\mathcal{D}(A^*)$ consists of all vector y in $\mathcal{H}^*$ for which the linear functional $\langle Ax, y \rangle$ is a continuous function of x on $\mathcal{D}(A)$. Observe that $\mathcal{N}(A^*) = \mathcal{R}(A)^{\perp}$, $\mathcal{N}(A) = \mathcal{R}(A^*)^{\perp}$. It is not difficult to show that $\|A\| = \|A^*\|$. Hence A^* is bounded if and only if A is bounded. Clearly A is the adjoint of A^*, that is, $(A^*)^* = A$.

The operator A has associated with it two further operators, namely, its pseudoinverse A^{-1} and its $*$-reciprocal A^{-*}. The operator A^{-*} is the adjoint of A^{-1} and the pseudoinverse of A^*. The pseudoinverse A^{-1} of A can be defined as follows. Observe that a vector x in $\mathcal{D}(A)$ is uniquely expressible in the form $x = \hat{x} + x_0$ with $\hat{x}$ in $\mathcal{C}(A)$ and x_0 in $\mathcal{N}(A)$. Since $Ax_0 = 0$ it follows that $Ax = A(\hat{x} + x_0) = A\hat{x}$. The range $\mathcal{R}(A)$ of A is therefore determined by the carrier $\mathcal{C}(A)$ of A. The transformation $\hat{y} = A\hat{x}$ from $\mathcal{C}(A)$ to $\mathcal{R}(A)$ is 1-1 and hence has an inverse $\hat{x} = A^{-1}\hat{y}$. Extend this transformation linearly to the larger domain $\mathcal{D}(A^{-1}) = \mathcal{R}(A) + \mathcal{R}(A)^{\perp}$ by setting $A^{-1}y_0 = 0$ for all vectors y_0 in $\mathcal{R}(A)^{\perp}$. If y is in $\mathcal{D}(A^{-1})$ there are unique vectors $\hat{y}$ in $\mathcal{R}(A)$ and y_0 in $\mathcal{R}(A)^{\perp}$ such that $y = \hat{y} + y_0$. By definition the vector

$$A^{-1}y = A^{-1}(\hat{y} + y_0) = A^{-1}\hat{y} + A^{-1}y_0 = A^{-1}\hat{y} = \hat{x}$$

is in $\mathcal{C}(A)$. The transformation A^{-1} from $\mathcal{V}^*$ to $\mathcal{V}$ obtained in this manner is the pseudoinverse of A. It is characterized by the relations

$$(2) \quad A = AA^{-1}A,\ A^{-1} = A^{-1}AA^{-1},\ \mathcal{N}(A) = \mathcal{R}(A^{-1})^{\perp},\ \mathcal{N}(A^{-1}) = \mathcal{R}(A)^{\perp}.$$

It is clear that A is the pseudoinverse of A^{-1}, that is, $(A^{-1})^{-1} = A$.

By virtue of (2), the pseudoinverse A^{-*} of A^* is characterized by the relations

$$(3) \qquad A^* = A^*A^{-*}A^*,\ A^{-*} = A^{-*}A^*A^{-*},$$

$$\mathcal{N}(A^*) = \mathcal{R}(A^{-*})^{\perp} = \mathcal{N}(A^{-1}),\ \mathcal{N}(A^{-*}) = \mathcal{R}(A^*)^{\perp} = \mathcal{N}(A).$$

The relations (2) and (3) are equivalent under the $*$-involution. Consequently A^{-*} is the adjoint of A^{-1}. Moreover $(A^{-*})^{-*} = A$, that is, A is the $*$-reciprocal of A^{-*}.

The relations between the operators A, A^*, A^{-1} and A^{-*} are summarized as follows

$$(4) \quad A = (A^*)^* = (A^{-1})^{-1} = (A^{-*})^{-*},\quad A^* = (A^{-*})^{-1} = (A^{-1})^{-*}$$

$$(5) \quad A^{-1} = (A^{-*})^* = (A^*)^{-*},\quad A^{-*} = (A^{-1})^* = (A^*)^{-1}.$$

(6) $A = AA^{-1}A,\ A^{*} = A^{*}A^{-*}A^{*},\ A^{-1} = A^{-1}AA^{-1},\ A^{-*} = A^{-*}A^{*}A^{-*}$

(7) $\eta(A) = \eta(A^{-*}) = \mathfrak{R}(A^{*})^{\perp} = \mathfrak{R}(A^{-1})^{\perp},\ \eta(A^{*}) = \eta(A^{-1}) = \mathfrak{R}(A)^{\perp} = \mathfrak{R}(A^{-*})^{\perp}.$

If A^{-1} or equivalently if A^{-*} is bounded, then A will be said to be reciprocally bounded. Clearly A is reciprocally bounded if and only if its range is closed.

3. Units and Relative Self-Adjointness

An operator U from $\mathfrak{H}$ to $\mathfrak{H}^{*}$ will be called a unit if $U^{-1} = U^{*}$ or equivalently if $U = U^{-*}$. A unit U is also characterized by the relation $U = UU^{*}U$. The zero operator $U = 0$ is a unit. If $U \neq 0$, then $\|U\| = 1$. A unit is a partial isometry. If U and V are units such that $UV^{*} = 0$ and $U^{*}V = 0$ then $W = U + V$ is also a unit. This follows because

$$WW^{*}W = (U+V)(U^{*}+V^{*})(U+V) = UU^{*}U + VV^{*}V = U+V = W .$$

If $\mathfrak{H}^{*} = \mathfrak{H}$ it is seen that any orthogonal projection is a unit. In particular the identity is a unit.

An operator A from $\mathfrak{H}$ to $\mathfrak{H}^{*}$ will be said to be self-adjoint relative to U if U is a unit from $\mathfrak{H}$ to $\mathfrak{H}^{*}$ such that

(8) $$A = UA^{*}U .$$

If $\mathfrak{H}^{*} = \mathfrak{H}$ and U is the identity this concept becomes the usual concept of self-adjointness. Suppose that A is self-adjoint relative to U. Then

(9) $A^{*} = U^{*}AU^{*},\quad A^{-1} = U^{*}A^{-*}U^{*},\quad A^{-*} = UA^{-1}U$

(10) $\eta(U) \subset \eta(A) = \eta(A^{-*}),\quad \eta(U^{*}) \subset \eta(A^{*}) = \eta(A^{-1})$.

Consequently A^{*} and A^{-1} are self-adjoint relative to U^{*} and A^{-*} is self-adjoint relative to U. Moreover, since $UU^{*}U = U$, we have

$$(11)\qquad A = UA^*U = UU^*UA^*U = UU^*A = UA^*UU^*U = AU^*U$$

and similar formulas involving A^{-*}, A^{-1} and A^*. From the relations

$$U^*A = U^*AU^*U = A^*U, \quad UA^* = UA^*UU^* = AU^*$$

it follows that the operators

$$(12)\qquad P = A^*U = UA^*, \quad Q = UA^* = AU^*$$

and their pseudoinverses P^{-1}, Q^{-1} are self-adjoint in the usual sense. In addition

$$P^2 = A^*UA^*U = A^*A, \; Q^2 = AA^*, \; P^{-2} = (P^{-1})^2 = A^{-1}A^{-*}, \; Q^{-2} = A^{-*}A^{-1}$$

$$\eta(P) = \eta(P^{-1}) = \eta(A) = \eta(A^{-*}), \; \eta(Q) = \eta(Q^{-1}) = \eta(A^*) = \eta(A^{-1}) \; .$$

It should be noted that $\eta(U^*) = \eta(A^*)$ whenever $\eta(U) = \eta(A)$. For if $A^*y = PU^*y = 0$, then $x = U^*y$ is in $\eta(P) = \eta(A)$. If $\eta(U) = \eta(A)$ we have $Ux = UU^*y = 0$ and hence $U^*y = 0$. Consequently $\eta(A^*) \subset \eta(U^*)$ so that $\eta(A^*) = \eta(U^*)$ by (10).

Given an operator A there is a unique unit U such that

$$(13)\qquad A = UA^*U, \quad \eta(U) = \eta(A)$$

and such that the self-adjoint operator $P = A^*U$ is nonnegative. Since $A = UP$ this can be viewed as the polar decomposition theorem. If we define P to be the square root

$$P = (A^*A)^{\frac{1}{2}}$$

then $U = (PA^{-1})^*$ is a unit of this type. Clearly $\eta(U) = \eta(A)$. If A is bounded and reciprocally bounded this can be seen by the computations

$$A^*U = A^*(PA^{-1})^* = A^*A^{-*}P = U^*A, \quad UA^*U = UU^*A = A$$

$$UU^*U = A^{-*}PPA^{-1}A^{-*}P = A^{-*}A^*AA^{-1}A^{-*}P = A^{-*}P = U \; .$$

Consequently (13) holds with $P = A^*U$ nonnegative. The proof in the unbounded case can be made in a similar manner by restricting our computations to the domain of A^*A. The uniqueness of U follows from the uniqueness of the square root of A^*A. The unit U obtained from A in this manner will be called its natural unit.

4. A Spectral Theory for Operators

The concept of relative self-adjoints enables us to develop a spectral theory for a general operator A from $\mathcal{H}$ to $\mathcal{H}^*$. This spectral theory is an extension of the principal (singular) value theory for matrices. It reduces to the usual eigenvalue theory when the operator is a positive self-adjoint operator in the usual sense.

In order to develop this spectral theory we introduce the concept of orthogonality of operators from $\mathcal{H}$ to $\mathcal{H}^*$. Two operators B and C will be said to be orthogonal if $\overline{BC^*} = 0$ and $\overline{B^*C} = 0$ or equivalently $\overline{CB^*} = 0$ and $\overline{C^*B} = 0$. If B and C are orthogonal and $A = B + C$ then $A^* = B^* + C^*$, $A^{-1} = B^{-1} + C^{-1}$ and $A^{-*} = B^{-*} + C^{-*}$. Moreover if B and C are units so also is A. An operator B will be called a section of an operator A if B is orthogonal to $C = A - B$. A section of a unit is a unit.

Two operators B and C from $\mathcal{H}$ to $\mathcal{H}^*$ will be said to permute if $BC^* = CB^*$ and $B^*C = C^*B$ on dense sets. Orthogonal operators permute. So also do sections of an operator. If two operators permute their natural units permute. If A is self-adjoint relative to a unit U then A and U permute. If A is self-adjoint relative to U, then U and A can be expressed as sums and differences of sections

$$U = U_+ + U_0 + U_-, \quad A = A_+ - A_-$$

such that U_+ and U_- are the natural units for A_+ and A_- respective. The unit $U_+ - U_-$ is the natural unit for A. The unit U_0 is orthogonal to A.

If U is the natural unit for operator A and λ is a positive number then A and U are expressible as sums of sections

$$A = A_\lambda + B_\lambda, \quad U = U_\lambda + V_\lambda$$

such that U_λ, V_λ are the natural units for A_λ and B_λ and such that

$$\|A_\lambda\| \leqq \lambda, \quad \|B_\lambda^{-1}\| \leqq 1/\lambda \ .$$

The spectrum of A consists of all numbers λ for which at least one of the relations $\|A_\lambda\| = \lambda$, $\|B_\lambda^{-1}\| = 1/\lambda$ holds. From this decomposition it can be shown that

$$A = \int_0^\infty \lambda\, dU_\lambda, \quad A^* = \int_0^\infty \lambda\, dU_\lambda^*$$

$$A^{-*} = \int_0^\infty \lambda^{-1} dU_\lambda, \quad A^{-1} = \int_0^\infty \lambda^{-1} dU_\lambda^*, \quad U = \int_0^\infty dU_\lambda \ .$$

If A is compact or reciprocally compact then these representations take the form

$$A = \sum_j \lambda_j U_j, \quad A^* = \sum_j \lambda_j U_j^*$$

$$A^{-*} = \sum_j \lambda_j^{-1} U_j, \quad A^{-1} = \sum_j \lambda_j^{-1} U_j^*, \quad U = \sum_j U_j \ .$$

The units $U_1, U_2, U_3, \ldots,$ are sections of U and hence are mutually orthogonal. The numbers $\lambda_1, \lambda_2, \lambda_3, \ldots,$ are called the principal or singular values of A.

Similar representations of A, A^*, A^{-1}, A^{-*} hold when U is not the natural unit for A provided that A is self-adjoint relative to U. However we shall not pause to discuss this case.

5. Examples

It is instructive to illustrate our results by simple examples.

Example 1. Let $\mathcal{K}$ and $\mathcal{K}^*$ be the real Hilbert space comprised of all square integrable functions $x : x(t)$ $(0 \leq t \leq \pi)$ on the interval $[0, \pi]$ with

$$\langle x, y \rangle = \int_0^\pi x(t)\, y(t) dt$$

as its inner product. Let $\mathcal{Q}$ denote the class of absolutely continuous functions $x: x(t)$ $(0 \leq t \leq \pi)$ with square integrable derivatives $\dot{x}(t)$. Let $A = \frac{d}{dt}$ be the differential operator whose domain $\mathcal{D}(A)$ consists of all functions x in $\mathcal{Q}$ having $x(0) = x(\pi) = 0$. It is a closed dense operator in $\mathcal{K}$. It is not self-adjoint in the usual sense. Its null space is $\eta(A) = 0$ and its range $\mathcal{R}(A)$ is comprised of all functions $y(t)$ in $\mathcal{K}^* = \mathcal{K}$ having

$$\int_0^\pi y(t) dt = 0 \quad .$$

Since $\mathcal{R}(A)$ is closed, its pseudoinverse A^{-1} is bounded. We have

$$A^{-1} y(t) = \int_0^t y(s) ds - \frac{t}{\pi} \int_0^\pi y(s) ds \quad .$$

The adjoint A^* of A is the operator $A^* = -\frac{d}{dt}$ with $\mathcal{D}(A^*) = \mathcal{Q}$ as its domain. Its null space consists of all constant functions on $(0, \pi)$. Its range is $\mathcal{R}(A^*) = \mathcal{K}$ and its pseudo-inverse is

$$A^{-*} x(t) = -\int_0^t x(s) ds + \frac{1}{\pi} \int_0^\pi \int_0^r x(s)\, ds\, dr \quad .$$

The domain of $A^*A = -\frac{d^2}{dt^2}$ consists of all functions in $\mathscr{D}(A)$ whose derivatives are in $\mathscr{a}$. The domain of $AA^* = -\frac{d^2}{dt^2}$ is the class of functions in $\mathscr{a}$ whose derivatives are in $\mathscr{D}(A)$ and hence have $\dot{x}(0) = \dot{x}(\pi) = 0$.

These operators can be represented in terms of Fourier series. To this end set

$$x_n(t) = \sqrt{\frac{2}{\pi}} \sin nt, \quad y_0(t) = \sqrt{\frac{2}{\pi}}, \quad y_n(t) = \sqrt{\frac{2}{\pi}} \cos nt.$$

The sequence $x_1, x_2, x_3, \ldots,$ forms a complete orthonormal sequence in $\mathscr{X}$. The same is true for the sequence $y_0, y_1, \ldots$. Hence every function x in $\mathscr{X}$ is expressible in the form

$$x = \sum_1^\infty a_n x_n, \qquad a_n = \langle x_n, x \rangle .$$

Similarly a vector y in $\mathscr{X}^* = \mathscr{X}$ is expressible in the form

$$y = \sum_0^\infty b_n y_n, \qquad b_n = \langle y_n, y \rangle .$$

We have

$$Ax = \sum_1^\infty n a_n y_n, \quad A^* y = \sum_1^\infty n b_n x_n$$

$$A^{-*} x = \sum_1^\infty \frac{1}{n} a_n y_n, \quad A^{-1} y = \sum_1^\infty \frac{1}{n} b_n x_n .$$

The unit U associated with A is defined by the first of the formulas

$$Ux = \sum_1^\infty a_n y_n, \quad U^* y = \sum_1^\infty b_n x_n .$$

We have

$$Px = A^* Ux = \sum_1^\infty n a_n x_n, \quad Qy = AU^* y = \sum_1^\infty \frac{1}{n} b_n y_n$$

$$U^* Ux = \sum_1^\infty a_n x_n, \qquad UU^* y = \sum_1^\infty b_n y_n$$

$$Ax = UA^* Ux = UPx = \sum_1^\infty n a_n y_n, \quad UU^* Ux = Ux = \sum_1^\infty a_n x_n .$$

If we define the operator U_n by the formula $U_n x = a_n y_n$, then $U_n^* y = b_n x_n$ and

$$U_j^* U_k = 0, \quad U_j U_k^* = 0 \quad (j \neq k)$$

$$U = \sum_1^\infty U_n, \quad A = \sum_1^\infty n U_n, \quad A^* = \sum_1^\infty n U_n^*, \quad A^{-1} = \sum_1^\infty \frac{1}{n} U_n^*,$$

$$A^{-*} = \sum_1^\infty \frac{1}{n} U_n .$$

In terms of the orthogonal projections

$$E_n = U_n^* U_n, \quad F_n = U_n U_n^*$$

we have

$$A^* A = \sum_1^\infty n^2 E_n, \quad AA^* = \sum_1^\infty n^2 F_n$$

From these results it is seen that the principal values of $A = \frac{d}{dt}$ are $\lambda_n = n$ $(n = 1, 2, 3, \ldots)$ while the principal values of $A^*A = -\frac{d^2}{dt^2}$ are $\lambda_n = n^2$ $(n = 1, 2, 3, \ldots)$, a fact that is well known.

If we introduce the Green's function

$$G(s, t) = \sum_{n=1}^{\infty} \frac{1}{n^2} x_n(s) x_n(t) = \frac{1}{\pi} \sin ns \sin nt = G(t, S)$$

for the one dimensional Laplacian $A^*A = -\frac{d^2}{dt^2}$ and set

$$K(s, t) = G_s(s, t) = G_t(t, s) = \frac{2}{\pi} \sum_{1}^{\infty} \cos ns \sin nt$$

it is easily verified that

$$A^{-1} y(t) = \int_0^{\pi} K(s, t)\, y(s) ds, \quad A^{-*} x(t) = \int_0^{\pi} K(t, s)\, x(s) ds \quad .$$

Example 2. Consider next the class $\mathscr{X}$ of all real valued functions

$$x: \qquad x(u, v) \qquad (0 \leqq u \leqq \pi, \quad 0 \leqq v \leqq \pi)$$

with

$$\langle x, y \rangle = \int_0^{\pi} \int_0^{\pi} x(u, v)\, y(u, v)\, du\, dv$$

as its inner product. Let $\mathscr{X}^* = \mathscr{X} \times \mathscr{X}$ be the cartesian product of $\mathscr{X}$ by itself. Let $\mathscr{A}$ be the class of Sobolev functions x in $\mathscr{X}$ whose gradient (x_u, x_v) is in $\mathscr{X}^*$. Such a function x has a representation $x(u, v)$ such that (1) $x(u, v)$ is absolutely continuous in u on $[0, \pi]$ for almost all v

in $[0, \pi]$ and is absolutely continuous in v on $[0, \pi]$ for almost all u on $[0, \pi]$; (2) the partial derivatives x_u and x_v which exist almost everywhere are in $\mathfrak{H}$. Using this representation the values $x(u, 0)$, $x(u, \pi)$, $x(0, v)$, $x(\pi, v)$ are the boundary values of x. If they are zero we say that x vanishes on the boundary.

Let A be the gradient operator, written $Ax = \text{grad } x$, whose domain $\mathcal{D}(A)$ consists of all functions x in $\mathcal{a}$ with zero boundary values. The operator A is a closed dense operator from $\mathfrak{H}$ to $\mathfrak{H}^*$. Its adjoint A^* is the negative divergence. In particular if $y(u, v) = (y^1(u, v), y^2(u, v))$ is an element in $\mathcal{D}(A^*)$ of class C' then

$$A^*y = -\left(\frac{\partial y^1}{\partial u} + \frac{\partial y^2}{\partial v}\right) .$$

Since the ranges of A and A^* are closed, the operators A and A^* are reciprocally bounded. The operator

$$A^*A = -\Delta = -\left[\frac{\partial^2}{\partial u^2} + \frac{\partial^2}{\partial v^2}\right]$$

is the negative Laplacian restricted to functions which vanish on the boundary.

The range $\mathcal{R}(A)$ of A consists of all gradients of functions that are orthogonal to the gradients of harmonic functions. The range of A^{-*} consists of all gradients in $\mathcal{R}(A)$ which possess divergences. It is dense in $\mathcal{R}(A)$.

In order to derive analogues of the expansions given in Example 1 set

$$x_{mn}(u, v) = \frac{2}{\pi} \sin mu \sin nv \quad (m, n = 1, 2, 3, \ldots) .$$

These functions form a complete orthornal system in $\mathfrak{H}$ and are the eigenfunctions of $A^*A = -\Delta$. If x is in $\mathfrak{H}$ we have

$$x = \sum_{m,n=1}^{\infty} a_{mn} x_{mn}, \quad a_{mn} = \langle x, x_{mn} \rangle .$$

The vectors y_{mn} in $\mathscr{Y}^*$ whose components are

$$y^1_{mn} = \frac{2m}{\pi\lambda_{mn}} \cos mu \sin nv, \quad y^2_{mn} = \frac{2n}{\pi\lambda_{mn}} \sin mu \cos nv$$

with $\lambda_{mn} = (m^2 + n^2)^{\frac{1}{2}}$ form a complete orthonormal system in the range $\mathscr{R}(A)$ of A. A vector y in $\mathscr{Y}^*$ is therefore expressible in the form

$$y = y_0 + \sum_{m,\,n=1}^{\infty} b_{mn} y_{mn}$$

where $b_{mn} = \langle y, y_{mn} \rangle$ and $A^* y_0 = 0$. We have

$$Ax = \sum_{m,\,n=1}^{\infty} \lambda_{mn} a_{mn} y_{mn}, \quad A^* y = \sum_{m,\,n=1}^{\infty} \lambda_{mn} b_{mn} x_{mn}$$

$$A^{-*}x = \sum_{m,\,n=1}^{\infty} \lambda^{-1}_{mn} a_{mn} y_{mn}, \quad A^{-1}y = \sum_{m,\,n=1}^{\infty} \lambda^{-1}_{mn} b_{mn} x_{mn}$$

$$A^*Ax = \sum_{m,\,n=1}^{\infty} \lambda^2_{mn} x_{mn}, \qquad AA^*y = \sum_{m,\,n=1}^{\infty} \lambda^2_{mn} b_{mn} y_{mn}$$

$$A^{-1}A^{-*}x = \sum_{m,\,n=1}^{\infty} \lambda^{-2}_{mn} x_{mn}, \qquad A^{-*}A^{-1}y = \sum_{m,n=1}^{\infty} \lambda^{-2}_{mn} b_{mn} y_{mn} .$$

The unit U associated with A is given by the formula

$$Ux = \sum_{m,\,n=1}^{\infty} a_{mn} x_{mn} .$$

It is easily verified that $UU^*U = U$ and that $A = UA^*U$. The operators A^{-1} and A^{-*} are expressible as integrals of derivatives of the Green's function for $A^*A = -\Delta$ but we shall not pause to describe this result.

Example 3. The gradient operator for a function $x(t_1, \ldots, t_m)$ on an m-dimensional Euclidean space has a natural spectral theory based upon the Fourier Transform. The details of this spectral theory can be found in the first reference given below.

6. A Ternary Algebra.

The ideas developed in the preceding pages leads us to the concept of a ternary algebra. The theory of the ternary algebras here considered is incomplete except in the finite dimensional case. The author is convinced that a theory is worth developing. The ternary algebra here considered can be viewed as a generalization of Hilbert space.

We motivate the concept of a ternary algebra by an example. Let $\mathcal{A}$ be the class of all bounded operators from a Hilbert space $\mathcal{H}$ to a second Hilbert space $\mathcal{H}^*$. The class $\mathcal{A}$ is a linear space over the reals or complexes according as $\mathcal{H}$ and $\mathcal{H}^*$ are real or complex Hilbert spaces. The triple product AB^*C has the following properties: Here A, B, C, D, E denotes element of $\mathcal{A}$ and α denotes a scalar

1) $(A+B)C^*D = AC^*D + BC^*D, \ DC^*(A+B) = DC^*A + DC^*B$,

$$A(B+C)^*D = AB^*D + AC^*D$$

2) $\alpha AB^*C = (\alpha A)B^*C = AB^*(\alpha C) = A(\alpha^*B)^*C$, where $\alpha^* = \alpha$ in the real case and $\alpha^* = \bar{\alpha}$ in the complex case.

3) $(AB^*C)D^*E = AB^*(CD^*E) = AB^*CD^*E = A(DC^*B)^*E$.

4) If $A \neq 0$ and $AA^*A = \alpha A$ then $\alpha > 0$.

5) $\|AB^*C\| \leqq \|A\| \|B\| \|C\|$, the equality holding if $A = B = C$.

Any linear space $\mathcal{a}$ with a ternary product AB^*C satisfying the conditions 1 - 4 will be called a ternary algebra. If in addition $\mathcal{a}$ is a normed linear space such that (5) holds then $\mathcal{a}$ is a normed ternary algebra. The ternary algebra of operators described above is a complete normed ternary algebra which has the following additional property.

6) The $*$-reciprocal A^{-*} of a reciprocally bounded operator A is characterized by the relations

$$A = BA^*A = AB^*A = AA^*B, \quad B = BB^*A = BA^*B = AB^*B$$

and in fact by the relations

$$A = BA^*A, \quad B = BB^*A .$$

These relations are used to define $*$-reciprocals in a ternary algebra. This concept plays a singificant role in the theory of ternary algebras.

Any inner product space $\mathcal{a}$ can be viewed as a ternary algebra by setting $AB^*C = \langle A, B\rangle C$ or by setting $AB^*C = A\langle C, B\rangle$. A right or left minimal ideal of a ternary algebra can be viewed as inner product spaces of this type. This suggests that a ternary algebra can be viewed as a generalization of a Hilbert space.

A ternary algebra $\mathcal{C}$ is said to be commutative if $AB^*C = CB^*A$ holds for all operators. In the complex case the study of commutative ternary algebras can be viewed as the theory of normal operators. A real ternary algebra $\mathcal{P}$ will be said to be permutative if the product AB^*C is invariant under permutations of A, B and C . The study of permutative algebras can be viewed as the theory of hermitian operators.

We shall not pursue the concept of ternary algebras further. The theory is incomplete, particularly in the infinitely dimensional case. Further research in this area should be rewarding.

7. Hilbert Spaces Associated with Operators

Let G be an operator from $\mathcal{H}$ to $\mathcal{H}^*$ whose null space is zero. The domain $\mathcal{D}(G)$ of G with

$$\langle x, y \rangle_G = \langle Gx, Gy \rangle$$

forms an inner product space. If G is reciprocally bounded this inner product space is complete and is accordingly a Hilbert space which will be designated by $\mathcal{H}_G$. If G is not recipracally bounded, then $\mathcal{H}_G$ denotes the completion of this inner product space. The space $\mathcal{H}_G$ will be called the Hilbert space associated with G. The norm of x in $\mathcal{H}_G$ is given by the formula

$$\|x\|_G = \|Gx\|$$

whenever x is in $\mathcal{D}(G)$. Observe that $\mathcal{H}_G$ is isometric with the Hilbert space defined by the closure $\mathcal{R}$ range of G. Similarly the Hilbert space $\mathcal{H}_{G^{-*}}$ associated with the $*$-reciprocal G^{-*} is isometric with $\overline{\mathcal{R}(G^{-*})} = \overline{\mathcal{R}(G)}$.

The significance of these concepts can be illustrated by the following example: Let T be an open set of Euclidean points $t = (t_1, \ldots, t_m)$. Let $\mathcal{H}$ be the class of square integrable functions x: $x(t)$ (t in T) with

$$\langle x, y \rangle = \int_T x(t)\, y(t)dt$$

as its inner product. Let $\mathcal{A}_k$ be the class of functions in $\mathcal{H}$ whose generalized derivatives of orders $\leqq k$ are in $\mathcal{H}$. The mapping G_k which maps a function x in $\mathcal{A}$ into itself and its derivatives of orders $\leqq k$ is a closed dense operator from $\mathcal{H}$ to a suitable large Cartesian product $\mathcal{H}^* = \mathcal{H} \times \mathcal{H} \times \ldots \times \mathcal{H}$ of $\mathcal{H}$ with itself. The space $\mathcal{A}$ with

$$\langle x, y \rangle = \langle G_k x, G_k y \rangle$$

forms a Hilbert space $\mathfrak{H}^k$. Its norm is

$$\|x\|_k = \|G_k x\| .$$

The functions in $\mathfrak{H}^k$ are called Sobolev functions of order k . The operator G_k is a reciprocally bounded on $\mathfrak{H}$ and accordingly has a bounded $*$-reciprocal G_k^{-*} . The Hilbert space $\mathfrak{H}^{-k} = \mathfrak{H}_{G_k^{-*}}$ plays a significant rôle in the theory of partial derivatives. Its norm $\|x\|_{-k}$ is called a "negative norm". If x is in $\mathfrak{H}$ then

$$\|x\|_{-k} = \|G_k^{-*} x\| .$$

This fact has been verified by E. Landesman.

Return now to the general nonsingular operator G . We assume that G is reciprocally bounded or equivalently that there is a positive number m such that

$$\|x\|_G = \|Gx\| \geqq m\|x\|$$

on $\mathfrak{D}(G)$. An operator K from $\mathfrak{H}$ to another Hilbert space $\mathfrak{H}'$ will be said to be <u>compact</u> relative to G if $\mathfrak{D}(K) \supset \mathfrak{D}(G)$ and K restricted to $\mathfrak{D}(G)$ is a compact operator in $\mathfrak{H}_G$. The operator G_j described above is compact relative to G_k if $j < k$. An operator A from $\mathfrak{H}$ to $\mathfrak{H}''$ will be said to be <u>similar</u> to G if $\mathfrak{D}(A) = \mathfrak{D}(G)$ and there is a compact operator K relative to G such that an inequality of the form

$$m\|G_k x\| \leqq \|Ax\| + \|Kx\| \leqq m\|G_k x\|$$

holds on $\mathfrak{D}_G$ for a suitable choice of positive constants m and n . If G is a closed dense restriction of the operator G_k described above, then an operator A that is similar to G is an elliptic operator. This viewpoint has been pursued by Landesman. Further studies in this direction should be fruitful.

REFERENCES

References to the material described in this paper can be found in the following articles.

1. Hestenes, M. R., Relatively self-adjoint operators in Hilbert space, Pacific Journal of Mathematics, Vol. 11, 1961, pp. 1315-1357.

2. Hestenes, M. R., A ternary algebra with applications to matrices and linear transformations, Archive for Rational Mechanics and Analysis, Vol. 11, 1962, pp. 138-194. An analysis of the associated ideal theory will appear shortly in Scripta Mathematica.

3. Landesman, E. M., Hilbert space methods in elliptic partial differential equations, dissertation, the University of California, Los Angeles, 1965. See also Pacific Journal of Mathematics, Vol. 21, 1967, pp. 113-131.

4. Hestenes, M. R., Quadratic variational theory, control theory and the calculus of variations, Academic Press, New York 1969.

The University of California at Los Angeles
IBM Research, Yorktown Heights, New York

The preparation of this paper was sponsored in part by the U. S. Army Research Office, Grant DA-31-124-ARO(D)-355, and the Office of Naval Research, Contract NONR 233(76). Reproduction in whole or in part is permitted for any purpose of the United States Government.

Aspects of Generalized Inverses in Analysis and Regularization

M. Zuhair Nashed

This paper provides glimpses into some aspects of generalized inverses in analysis and some aspects of regularization of ill-posed linear operator equations. The paper consists of three parts. Part I, which deals with linear analysis, is devoted primarily to generalized inverses of integral operators of the first and second kinds and to related topics. Part II provides some uses of generalized inverses in nonlinear analysis (nonlinear alternative problems, inverse mapping and open mapping theorems, nonlinear operator equations and constrained optimization problems). Part III deals briefly with regularization methods in which generalized inverses play a role.

Reference citations in this paper are to the Bibliography at the end of this volume. Also, the reader may refer to Section 2 of the paper on "Perturbations and Approximations..." in this volume for notation and results on generalized inverses that are used in the present paper.

I. ASPECTS OF GENERALIZED INVERSES IN LINEAR ANALYSIS

1. Singular Value Decomposition and the Generalized Inverse of a Compact Operator. Picard's Criteria for Solvability and Least-Squares Solvability of $Kx = y$

The existence theory of integral equations of the first kind with $\mathcal{L}_2$-kernels was first studied by Picard in 1910 in his little classic [1648]. This theory is based on the spectral decompositions of operators of the form K^*K and KK^*, where K is compact, as developed by E. Schmidt in three papers in 1907-1908. (For a relevant discussion of this theory, see

Smithies [1249]). Picard's theory generalizes immediately to the case of a compact operator on a Hilbert space.

Throughout this paper we let $\mathcal{H}_1$ and $\mathcal{H}_2$ be Hilbert spaces. Let $A: \mathcal{H}_1 \to \mathcal{H}_2$ be a compact (i.e. it maps bounded sets into totally bounded sets), symmetric $(A = A^*)$ and non-negative definite (i.e., $\langle Ax, x\rangle \geq 0$ for all $x \in \mathcal{H}_1$) linear operator. Then A has a finite or countably infinite number of eigenvalues λ_n; in the latter case $\lambda_n \to 0$ as $n \to \infty$. Each nonzero eigenvalue is of finite multiplicity and the λ_n's can be linearly ordered:

$$\lambda_1 \geq \lambda_2 \geq \dots \geq \lambda_n \geq \dots \geq 0$$

with corresponding orthonormal eigenvectors $\varphi_1, \varphi_2, \dots, \varphi_n, \dots$ We assume $A \neq 0$ so that $\lambda_1 > 0$. Let $\mathcal{E}$ be the set of subscripts for which $\lambda_n \neq 0$ and let $\sigma(A)$ denote the nonzero spectrum of A:

$$\sigma(A) = \{\lambda_i : A\varphi_i = \lambda_i \varphi_i,\ \lambda_i > 0,\ \varphi_i \neq 0\} = \{\lambda_n : n \in \mathcal{E}\} .$$

The set $\{\varphi_n : n \in \mathcal{E}\}$ is a Schauder basis for $\overline{\mathcal{R}(A)}$. (Recall that $\mathcal{B} \subset \mathcal{M}$ is a Schauder basis for a subspace $\mathcal{M} \subset \mathcal{H}$ if $\mathcal{B}$ is linearly independent and $\overline{\text{span}\, \mathcal{B}} = \mathcal{M}$, where $\text{span}\, \mathcal{B}$ is the set of all (finite) linear combination of elements of $\mathcal{B}$.) In particular, each element in $\mathcal{R}(A)$ can be expanded in terms of the eigenvectors φ_n in the form:

$$Ax = \sum_{n\in\mathcal{E}} \langle Ax, \varphi_n\rangle \varphi_n = \sum_{n\in\mathcal{E}} \lambda_n \langle x, \varphi_n\rangle \varphi_n$$

(see, for instance, Riesz and Nagy [1669]). Now each $x \in \mathcal{H}_1$ can be uniquely decomposed in the form $x = P_{\mathcal{N}} x + (I - P_{\mathcal{N}})x$, where $P_{\mathcal{N}}$ is the orthogonal projectors of $\mathcal{H}_1$ onto the null-space $\mathcal{N}(A)$. Note that $(I - P_{\mathcal{N}})x \in \mathcal{N}(A)^{\perp} = \overline{\mathcal{R}(A)}$ since A is symmetric. Thus,

$$(I - P_{\mathcal{N}})x = \sum_{k\in\mathcal{E}} \langle (I - P_{\mathcal{N}})x, \varphi_k\rangle \varphi_k ,$$

where by Bessel's inequality

$$\sum_{k\in\mathcal{E}} |\langle (I - P_{\eta})x, \varphi_k\rangle|^2 \leqq \|(I - P_{\eta})x\|^2 < \infty .$$

Hence,

$$x = P_{\eta} x + \sum_{k\in\mathcal{E}} c_k \varphi_k ,$$

where

$$\sum_{k\in\mathcal{E}} c_k^2 < \infty .$$

In particular, it follows that a necessary and sufficient condition that $\{\varphi_n : n \in \mathcal{E}\}$ form a complete orthonormal system for $\mathcal{V}_1$ is that $\eta(A) = \{0\}$; also $x \in \eta(A)$ if and only if $\langle x, \varphi_n\rangle = 0$ for all $n \in \mathcal{E}$, so that

$$\eta(A) = [\text{span}\{\varphi_n : n \in \mathcal{E}\}]^{\perp} .$$

In what follows we shall apply the preceding results to operators of the forms KK^* and K^*K, where $K: \mathcal{V}_1 \to \mathcal{V}_2$ is compact and K^* is the adjoint of K (it suffices to suppose that KK^* and K^*K are compact; however this implies that K is itself is compact, see p. 1260 of Dunford and Schwartz [350], Part II). Then clearly KK^* and K^*K are compact, symmetric, and nonnegative; it follows easily that $\sigma(KK^*) = \sigma(K^*K)$ and the multiplicities of the eigenvalues are the same in both spectra. Let

$$\sigma(KK^*) = \sigma(K^*K) = \{\lambda_n : n \in \mathcal{E}\} .$$

Singular functions and values. If $\phi \in \mathcal{V}_1$, $\psi \in \mathcal{V}_2$ are nonzero vectors which satisfy the equations

$$\phi = \mu K\psi \quad \text{and} \quad \psi = \mu K^* \phi ,$$

then they are said to be a pair of singular functions of K belonging to the singular value μ. It follows that μ is real and if (ϕ, ψ) is a pair of singular functions with singular value μ, then ϕ, ψ are eigenfunctions of KK^* and K^*K respectively with eigenvalue μ^{-2}, i.e.,

$$\psi = \mu^2 K^*K\psi \quad \text{and} \quad \phi = \mu^2 KK^* \phi .$$

Consequently if (ϕ_1, ψ_1) and (ϕ_2, ψ_2) correspond to different singular values, then ϕ_1 is orthogonal to ϕ_2, and ψ_1 to ψ_2.

Conversely let $\lambda > 0$ be an eigenvalue of K^*K with eigenfunction $v \in \mathcal{H}_2$. Then

$$KK^*(\lambda^{-\frac{1}{2}} Kv) = \lambda(\lambda^{-\frac{1}{2}} Kv) .$$

Let $u = \lambda^{-\frac{1}{2}} Kv$. Then $KK^*u = \lambda u$, so λ is an eigenvalue of KK^* with eigenfunction u. Similarly $v = \lambda^{-\frac{1}{2}} K^*u$; so that (u, v) is a pair of singular functions of K belonging to $\lambda^{-\frac{1}{2}}$.

Let $\sigma(K^*K) = \sigma(KK^*) = \{\lambda_n : n \in \mathcal{E}\}$,

$$\lambda_1 \geq \lambda_2 \geq \cdots \geq \lambda_n \geq \cdots > 0 .$$

Define $\mu_n = \lambda_n^{-\frac{1}{2}}$, $n \in \mathcal{E}$. Then

$$0 < \mu_1 \leq \mu_2 \leq \cdots \leq \mu_n \leq \cdots$$

with $\mu_n \to \infty$ if $\mathcal{E}$ is infinite. Let

$$\mathcal{U} = \{u_n : KK^* u_n = \lambda_n u_n, \quad n \in \mathcal{E}\}$$

$$\mathcal{V} = \{v_n : K^*K v_n = \lambda_n v_n, \quad n \in \mathcal{E}\} .$$

We have shown that (u_n, v_n) are singular functions belonging to the singular value $\mu_n = \lambda_n^{-\frac{1}{2}}$. We call $\{u_n, v_n; \mu_n\}$ a singular system for the compact operator K:

$$(1.1)\qquad \begin{cases} u_n = \mu_n K v_n \\ v_n = \mu_n K^* u_n \end{cases} .$$

Note that the "dual" relationships (1.1) establish a one-to-one correspondence between the sets $\mathcal{U}$ and $\mathcal{V}$; also,

$$\text{span}\,\mathcal{U} = K[\text{span}\,\mathcal{V}] \text{ and } \text{span}\,\mathcal{V} = K^*[\text{span}\,\mathcal{U}] .$$

The concept of a complete orthonormal system of pairs of singular functions is defined in the same way as a complete orthonormal system of eigenfunctions of a symmetric operator. If $\{u_n, v_n\}$ is a complete system of pairs of singular functions of K with singular values $\{\mu_n\}$, then $\{u_n\}$ is a C.O.N. system of eigenfunctions of KK^* with eigenvalues $\{\mu_n^{-\frac{1}{2}}\}$, and so on.

We summarize below properties of singular systems.

Theorem 1.1. Let $K: \mathcal{H}_1 \to \mathcal{H}_2$ be a compact linear operator. Then there exist orthogonal systems $\{u_n : n \in \mathcal{E}\}$, $\{v_n : n \in \mathcal{E}\}$ of eigenfunctions of KK^* and K^*K respectively, corresponding to $\sigma(KK^*) = \sigma(K^*K) = \{\lambda_n : n \in \mathcal{E}\}$, and $\{u_n, v_n; \lambda_n\}$ is a singular system for K. Furthermore,

$$\overline{\text{span}\,\mathcal{U}} = \overline{\mathcal{R}(KK^*)} = \overline{\mathcal{R}(K)} = \mathcal{N}(KK^*)^{\perp} = \mathcal{N}(K^*)^{\perp}$$

and (dually),

$$\overline{\text{span}\,\mathcal{V}} = \overline{\mathcal{R}(K^*K)} = \overline{\mathcal{R}(K^*)} = \mathcal{N}(K^*K)^{\perp} = \mathcal{N}(K)^{\perp} ,$$

so $\mathcal{U}$ is a Schauder basis for $\overline{\mathcal{R}(K)}$ and $\mathcal{V}$ for $\overline{\mathcal{R}(K^*)}$ and $\mathcal{V}$ for $\mathcal{R}(K^*)$.

Combining this with well-known orthogonal decomposition, we obtain,

$$\mathcal{H}_1 = \mathcal{N}(K) \oplus \mathcal{N}(K)^{\perp} = \mathcal{N}(K) \oplus \overline{\text{span}\,\mathcal{V}} ,$$

$$\mathcal{H}_2 = \mathcal{N}(K^*) \oplus \mathcal{N}(K^*)^{\perp} = \mathcal{N}(K^*) \oplus \overline{\text{span}\,\mathcal{U}} .$$

Picard's Criteria for $Kx = y$. Singular function analysis provides a useful tool for the study of existence, uniqueness and representation of solution or least-squares solution of equations of the first kind.

Theorem 1.2. Let $\{u_n, v_n; \mu_n\}$ be a singular system for the compact operator $K: \mathcal{H}_1 \to \mathcal{H}_2$. Let $y \in \mathcal{H}_2$ and let $K^\dagger$ denote the generalized inverse of K. Then the following statements are equivalent:

$$y \in \mathcal{D}(K^\dagger) = \mathcal{R}(K) + \mathcal{R}(K)^\perp \tag{1.2}$$

$$P_{\overline{\mathcal{R}(K)}}\, y \in \mathcal{R}(K) \tag{1.3}$$

$$\sum_{n \in \mathcal{E}} \mu_n^2 \, |\langle y, u_n\rangle|^2 < \infty . \tag{1.4}$$

For $y \in \mathcal{D}(K^\dagger)$,

$$K^\dagger y = \sum_{n \in \mathcal{E}} \mu_n \langle y, u_n\rangle v_n . \tag{1.5}$$

Proof. (1.2) $\Longleftrightarrow$ (1.3) is immediate. We show that (1.3) $\Longleftrightarrow$ (1.4). If (1.3) holds, then $\sum_{n \in \mathcal{E}} \langle y, u_n\rangle u_n = Kx$ for some $x \in \mathcal{H}_1$. But $x = \sum_{n \in \mathcal{E}} \langle x, v_n\rangle v_n + P_{\mathcal{N}(K)} x$ and thus $Kx = \sum_{n \in \mathcal{E}} \langle x, v_n\rangle \frac{u_n}{\mu_n} = \sum_{n \in \mathcal{E}} \langle y, u_n\rangle u_n$. Hence $\langle x, v_n\rangle = \mu_n \langle y, u_n\rangle$ and

$$\sum_{n \in \mathcal{E}} \mu_n^2 |\langle y, u_n\rangle|^2 = \sum_{n \in \mathcal{E}} |\langle x, v_n\rangle|^2 \leqq \|x\|^2 < \infty .$$

Conversely, if (1.4) holds, then $x = \sum_{n \in \mathcal{E}} \mu_n \langle y, u_n\rangle v_n \in \mathcal{H}_1$ and $Kx = \sum_{n \in \mathcal{E}} \langle y, u_n\rangle u_n = P_{\overline{\mathcal{R}(K)}}\, y$, which proves (1.3). Finally to prove (1.5) we note that $x = \sum_{n \in \mathcal{E}}' \mu_n \langle y, u_n \rangle v_n \in \mathcal{N}(K)^{\perp}$ and is a solution of the equation $Kx = P_{\overline{\mathcal{R}(K)}} y$; thus $x = K^{\dagger} y$. ■

Thus (1.4) is a necessary and sufficient condition for the existence of a least-squares solution of the equation $Kx = y$. If (1.4) is satisfied, then (1.5) gives the least-squares solution of minimal norm and the set of all least-squares solutions is given by

$$\mathcal{S}_y = \sum_{n \in \mathcal{E}} \mu_n \langle y, u_n \rangle v_n + [\operatorname{span} \mathcal{V}]^{\perp} .$$

Note that for any $y \in \mathcal{H}_2$,

$$\sum_{n \in \mathcal{E}} \mu_n^2 |\langle y, u_n \rangle|^2 = \sum_{n \in \mathcal{E}} \mu_n^2 |\langle P_{\overline{\mathcal{R}(K)}} y, u_n \rangle|^2$$

and that the sum on the left is divergent if and only if $P_{\overline{\mathcal{R}(K)}} y \in \overline{\mathcal{R}(K)} \setminus \mathcal{R}(K)$.

Condition (1.4) requires that the components of y along the vectors $\{u_n\}$ decrease fast enough to make the series convergent. Obviously the more rapidly the μ_n's increase, the more severely the y is restricted. Since u_n for large n represents higher frequency eigenfunctions of KK^*, (1.4) may be interpreted as a smoothness condition on y.

Theorem 1.2 extends to least-squares solvability Picard's criterion for solvability of $Kx = y$, which follows as a corollary: Let $y \in H_2$. Then $Kx = y$ has a solution if and only if (1.4) holds and $y \in \mathcal{N}(K^*)^{\perp}$. These two conditions can be combined into the singe condition

$$\sum_{n=0}^{\infty} \frac{|\langle y, u_n \rangle|^2}{\lambda_n} < \infty$$

where if $\lambda_n = 0$ we must have $\langle y, u_n \rangle = 0$ and agree to define the corresponding term in the series to be zero.

Note that (1.4) is equivalent to $y \in \mathcal{R}(K) + \mathcal{R}(K)^{\perp}$ so if y is known also to be in $\mathcal{N}(K^{*})^{\perp} = \overline{\mathcal{R}(K)}$, then it has no components in $\mathcal{R}(K)^{\perp}$ and it must be in $\mathcal{R}(K)$. If y satisfies Picard's criterion, then the solution with minimal norm of $Kx = y$ (which is the unique solution in $\mathcal{N}(K)^{\perp}$) is given by (1.5). The set of all solutions is

$$\sum \mu_i \langle y, u_i \rangle v_i + [\operatorname{span} \mathcal{V}]^{\perp} .$$

The solution is unique if and only if $\overline{\operatorname{span} \mathcal{V}} = \mathcal{H}_1$, i.e., $\mathcal{V}$ is a Schauder basis for $\mathcal{H}_1$.

Let K be the compact operator defined by

$$(1.6) \qquad (Kx)(s) := \int_0^1 k(s,t)\, x(t)dt, \qquad 0 \leq s \leq 1 ,$$

where $k \in \mathcal{L}_2([0,1] \times [0,1])$. Then the kernel $k(s,t)$ can be expanded (where convergence is in $\mathcal{L}_2$) in terms of the singular system $\{u_n, v_n; \mu_n\}$ of the compact operator K defined in (1.6):

$$(1.7) \qquad k(s,t) = \sum_{n \in \mathcal{E}} \frac{1}{\mu_i} u_i(s)\, v_i(t) .$$

As a corollary we have

$$\sum_{n \in \mathcal{E}} \frac{1}{\mu_i^2} = \int_0^1 \int_0^1 k^2(s,t)\, ds\, dt ,$$

and if the series in (1.7) converges in the ordinary sense almost everywhere, then its sum is equal to $k(s,t)$ for almost all s, t.

The singular system for K is not readily available in general, and hence it is of limited practical use for actual construction of solutions. Several procedures for approximating $K^{\dagger}y$ when no singular system is available are discussed in this paper and other papers in this volume. In the case of a symmetric operator, the singular system expansion reduces to the classical eigenfunction expansion.

2. $(I - \lambda K)^\dagger$ and Hurwitz's Pseudoresolvent

Let K be the integral operator defined in (1.6), and consider the integral equation of the second kind:

$$(I - \lambda K)x = y \ .$$

Recall that if the inverse of $I - \lambda K$ exists, then

$$(I - \lambda K)^{-1} = I + \lambda \Gamma$$

where Γ is the integral operator induced by the resolvent kernel $\Gamma(s,t;\lambda) \in \mathfrak{L}_2$. In this case for every $y \in \mathfrak{L}_2[0,1]$, the equation

$$x(s) - \lambda \int_0^1 k(s,t)\, x(t)dt = y(s) \tag{2.1}$$

has a unique solution given by $x = (I + \lambda\Gamma)y$. The resolvent kernel $\Gamma(s,t;\lambda)$ can be explicitly calculated from $k(s,t)$ and is uniquely determined by the equations

$$\Gamma = K + \lambda K\Gamma, \quad \Gamma = K + \lambda \Gamma K \ . \tag{2.2}$$

These results were first developed by Fredholm in his classic paper [414]. (Fredholm assumed all functions involved to be continuous). Hurwitz [604] reconsidered Fredholm's results and showed how to reduce the general case of solvability of (2.1) to the case when $(I - \lambda K)^{-1}$ exists, and expressed the general solution in terms of a pseudoresolvent. We consider from the generalized inverse viewpoint the essence of Hurwitz's construction.

According to Hurwitz, an $\mathfrak{L}_2$-kernel $\Gamma(s,t;\lambda)$ is called a pseudoresolvent of the kernel $k(s,t)$ if for any $y \in \mathfrak{R}(I-\lambda K)$, the function

$$x(s) = y(s) + \int_0^1 \Gamma(s,t;\lambda)\, y(t)\, dt$$

is a solution of (2.1). It then follows immediately that the operator $I + \lambda\Gamma$ is an inner inverse of $I - \lambda K$, i.e.,

$$(I - \lambda K)(I + \lambda \Gamma)(I - \lambda K) = I - \lambda K$$

(see Section 1 of the paper by Nashed and Votruba in this volume for an extensive treatment of inner inverses; in particular Proposition 1.4).

It is well-known that for any complex number λ,

$$\text{(2.3)} \qquad \dim \mathcal{N}(I - \lambda K) = \dim \mathcal{R}(I - \lambda K)^{\perp}, \quad \text{say } n(\lambda).$$

Let λ_0 be a characteristic value of K (i.e., $\lambda_0 Kx = x$, $x \neq 0$) and let $n(\lambda_0) = n$. Let $\{\varphi_1, \ldots, \varphi_n\}$ and $\{\psi_1, \ldots, \psi_n\}$ respectively be orthonormal bases for $\mathcal{N}(I - \bar{\lambda}_0 K^*) = \mathcal{R}(I - \lambda_0 K)^{\perp}$. The equation (2.1) is consistent if and only if $y \in \mathcal{N}(I - \lambda_0 K^*)^{\perp}$, in which case the general solution of (2.1) is given by

$$x = x_0 + \sum_{i=1}^{n} c_i \varphi_i ,$$

where x_0 is a particular solution of (2.1). We express the general solution in terms of pseudoresolvents. To this end define a new kernel

$$k_0(s,t) = k(s,t) - \frac{1}{\lambda_0} \sum_{i=1}^{n} \psi_i(s) \overline{\varphi_i(t)}$$

and let K_0 denote the corresponding integral operator. Then the operator $I - \lambda_0 K_0$ is one-to-one and onto. To prove this, it suffices to show in view of (2.3) that $\mathcal{N}(I - \lambda_0 K_0) = \{0\}$. Suppose $(I - \lambda_0 K_0)x = 0$. Then

$$x - \lambda_0 \int_0^1 k_0(s,t)\, x(t)dt = 0 ,$$

$$x - \lambda_0 \int_a^b k(s,t)\, x(t)dt = \sum_{i=1}^{n} \langle x, \varphi_i \rangle \psi_i(s) ,$$

i.e., $(I - \lambda_0 K)x = \sum_{i=1}^{n} \langle x, \varphi_i \rangle \psi_i(s)$. Now each $\psi_i \in \mathcal{R}(I - \lambda_0 K)^{\perp}$, so $(I - \lambda_0 K)x \in \mathcal{R}(I - \lambda_0 K)^{\perp}$. Thus $\sum_{i=1}^{n} \langle x, \varphi_i \rangle \psi_i = 0$, so $x \in \eta(I - \lambda_0 K)$. But by linear independence, we also have $\langle x, \varphi_i \rangle = 0$, $i = 1, \ldots, n$. That is, $x \in \eta(I - \lambda_0 K)^{\perp}$. Hence, $x = 0$.

Thus by the Fredholm theory of resolvents, the resolvent operator Γ_0 of K_0 exists and $(I - \lambda_0 K_0)^{-1} = I + \lambda_0 \Gamma_0$. Also,

$$(I + \lambda_0 \Gamma_0)\psi_i = \varphi_i, \qquad i = 1, \ldots, n ,$$

$$(I + \lambda_0 \Gamma_0)(I - \lambda_0 K)x = x \quad \text{for } x \in \eta(I - \lambda_0 K)^{\perp} ,$$

$$(I - \lambda_0 K)(I + \lambda_0 \Gamma_0)y = y \quad \text{for } y \in \mathcal{R}(I - \lambda_0 K) ;$$

thus $x_0 = (I + \lambda_0 \Gamma_0)y$ is a particular solution of (2.1) for each $y \in \mathcal{R}(I - \lambda_0 K)$ and $I + \lambda_0 \Gamma_0$ is an inner inverse to $I - \lambda_0 K$. (However, $I + \lambda_0 \Gamma_0$, being invertible, is not a generalized inverse. A discussion of invertible inner inverses will be given elsewhere).

For a given $\mathfrak{L}_2$-kernel $k(s,t)$, the pseudoresolvent is nonunique. Indeed, with k_0 as constructed above, it is easy to verify that the kernel

$$\Gamma_0(s,t) + \sum_{i,j=1}^{n} c_{ij}\, \varphi_i(t)\, \overline{\psi_j(s)}$$

is a pseudoresolvent kernel of $k(s,t)$ for any choice of the scalars c_{ij}. We also have the following characterization of all pseudoresolvents of a given kernel, which is an extension of the characterization (2.2) for resolvents: An $\mathfrak{L}_2$-kernel $\Gamma(s,t;\lambda)$ is a pseudoresolvent of $k(s,t)$ if and only if

$$\Gamma(s,t;\lambda) = k(s,t) + \lambda \int_0^1 k(s,r)\,\Gamma(r,t;\lambda)dr - \frac{1}{\lambda} \sum_{i=1}^{n} \alpha_i(t)\, \overline{\psi_i(s)}$$

$$\Gamma(s,t;\lambda) = k(s,t) + \lambda \int_0^1 k(s,r)\Gamma(r,t;\lambda)dr - \frac{1}{\lambda} \sum_{i=1}^{n} \varphi_i(t)\, \overline{\beta_i(s)} ,$$

where $\langle \alpha_i, \varphi_j \rangle = \delta_{ij}$ and $\langle \beta_i, \psi_j \rangle = \delta_{ij}$, $i, j = 1, \ldots, n$. The (Moore-Penrose) generalized inverse of $I - \lambda_0 K$ is given by

$$(I - \lambda_0 K)^\dagger = I + \lambda_0 \Gamma_0 - F,$$

where F is the finite rank integral operator whose kernel is $\sum_{i=1}^{n} \varphi_i \overline{\psi_j}$, with the corresponding pseudoresolvent kernel

$$\Gamma = \Gamma_0 - \frac{1}{\lambda_0} \sum_{i=1}^{n} \varphi_i \overline{\psi_i} .$$

Fredholm called the operator $I + \lambda_0 \Gamma_0$ a pseudoinverse of $I - \lambda_0 K$. The pseudoresolvent constructed by Fredholm (and extended by Hurwitz) is perhaps the first appearance of the concept of a generalized inverse.

The interplay between $(I - \lambda K)^\dagger$ in the spaces $\mathcal{L}_2[0,1]$ and $\mathcal{C}[0,1]$. Often the need arises to study integral equations in the space $\mathcal{C} = \mathcal{C}[0,1]$ equipped with the uniform norm with the best approximate solution taken in the sense of $\mathcal{L}_2[0,1]$. The integral operator is then regarded as both an operator on $\mathcal{C}$ and an operator $\widetilde{K}$ on $\mathcal{L}_2$; thus $K = \widetilde{K}|\mathcal{C}$. Relative to orthogonal projectors (in $\mathcal{L}_2$) we consider the generalized inverse $(I - \lambda\widetilde{K})^\dagger$. Under some mild assumptions, it follows that $(I - \lambda\widetilde{K})^\dagger|\mathcal{C}$, the restriction to $\mathcal{C}$ of $(I - \lambda\widetilde{K})^\dagger$, is indeed the (Banach space) generalized inverse of $I - \lambda K$ relative to the (restricted) orthogonal projectors. In particular for each $y \in \mathcal{C}[0,1]$, $(I - \lambda\widetilde{K})^\dagger y$ is a continuous function. For various aspects and applications of this interplay see Moore and Nashed [894; Section 4], Kammerer and Nashed [640; pp. 559-560]. A detailed study of the generalized inverse of $I - \lambda K$, where K is a (general) compact operator taking a Banach space X into itself is given in the paper by Nashed and Votruba in this volume. A canonical representation of $(I - \lambda K)^\dagger$ is also developed.

3. Some Contrasts Between Linear Integral and Operator Equations of the First and Second Kinds

In this section we draw some contrasts between integral

and operator equations of the first and second kinds, and highlight difficulties posed by equations of the first kind.

Let $\mathscr{H}_1$ and $\mathscr{H}_2$ be Hilbert spaces and K be a compact operator on $\mathscr{H}_1$ into $\mathscr{H}_2$. We consider the linear operator equations (of the first and second kinds respectively),

(3.1) $$Kx = y$$

(3.2) $$x - \lambda Kx = y$$

where y is a given element in $\mathscr{H}_2$ and λ is a scalar. A prototype of these equations is when K is the integral operator

(3.3) $$Kx = \int_0^1 k(\cdot, t)\, x(t)dt$$

when $k(s,t)$ is an $\mathfrak{L}_2$-kernel, and $\mathscr{H}_1 = \mathscr{H}_2 = \mathfrak{L}_2[0,1]$.

1. Both integral equations of the first and second kinds are old. Abel, in his paper on the tautochrone, was the first mathematician to pose and solve an integral equation. In volume 27 of Acta Mathematica, dedicated appropriately enough to Abel, Fredholm published in 1903 an account of his research on integral equations (of the second kind) which he began earlier. [Fredholm's paper was reproduced in "A Collection of Modern Mathematical Classics: Analysis" edited by Richard Bellman, Dover Publications, New York, 1961. The inclusion of the paper in a collection of thirteen papers of the late nineteenth and early twentieth centuries, was most appropriate since few papers have affected the future course of mathematics as Fredholm's paper]. The existence theory of integral equations of the first kind was developed by Picard in 1910. Fundamental contributions to integral equations of the second kind were advanced by Hilbert, F. Riesz, E. Schmidt, and others. In contrast, first kind equations have lain nearly dormant until quite recently when need arose for computing smooth approximate solutions to such equations, and the pathology and difficulties incurred in these equations became better understood. These equations are frequently encountered in physical, biological and engineering models. An extensive literature has developed over the past fifteen years on the theory, approximations, and regularization of first kind equations. These

equations have properties which are fundamentally and markedly different from second kind equations, and pose new problems. While several books have been devoted to the theory and applications of integral equations of the second kind, there is no book which is completely devoted to first kind equations (the author is preparing a manuscript, "Integral and Operator Equations of the First Kind: Theory and Approximation" which treats in detail some of the topics of this paper and others not considered here).

2. We should remark at the outset that this classification (i.e., first and second kinds) is a nineteenth century classification which reflects neither "genetic" nor "historic" aspects of such equations. A better classification could be based on the character of the spectral decompositions of the operators involved. From a modern viewpoint, key distinctions between the two types of equations reside in the operator-theoretic context:

(a) The range of $I - \lambda K$ is a closed subspace for any λ; whereas the range of K is nonclosed (unless $\mathcal{R}(K)$ is finite dimensional, which would correspond in the case of (3.3) to a separable kernel, a case of no interest for first kind equations).

(b) $(I - \lambda K)^{\dagger}$ is a bounded operator defined on all of $\mathcal{H}_2$; whereas $K^{\dagger}$ is unbounded and densely defined, with $\mathcal{D}(K^{\dagger}) = \mathcal{R}(K) + \mathcal{R}(K)^{\perp}$. (Statements (a) and (b) are equivalent).

(c) $\mathcal{N}(I - \lambda K)$ for each scalar λ is finite dimensional; whereas $\mathcal{N}(K)$ may be infinite dimensional.

(d) $\mathcal{H}_1 = \mathcal{N}(I - \lambda K) \oplus \mathcal{N}(I - \lambda K)^{\perp}$,

$\mathcal{H}_2 = \mathcal{R}(I - \lambda K) \oplus \mathcal{R}(I - \lambda K)^{\perp}$, and

$$\mathcal{R}(I - \lambda K) = \mathcal{N}(I - \bar{\lambda} K^{*})^{\perp} \; ; \tag{3.4}$$

whereas for K we have:

$$\mathcal{H}_1 = \mathcal{N}(K) \oplus \mathcal{N}(K)^{\perp} ,$$

$$\mathcal{H}_2 = \overline{\mathcal{R}(K)} \oplus \mathcal{R}(K)^{\perp}, \quad \text{and}$$

$$\overline{\mathcal{R}(K)} = \mathcal{N}(K^*)^{\perp} . \tag{3.5}$$

It follows from (3.4) that (3.2) has a solution if and only if $y \in \mathcal{N}(I - \lambda K^*)^{\perp}$; this is the Fredholm alternative criterion for solvability. In contrast, from (3.5), the condition $y \in \mathcal{N}(K^*)^{\perp}$ is necessary but not sufficient for the solvability of (3.1). The Fredholm alternative theorem does not hold for (3.1) and one needs an additional condition that $y \in \mathcal{N}(K^*)^{\perp}$ must satisfy in order for (3.1) to be solvable (this is the essence of Picard's criterion; see Section 1).

The preceding operator-theoretic properties cast important implications and manifestations in the approximation theory of these equations (see, e.g., [915], [937]).

3. Degenerate kernels. We consider (3.1) and (3.2) when K is the integral operator (3.3) with a separable kernel $k(s,t) = \sum_{i=1}^{n} \alpha_i(s)\, \beta_i(t)$. Then $\mathcal{R}(K)$ is finite dimensional; whereas $\mathcal{R}(I - \lambda K)$ is infinite dimensional. This places a severe limitation on solvability of integral equations of the first kind since $\mathcal{R}(K)$ is too "small" and $\mathcal{N}(K)$ is too "big". For example, the equation

$$\int_0^1 \sin s \, \sin t \; x(t)dt = y(s) \tag{3.6}$$

has a solution if and only if $y = C \sin s$. The determination of the general solution requires the knowledge of the null space, i.e., the set of all functions orthogonal to $\sin s$ on $[0,1]$. The class of functions that can be represented in the form (3.6) is extremely limited since $\mathcal{R}(K)$ in this case is a one-dimensional subspace spanned by $\sin s$. In contrast, a large class of $y \in \mathcal{L}_2[0,1]$ can be represented in the form $(I - \lambda K)x$ for some $x \in \mathcal{L}_2[0,1]$, namely, all y if $\lambda \neq 2$; and for $\lambda = 2$ all $y \in \mathcal{L}_2[0,1]$ which satisfy the compatability requirement

$\int_0^1 y(s) \sin s \, ds = 0$. The general solution of (3.2), which is unique in case $\lambda \neq 2$, can be easily obtained.

4. The problem (3.1) is ill-posed since $K^\dagger$ is unbounded and densely defined. The best approximate solution $K^\dagger y$ for $y \in \mathcal{R}(K) + \mathcal{R}(K)^\perp$ (or the unique solution if it exists) does not depend continuously on the data. In the case of (3.3), the ill-posedness also follows from elementary and intuitive considerations. The integral operator K has a smoothing effect on the function $x(t)$; the degree of smoothness depends on the kernel. The equation (3.1) does not have a solution for every $y \in \mathcal{L}_2[0,1]$; for if $k(s,t)$ has a certain degree of smoothness in s, then there can exist no function $x \in \mathcal{L}_2[0,1]$ satisfying (3.1) for any function $y \in \mathcal{L}_2[0,1]$ that has a lower degree of smoothness. Also, there is no continuous dependence on the data. For suppose x is the unique solution of (3.1) and that $k(s,t)$ is absolutely integrable as a function of t for each s . Let $\hat{x}(t) = x(t) + C \sin nt$; then for any given $\varepsilon > 0$ and arbitrary large C, we can choose n (according to the Riemann-Lebesgue Lemma) so that in the $\mathcal{L}_2$-norm:

$$\|y - K\hat{x}\| = \|Kx - K\hat{x}\| < \varepsilon \ .$$

The instability and ill-posedness of the first kind operator equations is also apparent from the singular function analysis and the representation (1.5) for $K^\dagger y$, since by choosing n sufficiently large, we can find a μ_n as large as we please.

In contrast, since $(I - \lambda K)^\dagger$ is everywhere defined and bounded, the best approximate solution $(I - \lambda K)^\dagger y$ of (3.2) for any $y \in \mathcal{L}_2$ depends continuously on the "data" y; thus the operator equation (3.2) is well-posed (in a narrow sense). If the pseudocondition number $\|(I - \lambda K)(I - \lambda K)^\dagger\|$ is very large, then the problem is numerically ill-conditioned. On the other hand, in a wide sense "data" could be interpreted to include y and the operator K (so that for the integral equation the data consist of y, the kernel, and the domain of integration). In a wide-sense, second kind equations are ill-posed: the map $(I - \lambda K) \to (I - \lambda K)^\dagger$ is not continuous (this is in contrast to the case of invertible operator, where the map $(I - \lambda K) \to (I - \lambda K)^{-1}$ is continuous).

5. The ill-posedness of the problem (3.1) also manifests itself seriously in the numerical instability of the solutions of the algebraic equations arising from discretization. Suppose the integral equation (3.1) is reduced to a system of linear algebraic equations using a quadrature approximation:

$$(3.7)\qquad \sum_{j=1}^{n} w_j k(s_i, t_j)\, x(t_j) = y(s_i), \qquad i = 1, \ldots, n .$$

The ill-posedness of (3.1) leads to an ill-conditioned matrix (i.e., $\|A\|\,\|A^\dagger\|$ is very large, where $A := (k(s_i, t_j))$, if n is sufficiently large). Since the matrix A is obtained from a compact integral operator, it has for large n a cluster of small eigenvalues, and the matrix is increasingly ill-conditioned with larger n. Any attempt to solve the system (3.7) directly, e.g. by Gaussian elimination, is likely to produce a widely oscillatory solution. On the other hand, if n is too small, then the system (3.7) is not likely to approximate (3.1) closely. The effects of round-off error together with error in $y(s_i)$ lead to highly oscillatory solutions which render the approximation meaningless. To obtain meaningful approximation, one has to resort to a "filtering" process or a regularization method in conjunction with the system (3.7); see also Part III of this paper.

In contrast, for integral equations of the second kind, the discretized algebraic system will be of the form

$$x(t_i) - \lambda \sum_{j=1}^{n} w_i\, k(s_i, t_j)\, x(t_j) = y(s_i), \qquad i = 1, \ldots, n$$

and unless λ is very large, the near singularity of A does not produce any difficulty.

6. Other aspects of the distinction between operator equations of the first and second kinds are reflected in theoretical numerical analysis and approximation. Some features related to the infinite-dimensionality which are not fully reflected in the matrix case are important in iterative and regularization methods. Convergence rates for several classes of iterative procedures for both types of equations are given in Kammerer and Nashed [640]; in particular convergence rates are geometric for second kind equations, and are as $1/n$ for

first kind equations. It should also be noted that many regularization methods (for finding stable numerical solutions) for first kind equations use a sequence of second kind equations. For the theory of best approximate solutions of Fredholm and Volterra integral equations of the first and second kinds, see [640].

4. Generalized Inverses of Fredholm and Semi-Fredholm Operators

In this section we cast in the context of generalized inverses some known results on Fredholm and semi-Fredholm operators (cf. Atkinson [67], Yood [1399], Gohberg [1555], [1556]; see also Schechter [1190], [1189], Gohberg and Krein [453], Goldberg [454] and Kato [655]).

Let $\mathcal{X}$ and $\mathcal{Y}$ be Banach spaces, let $\mathfrak{L}(\mathcal{X},\mathcal{Y})$ and $\mathcal{K}(\mathcal{X},\mathcal{Y})$ denote the spaces of all bounded, respectively compact, linear operators on $\mathcal{X}$ to $\mathcal{Y}$. An operator $A \in \mathfrak{L}(\mathcal{X},\mathcal{Y})$ is said to be a Fredholm operator if (i) $\mathcal{R}(A)$ is closed in $\mathcal{Y}$, (ii) $\alpha(A) := \dim \mathcal{N}(A)$ is finite, and (iii) $\beta(A) := \dim \mathcal{N}(A^*)$ is finite. The set of all Fredholm operators from $\mathcal{X}$ to $\mathcal{Y}$ is denoted by $\Phi(\mathcal{X},\mathcal{Y})$. The index of a Fredholm operator is defined by $i(A) := \alpha(A) - \beta(A)$. If $\mathcal{X} = \mathcal{Y}$ and $K \in \mathcal{K}(\mathcal{X}) := \mathcal{K}(\mathcal{X},\mathcal{X})$, then $I - K$ is a Fredholm operator and, by (2.3), $i(I-K) = 0$.

An operator $A \in \mathfrak{L}(\mathcal{X},\mathcal{Y})$ is called semi-Fredholm if $\mathcal{R}(A)$ is closed in $\mathcal{Y}$ and either $\alpha(A)$ or $\beta(A)$ is finite. Let $\Phi_+(\mathcal{X},\mathcal{Y})$ [resp. $\Phi_-(\mathcal{X},\mathcal{Y})$] denote the set of all $A \in \mathfrak{L}(\mathcal{X},\mathcal{Y})$ with $\mathcal{R}(A)$ closed in $\mathcal{Y}$ and $\alpha(A) < \infty$ [resp. $\beta(A) < \infty$].

Let $A \in \Phi(\mathcal{X},\mathcal{Y})$. Since every finite dimensional subspace of a Banach space has a topological complement (see the discussion in Sec. 3 of the paper by Nashed and Votruba in this volume, if necessary), there exists a closed subspace $\mathcal{M}$ such that

$$\mathcal{X} = \mathcal{N}(A) \oplus \mathcal{M} \quad . \tag{4.1}$$

Also if $\mathcal{R}$ is a closed subspace of $\mathcal{Y}$ such that $\mathcal{R}^\circ$, the set of all annihilators of $\mathcal{R}$, is of dimension n, then there is an n-dimensional subspace $\mathcal{S}$ of $\mathcal{Y}$ such that $\mathcal{Y} = \mathcal{R} \oplus \mathcal{S}$. For a Fredholm operator, $\mathcal{R}(A)^\circ = \mathcal{N}(A^*)$, which is finite dimensional. Hence,

(4.2) $$\mathcal{Y} = \mathcal{R}(A) \oplus \mathcal{S}, \quad \text{where } \dim \mathcal{S} = \beta(A) .$$

In view of (4.1) and (4.2), the generalized inverse of A exists (relative to these decompositions) and is bounded (see Nashed and Votruba).

Theorem 4.1. Let $A \in \Phi(\mathcal{X},\mathcal{Y})$. Then there is a closed subspace $\mathcal{M}$ of $\mathcal{X}$ such that (4.1) holds and a closed subspace $\mathcal{S}$ of $\mathcal{Y}$ of dimension $\beta(A)$ such that (4.2) holds. Relative to these decompositions there exists a unique generalized inverse $A^{\dagger} = A^{\dagger}_{\mathcal{S},\mathcal{M}}$ such that

$$\mathcal{N}(A^{\dagger}) = \mathcal{S}, \quad \mathcal{R}(A^{\dagger}) = \mathcal{M}$$

$$A^{\dagger}A = I \text{ on } \mathcal{M}, \quad AA^{\dagger} = I \text{ on } \mathcal{R}(A) .$$

In addition,

(4.3) $$A^{\dagger}A = I - F_1 \text{ on } \mathcal{X}$$

(4.4) $$AA^{\dagger} = I - F_2 \text{ on } \mathcal{Y}$$

where $F_1 \in \mathcal{B}(\mathcal{X})$ with $\mathcal{R}(F_1) = \mathcal{N}(A)$ and $F_2 \in \mathcal{B}(\mathcal{Y})$ with $\mathcal{R}(F_2) = \mathcal{S}$ (consequently F_1 and F_2 are of finite rank).

To prove (4.4), we note that $F_1 := I - A^{\dagger}A$ is equal to I on $\mathcal{N}(A)$ and vanishes on $\mathcal{M}$.

We also have the following converse (which together with (4.3) and (4.4) provide a characterization of Fredholm operators): Let $A \in \mathcal{B}(\mathcal{X},\mathcal{Y})$. Assume there are operators $A_1, A_2 \in \mathcal{B}(\mathcal{X},\mathcal{Y})$. Assume there are operators $A_1, A_2 \in \mathcal{B}(\mathcal{Y},\mathcal{X})$, $K_1 \in \mathcal{K}(\mathcal{X})$ and $K_2 \in \mathcal{K}(\mathcal{Y})$ such that

$$A_1 A = I - K_1 \text{ on } \mathcal{X} ,$$

$$AA_2 = I - K_2 \text{ on } \mathcal{Y} ,$$

then $A \in \Phi(\mathcal{X},\mathcal{Y})$. (See, e.g., Schechter [1190; Theorem 2.1]). Also it is well known that if $A \in \Phi(\mathcal{X},\mathcal{Y})$ and $B \in \Phi(\mathcal{Y},\mathcal{Z})$, then $BA \in \Phi(\mathcal{X}, \mathcal{Z})$ and $i(BA) = i(A) + i(B)$.

If $K \in \mathcal{K}(\mathcal{X})$, then $A = I - K$ is a Fredholm operator. If we add any compact operator to A, it remains a Fredholm operator. This is also true for any Fredholm operator: If $A \in \Phi(\mathcal{X}, \mathcal{Y})$ and $K \in \mathcal{K}(\mathcal{X}, \mathcal{Y})$, then $A + K \in \Phi(\mathcal{X}, \mathcal{Y})$ and $i(A + K) = i(A)$. That is, a Fredholm operator is preserved under a compact perturbation. It is also preserved under a perturbation with a bounded linear operator of small norm: Let $A \in \Phi(\mathcal{X}, \mathcal{Y})$. Then there is an $\varepsilon > 0$ such that for any $T \in \mathcal{L}(\mathcal{X}, \mathcal{Y})$ with $\|T\| < \varepsilon$, one has $A + T \in \Phi(\mathcal{X}, \mathcal{Y})$, $i(A+T) = i(A)$ and $\alpha(A+T) \leq \alpha(A)$. This result also follows as a special case of the general perturbation theory of [894; Corollary 1].

We now consider generalized inverses of perturbed semi-Fredholm operators. Following Atkinson [67], we say that $A \in \mathcal{L}(\mathcal{X}, \mathcal{Y})$ is *relatively regular* if there exists $B \in \mathcal{L}(\mathcal{Y}, \mathcal{X})$ such that $ABA = A$. This is equivalent to the existence of a *bounded generalized inverse*, for if we let $C = BAB$, then $ACA = A$ and $CAC = C$. In general the class of all relatively regular operators is not an open set in $\mathcal{L}(\mathcal{X}, \mathcal{Y})$; the generalized inverse of a perturbation $A + T$ of a relatively regular operator need not even exist. Goldman [1560] proved that for any operator $A \in \mathcal{L}(\mathcal{X}, \mathcal{Y})$ which has closed range and which does not have an index (i.e., if $\alpha(A)$ and $\beta(A)$ are both infinite), there is a compact operator $T \in \mathcal{L}(\mathcal{X}, \mathcal{Y})$ such that for $\lambda \neq 0$, $A + \lambda T$ has nonclosed range (a simpler construction is due to R. J. Whitley [1711]; see Goldberg [454], V.2.6). Thus the preceding perturbation results for Fredholm operators do not hold for general relatively regular operators, and it is necessary to approach the problem from a different viewpoint. An approach to generalized inverses of perturbations of relatively regular operators was developed recently by Moore and Nashed [894]. On the other hand, the sets $\Phi_+(\mathcal{X}, \mathcal{Y})$ and $\Phi_-(\mathcal{X}, \mathcal{Y})$ are open in $\mathcal{L}(\mathcal{X}, \mathcal{Y})$, and the class of all relatively regular semi-Fredholm operators (denoted by $\psi(\mathcal{X}, \mathcal{Y})$) has satisfactory stability properties under additive perturbations which are compact or of small norm.

Theorem 4.2. (i). Let $A \in \psi(\mathcal{X}, \mathcal{Y})$ and $K \in \mathcal{K}(\mathcal{X}, \mathcal{Y})$. Then $A + K \in \psi(\mathcal{X}, \mathcal{Y})$ and $A + K \in \Phi_+(\mathcal{X}, \mathcal{Y})$ if $A \in \Phi_+(\mathcal{X}, \mathcal{Y})$ (respectively, $A + K \in \Phi_-(\mathcal{X}, \mathcal{Y})$ if $A \in \Phi_-(\mathcal{X}, \mathcal{Y})$) .

(ii) Let $A \in \psi(\mathcal{X}, \mathcal{Y})$. There exists an $\varepsilon > 0$ such that if $T \in \mathcal{L}(\mathcal{X},\mathcal{Y})$ with $\|T\| < \varepsilon$, then $A + T \in \psi(\mathcal{X},\mathcal{Y})$ and $\alpha(A + T) \leq \alpha(A)$ and $\beta(A + T) \leq \beta(A)$.

We remark that there is a connection between relatively regular semi-Fredholm operators and the classes of right and left invertible elements in the quotient algebra $\mathcal{L}(\mathcal{X})/\mathcal{K}(\mathcal{X})$. This connection first appeared in Yood [1399]. The class $\mathcal{K}(\mathcal{X})$. The quotient algebra $\mathcal{L}(\mathcal{X})/\mathcal{K}(\mathcal{X})$ is called the Calkin algebra. Let π denote the canonical mapping $\mathcal{L}(\mathcal{X}) \to \mathcal{L}(\mathcal{X})$. Then we have the following characterizations: An operator $A \in \mathcal{L}(\mathcal{X})$ is relatively regular with $\alpha(A) < \infty$ (resp. $\beta(A) < \infty$) if and only if $\pi(A)$ is a left (resp. right) invertible element of the Calkin algebra $\mathcal{L}(\mathcal{X})/\mathcal{K}(\mathcal{X})$. In particular, $A \in \mathcal{L}(\mathcal{X})$ is Fredholm if and only if $\pi(A)$ is invertible in the Calkin algebra.

The results of this section can be extended to the case where A is a closed linear operator, except for the preceding characterizations since there is no direct analogue of the Calkin algebra for closed operators.

The results also apply to the case of a mermorphic function $A(\lambda)$ with values in $\psi(\mathcal{X},\mathcal{Y})$, i.e., for each λ in the domain of A, $A(\lambda)$ is a semi-Fredholm operator from $\mathcal{X}$ into $\mathcal{Y}$ with complemented null space and range. Then there exists $B(\lambda) \in \mathcal{L}(\mathcal{Y},\mathcal{X})$ such that

$$A(\lambda) = A(\lambda)\,B(\lambda)\,A(\lambda), \qquad B(\lambda) = B(\lambda)\,A(\lambda)\,B(\lambda) \ .$$

The study of the map $\lambda \to B(\lambda)$ is of a particular interest in the theory of operator-valued functions. Under certain conditions on the poles of A, the operators $B(\lambda)$ can be chosen in such a way that the map $\lambda \to B(\lambda)$ is meromorphic (see, e.g., Bart [1484; Chapter II] or [1485]).

Finally we remark that the adjoint of an operator $A \in \psi(\mathcal{X},\mathcal{Y})$ is in $\psi(\mathcal{Y},\mathcal{X})$, but the converse does not hold (in nonreflexive spaces). A counterexample was first constructed by Pietsch [1019; pp. 366, 367]; see also Nashed and Votruba in this volume. This shows that in general $\psi(\mathcal{X},\mathcal{Y})$ is a proper subspace of the class of all semi-Fredholm operators.

5. Generalized Inverses in Other Areas of Linear Analysis and Applications

There are several topics of generalized inverses in linear analysis which we could include in this part of the paper, but which we shall not consider. These include:

(a) Generalized inverses of differential operators and generalized Green's matrices.

(b) Aspects of generalized inverses in partial differential equations.

(c) Generalized inverses of random linear operators, and particularly questions of their measurability.

(d) Generalized inverses in control systems and quadratic optimization problems.

(e) Uses of generalized inverses in network and system theory.

(f) Uses of generalized inverses in filtering and linear estimation in infinite dimensional problems.

(g) Uses of generalized inverses in pattern recognition.

Aspects of (b), (d) and (f) are developed in other papers in this volume. Extensive references on each of these topics are given in the annotated bibliography at the end of this volume.

II. USES OF GENERALIZED INVERSES IN NONLINEAR ANALYSIS

6. Nonlinear Alternative Problems

Many problems in analysis and applied mathematics (e.g., bifurication theory in Hammerstein integral equations, differential equations under periodic or more general boundary conditions, etc.) can be reduced to the problem of finding an $x \in \mathcal{D}(A)$ such that

$$Ax = Nx \tag{6.1}$$

where:

$A: \mathcal{D}(A) \subset \mathcal{X} \to \mathcal{Y}$ is a linear operator with $\eta(A) \neq \{0\}$;

$N: \mathcal{X} \to \mathcal{Y}$ is a linear or nonlinear operator;

$\mathcal{X}$ and $\mathcal{Y}$ are topological vector spaces.

We describe the alternative method for solving (6.1). We assume that $\mathcal{N}(A)$ and $\mathcal{R}(A)$ admit (topological) complements, so

(6.2) $$\mathcal{X} = \mathcal{N}(A) \oplus \mathcal{M}, \quad \mathcal{Y} = \mathcal{R}(A) \oplus \mathcal{S} \ .$$

Equivalently, there exist (continuous) projectors U, E such that $\mathcal{R}(A) = E\mathcal{Y}$ and $\mathcal{N}(A) = U\mathcal{X}$. (Some technical modifications are necessary to extend the theory of this section to the case when $\overline{\mathcal{R}(A)}$ and $\mathcal{N}(A)$ admit topological complements, as well as for more general contexts in topological vector spaces under the assumption of domain decomposability in the sense of [934]; these will be discussed elsewhere).

Under these assumptions, A is 1-1 on $\mathcal{C}_A := \mathcal{M} \cap \mathcal{D}(A)$ the carrier of A, $A\mathcal{C}_A = \mathcal{R}(A)$, $A^{\dagger}_{E,U}$ exists and is onto $\mathcal{C}_A$, $A^{\dagger}A = I$ on $\mathcal{R}(A)$, $A^{\dagger}A = I - U$ on $\mathcal{D}(A)$, and $A^{\dagger}A = I$ on $\mathcal{C}_A$.

Obviously, the equation (6.2) is equivalent to the equation $Ax - AA^{\dagger}Nx = 0$, and is also equivalent to the system

$$(I - E)(A - N)x = 0 \quad \text{and} \quad E(A - N)x = 0 \ .$$

The following theorem follows easily from elementary properties of the generalized inverse.

Theorem 6.1. Let $\mathcal{X}$ and $\mathcal{Y}$ be topological vector spaces and let $A: \mathcal{X} \to \mathcal{Y}$ have a generalized inverse relative to projectors E of $\mathcal{Y}$ onto $\mathcal{R}(A)$ and U of $\mathcal{X}$ onto $\mathcal{N}(A)$. Then the following statements are equivalent for $x^* \in \mathcal{D}(N) \cap \mathcal{D}(A)$:

(i) x^* is a solution of (6.1).

(ii) x^* is a solution of the system

$$x - A^{\dagger} ENx = y$$
$$(I - AA^{\dagger})Nx = 0$$

for some $y \in \mathcal{N}(A)$.

(iii) $Ax^* - Nx^* \in \mathcal{N}(A^\dagger)$ and $Nx^* \in \mathcal{R}(A)$.

Letting $M := A^\dagger|\mathcal{R}(A)$, we can write (ii) in the form

$$(6.3) \qquad \begin{cases} \text{(a)} \quad x = Ux + MENx \\ \text{(b)} \quad (I - E)Nx = 0 \ . \end{cases}$$

The system (6.3), which is called an alternative problem for (6.1), is the basis for several methods for solving (6.1), particularly in the case of a "small" nonlinearity or "large" nonlinearity involving monotone operators. The idea is to choose an element $y \in \mathcal{N}(A)$ and solve

$$(6.4) \qquad x = y + MENx$$

for $x = x^*(y)$, then determine $y = y^*$ so that

$$(6.5) \qquad (I - E)Nx^*(y) = 0 \ .$$

Then $x^*(y^*)$ is a solution of (6.1). Usually (6.4) can be easily solved by some iterative method and is amenable to standard existence theorems based on contraction mappings or monotone operators. The crux of the matter is to solve (6.5), which is usually a finite system of nonlinear equations.

The alternative method is the underlying principle in several papers in nonlinear analysis and differential equations (see the survey paper of Hale [518] where various alternative methods for solving (6.1) in Banach space are discussed and an extensive bibliography is given. Extension of the alternative method will be given elsewhere).

The alternative method is also applicable to approximations of nonlinear (bifurication) problems involving a parameter. Let $\mathcal{S}$ be a ball in a parameter Banach space $\mathcal{B}$. Let $A(\lambda): \mathcal{X} \to \mathcal{Y}$ be a closed linear operator and for $\lambda \in \mathcal{S}$, assume there exist continuous projectors $U(\lambda)$, $E(\lambda)$ onto $\mathcal{N}(A(\lambda))$ and $\mathcal{R}(A(\lambda))$, respectively. Let $A^\dagger(\lambda)$ be the unique (bounded) generalized inverse of $A(\lambda)$ relative to these projectors, and let $M(\lambda) := A^\dagger(\lambda)|\mathcal{R}(A(\lambda))$, and assume that all of the above operators are continuous functions of $\lambda \in \mathcal{S}$. Let $N: \mathcal{S} \times \mathcal{X} \to \mathcal{Y}$ be a continuous, not necessarily linear, operator, and consider the problem of finding a solution $x^*(\lambda) \in \mathcal{X}$ of

(6.6) $$A(\lambda)x = N(\lambda, x)$$

depending continuously on λ in some subset of $\mathcal{g}$. The essence of the alternative method is to replace (6.6) by the equivalent problem of solving two simpler equations for $x = x^*(\lambda)$:

(6.7) $$x = x_0(\lambda) + M(\lambda)\, E(\lambda)\, N(\lambda, x)$$

and

(6.8) $$(I - E(\lambda))\, N(\lambda, x) = 0$$

where $x_0(\lambda) \in \mathcal{N}(A(\lambda))$, $\lambda \in \mathcal{g}$.

The alternative method generalizes to deal with the case of "large" nonlinearities, or with the situation where there is no continuous dependence on the parameter λ, by judiciously replacing $E(\lambda)$ by a new projector onto a subspace of $\mathfrak{R}(A(\lambda))$, leading to new alternative equations

(6.9) $$x = x_0(\lambda) + M(\lambda)\, Q(\lambda)\, N(\lambda, x)$$

(6.10) $$A(\lambda)\, x_0(\lambda) = (I - Q(\lambda))\, N(\lambda, x)$$

where $P(\lambda)$ and $Q(\lambda)$ are continuous projectors, replacing $U(\lambda)$ and $E(\lambda)$, respectively, satisfying the following conditions:

(6.11) $$\mathcal{N}(A(\lambda)) \subset \mathfrak{R}(P(\lambda)),\quad \mathfrak{R}(Q(\lambda)) \subset \mathfrak{R}(A(\lambda))\ ;$$

(6.12) $$Q(\lambda)\, A(\lambda)\, P(\lambda) = 0\ ;$$

(6.13) $$P(\lambda)\, M(\lambda)\, Q(\lambda) = 0\ .$$

Note that (6.12) accomodates for $\mathfrak{R}(P(\lambda)) \supsetneq \mathcal{N}(A(\lambda))$: $A(\lambda)$ does not annihilate $\mathfrak{R}(P(\lambda))$, but $Q(\lambda)\, A(\lambda)$ does.

The thrust of Sweet's paper [1288] is to adapt the alternative method in the setting of Banach spaces to approximate solutions of the nonlinear operator equation:

(6.14) $$F(x) = 0$$

where $F: \mathcal{X} \to \mathcal{Y}$ has, for each $x \in \mathcal{X}$, a Fréchet derivative which is assumed to be a closed operator with closed range and $\mathcal{R}(F'(x))$ admits a projector E and $\mathcal{N}(F'(x))$ admits a projector P. Assume x^* is a solution of (6.14) and let $\bar{x}$ be an approximate solution. Consider the problem of determining z such that

$$F(\bar{x} + z) = F'(\bar{x})z + F(\bar{x}) + N(\bar{x}, z) \tag{6.15}$$

when N is of higher order in z when z is small. Equation (6.15) is of the form of the alternative method discussed above, where now the approximate solution $\bar{x}$ plays the role of the parameter λ. It is not hard to show the existence of projectors which depend continuously on $\bar{x}$ with the structure permitting the use of the alternative method.

A general framework for perturbation and approximation of generalized inverses in Banach space is given in Moore and Nashed [894]. An important motivation for [894] is its connection with some aspects of the alternative method. Some of the estimates and convergence results of [894], as indicated therein, can be used to strengthen results on approximations of alternative problems and provide new results on perturbations of these problems.

7. Inverse Mapping Theorem and Open Mapping Theorem for Differentiable Operators

Let $\mathcal{X}$ and $\mathcal{Y}$ be Banach spaces and let $\mathcal{U}$ be a neighborhood of the origin in $\mathcal{X}$. Let $F: \mathcal{U} \to \mathcal{Y}$ with $F(0) = 0$ In the classical version of the inverse mapping theorem (e.g. Hildebrandt and Graves [1581]) one usually assumes that the Fréchet derivative $F'(x)$ exists for all x in a neighborhood of the origin and is continuous at 0 and $F'(0)$ is invertible. Then the mapping F is open and uniquely invertible near the origin.

Several variants and extensions of the inverse mapping theorem have been given. We consider extensions of this theorem in two directions. The first of such includes a stronger version (due to Leach [741]) under a stronger version of differentiability at the origin only. In the second direction, we relax the assumptions of the inverse mappings theorem (Graves [1569], Bartle [1488]). In both cases the assumption of the invertibility of $F'(0)$ is replaced by some generalized invertibility.

The operator F is said to be strongly differentiable at $x_0 \in \mathcal{X}$ if there exists $A \in \mathcal{L}(\mathcal{X},\mathcal{Y})$ such that for each $\varepsilon > 0$, there is $\delta > 0$ such that

$$\|F(x_2) - F(x_1) - A(x_2 - x_1)\| \leq \varepsilon \|x_2 - x_1\| \text{ whenever } \|x_2 - x_1\| < \delta .$$

Clearly strong differentiability implies Fréchet differentiability. The operator A is called the strong derivative of F at x_0. For properties of strong differentiability and other types of differentiations, see Nashed [916].

Theorem 7.1 (Inverse Mapping Theorem). Let $F: \mathcal{U} \to \mathcal{Y}$ with $F(0) = 0$ have a strong differential at the origin and that the strong derivative $A = F'(0)$ has a bounded linear outer inverse, i.e., there exists $B \in \mathcal{L}(\mathcal{Y},\mathcal{X})$ such that $BAB = B$. Then there exists an operator $G(0) = 0$ and G is strongly differentiable at $y_0 = 0$ and its derivative is B. Moreover,

$$F(G(y)) - y \in \mathcal{N}(B) \text{ for } \|y\| \text{ sufficiently small.}$$

and

$$G(F(x)) = x \text{ for all } x \in \mathcal{R}(B) .$$

Note that if A is invertible (so $B = A^{-1}$), then $F(G(y)) = y$ as in the classical inverse mapping theorem. For the proof we use Newton's method to define the sequence $x_{n+1} = x_n + B(y - F(x_n))$. It is not hard to show that for all sufficiently small y, $\lim_{n\to\infty} x_n$ exists. Define $G(y) = \lim_{n\to\infty} x_n$ whenever this limit exists and $G(y) = 0$ otherwise. The G satisfies the conclusions of the theorem.

A "soft" implicit function version of Theorem 7.1 can also be easily given (see [741], [114]).

Theorem 7.2 (Open Mapping Theorem). Let $F: \mathcal{U} \subset \mathcal{X} \to \mathcal{Y}$ with $F(0) = 0$. Suppose F is Fréchet differentiable for each $x \in \mathcal{B}_r := \{x: \|x\| < r\} \subset \mathcal{U}$ for some $r > 0$, and that the Fréchet derivative $F'(0)$ at the origin maps $\mathcal{X}$ onto all of $\mathcal{Y}$. Let φ be the monotone function defined for $0 \leq \rho < r$ by

$$\varphi(\rho) := \sup\{\|F'(x) - F'(0)\|: \|x\| \leq \rho\} .$$

Assume in addition that

(7.1) $$\varphi(0^+) := \limsup_{x \to 0} \|F'(x) - F'(0)\| < 1/\|[F'(0)]^\dagger\|$$

and let

$$\hat{\mathcal{B}} := \sup\{\rho: \varphi(\rho) < 1/\|[F'(0)]^\dagger\|\} .$$

Then the image under F of any open set contained in $\hat{\mathcal{B}}$ is an open set.

Condition (7.1) involves the minimum modulus of the operator $F'(0)$ and is satisfied in particular in the case that the map $x \to F'(x)$ is continuous at $x = 0$.

8. Least Squares Solutions of Nonlinear Operator Equations

Let $F: \mathcal{H}_1 \to \mathcal{H}_2$, where $\mathcal{H}_1$ and $\mathcal{H}_2$ are real Hilbert spaces and consider the problem of minimizing the functional

(8.1) $$\Phi(x) = \frac{1}{2}\|F(x)\|^2 = \frac{1}{2}\langle F(x), F(x)\rangle$$

without the assumption that $F(x) = 0$ has a solution. In other words, we seek a "solution" $\hat{x}$ which minimizes the Hilbert space norm of the residuals. Any such solution $\hat{x}$ must satisfy grad $\Phi(\hat{x}) = 0$, i.e.,

(8.2) $$G(\hat{x}) := [F'(\hat{x})]^* F(\hat{x}) = 0$$

where $F'(x)$ is the Fréchet derivative of F at x and $*$ denotes the adjoint. We assume that $\mathcal{R}(F'(x))$ is closed in $\mathcal{H}_2$ and consider the case where $F'(x)$ is not necessarily invertible. (In the case of nonlinear equations on $\mathbb{R}^n$, $F'(x)$ is simply the Jacobian matrix.)

To motivate some algorithms for solving (8.2), we consider first the case of linear operator equations. Let $F(x) = Ax - b$, where A is a bounded linear operator with closed range. Theoretically, a least squares solution of $Ax = b$ can be obtained in one step starting from any $x_0 \in \mathcal{H}_1$ by taking $x_1 = A^\dagger b + (I - A^\dagger A)x_0$, which we rewrite in the form:

(8.3) $$x_1 = x_0 - A^\dagger(Ax_0 - b) = x_0 - A^\dagger F(x_0) \ .$$

(The weakness of this, however, is that it poses still the same problem, that of computing $A^\dagger$. In particular, if A possesses an inverse, then $x_1 = A^{-1}b$). Noting that for the case $F(x) = Ax - b$, we have $F'(x) = A$, (8.3) suggests that a suitable iterative procedure for minimizing $\Phi(x)$ in (8.1) is defined by:

(8.4) $$x_{n+1} = x_n - \alpha_n[F'(x_n)]^\dagger F(x_n) \ ,$$

or equivalently, since $A^\dagger = (A^*A)^\dagger A^*$,

(8.5) $$x_{n+1} = x_n - \alpha_n[(F'(x_n))^* F'(x_n)]^\dagger [F(x_n)]^* F(x_n) \ ,$$

where the sequence $\{\alpha_n\}$ is chosen so that $\Phi(x)$ does not increase (or is optimally chosen to minimize $\Phi(x)$ in the direction of $h_n = x_{n+1} - x_n$ at each iteration).

The functional $\Phi(x)$ in (8.1) is decreasing along the directions

$$h = -\alpha[F'(x)]^\dagger F(x), \quad \alpha > 0$$

unless x_0 is a stationary point. To show this we note that

(8.6) $$\langle \operatorname{grad} \Phi(x), h \rangle = -\alpha \langle F(x), F'(x)[F'(x)]^\dagger F(x) \rangle < 0$$

since $F'(x)[F'(x)]^\dagger$ is a projector onto $\mathcal{R}(F'(x))$.

Convergence of the sequence, together with the modified Newton's method, can be established under some conditions. The analysis is similar to the nonsingular case. The gist of convergence results is that if the generalized inverse of $F'(x)$ is continuous in some neighborhood of the iterates, then a local convergence proof holds for a stationary point of $\Phi(x)$. Under similar hypotheses, global convergence results are known for certain classes of the step size α_n. Other gradient methods, including the conjugate gradient methods, can also be adapted to this setting.

The preceding techniques are of particular interest in the case of a system of nonlinear equations in $\mathbb{R}^n$:

$$(8.7) \qquad f_i(x) = 0, \quad x = (x^1, \ldots, x^n), \qquad i = 1, \ldots, m .$$

If $m = n$ and the Jacobian $J(x) = F'(x)$ of $F = (f_1, \ldots, f_m)^t$ is nonsingular, then (8.4) is Newton's method modified by the factor α_n. If $J(x)$ is singular, if the system (8.7) is overdetermined $(m > n)$, or if it underdetermined $(m < n)$, then the classical Newton's method is not applicable, and the system (8.7) may not even possess a solution. An alternate way in these cases is seek a least squares solution of (8.7), i.e., a solution which minimizes the sum of the squares of the residuals, and to apply the generalized Newton's method (8.4). In this connection we remark that the "downhill" property of the direction h in (8.6) assumes exact arithmetic and is not necessarily valid if the rank of the Jacobian matrix becomes indeterminate due to rounding errors. Thus while the generalized Newton's method theoretically applies to the general case of a system of equations, and does not depend (in theory) upon knowing a priori the rank of J, the effect of applying the method to problems with rank-deficient Jacobians may present numerical difficulties when rounding errors are present. The method (8.5) reduces in the case when rank $J = n$ and $\alpha_n = 1$ to the classical Gauss-Newton method:

$$x_{n+1} = x_n - [(F'(x_n))^* F'(x_n)]^{-1} [F(x_n)]^* F(x_n) .$$

Approximations to $J^\dagger$ can be updated by differences as in the classical Newton or Gauss-Newton method in case when the derivatives are not explicitly available.

For aspects of least squares solutions of nonlinear equations, Newton's method for the case of a singular derivative, and related topics, see [109], [114], [398], [399], [1004] and related references cited therein.

In some problems, the functional $\Phi(x)$ cannot be reduced appreciably at each iteration of the algorithm (8.5). In order to overcome this practical drawback, several authors (Levenberg, Hoerl, Morrison, Marquardt and others) have suggested in different contexts the use of directions $h_n(\lambda_n)$ defined by

$$(8.8) \quad \{[F'(x_n)]^* F'(x_n) + \lambda_n I\}h_n = -[F'(x_n)]^* F(x_n), \quad \lambda_n > 0 .$$

The use of this idea in linearization of nonlinear problems is often called Marquardt's method. Note that the operator on the left side of (8. 8) is always invertible for $\lambda_n > 0$, so h_n can always be uniquely solved for. As $\lambda_n \to 0$ in (8. 8) we get the classical Gauss-Newton method, while for large λ_n the direction is asymptotically the gradient direction. Thus Marquardt's method is an interpolation or a compromise between the Gauss-Newton method and the steepest descent method. From a theoretical viewpoint, the method also may be interpreted as a steepest descent method relative to a family of semidefinite inner products defined by

$$[x,y]_n = \langle \{[F'(x_n)]^* F'(x_n) + \lambda_n I\} x, y \rangle, \qquad \lambda_n > 0 .$$

However, this does not contribute to the practical aspects of the method since the main point of the method is the choice of the adjustable parameter λ, which controls the length and direction of the correction vector used to update the current iterate. The problem is intimately connected with ridge analysis (cf., e.g. [802]) where the study of $h(\lambda)$ as a function of λ is important. Various algorithms and programs for choosing λ at each iteration have been proposed. For some aspects of practical implementation of the Marquardt algorithm using matrix decompositions and related computational method for least squares problems, see Golub and Saunders [475], Lawson and Hanson [735] and references and programs cited therein.

Finally we remark that other modifications are also possible by replacing the identity operator I in (8. 8) by a positive definite operator. This is closely connected with some regularization methods for ill-posed linear operator equations (see [920]).

9. Constrained Nonlinear Optimization Problems

We give an indication of the use of generalized inverses in the construction of a suitable penalty function for constrained nonlinear optimization problems. The discussion is patterned on that of Fletcher [1541]. Consider the optimization problem:

$$(9.1) \quad \begin{cases} \text{Minimize } f(x) & x \in \mathbb{R}^n \\ \text{Subject to } g_i(x) = 0, & i = 1, 2, \ldots, k \le n \end{cases}$$

with solution $x = \xi$. Let $F(x) = \nabla f(x)$ and denote by $H(x)$ the Hessian matrix of second partial derivatives of f . With t denoting transpose, let $g = (g_1, g_2, \ldots, g_k)^t$ and let N denote the $n \times k$ matrix whose columns are $\nabla g_i(x)$, $i = 1, \ldots, k$. Assume N is of full rank k . Then $N^\dagger = (N^t N)^{-1} N^t$. A class of methods for the problem (9.1) is based on consideration of the function (see [1541] for motivation)

$$\psi := f - g^t N^\dagger F \quad . \tag{9.2}$$

Then

$$\nabla\psi = F - NN^\dagger F - \nabla(N^\dagger F)g \quad .$$

The matrix $P := NN^\dagger$ projects vectors into $\mathcal{R}(N)$, i.e. into the subspace spanned by the normals to the constraints, and $I - P$ projects vectors so that their direction lies in the space formed by the intersection of the constraints. The solution ξ to (9.2) must satisfy the conditions: $(I - P)F = 0$ and $g = 0$, i.e. the component of the gradient of f in the constraint manifold is zero. Using these conditions in (9.2) we see that $\nabla\psi = 0$, so ξ is a stationary point of ψ . For ψ to be a suitable pentality function, it is important to know that the stationary point is a minimum. This is not always the case for the function ψ . To see this, consider the second derivative of ψ in the case when f is quadratic and g is linear. Then N, $N^\dagger$ and H are all constant matrices and

$$\nabla^2\psi = (I - P)\, H(I - P) - PHP \quad .$$

Thus ψ only has a minimum at ξ if PHP is negative semi-definite and of rank k . A suitable penalty function can be obtained from (9.2) by adding a sufficiently "large" positive definite form, i.e. by replacing ψ in (9.2) by

$$\phi = F - NN^\dagger F + g^t Q g \, . \tag{9.3}$$

Two simple choices for Q are $Q = cI$ and $Q = c\, N^\dagger N^{\dagger t}$, where c is a sufficiently large positive number. The following result is due to Fletcher [1541].

Theorem 9.1. Let f and g be twice differentiable at $x = \xi$, let $N(\xi)$ be of full rank and let $N^\dagger(x)$ be a differentiable function at ξ. Let $h(x, \lambda) := f - g^t\lambda$. If

$$\nabla_x h(x, \lambda) = 0, \quad y^t \nabla_x^2 h\, y > 0$$

at $x = \xi$, where $y \in \eta(N^t)$, $y \neq 0$, then there exists a positive number α such that for all $c > \alpha$, the penalty function

$$\phi(x) = F - NN^\dagger F + c\, g^t N^\dagger N^{\dagger t} g$$

satisfies the following properties:

$$\nabla\phi(\xi) = 0 \quad \text{and} \quad v^t \nabla^2 \phi(\xi) v > 0 \quad \text{for} \quad v \neq 0 .$$

The preceding conditions are not very restrictive and the penalty function (9. 3) with the indicated choices of Q are of quite general applicability. Practical approaches and recommendation for the best computational scheme for minimizing the penalty function (9. 3) are given in Fletcher and Lill [1544] and Lill [1616], where numerical results are also given. It should be noted that (9. 3) is an "exact" penalty method for equality-constraint nonlinear programming. The method is also adaptable to inequality constraints. The method is suitable for problems in which the objective functions are differentiable but nonlinear and nonseparable. The method has strong interface with Lagrangian and penalty methods. It has an advantage over both in that a sequence of unconstrained minimization problems is not required (we have a single unconstrained problem (9. 3)) and that the penalty function is well-conditioned.

III. ASPECTS OF GENERALIZED INVERSES IN REGULARIZATION AND APPROXIMATION OF ILL-POSED LINEAR OPERATOR EQUATIONS

10. Ill-Posed Problems and Regularization Methods

An operator equation $Ax = y$, where $A: \mathcal{X} \to \mathcal{Y}$ is said to be well-posed if for each $y \in \mathcal{Y}$ a unique "solution" exists which depends continuously on the "data"; otherwise the equation is said to be ill-posed (improperly-posed). Clearly the notion of well-posedness depends on the adopted notion of "solution", the admissible data and the "measure" of continuous dependence.

In a narrow sense, continuous dependence on the "data" is often interpreted as continuous dependence on $y \in \mathcal{Y}$. However in a wide sense, the "data" should be interpreted to include y and the operator A. In the case of a differential equation, for example, one must regard as "data" any initial or boundary values, prescribed values of the operator, other coefficients or parameters in the equation, and the geometry of the domain of definition of the operator. Similarly for an integral equation, one must regard as data, the kernel, the domain of the kernel, the prescribed function y, and other parameters. In this wide sense, ill-posed problems encompass large classes of problems of analysis and applications (in addition to the variety of problems which are ill-posed in the narrow sense). This includes, for example, forced vibration problems, bifurication problems, least-squares problems for integral equations of the second kind, and even linear least squares problems in finite dimensional spaces (since the generalized inverse of a matrix does not depend continuously on perturbations of the matrix which alter its rank).

A lack of continuous dependence of the solution on the data makes direct investigation (and particularly approximation) of ill-posed problems difficult and often meaningless, and manifests itself most seriously in the numerical analysis of such problems.

"Regularizations of ill-posed problems" is a phrase that is used for various approaches to circumvent lack of continuous dependence (as well as to impart existence and uniqueness if necessary). Roughly speaking, a regularization method entails an analysis of an ill-posed problem via an analysis of an associated well-posed problem, or a family (usually a sequence or a filter) of well-posed problem, provided this analysis yields meaningful answers and/or approximations to the given ill-posed problem.

Review and survey of various aspects of ill-posed problems and regularization methods, with extensive bibliographies, are to be found in Payne [991], [990], Tikhonov [1695], Nashed [918], Turchin, Kozlov and Malkevich [1699], Nedelkov [1636],

Hilgers [583], and Medgyessy [1627]. For recent results on approximate regularized solutions to improperly posed linear integral and operator equations, see [920]. Operator-theoretic and approximations aspects of ill-posed problems are examined in a broader framework in [927].

Numerous methods have been proposed for treating and regularizing various types of ill-posed problems. For classifications of the various approaches see Nashed [920] and Payne [991], where the emphases are respectively ill-posed operator equations and Cauchy problems. The various approaches to regularization involve essentially one or more of the following intuitive ideas:

(a) a change of the concept of a solution;

(b) a restriction of the data;

(c) a change of the spaces and/or topologies;

(d) a modification of the operator itself;

(e) the concept of regularization operators;

(f) probabilistic methods or well-posed stochastic extensions of ill-posed problems.

The various approaches to regularization overlap in many aspects. Usually several of the above ideas manifest themselves in any particular approach to regularization.

The discussion in the remaining part of this paper (which partly overlaps with a lecture on ill-posed problems presented at another conference [920]) is limited primarily to classes of ill-posed linear operator equations in Hilbert space with emphasis on regularization methods in which generalized inverses play a role. These regularization methods include:

1. The use of singular values decomposition for a compact operator and its interpretation in reproducing kernel Hilbert spaces.

2. Constrained minimization methods for stabilization of ill-posed linear operator equations.

3. Method of generalized inverses in reproducing kernel Hilbert spaces (RKHS).

4. Numerically filtered iterative methods.

Let $\mathcal{X}$ and $\mathcal{Y}$ be two Hilbert spaces over the same (real or complex) scalars and let A be a linear operator on $\mathcal{D}(A) \subset \mathcal{X}$ into $\mathcal{Y}$. We consider the equation

$$Ax = y \quad \text{where} \quad y \in \mathcal{Y} . \tag{10.1}$$

<u>Definition 10.1.</u> An element $u \in \mathcal{X}$ is said to be a <u>least squares</u> solution of (10.1) if $\inf\{\|Ax - y\| : x \in \mathcal{X}\} = \|Au - y\|$. A <u>pseudosolution</u> of (10.1) is a least squares solution of minimal norm.

<u>Definition 10.2.</u> The operator equation (10.1) is said to be <u>well-posed</u> (relative to the spaces $\mathcal{X}$ and $\mathcal{Y}$) if for each $y \in \mathcal{Y}$, (10.1) has a unique pseudosolution which depends continuously on y; otherwise the equation is said to be <u>ill-posed</u>

Clearly (10.1) has a least squares solution for a given $y \in \mathcal{Y}$ if and only if there exists an element $w \in \mathcal{R}(A)$ which is closest to y, i.e., if and only if the orthogonal projection of y onto $\overline{\mathcal{R}(A)}$ is in $\mathcal{R}(A)$, or equivalently if $y \in \mathcal{R}(A)^{\perp} + \mathcal{R}(A)$. For such y it is easy to see that the set $\mathcal{S}_y$ of all least squares solutions has a unique element of minimal norm if and only if the orthogonal projection of a vector $u \in \mathcal{S}_y$ onto $\overline{\mathcal{N}(A)}$ is in $\mathcal{N}(A)$. It then follows that a pseudosolution of (10.1) exists if and only if

$$y \in A[\mathcal{D}(A) \cap \mathcal{N}(A)^{\perp}] + \mathcal{R}(A)^{\perp} . \tag{10.2}$$

In what follows we shall be particularly interested in the cases when A is a <u>closed</u> linear operator on a dense domain $\mathcal{D}(A) \subset \mathcal{X}$, or when A is a <u>bounded</u> linear operator on $\mathcal{X}$. In either of these cases condition (10.2) reduces to the condition

$$y \in \mathcal{R}(A) + \mathcal{R}(A)^{\perp} . \tag{10.3}$$

The (linear) map which associates with y satisfying (10.2) a unique pseudosolution defines the generalized inverse $A^\dagger$ of A. The following proposition is an immediate consequence of the closed graph theorem.

Proposition 10.1. Let $A: \mathcal{D}(A) \subset \mathcal{X} \to \mathcal{Y}$ be a bounded or densely defined closed linear operator. Then the following statements are equivalent:

(a) The operator equation (10.1) is well-posed in $(\mathcal{X}, \mathcal{Y})$;

(b) A has a closed range in $\mathcal{Y}$;

(c) $A^\dagger$ is a bounded operator on $\mathcal{Y}$ into $\mathcal{X}$. (When $\mathcal{R}(A)$ is not closed, $A^\dagger$ is an unbounded densely defined operator.)

An ill-posed problem relative to $(\mathcal{X}, \mathcal{Y})$ may be recast in some cases as a well-posed problem relative to new spaces $\mathcal{X}' \subset \mathcal{X}$ and $\mathcal{Y}' \subset \mathcal{Y}$, with topologies on $\mathcal{X}'$ and $\mathcal{Y}'$ which are different from the (original) topologies on $\mathcal{X}$ and $\mathcal{Y}$. To provide regularization, the topologies on $\mathcal{X}'$ and $\mathcal{Y}'$ should not be restrictive and must lend themselves to the original problem. This is the central idea in the method of generalized inverses in reproducing kernel Hilbert spaces (Nashed and Wahba [938], [939]).

11. Singular Values Decomposition, Revisited: Implications for Regularization and Reproducing Kernel Hilbert Spaces

1. Let K be the compact operator defined in (1.6). Let $\{\lambda_n\}$, $\{u_n\}$ be the eigenvalues and eigenvectors of KK^*. Define

$$\mathcal{H}_Q := \{g \in \mathcal{L}_2[0,1]: \sum_n \frac{|\langle g, u_n \rangle|^2}{\lambda_n} < \infty\},$$

where if $\lambda_n = 0$, we must have $\langle g, u_n \rangle = 0$, and the corresponding term in the sum is taken to be zero. By Picard's criterion (see Theorem 1.2), the vector space $\mathcal{H}_Q$ is precisely the set of all functions in $\mathcal{L}_2[0,1]$ for which the equation $Kx = y$ has a solution. $\mathcal{H}_Q$ is a proper dense subspace of $\mathcal{L}_2[0,1]$. If we define a new inner product in $\mathcal{H}_Q$ by

$$\langle f, g \rangle_Q = \sum_n \frac{1}{\lambda_n} \langle f, u_n \rangle \langle g, u_n \rangle ,$$

then $\mathcal{H}_Q$ becomes a Hilbert space. Letting

$$Q(s, s') := \int_0^1 k(s,t)\, k(s',t) dt$$

and $Q_s(s') := Q(s, s')$, it follows also that

$$\langle Q_s, z \rangle_Q = z(s), \quad z \in \mathcal{H}_Q \quad \text{for} \quad 0 \le s \le 1 .$$

That is, $\mathcal{H}_Q$ is a reproducing kernel Hilbert space (RKHS) with reproducing kernel Q. (For details see [937] where appropriate references to properties of RKHS are also given. A Hilbert space $\mathcal{H}$ of real-valued functions on $[0,1]$ is said to be an RKHS if all the evaluation functionals $f \to f(s)$ for $f \in \mathcal{H}$, $0 \le s \le 1$ are continuous.) Furthermore, it is shown that the spaces $\mathcal{L}_2$ and $\mathcal{H}_Q$ and their respective norms are related by the following relations:

$$\mathcal{H}_Q = Q^{\frac{1}{2}} \mathcal{L}_2$$

$$\|z\| = \inf\{\|P\|_{\mathcal{L}_2} : Q^{\frac{1}{2}}P = z\} = \|Q^{-\frac{1}{2}}\|_{\mathcal{L}_2} ,$$

where Q is the Fredholm integral operator whose kernel is $Q(s, s')$ and $Q^{-\frac{1}{2}} := (Q^{\frac{1}{2}})^{\dagger}$.

Thus if the operator K is regarded as a map $K: \mathcal{L}_2 \to \mathcal{H}_Q$, then it becomes onto. Hence for each $y \in \mathcal{H}_Q$, the equation $Kx = y$ has a unique least squares solution of minimal norm and this solution depends continuously on y (in the topology of $\mathcal{H}_Q$), i.e., the generalized inverse $K^{\dagger}: \mathcal{H}_Q \to \mathcal{L}_2$ is bounded.

The introduction of the RKHS $\mathcal{H}_Q$ enabled us to cast Picard's criteria for $Kx = y$ (see Theorem 1.2) in the equivalent form: $y \in \mathcal{H}_Q$. While this casting may appear as a mere formality, it has several important ramifications. In the above process we have in fact <u>shrunk</u> the image space $\mathcal{L}_2$ to the range of K and endowed the latter with a <u>new topology</u> relative to which the range is closed. (This is typical of the more general situations where continuous dependence is imparted by a change in the topologies of the spaces.) By a simple computation using (1.5) one can verify that $\|K\|_Q \|K^\dagger\|_Q = 1$, so the equation $Kx = y$ is optimally conditioned in the RKHS setting.

We may also narrow the domain of K and endow it with a different topology in order to provide higher orders of smoothness in the solution $K^\dagger y$. For example consider $K: \mathcal{H}_P \to \mathcal{H}_Q$, where $\mathcal{H}_P$ is the space of all absolutely continuous functions with derivative in $\mathcal{L}_2[0,1]$, with the inner product on $\mathcal{H}_P$ defined by

$$\langle u, v \rangle = u(0)\, v(0) + \int_0^1 u'(t)\, v'(t) dt \ .$$

($\mathcal{H}_P$ is an RKHS with reproducing kernel $P(s,t) = \min(s,t)$).

The preceding idea provide a preview of one aspect of the regularization method of generalized inverses in RKHS developed by Nashed and Wahba in [939], [935] and summarized in Section 13.

2. A close look at the singular value decomposition of a compact operator and the expression for $K^\dagger y$ (developed in Section 1) reveals the need for filtering high frequency components (in order to remedy the instability) and suggests one such method.

In physical considerations, we may view the operator K as some measuring device which produces data y when applied to some unknown entity x. We wish to estimate (or preferably determine if possible) x from the observation y. Often (as in the example of an integral operator) K gives a weighted average. The physical sensing device has limitations which results in a smearing of the sharp features of the unknown and a loss of some fine details of x. The singular system (1.1) reveals that "high frequency" components become attenuated. Ideally, small perturbations on y should be treated as noise, rather than the result of large highly oscillatory components in x. This is not easily implemented since it reflects the inherent instability of the problem. One way of remedying this situation is to restrict the class of admissible solutions, e.g. by imposing a smoothness requirement which would have the effect of filtering out undesirable contributions from the high frequency singular functions.

From (1. 6) we have the singular function expansion:

(11. 1) $$K^{\dagger}y = \sum_{n \in \mathcal{E}} \mu_n \langle y, u_n \rangle v_n \ .$$

One method of filtering is to calculate to successive terms in the expansion (11. 1) and terminate at some point before excessive contributions from the high frequency singular functions begin to creep in. The use of truncated eigenfunction expansions as a regularization method was proposed by Baker, Fox, Mayer and Wright [1478] for the case of a symmetric matrix. Extensions to the general case of K using truncated singular function expansions have been considered by several authors ([534], [1701], [1702], [285], [1339], [1462]).

For example, we may calculate, using discretization and matrix methods, approximations to v_n and $\langle y, u_n \rangle$ and form the sequence of partial sums:

$$x^{(m)} = \sum_{n=1}^{m} \mu_n \langle y, u_n \rangle v_n, \qquad m = 1, 2, \ldots \ .$$

To control the error, we also compute the residual:

$$r^{(m)} = \|Kx^{(m)} - y\|^2 = \|y\|^2 - \sum_{n=1}^{m} |\langle y, u_n \rangle|^2 \ .$$

The method has practical drawbacks due to the difficulty of computing higher singular functions accurately. The problem of compromise between stability and higher accuracy has to be resolved by other modifications.

3. Another approach to the regularization of the equation $Kx = y$ is to replace this equation by a suitable equation of the second kind, e.g.,

(11. 2) $$(K^*K + \alpha I)x = K^*y, \quad \alpha > 0 \ .$$

The operator $K^*K + \alpha I$ (for $\alpha > 0$) is a one-to-one onto; hence its inverse is continuous. Let $x_\alpha = (K^*K + \alpha I)^{-1}K^*y$. Then as $\alpha \to 0$, $x_\alpha \to K^{\dagger}y$.

The problem (11.2) is equivalent to minimizing the functional

$$J(x; \alpha, L_m) = \|Kx - y\|^2 + \alpha \|L_m x\|^2, \quad \alpha > 0 \tag{11.3}$$

where $L_m = I$. In general the operator L_m in (11.3) is suitably chosen to provide smoothing and stabilization, and is usually taken to be an mth order linear differential operator. Common choices for L_m include $L_0 = I$, $L_1 x = x'$, $L_2 x = x''$.

Using singular function expansions we solve (11.2) for x_α:

$$x_\alpha = \sum_{n \in \mathcal{E}} \frac{\mu_n \langle y, u_n \rangle}{1 + \alpha \mu_n^2} v_n . \tag{11.4}$$

Comparing (11.4) with (11.1) we note that the net effect has been to insert a filter factor $(1 + \alpha\mu_n^2)^{-1}$. If error is introduced in y, so $\tilde{y} = y + \delta y$, then

$$x_\alpha = \sum_{n \in \mathcal{E}} \frac{\mu_n \langle y, u_n \rangle}{1 + \alpha \mu_n^2} v_n + \sum_{n \in \mathcal{E}} \frac{\mu_n \langle \delta y, u_n \rangle}{1 + \alpha \mu_n^2} v_n . \tag{11.5}$$

Since the second term on the right side consists of error (due to measurement or roundoff), we would like to make the term small by taking α large. On the other hand, we should take α small in the first term of (11.5) so that x_α is close to $K^\dagger y$. Thus some compromise has to be made to resolve the conflict of choices. Several authors have discussed the choice of optimum α, but there is no universal rule for general operator equations.

12. Constrained Quadratic Minimization Problems in Hilbert Space

We mention briefly several minimization problems which have close interface with regularization methods and whose solutions can be characterized by (or related to) appropriate generalized inverses. A unified approach to various multi-stage minimization problems for regularization in the framework of generalized inverses and their approximations will be the subject of another paper.

Let $\mathcal{X}, \mathcal{Y}, \mathcal{Z}$ be (real or complex) Hilbert spaces, let $A: \mathcal{X} \to \mathcal{Y}$ and $L: \mathcal{X} \to \mathcal{Z}$ be bounded linear operators. We assume that $\mathcal{R}(L)$ is closed, but $\mathcal{R}(A)$ is not necessarily closed in $\mathcal{Y}$. We also assume that $\mathcal{N}(A)$ and $\mathcal{N}(L)$ are nontrivial subspaces. For $y \in \mathcal{D}(A^\dagger)$, let

$$\mathcal{S}_y = \{u \in \mathcal{X}: \|Au - y\| = \inf\|Ax - y\|, \quad x \in \mathcal{X}\} .$$

Then $\mathcal{S}_y$ is a nonempty closed convex subset and

$$u \in \mathcal{S}_y \iff u = A^\dagger y + (I - A^\dagger A)v \quad \text{for some} \quad v \in \mathcal{X} .$$

We consider the following minimization problems:

(I) Find $w \in \mathcal{S}_y$ such that

$$\|Lw\| = \inf\{\|Lu\|: u \in \mathcal{S}_y\} .$$

(II) For $z \in \mathcal{R}(L)$, let $\Omega_z = \{x \in \mathcal{X}: Lx = z\}$. Find $w \in \Omega$ such that

$$\|Aw - y\| = \inf\{\|Ax - y\|: x \in \Omega_z\} .$$

(III) Among all least squares solutions of $Lx = z$, find an element w which minimizes $\|Ax - y\|$.

(IV) Let $\alpha > 0$. Find $x_\alpha \in \mathcal{X}$ which minimizes the functional

$$J_\alpha(x) = \|Ax - y\|^2 + \alpha\|Lx\|^2 .$$

Theorem 12.1. An element $w \in \mathcal{X}$ is a solution to proble

(I) $\iff A^*Aw = A^*y$ and $L^*Lw \in \mathcal{N}(A)^\perp$;

(II) $\iff A^*Aw - A^*y \in \mathcal{N}(L)^\perp$ and $Lw = z$;

$$\text{(III)} \iff A^*Aw - A^*y \in \mathcal{N}(L)^{\perp} \text{ and } L^*Lw = L^*z \ ;$$

$$\text{(IV)} \iff (A^*A + \alpha L^*L)x = A^*y \ .$$

In particular if $\mathcal{R}(A)$ is closed, then a solution to problem (I) exists if and only if there is a solution to the system

$$\begin{pmatrix} A^*A & 0 \\ L^*L & -A^* \end{pmatrix} \begin{pmatrix} w \\ v \end{pmatrix} = \begin{pmatrix} A^*y \\ 0 \end{pmatrix} .$$

If $\mathcal{R}(L)$ is closed, then a solution to problem (II) exists if and only if there is a solution to the system

$$\begin{pmatrix} A^*A & -L^* \\ L & 0 \end{pmatrix} \begin{pmatrix} w \\ v \end{pmatrix} = \begin{pmatrix} A^*y \\ 0 \end{pmatrix} .$$

Each of the minimization problems (I) - (IV) can be reduced to a standard minimum norm problem and the solutions can be expressed in terms of an appropriate generalized inverse. For example, (I) is equivalent to

$$\inf_{x \in \mathcal{S}_y} \|Lx\| = \inf_{v \in \mathcal{N}(A)} \|L(A^{\dagger}y + v)\| = \inf_{u \in L\mathcal{S}_y} \|u\| .$$

Note that $L\mathcal{S}_y$ is a translate of the subspace $L\mathcal{N}(A)$. Thus the problem has a solution for every y if and only if $L\mathcal{N}(A)$ is closed. The solution is unique if and only if $\mathcal{N}(A) \cap \mathcal{N}(L) = \{0\}$. The following statements can be shown to be equivalent:

(i) $L\mathcal{N}(A)$ is closed;

(ii) $\mathcal{N}(A) + \mathcal{N}(L)$ is closed;

(iii) $\mathcal{N}(A)$ and $\mathcal{N}(L)$ make a positive angle modulo $\mathcal{N}(A) \cap \mathcal{N}(L)$; in particular if $\mathcal{N}(A) \cap \mathcal{N}(L) = \{0\}$, there exists $\delta > 0$ such that for all $x \in \mathcal{N}(A)$, $y \in \mathcal{N}(L)$,

$$|\langle x, y\rangle| \leqq \|x\| \, \|y\| \, (1-\delta) \ .$$

Assume that the conditions for existence and uniqueness of solutions of (I) are satisfied and let w_y denote the solution of (I) for a given y. Let $A_L^\dagger$ denote the (linear) map induced by $y \to w_y$. We note that the subspace

$$\mathcal{M} = \{w \in \mathcal{X}: \ L^*Lw \in \mathcal{N}(A)^\perp\}$$

is closed in $\mathcal{X}$ and is a topological complement to $\mathcal{N}(A)$. Thus $A_L^\dagger$ is a generalized inverse of A relative to the compositions: $\mathcal{X} = \mathcal{N}(A) \oplus \mathcal{M}$, $\mathcal{Y} = \overline{\mathcal{R}(A)} \oplus \mathcal{R}(A)^\perp$, and we have $A_L^\dagger = (2I - A^\dagger A - U)A^\dagger$, where $A^\dagger$ is the Moore-Penrose inverse (i.e., corresponding to $L = I$) and U is the projector of $\mathcal{X}$ onto $\mathcal{N}(A)$ along $\mathcal{M}$ (see Section 2 of [924]).

Projection methods and approximation methods applicable to these minimization problems and hence to regularization are considered in another paper in this volume.

13. Regularization Method of Generalized Inverses in Reproducing Kernel Hilbert Spaces

In this section, which is based on Nashed and Wahba [939], [935], we describe the regularization method of generalized inverses in RKHS. The results apply to a large class of operator equations that includes two-point boundary value problems, Fredholm integral equations of the first kind and mixed integrodifferential equations. Other aspects of the regularization method in RKHS and comparisons with Tikhonov's regularization method are given in Hilgers [583].

We recall that a Hilbert space $\mathcal{H}$ of real-valued function defined on a set S is said to be a reproducing kernel Hilbert space (RKHS) if all the evaluation functionals $f \to f(s)$ for $f \in \mathcal{H}$, $s \in S$, are continuous. Then by the Riesz theorem, there exists a unique element in $\mathcal{H}$, call it Q_s, such that $f(s) = \langle Q_s, f\rangle$, $f \in \mathcal{H}$. (Here $\langle \cdot , \cdot \rangle$ is the inner product in $\mathcal{H}$). The reproducing kernel (RK) is defined by $Q(s, s') := \langle Q_s, Q_{s'}\rangle$, $s, s' \in S$. Let $\mathcal{H}_Q$ denote the RKHS with reproducing kernel Q, and denote the inner product and norm in $\mathcal{H}_Q$ by $\langle \cdot , \cdot \rangle_Q$ and $\|\cdot\|_Q$ respectively. For properties of RKHS, see N. Aronszajn, Trans. Amer. Math. Soc., 68(1950), 337-404.

We take S to be a bounded interval and assume $Q(s, s')$ is continuous on $S \times S$. Then it is easy to show that $\mathcal{H}_Q$ is a space of continuous functions. (Note that $L_2[S]$, the space of square integrable real functions on S is <u>not</u> an RKHS since the evaluation functionals are not continuous).

An RKHS with RK Q induces a symmetric Hilbert-Schmidt operator (also denoted by Q) on $L_2[S]$ into $L_2[S]$ by

(13.1) $$(Qf)(s) = \int_S Q(s,t)\, f(t)dt \quad .$$

The operator Q has an $L_2[S]$ - complete orthonormal system of eigenfunctions $\{\phi_i\}_{i=1}^{\infty}$ and corresponding eigenvalues $\{\lambda_i\}_{i=1}^{\infty}$ with $\lambda_i \geq 0$ and $\sum_{i=1}^{\infty} \lambda_i < \infty$. We also have the following uniformly convergent Fourier expansions:

$$Q(s, s') = \sum_{i=1}^{\infty} \lambda_i\, \phi_i(s)\, \phi_i(s') \quad ,$$

$$Qf = \sum_{i=1}^{\infty} \lambda_i\, (f, \phi_i)_{L_2}\, \phi_i \quad .$$

Furthermore

$$\mathcal{H}_Q = \{f \in L_2[S]: \sum_{i=1}^{\infty} \lambda_i^{-1}(f, \phi_i)_{L_2}^2 < \infty\} \quad ,$$

where the notational convention $\frac{0}{0} = 0$ is being adopted, and

$$\langle f, g \rangle_Q = \sum_{i=1}^{\infty} \lambda_i^{-1}(f, \phi_i)_{L_2}\, (g, \phi_i)_{L_2} \quad .$$

The symmetric square root of the operator Q is given by

$$Q^{\frac{1}{2}} f = \sum_{i=1}^{\infty} \sqrt{\lambda_i}\; (f, \phi_i)_{L_2}\, \phi_i \quad ,$$

and since $\eta(Q) = \eta(Q^{\frac{1}{2}})$, we get

$$\mathcal{H}_Q = Q^{\frac{1}{2}} L_2[S] = Q^{\frac{1}{2}}(L_2[S] \ominus \eta(Q)) .$$

$(Q^{\frac{1}{2}})^{\dagger}$ has the representation

$$(Q^{\frac{1}{2}})^{\dagger} f = \sum_{i=1}^{\infty} (\sqrt{\lambda_i})^{\dagger} (f, \phi_i)_{L_2} \phi_i$$

on $\mathcal{H}_Q + \mathcal{H}_Q^{\perp}$ ($\perp$ is in L_2), where for α a real number, $\alpha^{\dagger} = \alpha^{-1}$ if $\alpha \neq 0$ and $0^{\dagger} = 0$. Similarly

$$Q^{\dagger} f = \sum_{i=1}^{\infty} \lambda_i^{\dagger} (f, \phi_i)_{L_2} \phi_i$$

on its domain. We adopt the notational conventions

$$Q^{-\frac{1}{2}} := (Q^{\frac{1}{2}})^{\dagger} \text{ and } Q^{-1} := Q^{\dagger} .$$

We have the following relations for $f, g \in \mathcal{H}_Q$:

$$\|f\|_Q = \inf \{ \|p\|_{L_2} : p \in L_2[S], \ Q^{\frac{1}{2}} p = f \}$$

$$\langle f, g \rangle_Q = (Q^{-\frac{1}{2}} f, Q^{-\frac{1}{2}} g)_{L_2} .$$

Let $\mathcal{H}_Q$ and $\mathcal{H}_P$ be RKHS with norms $\|\cdot\|_Q$ and $\|\cdot\|_P$ respectively. Let A be a linear operator with a dense domain in $\mathcal{X} = L_2[S]$ into $\mathcal{Y} = L_2[T]$, where S, T are bounded intervals. We assume throughout that $\mathcal{H}_Q$ is chosen so that $A[\mathcal{H}_Q] = \mathcal{H}_{R}$, where $\mathcal{H}_R$ is some RKHS contained (as a set) in $L_2[T]$, and $\mathcal{H}_P$ is a subset of $L_2[T]$.

Definition 13.1. A regularized pseudosolution (in RKHS) of the equation

(13.2) $$Af = g$$

is a solution to the variational problem: Find $f_\lambda \in \mathcal{H}_Q$ to minimize

$$(13.3) \qquad \phi(f;g,\lambda) = \|Af - g\|_P^2 + \lambda\|f\|_P^2 + \lambda\|f\|_Q^2, \quad \lambda > 0$$

($\phi(f;g,\lambda)$ will be assigned $+\infty$ if $Af - g \notin \mathcal{H}_P$) .

For $\lambda > 0$, let $\mathcal{H}_{\lambda P}$ be the RKHS with kernel $\lambda P(t, t')$, where $P(t, t')$ is the (continuous) RK of $\mathcal{H}_P$. Then $\mathcal{H}_P = \mathcal{H}_{\lambda P}$ and $\|\cdot\|_P^2 = \lambda\|\cdot\|_{\lambda P}^2$. Let $R(\lambda) = R + \lambda P$, where the operators R and P are induced as in (13.1), and let $\mathcal{H}_{R(\lambda)}$ be the RKHS with kernel $R(\lambda; t, t')$. Then (see Aronszajn, p. 352) $\mathcal{H}_{R(\lambda)}$ is the Hilbert space of all functions of the form $g = g_0 + \xi$, where $g_0 \in \mathcal{H}_R$ and $\xi \in \mathcal{H}_P$. The norm on $\mathcal{H}_{R(\lambda)}$ is given by

$$\|g\|_{R(\lambda)}^2 = \min\{\|g_0\|_R^2 + \|\xi\|_{\lambda P}^2 : g_0 \in \mathcal{H}_p, \ g_0 + \xi = g\} .$$

Following Aronszajn, we note that the decomposition $g = g_0 + \xi$ is not unique unless $\mathcal{H}_P \cap \mathcal{H}_R = \{0\}$. However, the g_0 and ξ attaining the minimum in the above expression are easily shown to be unique by the strict convexity of the norm.

Theorem 13.2. Suppose $\mathcal{D}(A^*)$, the domain of the adjoint (in L_2) of A, is dense in $\mathcal{Y}$, $\mathcal{H}_Q \subset \mathcal{D}(A)$, and A and $\mathcal{H}_Q$ are such that the linear functionals E_t defined by $E_t f = (Af)(t)$ are continuous in $\mathcal{H}_Q$. Suppose $\mathcal{H}_Q$, $\mathcal{H}_R$ and $\mathcal{H}_P \subset \mathcal{Y}$ all have continuous kernels. Then for $q \in \mathcal{H}_{R(\lambda)}$, the unique minimizing element $f_\lambda \in \mathcal{H}_Q$ of the functional $\phi(f; g, \lambda)$, defined by (13.3) is given by

$$f_\lambda(s) = \langle AQ_s, g\rangle_{R(\lambda)} = QA^*(AQA^* + \lambda P)^\dagger g(s) .$$

Theorem 13.3. In addition to the assumptions of Theorem 13.2 suppose that $\mathcal{H}_P \cap \mathcal{H}_R = \{0\}$. Then the minimizing element f_λ of (13.3) is the solution to the least-square problem: Find $f \in \Omega$ to minimize $\|f\|_Q$, where

$$\Omega = \{f \in \mathscr{H}_Q : \|Af - g\|_{R(\lambda)} = \inf_{h \in \mathscr{H}_Q} \|Ah - g\|_{R(\lambda)}\} .$$

For proofs of Theorems 13.2 and 13.3 see [939]. We note that if $\mathscr{H}_R \cap \mathscr{H}_P = \{0\}$, then $\mathscr{H}_R$ is a closed subspace of $\mathscr{H}_{R+\lambda P}$; in this case Theorem 13.3 says that regularization operator is a generalized inverse in an appropriate RKHS. However, the topology in $\mathscr{H}_R$ is not, in general, the restriction of the topology of $\mathscr{H}_{R+\lambda P}$, with the notable exception of the case $\mathscr{H}_R \cap \mathscr{H}_P = \{0\}$.

We now describe a procedure for obtaining <u>computable approximations</u> $\{f_{\lambda,n}\}$ to f_λ using moment discretization, which are uniformly <u>pointwise</u> convergent. Let $T_n = \{t_1, t_2, \ldots, t_n\}$, where $t_i \in T$, $t_1 < t_2 < \ldots < t_n$. For a generic function h on T, let $h_n := (h(t_1), \ldots, h(t_n))$. Let P_n be the $n \times n$ matrix whose ijth entry is $P(t_i, t_j)$, and define

$$\|h_n\|_{P_n} = \min\{\|e\| : e \in R^n, \ P_n^{\frac{1}{2}} e = h_n\} .$$

Let $f_{\lambda,n}$ be the minimizing element in $\mathscr{H}_Q$ of the functional

(13.4) $$J_n(f) = \|(Af)_n - g_n\|_{P_n}^2 + \lambda \|f\|_Q^2, \qquad \lambda > 0 .$$

If the matrix $R_n(\lambda) = R_n + \lambda P_n$ is nonsingular, an explicit expression for $f_{\lambda,n}$ is given by (see [935])

(13.5) $$f_{\lambda,n}(s) = (\eta_{t_1}(s), \ldots, \eta_{t_n}(s))\, R_n^{-1}(g(t_1), \ldots, g(t_n)) ,$$

where η_t is defined by

(13.6) $$E_t f := (Af)(t) = \langle \eta_t, f \rangle_Q .$$

(Such a representation exists by Riesz's theorem since E_f by assumption is a continuous functional in $\mathscr{H}_Q$). Define η_s^* by $\eta_s^*(t) = \eta_t(s)$, and let $P_{T_n}(\lambda)$ be the orthogonal projector in

$\mathcal{H}_{R(\lambda)}$ onto the subspace spanned by $\{R_t(\lambda): t = t_1, \ldots, t_n\}$. Here $R_t(\lambda)$ is the representer of the evaluation functional at t in $\mathcal{H}_t(\lambda)$, i.e. $R_t(\lambda)(t') = R(\lambda; t, t')$. It is easy to show that

$$f_{\lambda,n}(s) = \langle P_{T_n}(\lambda)\, \eta_s^*, P_{T_n}(\lambda) g \rangle_{R(\lambda)} ,$$

whereas

$$f_\lambda(s) = \langle \eta_s^*, g \rangle_{R(\lambda)} . \tag{13.7}$$

Theorem 5.4 [935]. Let A be a linear operator from $\mathcal{H}_Q$ into $\mathcal{H}_R$ which satisfies the assumptions of Theorem 5.2. For $g \in \mathcal{H}_{R(\lambda)}$, let $f_\lambda(s)$ and $f_{\lambda,n}(s)$ be given by (13.7) and (13.5) respectively. Suppose that $R(\lambda; t, t') = R(t, t') + \lambda P(t, t')$ satisfies a smoothness condition of order m (see [935] for more details).

Suppose further that $g \in \mathcal{H}_{R(\lambda)}$ satisfies

(a) $g = (R + \lambda P)\rho$ for some bounded ρ, or

(b) $\eta_s^* = AQ(\cdot, s) = (R + \lambda P)\rho_s$ for some bounded ρ_s.

Then

$$|f_\lambda(s) - f_{\lambda,n}(s)| = O(\Delta_n^m) ,$$

where

$$\Delta_n := \sup_{t \in T} (\inf_{t_i \in T_n} |t - t_i|) .$$

If g satisfies both (a) and (b), then

$$|f_\lambda(s) - f_{\lambda,n}(s)| = O(\Delta_n^{2m}) .$$

In the case $\mathcal{H}_P \cap \mathcal{H}_R = \{0\}$, $f_{\lambda,n}$ converges uniformly to $A^\dagger g$.

The techniques of this section can be extended to restricted generalized inverses in RKHS and to minimization problems subject to a family of linear constraints (see [1348]).

The usefulness of RKHS in regularization problems (and more generally in numerical analysis and approximation theory) is partly due to the following observations:

1. Point functionals are the only functionals that can be evaluated directly from the computational point of view.

2. Strong convergence in RKHS implies pointwise convergence.

3. The choice of the norm is significant in the analysis and approximation of operator equations. In an increasing number of problems, it is more effective if this norm is chosen a posteriori to take into consideration properties of the operator. RKHSs can be constructed in this manner, and provide instances of tailoring the spaces to the problems in questions, where the operator itself is used in defining the norm or inner product. Another familiar instance is provided by the graph norm of a closed linear operator.

4. Satisfactory numerical results for linear operator equations can only be obtained really in the case when the range of the operator is closed. Thus a natural and satisfactory approach to ill-posed problems is to recast them relative to new spaces in which the operator has a closed range. This puts restrictions on the class of "noise" or "error" admitted. RKHSs provide a natural setting for such problems if the range of the operator can be viewed as an RKHS .

It should also be observed that even the operators admitted in the setting of this section are not defined a priori by specific properties; rather they are required to satisfy certain representations in the RKHS in question.

14. Regularization Methods Based on Iterative Procedures and Filtering

Theoretical convergence results for iterative methods for ill-posed problems (specifically, bounded linear operator equations with nonclosed range) are given in another paper in this volume [924]. In this section we comment on iterative methods from the viewpoint of filtering and regularization.

The iterative method

$$x_{n+1} = x_n + \beta K^*(y - Kx_n), \qquad n = 0, 1, 2, \ldots \tag{14.1}$$

for the solution of the integral equation of the first kind, $Kx = y$, was proposed by Landweber [722], Fridman [416], and others. The iteration (14.1) is the same as the (Picard) method of successive approximations applied to the normal equation $K^*Kx = K^*y$ or, equivalently, to the equation $Fx = x$, where $Fx = x - \beta K^*(y - Kx)$, $\beta \neq 0$. Convergence proofs are given by Landweber [722], Fridman [416], Bialy [145] and others. Bakušinskii [1480] showed that the iteration (14.1) may be regarded as a regularization method with the parameter $\alpha = \frac{1}{n}$. See also Bakušinskii and Strahov [79].

Various contributions to theoretical and practical aspects of extensions of (14.1) have been made recently. See [915], [924]. The method (14.1) also has been generalized in other directions, e.g., by using the iteration

$$x_{n+1} = x_n + DK^*(y - Kx_n), \tag{14.2}$$

where D is a suitable operator.

Using singular values and functions in (14.1), it is not difficult to show by repeated iterations that

$$x_n = \sum_i \{1 - (1 - \frac{\beta}{\mu_i^2})^n\} \langle y, u_i \rangle \mu_i v_i \,. \tag{14.3}$$

From this it follows readily that the sequence $\{x_n\}$ converges to $K^\dagger y$ provided $0 < \beta < 2\mu_1^2$.

The net effect of various modifications of iterative procedures for the regularization method (of minimizing the functional $\|Kx - y\|^2 + \alpha\|Lx\|^2$, where $Lx = x - \tilde{x}$, and $\tilde{x}$ is an estimate of x) is to provide some filtering in (14.3). Strand [1284] extended the theory of singular functions expansion for the study of the iteration (14.2) and derived a quantitative method of choosing the operator D to shape the response to singular functions of K.

School of Mathematics
Georgia Institute of Technology
Atlanta, Georgia 30332

Methods for Computing the Moorse-Penrose Generalized Inverse, and Related Matters

B. Noble

1. Introduction

This paper gives an elementary and expository account of various topics connected with computing generalized inverses of matrices. A large number of algorithms have been suggested in recent years. We wish to understand why some will give more satisfactory results than others, when implemented on digital computers.

Many of the ideas are generalizations of those encountered when inverting $n \times n$ nonsingular matrices, but there are significant additional points that must be taken into consideration. Thus in the present context a matrix is ill-conditioned if small changes in the original matrix that do not increase the rank of the matrix produce large changes in the generalized inverse. (But a small change that increases the rank of the matrix will produce a large change in the generalized inverse — see the discussion in §3 below). Two points will recur throughout this paper:

(a) We deal with methods for calculating generalized inverses of matrices that involve an explicit decision regarding the rank of the matrix at some point in the calculation. (See Appendix B, Additional Note (a).)

(b) If we are not careful, the computing problem will involve the inversion of matrices or the solving of linear systems whose condition is worse than it need be. If the original problem is ill-conditioned in the sense explained above, it is essential to avoid <u>unnecessary</u> worsening of condition.

We illustrate these points by considering the generalized inverse of

$$A = \begin{bmatrix} 1 & 1 \\ 1 & 1+\epsilon \\ 1 & 1-\epsilon \end{bmatrix} \tag{1.1}$$

If $\epsilon \neq 0$, the rank of A is 2, and the generalized inverse is given by

$$A^{\dagger} = (A^TA)^{-1} A^T \tag{1.2}$$

$$= \begin{bmatrix} \frac{1}{3} & \frac{1}{3} - \frac{1}{2\epsilon} & \frac{1}{3} + \frac{1}{2\epsilon} \\ 0 & \frac{1}{2\epsilon} & -\frac{1}{2\epsilon} \end{bmatrix} \tag{1.3}$$

Suppose that ϵ is a small number $(|\epsilon| << 1)$ so that the rows of A are "nearly" linearly dependent and the rank of A is "nearly" 1. Suppose also that we are computing $A^{\dagger}$ on a computer whose word-length is such that $1+\epsilon$ is represented reasonably accurately, but $1 + 2\epsilon + \epsilon^2$ is rounded to $1 + 2\epsilon$. If we try to compute $A^{\dagger}$ from (1.1) we first compute A^TA, and we find

$$(A^TA)_{\text{computed}} = \begin{bmatrix} 3 & 3 \\ 3 & 3 \end{bmatrix}.$$

When the machine attempts to invert A^TA, it will tell us that A^TA is singular, i.e. A has rank 1.

We do not wish to imply that (1.2) should never be used when computing $A^{\dagger}$. The form (1.2) would be perfectly satisfactory if the columns of A are nearly orthogonal, for example. Also the difficulty could have been avoided by working in double precision which is efficient on some machines, and it may be sensible to do this rather than use the more sophisticated methods described later. Nevertheless it is also true that direct computation using (1.2) introduces the rank difficulty unnecessarily since there are stable numerical methods that will give a satisfactory generalized inverse via single-precision computation. For example, if we decompose A into the product of a lower trapezoidal and upper triangular matrix, as described in §6 below, we find, using single precision,

$$A = LU = \begin{bmatrix} 1 & 0 \\ 1 & 1 \\ 1 & -1 \end{bmatrix} \begin{bmatrix} 1 & 1 \\ 0 & \epsilon \end{bmatrix} \tag{1.4}$$

Then

$$A^{\dagger} = (A^T A)^{-1} A^T = (U^T L^T L U)^{-1} U^T L^T \tag{1.5}$$

$$= U^{-1} (L^T L)^{-1} L^T \tag{1.6}$$

and this last form gives (1.3), again using single-precision. The stabilization has come from the cancellation involving $(U^T)^{-1}$ and U^T in going from (1.5) to (1.6).

The plan of this paper is the following. Some relevant mathematics is discussed in §2, including the key idea which is exploited numerically, namely the decomposition of the $m \times n$ matrix A of rank k (with columns permuted if necessary so that the first k columns are linearly independent) into the product of $m \times k$ and $k \times n$ matrices, each of rank k. The continuity of the Moore-Penrose inverse is discussed in §3, and the question of the reduction of rank by perturbation in the 2-norm. Generalization of standard square matrix perturbation theory is discussed in §4. Some special results for the full-rank case are considered in §5.

Methods generalizing Gauss elimination are described in §6. Householder transformations, which are very stable numerically, are discussed in §7. A numerically stable modification of the classical Gram-Schmidt orthogonalization procedure is discussed in §8. The paper concludes with a discussion of the advantages and disadvantages of the methods available for computing generalized inverses. This can hardly claim to be definitive, but it should clarify the issues involved.

Some (incomplete) comments on the literature are made in §12.

2. Mathematical preliminaries

From the point of view adopted in this paper, a generalized inverse of an $m \times n$ real matrix A, rank k, is a convenient tool for handling numerically the system $Ax = b$ of m linear equations in n unknowns. Any numerical procedure for solving such systems will have to distinguish four cases. Suppose that the rank of $[A, b]$ is k'.

(1) $k = k'$. Then the equations have one or more solutions

(i) If $k = n \leq m$, there is a unique solution

(ii) If $k < n$, the equations have an infinite number of solutions that can be written in the form $x = x_0 + u$, where x_0 is a particular solution, and $Au = 0$, i.e., $u \in N(A)$, the null space of A, where N(A) is a vector space of dimension $n - k$.

(2) $k < k'$. The equations are inconsistent. In this case it makes sense in many applications to look for a least-squares solution, i.e., the solution that minimizes $\|Ax - b\|_2$, the size of the residual $r = b - Ax$ in the 2-norm. We again distinguish two cases

(i) If $k = n$ there exists a unique least-squares solution.

(ii) If $k < n$, the residual is minimized by a family of solutions $x = x_0 + u$, where x_0 is a particular least squares solution, and $u \in N(A)$.

From our point of view, the generalized inverse is motivated by the desire to cover all of these cases by one solution formula $x = x_0 + u$, $x_0 = Gb$, $u \in N(A)$ where G and $N(A)$ are defined uniquely, so that different persons solving $Ax = b$ will come up with directly comparable answers. We require the following definitions and results.

Definition 2.1. A g-inverse (also called a partial or inner or 1-inverse in these proceedings) of an $m \times n$ matrix A is any matrix G such that

$$AGA = A .$$

A symmetric g-inverse satisfies

$$AGA = A, \quad (AG)^T = AG \tag{2.1}$$

Definition 2.2. A least-squares solution of $Ax = b$ is a solution that minimizes $\|Ax - b\|_2$. A minimal least-squares solution is one that minimizes $\|Ax - b\|_2$ and $\|x\|_2$.

Theorem 2.1. *A necessary and sufficient condition for $x = Gb$ to be a least-squares solution of $Ax = b$ is that G be a symmetric g-inverse.*

Theorem 2.2. *If G is a symmetric g-inverse, necessary and sufficient conditions for $x = Gb$ to be a minimal least-squares solution of $Ax = b$ is that G also satisfy:*

$$GAG = G, \quad (GA)^T = GA . \tag{2.2}$$

Definition 2.3. The Moore-Penrose (generalized) inverse of A is a matrix $G = A^{\dagger}$ that satisfies the four conditions in (2.1), (2.2).

Theorem 2.3. *The Moore-Penrose inverse is unique.*

The motivation for these definitions is the following. If $Ax = b$ are consistent, any g-inverse G gives a solution

$x = Gb$. To obtain, in addition, a least-squares solution of inconsistent systems, we need at least a symmetric g-inverse. To obtain uniqueness, it is convenient to also impose the conditions (2.2) necessary for a minimal least-squares solution.

We have spelled out these definitions and theorems partly to be quite specific about the role of the Moore-Penrose inverse in numerical work in giving the unique minimal least-squares solution, and partly to emphasize the central role of the 2-norm, which permeates other aspects of the theory.

The methods for computing inverses that we shall discuss are based on factorizing the $m \times n$ matrix A of rank k in the form (assuming that the first k columns of A are linearly independent):

$$A = BC \tag{2.3}$$

where B is $m \times k$, C is $k \times n$, and B, C are each of rank k . Then

$$A^\dagger = C^\dagger B^\dagger , \tag{2.4}$$

where

$$C^\dagger = C^T(CC^T)^{-1}, \quad B^\dagger = (B^TB)^{-1}B^T . \tag{2.5}$$

We shall consider essentially three different types of factorization. (There are many variants in detail, some of which will be indicated later.)

(i) $$A = LU , \tag{2.6}$$

where L is $m \times k$ lower trapezoidal, with units on the principal diagonal and zeros above the principal diagonal, and U is upper trapezoidal. This is the subject of §6.

$$A^\dagger = U^\dagger L^\dagger = U^T(UU^T)^{-1}(L^TL)^{-1}L^T . \tag{2.7}$$

(ii) $$A = QS \tag{2.8}$$

where Q is $m \times k$ with orthonormal columns, i.e. $Q^TQ = I_k$, and S is upper triangular. It is easy to verify from (2.4), (2.5) that

$$A^\dagger = S^T(SS^T)^{-1}Q^T \; . \tag{2.9}$$

The standard numerical procedures for finding Q and S use either Householder transformations, described in §7, or the modified Gram-Schmidt orthogonalization procedure, discussed in §8.

(iii) $$A = V \Sigma U^T \; , \tag{2.10}$$

where V, U, Σ are $m \times k$, $n \times k$, $k \times k$. The columns of V are orthonormal, and similarly for the columns of U. Σ is a diagonal matrix whose diagonal elements are the singular values. This is the singular value decomposition, discussed briefly in Appendix A. A practical method for finding the singular value decomposition, based on Householder transformations and the QR algorithm, is described briefly in §7. From (2.5), (2.6) we deduce

$$A^\dagger = U \Sigma^{-1} V^T \; . \tag{2.11}$$

Before we discuss these factorizations in §§ 6-8, we require some understanding of the effects of perturbations. This will emphasize the importance of building some method for determining the rank into the computing procedure.

3. Continuity of the generalized inverse, and the reduction of rank by perturbation

It is a standard result for square matrices that if A, E are $n \times n$, A nonsingular, and

$$\Delta = \|A^{-1}\| \, \|E\| < 1 \; ,$$

then $A + E$ is nonsingular, and

$$\|(A+E)^{-1} - A^{-1}\| \le \frac{\|A^{-1}\|\Delta}{1-\Delta} .$$

This indicates that if $\|E\|$ is small enough, then $(A+E)^{-1}$ is close to A^{-1}. Simple examples show that this is not true for generalized inverses. Consider

$$A = \begin{bmatrix} 1 & 0 \\ 0 & 0 \end{bmatrix}, \quad F = \begin{bmatrix} 0 & 1 \\ 0 & 0 \end{bmatrix}, \quad G = \begin{bmatrix} 0 & 0 \\ 0 & 1 \end{bmatrix} .$$

We have $A^{\dagger} = A$, and

$$(A+\epsilon F)^{\dagger} = \frac{1}{1+\epsilon^2}\begin{bmatrix} 1 & 0 \\ \epsilon & 0 \end{bmatrix}, \quad (A+\epsilon G)^{\dagger} = \begin{bmatrix} 1 & 0 \\ 0 & 1/\epsilon \end{bmatrix},$$

so that

$$\lim_{\epsilon \to 0} (A+\epsilon F)^{\dagger} = A^{\dagger}, \quad \lim_{\epsilon \to 0} (A+\epsilon G)^{\dagger} \text{ does not exist} .$$

Theorems 3.1, 4.2 indicate that the key difference between $A+\epsilon F$ and $A+\epsilon G$ is that the rank of $A+\epsilon F$ is the same as the rank of A, whereas the rank of $A+\epsilon G$ is greater than the rank of A.

Theorem 3.1. Let A be an $m \times n$ matrix of rank k, and $B = A + E$.

(i) If rank $(B) >$ rank (A), then

$$\|B^{\dagger}\|_2 \ge \frac{1}{\|E\|_2} \tag{3.1}$$

(ii) If $\|A^{\dagger}\|_2 \|E\|_2 < 1$, then rank $(B) \ge$ rank (A).

Proof: We use Theorem A. 3 in Appendix A. If $\text{rank}(B) = p > k$, then $\sigma_p(A) = 0$. Choose $j = p$, $G = B$, $K = A$ in (A.11). Then

$$\sigma_p(B) \leq \|E\|_2 \ .$$

Using (A. 8), the result (3. 1) follows immediately. We give a separate proof of part (ii) even though it is the contrapositive of (i). Choose $G = -A$, $K = -B$ in (A. 11). Then

$$\sigma_j(B) \geq \sigma_j(A) - \|E\|_2, \qquad j = 1, \ldots, n \ .$$

Since the eigenvalues are arranged in descending order, and using (A. 8),

$$\sigma_j(A) - \|E\|_2 \geq \sigma_k(A) - \|E\|_2 = \frac{1}{\|A^\dagger\|_2} - \|E\|_2 > 0 \ ,$$

because of the hypothesis $\|A^\dagger\|_2 \|E\|_2 < 1$. Hence $\sigma_j(B) > 0$ for $j = 1, \ldots, k$ and $\text{rank}(B) \geq k = \text{rank}(A)$.

The first part of this theorem is remarkable and instructive. It states that if a matrix $B = A + E$ is near A, but has rank greater than A, its generalized inverse can be larger and completely different from $A^\dagger$, and the smaller E, the worse the trouble can be. The moral from a numerical point of view is inescapable. Before we can compute a generalized inverse with confidence, we must know the rank of A; rounding errors in the computer will mean that we actually compute the inverse of some matrix $A + E$, and we must make sure that the rank assumed for $A + E$ is the same as the rank of A.

A question related to the results in the above theorem is the following: what is the "smallest" matrix E such that $\text{rank}(A + E) < \text{rank}(A)$? This is answered in the 2-norm, by the following (cf. L. Minsky, Quart. J. Math. 11(1960), 50-59):

Theorem 3.2. If A is $m \times n$, rank k, there is a matrix E with $\|E\|_2 = 1/\|A^\dagger\|_2$ such that $A + E$ is of rank $k - 1$. There is no E with $\|E\|_2 < 1/\|A^\dagger\|_2$ such that $A + E$ is of rank less than k.

Proof: We require some background from Appendix A, in

particular the singular value decomposition in Theorem A.1, and the result that $\|A^\dagger\|_2 = 1/\sigma_k(A)$, where $\sigma_k(A)$ is the smallest singular value of A. Using the notation in Equation (A.7), choose E so that

$$E = -VDU^T ,$$

where D is a zero matrix except for the kth element on the principal diagonal which is chosen to be $\sigma_k(A)$. Then

$$A + E = V \begin{bmatrix} \Sigma' & 0 \\ 0 & 0 \end{bmatrix} U^T ,$$

where Σ' is a diagonal matrix of order $k - 1$ with diagonal elements $\sigma_i(A)$. The rank of $A + E$ is $k - 1$ and $\|E\|_2 = \sigma_k(A)$, so the first part of the theorem is proved. The second part follows directly from Theorem 3.1(ii).

4. Perturbation theory

Certain results connected with perturbation theory for nonsingular square matrices can be deduced conveniently from the identity (A, B nonsingular, $B = A + E$)

$$B^{-1} - A^{-1} = -B^{-1}EA^{-1} . \qquad (4.1)$$

We obtain a generalization for $m \times n$ matrices as follows:

$$\begin{aligned} B^\dagger - A^\dagger &= -B^\dagger EA^\dagger + (B^\dagger - A^\dagger + B^\dagger(B-A)A^\dagger) \\ &= -B^\dagger EA^\dagger + B^\dagger(I - AA^\dagger) - (I - B^\dagger B)A^\dagger . \qquad (4.2) \end{aligned}$$

This can be transformed into a more useful form by noting that:

$$A^T(I - AA^\dagger) = (A - AA^\dagger A)^T = 0 \ .$$

Hence

$$B^\dagger(I - AA^\dagger) = B^\dagger B^{\dagger T} B^T (I - AA^\dagger)$$

$$= B^\dagger B^{\dagger T} E^T (I - AA^\dagger) \ .$$

The last term in (4.1) can be dealt with similarly, and we obtain (see W. Kahan, IFIPS Congress '71, Ed. C. V. Freiman, North-Holland (1972)).

Theorem 4.1. _If_ A, B _are general_ $m \times n$ _matrices and_ B = A + E, _then_

$$B^\dagger - A^\dagger = -B^\dagger E A^\dagger + B^\dagger B^{\dagger T} E^T (I - AA^\dagger) + (I - B^\dagger B) E^T A^{\dagger T} A^\dagger \ . \tag{4.3}$$

A similar formula holds when A and B are interchanged. The results are slightly simpler in the full-rank case since then the second or third terms can vanish.

Since $I - AA^\dagger$ is a symmetric idempotent (i.e., projection) operator, we have $\|I - AA^\dagger\|_2 = 1$ if $AA^\dagger \neq I$. Similarly for $I - B^\dagger B$, etc. This indicates that we should work in the 2-norm, and (4.3) gives, for general A, B ,

$$\|B^\dagger - A^\dagger\|_2 \leq (\|A^\dagger\|_2 \|B^\dagger\|_2 + \|B^\dagger\|_2^2 + \|A^\dagger\|_2^2)\|E\|_2 \ . \tag{4.4}$$

If A and B are square and nonsingular, (4.1) gives

$$\|B^{-1} - A^{-1}\| \leq \|B^{-1}\| \|E\| \|A^{-1}\| \ . \tag{4.5}$$

If A is known and B is unknown, and $\|E\|$ is sufficiently small, we eliminate B^{-1} from the right of this inequality by noting that:

$$B = A + E = A(I + A^{-1}E) \tag{4.6}$$

We now use the Banach lemma: If S is square and $\|S\| < 1$, where the norm is such that $\|I\| = 1$, then $I + S$ has an inverse, and

$$\|(1 + S)^{-1}\| \leq \frac{1}{1 - \|S\|} .$$

This tells us that if $\|A^{-1}\| \|E\| < 1$, then $I + A^{-1}E$ is nonsingular and we deduce from (4.6) that B is nonsingular, and

$$\|B^{-1}\| \leq \frac{\|A^{-1}\|}{1 - \|A^{-1}\| \|E\|} . \tag{4.7}$$

Combining (4.5) and (4.7), we see that if A is nonsingular and $\Delta = \|A^{-1}\| \|E\| < 1$, then

$$\|B^{-1} - A^{-1}\| \leq \frac{\|A^{-1}\|\Delta}{1 - \Delta} . \tag{4.8}$$

It seems to be possible to obtain formulae analogous to (4.7), (4.8) for the Moore-Penrose inverses of arbitrary $m \times n$ matrices if we work in the 2-norm, but not in the 1- or ∞-norms. Perhaps this is not too surprising because of the connection of the Moore-Penrose inverse with least squares. In contrast to the elementary derivation of (4.7), it seems to be necessary to use the minimax theorem to generalize (4.7):

Theorem 4.2. If A is an $m \times n$ matrix, rank k, and $B = A + E$, rank $(B) =$ rank (A), and $\|A^{\dagger}\|_2 \|E\|_2 < 1$, then

$$\|B^{\dagger}\|_2 \leq \frac{\|A^{\dagger}\|_2}{1 - \|A^{\dagger}\|_2 \|E\|_2} . \tag{4.9}$$

Proof: In (A.11) choose $G = -A$, $K = -B$, $j = k$, which gives

$$\sigma_k(B) \geq \sigma_k(A) - \|E\|_2 .$$

Hence, using (A.8),

$$\frac{1}{\|B^{\dagger}\|_2} \geq \frac{1}{\|A^{\dagger}\|_2} - \|E\|_2$$

which is (4.9).

Combining (4.4), (4.9), if $B = A + E$, rank $(B) =$ rank (A), then

$$\|B^{\dagger} - A^{\dagger}\|_2 \leq [1 + \frac{1}{1-\Delta} + \frac{1}{(1-\Delta)^2}] \|A^{\dagger}\|_2 \Delta , \tag{4.10}$$

where $$\Delta = \|A^{\dagger}\|_2 \|E\|_2 .$$

This formula indicates the essential fact that a small perturbation in A produces a small perturbation in $A^{\dagger}$, provided that the perturbation does not change the rank of A.

The multiplier in brackets in (4.10) is not the best possible. A somewhat smaller multiplier can be obtained by realizing that $\|B^{\dagger}\|_2^2$ in (4.4) can be replaced by $\|A^{\dagger}\|_2 \|B^{\dagger}\|_2$. A result that can be used to derive this from (4.3) is

$$\|BB^\dagger(I - AA^\dagger)\|_2 = \|AA^\dagger(I - BB^\dagger)\|_2 .$$

Details can be found in Wedin [1] and Lawson and Hanson [2], Chapter 8. An elegant improvement on (4.10) (compare (4.8) – the result is again not best possible) is:

$$\|B^\dagger - A^\dagger\|_2 \leq \frac{3\|A^\dagger\|_2 \Delta}{1-\Delta}.$$

It is not altogether clear whether this part of the theory is in definitive form.

Results of the above type can be used to examine the effect of perturbations of A and b on the minimal least-squares solution of $Ax = b$. One result is the following:

Theorem 4.3. *If* A *is* $m \times n$, *rank* k, *and* x, $x + z$ *are the minimal least-squares solutions of*

$$Ax = b, \qquad (A + E)(x+z) = b + d ,$$

where $\Delta = \|A^\dagger\|_2 \|E\|_2 < 1$ *and* $\mathrm{rank}(A + E) = \mathrm{rank}(A)$, *then*

$$\|z\|_2 \leq \frac{\|A^\dagger\|_2}{1-\Delta} \{2\|E\|_2 \|x\|_2 + \|d\|_2 + \frac{\Delta \|r\|_2}{1-\Delta}\} , \quad (4.11)$$

where $r = b - Ax$.

Proof: $$x = A^\dagger b, \qquad x + z = (A+E)^\dagger (b+d) ,$$

$$z = \{(A+E)^\dagger - A^\dagger\}b + (A+E)^\dagger d$$

$$= -B^\dagger Ex + B^\dagger B^{\dagger T} E^T r + (I - B^\dagger B)E^T A^{\dagger T} x + B^\dagger d ,$$

where $B = A+E$. The result (4.11) follows immediately on taking norms and using (4.9).

Again the result is not the best possible. However the formula does show the essential point, that the perturbation in the solution may be large if the residual is large. This is a feature of least-squares problems. (See Bjorck [2], Bjorck and Golub [1], Golub and Wilkinson [1].)

5. Special results for the full rank case

A feature of perturbation theory for a nonsingular (square) matrix compared with the theory for a general $m \times n$ matrix, rank k, is that the results for the nonsingular matrix hold for a quite general norm, whereas the results for the $m \times n$ matrix are valid only for a restricted class of norm, in particular the 2-norm. This is connected with the fact that the proofs for the nonsingular matrix require little more than basic properties of norms, whereas the proofs for the $m \times n$ case require the minimax theorem. It is a tantalizing pastime to try to extend the results for the $m \times n$ case to the 1- or ∞-norms because limited progress is possible in the full-rank case, and the generalizations we would like always seem to be just around the corner. We present some special results connected with the full-rank case. Carl de Boor has made the comment that a more profitable point of view might be to try to find the correct definition of generalized inverse for which results are valid in norms other than the 2-norm. (The Moore-Penrose inverse is closely tied to the 2-norm by its definition.)

We first give a variant of Theorem 3.1(i). (A variant of Theorem 3.1(ii) is given in Theorem 5.2 below.)

Theorem 5.1. *Let* A, B *be* $m \times n$, $B = A + E$. *If* $\mathrm{rank}(B) = n$, $\mathrm{rank}(A) < n$, *then*

$$\|B^\dagger\| \geq \frac{1}{\|E\|} \tag{5.1}$$

Proof: Since $\mathrm{rank}(A) < n$, there exists $x \neq 0$ such that $Ax = 0$. Since $\mathrm{rank}(B) = n$, $(A+E)x = y \neq 0$, where this defines y. This set of linear equations is consistent and has the unique solution $x = (A+E)^\dagger y$. Also $Ex = y$ so that

$$x = (A+E)^{\dagger} Ex \ .$$

The result (5.1) follows on taking norms.

Theorem 3.2 is not true in a general norm. Thus consider

$$A = \begin{bmatrix} a \\ b \end{bmatrix}, \qquad \|A\|_{\infty} = \max(|a|, \ |b|) \ ,$$

$$A^{\dagger} = \frac{1}{a^2+b^2}[a, b], \quad \|A^{\dagger}\|_{\infty} = \frac{|a| + |b|}{a^2+b^2} \ .$$

The only matrix such that the rank of $A + E$ is zero is $E = -A$, and clearly $\|A^{\dagger}\|_{\infty} \|E\|_{\infty} > 1$ if $a \neq b$. However Theorem 3.2 is true for square matrices for a general norm as stated in Conte and de Boor [1], p. 150, without proof. Since this does not seem to be widely known,* we sketch C. de Boor's proof, specialized to the ∞-norm. (In his general setting, the vector u^T below is replaced by a linear functional in the dual space.) Let A be a nonsingular matrix. Write

$$A + E = (I + EA^{-1})A \ . \tag{5.2}$$

We shall choose E so that $\|E\|_{\infty} = 1/\|A^{-1}\|_{\infty}$, and $I + EA^{-1}$ is singular, from which it will follow from (5.2) that the rank of $A + E$ is less than n. If $A^{-1} = [\alpha_{ij}]$ we have

$$\|A^{-1}\|_{\infty} = \max_{i} \sum_{j=1}^{n} |\alpha_{ij}| \ .$$

Suppose that the maximum is attained for $i = k$. If we define $z_j = +1(\alpha_{kj} > 0)$, $-1(\alpha_{kj} < 0)$, then for this choice of z we have

* It is attributed to N. Gastinel in W. Kahan, Can. Math. Bull. 9 (1966), p. 775.

$$(A^{-1}z)_k = \sum_{j=1}^{n} |\alpha_{kj}|, \quad \|A^{-1}z\|_\infty = \|A^{-1}\|_\infty .$$

Now choose

$$E = -zu^T ,$$

where u^T is a vector which is at our disposal. This gives

$$(I + EA^{-1})z = (I - zu^T A^{-1})z = (1 - u^T A^{-1} z)z .$$

It is possible to pick u so that $u^T A^{-1} z = 1$, $\|E\|_\infty = 1/\|A^{-1}\|_\infty$:

$$u_j = \begin{cases} 1/\|A^{-1}\|_\infty, & i = k , \\ 0, & i \neq k . \end{cases}$$

For this choice of E, $(I + EA^{-1})z = 0$, $z \neq 0$, so that $I + EA^{-1}$ is singular, and this completes the proof.

We next consider a generalization of Theorem 4.2 for the full-rank case.

Theorem 5.2. If A is $m \times n$, rank n, and $B = A + E$ where $\Delta = \|A^\dagger\|\|E\| < 1$ and the norm is such that $\|I\| = 1$, then B is of rank n and

$$\|B^\dagger\| \leq \frac{\|A^\dagger\|\|BB^\dagger\|}{1 - \Delta} . \tag{5.3}$$

First proof: Since $A^\dagger A = I$,

$$B = A + E = A + EA^\dagger A = (I + EA^\dagger)A .$$

Since $\|EA^{\dagger}\| \leq \Delta < 1$, $I + EA^{\dagger}$ is nonsingular, and B has rank n . Hence

$$A = (I + EA^{\dagger})^{-1}B$$

Multiply on the left by $A^{\dagger}$ and on the right by $B^{\dagger}$:

$$B^{\dagger} = A^{\dagger}(I + EA^{\dagger})^{-1}BB^{\dagger} . \tag{5.4}$$

The result (5.3) is obtained on taking norms.

<u>Second proof:</u> Multiply $B = A + E$ on the left by $A^{\dagger}$:

$$A^{\dagger}B = A^{\dagger}A + A^{\dagger}E = I + A^{\dagger}E .$$

Hence

$$B^{\dagger} = (I + A^{\dagger}E)^{-1} A^{\dagger}BB^{\dagger} , \tag{5.5}$$

and (5. 3) again follows, on taking norms.

This concludes our discussion of perturbation theory.

6. Methods based on Gauss elimination

We turn our attention to methods for computing generalized inverses. Standard general-purpose computer algorithms for solving the linear system $Ax = b$, where A is square, are usually based on Gauss elimination with partial pivoting. It is natural to extend this method to solve $Ax = b$ when A is $m \times n$, rank k . The pivoting strategy will be important since in general the rank, and a set of linearly independent columns, will not be known beforehand, and partial pivoting alone will not be sufficient.

We describe in detail one method for finding the generalized inverse, based on Gauss elimination with complete pivoting. The matrix $A = A_0$ is transformed in succession to matrices $A_1, A_2, \ldots, A_k$. The matrix A_r is $m \times n$ and has the form

$$A_r = \begin{bmatrix} U_r & V_r \\ 0 & W_r \end{bmatrix} \qquad (6.1)$$

where U_r is $r \times r$ upper triangular. Denote the (i, j) element in W_r by w_{ij}. We determine the element w_{ij} with largest modulus. Suppose that this is w_{pq}. Interchange the pth and (r+1)th columns in the complete $m \times n$ matrix A_r, and the qth and (r+1)th rows, so that w_{pq} is now in the (r+1, r+1) position. Form

$$\ell_{i, r+1} = w_{i, r+1}/w_{r+1, r+1}, \quad i = r+2, \ldots, m . \qquad (6.2)$$

Subtract $\ell_{i, r+1}$ times the (r+1)th row from the ith row to give A_{r+1} which has exactly the same form as A_r:

$$A_{r+1} = \begin{bmatrix} U_{r+1} & V_{r+1} \\ 0 & W_{r+1} \end{bmatrix},$$

where now U_{r+1} is an upper triangular matrix. With exact computation, the process will terminate with A_k.

$$A_k = \begin{bmatrix} U_k & V_k \\ 0 & 0 \end{bmatrix} \qquad (6.3)$$

where U_k is a $k \times k$ upper triangular matrix, and $W_k = 0$, since A is of rank k, and so far we have assumed exact calculation.

It is convenient to summarize the row and column interchanges by saying that permutation matrices P, Q exist such that the form (6.3) is obtained by transforming the matrix PAQ as described, no row and column interchanges now being necessary. The transformations to the form (6.3) can be written

$$M_k M_{k-1} \ldots M_1 PAQ = \begin{bmatrix} U \\ 0 \end{bmatrix},$$

where M_r is a unit matrix plus elements $-\ell'_{ir}$, $i = r+1, \ldots, m$ below the diagonal in the rth column. (The ℓ'_{ir} are row permutations of the ℓ_{ir} defined in (6.2).) Generalizing the situation for Gauss elimination applied to a square matrix, we deduce that PAQ has been factored in the form

PAQ = LU,

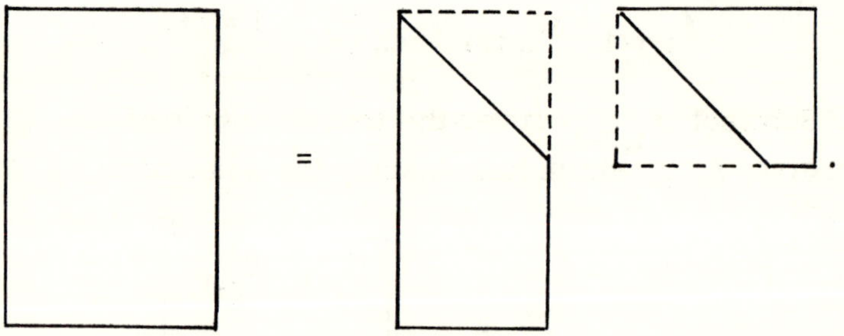

Here L is $m \times k$ lower trapezoidal, with zeros above the principal diagonal, units on the principal diagonal, and elements ℓ'_{ij} below the principal diagonal. (L is the first k columns of $M_1^{-1} M_2^{-1} \ldots M_k^{-1}$.) Also U is $k \times n$ upper trapezoidal. ($U = [U_k \; V_k]$ in the notation of (6.3).)

The corresponding generalized inverse is

$$A^{\dagger} = QU^{\dagger} L^{\dagger} P ,$$

where

$$U^{\dagger} = U^T(UU^T)^{-1}, \quad L^{\dagger} = (L^T L)^{-1} L^T .$$

In the above description it was assumed that the rank was exactly k, and that exact computation was used. In practice, the rank may not be known, and rounding errors will occur. We have to decide, at the rth stage of the reduction, when A is reduced to A_r of the form (6.1), whether the elements of W_r should be regarded as zero or not. In many practical situations it is sufficient to regard numbers less than an arbitrary number ϵ as zero. Then A has a "computed ϵ-rank" equal to k if all the elements of the computed W_k has magnitude less than or equal to ϵ, and at least one element of W_{k-1} has magnitude greater than ϵ. The theoretical motivation for this lies in the following result.

Theorem 6.1. If exact computation is used in the LU decomposition, there is a matrix A + E, where the elements e_{ij} of E are bounded by the largest element in W_r in (6.1), such that A + E is of rank r, $1 \leq r \leq k$.

Proof: Without loss of generality we can assume that no row or column changes are involved in transforming A to A_r. Choose

$$E = -\begin{bmatrix} 0 & 0 \\ 0 & W_r \end{bmatrix}.$$

Exactly the same transformations as before, applied to A+E, will now give, in the notation of (6.1),

$$(A+E)_r = \begin{bmatrix} U_r & V_r \\ 0 & 0 \end{bmatrix}$$

which gives the required result.

Of course the matrix E in this theorem may have elements which are much larger than necessary to reduce the rank of A + E to r. The classic example (due to Wilkinson) is

$$A = \begin{bmatrix} 1 & -1 & -1 & -1 & \cdots \\ 0 & 1 & -1 & -1 & \cdots \\ 0 & 0 & 1 & -1 & \cdots \\ & & \cdots & & \end{bmatrix}.$$

If A is a square matrix of order n, and ϵ is added to each element in the first column below the diagonal, the resulting matrix is singular for $\epsilon = -1/(2^{n-1} - 1)$. The LU decomposition gives no indication of trouble $(L = I,\ U = A)$. However trouble would be indicated if we computed the inverse, since this would have some large elements (and rows and columns of widely varying magnitude, which is an indication of possible faulty scaling).

Instead of using complete pivoting, it will usually be sufficient to use partial pivoting with a linear independence check. In (6.1) we check for the largest element in the first column of W_r, say $w_{p,r+1}$; if $|w_{p,r+1}| \leq \epsilon$, we decide that the $(r+1)$th column is linearly dependent on the first $(r+1)$ columns, and go on to the $(r+2)$th column; if $|w_{p,r+1}| > \epsilon$, interchange the $(r+1)$th and pth rows of $p \neq r+1$, and pivot on the resulting $w_{r+1,r+1}$ element, as before.

If no interchange of rows and columns is involved we have the usual formulae

$$u_{pj} = a_{pj} - \sum_{k=1}^{p-1} \ell_{pk} u_{kj}, \qquad j = p, p+1, \ldots, n ,$$

$$\ell_{iq} = (a_{iq} - \sum_{k=1}^{q-1} \ell_{ik} u_{kq})/u_{qq}, \qquad i = q+1, \ldots, m .$$

Here p and q run from 1 to k. If partial pivoting with a linear independence check is used, these can be programmed by a direct extension of the method used for non-singular (square) systems, accumulating inner products in double precision.

Returning to the question of the determination of rank, it is clear that the scaling of the matrix is important. In practice, the best that we can do is usually to scale so that the largest element in each row and column is of order unity (or, equivalently, use scaled pivoting). However in least squares problems we have to be careful, since scaling of rows corresponds to weighting the equations.

Enthusiasts for singular-value methods and orthogonal transformations sometimes tend to exaggerate the difficulties involved in deciding the rank. In our experience when using Gauss elimination with the matrix scaled as described, one of two situations usually arises when the original matrix is known exactly:

(a) The pivots w_{rr} stay well above 10^{-t} on a t-decimal machine for $r = 1, 2, \ldots$ to some value k, and then $w_{k+1,k+1}$ drops abruptly to some small multiple of 10^{-t}. In this case the rank should be taken as k.

(b) The pivots w_{rr} decrease gradually like $1, p, p^2, \ldots$ until r is so large that p^r is comparable with 10^{-t}, after which the pivots remain comparable in size (e.g. for the Hilbert matrix $p \approx 0.1$). This is a typical ill-conditioned situation. In this case the rank cannot be determined by the method we are using, and it is probably meaningless to ask what the rank of the matrix stored in the machine is, since changes in the original matrix comparable with the rounding errors will change the rank of the resulting matrix.

Similar remarks apply when the elements of the original matrix are uncertain – instead of 10^{-t} we must use some measure of uncertainty of the elements.

We make two further comments on the LU decomposition

(1) The number of operations involved (an operation = an addition plus a multiplication) if $m, n, k >> 1$ is of the order of

$$k[mn - \frac{1}{2}(m+n)k + \frac{1}{3}k^2] . \qquad (6.4)$$

(2) Wilkinson backward error analysis can be applied to show that if $\hat{L}$, $\hat{U}$ are the computed L, U, then

$$\hat{L}\hat{U} = A + \Delta ,$$

where Δ is "small". In particular, if partial pivoting with a linear independence check is used, accumulating products in double precision and neglecting higher order terms, if $\Delta = [\delta_{ij}]$, then

$$\delta_{1j} = 0, \qquad j = 1 \text{ to } n,$$

$$|\delta_{ij}| \le |\hat{\ell}_{ij}\, \hat{u}_{jj}|e, \qquad j = 1 \text{ to } k, \quad i = j+1 \text{ to } m.$$

$$|\delta_{ij}| \le |\hat{u}_{ij}|e, \qquad i = 2 \text{ to } k, \quad j = i \text{ to } n,$$

$$|\delta_{ij}| \le \delta, \qquad i = k+1 \text{ to } m, \quad j = k+1 \text{ to } n,$$

where the rounding rule is $fl(x) = x(1+\epsilon)$, $|\epsilon| \le e$, and the elements in the computed W_k have magnitude less than δ. (We are using the notation of (6.1) – these are the elements that are neglected to give the "computed δ-rank" equal to k.)

7. Householder transformations

Householder transformations can be motivated by the following argument. Suppose that we wish to transform a real arbitrary vector u into a second real vector v of the same length (i.e. $\|u\| = \|v\|$) by means of an equation

$$v = Pu, \tag{7.1}$$

where P is a square matrix that will depend on u and v. This can be done by noting that (Fig. 1):

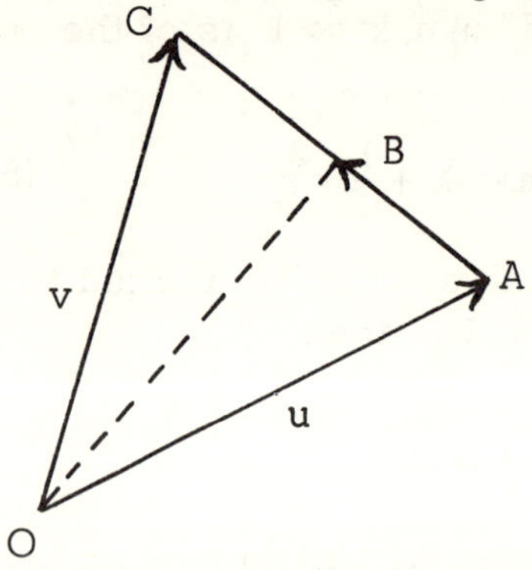

Fig. 1. Geometrical interpretation of Householder transformation.

$$\vec{OC} = \vec{OA} + 2\vec{AB},$$

where B is the midpoint of AB, so that OB is perpendicular to AC, and $\vec{AB}$ is minus the projection of OA on AB. If we introduce

$$\omega = \frac{v-u}{\|v-u\|},$$

so that ω is a unit vector along AC, then, remembering the minus sign,

$$v = u - 2(\omega^T u)\omega \quad .$$

Since $(\omega^T u)\omega = (\omega\omega^T)u$, this gives

$$v = (I - 2\omega\omega^T)u \quad ,$$

which gives the required result, namely that u can be transformed into v by a transformation of the form (7.1), where

$$P = I - 2\omega\omega^T, \quad \|\omega\| = 1 \quad .$$

We have derived this result by a geometric argument, but the result is of course an algebraic identity that can be verified directly. This geometric derivation of P shows that the transformation represents a rotation from u to v. We therefore expect that P should be orthogonal, and it is easy to verify that this is true.

Suppose now that we wish to transform u into $v = \alpha e_1$, a multiple of e_1, the unit vector whose first element is unity and whose remaining elements are zero. The condition $\|u\| = \|v\|$ gives $\alpha = \pm\|u\|$ so that

$$v - u = \pm\|u\|e_1 - u \quad .$$

We choose the sign that avoids cancellation when computing, i.e., if u_1 denotes the first element of u,

$$\alpha = -\|u\|, \quad u_1 > 0 : \alpha = \|u\|, \quad u_1 < 0 \quad .$$

For definiteness we choose $\alpha = -\|u\|$ if $u_1 = 0$. It is easy to verify that

$$\|v - u\| = 2\|u\|(|u_1| + \|u\|) \quad .$$

This leads to the following lemma: A given vector $u \neq 0$, can be transformed into a multiple of the unit vector e_1 by means of the following Householder transformation:

$$v = Pu = -\sigma \|u\| e_1$$

where

$$P = I - \frac{2zz^T}{z^T z},$$

$$z = u + \sigma \|u\| e_1, \qquad z^T z = 2\|u\|(|u_1| + \|u\|),$$

$$\sigma = +1(u_1 \geq 0), \ -1(u_1 < 0) .$$

We use these ideas to show that we can premultiply an arbitrary real $m \times n$ matrix A by a real orthogonal matrix P to produce a matrix B in which all the elements in the first column are zero except the first:

$$B = PA = \begin{bmatrix} b_{11} & \beta \\ 0 & B_1 \end{bmatrix}.$$

We do this by means of a Householder transformation. If the first columns of A and B are denoted by a_1 and b_1, we must have $a_1 = Pb_1$, $\|a_1\| = \|b_1\|$. This last condition gives

$$b_{11} = \pm \|a_1\| .$$

Also

$$\omega = \frac{b_1 - a_1}{\|b_1 - a_1\|} .$$

The first element of ω is proportional to

$$b_{11} - a_{11} = \pm \|a_1\| - a_{11} .$$

We can deduce:

Theorem 7.1. Let A be an $m \times n$ matrix of rank k. There is an $m \times n$ orthogonal matrix Q and an $n \times n$ permutation matrix P such that

$$QAP = \begin{bmatrix} U & V \\ 0 & 0 \end{bmatrix},$$

where U is $k \times k$ upper triangular, with nonzero diagonal elements. The appropriate null matrices do not exist if $k = m$ or n.

Proof: Choose P so that the first k columns of A are linearly independent. Apply a Householder transformation Q_1 to AP such that the first column of the resulting matrix is a multiple of the unit vector:

$$Q_1AP = \begin{bmatrix} u_{11} & v_1 \\ 0 & W_1 \end{bmatrix}. \tag{7.2}$$

Let Q_2 be the Householder transformation of rank $m-1$ that transforms the first column of W_1 to a multiple of an $(m-1) \times 1$ unit column vector e_1. Then

$$\begin{bmatrix} 1 & 0 \\ 0 & Q_2 \end{bmatrix} Q_1AP = \begin{bmatrix} U_2 & V_2 \\ 0 & W_2 \end{bmatrix}$$

where U_2 is 2×2 upper triangular. Proceeding in this way, we find an $m \times m$ matrix Q which is a product of k Householder transformations such that

$$QAP = \begin{bmatrix} U & V \\ 0 & W \end{bmatrix}$$

where U is $k \times k$ upper triangular. Since the first k columns of AP are linearly independent, the same must be true of the columns of U, i.e. the diagonal elements of U must be nonzero. The matrix W must be zero since otherwise the rank of QAP would be greater than k, which would contradict the fact that the rank of A is k.

Suppose next that the rank of A and a set of linearly independent columns are unknown, but have to be determined in the course of the computation. Suppose that at the rth stage of the computation we have determined a permutation matrix P_r and an orthogonal matrix Q_r such that

$$Q_r AP_r = \begin{bmatrix} U_r & V_r \\ 0 & W_r \end{bmatrix} \qquad (7.3)$$

where U_r is upper triangular. In exact computation, if the first column of W_r were zero, this would mean that the (r+1)th column in (7.3) is linearly dependent on the first r columns. When computing, a convenient strategy is the following. Select the column of W_r whose sum of squares is the greatest, call this ω, and suppose that it is in the pth column of $Q_r AP_r$. If $\|\omega\|_2 < \delta$, where δ is a number assigned by the programmer, set $W_r = 0$, and take the rank of A to be r. Otherwise interchange the pth and (r+1)th columns of $Q_r AP_r$, thereby obtaining $Q_r AP_{r+1}$, and perform a Householder transformation to obtain $Q_{r+1} AP_{r+1}$. A variant of this strategy is to stop when $|w_{rr}| < \delta$.

Important features of Householder transformations are:

(a) The vectors involved in the above transformation can be stored efficiently in the space occupied by A.

(b) Householder transformations are very stable with respect to rounding errors.

However a detailed discussion of these matters lies outside the scope of this paper, and the reader is referred to the discussion in Lawson and Hanson [1].

We turn our attention to a decomposition that is the basis for an efficient method for computing the singular value decomposition (and could also be used to compute $A^{\dagger}$ directly):

Theorem 7.2. Let A be an $m \times n$ matrix of rank k. There are orthogonal matrices $Q(m \times m)$ and $R(n \times n)$, and a permutation matrix $P(n \times n)$ such that

$$QAPR = \begin{bmatrix} U & 0 \\ 0 & 0 \end{bmatrix}$$

where U is $k \times k$ upper bidiagonal with nonzero diagonal elements. The appropriate null matrices do not exist if $k = m$ or n.

Proof: Choose P and Q_1 as in the last theorem to obtain (7.2). Then postmultiply by a matrix R_1 of the following form:

$$Q_1APR_1 = \begin{bmatrix} u_{11} & v_1 \\ 0 & W_1 \end{bmatrix} \begin{bmatrix} 1 & 0 \\ 0 & S_2 \end{bmatrix} = \begin{bmatrix} u_{11} & v_1S_2 \\ 0 & W_1S_2 \end{bmatrix}.$$

Choose S_2 to be a Householder transformation such that v_1S_2 is a multiple of a unit row vector with a nonzero element in its first position. Then

$$Q_1APR_1 = \begin{bmatrix} u_{11} & u_{12} & 0 \\ 0 & W_2 & X_2 \end{bmatrix}.$$

Continue in this way with pre- and post-Householder transformations until we obtain

$$Q_kQ_{k-1} \cdots Q_1APR_1R_2 \cdots R_{k-1} = \begin{bmatrix} U & V \\ 0 & W \end{bmatrix}$$

where U is $k \times k$ upper triangular, W must be zero since A is of rank k, and U is zero except for its last row. The last row of V can then be reduced to zero by a post-Householder transformation, and this completes the proof.

The singular values of A are the square roots of the eigenvalues of $U^T U$ which is a tridiagonal matrix. These can be computed efficiently by the QR algorithm.

For further information the reader is referred to the detailed account in Lawson and Hanson [1]. From our point of view, the important thing about the singular value decomposition is that we have the very remarkable Theorem 3.2 which says that if A is of rank k, we can construct a matrix $A + E$ with $\|E\|_2 = 1/\|A^\dagger\|_2 = \sigma_k(A)$ such that $A + E$ has rank k-1, and this is the "smallest" matrix E (in the 2-norm) that will reduce the rank. No comparable "best-possible" theorem is known for the LU decomposition of §6, or the decomposition $A = QU$ (where the columns of Q are orthonormal and U is upper trapezoidal) based on Householder transformations or the modified Gram–Schmidt method of §8.

8. Modified Gram-Schmidt orthogonalization

Suppose first of all that A is an $m \times n$ matrix of rank n, i.e. the columns of A are linearly independent. Let a_i denote the ith column of A:

$$A = [a_1, a_2, \ldots, a_n].$$

We wish to produce from A a matrix Q :

$$Q = [q_1, q_2, \ldots, q_n]$$

whose columns are orthogonal, such that q_p is a linear combination of $a_1, \ldots, a_p$. This is equivalent to finding an upper triangular nonsingular square matrix S of order n such that $AS = Q$, which in turn is equivalent to finding a decomposition of A in the form

$$A = QR \ ,$$

where Q is orthogonal and $R(=S^{-1})$ is upper triangular.

The columns of Q can be obtained by using the classical Gram-Schmidt orthogonalization procedure. In the present context this means computing in succession $A_1, A_2, \ldots, A_n$, where

$$A_p = [q_1, \ldots, q_p, \ a_{p+1}, \ldots, a_n] \ , \qquad q_r^T q_s = 0, \qquad r \neq s \ , \tag{8.1}$$

and q_p is computed from

$$q_p = a_p - \sum_{j=1}^{p-1} r_{jp} q_j \ ,$$

$$r_{jp} = (q_j^T a_p)/q_j^T a_j, \qquad j = 1, \ldots, p-1 \ .$$

This process cannot break down since if $q_p = 0$ at any stage this would mean that a_p is linearly dependent on $a_1, \ldots, a_{p-1}$, which is not possible since A has been assumed to be of rank n.

It is found that this process is not very stable numerically. The computed vectors are often far from orthogonal, and in fact, before the superior numerical stability of the variant to be described presently was appreciated, it was often recommended that the vectors found by the classical Gram-Schmidt procedure be reorthogonalized, i.e. that the process be applied twice.

Nowadays one would recommend universal use of a stable (mathematically equivalent) variant known as the modified Gram-Schmidt procedure. This computes in succession $A_1, A_2, \ldots, A_k$, where

$$A_p = [q_1, \ldots, q_p, \ a_{p+1}^{(p)}, \ldots, a_n^{(p)}] \ , \tag{8.2}$$

$$a_j^{(p)} = a_j^{(p-1)} - r_{p-1,j} q_{p-1}, \quad q_p = a_p^{(p)} , \tag{8.3}$$

$$r_{p-1,j} = q_{p-1}^T a_j^{(p-1)} / q_{p-1}^T q_{p-1}, \quad p \le j \le n .$$

To illustrate the different behavior of the two procedures numerically, consider

$$A = \begin{bmatrix} 1 & 1 & 1 \\ 1+\epsilon & 1 & 1 \\ 1 & 1+\epsilon & 1 \\ 1 & 1 & 1+\epsilon \end{bmatrix} .$$

Here ϵ is assumed to be a small number $(|\epsilon| << 1)$ such $1 + \epsilon$ is represented accurately on our computer, but $1 + 2\epsilon + \epsilon^2$ is rounded to $1 + 2\epsilon$. The classical Gram-Schmidt procedure gives, with this rounding rule,

$$q_1 = \begin{bmatrix} 1 \\ 1+\epsilon \\ 1 \\ 1 \end{bmatrix}, \quad q_2 = \begin{bmatrix} 0 \\ -\epsilon \\ \epsilon \\ 0 \end{bmatrix}, \quad q_3 = \begin{bmatrix} 0 \\ -\epsilon \\ 0 \\ \epsilon \end{bmatrix} .$$

Note that

$$\frac{q_2^T q_3}{\|q_2\| \, \|q_3\|} = \frac{1}{2} ,$$

so that q_2, q_3 are by no means orthogonal. The modified Gram-Schmidt, with the above rounding rule, gives

$$q_1 = \begin{bmatrix} 1 \\ 1+\epsilon \\ 1 \\ 1 \end{bmatrix}, \quad q_2 = \begin{bmatrix} 0 \\ -\epsilon \\ \epsilon \\ 0 \end{bmatrix}, \quad q_3 = \begin{bmatrix} 0 \\ -\frac{1}{2}\epsilon \\ -\frac{1}{2}\epsilon \\ \epsilon \end{bmatrix},$$

and now $q_2^T q_3 = 0$.

What has happened is that cancellation due to subtraction of nearly equal numbers (the usual source of trouble in numerical work) occurred when forming q_2 and q_3 in the classical Gram-Schmidt, rendering q_3 almost meaningless. Cancellation occurs in the modified Gram-Schmidt when forming q_2 and $a_3^{(2)}$, but we force orthogonality of q_2 and q_3, and the resulting q_3 turns out to be reasonably accurate.

The superiority of the modified procedure seems to have been recognized explicitly for the first time in print by J. Rice [1]. An error analysis has been given by A. Bjorck [1], but the details are complicated. The reader is referred to this paper for further discussion. We content ourselves with the following observations.

(a) If, using the notation (8.2), we find $q_p = 0$, this means that the pth column of A is a linear combination of the first p-1 columns. In practice, to take account of rounding errors and to give numerical stability, rather than (8.2) we should compute

$$A_p = [q_1, \ldots, a_p^{(p)}, \; a_{p+1}^{(p)}, \ldots, a_n^{(p)}]$$

and then compute $\|a_i^{(p)}\|$ for $i = p, p+1, \ldots, n$. Let the maximum occur for $i = q$. If this maximum is less than δ (prescribed by the program), all the columns $a_i^{(p)}$ are regarded as zero, the reduction is complete, and rank(A) = p-1 . Otherwise interchange columns q and p, and choose q_p equal to the resulting pth column.

(b) The number of arithmetic operations required is slightly greater for the modified Gram-Schmidt compared with Householder, since Gram-Schmidt always deals with vector of length m, whereas Householder deals with successively shorter columns. This also means that modified Gram-Schmidt needs somewhat more storage than Householder.

(c) Modified Gram-Schmidt and Householder (with appropriate column interchanges) are broadly speaking comparable in stability, accuracy and general utility for the same amount of work.

9. Computing the generalized inverse of full-rank matrices

Having developed the necessary background, we next consider computation of the generalized inverse of an $m \times k$ matrix A whose rank is known to be k. The most straightforward procedure is to use the formula

$$A^{\dagger} = (A^T A)^{-1} A^T$$

by solving the following set of k equations in k unknowns, with m different right-hand sides:

$$(A^T A)X = A^T \quad . \tag{9.1}$$

Since $A^T A$ is positive definite, no pivoting is needed.

The number of operations (1 operation = 1 addition + 1 multiplication) to form $A^T A$ (which is symmetric) for $k \gg 1$, is of order $\frac{1}{2} mk^2$, the LU decomposition (again exploiting symmetry) takes approximately $\frac{1}{6} k^3$, and the operations involving the m right-hand sides require approximately mk^2, a total of

$$k^2 (\frac{3}{2} m + \frac{1}{6} k) \text{ operations} .$$

If A is $k \times n$, rank k, we can either compute the generalized inverse of A^T by the above procedure, and transpose the result; or compute directly $A^T(AA^T)^{-1}$ which takes slightly more operations, namely $k^2(\frac{3}{2} m + \frac{1}{2} k)$.

If (9.1) are solved by a well-constructed subroutine using Gauss elimination, ill-conditioning will be indicated by two warnings (a) Iterative correction needs an excessive number of iterations, (b) Small pivots occur. If ill-conditioning is indicated, the simplest way of coping with this is to repeat the computation in double-precision throughout, i.e. in the formation of $A^T A$ and the solution of (9.1), assuming that double-precision working is reasonably efficient. However we will still have to worry about the degree of ill-conditioning in the original problem.

If the least-squares problem $Ax = b$ is ill-conditioned, the computing problem

$$(A^T A)x = A^T b \tag{9.2}$$

worsens the condition <u>unnecessarily</u>. To try to explain this, we prove:

<u>Theorem 9.1</u>. <u>If</u> A <u>is</u> $m \times k$, <u>rank</u> k, <u>and we define</u>

$$\text{cond}(A) = \|A\|_2 \, \|A^+\|_2 \, ,$$

<u>then</u>

$$\text{cond}(A^T A) = [\text{cond}(A)]^2 \, .$$

<u>Proof</u>: If $\sigma_j(A)$ denotes the jth singular value of A, the singular values of $A^T A$ are $[\sigma_j(A)]^2$, and the result in the theorem follows immediately from (A. 8).

If A were square, a large value of cond (A), assuming reasonable scaling of A, would indicate that A was ill-conditioned. The theorem indicates that if cond(A) is large, then $\text{cond}(A^T A)$ is much larger. The perturbation results presented earlier (e.g. Theorem 4. 3) indicate that if $\|r\|_2 = \|b - Ax\|_2$ is small, the condition of the least-squares problem $Ax = b$ is governed by cond(A), and a stable computing procedure should therefore have a sensitivity to rounding errors governed by cond(A), not the square of this as is the case with (9.1).

To illustrate how we can avoid this unnecessary worsening of condition in (8.2), suppose that we can partition A in the form

$$A = \begin{bmatrix} A_1 \\ A_2 \end{bmatrix} = \begin{bmatrix} I \\ P \end{bmatrix} A_1 , \qquad (9.3)$$

where $P = A_2 A_1^{-1}$, and A_1 has been chosen to be "as well-conditioned as possible". For example, using the LU decomposition method of §6,

$$A = LU = \begin{bmatrix} L_1 \\ L_2 \end{bmatrix} U = \begin{bmatrix} I \\ P \end{bmatrix} L_1 U ,$$

and partial pivoting will avoid choices such that $\det(L_1 U)$ is unnecessarily small.

Substituting (9.4) in the equation $Ax = b$, $b^T = [b_1 b_2]$, we find

$$A_1^T [I + P^T P] A_1 x = A_1^T (b_1 + P^T b_2) .$$

Since A_1^T is nonsingular it can be cancelled, and this is the key step that improves the condition:

$$(I + P^T P) A_1 x = b_1 + P^T b_2 . \qquad (9.4)$$

The matrix $I + P^T P$ is well-conditioned. (Its eigenvalues and determinant are greater than 1.) The condition of the problem is governed by the condition of A_1, i.e., crudely speaking, the "best" submatrix then can be chosen from the rows of A. It is instructive to rearrange (9.4) in the form

$$A_1 x = b_1 + (I + P^T P)^{-1} P^T (b_2 - P b_1) . \qquad (9.5)$$

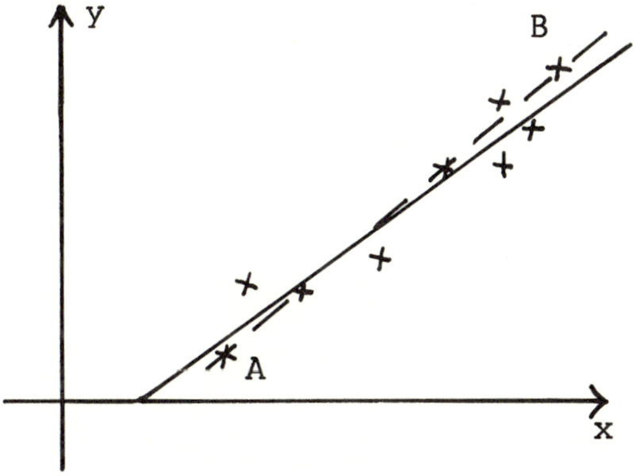

Figure 2. Least-squares fitting of straight line

If $b_2 = Pb_1$, i.e. the equations are consistent, we simply have to solve the square system $A_1x = b_1$. If the equations are "nearly" consistent, the second term on the right of (9.5) will be a small correction to the first. Geometrically in Fig. 2, we choose a first approximation (corresponding to $A_1x = b_1$) by drawing a straight line through AB, and then make a small correction to obtain the final least-squares line.

Related considerations apply to other methods that avoid unnecessary worsening of condition.

When we come to list specific methods for avoiding the unnecessary worsening of condition implicit in (9.2), we have a Pandora's box of possibilities based on the methods developed in §§6-8 (and a second box involving methods we shall not consider). For simplicity ignore any row or column interchanges that occur in implementation of the methods. We discuss four possibilities:

(1) Decompose A in the form $A = LU$ where L is lower trapezoidal with units on the diagonal, and U is upper triangular. Then

$$A^{\dagger} = U^{-1}(L^TL)^{-1} L^T .$$

In practice we would probably first do the triangular decomposition part of solving

$$L^T L Y = L^T .$$

If $L^T L = L_1 U_1$, this gives

$$U_1 Y = L_1^{-1} L^T . \qquad (9.6)$$

We now set $Y = UX$, form $U_1 U$, and solve

$$(U_1 U) X = L_1^{-1} L^T , \quad (X = A^\dagger) . \qquad (9.7)$$

This involves $k^2(\frac{1}{2}m - \frac{1}{6}k)$ operations to form $L, U, \frac{1}{2}mk^2$ to form $L^T L$, $\frac{1}{6}k^3$ to form L_1, U_1, $\frac{1}{2}mk^2$ to form $L_1^{-1}L^T$, $\frac{1}{6}k^3$ to form $U_1 U$, and $\frac{1}{2}mk^2$ to solve (9.7), i.e., in all,

$$k^2(2m + \frac{1}{6}k) \text{ operations.} \qquad (9.8)$$

(2) By Householder transformations or the modified Gram-Schmidt procedure, we can decompose A in the form

$$A = QU ,$$

where the columns of A are orthonormal, and U is upper triangular. Then

$$A^\dagger = U^{-1} Q^T . \qquad (9.9)$$

The number of operations required to find Q is $k^2(m - \frac{1}{3}k)$ (see (10.2) below), and the number required to solve $UX = Q^T$ is $\frac{1}{2}mk^2$ so the total is approximately

$$k^2\left(\frac{3}{2}m - \frac{1}{3}k\right) .$$

This method is both efficient and stable. (The stability comes from the facts that (a) the decomposition by Householder transformations is stable (b) the only computed inverse is U^{-1} in (9.9).)

(3) Decompose A in the form $A = LU$ as in (1), and then decompose further:

$$L = \begin{bmatrix} L_1 \\ L_2 \end{bmatrix} = \begin{bmatrix} I \\ P \end{bmatrix} L_1 ,$$

where L_1 is lower triangular, giving

$$A^\dagger = U^{-1}L_1^{-1}(I + P^TP)^{-1} [I, P^T] . \tag{9.10}$$

This is related to Jordan's method for finding the inverse of a matrix. Write

$$A = \begin{bmatrix} A_1 \\ A_2 \end{bmatrix} = \begin{bmatrix} I \\ P \end{bmatrix} A_1 , \quad A^\dagger = A_1^{-1} \begin{bmatrix} I \\ P \end{bmatrix}^\dagger , \tag{9.11}$$

where this last formula is (9.10). To find A_1^{-1} and P, use elementary row operations to perform the following transformations:

$$\begin{bmatrix} A_1 & I \\ A_2 & 0 \end{bmatrix} \rightarrow \begin{bmatrix} I & A_1^{-1} \\ 0 & -P \end{bmatrix} .$$

This type of procedure takes somewhat more operations than that in (1), and there does not seem to be any particular advantage in stability.

(4) In the decomposition

$$A = \begin{bmatrix} I \\ P \end{bmatrix} A_1$$

of (9.11), however derived, if we use Householder transformations or modified Gram-Schmidt to perform the decomposition

$$\begin{bmatrix} I \\ P \end{bmatrix} = QU = \begin{bmatrix} Q_1 \\ Q_2 \end{bmatrix} U$$

where the columns of Q are orthogonal, then

$$\begin{bmatrix} I \\ P \end{bmatrix}^{\dagger} = Q_1 Q^T .$$

Although this is an elegant formula, there seems to be no particular advantage in speed or stability over the methods (1), (2).

To sum up, if I had to write from scratch a program to invert a small number of full-rank matrices, I would simply use (9.1) in single precision, re-running in double precision if ill-conditioning were indicated. (It is important to understand that this is to avoid unnecessary worsening of condition introduced by (9.1). If we have to use double precision, we have to worry about the inherent ill-condition of the original problem.) However there would seem to be every advantage in using well-constructed least-squares subroutines based on method (2) above, using Householder transformations. These are available in, for example, Lawson and Hanson [1], and Wilkinson and Reinsch [1].

10. Computing the generalized inverse of an $m \times n$ matrix, rank k

We have already summarized at the end of §2 the methods that we wish to consider. It will facilitate the discussion, without loss of generality, if we ignore any interchanges of rows and columns required to determine the rank of A and make the computation stable.

The first class of method depends on the decomposition

$$A = LU$$

where L, U are $m \times k$, $k \times n$ upper and lower trapezoidal, and L has units on the diagonal. If $k >> 1$ this takes approximately

$$k[mn - \frac{1}{2}(m+n)k + \frac{1}{3}k^2] \text{ operations .} \qquad (10.1)$$

The second class of method depends on the decomposition

$$A = QS \quad ,$$

where the columns of Q are orthogonal and S is upper trapezoidal. If this decomposition is performed by Householder transformations we require approximately

$$2k[mn - \frac{1}{2}(m+n) + \frac{1}{3}k^2] \text{ operations.} \qquad (10.2)$$

If it is performed by Gram-Schmidt, we require approximately

$$2km(n - \frac{1}{2}k) \text{ operations .} \qquad (10.3)$$

Thus the LU decomposition takes roughly half the number of operations required for the QS decomposition. However this advantage is lost when we come to complete the computation of $A^\dagger$.

Using the LU decomposition,

$$A = U^{\dagger}L^{\dagger} = U^{T}(UU^{T})^{-1}(L^{T}L)^{-1}L^{T} .$$

A stable method of computation is to solve, in turn,

$$L^{T}LX = L^{T}, \quad UU^{T}Y = X , \tag{10.4}$$

and then form $U^{T}Y$. The coefficient matrices are positive definite, and we pivot along the principal diagonal. No scaling is needed since we know how to choose the pivots. (The effect of scaling is to modify the choice of pivots in Gauss elimination.) However (10.4) involves the solution of two sets of equations with m different right-hand sides. If we wish to solve only one set of equations we can set $U = DU_1$, where D is diagonal and U_1 has units on the diagonal, then solve

$$(L^{T}L\,DU_1U_1^{T})Y = L^{T} , \tag{10.5}$$

and form $U^{T}Y$. The matrix of coefficients is not symmetric and partial pivoting is needed. The rescaling involved in writing $U = DU_1$ is essential to preserve stability. The total number of operations required to compute $A^{\dagger}$ from A via (10.5) is

$$k[2mn + \frac{1}{2}mk + k^{2}] . \tag{10.6}$$

Other arrangements seem to require a comparable amount of calculation.

Using the QS decomposition,

$$A^{\dagger} = S^{T}(SS^{T})^{-1}Q^{T} .$$

If Householder transformations are used to find Q, S and then we solve $(SS^{T})X = Q^{T}$, and finally form $S^{T}X$, this will take a total of approximately

$$k[2mn - \frac{1}{2}mk - nk + \frac{4}{3}k^2] \text{ operations.} \qquad (10.7)$$

A comparison of (10.7) and (10.6) illustrates the remark made earlier, that the advantage of the first class of method over the second in terms of numbers of operations to perform the basic decomposition is to a large extent lost when we have to complete the calculation to find $A^\dagger$.

There are many other aspects of these methods that would be covered in a comprehensive treatment:

(1) When dealing with square matrices we very seldom compute and store the inverse of A – we compute and store the triangular factors L, U . Related remarks are true in connection with the generalized inverse.

(2) As is true for square matrices, the storage space required to compute and store $A^\dagger$ (or the essential factors like L, U or Q, S) is little more than the original storage space occupied by A .

(3) Iterative correction is usually incorporated in good subroutines for solving $Ax = b$, nonsingular A . It provides a valuable check on the condition of the problem and it is cheap. Corresponding techniques for the least squares problem have been discussed by Bjorck [2], Bjorck and Golub [1], Golub and Wilkinson [1]. The situation is more subtle than for nonsingular systems since the size of the residual is important (cf. Theorem 4.3).

(4) The elimination and orthogonalization techniques described can be used to solve other problems, for instance, find the least-squares solution of $Ax = b$ subject to $Cx = d$. See Lawson and Hanson [1].

(5) From (10.6), (10.7), we see that if the corresponding methods are used to compute the inverse of an $n \times n$ matrix A, this would seem to involve about $3n^3$ operations instead of the n^3 required by the usual Gauss elimination procedure. One would like to have a method for computing

the generalized inverse of an $(n+1) \times n$ matrix that requires only slightly more than n^3 operations. This can be achieved by using updating methods discussed in Gill et al. [1], also Bennett [1], or by using the formula:

$$(A + BDC)^{-1} = A^{-1} - A^{-1}B(D^{-1} + CA^{-1}B)^{-1} CA^{-1} ,$$

where A, B, C, D are $n \times n$, $n \times p$, $p \times p$, $p \times n$. A related question is the problem of "updating", i.e., adding or removing or modifying a single row or column of A . See the above references, and Lawson and Hanson [1], Chapter 24.

(6) We have concentrated on the Moore-Penrose inverse, but in practice it may be sufficient to compute an inverse satisfying only some of the four defining conditions in (2.1), (2.2). For instance, if the LU decomposition of an $m \times n$, rank k matrix A is

$$A = \begin{bmatrix} L_1 \\ L_2 \end{bmatrix} [U_1 \; U_2] ,$$

then the general solution of a <u>consistent</u> system $Ax = b$ is given by $x = Gb + u$, where

$$G = \begin{bmatrix} U_1^{-1} L_1^{-1} & 0 \\ 0 & 0 \end{bmatrix}, \qquad u = \begin{bmatrix} -U_1^{-1} U_2 \\ I \end{bmatrix} z ,$$

where z is an arbitrary $(n-k) \times 1$ vector. However the matrix G is not unique, and it is not particularly useful in most least-squares problems.

Very many iterative methods for computing generalized inverses have been proposed in the literature. We have ignored these because they seem to present no advantages over direct methods for full matrices of reasonable size (say 500×100). In particular the explicit determination of rank is usually ignored. The impression is left with the reader that we iterate to convergence and the generalized inverse appears. The situation is much more complicated than this. We need do no more than refer the reader to the excellent discussion in Soderstrom and Stewart [1] which should be used as a guide by proposers of iterative methods, to illustrate the questions that must be answered before a proposed iterative method can be taken seriously.

Similar remarks apply to other definitions of the generalized inverse that do not mention rank explicitly, and therefore might seem attractive for numerical purposes, for example,

$$A^{\dagger} = \lim_{\delta \to 0} \{(A^T A + \delta^2 I)^{-1} A^T\} .$$

We have stressed throughout that the determination of rank is crucial. A related question is whether the problem is ill-conditioned. In the methods based on the decompositions in §§6-8, we have to watch a sequence of numbers $p_1, p_2, \ldots$ such that (assuming that scaling has been taken into account), if $p_1, \ldots, p_k$ are of order of magnitude unity and p_{k+1} drops suddenly to a small multiple of 10^{-t} on a t-decimal machine, we can assume that the rank is k; if the numbers decrease gradually to a small multiple of 10^{-t}, the problem is ill-conditioned. In Gauss elimination the p's are the pivots. In orthogonalization methods the p's are the sizes of successive orthogonal vectors, assuming that these are not normalized. I have found this perfectly satisfactory in practical situations. There are usually methods for cross-checking on rank-determination in physical situations. Ill-conditioning can be remedied usually only by going back and reformulating the problem; there is no magic way of removing it from a given matrix.

If the elements of the generalized inverse are large, this indicates either ill-conditioning or faulty determination of rank. One can always carry out a posteriori checks on the reliability of the computed inverse by seeing whether small changes in the original matrix produce unacceptably large changes in the generalized inverse, and answers derived from it.

If the reader does not find these methods of dealing with rank and condition convincing, his best recourse is to go to the singular value decomposition described briefly at the end of §2. At least he has then Theorem 3.2 which gives him a precise statement regarding the smallest perturbation that will change A to a matrix of any required rank. A detailed discussion of various facets of the singular value method is given in Lawson and Hanson [1].

To conclude this section we remark that it is very difficult to decide which method is "best" by numerical studies. (It can even be dangerous to conclude that a method is "bad" – because this may depend on the way the program is written.) There are just too many factors and too many imponderables. One study concluded that elimination and Gram-Schmidt were more accurate than Householder and singular value decomposition. Such comparisons boil down, in the end, to saying that programs A and B performed better than C and D when used on examples E, F on machines G and H. The evidence so far available seems to indicate that elimination and orthogonalization and singular values give comparable accuracy in comparable circumstances. The best paper I know on a comparative study is the one by Peters and Wilkinson [1].

11. Conclusions on computing methods

A "good" numerical method for computing generalized inverses will satisfy the following criteria:

(1) STABILITY. In particular, the method must avoid squaring the condition number.

(2) EXPLICIT DETERMINATION OF RANK, together with some basis for deciding whether the problem is ill-conditioned. (On the question of rank see also Appendix B(a).)

(3) EFFICIENCY. This can be measured in several ways. The numbers of additions and multiplications, and the storage space required, are important.

(4) FLEXIBILITY. We may want to find the null space of A, solve sets of equations with linear constraints, compute sequentially (e.g. add rows to A one at a time), and so on.

The situation is complicated by the fact that many variations in detail are possible when implementing any given algorithm, and tradeoffs are possible between execution time, accuracy, information generated, and so on.

The present "state of the art" in connection with the Moore-Penrose inverse seems to be the following. (These comments summarize our previous discussions.)

(a) If A is $m \times n$ $(m \neq n)$ and it is known to be of full rank, the most straightforward method for computing $A^\dagger$ is via

$$A^\dagger = (A^TA)^{-1}A^T \quad (m > n), \quad \text{or} \quad A^T(AA^T)^{-1} \quad (m < n), \qquad (11.1)$$

using double precision if A^TA is badly conditioned. Double precision can be avoided by using the Gauss elimination, Householder transformation, or Gram-Schmidt methods of §§6-8, as discussed in §9. Compared with the double-precision solution of (11.1), these offer a saving of storage space and an increase in efficiency, depending on how efficient double-precision working is on the machine used.

(b) If A is of less than full rank, but its rank can be determined by the Gauss elimination, Householder transformation, or Gram-Schmidt methods of §§6-8, then these methods, or variants based on them, can give satisfying results of comparable accuracy, with comparable stability, efficiency, flexibility, and reliability of determination of rank.

(c) If there is real difficulty in rank-determination, the safest procedure is to use the singular-value decomposition and formula (2.11), but this is more expensive than the methods in (b).

(d) Well-constructed programs for methods used in Householder transformations and the singular value decomposition are readily available in Lawson and Hanson [1] and Wilkinson and Reinsch [1]. Advice to the general user is simply to use these published programs. (See Appendix B(c).)

As remarked earlier, we do not claim that these conclusions are definitive, but we hope that the issues involved have been clarified.

12. Concluding comments

First consider perturbation theory, treated in §§3-5. We have simply tried to emphasize some of the more important results in as transparent a fashion as possible. As remarked earlier, it is not at all clear to us how definitive the existing theory is. Most of our results are stated in the 2-norm, and many of these can be generalized to unitarily invariant norms, but not to 1- or ∞-norms. To get the "right" kind of theorems in other norms it may be necessary to use inverses other than Moore-Penrose. C. de Boor has pointed out to me that the paper of M. Z. Nashed and G. F. Votruba in this volume may be relevant. One would certainly like to have a generalization of Theorem 3.2 in an arbitrary norm: If A is of rank k, what is the "smallest" matrix E such that $A + E$ is of rank $k - 1$?

The most definitive treatment of perturbation theory to date is the excellent article by Wedin [1] which is mainly publication of his University of Lund report of 1969. Several of his results have been discovered independently by W. Kahan, and announced at the Gatlinburg conference of 1972 and elsewhere. Several special results were announced earlier by Ben-Israel [1], Hanson and Lawson [1], Pereyra [1] and Stewart [1].

In dealing with computational methods in §§6-11, it may seem that I have devoted inordinate space to methods based on Gauss elimination, and not enough to orthogonal transformations. In part this has been a deliberate attempt to redress the balance in the literature, where orthogonal transformations are emphasized, and Gauss elimination neglected (except in the excellent survey paper of Peters and Wilkinson [1]). My reaction when I first became interested in computing generalized inverses almost ten years ago was: Since Gauss elimination is the "best" method for solving square systems, it is plausible that extensions of Gauss will be "best" for arbitrary systems. Somewhat against my will, the writing of this paper has led me to a more balanced appreciation of the virtues of Householder transformations. Methods based on Gauss elimination will not be competitive with orthogonal transformations until another Golub (or Wilkinson) writes standard subroutines and builds up the necessary body of supporting theory and practice, comparable to that in Lawson and Hanson [1] for orthogonalization methods.

I am indebted to Carl de Boor for several stimulating conversations during the (belated) writing of this paper. Anyone writing on this subject must express a debt of gratitude to Gene Golub. Although his name does not appear explicitly in connection with the methods described, he is the father of many of the developments in this subject.

* * *

Appendix A: Singular values and singular vectors

Various results in the main part of the paper rely heavily on singular values, and we summarize the required background:

Definition A.1. If A is a general $m \times n$ matrix, σ a non-negative number, and u, v vectors, such that

$$Au = \sigma v, \quad A^T v = \sigma u \; , \tag{A.1}$$

then σ is called a singular value of A, and u, v are called singular vectors.

Note that

$$A^T A u = \sigma^2 u, \quad A A^T v = \sigma^2 v \ , \tag{A.2}$$

Theorem A.1. If A is $m \times n$, rank k, there are exactly k nonzero (real) singular values σ_i, with corresponding pairs of singular vectors u_i, v_i, $i = 1, \ldots, k$. The u_i can be chosen to be orthonormal, and also the v_i. If

$$U = [u_1, \ldots, u_k], \qquad V = [v_1, \ldots, v_k] \tag{A.3}$$

then

$$V^T A U = \Sigma \ , \qquad \Sigma = \begin{bmatrix} \sigma_1 & \ldots & 0 \\ & & \\ 0 & \ldots & \sigma_k \end{bmatrix} \tag{A.4}$$

$$A = V \Sigma U^T, \quad A^\dagger = U^T \Sigma^{-1} V \ . \tag{A.5}$$

Note that $U^T U = V^T V = I_k$. It is sometimes convenient to find $n - k$ vectors u_i such that $Au_i = 0$, $i = k+1, \ldots, n$, and $m - k$ vectors v_i such that $Av_i = 0$, $i = k+1, \ldots, m$, such that

$$U = [u_1, \ldots, u_n], \quad V = [v_1, \ldots, v_n] \tag{A.6}$$

are orthogonal matrices. In this case

$$A = V \begin{bmatrix} \Sigma & 0 \\ 0 & 0 \end{bmatrix} U^T, \qquad A^\dagger = U^T \begin{bmatrix} \Sigma^{-1} & 0 \\ 0 & 0 \end{bmatrix} V \ . \tag{A.7}$$

If the singular values are arranged in decreasing order:

$$\sigma_1(A) \geq \dots \geq \sigma_k(A) \quad ,$$

then

$$\|A\|_2 = \sigma_1(A), \quad \|A^\dagger\|_2 = 1/\sigma_k(A) \quad . \tag{A. 8}$$

Theorem A. 2. If Q is hermitian with eigenvalues $\lambda_1 \geq \dots \geq \lambda_n$, with corresponding eigenvectors $x_1, \dots, x_n$, and x is a nonzero vector such that $x_i^T x = 0$, $i = 1, \dots, j-1$, then

(i)
$$\lambda_j = \max_x \frac{x^T Q x}{x^T x} \quad . \tag{A. 9}$$

(ii)
$$\lambda_j = \min \left\{ \max \frac{x^T Q x}{x^T x} \right\} \tag{A. 10}$$

where (a) first take the max over all $x \neq 0$ subject to $p_i^T x = 0$, $i = 1, \dots, j-1$, where the p_i are given (arbitrary, but fixed when taking the max over x).

(b) then take the min over all p_i .

So far we have quoted standard theorems, proofs of which can be found in standard texts. We shall also need the following, which is not quite so well known. (K. Fan, Proc. Nat. Acad. Sc. 37 (1951), 764.)

Theorem A. 3. If $\sigma_j(H)$ denotes the jth singular value of H, $\sigma_1 \geq \dots \geq \sigma_n$, and $G = K + E$, then

$$\sigma_j(G) \leq \sigma_j(K) + \|E\|_2, \quad j = 1, \dots, n \quad . \tag{A. 11}$$

Proof: We use 2-norms throughout this proof. From (A. 10),

$$\sigma_j(G) = \min \left\{ \max \frac{\|Gx\|}{\|x\|} \right\},$$

where the max is taken over all $x \neq 0$, $p_i^T x = 0$, $i = 1, \ldots,$ $j-1$ for given p_i, and the min is taken over all p_i. Let p_i be the set S of singular vectors u_i of K corresponding to the $j-1$ largest singular values. Then

$$\sigma_j(G) \leq \max_S \frac{\|Gx\|}{\|x\|}$$

$$\leq \max_S \left\{ \frac{\|Kx\|}{\|x\|} + \frac{\|Ex\|}{\|x\|} \right\}$$

$$\leq \max_S \frac{\|Kx\|}{\|x\|} + \|E\| = \sigma_j(K) + \|E\| ,$$

where we have used (A.10). This completes the proof.

Appendix B: Additional Notes

(a) In this paper we have discussed methods for computing generalized inverses of matrices that make a quite explicit decision regarding the rank of the matrix at some point in the calculation. It is clear from the literature and from remarks of a reviewer of this paper that some people feel that methods might exist that would allow one to avoid making an explicit determination of rank.

This feeling is reinforced by the literature on generalized inverses of infinite-dimensional linear operators where we can no longer compare operators on the basis of rank. (See, for instance, the paper by M. Z. Nashed on "Perturbations and Approximations..." in this volume, and an earlier paper by R. H. Moore and M. Z. Nashed, "Approximations to generalized inverses of linear operators", SIAM J. Applied Math., 27 (1974), 1-16.) Suppose that we have matrices A and A_ϵ such that $A_\epsilon^\dagger$ does not converge to $A^\dagger$ as $\epsilon \to 0$. The proposal is that we define a new generalized inverse A_ϵ^ϕ such that A_ϵ^ϕ tends to $A^\dagger$ as $\epsilon \to 0$. The difficulty in practice in applying this idea in the context of the present paper is that we are dealing with situations where we are not given A; we are given an approximation A_δ to A. We do not know A, so that we cannot know

$A^\dagger$, and there is no means of saying with certainty what A_ϵ^ϕ should converge to. In practice we make a decision concerning the rank of A_δ, and assume that this is the rank of A. If we wanted to use the A_ϵ^ϕ idea, this would give us a handle on what A_ϵ^ϕ should converge to. In the context of the present paper, we replace A_δ by A_Δ, where A_Δ has the rank we have decided to attribute to A, and use $A_\Delta^\dagger$ as an approximation to $A^\dagger$.

I would go so far as to claim that if a method for computing generalized inverses of a matrix does not explicitly make a decision about the rank of the matrix at some point in the calculation, then such a decision is built in implicitly somewhere in the procedure, with the corollary that it is better to use methods that make decisions explicitly rather than implicitly. I had intended originally to include a section illustrating this thesis, but the paper of Soderstrom and Stewart [1] appeared that illustrates the point better than I could have done. A reviewer of the present paper made the remark that Soderstrom and Stewart do not mention rank much. From my point of view this is a matter of terminology rather than substance. If one is using a singular value decomposition, it is natural to discuss the situation in terms of the sizes of the singular values. A statement that singular values of A less than ϵ will be set equal to zero is a statement that we replace A by a matrix A_ϵ of rank equal to the number of nonzero singular values that we retain.

I have not intended to imply that iterative methods should never be used for computing generalized inverses. They may be useful for large sparse matrix problems, for instance.

(b) I should have said somewhere that if we wish to find the least-squares solution $x = A^\dagger b$ of $Ax = b$, it is usually unnecessary and inefficient to first calculate $A^\dagger$ and then form $A^\dagger b$. However most of our discussion is relevant to this problem and the point is well-known in connection with nonsingular matrices.

(c) This paper has not been concerned with computer codes for solving least squares problems or finding generalized inverses. The interested reader should consult Lawson and Hanson [1] for FORTRAN programs for methods depending on Householder transformations and singular value decompositions. Similar programs are available in the following "packages": IMSL (International Mathematical and Statistical Library),

NAG (Numerical Algorithms Group, Oxford), EISPACK and a forthcoming package LINPACK (Applied Mathematics Division, Argonne). ALGOL programs related to those in Lawson and Hanson are given in Wilkinson and Reinsch [1]. ALGOL programs for Gram-Schmidt methods are discussed in Bjorck [1], [2]. As far as I know there are no standard packages available using Gauss-elimination type methods. These are relatively easy to write in principle, but one would have to be careful about several matters including scaling, which we have not discussed in detail. (We note in passing that there is almost no discussion in the literature of the effect of scaling on the practical computation of generalized inverses. This is a definite gap in the literature. Even the discussion of scaling when inverting nonsingular matrices is somewhat unsatisfactory and incomplete.)

Advice to the general user is to use probrams in the well-tested packages mentioned earlier.

REFERENCES

N. N. ABDELMALEK [1], Round-off error analysis for Gram-Schmidt method and solutions of linear least squares problems, BIT 11 (1971), 345-367.

S. M. AFRIAT [1], Orthogonal and oblique projectors and the characteristics of pairs of vector spaces, Proc. Camb. Phil. Soc. 53 (1957), 800-816.

A. ALBERT [1], Regression and the Moore-Penrose Inverse, Academic Press (1972).

F. L. BAUER [1], Elimination with weighted row combinations for solving linear equations and least squares problems, Numer. Math. 7 (1965), 338-352, reprinted in Wilkinson and Reinsch [1].

A. BEN-ISRAEL [1], On error bounds for the generalized inverse, SIAM J. Numer. Anal. 3 (1966), 585-592.

A. BEN-ISRAEL and T. N. E. GREVILLE [1], Generalized Inverses, Wiley (1974).

J. M. BENNETT [1], Triangular factors of modified matrices, Numer. Math. 7 (1965), 217-221.

A. BJORCK [1], Solving linear least squares problems by Gram-Schmidt orthogonalization, BIT 7 (1967), 1-21.

[2] Iterative refinement of linear least squares solutions I, BIT 7 (1967), 257-278, II BIT 8 (1968), 8-30.

A. BJORCK and G. H. GOLUB [1], Iterative refinement of linear least squares solutions by Householder transformations, BIT 7 (1967), 322-337.

P. BUSINGER and G. GOLUB [1], Linear least squares by Householder transformations, Numer. Math. 7 (1965), 269-276.

S. D. CONTE and C. de BOOR [1], Elementary Numerical Analysis, McGraw Hill, 2nd Edn. (1972).

C. ECKART and G. YOUNG [1], The approximation of one matrix by another of lower rank, Psychometrika, 1(1936), 211-218.

[2] A principal axis transformation for non-hermitian matrices, Bull. Amer. Math. Soc. 45(1939), 118-121.

P. E. GILL, G. H. GOLUB, W. MURRAY, M. A. SAUNDERS [1], Matrix updating methods, Math. Comp. 28 (1974), 505-535.

G. H. GOLUB [1], Numerical methods for solving linear least squares problems, Numer. Math. 7 (1965), 206-216.

[2] Least squares, singular values, and matrix approximations, Apl. Mat. 13 (1968), 44-51.

G. H. GOLUB and W. KAHAN [1], Calculating the singular values and pseudo-inverse of a matrix, SIAM J. Numer. Anal. 2 (1965), 205-224.

G. H. GOLUB and V. PEREYRA [1]. The differentiation of pseudo-inverses and nonlinear least squares problems whose variables separate, SIAM J. Numer. Anal. 10 (1973), 413-432.

G. H. GOLUB and J. H. WILKINSON [1], Note on the iterative refinement of least squares solutions. Numer. Math. 9 (1966), 139-148.

R. J. HANSON and C. L. LAWSON [1], Extensions and applications of the Householder algorithm for solving linear least-squares problems, Math. Comp. 27 (1969), 787-812.

M. R. HESTENES [1], Inversion of matrices by biothogonalization and related results, SIAM J. 6 (1958), 51-90.

T. L. JORDAN [1], Experiments in error growth associated with some linear least-squares procedures, Math. Comp. 21 (1966), 579-588.

V. N. KUBLANOVSKAYA [1], On the calculations of generalized inverses and projections, Z. Vycisl. Mat. i Mat. Fiz. 6 (1966), 326-332,

P. LAUCHLI [1], Jordan-Elimination und Ausgleichung nach kleinsten Quadraten, Numer. Math. 3 (1961), 226-240.

C. L. LAWSON and R. J. HANSON [1], Solving Least Squares Problems, Prentice-Hall (1974).

E. E. OSBORNE [1], On least squares solution of linear equations, JACM 8 (1961), 628-636.

[2] Smallest least squares solutions of linear equations, SIAM J. Numer. Anal. 2 (1965), 300-307.

V. PEREYRA [1], Stability of general systems of linear equations, Aeq. Math. 2 (1969), 194-206.

G. PETERS and J. H. WILKINSON [1], The least-squares problem and pseudo-inverses, Comp. J. 13 (1970), 309-316.

J. R. RICE [1], Experiments on Gram-Schmidt orthogonalization, Math. Comp. 20 (1966), 325-328.

B. W. RUST, W. R. BURRUS, and C. SCHNEEBERGER [1], A simple algorithm for computing the generalized inverse of a matrix, Comm. ACM 9 (1966), 381-386.

N. SHINOZAKI, M. SIBUYA, and K. TANABE [1], Numerical algorithms for the Moore-Penrose inverse of a matrix, Direct methods, Ann. Inst. Statist. Math. 24 (1972), 193-203.

T. SODERSTROM and G. W. STEWART, On the numerical properties of an iterative method for computing the Moore-Penrose generalized inverse, SIAM J. Numer. Analysis, 11 (1974), 61-74.

R. P. TEWARSON [1], A direct method for generalized matrix inversion, SIAM J. Numer. Anal. 4 (1967), 499-507.

[2] A computational method for evaluating generalized inverses, Comp. J. 10 (1968), 411-413.

[3] On computing generalized inverses, Computing 4 (1969), 139-151.

[4] On two direct methods for computing generalized inverses, Computing 7 (1971), 236-239.

P-A.WEDIN [1], Perturbation theory for pseudo-inverses, BIT 13 (1973), 217-232.

J. H. WILKINSON and C. REINSCH [1], Handbook for Automatic Computation II Linear Algebra, Springer (1971).

Mathematics Research Center, University of Wisconsin

Differentiation of Pseudoinverses, Separable Nonlinear Least Square Problems and Other Tales

G.H. Golub

V. Pereysa

1. Introduction

We present in this paper two main themes and some variations. After some preliminary results and notation on matrix decompositions relevant to the study of pseudoinverses and orthogonal projectors, we study the differentiation of matrix functions of several variables, under the hypothesis of locally constant rank.

In Section 4 these concepts are applied to the least squares fitting of models which are linear combinations of nonlinear functions. The method of variable projections permits the modification of the model in such a way that only the nonlinear parameters must be estimated. This is described both in the unconstrained and constrained cases.

A comprehensive set of references is given. In the main body of the paper, most references are collected in historical paragraphs at the end of Sections 3 and 4.

2. Preliminary Results

Orthogonal decompositions and Householder transformations

Every $m \times n$ matrix A of rank r, $r \leq n \leq m$, can be orthogonally transformed into "trapezoidal" form. That is, there exist orthogonal matrices Q, S $(QQ^T = SS^T = I)$ such that

(2.1)
$$QAS = \left[\begin{array}{c|c} T_{11} & T_{12} \\ \hline 0 & 0 \end{array}\right] \begin{array}{l} \} \; r \\ \} \; m-r \end{array} \equiv T_0$$
$$\underbrace{\quad}_{r} \quad \underbrace{\quad}_{n-r}$$

where T_{11} is an $r \times r$ non-singular, upper triangular matrix. Various procedures are available for constructing decomposition (2.1). The most popular one is based on Householder transformations (or elementary reflections). Householder transformations are orthogonal matrices of the form

(2.2)
$$U = I - 2\underset{\sim}{u}\,\underset{\sim}{u}^T$$

where the vector $\underset{\sim}{u}$ has Euclidean length equal to one.

In order to obtain (2.1) the matrix A is successively premultiplied by r Householder transformations Q_i, permuting the columns if necessary (thus the matrix S). The decomposition is not unique since the permutation matrix S is not unique.

<u>Generalized inverses</u>

Every $n \times m$ matrix A^- satisfying

(2.3)
$$AA^-A = A$$

is called a <u>generalized inverse</u>† of A. If it also satisfies

(2.4)
$$(AA^-)^T = AA^- \quad ;$$

it will be called a <u>symmetric g-inverse.</u>

From the decomposition (2.1), it is possible to construct the g-inverses for A. In fact, for any choice of matrices X, Y:

† In other papers in this volume, A^- is also called a partial inverse or a 1-inverse.

(2.5)
$$A^{-} = S \left[\begin{array}{c|c} T_{11}^{-1} & Y \\ \hline X & 0 \end{array}\right] Q ,$$

is a g-inverse of A provided that $T_{12}X = 0$. If we put $Y = 0$ then the resulting g-inverse is symmetric. Finally, if we also put $X = 0$ then the matrix

(2.6)
$$A^{B} \equiv S \left[\begin{array}{c|c} T_{11}^{-1} & 0 \\ \hline 0 & 0 \end{array}\right] Q$$

satisfies the additional condition

(2.7)
$$A^{B}AA^{B} = A^{B} .$$

Orthogonal projectors

The decomposition (2.1) is also useful for representing the associated orthogonal projector. In fact, if we call P_A the orthogonal projector on the column space of the matrix A, it is easy to see that for any symmetric g-inverse A^{-} : $P_A \equiv AA^{-}$. But, from (2.1) and (2.5) (with $Y = 0$) we get:

(2.8)
$$P_A = AA^{-} = Q^{T} \left[\begin{array}{c|c} I_r & 0 \\ \hline 0 & 0 \end{array}\right] Q ,$$

where I_r is the $r \times r$ identity matrix. The projector on the orthogonal complement $P_A^{\perp} = I - AA^{-}$ then becomes:

(2.9)
$$P_A^{\perp} = Q^T \left[\begin{array}{c|c} 0 & 0 \\ \hline 0 & I_{m-r} \end{array}\right] Q \ .$$

<u>Complete orthogonal decomposition and the Moore-Penrose pseudoinverse</u>

Applying additional unitary transformations on the right in (2.1), it is possible to obtain an even more compact representation that we shall call the <u>complete orthogonal decomposition</u> of A:

(2.10)
$$QAT = \left[\begin{array}{c|c} R & 0 \\ \hline 0 & 0 \end{array}\right] ,$$

where all blocks have the same dimensions as before, R is an $r \times r$ non-singular, upper triangular matrix and Q and T are orthogonal. With this new decomposition we can define another g-inverse as follows:

(2.11)
$$A^+ = T \left[\begin{array}{c|c} R^{-1} & 0 \\ \hline 0 & 0 \end{array}\right] Q$$

which has the three properties (2.3), (2.4), (2.7) and the additional one:

(2.12)
$$(A^+A)^T = (A^+A) \ .$$

Since for any A there is only one matrix A^+ having these four properties, (2.11) is a representation of the

Moore-Penrose pseudoinverse of A. With A^+ we can represent the orthogonal projector on the row space of A, ${}_AP \equiv A^+A$, and also the projector on its orthogonal complement ${}_AP^{\perp} \equiv I - A^+A$:

$$(2.13)\quad {}_AP = T\begin{bmatrix} I_r & 0 \\ \hline 0 & 0 \end{bmatrix}T^T, \quad {}_AP^{\perp} = T\begin{bmatrix} 0 & 0 \\ \hline 0 & I_{m-r} \end{bmatrix}T^T .$$

Linear least squares problems

Generalized inverses provide a natural language for dealing with linear subspaces and orthogonal projections. Thus, it comes as no surprise the role they play in the solution of linear least squares problems.

For a given $m \times n$ matrix A and m-vector $\underset{\sim}{b}$ the linear least squares problem consists of finding n-vectors $\hat{\underset{\sim}{a}}$ that satisfy

$$(2.14)\qquad \min_{\underset{\sim}{a}} \| \underset{\sim}{b} - A\underset{\sim}{a} \|^2 .$$

If $\text{rank}(A) = r < n$, this problem has infinitely many solutions: exactly those vectors which are solutions of the linear system of equations:

$$(2.15)\qquad (A^TA)\underset{\sim}{a} = A^T\underset{\sim}{b} .$$

For any symmetric g-inverse $\hat{\underset{\sim}{a}} = A^-\underset{\sim}{b}$ is a least squares solution. It is enough to verify that any such vector $\hat{a}$ satisfies (2.15), which follows immediately from the identity

$$A^T = (AA^-A)^T = A^T(AA^-)^T = A^TAA^- .$$

The pseudoinverse A^+ singles out the least squares solution of minimum length. Thus $\hat{a} = A^+ b$ will be called the <u>minimal</u> least squares solution of (2.14).

Computing $\hat{a}$ requires more operations than computing

$$a^B \equiv A^B b \quad .$$

Furthermore, the vector a^B has at most r non-zero elements. Since $\hat{a}$ is the unique minimal length solution, $\|a^B\| \geq \|\hat{a}\|$. We now give an upper bound for $\|a^B\|$ in terms of $\|\hat{a}\|$.

Using the representation (2.1), we see that

$$\|b - Aa\|^2 = \|Qb - \begin{bmatrix} T_{11} & T_{12} \\ 0 & 0 \end{bmatrix} S^T a\|^2 \quad .$$

Let us write

$$\begin{pmatrix} c \\ d \end{pmatrix} \begin{matrix} \}\, r \\ \}\, m-r \end{matrix} \equiv Qb \quad \text{and} \quad \begin{pmatrix} u \\ v \end{pmatrix} \begin{matrix} \}\, r \\ \}\, n-r \end{matrix} \equiv S^T a \quad .$$

Then

$$\|b - Aa\|^2 = \text{min. when } T_{11} u + T_{12} v = c \quad .$$

Hence

$$u = T_{11}^{-1} c - T_{11}^{-1} T_{12} v \equiv \ell + Kv \quad .$$

The vector ℓ augmented by n-r zeroes yields the vector a^B, and the vector $\hat{a}$ is determined from the equation

$$\|\hat{a}\|^2 = \|u\|^2 + \|v\|^2 = \text{min.}$$

This condition implies that

$$\underset{\sim}{u} = (I - K(K^TK+I)^{-1}K^T)\underset{\sim}{\ell}, \qquad \underset{\sim}{v} = -(K^TK+I)^{-1} K^T \underset{\sim}{\ell} .$$

Since $\|\hat{\underset{\sim}{a}}\|^2 = \|\underset{\sim}{u}\|^2 + \|\underset{\sim}{v}\|^2$, we have

$$\|\hat{\underset{\sim}{a}}\|^2 = \underset{\sim}{\ell}^T(I - K(K^TK + I)^{-1} K^T)\underset{\sim}{\ell} .$$

Noting that $\|\underset{\sim}{a}^B\|^2 = \|\underset{\sim}{\ell}\|^2$, we see that

$$(1+\sigma^2_{max}(K))^{-1} \leq (\|\hat{\underset{\sim}{a}}\|^2/\|\underset{\sim}{a}^B\|^2) \leq 1$$

where $\sigma_{max}(K)$ is the largest singular value of K. The lower bound is easy to calculate by noting that $\sigma_{max}(K) \leq \|K\|_1 \|K\|_\infty$.

Vector calculus

We consider now $m \times n$ matrices $A(\underset{\sim}{\alpha})$ whose elements are functions of the k variables $(\alpha_1, \ldots, \alpha_k)^T \equiv \underset{\sim}{\alpha}$. If the elements of $A(\underset{\sim}{\alpha})$, $a_{ij}(\underset{\sim}{\alpha})$, are differentiable functions at a point $\underset{\sim}{\alpha}^0$, then the derivative at $\underset{\sim}{\alpha}^0$ (in the sense of Fréchet) of the mapping $A(\underset{\sim}{\alpha})$ is the tridimensional tensor

$$(2.16) \quad DA(\underset{\sim}{\alpha}^0) \equiv \left(\frac{\partial a_{ij}(\underset{\sim}{\alpha}^0)}{\partial \alpha_s} \right) \quad (i=1, \ldots, m; \quad s = 1, \ldots, k) .$$

There are several elementary properties in this matrix calculus that we pause to list now and that will be used later.

(a) If A is a constant matrix then:

$$(2.17) \qquad DA \equiv 0; \quad D[A\underset{\sim}{\alpha}] \equiv A \qquad (k = n \text{ in this case}) .$$

(b) If we have two matrix functions $A(\underset{\sim}{\alpha})$, $B(\underset{\sim}{\alpha})$ such that the matrix product $A(\underset{\sim}{\alpha}) \cdot B(\underset{\sim}{\alpha})$ is defined then:

$$D[A(\underset{\sim}{\alpha})\, B(\underset{\sim}{\alpha})] = DA(\underset{\sim}{\alpha}) \cdot B(\underset{\sim}{\alpha}) + A(\underset{\sim}{\alpha}) \cdot DB(\underset{\sim}{\alpha}) \;. \tag{2.18}$$

In components, we have

$$[D(A(\underset{\sim}{\alpha})\, B(\underset{\sim}{\alpha})))]_{ijs} = \sum_{t=1}^{n} \left(\frac{\partial a_{it}}{\partial \alpha_s} \cdot b_{tj} + a_{it} \cdot \frac{\partial b_{tj}}{\partial \alpha_s} \right) .$$

(c) Taylor's formula is valid if A is twice differentiable:

$$A(\underset{\sim}{\alpha} + \underset{\sim}{h}) = A(\underset{\sim}{\alpha}) + DA(\underset{\sim}{\alpha})\underset{\sim}{h} + O(\|\underset{\sim}{h}\|^2) \;. \tag{2.19}$$

(d) If $A(\underset{\sim}{\alpha})$ is $n \times n$, differentiable and nonsingular at $\underset{\sim}{\alpha}_0$, then

$$D[A^{-1}]_{\underset{\sim}{\alpha}=\underset{\sim}{\alpha}_0} = -A^{-1}(\underset{\sim}{\alpha}_0)\, DA(\underset{\sim}{\alpha}_0) A^{-1}(\underset{\sim}{\alpha}_0) \;. \tag{2.20}$$

The scope of the differentiation operator D will always be the symbol (letter, letters forming a name, or parenthetical expression) immediately following the differentiation operator.

3. Differentiation of Projectors and Pseudoinverses

It is natural to ask if there exists a closed formula for the derivative of the pseudoinverse of a matrix function, as in the case of a square nonsingular matrix. The answer is yes, but it is considerably more complicated. Because it has intrinsic interest, and because we shall need it later we first derive the derivatives of projectors.

Let then $A(\underset{\sim}{\alpha})$ be a differentiable $m \times n$ matrix function of the k variables $(\alpha_1, \ldots, \alpha_k)$ on an open set $\Omega \subset R^k$. An important assumption throughout this section is that $A(\underset{\sim}{\alpha})$

is of local constant rank at any point α^0 in which differentiation is to be performed. By this we mean that there exists an open neighborhood $S(\underset{\sim}{\alpha}^0) \subset \Omega$ such that $A(\underset{\sim}{\alpha})$ has constant rank for every $\underset{\sim}{\alpha} \in S(\underset{\sim}{\alpha}^0)$. This is a natural assumption since otherwise the pseudoinverse of $A(\underset{\sim}{\alpha})$ is not even a continuous function of $\underset{\sim}{\alpha}$. In what follows we omit the argument $\underset{\sim}{\alpha}$ in all the functions whenever this is possible without obscuring the meaning.

Recalling that for a symmetric generalized inverse A^- the orthogonal projector P_A is represented as AA^-, we have the following lemma.

Lemma 3.1. Let $A^-(\underset{\sim}{\alpha})$ be a symmetric generalized inverse of $A(\underset{\sim}{\alpha})$ for each $\underset{\sim}{a} \in \Omega$. Then

$$DP_A = P_A^{\perp} DAA^- + (P_A^{\perp} DAA^-)^T \quad . \tag{3.1}$$

Proof: Since $P_A A = A$ then,

$$D(P_A A) = DP_A A + P_A DA = DA \quad ,$$

and therefore

$$DP_A A = P_A^{\perp} DA \quad . \tag{3.2}$$

Since $P_A = AA^-$ it follows from (3.2) that

$$DP_A P_A = DP_A AA^- = P_A^{\perp} DAA^- \quad . \tag{3.3}$$

But P_A and DP_A are symmetrical, and therefore

$$(DP_A P_A)^T = P_A DP_A \tag{3.4}$$

(transposition in the tensor DP_A is performed on each

"plane" $\frac{\partial}{\partial \alpha_i} P_A$). Finally, since P_A is idempotent, it turns out that:

$$DP_A = D(P_A^2) = DP_A P_A + P_A DP_A \tag{3.5}$$

and (3.3), (3.4), (3.5) together give (3.1). □

Unfortunately, the projector on the row space of A cannot be represented with A^-, and therefore we must resort to A^+ in order to obtain

<u>Corollary 3.2.</u> <u>Let</u> ${}_A P = A^+ A$. <u>Then</u>

$$D_A P = A^+ DA_A P^\perp + (A^+ DA_A P^\perp)^T \ . \tag{3.6}$$

<u>Proof</u>. Since $(P_{A^T})^T = (A^T A^{+T})^T = A^+ A = {}_A P$, (3.6) follows from (3.1) with A replaced by A^T and A^- replaced by A^+.

□

Our interest in using A^- whenever possible stems from the fact that A^- is considerably cheaper to evaluate than A^+. With these two results we can obtain the following main theorem, which we give here without proof:

<u>Theorem 3.3.</u> <u>If</u> A <u>has local constant rank then</u>

$$DA^+(\alpha) = -A^+ DAA^+ + A^+ A^{+T} DA^T P_A^\perp + {}_A P^\perp DA^T A^{+T} A^+ \ . \tag{3.7}$$

□

For more details and additional results we refer the reader to [6].

We observe that if A has full column rank then $P_A \equiv 0$, and the second term on the right of (3.7) vanishes; if A has full row rank then it is the third term that vanishes, while if A is square and nonsingular we retrieve the usual formula (2.20).

A different proof of Theorem 3.3 can be obtained from the following incremental identity due to Wedin [24]:

$$B^+ - A^+ = -B^+(B-A)A^+ + {}_BP^{\perp}(B-A)^T A^{+T}A^+ + B^+B^{+T}(B-A)^T P_A^{\perp} . \tag{3.8}$$

In fact, if we put $A \equiv A(\underset{\sim}{\alpha}^0)$, $B \equiv A(\underset{\sim}{\alpha}^0 + \underset{\sim}{h})$ and take limits for $h \to 0$ then (3.7) is obtained.

Historical comments

Despite the fact that generalized inverses have been around for a long time we have not been able to find references related to the problem of differentiating the pseudoinverse of a matrix function before 1967. M. Pavel-Parvu and A. Korganoff seem to have been the first in discussing this problem in a paper given at the Symposium on "Constructive aspects of the fundamental theorem of algebra" [17] and published in 1969.†

Although the beginning of the modern interest in generalized inverses (après Moore) is usually ascribed to Penrose [18], or by more risqué authors [15] to Yu. Ya. Tseng [22], we have been able to find another early reference to the subject by Bergmann, Penfield, Schiller, and Zatzkis [2], which has probably passed unseen firstly for its main subject, and secondly because the pertinent section is headed: "3. The quasi-universe" with an obvious misprint.

By decomposing the matrix A as:

$$A = U\Gamma^{-1}V , \tag{3.9}$$

where U and V are bases for the column and row subspaces of A, and where Γ is a nonsingular matrix having the rank of A, Pavel-Parvu and Korganoff prove the following formula for DA^+:

$$DA^+ = -A^+DAA^+ + A^+U^{+T}DU^T P_A^{\perp} + {}_AP^{\perp}DV^T V^{+T}A^+ . \tag{3.10}$$

† We thank Professor Korganoff for calling our attention to this reference.

Observe the similarity between this formula and (3.7).

In 1969, P. A. Wedin [24] proved both the incremental formula (3.8), and pointed out that for matrix functions of one independent variable one obtains (3.7) readily from (3.8) A bit earlier, Hearon and Evans [8] obtained several differentiability results, also for matrix functions of one independent variable. They proved, for example, the following formulas. For any differentiable projection:

$$PDPP = 0 \quad , \tag{3.11}$$

which follows readily from Lemma 3.1 and Corollary 3.2. Also they proved

$$P_A DA^+ P_A = -A^+ DAA^+ \tag{3.12}$$

which follows from our general formula (3.7).

Formula (3.7) in the full rank case has been rediscovered a number of times recently [5, 19]. In [19] the general case is considered also, though the final formula uses a decomposition as in [17].

Formula (3.7) in its full generality has been proved rigorously for the first time (as far as we know) in our paper [6].

4. Variable Projections for Separable Nonlinear Least Squares

In recent times, separable nonlinear least squares problems have come into the limelight as new powerful techniques for their solution have appeared. The problem is the following:

Given $\underset{\sim}{t}, \underset{\sim}{b} \in \mathcal{R}^m$ and n functions $\varphi_j(\underset{\sim}{\alpha};t)$, $j = 1, \ldots, n$, of the (k+1) variables $(\alpha_1, \ldots, \alpha_k;t)$, determine optimal parameters $\hat{\underset{\sim}{a}}, \hat{\underset{\sim}{\alpha}}$ satisfying:

$$\min_{\underset{\sim}{a}, \underset{\sim}{\alpha}} \sum_{i=1}^{n} (b_i - \sum_{j=1}^{n} a_j \varphi_j(\underset{\sim}{\alpha};t_i))^2 \quad . \tag{4.1}$$

In a typical application, b_i is an observation corresponding to the independent variable t_i, while

(4.2) $$\mu(\underset{\sim}{a}, \underset{\sim}{\alpha}; t) = \sum_{j=1}^{n} a_j \varphi_j(\underset{\sim}{\alpha}; t)$$

is the nonlinear analytic model chosen to fit the given observations in the least squares sense. Models of the form (4.2) are in a way, more general than the usual nonlinear models, which would consist of only one function $\varphi(\underset{\sim}{\alpha}; t)$. The fact that there are two different sets of parameters $\underset{\sim}{a}$ and $\underset{\sim}{\alpha}$, and that the parameters $\underset{\sim}{a}$ appear linearly (for fixed $\underset{\sim}{\alpha}$), is what makes the model "separable".

In what follows we shall denote by $\underset{\sim}{\varphi}_j(\underset{\sim}{\alpha})$ the vector $(\varphi_j(\underset{\sim}{\alpha}; t_1), \ldots, \varphi_j(\underset{\sim}{\alpha}; t_m))^T$, and do similarly with other functions of the independent variable t. With the column vectors $\underset{\sim}{\varphi}_j(\underset{\sim}{\alpha})$ we form the matrix $\Phi(\alpha) = [\underset{\sim}{\varphi}_1, \ldots, \underset{\sim}{\varphi}_n]$, and therefore (4.1) can be written as:

(4.3) $$\min_{\underset{\sim}{a}, \underset{\sim}{\alpha}} r(\underset{\sim}{a}, \underset{\sim}{\alpha}) \equiv \min_{\underset{\sim}{a}, \underset{\sim}{\alpha}} \| \underset{\sim}{b} - \Phi(\underset{\sim}{\alpha})\underset{\sim}{a} \|^2 ,$$

where $\|\cdot\|$ is the Euclidean norm. We observe that for fixed $\underset{\sim}{\alpha}$, the problem of minimizing $r(\underset{\sim}{a}, \underset{\sim}{\alpha})$ with respect to the parameters $\underset{\sim}{a}$ is a <u>linear</u> least squares problem which has the minimal solution

(4.4) $$\underset{\sim}{a} = \Phi(\underset{\sim}{\alpha})^+ \underset{\sim}{b} .$$

Replacing this value in (4.3) we obtain the reduced functional:

(4.5) $$r_2(\underset{\sim}{\alpha}) = \| \underset{\sim}{b} - \Phi(\underset{\sim}{\alpha})\Phi^+(\underset{\sim}{\alpha})\underset{\sim}{b} \|^2 = \| P^{\perp}_{\Phi(\underset{\sim}{\alpha})} \underset{\sim}{b} \|^2 ,$$

which will be named (for obvious reasons) "the variable projection" functional.

The <u>variable projection</u> method consists of minimizing first $r_2(\alpha)$, obtaining an optimal vector $\hat{\alpha}$, and then computing $\hat{a} = \Phi(\hat{\alpha})^+ b$. In [6] it has been proved that under suitable conditions this process yields an optimal solution of the original problem (4.3).

Many methods for minimizing $r_2(\alpha)$ require either its gradient or the Jacobian of the residual vector function $r(\alpha) = P^\perp_{\Phi(\alpha)} b$. It is here where the results of Section 3 are used. In fact, from (3.1) we obtain, by putting $C \equiv P^\perp_\Phi D\Phi\Phi^+$

$$\frac{1}{2}\nabla r_2(\alpha) = \frac{1}{2}\nabla(b^T P^\perp_\Phi b)$$

$$= -\frac{1}{2} b^T D P_\Phi b$$

$$= -\frac{1}{2} b^T P^\perp_\Phi (C + C^T) b$$

$$= -b^T P^\perp_\Phi D\Phi\Phi^+ b$$

or

$$\frac{1}{2}\nabla r_2(\alpha) = -r^T D\Phi\, a \, , \tag{4.6}$$

where a is defined as in (4.4).

Separable problems with separable nonlinear constraints

The technique of Section 4.1 can be extended to problems of the form (4.3) subject to the constraints

$$H(\alpha)a = g(\alpha) \, , \tag{4.7}$$

where H is a $p \times n$ matrix function, g is a p-vector function, and all their elements are differentiable functions of α. Furthermore, it is assumed that no row of H vanishes. For fixed α, the set of all solutions of the linear equations (4.7) is given by:

(4.8) $$\underset{\sim}{a} = H(\underset{\sim}{\alpha})^{+} \underset{\sim}{g}(\underset{\sim}{\alpha}) + {}_{H}P^{\perp}\underset{\sim}{z}, \quad z \in \mathcal{R}^{n} .$$

Thus minimizing $r(\underset{\sim}{a}, \underset{\sim}{\alpha})$ subject to (4.7) is equivalent to minimizing the unconstrained functional:

(4.9) $$\min_{\underset{\sim}{z}, \underset{\sim}{\alpha}} s(\underset{\sim}{z}, \underset{\sim}{\alpha}) = \min \| \underset{\sim}{b} - \Phi(\underset{\sim}{\alpha})(H^{+}(\underset{\sim}{\alpha})\underset{\sim}{g}(\underset{\sim}{\alpha}) + {}_{H}P^{\perp}\underset{\sim}{z})\|^{2} .$$

By putting $\underset{\sim}{c}(\underset{\sim}{\alpha}) \equiv \underset{\sim}{b} - \Phi(\underset{\sim}{\alpha})H^{+}(\underset{\sim}{\alpha})\underset{\sim}{g}(\underset{\sim}{\alpha})$, and $G(\underset{\sim}{\alpha}) = \Phi(\underset{\sim}{\alpha})_{H}P^{\perp}$, we can write:

(4.10) $$s(\underset{\sim}{z}, \underset{\sim}{\alpha}) = \|\underset{\sim}{c}(\underset{\sim}{\alpha}) - G(\underset{\sim}{\alpha})\underset{\sim}{z}\|^{2}$$

which has a striking similarity with the unconstrained functional of problem (4.3). The only difference is in fact that the "vector of observations" $\underset{\sim}{c}$ depends upon the nonlinear parameters $\underset{\sim}{\alpha}$. However that is no obstacle to employing the technique of separation of variables, which, by setting $\hat{\underset{\sim}{z}} = G^{+}(\underset{\sim}{\alpha})\underset{\sim}{c}(\underset{\sim}{\alpha})$, leads to the reduced functional

(4.11) $$s_{2}(\underset{\sim}{\alpha}) = \| P^{\perp}_{G(\underset{\sim}{\alpha})} \underset{\sim}{c}(\underset{\sim}{\alpha})\|^{2} .$$

If $\hat{\underset{\sim}{\alpha}}$ is a minimum of the functional $s_{2}(\underset{\sim}{\alpha})$ then

(4.12) $$\hat{\underset{\sim}{a}} = H(\hat{\underset{\sim}{\alpha}})^{+} \underset{\sim}{g}(\hat{\underset{\sim}{\alpha}}) + {}_{H(\hat{\underset{\sim}{\alpha}})}P^{\perp}G^{+}(\hat{\underset{\sim}{\alpha}})(\underset{\sim}{b} - \Phi(\hat{\underset{\sim}{\alpha}})H^{+}(\hat{\underset{\sim}{\alpha}})\underset{\sim}{g}(\hat{\underset{\sim}{\alpha}}))$$

is such that the pair $(\hat{\underset{\sim}{a}}, \hat{\underset{\sim}{\alpha}})$ is optimal for the original constrained problem (4.3), (4.7).

Naturally, minimizing the functional $s_{2}(\underset{\sim}{\alpha})$ and obtaining $\hat{\underset{\sim}{a}}$ is now a more complicated task than in the unconstrained case. However L. Kaufman [10] has shown that if an appropriate algorithm is used, the work for computing $DP^{\perp}_{G(\underset{\sim}{\alpha})} \underset{\sim}{c}(\underset{\sim}{\alpha})$ is only about twice that necessary for differentiating the simple unconstrained functional. Of course, in analytically differentiating $P^{\perp}_{G(\alpha)} c(\alpha)$ heavy use is made of the formulas of Section 3. L. Kaufman has also devised an algorithm for the case of linear inequality constraints on the nonlinear variables, where (4.7) is replaced by

$$H^T \alpha \geq d \quad . \tag{4.13}$$

She proves a theorem similar to the one in Golub and Pereyra [6, Th. 2.1], establishing the validity of the variable projections approach. Thus, problem (4.3) subject to (4.13) is equivalent to:

$$\min r_2(\alpha) \quad \text{subject to} \quad H^T \alpha \geq d \tag{4.14}$$

and $\hat{a} = \Phi^+(\hat{\alpha})b$.

<u>Separable problems in a variable metric</u>

In some statistical problems, it is desirable to determine

$$\min_{a, \alpha, \rho} t(a, \alpha, \rho) \equiv \min_{a, \alpha, \rho} (b - \Phi(\alpha)a)^T \Sigma(\rho)^{-1} (b - \Phi(\alpha)a) \tag{4.15}$$

Here it is assumed that $\Sigma(\rho)$ is a symmetric positive definite matrix for $\rho \in P$. We denote an optimal set of parameters as (a, α, ρ) . Since $\Sigma(\rho)$ is a positive symmetric matrix, it is possible to determine a lower triangular matrix $F(\rho)$ such that

$$F(\rho)\Sigma(\rho) F^T(\rho) = I \quad . \tag{4.16}$$

The matrix $F(\rho)$ is the Cholesky factor of $\Sigma(\rho)$ and can be easily computed directly from $\Sigma(\underset{\sim}{\rho})$. Substituting (4.16) into (4.15), we see that finding $\min_{\underset{\sim}{a}, \underset{\sim}{\alpha}, \underset{\sim}{\rho}} t(\underset{\sim}{a}, \underset{\sim}{\alpha}, \underset{\sim}{\rho})$ is equivalent to determining

$$\min_{\underset{\sim}{a}, \underset{\sim}{\alpha}, \underset{\sim}{\rho}} \| F(\underset{\sim}{\rho})\underset{\sim}{b} - F(\underset{\sim}{\rho})\Phi(\underset{\sim}{\alpha})\underset{\sim}{a} \|^2 .$$

which is precisely of the same form as (4.10).

In order to use the separation of variables technique we need to determine

$$D\, P^{\perp}_{K(\underset{\sim}{\rho}, \underset{\sim}{\alpha})} \tag{4.17}$$

where $K(\underset{\sim}{\rho}, \underset{\sim}{\alpha}) \equiv F(\underset{\sim}{\rho})\Phi(\underset{\sim}{\alpha})$, which requires the calculation of $D\,F(\underset{\sim}{\rho})$. Using the rules of Fréchet differentiation and (4.16), we see that (omitting the argument)

$$F\,\Sigma\, DF^T + DF\,\Sigma\, F^T = -FD\,\Sigma\, F^T$$

or

$$F^{-T}\, DF^T + DF\, F^{-1} = -FD\,\Sigma\, F^T . \tag{4.18}$$

Since F is a lower triangular matrix, F^{-1} and DF are lower triangular. Thus, it is a simple matter to solve the Riccatti equation, (4.18).

Historical account of developments and applications

It is well to observe that a large number of nonlinear least squares problems are separable. Among the most usual applications we have: exponential fitting, or more generally, least squares fitting by γ-polynomials [20], with the model:

$$\mu(\underset{\sim}{a}, \underset{\sim}{\alpha}; t) = \sum_{j=1}^{n} a_j\, \varphi(\alpha_j; t) , \tag{4.19}$$

where we have only one function $\varphi(\alpha;t)$, of two variables. In the exponential fitting case we have $\varphi(\alpha;t) = e^{\alpha t}$. An important application is the fitting by splines with variable knots [3, 9]. Mathon [13] has considered γ-polynomials with more than one parameter $\sum_{j=1}^{n} a_j \varphi(\underset{\sim}{\alpha};\underset{\sim}{t})$. He was interested in the solution of boundary value problems for elliptic second order equations in m-dimensions, and the $\varphi(\underset{\sim}{\alpha};\underset{\sim}{t})$ were normalized fundamental solutions with singularities (at $\underset{\sim}{\alpha}$) located outside of the region in which a solution was sought.

Model (4.2) is considerably more general and can be used in other types of applications. In [6], we have exemplified the use of variable projections in the least squares fitting of: (a) linear combinations of Gaussians with an exponent background, as the type of model appearing when data is collected by a multi-channel analyzer [14]; (b) Iron Mössbauer spectrums with two sites of different electric field gradient and one single line. This last application is interesting, because a fairly general program can be devised for routinely handling cases with p doublets and q single lines, which usually are very troublesome with standard procedures.

The idea of separating variables in nonlinear least squares problems has been around for a while, and surely many practitioners have used it in one way or another. In the printed literature, we have been able to find some references to these early users of the technique [1, 11, 12, 16, 23]. All these papers arrive more or less at the reduced functional (4.5) but then beg out when differentiating time arrives. It seems to have been Scolnik [21] who was the first to pursue the differentiation of $P^{\perp}_{\Phi(\underset{\sim}{\alpha})}\underset{\sim}{b}$ in the simple γ-polynomial case $\varphi(\alpha;t) = t^{\alpha}$. This was achieved in a direct painstaking fashion, which was generalized to a one-dimensional version of (4.2) in [7], and further extended to the case (4.2) in [19]. Although formulas for the differentiation of the pseudoinverse were developed in [19], the authors did not use them for obtaining $DP^{\perp}_{\Phi(\underset{\sim}{\alpha})}\underset{\sim}{b}$, as was done in [6].

In [7], Guttman, Pereyra and Scolnik first proved that the variable projection approach solved the original minimization problem; in [6] that result was extended to the complete case (4.1), and finally, L. Kaufman [10] has extended the method to the case of inequality constraints. See also Jupp [9].

REFERENCES

[1] R. H. Barham and W. Drane, "An algorithm for least squares estimation of nonlinear parameters when some of the parameters are linear," Techn. 14 (1972), 757-766.

[2] P. G. Bergmann, R. Penfield, R. Schiller and H. Zatzkis, "The Hamiltonian of the general theory of relativity with electromagnetic field," Phys. Rev. 80 (1950), 81-88.

[3] Carl de Boor and J. R. Rice, "Least square cubic spline approximation II: variable knots," Techn. Rep. CSD TR 21, Purdue Univ. (1968).

[4] K. M. Brown, "Computer oriented methods for fitting tabular data in the linear and nonlinear least squares sense," manuscript (1972).

[5] R. Fletcher and S. A. Lill, "A class of methods for nonlinear programming II: Computational experience," Nonlinear Programming (edit. by J. B. Rosen, O. L. Mangasarian and K. Ritter), Academic Press, New York (1970), 67-92.

[6] G. H. Golub and V. Pereyra, "The differentiation of pseudoinverses and nonlinear least squares problems whose variables separate," SIAM J. Numer. Anal. 10 (1973), 413-432.

[7] I. Guttman, V. Pereyra and H. D. Scolnik, "Least squares estimation for a class of nonlinear models," Techn. 15 (1973), 209-218.

[8] John Z. Hearon and J. W. Evans, "Differentiable generalized inverses," J. Res. N. B. S. 72B (1968), 109-113.

[9] D. Jupp, "Nonlinear least square spline approximation," Techn. Rep., The Flinders Univ. of South Australia (1971).

[10] L. Kaufman, Manuscript (1973).

[11] W. H. Lawton and E. A. Sylvestre, "Elimination of linear parameters in nonlinear regression," Techn. 13 (1971), 461-478.

[12] W. H. Lawton, E. A. Sylvestre and M. S. Maggio, "Self modeling nonlinear regression," Techn. 14 (1972), 513-532.

[13] R. Mathon, "On the approximation of elliptic boundary value problems by fundamental solutions," Techn. Rep. No. 49, Computer Science Dept., Univ. of Toronto (1973).

[14] R. H. Moore and R. K. Zeigler, "The solution of general least squares problem with special reference to high-speed computers," Techn. Rep. LA-2367, Univ. of California, Los Alamos (1959).

[15] M. Z. Nashed, "Generalized inverses, normal solvability, and iteration for singular operator equations," Nonlinear Functional Analysis and Applications (ed. L. B. Rall), Academic Press, New York (1971).

[16] D. L. Nelson and T. O. Lewis, "A method for the solution of nonlinear least squares problem when some of the parameters are linear," The Texas J. Sc. 21 (1970), 480.

[17] M. Pavel-Parvu and A. Korganoff, "Iteration functions for solving polynomial equations," Constructive Aspects of the Fundamental Theorem of Algebra (ed. B. Dejon and P. Henrici), Wiley, New York (1969), 225-280.

[18] R. Penrose, "A generalized inverse for matrices," Proc. Cambridge Philos. Soc. 51 (1955), 406-413.

[19] A Pérez and H. D. Scolnik, "Derivatives of pseudoinverses and constrained nonlinear regression problems," Manuscript (1972).

[20] J. R. Rice, "The Approximation of Functions," Vol. 2, Multilineal and Multivariate Theory, Addison-Wesley, Reading, Mass. (1969).

[21] H. D. Scolnik, "On the solution of nonlinear least squares problems," Proc. IFIP-71, Numerical Math. (1971), 18-23.

[22] Yu. Ya. Tseng, "Sur les solutions des équations opératrices fonctionelles entre les espaces unitaires. Solutions extrémales. Solutions virtuelles," C. R. Acad. Sci. Paris 228 (1949), 640-641.

[23] D. Walling, "Nonlinear least squares curve fitting when some parameters are linear," The Texas J. Sci., 20 (1968), 119-124.

[24] P. A. Wedin, "On pseudoinverses of perturbed matrices," Techn. Rep. Computer Science Dept., Lund Univ., Sweden (1969).

G. H. Golub, Computer Science Department, Stanford University. The work of this author was supported by the NSF and AEC .

V. Pereyra, Departamento de Computacion, Fac. Ciencias, Univ. Central de Venezuela.

Perturbations and Approximations for Generalized Inverses and Linear Operator Equations

M. Zuhair Nashed

"All exact science is dominated by the idea of approximation."

Bertrand Russell

"So how do you go about teaching them something new? By mixing what they know with what they don't. Then, when they see in their fog something they recognize they think, "Ah, I know that!" And then it is just one more step to "Ah, I know the whole thing." And their mind thrusts forward into the unknown and they begin to recognize what they didn't know before, and they increase their power of understanding."

Picasso (From F. Gilot and C. Lake, Life with Picasso, McGraw-Hill Book Co., New York, 1964).

CONTENTS

1. Introduction

This paper addresses itself to problems of perturbations and approximations of generalized inverses of linear operators. The approximation methods include projectional and iterative methods; and collectively-compact operator approximations. The Table of Contents gives a general idea of the topics treated.

The papers in this volume by Golub and Pereyra, and Noble, address questions of approximations and numerical analysis of generalized inverse matrices and least-squares problems. The paper by Beutler and Root deals with some aspects of approximations of generalized inverses of closed linear operators in Hilbert space. We have omitted several topics (which could have been included here) and considered others only briefly in the context of linear operators to minimize overlap with these papers. On the other hand, we do not restrict ourselves to the Moore-Penrose inverse and the geometry of Hilbert spaces. Our setting includes generalized inverses in Banach spaces.

The existence of an inverse of a matrix is not tied with "structural" properties of the space R^n or with the norm used. A choice of a norm for error analysis, for example, can be made in this case a posteriori. In contrast, generalized inverses in analysis (see also Sections 2 and 3 of the paper by Nashed and Votruba in this volume) are intimately connected with properties of the spaces involved. The Moore-Penrose inverse, for example, is connected with inner product norms and orthogonal projectors. In some settings these are not used, and it is desirable to obtain error bounds in other norms. In such cases one should not use the Moore-Penrose inverse, but rather a generalized inverse in an appropriate (Banach space) geometry relative to some suitable decompositions of the spaces. Generalized inverses in analysis involve the geometry of the (Banach) spaces and the existence of certain topological complements, whereas these issues do not arise for the inverse. Thus attempts to extend verbatim error and perturbation results for operator or matrix inverse approximations to generalized inverse settings (without taking these factors into consideration) do not lead to fruitful results. One cannot simply mimic the theory of inverse operator approximations; the questions that should be asked are not always the same. We believe in an ancient dictum that to obtain a satisfactory solution to a problem, one should first formulate the proper question.

For example, given a perturbation $A+\Delta$ of a given linear operator A whose generalized inverse $A^{\dagger}$ (relative to some suitable projectors) can be computed with confidence, what should we consider (and compute) as a reasonable approximation to $(A+\Delta)^{\dagger}$? In some cases $(A+\Delta)^{\dagger}$ itself may work; but in general we should seek a variant of the generalized inverse of $A+\Delta$. Because of the lack of continuity of the map $A \to A^{\dagger}$, a markedly different approach is called for.

The approximation theory of generalized inverses of linear operators has many subtle points involving several modes of convergence, analytic and computational tractability, and techniques which are not merely extensions of those used in the matrix case.

Often the study of approximations for a given mathematical object leads to a deeper understanding of the properties of that object, as suggested by Bertrand Russell. For example this holds for integral equations of the first kind. Resolution of the difficulties arising in the approximations leads to sharper insight into the properties of the exact object.

This paper is for the most part an expository survey of the topics treated. (Some of the results and observations in Sections 3 and 5 are new.) Following the contention of Picasso (there are some who believe that mathematics is also an art), and in the hope of providing a proper perspective, we proceed in our exposition from known approximation and perturbation results when the operator is one-to-one and onto, to the general case when it is neither. Similarities and contrasted difficulties are highlighted.

2. Generalized Inverses of Linear Operators in Banach and Hilbert Spaces

We provide in this section a synopsis of results on generalized inverses which will be essential to the main results in the remaining sections; the generality is confined to the context that will be needed. For details, see [40], [49].

A projector P on a Banach space $\mathcal{X}$ is a continuous linear idempotent (i.e., $P^2 = P$) operator. A subspace $\mathcal{M}$ of $\mathcal{X}$ is said to have a topological complement if and only if there exists a subspace $\mathcal{N}$ such that $\mathcal{X} = \mathcal{M} \oplus \mathcal{N}$ (topological direct sum). In this case $\mathcal{M}$ and $\mathcal{N}$ are closed; however a closed subspace need not necessarily have a topological complement. A projector P on $\mathcal{X}$ induces a decomposition of $\mathcal{X}$ into two topological complements $P\mathcal{X}$ and $(I-P)\mathcal{X}$. Conversely, a subspace $\mathcal{M}$ of $\mathcal{X}$ has a topological complement if and only if there exists a projector P of $\mathcal{X}$ onto $\mathcal{M}$.

Let $\mathcal{X}$ and $\mathcal{Y}$ be Banach spaces and let $A : \mathcal{X} \to \mathcal{Y}$ be a linear operator with null space $\mathcal{N}(A)$ and range $\mathcal{R}(A)$, and let $\overline{\mathcal{R}(A)}$ denote the closure of $\mathcal{R}(A)$. Assume that $\mathcal{N}(A)$ has a topological complement $\mathcal{M}$ and that $\overline{\mathcal{R}(A)}$ has a topological complement $\mathcal{S}$, i.e. there exist projectors U, E with $U\mathcal{X} = \mathcal{N}(A)$ and $E\mathcal{Y} = \overline{\mathcal{R}(A)}$ so that

$$\mathcal{X} = \mathcal{N} \oplus \mathcal{M}, \quad \mathcal{Y} = \overline{\mathcal{R}(A)} \oplus \mathcal{S}. \tag{2.1}$$

Relative to these projectors, A has a unique generalized inverse $A^\dagger$ with domain $\mathcal{D}(A^\dagger) = \mathcal{R}(A) + \mathcal{S}$, $\mathcal{R}(A^\dagger) = \mathcal{M}$, and $\mathcal{N}(A^\dagger) = \mathcal{S}$ such that

$$(2.2)\quad \begin{cases} \text{(a)}\ A^{\dagger}A = I-U, & \text{(b)}\ A^{\dagger}AA^{\dagger} = A^{\dagger}\ \text{on}\ \mathcal{D}(A^{\dagger}), \\ \text{(c)}\ AA^{\dagger} = E\ \text{on}\ \mathcal{D}(A^{\dagger}), & \text{(d)}\ AA^{\dagger}A = A. \\ \text{and (hence)} & \end{cases}$$

Clearly $A^{\dagger}$ depends on the projectors E, U (or equivalently on the choice of topological complements to $\mathcal{N}(A)$ and $\overline{\mathcal{R}(A)}$); thus we shall use the notation $A^{\dagger}_{U,E}$. Suppose now that new projectors U', E' are chosen, where $\mathcal{R}(U') = \mathcal{N}(A)$ and $\mathcal{R}(E') = \overline{\mathcal{R}(A)}$. Then the restrictions of $A^{\dagger}_{U',E'}$, $A^{\dagger}_{U,E}$ to $\mathcal{R}(A)$ are related as follows [48 ; Theorem 1.3(a)], [49; Theorem 1.20]

$$(2.3)\qquad A^{\dagger}_{U',E'} = (2I - A^{\dagger}_{U,E}A + U')A^{\dagger}_{U,E}(I - AA^{\dagger}_{U,E} + E') .$$

We now consider the important case when $\mathcal{X}$ and $\mathcal{Y}$ are Hilbert spaces, and A is a bounded or a densely defined closed linear operator. In either case, $\mathcal{N}(A)$ is closed. Thus $\mathcal{N}(A)$ and $\overline{\mathcal{R}(A)}$ have topological complements. In particular we may choose $\mathcal{M} = \mathcal{N}^{\perp}$ and $\mathcal{S} = \mathcal{R}(A)^{\perp}$ in (2.1). The generalized inverse corresponding to this choice of $\mathcal{M}$ and $\mathcal{S}$ (equivalently the choice of orthogonal projectors) is the operator version of the Moore-Penrose inverse. It possesses (and is equivalently characterized by) the following important "best approximation" property. For any $y \in \mathcal{D}(A^{\dagger}) = \mathcal{R}(A) \oplus \mathcal{R}(A)^{\perp}$, the vector $A^{\dagger}y$ minimizes $\|Ax-y\|$ over $\mathcal{D}(A)$ and $\|A^{\dagger}y\|$ is smaller than the norm of any other vector which minimizes $\|Ax-y\|$.

For a given $y \in \mathcal{Y}$, let $\mathcal{S}_y$ denote the set of all least squares solutions of the equation

$$(2.4)\qquad Ax = y.$$

That is,

(2.5) $$\mathcal{S}_y = \{u : \inf_{x \in \mathcal{X}} \|Ax-y\| = \|Au-y\|\} .$$

It is easy to show that the following characterizations of $\mathcal{S}_y$ hold (see [40; Sect. 2]):

(2.6) $$\mathcal{S}_y = \{u : A^*(Au - y) = 0\}$$

(2.7) $$\mathcal{S}_y = \{u : Au = Qy\} ,$$

where A^* denotes the adjoint of A, and Q is the orthogonal projector of $\mathcal{Y}$ onto $\overline{\mathcal{R}(A)}$.

For a given $y \in \mathcal{Y}$, $\mathcal{S}_y$ is nonempty if and only if it satisfies any (and hence all) of the following conditions: $Qy \in \mathcal{R}(A)$, $A^*y \in \mathcal{R}(A^*A)$, $y \in \mathcal{R}(A) + \mathcal{R}(A)^{\perp}$. That is $\mathcal{S}_y$ is nonempty if and only if $y \in \mathcal{D}(A^{\dagger})$. For any $y \in \mathcal{D}(A^{\dagger})$, $\mathcal{S}_y$ is a closed convex set and $\mathcal{S}_y = A^{\dagger}y + \mathcal{N}(A)$. Define the closest-point map $P_{\mathcal{S}} : \mathcal{X} \to \mathcal{S}_y$ by $P_{\mathcal{S}}x = v$, where

$$\|x - v\| = \inf\{ \|x - u\| : u \in \mathcal{S}_y\} .$$

Clearly,

$$A^*A(x - P_{\mathcal{S}}x) = A^*(Ax - y), \quad x \in \mathcal{X} ,$$

$$x - P_{\mathcal{S}}x \in \mathcal{N}(A)^{\perp} ,$$

and

$$P_{\mathcal{S}}x = A^{\dagger}y + P_{\mathcal{N}}x$$

where $P_{\mathcal{N}}$ is the orthogonal projector of $\mathcal{X}$ onto $\mathcal{N}(A)$.

In the case of Hilbert spaces, one may also use generalized inverses relative to non-orthogonal projectors. For various properties and characterizations of $A^\dagger$, see Sections 4, 5.2, and 5.3 in the paper by Nashed and Votruba in this volume.

We remark that in the case of matrices, every algebraic generalized inverse (i.e., 1, 2-inverse) B:

$$ABA = A \quad \text{and} \quad BAB = B \quad , \tag{2.8}$$

is a Moore-Penrose inverse relative to some new inner products. That is, if we consider $A: \mathbb{R}^n \to \mathbb{R}^m$, then there exist inner products $[\cdot\,,\cdot\,]$ and $\langle\cdot\,,\cdot\rangle$ on $\mathbb{R}^n$ and $\mathbb{R}^m$, respectively, such that

$$\langle ABy_1, y_2\rangle = \langle y_1, ABy_2\rangle, \quad y_1, y_2 \in \mathbb{R}^m$$

and

$$[BAx_1, x_2] = [x_1, BAx_2], \quad x_1, x_2 \in \mathbb{R}^n \quad .$$

This follows from properties of the weighted and oblique generalized inverses and a known characterization of all inner products on finite dimensional spaces. This result is a useful way of viewing a generalized inverse satisfying (2.8) which seems to have been unnoted before.

We conclude this section with additional notation: $\mathfrak{L}(\mathcal{X},\mathcal{Y})$ or $[\mathcal{X},\mathcal{Y}]$ is the space of all bounded linear operators on $\mathcal{X}$ into $\mathcal{Y}$; $[\mathcal{X},\mathcal{X}]$ is abreviated as $[\mathcal{X}]$. The norm on $\mathfrak{L}[\mathcal{X},\mathcal{Y}]$ is as usual: $\|A\| = \sup\{\|Ax\| : x \in \mathcal{X}, \|x\| = 1\}$. The identity operator on a particular space is denoted by I, and the adjoint of A is denoted by A^* .

The results and techniques of this paper can also be adapted to random linear operator equations (see [47], [46]). The necessary technicalities will be presented elsewhere.

3. Perturbations and Continuity of Generalized Inverses of Linear Operators

3.1. Inverse operator approximations

Before we consider perturbations and continuity of generalized inverses of linear operators on Banach space, we review briefly, as for example in [2], the elementary theory of approximations of inverses based on operator norm convergence. This will also provide motivation and comparison for what follows, and show the limitations of the theory.

Proposition 3.1 (Banach's lemma). Let $L \in [\mathcal{X}]$. If $\|L\| < 1$, then $(I - L)^{-1}$ exists and is bounded,

$$(I - L)^{-1} = \sum_{m=0}^{\infty} L^m ,$$

where the (Neumann) series converges in operator norm, and

$$\|(I - L)^{-1}\| \leq \frac{1}{1 - \|L\|} .$$

Proposition 3.2. Let $T: \mathcal{X} \to \mathcal{Y}$, $S: \mathcal{Y} \to \mathcal{X}$ be bounded linear operators. If $ST = I - L$, with $\|L\| < 1$, then T^{-1} exists (on $T\mathcal{X}$), and T^{-1} is bounded.

$$T^{-1} = (I - L)^{-1} S, \quad T^{-1} - S = (I - L)^{-1} LS$$

$$\|T^{-1}\| \leqq \frac{\|S\|}{1 - \|L\|}, \quad \|T^{-1} - S\| \leqq \frac{\|L\| \|S\|}{1 - \|L\|} .$$

Thus, if T has a sufficiently good approximate left inverse S, then T^{-1} exists (on $T\mathcal{X}$) and is near S. Similarly, if $TS = I - M$, with $\|M\| < 1$, then $T\mathcal{X} = \mathcal{Y}$.

Proposition 3.3. Let $A, B \in \mathcal{L}(\mathcal{X},\mathcal{Y})$. Assume $A^{-1} \in \mathcal{L}(\mathcal{Y},\mathcal{X})$ exists and

$$\Delta := \|A^{-1}\| \, \|A - B\| < 1 .$$

Then there exists $B^{-1} \in \mathcal{L}(\mathcal{Y},\mathcal{X})$ and

$$\|B^{-1}\| \leqq \frac{\|A^{-1}\|}{1-\Delta} \tag{3.1}$$

$$\|B^{-1} - A^{-1}\| \leqq \frac{\Delta \|A^{-1}\|}{1-\Delta} . \tag{3.2}$$

Proposition 3.3 follows from Proposition 3.2 by letting $S = A^{-1}$. The traditional proof is based on the decompositions:

$$B^{-1} - A^{-1} = B^{-1}(A-B)A^{-1} ,$$

$$B = A[I - A^{-1}(A-B)] ,$$

and Proposition 3.1. For since $\|A^{-1}(A-B)\| < 1$ and A^{-1} exists, B^{-1} exists and

$$\|B^{-1}\| \leqq \|A^{-1}\| \, \|I - A^{-1}(A-B)\| \leqq \frac{\|A^{-1}\|}{1-\Delta} .$$

It is also convenient to write (3.2) in the form

$$\frac{\|(A+T)^{-1} - A^{-1}\|}{\|A^{-1}\|} \leqq \frac{\kappa(A)\,\|T\|/\|A\|}{1-\kappa(A)\,\|T\|/\|A\|} , \tag{3.3}$$

where $T = B - A$ and

$$\kappa(A) := \|A\| \, \|A^{-1}\| \tag{3.4}$$

is called the condition number of the operator A. It reflects how perturbations in the operator may be magnified in its inverse.

Proposition 3.3 implies that the operators $T \in \mathcal{L}(\mathcal{X},\mathcal{Y})$ with $T^{-1} \in \mathcal{L}(\mathcal{Y},\mathcal{X})$ form an open set in $\mathcal{L}(\mathcal{X},\mathcal{Y})$, and that the map $T \to T^{-1}$ is continuous. The set of all left (resp. right) invertible operators in $\mathcal{L}(\mathcal{X},\mathcal{Y})$ is an open set. This also follows as a corollary of some perturbation results for generalized inverses (see Section 3.3).

In Proposition 3.3 replace B and A by A_n and A in both orders to obtain the following criterion for the invertibility of A in terms of information on invertibility of the approximents A_n.

Proposition 3.4. Let $A, A_n \in \mathcal{L}(\mathcal{X},\mathcal{Y})$ and $\|A_n - A\| \to 0$. Then there exists $A^{-1} \in \mathcal{L}(\mathcal{Y},\mathcal{X})$ if and only if for some N and all $n \geq N$ there exist uniformly bounded $A_n^{-1} \in \mathcal{L}(\mathcal{Y},\mathcal{X})$, in which case $\|A_n^{-1} - A^{-1}\| \to 0$.

It is natural to seek extension of the preceding results on operator inverses to the framework of generalized inverses. Some of the above properties fail for generalized inverses (e.g. the map $A \to A^{\dagger}$ is discontinuous in $\mathcal{L}(\mathcal{X},\mathcal{Y})$). Also, some derivations of exactly analogous bounds for generalized inverses turn out to be either too restrictive or too complicated to be useful. Various perturbation results have been obtained, but the complete theory is not yet in a definitive form. This section is a brief excursion along a road with only a few landmarks as yet. In part of this section and in Section 4, we stress, however, a new point of view which circumvents some of the difficulties and leads to more appropriate questions that one should ask in connection with problems of perturbation, approximation, and error bounds for generalized inverses. We also provide some answers to these questions.

3.2. On the continuity of the map $A \to A^{\dagger}$

The Moore-Penrose inverse of a matrix is not necessarily a continuous function of the matrix. For example, let

$$A = \begin{pmatrix} 1 & 1 \\ 1 & 1 \end{pmatrix}, \quad A_\varepsilon = \begin{pmatrix} 1+\varepsilon & 1 \\ 1 & 1 \end{pmatrix}, \quad \varepsilon \neq 0 .$$

Then,

$$A^\dagger = \begin{pmatrix} \frac{1}{4} & \frac{1}{4} \\ \frac{1}{4} & \frac{1}{4} \end{pmatrix} \quad \text{and} \quad A_\varepsilon^\dagger = A^{-1} = \begin{pmatrix} \varepsilon^{-1} & -\varepsilon^{-1} \\ -\varepsilon^{-1} & 1+\varepsilon^{-1} \end{pmatrix}, \quad \varepsilon \neq 0 .$$

Thus it is not true that $\lim_{\varepsilon \to 0} A_\varepsilon^\dagger = A^\dagger$ although $\lim_{\varepsilon \to 0} A_\varepsilon = A$. Note that in this example $\operatorname{rank} A \neq \operatorname{rank} A_\varepsilon$, $\varepsilon \neq 0$. The example dramatizes the extent to which the generalized inverse could be distorted if the rank is changed; the generalized inverse of a slightly perturbed matrix might bear no relation to the generalized inverse of the original matrix. None of the entries of A_ε converge as $\varepsilon \to 0$.

The object of Stewart's paper [66] is to show that for a sequence of matrices T_n satisfying

$$\|A^\dagger\| \, \|T_n\| < 1 \tag{3.5}$$

and

$$\lim_{n \to \infty} T_n = 0 , \tag{3.6}$$

a necessary and sufficient condition that

$$\lim_{n \to \infty} (A + T_n)^\dagger = A^\dagger$$

is that for all sufficiently large n,

$$\operatorname{rank}(A + T_n) = \operatorname{rank} A .$$

In [66] this was done by exhibiting first explicit error bound, for $\|(A+T)^{\dagger} - A^{\dagger}\|$. However, because the columns (or rows) of A and $A+T$ may span different spaces, the form (and derivation) is considerably more complicated than the right-hand side of (3.3).

The above condition for continuity of the map $A \to A^{\dagger}$ can be obtained more easily via other routes and also follow as a corollary of the basic approximation setting of [37], which provides a sharper insight to generalized inverse approximations of perturbed matrices. The fact that $A^{\dagger}$ is a continuous function of A if the rank of A is kept fixed was observed first by Penrose [54, p. 408].

Theorem 3.5. Let A_n and A be $m \times n$ matrices over $\mathbb{C}$. If $A_n \to A$, then the following conditions are equivalent:

(a) $\operatorname{rank} A_n = \operatorname{rank} A$ for all sufficiently large n.

(b) $A_n^{\dagger} \to A^{\dagger}$.

Proof. We use the Frobenius norm:

$$\|A\|_F^2 = \operatorname{trace}(A^*A) = \sum_{i,j} |a_{ij}|^2 .$$

Using the fact $\mathcal{R}(A^{\dagger}) = \mathcal{N}(A)^{\perp} = \mathcal{R}(A^*)$ and well known properties of ranks of matrices, it follows that A, A^*A, $A^{\dagger}$ and $A^{\dagger}A$ all have equal rank. Also $A^{\dagger}A$, being idempotent, has rank equal to its trace.

(a) $\Longrightarrow$ (b): Since A^*A, $(A^*A)^2, \ldots$ cannot all be linearly independent, there exist scalars $c_1, c_2, \ldots, c_k$, not all zero, such that

$$(3.7) \qquad c_1 A^*A + c_2 (A^*A)^2 + \ldots + c_k (A^*A)^k = 0 .$$

We assume A is singular. The polynomial in (3.7) could then be taken to be the characteristic polynomial of A^*A. Let c_r be the first nonzero c in (3.7) and put

$$B := -c_r^{-1}\{c_{r+1}I + c_{r+2}A^*A + \dots + c_k(A^*A)^{k-r-1}\} .$$

Then it can be easily verified that $B(A^*A)^{r+1} = (A^*A)^r$ and

$$A^\dagger = BA^* . \tag{3.8}$$

Now

$$r = \dim \mathcal{R}(A^*A)^\perp = \dim \eta(A^*A) = \dim \eta(A) = n - \operatorname{rank} A .$$

Thus r is determined by the rank of A. Continuity of $A^\dagger$ under condition (a) then follows using (3.8) and the above properties of ranks.

It also follows that $A^\dagger$ varies discontinuously when the rank of A changes since $\operatorname{rank} A = \operatorname{trace} A^\dagger A$.

$$\text{(b)} \Longrightarrow \text{(a)}: \quad P_{\mathcal{R}(A_n)} = A_n A_n^\dagger \to AA^\dagger = P_{\mathcal{R}(A)} .$$

But,

$$\operatorname{rank} A_n = \operatorname{rank} A_n A_n^\dagger$$

$$= \operatorname{trace} P_{\mathcal{R}(A_n)} \to \operatorname{trace} P_{\mathcal{R}(A)} = \operatorname{rank} A . \qquad \blacksquare$$

Remark 3.6. The investigation of continuity properties of the map $A \to A^\dagger_{P,Q}$ in Banach spaces leads to more subtle problems. This investigation was first undertaken by Moore and Nashed [37], as a by-product of a general setting for approximations of generalized inverse operators. It should be noted that in normed spaces the generalized inverse of a perturbed operator need not even exist. The setting of [37] also leads to necessary and sufficient conditions for the continuity of the generalized inverse of a matrix relative to nonorthogonal projectors and other generalized inverses (e.g., the group inverse, Drazin's inverse, etc.). We summarize these results in Section 4.

Remark 3.7. Let A, B be $m \times n$ matrices and let A be of rank r. If rank $B >$ rank A, then

$$\|B^{\dagger}\|_2 \geqq \|B - A\|_2^{-1} .$$

(see e.g. Stewart [66 ; Theorem 5.2] and Wedin [72]). This bound also dramatizes the extent to which things can go wrong in generalized inverses if the rank is perturbed. For other bounds see Section 3.3, [51], [34] and [71].

Remark 3.8. The moral of Theorem 3.5 from the point of view of computation is evident: Never compute a generalized inverse of a matrix whose rank can change under small perturbation. This moral has dominated the literature and practice of generalized inverse computations. But there is also the second side of the coin which says that perhaps we are not asking the right question in this case. That is, if B approximates A, but rank $B \neq$ rank A, then $B^{\dagger}$ is not the matrix to seek as an approximation to $A^{\dagger}$. Instead we should develop a procedure to compute a variant B^{ϕ} (of $B^{\dagger}$) which is a good approximation to the true $A^{\dagger}$ and which has in an approximate sense the properties of a true generalized inverse. If we adopt this point of view, then several questions arise:

(i) What is such a variant B^{ϕ} ?

(ii) Can this variant be interpreted as the true generalized inverse of a variant of B, in case rank $B \neq$ rank A ?

(iii) Does B^{ϕ} coincide with $B^{\dagger}$ in case rank $A =$ rank B ?

This point of view is extremely important in the case of

infinite dimensional spaces, where we no longer have the possibility of equality of rank. We address these questions in Section 4.

3.3. A perturbation theory and error bounds under some restrictions

A perturbation theory and bounds for $B^\dagger - A^\dagger$ can be obtained under some restrictions as corollaries of the general setting of [37] (see also Section 4 below). For the sake of simplicity we prove a special case directly.

Theorem 3.9. Let $A \in \mathcal{L}(\mathcal{X},\mathcal{Y})$ and suppose $\mathcal{X} = \mathcal{N}(A) \oplus \mathcal{M}$ and $\mathcal{Y} = \mathcal{R}(A) \oplus \mathcal{S}$. Let $A^\dagger_{\mathcal{M},\mathcal{S}}$ denote the generalized inverse of A relative to these decompositions. Let $T \in \mathcal{L}(\mathcal{X},\mathcal{Y})$ and $B := A + T$. Suppose that

$$\|TA^\dagger_{\mathcal{M},\mathcal{N}}\| < 1 \tag{3.9}$$

and

$$(I + TA^\dagger)B \text{ maps } \mathcal{N}(A) \text{ into } \mathcal{R}(A)\ . \tag{3.10}$$

Then $B^\dagger := B^\dagger_{\mathcal{R}(A^\dagger),\mathcal{N}(A^\dagger)}$ exists and

$$B^\dagger_{\mathcal{R}(A^\dagger),\mathcal{N}(A^\dagger)} = A^\dagger(I+TA^\dagger)^{-1} = (I+A^\dagger T)^{-1}A^\dagger\ . \tag{3.11}$$

Moreover,

$$\|B^\dagger\| \le \frac{\|A^\dagger\|}{1 - \|TA^\dagger\|} \tag{3.12}$$

and

(3.13) $$\|B^{\dagger} - A^{\dagger}\| \leq \frac{\|A^{\dagger}\| \, \|TA^{\dagger}\|}{1 - \|TA^{\dagger}\|} .$$

If in addition, $\|T\| < \|A^{\dagger}\|^{-1}$, then

(3.14) $$\frac{\|(A+T)^{\dagger} - A^{\dagger}\|}{\|A^{\dagger}\|} \leq \frac{\kappa(A) \, \|T\| / \|A\|}{1 - \kappa(A) \|T\| / \|A\|}$$

where

$$\kappa(A) = \|A\| \, \|A^{\dagger}\|$$

is the (pseudo) condition number of A.

Proof. In view of (3.9) and Banach's lemma (Proposition 3.1), $(I + TA^{\dagger})^{-1}$ exists. Using the Neumann series it also follows that

$$A^{\dagger}(I + TA^{\dagger})^{-1} = (I + A^{\dagger}T)^{-1} A^{\dagger} .$$

Let $C := A^{\dagger}(I + TA^{\dagger})^{-1}$. Then

$$\begin{aligned} B - BCB &= \{I - (A+T)A^{\dagger} (I + TA^{\dagger})^{-1}\} (A + T) \\ &= \{I + TA^{\dagger} - (A + T)A^{\dagger}\} (I + TA^{\dagger})^{-1} (A + T) \\ &= (I - AA^{\dagger})(I + TA^{\dagger})^{-1} (A + T) \\ &= (I - AA^{\dagger})(A + TA^{\dagger})^{-1} \{(I + TA^{\dagger})A + T(I - A^{\dagger}A)\} \\ &= (I - Q)(I + TA^{\dagger})^{-1} TP = 0, \text{ by assumption (3.10).} \end{aligned}$$

Similarly,

$$CBC - C = A^{\dagger}(I+TA^{\dagger})^{-1}\{(A+T)A^{\dagger}(I+TA^{\dagger})^{-1} - I\}$$

$$= A^{\dagger}(I+TA^{\dagger})^{-1}(AA^{\dagger} - I)(I+TA^{\dagger})^{-1} = 0$$

since

$$\mathcal{N}(C) = \mathcal{N}[A^{\dagger}(I+TA^{\dagger})^{-1}] = \mathcal{N}[(I+A^{\dagger}T)^{-1}A^{\dagger}] = \mathcal{N}(A^{\dagger}) .$$

Note also that

$$\mathcal{R}(C) = \mathcal{R}[A^{\dagger}(I+TA^{\dagger})^{-1}] = \mathcal{R}(A^{\dagger}) .$$

Thus

$$\mathcal{X} = \mathcal{N}(A) \oplus \mathcal{M} = \mathcal{N}(A) \oplus \mathcal{R}(A^{\dagger}) = \mathcal{N}(A) \oplus \mathcal{R}(B^{\dagger}) ,$$

$$\mathcal{Y} = \mathcal{R}(A) \oplus \mathcal{S} = \mathcal{R}(A) \oplus \mathcal{N}(A^{\dagger}) = \mathcal{R}(A) \oplus \mathcal{N}(B^{\dagger}) ;$$

also

$$\mathcal{X} = \mathcal{N}(B) \oplus \mathcal{R}(A^{\dagger}) \quad \text{and} \quad \mathcal{Y} = \mathcal{R}(B) \oplus \mathcal{N}(A^{\dagger})$$

(see [49, Corollary 5.3(b)] if necessary). This proves (3.11).

The bounds can be easily obtained using the Neumann series:

$$\|B^{\dagger}\| = \|A^{\dagger}(I+TA^{\dagger})^{-1}\| \leq \frac{\|A^{\dagger}\|}{1 - \|TA^{\dagger}\|} ,$$

and

$$\|B^{\dagger} - A^{\dagger}\| = \|A^{\dagger}[(I+TA^{\dagger})^{-1} - I]\| \leq \frac{\|A^{\dagger}\| \, \|TA^{\dagger}\|}{1 - \|TA^{\dagger}\|} . \quad \blacksquare$$

Corollaries and remarks. (i) Theorem 3.9 shows that the generalized inverse $A^{\dagger}$ is a continuous function of A

in the class of mapping satisfying the hypotheses of the theorem.

(ii) Suppose (3.9) holds. Then (3.10) is satisfied in particular if either $\mathcal{N}(B) \supseteq \mathcal{N}(A)$ (equivalently $\mathcal{N}(T) \supseteq \mathcal{N}(A)$) or $\mathcal{R}(B) \subseteq \mathcal{R}(A)$ (equivalently $\mathcal{R}(T) \subseteq \mathcal{R}(A)$). For if $\mathcal{N}(B) \supseteq \mathcal{N}(A)$, then $(I + TA^{\dagger})^{-1}B$ maps $\mathcal{N}(A)$ into $\{0\}$. Similarly if $\mathcal{R}(T) \subseteq \mathcal{R}(A)$, then on $\mathcal{N}(A)$, $(I + TA^{\dagger})^{-1}B = (I + TA^{\dagger})^{-1}T = T(I + A^{\dagger}T)^{-1}$ and its range is contained in $\mathcal{R}(T)$, and hence in $\mathcal{R}(A)$.

(iii) Theorem 3.9 was established for the Moore-Penrose inverse of a matrix in [5], [66], [72]. In [5] and [66], the assumptions

$$P_{\mathcal{R}(A)}T = T \quad \text{and} \quad TP_{\mathcal{R}(A^*)} = T$$

(equivalently $P_{\mathcal{R}(A)}B = B$ and $BP_{\mathcal{R}(A^*)} = B$) were used instead of (3.10). In view of (ii) above, we see that one of these two conditions is superflous. The conclusions of Theorem 3.9 holds for matrices if rank A = rank B and (3.9) is satisfied. See also Section 4.

(iv) Recall that $T \in \mathcal{L}(\mathcal{X},\mathcal{Y})$ is left invertible if there exists $S \in \mathcal{L}(\mathcal{Y},\mathcal{X})$ such that $ST = I_{\mathcal{X}}$, and is right invertible if there exists $S \in \mathcal{L}(\mathcal{X},\mathcal{Y})$ such that $TS = I_{\mathcal{Y}}$. Also $T \in \mathcal{L}(\mathcal{X},\mathcal{Y})$ is said to be relatively regular if T has a bounded inner inverse, i.e. if there exists $S \in \mathcal{L}(\mathcal{Y},\mathcal{X})$ such that $TST = T$. It then follows that the set of all left (right) invertible operators in $\mathcal{L}(\mathcal{X},\mathcal{Y})$ is the same as the set of all relatively regular operators for which $\mathcal{N}(T) = \{0\}$ (resp. $\mathcal{R}(T) = \mathcal{Y}$). In view of (ii) both of these sets are open in $\mathcal{L}(\mathcal{X},\mathcal{Y})$.

(v) We remark that the bounds (3.12), (3.13) and (3.14) for the generalized inverse are perfectly analogous to the corresponding bounds (3.1), (3.2) and (3.3) for the ordinary inverse. They are valid for the operator norm on $\mathcal{L}(\mathcal{X}, \mathcal{Y})$ and are not restricted to 2-norms. Furthermore, the derivation is almost as elementary as the derivation for inverses. These bounds can be used to study the effect of perturbations on best approximate solutions of operator equations as was done for matrices for example in [34], [66] and [51].

Various more explicit error and perturbation bounds can be derived using results on bounds for the product of two projectors and singular value decompositions. See for example Wedin [71], Lawson and Hanson [34; Ch. 8] and Stewart [66]. Instead of pursuing this line of perturbation theory, we provide in Section 4 a synoposis of a fruitful approach to approximations of generalized inverses of perturbed operators (or matrices), which represents a departure from other perturbation theories that try to mimic the corresponding theory for invertible operators.

It should be noted, moreover, that (3.12), (3.13) and (3.14) were derived under some stringent conditions, and that the standard approaches to perturbation theory of generalized inverses do not completely indicate the subtle features of the general case, or more precisely what to do if a condition such as (3.10) or similar conditions are not satisfied. The alternative approach of Section 4 considers this situation.

3.4. Decompositions for $B^{\dagger}_{P', Q'} - A^{\dagger}_{P, Q}$ and associated error bounds

Some results on perturbation theory for invertible linear operators and associated error bounds can be easily deduced, as noted in Section 3.1, from the identity $B^{-1} - A^{-1} = -B^{-1}TA^{-1}$, where A, B are invertible and $T = B - A$. Extensions have also been obtained for generalized inverses of $m \times n$ matrices (see, e.g., [72]). We show that similar generalizations hold in Banach spaces.

Let $\mathcal{X}$ and $\mathcal{Y}$ be Banach spaces and let $A, B \in \mathcal{L}(\mathcal{X}, \mathcal{Y})$. Assume that

(3.15) $$\mathcal{X} = \mathcal{N}(A) \oplus \mathcal{M}, \quad \mathcal{Y} = \mathcal{R}(A) \oplus \mathcal{S}$$

(3.16) $$\mathcal{X} = \mathcal{N}(B) \oplus \mathcal{M}', \quad \mathcal{Y} = \mathcal{R}(B) \oplus \mathcal{S}' .$$

Let

P be the projector of $\mathcal{X}$ onto $\mathcal{N}(A)$ along $\mathcal{M}$,

P' be the projector of $\mathcal{X}$ onto $\mathcal{N}(B)$ along $\mathcal{M}'$,

Q be the projector of $\mathcal{Y}$ onto $\mathcal{R}(A)$ along $\mathcal{S}$,

and

Q' be the projector of $\mathcal{Y}$ onto $\mathcal{R}(B)$ along $\mathcal{S}'$.

Let $A^{\dagger} := A^{\dagger}_{P,Q}$ and $B^{\dagger} := B^{\dagger}_{P',Q'}$.

Theorem 3.10. Under assumptions (3.15) and (3.16) we have:

(3.17) $$B^{\dagger} - A^{\dagger} = -B^{\dagger}(B-A)A^{\dagger} + B^{\dagger}(I-Q) - P' A^{\dagger}$$

(3.18) $$B^{\dagger} - A^{\dagger} = -A^{\dagger}(B-A)B^{\dagger} + PB^{\dagger} - A^{\dagger}(I - Q') .$$

In the case of Hilbert spaces we also have the following decompositions for the Moore-Penrose inverse:

(3.19) $$B^{\dagger} - A^{\dagger} = -B^{\dagger}(B-A)A^{\dagger} + B^{\dagger}B^{\dagger *}(B-A)^{*} (I - Q) - P'(B - A)^{*} A^{\dagger *} A^{\dagger} .$$

(3.20) $$B^{\dagger} - A^{\dagger} = -A^{\dagger}(B-A)B^{\dagger} + P(B-A)^{*} B^{*\dagger} - A^{\dagger} A^{*\dagger}(B-A)^{*} (I - Q') .$$

Proof. We use the identity

$$B^{\dagger} - A^{\dagger} = (P' + I - P')(B^{\dagger} - A^{\dagger})(Q + I - Q) .$$

Noting that

$$A^{\dagger} = A^{\dagger}Q, \quad B^{\dagger} = B^{\dagger}Q',$$

$$A^{\dagger}\mathfrak{s} = \{0\}, \quad P'B^{\dagger} = 0 ,$$

we obtain

$$B^{\dagger} - A^{\dagger} = (I-P')(B^{\dagger}-A^{\dagger})Q + (I-P')B^{\dagger}(I-Q) - P'A^{\dagger}Q .$$

Now since $(I - P')B^{\dagger} = B^{\dagger}$,

$$\begin{aligned}(I - P')(B^{\dagger} - A^{\dagger})Q &= (I - P')B^{\dagger}Q - (I - P')A^{\dagger}Q \\ &= B^{\dagger}Q - (I - P')A^{\dagger} \\ &= B^{\dagger}AA^{\dagger} - B^{\dagger}BA^{\dagger} = -B^{\dagger}(B - A)A^{\dagger} .\end{aligned}$$

Inserting this in the above expression for $B^{\dagger} - A^{\dagger}$ we obtain (3.17). Interchanging A and B in (3.17) we obtain (3.18).

To prove (3.19) we note that for the Moore-Penrose inverse the projectors are orthogonal (equivalently self-adjoint). Hence,

$$\begin{aligned}B^{\dagger}(I - Q) &= B^{\dagger}Q'(I - Q) = B^{\dagger}[(I-Q)Q']^{*} \\ &= B^{\dagger}[(I - Q)BB^{\dagger}]^{*} = B^{\dagger}[(I - Q)(B - A)B^{\dagger}]^{*} \\ &= B^{\dagger}B^{\dagger *}(B-A)^{*}(I - Q) .\end{aligned}$$

Similarly to prove (3.20) we show that

$$P'A^{\dagger} = [(I-P)P']^{*}A^{\dagger} = -[A^{\dagger}(B - A)P']^{*} = -P'(B-A)A^{\dagger *}A^{\dagger} . \blacksquare$$

We remark that (3.17) and (3.18) can also be verified very easily using the identity:

$$B^{\dagger} - A^{\dagger} = -B^{\dagger}(B - A)A^{\dagger} + B^{\dagger} - A^{\dagger} + B^{\dagger}(B - A)A^{\dagger} .$$

The preceding decomposition results for $B^{\dagger} - A^{\dagger}$ shed more light on continuity properties of $A^{\dagger}$. Note that when A and B are invertible, the second and third terms on the right-hand side of (3.17) vanish. However for noninvertible operators, these terms are not necessarily small if $T = B-A$ is small. For the example of Section 3.2 we have, letting $B = A_{\varepsilon}$,

$$T = B - A = \begin{bmatrix} \varepsilon & 0 \\ 0 & 0 \end{bmatrix},$$

while,

$$B^{\dagger}(I - Q) = B^{\dagger}(I - AA^{\dagger}) = \begin{bmatrix} \varepsilon^{-1} & -\varepsilon^{-1} \\ -\varepsilon^{-1} - \frac{1}{2} & \varepsilon^{-1} + \frac{1}{2} \end{bmatrix}.$$

In fact the trouble can be worse with smaller T, as the example shows. It should also be mentioned that not even $(I - P)(B^{\dagger} - A^{\dagger})Q$ is small.

An additional complexity in the perturbation theory of generalized inverses in Banach spaces arises from the fact that the generalized inverse of a perturbed operator need not necessarily exist. If $\mathcal{N}(A)$ is complemented in $\mathcal{X}$ and $\mathcal{R}(\mathcal{Y})$ is complemented in $\mathcal{Y}$, then $\mathcal{N}(A + T)$ and $\mathcal{R}(A + T)$ need not have complements in $\mathcal{X}$ and $\mathcal{Y}$ respectively. We have avoided this difficulty in Theorem 3.10 by imposing a priori that such complements exists. Another approach is to arrive at the existence of these complements a posteriori by imposing some other conditions. This was illustrated in Section 3.3.

Alternate decomposition results for $B^{\dagger} - A^{\dagger}$ may be derived under additional comparison conditions of the complementary subspaces using results on pairs of projectors from, e.g., [1] and [26].

3.5. Error bounds for generalized inverses under change of projectors

Theorem 3.10 does not provide useful identities for the case of generalized inverses of the same operator relative to different projectors, i.e. for $A^{\dagger}_{P',Q'} - A^{\dagger}_{P,Q}$. However this was considered in [49] and we have

$$A^{\dagger}_{P',Q'} = (I + P - P')\, A^{\dagger}_{P,Q}(I - Q + Q') \ ,$$

and

$$A^{\dagger}_{P',Q'} - A^{\dagger}_{P,Q} = (P - P')A^{\dagger}_{P,Q} + (P-P')A^{\dagger}_{P,Q}(Q'-Q) + A^{\dagger}_{P,Q}(Q' - Q) \ .$$

Thus

$$\| A^{\dagger}_{P',Q'} - A^{\dagger}_{P,Q} \| \leq (\alpha + \beta + \alpha\beta) \| A^{\dagger}_{P,Q} \| \ ,$$

where

$$\alpha = \| P - P' \| \quad \text{and} \quad \beta = \| Q - Q' \| \ .$$

We obtain next estimates for α and β.

Let $\mathcal{M}_1$ and $\mathcal{M}_2$ be two closed subspaces of $\mathcal{X}$, and define

$$\delta(\mathcal{M}_1, \mathcal{M}_2) = \sup \{ \mathrm{dist}(x, \mathcal{M}_2) : \ x \in \mathcal{M}_1, \ \|x\| = 1 \} \ .$$

Note that in general $\delta(\mathcal{M}_1, \mathcal{M}_2) \neq \delta(\mathcal{M}_2, \mathcal{M}_1)$. The gap between $\mathcal{M}_1$ and $\mathcal{M}_2$ is defined by

$$g(\mathcal{M}_1,\mathcal{M}_2) = \max\{\delta(\mathcal{M}_1,\mathcal{M}_2),\ \delta(\mathcal{M}_2,\mathcal{M}_1)\} .$$

The "gap" between two different complements to a given subspace may be used to obtain estimates for the difference of the associated projectors onto this subspace.

Proposition 3.11. Let $\mathcal{N}$, $\mathcal{M}$ and $\mathcal{M}'$ be closed subspaces of $\mathcal{X}$ such that

$$\mathcal{X} = \mathcal{N} \oplus \mathcal{M} = \mathcal{N} \oplus \mathcal{M}' ,$$

and let P and P' be the projectors of $\mathcal{X}$ onto $\mathcal{N}$ along $\mathcal{M}$ and $\mathcal{M}'$, respectively. If $\|P'\|\,\delta(\mathcal{M},\mathcal{M}') < 1$, then

$$\|P - P'\| \le \frac{\|I - P'\|\,\|P'\|\,\delta(\mathcal{M},\mathcal{M}')}{1 - \|P'\|\,\delta(\mathcal{M},\mathcal{M}')} \tag{3.21}$$

and hence,

$$\|P-P'\| \le \min\left\{\frac{\|I-P\|\,\|P\|\,\delta(\mathcal{M}',\mathcal{M})}{1 - \|P\|\,\delta(\mathcal{M}',\mathcal{M})},\ \frac{\|I-P'\|\,\|P'\|\,\delta(\mathcal{M},\mathcal{M}')}{1 - \|P'\|\,\delta(\mathcal{M},\mathcal{M}')}\right\}.$$

Proof. Let $\delta_1 > \delta(\mathcal{M},\mathcal{M}')$, $x \in \mathcal{M}$, $\|x\| = 1$. Then there exists $u \in \mathcal{M}'$ (i.e. $P'u = 0$) such that $\|x-u\| < \delta_1$. Thus,

$$\|P'x\| = \|P'(x-u)\| \le \|P'\|\,\delta_1$$

and hence,

$$\rho := \sup\{\|P'x\|: x \in \mathcal{M},\ \|x\| = 1\} \le \|P'\|\,\delta(\mathcal{M},\mathcal{M}') \le 1 .$$

Now $P = P'P$ since $\mathcal{R}(P) = \mathcal{N}(I - P')$. Thus,

$$\|(I-P)x\| \leq \|(P'-P'P)x\| + \|(I-P')x\|$$

$$\leq \rho\|(I-P)x\| + \|(I-P')x\| ,$$

or

$$\|(I-P)x\| \leq (1-\rho)^{-1} \|(I-P')x\| .$$

Therefore,

$$\|P-P'\| \leq \frac{\|P'\| \, \delta(\mathcal{M},\mathcal{M}')\|I-P'\|}{1-\|P'\| \, \delta(\mathcal{M},\mathcal{M}')} . \qquad \blacksquare$$

In the case of Hilbert spaces, an inequality for $\|P-P'\|$ that cannot be improved may be used to obtain error bounds for the norm of the difference of the Moore-Penrose inverse and $A^{\dagger}_{P,Q}$.

<u>Proposition 3.12.</u> Let $\mathcal{X}$ be a Hilbert space with respect to each of two equivalent inner products $\langle \cdot , \cdot \rangle$ and $[\cdot , \cdot]$. Let $\mathcal{N}$ be a closed subspace of $\mathcal{X}$ and let $\mathcal{M}$ and $\mathcal{M}'$ be the orthogonal complements of $\mathcal{N}$ with respect to the inner products $\langle \cdot , \cdot \rangle$ and $[\cdot , \cdot]$ respectively. Let

$$m_1 = \inf\left\{\frac{[x, x]}{\langle x, x \rangle} : x \in \mathcal{X}, \quad x \neq 0\right\}$$

and

$$m_2 = \sup\left\{\frac{[x, x]}{\langle x, x \rangle} : x \in \mathcal{X}, \quad x \neq 0\right\} .$$

Then

$$\delta(\mathcal{M},\mathcal{M}') = \left(\frac{m_2 - m_1}{m_2 + m_1}\right)^2 .$$

Let P and P' be the orthogonal projectors of $\mathcal{X}$ onto $\mathcal{N}$ along $\mathcal{M}$ and $\mathcal{M}'$ respectively. Then

$$\|P - P'\| \leq \frac{(m_1 - m_2)^2}{4m_1 m_2} \|I - P'\| .$$

3.6. Generalized inverses of modified matrices

For a square, nonsingular matrix A and two column vectors, c and d, it is well known that when A is modified by the matrix cd^* where d^* is the conjugate transpose of d, and if the resulting matrix $M = A + cd^*$ is nonsingular, then

$$M^{-1} = A^{-1} - \frac{1}{\beta} A^{-1} cd^* A^{-1} , \tag{3.22}$$

where $\beta = 1 + d^* A^{-1} c$. Note that in this case A is modified by a matrix of rank 1. By means of (3.22), one can alter one or more of the elements of A and still use A^{-1} to invert the modified matrix. (For the literature on this technique and inversion schemes which are based on (3.22), see the references in [36]).

For singular matrices, it is not always possible to obtain the Moore-Penrose inverse of the modified matrix $A + cd^*$ using (3.22) with A^{-1} replaced by $A^\dagger$. Some results on representations for the generalized inverse of sums of matrices which were obtained by Cline [10], allow one to compute $(A + cd^*)^\dagger$ in terms of $A^\dagger$ in the special cases when $Ad^*c = 0$ or else when $c = d$ and $A = SS^*$ for some S.

Recently, Meyer [36] obtained expressions for $(A + cd^*)^\dagger$ which cover all possible cases:

(i) $c \notin \mathcal{R}(A)$ and $d \notin \mathcal{R}(A^*)$;

(ii) $c \in \mathcal{R}(A)$ and $d \notin \mathcal{R}(A^*)$ and $\beta = 0$;

(iii) $c \in \mathcal{R}(A)$ and d arbitrary and $\beta \neq 0$;

(iv) c is arbitrary and $d \in \mathcal{R}(A^*)$ and $\beta \neq 0$;

(v) $c \notin \mathcal{R}(A)$ and $d \in \mathcal{R}(A^*)$ and $\beta = 0$.

(vi) $c \in \mathcal{R}(A)$ and $d \in \mathcal{R}(A^*)$ and $\beta = 0$.

Each of the expressions for $(A + cd^*)^\dagger$ is of the form

$$(A + cd^*)^\dagger = A^\dagger + G ,$$

where G is a matrix obtained from only sums and products of A, $A^\dagger$, c, d, and their conjugate transposes.

4. A New Approach to Approximations and Perturbation Bounds for Generalized Inverses. Collectively Compact Pointwise Convergent Approximations

4.1. The setting

In this section we consider a general approximation theory for generalized inverses alluded to in Section 3 (Remarks 3.6 and 3.8), and address the questions raised in Remark 3.8. The proofs of all the results of this section are given in Moore and Nashed [37]. We confine our exposition to a summary and interpretation of the main results.

The setting is a (real or complex) Banach space $\mathcal{X}$. Consider $A \in [\mathcal{X}]$ with null space $\mathcal{N} = \mathcal{N}(A)$ and range $\mathcal{R} = A\mathcal{X}$. Assume there exist projectors $U, E \in [\mathcal{X}]$ with $U\mathcal{X} = \mathcal{N}$ and $E\mathcal{X} = \mathcal{R}$. Let $P = I - U$ and $\mathcal{M} = P\mathcal{X}$ so that

$$\mathcal{X} = \mathcal{M} \oplus \mathcal{N} = P\mathcal{X} \oplus U\mathcal{X} .$$

Also

$$\mathcal{X} = E\mathcal{X} \oplus (I - E)\mathcal{X} .$$

Relative to these projectors, A has a unique generalized inverse $A^\dagger = A^\dagger_{E,U}$ such that

$$(4.1)\qquad \begin{cases} \text{(a)}\ A^\dagger A = P & \text{(b)}\ A^\dagger A A^\dagger = A^\dagger \\ \text{(c)}\ AA^\dagger = E & \text{(d)}\ AA^\dagger A = A \end{cases}$$

In this section the projectors E, U will remain fixed, so the subscript in $A^\dagger_{E,U}$ will be omitted. More general results allowing P and E to vary can be easily developed by combining the results of this section with those of Section 3, and [49].

The thrust of [37] is to show that if $B \in [\mathcal{X}]$ approximates A <u>either in the operator norm or pointwise</u>, then while $B^\dagger$ <u>need not</u> approximate $A^\dagger$ still there is $B^\phi \in [\mathcal{X}]$, with $\mathcal{R}(B^\phi) = \mathcal{M}$, which approximates $A^\dagger$ and corresponds to projectors satisfying conditions similar to (4.1). Theorem 4.1 and 4.5 state these results precisely. We denote the restriction of an operator $T \in [\mathcal{X}]$ to a subspace $\mathcal{M}$ of $\mathcal{X}$ by $T_{\mathcal{M}}$. All statements concerning dimension or codimension will be understood to apply only when these are finite. Let $p(\alpha)$ denote a polynomial in α such that $p(0) = 1$, so

$$(4.2)\qquad p(\alpha) = 1 - \alpha q(\alpha) .$$

4.2. <u>A general theorem for approximations of generalized inverses</u>

<u>Theorem 4.1.</u> Let $A \in [\mathcal{X}]$ with associated projectiors E, U and P, and with $\mathcal{M} = P\mathcal{X}$, as described above, and let $p(\alpha)$ be a polynomial satisfying (4.2). Let $B \in [\mathcal{X}]$, and let

$$(4.3)\qquad H := A^\dagger(A - B)\, p(EB) .$$

If $\delta := \|H_{\mathcal{M}}\| < 1$, for some polynomial satisfying (4.2), then there exists a continuous linear operator B^{ϕ} from $\mathcal{X}$ onto $\mathcal{M}$, defined by

(4.4) $$B^{\phi} = (I - H)^{-1} A^{\dagger}[I + (A-B)\, q(EB)E]$$

such that

(a) $B^{\phi} B_{\mathcal{M}} = I_{\mathcal{M}}$ so $B^{\phi} B B^{\phi} = B^{\phi}$ and $B B^{\phi} B_{\mathcal{M}} = B_{\mathcal{M}}$;

(b) $F := BB^{\phi}$, $Q := B^{\phi}B$ and $V := I - Q$ are projectors;

(c) $Q\mathcal{X} = \mathcal{M}$, $F\mathcal{X} = B\mathcal{M}$;

(d) $B_{\mathcal{M}}$ is one-to-one, $(B_{\mathcal{M}})^{-1} = B^{\phi}\big|_{F\mathcal{X}}$;

(e) $\eta(B) \subset \eta(FB) = V\mathcal{X}$;

(f) $\dim \eta(B) \leqq \dim \eta(A)$;

(g) $B^{\phi}(I - E) = 0$ and $\eta(A^{\dagger}) = (I-E)\mathcal{X} \subset (I-F)\mathcal{X} = \eta(B^{\phi})$.

Furthermore, the following estimates hold:

(h) $\|Q - P\| = \|V - U\| \leqq \frac{1}{1-\delta} \|HU\| \leqq \frac{\delta}{1-\delta} \|U\|$.

(i) $\|B^{\phi}x - A^{\dagger}x\| \leqq \frac{\|A^{\dagger}\|}{1-\delta} (\delta \|x\| + \|(A-B)\, q(EB)Ex\|)$, $x \in \mathcal{X}$;

(j) $\|Fx - Ex\| \leqq \|B\|\, \|B^{\phi}x - A^{\dagger}x\| + \|(B-A)A^{\dagger}x\|$, $x \in \mathcal{X}$.

The proof is not hard. The main point is to arrive at the right setting and an appropriate approximation B^{ϕ}. Let $J := P - H$. Since $J_{\mathcal{M}} = I_{\mathcal{M}} - H_{\mathcal{M}}$ and $\delta := \|H_{\mathcal{M}}\| < 1$, there exists $J_{\mathcal{M}}^{-1} \in [\mathcal{M}]$. Also it follows that $J = A^{\dagger}[I + (A-B)\, q(EB)E]B$. Let $B^{\phi} := J_{\mathcal{M}}^{-1} A^{\dagger}[I + (A-B)\, q(EB)E]$. Then $B^{\phi} \in [\mathcal{X}]$

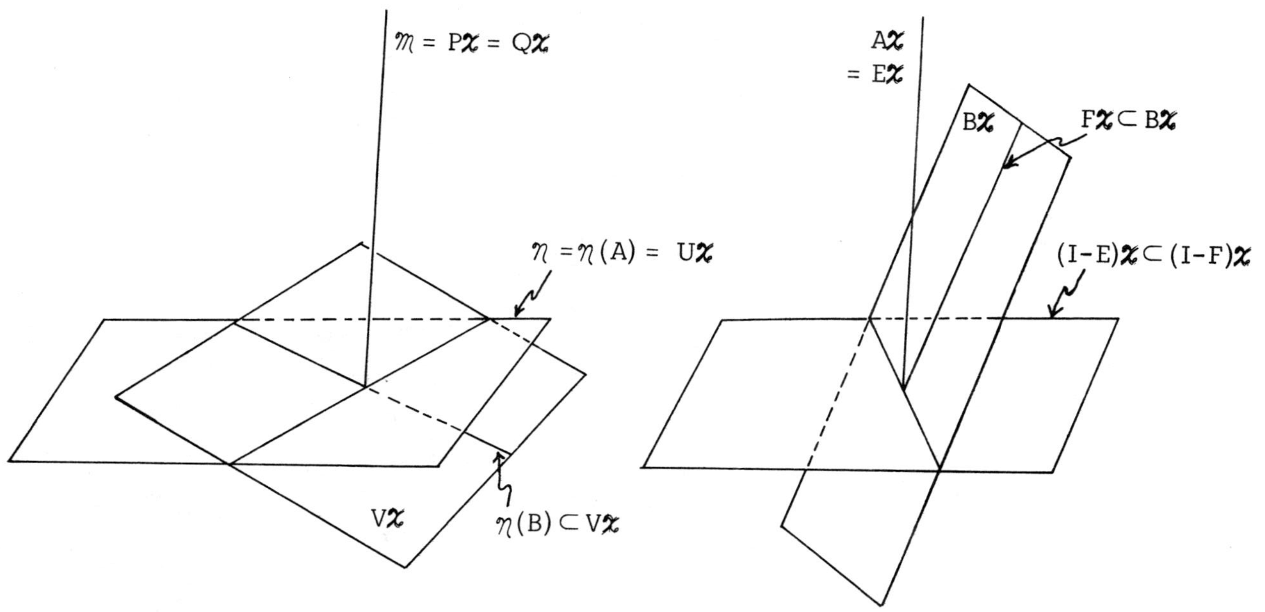
$\mathfrak{M} = P\mathfrak{X} = Q\mathfrak{X}$
$\mathfrak{N} = \mathfrak{N}(A) = U\mathfrak{X}$
$V\mathfrak{X}$
$\mathfrak{N}(B) \subset V\mathfrak{X}$
$A\mathfrak{X}$
$= E\mathfrak{X}$
$B\mathfrak{X}$
$F\mathfrak{X} \subset B\mathfrak{X}$
$(I-E)\mathfrak{X} \subset (I-F)\mathfrak{X}$

Fig. 1

and some computations show that it coincides with (4.4) and satisfies all the conclusions of the theorem. In view of (4.4), one may be tempted to define $J = I - H$. However, then B could not be obtained as a factor in J as above, as needed for (a), and $\|Q - P\|$ would not be shown to be small when $\|H\|$ is small.

An important interpretation of what is actually constructed in Theorem 4.1 is provided in the answer to the following question:

Can B^{ϕ} be viewed as the generalized inverse of a variant of the operator B, relative to some projectors?

Let $\hat{B} = BP$, so $\hat{B}x = Bx$ for $x \in \mathfrak{M}$ and $\hat{B}x = 0$ for $x \in \mathfrak{N}$. *Then B^{ϕ} is the true generalized inverse of $\hat{B}$ relative to F and P,* i.e. $B^{\phi} = \hat{B}^{\dagger}$. To prove this, we note that by (a) $B^{\phi} B_{\mathfrak{M}} = I_{\mathfrak{M}}$ so $B^{\phi}BP = P$, i.e. (i) $B^{\phi}\hat{B} = P$. Since B^{ϕ} maps $\mathfrak{X}$ onto $\mathfrak{M}$, one has $PB^{\phi} = B^{\phi}$ so (ii) $\hat{B}B^{\phi} = F$. By (a) $B^{\phi}BB^{\phi} = B^{\phi}$ so (ii) yields (iii) $B^{\phi}\hat{B}B^{\phi} = B^{\phi}$. Again by (a), $BB^{\phi}B_{\mathfrak{M}} = B_{\mathfrak{M}}$ so (ii) also yields (iv) $\hat{B}B^{\phi}\hat{B} = \hat{B}$. Thus (i)-(iv) show that equations of the form (1.1) hold for $\hat{B}$ and B^{ϕ}. Now $\hat{B} = BP$ while B is one-to-one on $\mathfrak{M} = P\mathfrak{X}$, $\mathfrak{N}(\hat{B}) = \mathfrak{N}(P)$; hence P is a projector onto a complement of $\mathfrak{N}(\hat{B})$. Since by (c) $F\mathfrak{X} = B\mathfrak{M} = \hat{B}\mathfrak{X}$, F is a projector onto the range of $\hat{B}$; and $\mathfrak{N}(B^{\phi}) = (I - F)\mathfrak{X}$ is a complementary subspace to $F\mathfrak{X}$. From the uniqueness of the generalized inverse relative to given projectors satisfying (4.1) it follows that B^{ϕ} is $\hat{B}_{F,P}$. The interpretation of this result in the case of matrices is evident.

Examples (In Theorem 4.1 take $p(\alpha) \equiv 1$ so $q(\alpha) \equiv 0$, and compute B^{ϕ} using (4.4)).

(i) $A = \begin{pmatrix} 2 & 0 \\ 0 & 0 \end{pmatrix}$, $B = \begin{pmatrix} 2 & 0 \\ 0 & \varepsilon \end{pmatrix}$.

Then $\mathfrak{N}(A)^{\perp} = \{(a, 0): a \in \mathbb{R}\}$, $P = \begin{pmatrix} 1 & 0 \\ 0 & 0 \end{pmatrix}$,

$$\hat{B} = BP = \begin{pmatrix} 2 & 0 \\ 0 & 0 \end{pmatrix} \quad \text{and} \quad B^{\phi} = \begin{pmatrix} \frac{1}{2} & 0 \\ 0 & 0 \end{pmatrix} = A^{\dagger} \quad .$$

(ii) $$A = \begin{pmatrix} \sigma & 0 \\ 0 & 0 \end{pmatrix}, \quad B = \begin{pmatrix} \sigma & \varepsilon \\ \varepsilon & 0 \end{pmatrix}, \quad \frac{\varepsilon}{\sigma} < 1 \ .$$

Then $$P = \begin{pmatrix} 1 & 0 \\ 0 & 0 \end{pmatrix}, \quad \hat{B} = BP = \begin{pmatrix} \sigma & 0 \\ \varepsilon & 0 \end{pmatrix}, \quad A^{\dagger} = \begin{pmatrix} \frac{1}{\sigma} & 0 \\ 0 & 0 \end{pmatrix},$$

$$H = A^{\dagger}(A - B) = \begin{pmatrix} 0 & \frac{\varepsilon}{\sigma} \\ 0 & 0 \end{pmatrix} ,$$

$$B^{\phi} = \begin{pmatrix} 1 & \frac{\varepsilon}{\sigma} \\ 0 & 1 \end{pmatrix}^{-1} \begin{pmatrix} \frac{1}{\sigma} & 0 \\ 0 & 0 \end{pmatrix} = \begin{pmatrix} 1 & \frac{\varepsilon}{\sigma} \\ 0 & 1 \end{pmatrix} \begin{pmatrix} \frac{1}{\sigma} & 0 \\ 0 & 0 \end{pmatrix} = \begin{pmatrix} \frac{1}{\sigma} & 0 \\ 0 & 0 \end{pmatrix} .$$

$$(B^{\dagger} - A^{\dagger}) P_{\Re(A)} = \begin{pmatrix} -\frac{1}{\sigma} & 0 \\ \frac{1}{\varepsilon} & 0 \end{pmatrix}, \quad \text{and}$$

$$P_{N(A)^{\perp}} (B^{\dagger} - A^{\dagger}) P_{\Re(A)} = \begin{pmatrix} -\frac{1}{\sigma} & 0 \\ 0 & 0 \end{pmatrix} .$$

On the other hand,

$$(B^{\phi} - A^{\dagger}) P_{\Re(A)} = \begin{pmatrix} 0 & 0 \\ 0 & 0 \end{pmatrix} .$$

(iii) $$A = \begin{pmatrix} 1 & 0 \\ 2 & 0 \end{pmatrix}, \quad B = \begin{pmatrix} 1 & \varepsilon \\ 2 & 2\varepsilon \end{pmatrix}.$$

Then

$$B^{\phi} = \begin{pmatrix} \frac{1}{5} & \frac{2}{5} \\ 0 & 0 \end{pmatrix}.$$

4.3. The case when $B^{\phi} = B^{\dagger}$. Perturbation bounds for $B^{\dagger} - A^{\dagger}$

In the standard perturbation theory for generalized inverse, the condition $\|A^{\dagger}(A-B)\| < 1$ is usually imposed. We obtain stronger results from Theorem 4.1 for perturbation bounds as a byproduct of the answer of the following question:

When is $B^{\phi} = B^{\dagger} = (B^{\dagger}_{F,Q})$, given that $\delta < 1$?

According to the definition of the generalized inverse it is necessary and sufficient that $Q\mathcal{X} = \mathcal{M}$ be complementary to $\mathcal{N}(B)$, and that $F\mathcal{X} = B\mathcal{X}$. Theorem 4.1 then yields the following corollary.

Corollary 4.2. In addition to the assumption of Theorem 4.1 suppose that

(4.5) $Q\mathcal{X} = \mathcal{M}$ is complementary to $\mathcal{N}(B)$

and

(4.6) $F\mathcal{X} = B\mathcal{X}$.

Then

$$B^{\dagger} := B^{\dagger}_{F,Q} = (I - H)^{-1} A^{\dagger}[I + (A-B)\, q(EB)E]$$

is a generalized inverse of B and

$$\|B^{\dagger} - A^{\dagger}\| \leq \frac{\|A^{\dagger}\|}{1-\delta} (\delta + \|(A-B)\, q(EB)E\|)$$

$$\|Q - P\| = \|V - U\| \leq \frac{1}{1-\delta} \|HU\| \leq \frac{\delta}{1-\delta} \|U\|$$

$$\|F - E\| \leq \|B\| \, \|B^{\dagger} - A^{\dagger}\| + \|(B - A)A^{\dagger}\| \; .$$

Various sufficient conditions under which (4.5) and (4.6) hold, generalizing for example (3.10), can be stated. We consider instead a more useful condition of the preceding criterion for a special case (cf. [37 , Corollary 1]).

Corollary 4.3. Suppose $\|A^{\dagger}(A-B)\| =: \delta_0 < 1$. Then $B^{\Phi} = B^{\dagger} \iff \dim \eta(B) = \dim \eta(A)$ and $\operatorname{codim} B\mathfrak{X} = \operatorname{codim} A\mathfrak{X}$.

It is interesting to note, in comparison with Corollary 4.3, that for the case $p(\alpha) \equiv 1$, so $q(\alpha) \equiv 0$, one obtains, as an immediate consequence of Theorem 4.1, a result which imposes no dimensionality conditions, namely:

Corollary 4.4. If

$$\|A^{\dagger}(A - B)\| =: \delta_0 < 1 \tag{4.7}$$

then

$$B^{\Phi} := [I - A^{\dagger}(A - B)]^{-1} A^{\dagger} \tag{4.8}$$

satisfies (a)-(h) in Theorem 4.1. Furthermore

(i_0) $$\|B^{\phi} - A^{\dagger}\| \leq \frac{\delta_0}{1-\delta_0} \|A^{\dagger}\|$$

(j_0) $$\|F - E\| \leq (1 + \frac{1}{1-\delta_0} \|B\| \|A^{\dagger}\|)\delta_0 \ .$$

If B is a varying operator converging to A in norm, $\|B - A\| \to 0$, then

$$\|B^{\phi} - A^{\dagger}\| \to 0, \quad \|Q - P\| \to 0, \quad \text{and } \|F - E\| \to 0 \ .$$

It should be noted that Corollary 4.4 is the exact analog of Proposition 3.3, with no additional assumptions, while in contrast, for example, Theorem 3.9 requires an additional assumption. Note from Corollary 4.4 that (4.7) with no added dimensionality conditions, implies that B^{ϕ}, given by (4.8), satisfies (i_0) (even though $B^{\dagger}$ might not); also, as B converges to A, B^{ϕ} converges to $A^{\dagger}$ (even though $B^{\dagger}$ might not). The moral of Corollary 4.4 is that we should not use $B^{\dagger}$, but rather a variant B^{ϕ} of $B^{\dagger}$, which has nicer properties than $B^{\dagger}$, and which reduces to $B^{\dagger}$ if the dimensionality conditions of Corollary 4.3 are satisfied. Thus Corollary 4.4 covers the general case both for matrices as well as for operators. Theorem 4.1 of course leads to the same conclusions, under a more relaxed condition, i.e. with $p(\alpha) \equiv 1$ replaced by a general polynomial $p(\alpha)$ with $p(0) = 1$. This latter generalization is important in the approximation theory of equations involving compact operators.

The setting of this section includes as special cases results on continuity and perturbation of generalized inverses of matrices (e.g. [5], [34], [66]). However, the techniques are not extensions of those in the matrix case. Other specializations of Theorem 4.1 yields standard and recent results on inverse operator approximations (e.g. [2], [3], [52]).

Finally, we note from (4.4), in the general case, that the operator

(4.9) $$C := A^{\dagger}[I + (A-B)\, q(EB)E]$$

serves as a computable approximation to B^{ϕ}, and

$$\|B^{\phi} - C\| \leq \frac{\delta}{1-\delta} \|C\| .$$

4.4. Collectively compact pointwise convergent approximations to $(I - K)^{\dagger}$

We now consider the case when the operator appximations converge pointwise rather than in operator norm, i.e. $A_n x \to Ax$ for each $x \in \mathcal{X}$, but $\|A_n - A\| \not\to 0$. The classical and elementary analysis of inverse operator approximations with operator norm convergence (Section 3.1) has been extended to the case of pointwise convergence using a collective compactness hypothesis which effectively compensates for the discrepancy between pointwise and operator norm convergence. For a thorough exposition of the theory of collectively compact operators for inverse operator approximations, see Anselone [2]. We consider in this section an extension of this theory for generalized inverse operator approximations. Convergence theorems and error bounds for operator generalized inverses are obtained.

We consider the case where $A = I - K$, $A_n = I - K_n$, where $\|K_n x - Kx\| \to 0$ but $\|K_n - K\| \not\to 0$, and $\{K_n\}$ is collectively compact, i.e., $\bigcup_n K_n \mathcal{B}$ is compact, where $\mathcal{B} = \{x: \|x\| \leq 1\}$. On the basis of Theorem 4.1 two types of theorems were established in [37], generalizing settings of Kantorovich and Akilov [25 ; Ch. XIV, p. 542] and Anselone [2 , §1.8, p. 10]:

(a) Given A, $A^{\dagger}$, and A_n, obtain A_n^{ϕ} and relate to $A^{\dagger}$.

(b) Given A, A_n, and $A_n^{\dagger}$, obtain a related operator $A^{\phi, n}$ as a variant of $A^{\dagger}$.

Theorem 4.5. Let $A = I - K$, $A_n = I - K_n$, $n = 1, 2, \dots$, where $\{K_n : n = 1, 2, \dots\}$ is a collectively compact family of operators, $K_n x \to Kx$ as $n \to \infty$ for all $x \in \mathcal{X}$. Let E and U be projectors on $\mathcal{R} = A\mathcal{X}$ and $\eta = \eta(A)$, respectively, and let $A^\dagger$ denote the generalized inverse of A relative to U and E. Let p be a polynomial such that $p(0) = 1$ and $p(1) = 0$, so that

$$(4.10) \qquad p(\alpha) = 1 - \alpha q(\alpha) = (1-\alpha)\, r(\alpha) .$$

Let

$$H_n := A^\dagger (A_n - A)\, p(EA_n) .$$

Then for n sufficiently large one has $\theta := \|H_n\| < 1$ and the conclusions of Theorem 4.1 hold with

$$B, H, \delta, B^\phi \text{ replaced by } A_n, H_n, \theta, A_n^\dagger, \text{ resp.}$$

and

$$F, Q, V \text{ replaced by } F_n, Q_n, V_n, \text{ resp.}$$

where

$$F_n = A_n A_n^\phi, \quad Q_n = A_n^\phi A_n, \quad V_n = I - Q_n .$$

In particular, $\dim \eta(A_n) \le \dim \eta(A)$ and

(h') $\|Q_n - P\| = \|V_n - U\| \le \frac{1}{1-\theta} \|H_n U\| \le \frac{\theta}{1-\theta} \|U\| \to 0, \ n \to \infty$,

(i') $\|A_n^\phi x - A^\dagger x\| \le \frac{\|A^\dagger\|}{1-\theta} (\theta \|x\| + \|(K - K_n)\, q(EA_n)\, Ex\| \to 0$,

(j') $\|F_n x - Ex\| \le \|A_n\| \, \|A_n^\phi x - A^\dagger x\| + \|(K_n - K)\, A^\dagger x\| \to 0$,

as $n \to \infty$ for each $x \in \mathcal{X}$.

The key estimate in the proof of Theorem 4.5 is to show that $\|(A_n - A)\, p(EA_n)\| \to 0$ as $n \to \infty$ even when $\|A_n - A\| \not\to 0$. It should be noted that the convergence in (i') and (j') is pointwise and not uniform, in contrast with the estimates (i) and (j) which imply uniform convergence if B is a varying operator converging in norm to A.

Now consider the reverse situation to that above: A_n is again one of a sequence of operators approximating A, however the operators A and B of Theorem 4.1 are now replaced by A_n and A, respectively. Using $A_n^\dagger$, we obtain a related operator $A^{\phi, n}$ as a variant of $A^\dagger$.

Theorem 4.6. Let $A = I - K$, $A_n = I - K_n$, where $K, K_n \in [\mathcal{X}]$ and $\{K_n\colon n = 1, 2, \ldots\}$ is a collectively compact family of operators. Suppose $K_n x \to Kx$ as $n \to \infty$ for all $x \in \mathcal{X}$. Let E_n and U_n be projectors on $\mathcal{R}_n = A_n\mathcal{X}$ and $\eta_n = \eta(A_n)$, respectively, $n = 1, 2, \ldots$; and $A_n^\dagger$ denote the generalized inverse of A_n relative to U_n and E_n. Suppose $\{I - E_n\colon n = 1, 2, \ldots\}$ and $\{E_n K\colon n = 1, 2, \ldots\}$ are collectively compact families of operators. Let p be a polynomial satisfying (4.10), and let

$$G_n := A_n^\dagger (A - A_n)\, p(E_n A) .$$

Suppose for a particular polynomial satisfying (4.10) and a particular n, $\gamma_n := \|G_n\| < 1$. Then the conclusions of Theorem 4.1 hold with

$$A, B, H, \delta, B^\phi \text{ replaced by } A_n, A, G_n, \gamma_n, A^{\phi,n}, \text{ resp.},$$

and with E, P, U, F, Q, V replaced by

$$E_n = A_n A_n^\dagger, \quad P_n = A_n^\dagger A_n, \quad U_n = I - P_n$$

$$F_n = AA^{\phi, n}, \quad Q_n = A^{\phi, n} A, \quad V_n = I - Q_n .$$

In particular $\dim \eta(A) \leq \dim \eta(A_n)$ and

(h") $$\|Q_n - P_n\| = \|V_n - U_n\| \leq \frac{1}{1-\gamma_n} \|G_n U_n\| ,$$

(i") $$\|A^{\phi, n}x - A_n^{\dagger}x\| \leq \frac{\|A_n^{\dagger}\|}{1-\gamma_n}(\gamma_n \|x\| + \|(A_n - A)q(E_n A)E_n x\|)$$

(j") $$\|F_n x - E_n x\| \leq \|A\| \|A^{\phi, n}x - A_n^{\dagger}x\| + \|(A-A_n)A_n^{\dagger}x\| .$$

Furthermore, if the operators $A_n^{\dagger}$, $n = 1, 2, \ldots$, are uniformly bounded, then $\gamma_n < 1$ for all sufficiently large n, and the right members of the inequalities (h"), (i"), (j") tend to zero as $N \to \infty$.

For details and other key observations on collectively compact pointwise convergent approximations to generalized inverses, we refer to [37]. The results are also applied in [37] to least squares solutions of Fredholm equations of the second kind, and to consideration of convergence of approximtions obtained by means of numerical quadrature in the space of continuous functions. The analysis is more subtle than the corresponding analysis in the case of a unique solution, and involves an interplay between the generalized inverses of $I - K$ in the spaces $\mathcal{L}_2[a, b]$ and $C[a, b]$.

Collectively compact operator approximation theory has found applications and extensions to nonlinear operators with the aid of the Fréchet derivative; approximation theory of nonlinear operator equations via Newton's method; projection, Galerkin, and collocation methods for operator and integral equations of the second kind; finite element and spline approximations for integral equations of the first and second kinds; and boundary value problems reformulated in terms of integral operators; and many other areas. The advance of the collectively compact approximation theory to generalized inverse as made in [37] and outlined in this section promises in a similar manner a broad range of applications. These incluce alternative problems for nonlinear operator equations with singular Fréchet derivative via Newton's method and corresponding discretization procedures, projectional and collocation methods for best approximate solutions of linear and integral equations of the first and second kinds, finite-element and regularization methods for equations of the first kind, as well as other possibilities. Some of these extensions will be reported in [38]. Some preliminary results on the extensions of the theory of generalized inverses to alternative problems in topological vector spaces have also been obtained.

5. Projection Methods for Generalized Inverses and Best Approximate Solutions of Linear Operator Equations of the First and Second Kinds

5.1. Preliminaries and a framework for some projection methods

We review first a general setting for projection methods for operator equations (see, e.g., Petryshyn [57]). Let X and Y be real or complex Banach spaces with the property that there exists a sequence $\{X_n\}$ of finite dimensional subspaces X_n of X, a sequence of linear projectors $\{P_n\}$ and a constant $C > 0$ such that

$$(5.1)\qquad P_n X = X_n, \quad X_n \subset X_{n+1}, \quad n = 1, 2, \ldots, \qquad \overline{\bigcup_{n=1}^{\infty} X_n} = X .$$

$$(5.2)\qquad \|P_n\| \leq C, \quad n = 1, 2, \ldots, \quad P_n P_j = P_j \quad \text{for } n \geq j.$$

Assume that Y has similar properties with subspaces $\{Y_n\}$ and projectors $\{Q_n\}$.

The quadruple of sequences $\Gamma_n = (\{X_n\}, \{Y_n\}, \{P_n\}, \{Q_n\})$ is said to form a complete projectional scheme.

These conditions are realized if X and Y are Banach spaces with Schauder bases $\{u_n\}$ and $\{v_n\}$ respectively, and X_n and Y_n are taken to be the n-dimensional subspaces spanned by $u_1, u_2, \ldots, u_n$ and $v_1, v_2, \ldots, v_n$ respectively. Let P_n be the linear projector of X onto X_n and Q_n of Y onto Y_n given by

$$P_n x = \sum_{i=1}^{n} \langle f_i, x\rangle u_i, \quad Q_n y = \sum_{i=1}^{n} \langle g_i, y\rangle v_i, \quad x \in X, \ y \in Y,$$

where $\langle f, x\rangle$ denotes the value of the functional $f \in X^*$ at $x \in X$, and the sequences $\{f_i\}$ and $\{g_i\}$ in the dual spaces X^* and Y^* respectively satisfy the biorthogonality relations

$$\langle f_i, u_i\rangle = \delta_{ij}, \quad \langle g_i, v_j\rangle = \delta_{ij}, \quad i, j = 1, 2, \ldots$$

It follows easily that $P_m P_n = P_n$ and $Q_m Q_n = Q_n$ for $m \geq n$; $\|P_n\| \leq a$, $\|Q_n\| \leq b$ for some constants $a \geq 1$ and $b \geq 1$, and $P_n x \to x$ and $Q_n y \to y$. Also,

$$X = X_n \oplus \mathcal{R}(I - P_n) \text{ and } Y = Y_n \oplus \mathcal{R}(I - Q_n) .$$

If X is a separable Hilbert space, then (5.2) with $C = 1$ follows from (5.1) alone.

Let A be a mapping of X to Y and $\{A_n\}$ the sequence of mappings of X_n into Y_n defined by

$$A_n x = Q_n A P_n x \text{ for } x \in X_n, \quad n = 1, 2, \ldots .$$

A general class of projection methods is concerned with obtaining solutions of the "exact" equation

$$(5.3) \qquad Ax = y \text{ for } y \in Y$$

as the strong limit of solutions $x_n \in X_n$ of "approximate" equations

$$A_n x_n = Q_n y \quad (5.4)$$

Equation (5.3) is said to be (uniquely) projectionally solvable with respect to the scheme Γ_n if there exist a positive integer N such that for each $n \geq N$ and each $y \in Y$, (5.4) has a unique solution $x_n \in X_n$ such that $\{x_n\}$ converges to x strongly and x is a (unique) solution of (5.3). One also says that the projectional scheme Γ_n is applicable to A if $(Q_n A P_n)^{-1}$ converges strongly to A^{-1} as $n \to \infty$, so that (5.3) is uniquely projectionally solvable with respect to Γ_n. This entails that $Q_n A P_n$ as an operator from $P_n X$ to $Q_n y$ be invertible for sufficiently large n and that $(Q_n A P_n)^{-1} Q_n$ converges strongly as $n \to \infty$. (Then A is necessarily invertible and the limit is A^{-1}).

Let A be a bounded linear operator of X to Y, and let Γ_n be the projection scheme for (5.3) introduced above. Then (5.3) is uniquely projectionally-solvable if and only if A maps X onto Y and satisfies the <u>Polsky condition</u>: For some $c > 0$ and $N > 0$,

$$\|A_n x\| \geq c\|x\| \text{ for all } x \in X_n \text{ and } n \geq N . \quad (5.5)$$

If X or Y is reflexive (in particular if either is a Hilbert space), then this assertion remains valid without the additional assumption that A maps X onto Y. See [58], [60], [19], [57] for various aspects of projection methods in the context of functional analysis and for related references.

In this section we consider projection methods within this framework for best approximate solutions of linear operator equations when $y \notin \mathcal{R}(A)$. Operators with closed range as well as operators with nonclosed range are considered. The results of this section and Section 6 apply in particular to integral and operator equations of the first and second kinds. We also consider other projection-type methods which are not viewed in this framework, i.e. for which P_n, Q_n, X_n, Y_n are no longer required to satisfy (5.1) and (5.2).

5.2. Projection methods for best approximate solutions of linear operator equations

Let X, Y be Hilbert spaces and let A be a bounded linear operator on X into Y with $\mathcal{R}(A)$ not necessarily closed. We consider projection methods for best approximate solutions of the operator equation

$$Ax = y, \quad y \in Y. \tag{5.6}$$

Let P be the orthogonal projectior of X onto $\mathcal{N}(A)^{\perp}$ and Q the orthogonal projector of Y onto $\overline{\mathcal{R}(A)}$. Consider the operator $\hat{A} : \mathcal{N}(A)^{\perp} \to \overline{\mathcal{R}(A)}$. Then $\hat{A}$ is one-to-one; also $\hat{A}^{-1}$ is bounded and defined on all $\mathcal{R}(A)$ if and only if $\mathcal{R}(A)$ is closed. For $y \in \mathcal{D}(A^{\dagger})$, $A^{\dagger}y$ is the unique solution of the equation

$$\hat{A}x = Qy, \tag{5.7}$$

or equivalently the unique solution of $Ax = Qy$ in $\mathcal{N}(A)^{\perp}$.

Thus projection methods for equations with unique solution carry over immediately to the approximation of $A^{\dagger}y$ once a suitable sequence of finite dimensional subspaces of the Hilbert space $\mathcal{N}(A)^{\perp}$ are found to satisfy the conditions of Section 5.1. Usually one has such a sequence for the space X or Y; and it is only necessary to "filter out" this sequence to obtain a corresponding sequence in $\mathcal{N}(A)^{\perp}$. Such an end is accomplished by the following proposition:

Proposition 5.1. (a) Let $\{Y'_n\}$ be a sequence of finite dimensional subspaces of Y satisfying conditions analogous to (5.1) and (5.2). Let $X_n := A^*Y'_n$. Then $\{X_n\}$ is a sequence of finite dimensional subspaces which satisfy (5.1) and (5.2) with X replaced by $\mathcal{N}(A)^\perp$.

(b) Let $\{X'_n\}$ be a sequence of finite dimensional subspaces of X satisfying (5.1) and (5.2). Let $X_n := A^*AX'_n$. Then the conclusion of part (a) holds for $\{X_n\}$.

(c) Let $\{\varphi_1, \varphi_2, \ldots, \varphi_n\}$ be a linearly independent subset of X. Denote $A^*A\varphi_i$ by ψ_i and assume that $\{\psi_1, \ldots, \psi_N\}$ is a maximal linearly independent subset of $\{A^*A\varphi_i: i = 1, 2, \ldots, n\}$. Let x_N denote the span of $\{\psi_1, \ldots, \psi_N\}$. Then the best approximation of $A^\dagger y$ by elements of X_N is given by $x_N^* = \sum_{i=1}^{N} c_i\psi_i$, where the c_i's are the solutions of the system

$$(5.8) \qquad \sum_{j=1}^{N} \langle \psi_i, \psi_j \rangle c_j = \langle A^* y, \varphi_i \rangle, \quad i = 1, \ldots, N.$$

(d) Let $\{\varphi_1, \varphi_2, \ldots, \varphi_n\}$ be a linearly independent subset of Y. Denote $A^*\varphi_i$ by ψ_i and assume that $\{\psi_1, \ldots, \psi_N\}$ is a maximal linearly independent subset of $\{A^*\varphi_i : i = 1, 2, \ldots, n\}$. Let X_N be the span of $\{\psi_1, \ldots, \psi_N\}$. Then the best approximation of $A^\dagger y$ by elements of X_N is given by $x_N^* = \sum_{i=1}^{N} c_i\psi_i$, where the c_i's are the solutions of the system

$$(5.9) \qquad \sum_{j=1}^{N} \langle \psi_i, \psi_j \rangle c_j = \langle Qy, \varphi_i \rangle, \quad i = 1, \ldots, N.$$

In particular, if the φ_i's are chosen to be in $\overline{\mathcal{R}(A)}$ or if $y \in \overline{\mathcal{R}(A)}$ then the right hand side of (5.9) may be replaced by $\langle y, \phi_i \rangle$.

(e) If A is self-adjoint, then in parts (b) and (c), $X_n := A^*AX'_n$ and $\psi_i := A^*A\varphi_i$ may be replaced by $X_n := AX'_n$

and $\psi_i := A\varphi_i$, respectively.

Proof. (a) $A^*Y = A^*[\overline{UY'_n}] \subset \overline{A^*[UY'_n]} = \overline{UA^*Y'_n}$. Thus $\mathcal{N}(A)^\perp = \overline{\mathcal{R}(A^*)} \subseteq \overline{UA^*Y'_n} = \overline{UX_n}$. Also, $X_n = A^*Y'_n \subset \overline{\mathcal{R}(A^*)} = \mathcal{N}(A)^\perp$. Thus $\mathcal{N}(A)^\perp = \overline{UA^*Y'_n}$. The remaining properties are obviously satisfied.

(b) The proof is similar recalling that $\overline{\mathcal{R}(A^*A)} = \overline{\mathcal{R}(A^*)} = \mathcal{N}(A)^\perp$.

(c) Let $\mathcal{X}_n$ be the finite dimensional subspace of X spanned by the linearly independent set $\{\varphi_1, \ldots, \varphi_n\}$. Then $A^*A[\mathcal{X}_n] = \mathrm{span}\{A^*A\varphi_i : i = 1, 2, \ldots, n\}$. Note that $X_N = A^*A[\mathcal{X}_n]$ is a finite dimensional subspace of $\mathcal{R}(A^*)$ and that $A^\dagger y \in \mathcal{R}(A^*)$. Thus the problem of best approximation of $A^\dagger y$ by elements of X_N reduces to finding the orthogonal projection of $A^\dagger y$ onto X_N. Let

$$\|A^\dagger y - x_N^*\| = \min\{\|A^\dagger y - x_N\| : x_N \in X_N\},$$

where

$$x_N^* = \sum_{j=1}^{N} c_j \psi_j .$$

The c_i's are easily shown to be solutions of the system

$$\sum_{j=1}^{N} \langle \psi_i, \psi_j \rangle c_j = \langle A^\dagger y, \psi_i \rangle, \quad i = 1, 2, \ldots, N .$$

But

$$\langle A^\dagger y, \psi_i \rangle = \langle A^\dagger y, A^*A\varphi_i \rangle = \langle A^*AA^\dagger y, \varphi_i \rangle$$

$$= \langle A^*Qy, \varphi_i \rangle = \langle A^*y, \varphi_i \rangle .$$

This proves (5.8).

(d) The proof is the same noting that in this case,

$$\langle A^{\dagger}y, \psi_j\rangle = \langle A^{\dagger}y, A^{*}\varphi_i\rangle = \langle AA^{\dagger}y, \varphi_i\rangle = \langle Qy, \varphi_i\rangle .$$

(e) If $A = A^{*}$, then $\overline{\mathcal{R}(A^{*})} = \overline{\mathcal{R}(A)} = \mathcal{N}(A)^{\perp}$ and the result follows. ∎

Note that in all cases we are producing approximations to $A^{\dagger}y \in \mathcal{R}(A^{*})$ by elements of $\mathcal{R}(A^{*})$. In general, the systems of linear equations (5.8) and (5.9) are ill-conditioned and could be replaced by the regularized systems

$$\sum_{j=1}^{N} [\langle A^{*}A\varphi_i, A^{*}A\varphi_j\rangle + \alpha\delta_{ij}]c_j = \langle A^{*}y, \varphi_i\rangle$$

and

$$\sum_{j=1}^{N} [\langle A^{*}\varphi_i, A^{*}\varphi_j\rangle + \alpha\delta_{ij}]c_j = \langle Qy, \varphi_i\rangle ,$$

respectively, where $\alpha > 0$ and δ_{ij} is the Kronecker symbol.

With the aid of Proposition 5.1, one can without difficulty adapt convergence results on projectional solvability of operator equations with unique solutions to the setting of best approximate solutions of (5.6). We say that equation (5.6) is projectionally solvable in the least-squares sense with respect to the scheme $\Gamma_n = (\{X_n\}, \{Y_n\}, \{P_n\}, \{Q_n\})$ if there exist a positive integer N such that for each $n \geq N$ and each $y \in \mathcal{D}(A^{\dagger})$, (5.4) has a unique solution $x_n \in X_n$ such that $x_n \to x$.

Approximations $\{x_n^{*}\}$ to $A^{\dagger}y$ with respect to the scheme Γ_n are given in Lemma 6.1. Explicit error bounds for $\|A^{\dagger}y - x_n^{*}\|$ can be obtained depending on the choice of the spaces $\{X_n\}$. Crucial to the determination of convergence rates and error bounds are estimates for $\|x - P_n x\|$, where P_n is the orthogonal projector of X onto X_n. Various such estimates, depending on the choice of the spaces $\{X_n\}$, are known in approximation theory.

Given Setting for $Ax=y$

$$X \xrightarrow{A} Y$$
$$P'_n \downarrow \qquad Q'_n \downarrow$$
$$X'_n \qquad Y'_n$$

Constructed Setting (Lemma 5.1) for $Ax = Qy$

$$X \underset{A^\dagger_{P,Q}}{\overset{A}{\rightleftarrows}} Y$$

$$P \downarrow \qquad Q \downarrow$$

$$\eta(A)^\perp \xrightarrow[1\text{-}1]{\hat{A}} \overline{\Re(A)}$$

$$P_n \downarrow \qquad Q_n \downarrow$$

$$X_n \underset{A_n^\dagger = A_n^{-1}}{\overset{A_n}{\rightleftarrows}} Y_n$$

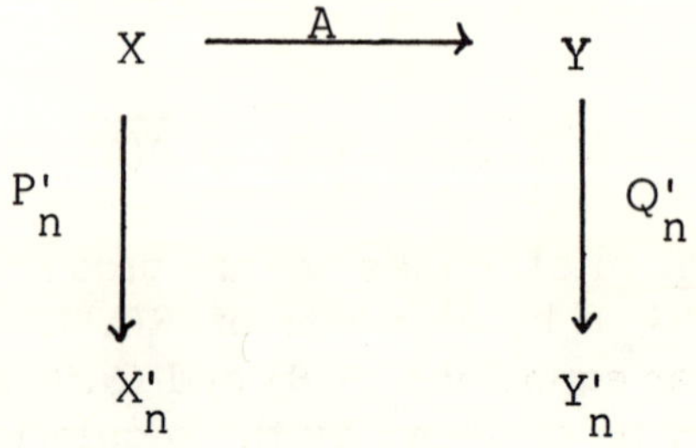

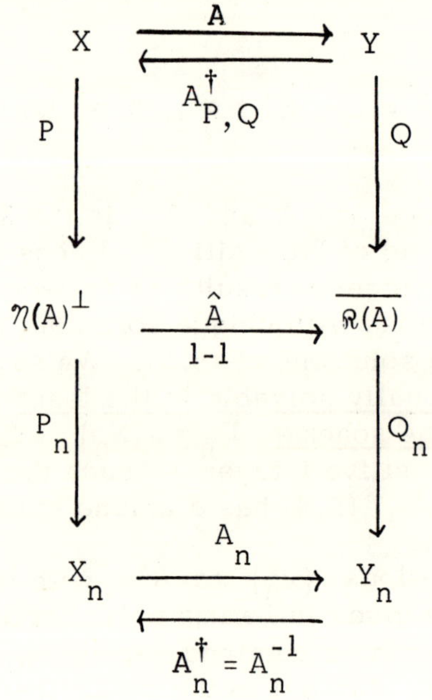

Figure 2

In the above construction (see Figure 2), the sequences of orthogonal projectors $\{P_n\}$, $\{Q_n\}$ converge strongly to the orthogonal projectors P, Q, respectively. Also, if $A = A^*$, then $\{P_n\}$ converges to Q.

In the case of a bounded linear operator A with closed range, we have for some constant $c > 0$, $\|Ax\| > c\|x\|$ for $x \in \mathcal{N}(A)^{\perp}$, which implies that $\|Q_n Ax\| \geq c\|x\|$ for $x \in X_n$ since $Q_n A = A$ on X_n. Thus in this case the equation $Q_n A x_n = Q_n y$ is uniquely solvable for each $y \in Y$ and $x_n \to A^{\dagger}y$ as $n \to \infty$ for any $y \in Y$.

In the case when A maps X into X it is easier to use for approximations the solutions of the equations

$$P_n A x_n = P_n y, \quad x_n \in X_n \quad \text{(Galerkin's method)} \tag{5.10}$$

One obtains convergence of the solutions x_n of (5.10) to $A^{\dagger}y$ if Polsky's condition

$$\begin{cases} \|P_n Ax\| \geq c\|x\| \\ \qquad \text{for all } x \in X_n \text{ and some constant } c > 0 \end{cases} \tag{5.11}$$

is satisfied. To establish convergence, we write

$$\|A^{\dagger}y - x_n\| \leq \|A^{\dagger}y - P_n A^{\dagger}y\| + \|x_n - P_n A^{\dagger}y\| .$$

Note that $\|A^{\dagger}y - P_n A^{\dagger}y\| \to 0$ since $A^{\dagger}y \in \mathcal{N}(A)^{\perp}$. Thus using (5.11),

$$c\|x_n - P_nA^\dagger y\| \leq \|P_nAx_n - P_nAP_nA^\dagger y\|$$

$$= \|P_ny - P_nAP_nA^\dagger y\|$$

$$= \|P_nAA^\dagger y - P_nAP_nA^\dagger y\|$$

$$\leq \|P_nA\| \; \|A^\dagger y - P_nA^\dagger y\| \to 0 \quad \text{as} \quad n \to \infty.$$

We remark that condition (5.11) is satisfied in particular if $A = I + K$, where K is a compact self-adjoint operator, or if $A = B^*B$ where B is a bounded linear operator with closed range; see also [19], [60] for several classes of operators which satisfy (5.11).

For operators with nonclosed range, the derivation of the rate of convergence is made by considering the two cases: $A^\dagger y \in \mathcal{R}(A^*)$ and $A^\dagger y \in \overline{\mathcal{R}(A^*)} \setminus \mathcal{R}(A^*)$. In both cases error bounds associated with the projection schemes of Lemma 5.1 are reduced to estimating $\|P_Nx - x\|$. An alternate approach to projection methods for linear operator equations in the case of nonclosed range is to use projection methods for approximate minimization of the functional

$$\|Ax - y\|^2 + \alpha\|x\|^2, \quad \alpha > 0 .$$

This would be a regularization-approximation scheme.

Finally we remark that other settings of approximation-projection methods which have extensions to generalized inverse approximations in Banach and Hilbert spaces include Kantorovich's setting for general theory of approximate method for linear operator equations of the second kind (see [25], Chapter 14, [62], and [14]) and Stummel's setting for discrete convergence of linear operators (see [68] and other papers by Stummel). Details of other aspects of projection methods for ill-posed linear operator equations will be given elsewhere [43].

5.3. Convergence rates of approximate least squares solutions of linear operator equations obtained by moment discretization

Let X be an arbitrary Hilbert space with inner product $\langle \cdot, \cdot \rangle_X$ and let A be a linear operator mapping X into the space of real-valued functions on an interval T, with the property that

$$|(Ax)(t)| \leqq M_t \|x\|_X \quad t \in T, \quad x \in X .$$

Then by Riesz's representation theorem, there exists a family $\{a_t : t \in T\}$ of elements in X with the property that

$$(Ax)(t) = \langle a_t, x \rangle_X, \quad t \in T, \quad x \in X .$$

Let $Q(t, t') := \langle a_t, a_{t'} \rangle_X$, $t, t' \in T$. Then Q is a nonnegative definite symmetric kernel on $T \times T$ and one can show that AX is a reproducing kernel Hilbert space (RKHS) $\mathcal{H}_Q$ with reproducing kernel $Q(t, t')$; see [50].

We consider approximate solutions of the operator equation

$$Ax = y \tag{5.12}$$

via moment discretization using n values $y(t_i)$ of $y(t)$. We consider first the case when $y \in \mathcal{R}(A)$, i.e. $y \in \mathcal{H}_Q$, but A is not necessarily one-to-one. Suppose that $y(t)$ is known on the set $T_n = \{t_1, \ldots, t_n\}$, $t_i \in T$, and consider the set of linear equations obtained by moment discretization on T_n of (5.12):

(5.13) $$(Ax_n)(t) = y(t), \quad t \in T_n .$$

Let x_n be that element in X of minimal norm which satisfies (5.13). (Note that x_n exists and is unique for each given y since the operator induced by the system (5.13) has a finite-dimensional range). It is easy to show that x_n is given explicitly by

(5.14) $$x_n(\cdot) = (y_1, y_2, \ldots, y_n)Q_n^{-1}(a_{t_1}, a_{t_2}, \ldots, a_{t_n})',$$

where the prime denotes transpose, $y_i := y(t_i)$, Q_n is the $n \times n$ matrix whose ij^{th} entry is given by $Q(t_i, t_j) = \langle a_{t_i}, a_{t_j} \rangle$, and the set $\{a_t : t \in T_n\}$ is assumed to be linearly independent (otherwise Q_n^{-1} is to be replaced by $Q_n^\dagger$ in (5.14)). Let

(5.15) $$\Delta_n := \sup_{t \in T} \left(\inf_{t_i \in T_n} |t-t_i| \right).$$

We now state a theorem on the convergence of $\{x_n\}$ to the solution of (5.12) of minimal X-norm (see [50] for a proof and other details).

<u>Theorem 5.2.</u> (a) If $Q(t, t')$ is continuous on $T \times T$ and if $y \in \mathcal{R}(A)$, then

$$\lim_{\Delta_n \to 0} \|x_n - A^\dagger y\| = 0$$

where Δ_n is defined by (5.15).

(b) If $y \in \mathcal{R}(A)$, $y \in \mathfrak{Q}(\mathcal{L}_2(T))$, where $\mathfrak{Q}$ is the Fredholm integral operator with kernel $Q(t, t')$ and $Q(t, t')$

satisfies the following smoothness properties:

(i) $\frac{\partial^{\ell}}{\partial t^{\ell}} Q(t,t')$ exists and is continuous on $T \times T$ for $t \neq t'$, $\ell = 0, 1, 2, \ldots, 2m$ for some $m \geq 2$ $\frac{\partial^{\ell}}{\partial t^{\ell}} Q(t,t')$ exists and is continuous on $T \times T$ for $\ell = 0, 1, \ldots, 2m-2$,

(ii) $\lim_{t \nearrow t'} \frac{\partial^{2m-1}}{\partial t^{2m-1}} Q(t,t')$ and $\lim_{t \searrow t'} \frac{\partial^{2m-1}}{\partial t^{2m-1}} Q(t,t')$

exist and are bounded for all $t' \in T$, then

$$\|x_n - A^{\dagger}y\| = O(\Delta_n^m) .$$

(The smoothness properties assumed on $Q(t, t')$ are tantamount to smoothness assumptions on $A^{\dagger}y$; in particular they imply that $A^{\dagger}y \in A^{*}\mathfrak{L}_2(T)$).

(c) If y does not coincide with some $y_0 \in \mathfrak{R}(A)$ on $\bigcup_n T_n$, then $\|x_n\| \to \infty$.

Now we consider the case $y \notin \mathfrak{R}(A)$. Suppose that X and Y are Hilbert spaces of real-valued functions on intervals S and T respectively. Let A be a linear operator on X into Y such that $A^{*}A$ exists and has the property that

$$|(A^{*}Ax)(s)| \leq M_s \|x\|_X, \quad x \in X, \quad s \in S .$$

Then there exists a family $\{p_s : s \in S\}$ in X such that

$$(A^{*}Ax)(s) = \langle p_s, x\rangle_X .$$

Let $P(s, s') = \langle p_s, p_{s'} \rangle_X$ and let P_n be $n \times n$ matrix whose ij^{th} element is given by $P(s_i, s_j)$. Let x_n be the element of minimal X-norm which satisfies

$$(A^* Ax)(s) = (A^* y)(s), \quad s \in S_n = \{s_1, s_2, \ldots, s_n\}.$$

Then a theorem analogous to Theorem 6.2 holds with the obvious modifications.

Commutativity of discretization and least-squares operations for linear integral equations was established in [42].

6. Iterative Methods for Generalized Inverses and Linear Operator Equations. Series and Integral Representations of Generalized Inverses

Several results on the convergence of some iterative methods for best approximate solutions of linear operator equations were given in [40; pp. 334-345]. In this section we make additional comments on these results and summarize briefly some of the recent contributions to this topic.

Let X and Y be Hilbert spaces and let A be a bounded linear operator on X into Y. We consider iterative methods for least-squares solutions of the operator equation:

$$(6.1) \qquad Ax = y, \quad y \in Y.$$

Let $\mathcal{S}_y$ denote the set of all least-squares solutions of (6.1) for a given $y \in Y$, and recall that $\mathcal{S}_y$ is nonempty if and only if $y \in \mathcal{D}(A^\dagger) = \mathcal{R}(A) + \mathcal{R}(A)^\perp$. Let $P_{\mathcal{S}}$ be the closest-point projector of X onto $\mathcal{S}_y$, i.e.

$$\|x - P_{\mathcal{S}} x\| = \inf\{ \|x-w\| : w \in \mathcal{S}_y\} .$$

Then

$$x - P_{\mathcal{S}} x \in \mathcal{N}(A)^{\perp} \quad \text{and} \quad P_{\mathcal{S}} x = A^{\dagger} y + P_{\mathcal{N}} x \ .$$

where $P_{\mathcal{N}}$ is the orthogonal projector of X onto $\mathcal{N}(A)$.

In the study of iterative methods, one is concerned with the domain of convergence (the set of $y \in Y$ for which an iterative method converges to a least-squares solution, and conditions on the initial approximation if any), rates of convergence and error bounds.

6.1. General properties of some iterative methods

Classical iterative methods (successive approximation method with a fixed parameter, steepest descent and conjugate gradient methods, etc.) have the following properties when applied to the equation (6.1) where y is not necessarily in the range of A, and the range of A is not necessarily closed:

(i) The iterative methods converge for each $y \in \mathcal{D}(A^{\dagger})$: the domain of convergence coincides with the domain of least-squares solvability.

(ii) For any initial approximation $x_0 \in X$, the iterative methods converge monotonically to the element in $\mathcal{S}_y$ closest to x_0, i.e. $\{x_n\}$ converges to $u = P_{\mathcal{S}} x_0 = A^{\dagger} y + P_{\mathcal{N}} x_0$. (See also Figure 3.) In particular a necessary and sufficient condition for $\{x_n\}$ to converge to $A^{\dagger} y$ is that $x_0 \in \mathcal{N}(A)^{\perp}$. (We shall assume thoughout that $x_0 \notin \mathcal{S}_y$, to avoid trivialities).

(iii) Convergence rates of iterative methods for least-squares solutions are markedly different for the cases of linear operators with closed or arbitrary ranges. To obtain specific convergence rates, it is necessary to treat the cases of operators with closed or nonclosed range separately. For all the iterative methods mentioned above, the rate of convergence for the case when $\mathcal{R}(A)$ is closed is at least geometric, while for the case when $\mathcal{R}(A)$ is nonclosed, the rate is asymptotic to $\frac{1}{n}$ (see also Table 1).

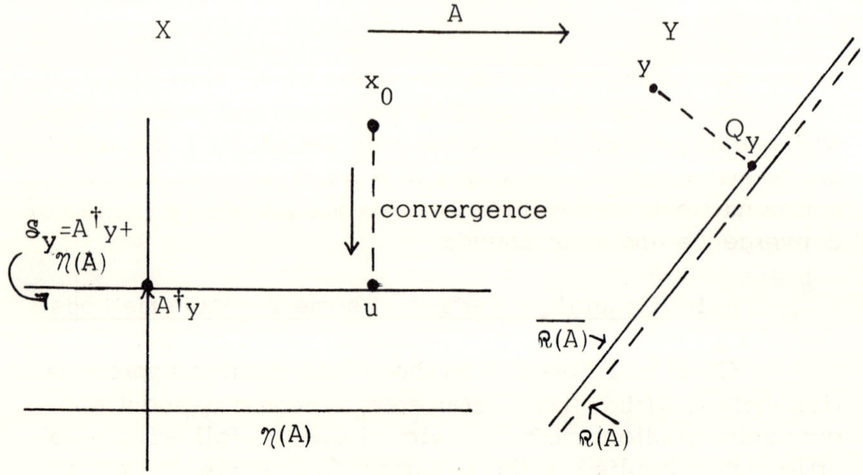

Q := orthogonal projector of Y onto $\overline{\mathcal{R}(A)}$

$P_{\mathcal{N}}$:= orthogonal projector of X onto $\mathcal{N}(A)$

Figure 3

6.2. Gradient methods

With gradient methods for least-squares solutions of (6.1) we seek to minimize the quadratic functional:

(6.1) $$J(x) = \frac{1}{2} \|Ax - y\|^2 .$$

The method of steepest descent for minimizing $J(x)$ is defined by the sequence

$$x_{n+1} = x_n - \alpha_n \operatorname{grad} J(x_n),$$

TABLE 1

Method \ Case	$\mathfrak{R}(A)$ is closed	$\mathfrak{R}(A)$ is nonclosed
Successive Approximations (with a fixed averaging parameter)	$\lVert A^{\dagger}y + P_{\eta}x_0 - x_n \rVert \leqq \left(\frac{\kappa^2 - 1}{\kappa^2 + 1}\right)^n \lVert x_0 - A^{\dagger}y \rVert$	$\lVert A^{\dagger}y - x_n \rVert^2 \leqq \frac{\lVert A^{\dagger}y \rVert^2 \lVert (AA^*)^{\dagger}y \rVert^2}{\lVert (AA^*)^{\dagger}y + n_{\nwarrow}\alpha(2-\alpha\lVert A \rVert^2) \lVert A^{\dagger}y \rVert^2}$ $x_0 = 0$, $Qy \in \mathfrak{R}(AA^*)$, $0 < \alpha < \frac{2}{\lVert A \rVert^2}$
Steepest Descent	$\lVert A^{\dagger}y + P_{\eta}x_0 - x_n \rVert \leqq C_1 \left(\frac{\kappa^2 - 1}{\kappa^2 + 1}\right)^n$	$\lVert A^{\dagger}y - x_n \rVert^2 \leqq \frac{\lVert A \rVert^2 \lVert A^{\dagger}y \rVert^2 \lVert (AA^*)^{\dagger}y \rVert^2}{\lVert A \rVert^2 \lVert (AA^*)^{\dagger}y \rVert^2 + n_{\nwarrow} \lVert A^{\dagger}y \rVert^2}$ $x_0 = 0$, $Qy \in \mathfrak{R}(AA^*)$.
Conjugate Gradient	$\lVert A^{\dagger}y + P_{\eta}x_0 - x_n \rVert \leqq C_2 \left(\frac{\kappa^2 - 1}{\kappa^2 + 1}\right)^n$	$\lVert A^{\dagger}y - x_n \rVert^2 \leqq \frac{\lVert A \rVert^2 \lVert A^{\dagger}y \rVert^2 \lVert (AA^*)^{\dagger}y \rVert^2}{\lVert A \rVert^2 \lVert (AA^*)^{\dagger}y \rVert^2 + n_{\nwarrow} \lVert A^{\dagger}y \rVert^2}$ $x_0 = 0$, $Qy \in \mathfrak{R}(AA^*A)$.
Constants	κ = pseudocondition no. = $\lVert A \rVert \; \lVert A^{\dagger} \rVert$; C_1, C_2 are constants.	

where α_n is chosen to minimize $J(x_{n+1})$. Noting that grad $J(x) = A^*Ax - A^*y$, the steepest descent sequence becomes

$$x_{n+1} = x_n - \alpha_n r_n \qquad n = 0, 1, 2, \ldots .$$

where

$$r_n = A^*(Ax-y),$$

x_0 is any initial approximation, and

$$\alpha_n = \frac{\|r_n\|^2}{\|Ar_n\|^2} \quad .$$

Convergence rates of this method for the cases of closed and nonclosed range are displayed in Table 1. Proofs and related results are given in Nashed [39] and Kammerer and Nashed [20]. (For simplicity we have taken $x_0 = 0$ in the case of operators with nonclosed range).

In the conjugate gradient method of Hestenes and Stiefel for minimizing the functional $J(x)$ in (6.1), we choose an initial approximation $x_0 \in X$ and compute $r_0 = p_0 = A^*(Ax-y)$. If $p_0 \neq 0$, we compute $x_1 = x_0 - \alpha_0 p_0$, where $\alpha_0 = \|r_0\|^2/\|Ap_0\|^2$. Then for $n = 1, 2, \ldots$ we compute

$$r_n = A^*(Ax_n - y) = r_{n-1} - \alpha_{n-1} A^* Ap_{n-1},$$

where

$$r_{n-1} = \frac{\langle r_{n-1}, p_{n-1}\rangle}{\|Ap_{n-1}\|^2}$$

and if $r_n \neq 0$, then we compute

$$p_n = r_n - \beta_{n-1} p_{n-1}, \quad \text{where } \beta_{n-1} = \frac{\langle r_n, A^* A p_{n-1} \rangle}{\|A p_{n-1}\|^2} .$$

and define the conjugate gradient sequence by

$$x_{n+1} = x_n - \alpha_n p_n .$$

Convergence of this method for the general case was investigated by Kammerer and Nashed [22]. Convergence rates are displayed in Table 1.

A class of gradient methods for minimizing $J(x)$ in (6.1) was also recently considered by McCormick and Rodigue [35]. This class is characterized by the step-size used in the iteration

$$(6.2) \qquad x_{n+1} = x_n - s(x_n) A^* (Ax_n - y),$$

where $s(x)$ is a real-valued function defined on $X - \mathcal{S}_y$. For $x \notin \mathcal{S}_y$, let $\Delta x = x - P_{\mathcal{S}_y} x$. The class of gradient methods in [35] is defined by (6.2) where s is assumed to satisfy the condition:

$$(6.3) \qquad \frac{1}{\|A\|^2} \leqq s(x) \leqq \frac{\|A\Delta x\|^2}{\|A^* A \Delta x\|^2}, \quad x \notin \mathcal{S}_y .$$

Step-size functionals that satisfy (6.3) include the method of steepest descent [39] the successive approximation method with a fixed parameter [40], [56] and for the case $y \in \mathcal{R}(A)$ the step size [13]:

$$s_0(x) = \frac{\|Ax-y\|^2}{\|A^*(Ax-y)\|^2} .$$

Suppose s satisfies (6.3). Then for any $x_0 \in X$ and $0 < \alpha < 2$, the sequence $\{x_n\}$ generated by (6.2) converges to $P_{\mathfrak{S}} x_0$. Convergence rates are established using some of the techniques in [20], [39]. These rates are similar to those in Table 1.

Groetsch [15] showed that each weak limit point of the steepest descent sequence $\{x_n\}$ is a least-squares solution of (6.1). As a corollary, (6.1) has a least-squares solution if and only if the steepest descent is bounded for any $x_0 \in X$. In [20], the blanket assumption that the projection of y onto $\overline{\mathfrak{R}(A)}$ is in $\mathfrak{R}(A^*A)$ was used to obtain a rate of convergence in the case when $\mathfrak{R}(A)$ is not closed. Convergence (but not the rate) holds without this assumption. Thus if steepest descent sequence $\{x_n\}$ is bounded, then it converges to a least-squares solution. This is also valid for the class of gradient methods considered above and for successive averaging methods.

A crucial point for obtaining geometric rates of convergence for gradient and other methods in the case when the operator A has a closed range is the fact that A^* has a bounded inverse on $\mathfrak{R}(A)$.

The practical usefulness of iterative methods for a best approximate solution of (6.1) in the case of a non-closed range is limited by sensitivity to generated inaccuracies since $A^\dagger$ in this case is unbounded and its domain is dense in Y. For such problems one resorts usually to some regularization methods as discussed in another paper in these proceedings, or uses a "filtering" scheme in conjunction with the iterative methods.

6.3. Series and integral representations of generalized inverses

The spectral theory for a bounded or densely defined closed linear operator in a Hilbert space is a particularly useful tool for convergence proofs of iterative processes, and series and integral representations of generalized inverses. (The original proof of the convergence of the method of steepest descent for positive definite bounded linear operators, due to Kantorovich, was based on the spectral theorem. Subsequently, easier proofs have been given which do not use the spectral theory.) Beutler and Root, in their paper in this volume, use the spectral theory to extend some known generalized matrix approximations to the context of densely defined closed linear operators in Hilbert space.

The spectral theory and functional calculus have been used by several authors in context of iterative methods, and/or general inverse approximations and representations (see, e.g. Bialy [9], Fridman [13], Groetsch [16], Koliha [28], Lardy [33], Nashed [45], and Patterson [53]).

Let X and Y be two Hilbert spaces over the same scalars and let A be a bounded linear operator on X into Y with closed range. Let $T = A^*A \mid \mathcal{R}(A^*)$. Let $\{B_\alpha(x)\}$ be a net of continuous real functions on $[0, \|A\|^2]$ such that $\{x B_\alpha(x)\}$ is uniformly bounded and

$$\lim_\alpha B_\alpha(x) = \frac{1}{x} \quad \text{uniformly.} \tag{6.4}$$

Then

$$A^\dagger = \lim_\alpha B_\alpha(T) A^* \tag{6.5}$$

holds in the uniform operator topology (see Groetsch [16] for a variant of this result).

An analogous formula can also be given in the form

$$A^\dagger = \lim_\alpha A\, C_\alpha(AA^*) \ .$$

Several series and integral representations (some old, some new) of $A^\dagger$ follow from this result and using various well-known classical summability transforms of the series $1 + x + x^2 + \ldots$. For example:

$$A^\dagger = \sum_{n=0}^{\infty} \beta[I-\beta A^*A]^n A^* \qquad (0 < \beta < 2\|A\|^{-2}),$$

$$A^\dagger = \int_0^\infty e^{-A^*As} A^* ds,$$

$$A^\dagger = \lim_{\epsilon \to 0^+} (A^*A + \epsilon I)^{-1} A^* = \lim_{\epsilon \to 0^+} A(AA^* + \epsilon I)^{-1}$$

$$A^\dagger = \sum_{n=1}^{\infty} A^*(I+AA^*)^{-n}$$

$$A^\dagger = \sum_{n=0}^{\infty} \frac{1}{n+2} \{\prod_{j=1}^{n} [I - \frac{1}{j+1} A^*A]\}A^*,$$

$$A^\dagger = \lim_{t \to 0^+} \sum_{n=0}^{\infty} a_n(t)[I-A^*A]^n A^*,$$

where $\quad a_0(t) \equiv 1$ and $a_n(t) = e^{-t\, n \log n} \quad (n \geq 1)$,

or, $\quad a_n(t) = [\Gamma(1 + tn)]^{-1}, \quad n \geq 1$

or, $\quad a_n(t) = \dfrac{\Gamma(1 + (1-t)n)}{\Gamma(1+n)}, \quad n \geq 1$

where Γ is the Gamma function.

By invoking the spectral theory, similar results and representations can be established for bounded linear operators with arbitrary range and for closed densely defined linear operators. However, for the case of arbitrary range, the requirement (6.4) would be pointwise, $T = A^*A|\overline{\mathcal{R}(A^*)}$, and the representations and limits hold only pointwise. The representations hold uniformly if and only if A has a closed range. See also Beutler and Root [8], Lardy [33], Nashed [40;

pp. 338-345] [45], Showalter and Ben-Israel [60], and Koliha [28]. A unified view of uses of the spectral theory in convergence proofs of iterative methods and series and integral representations is given in [45].

6.4. Successive approximations methods

Some successive approximations methods for generalized inverses and best approximate solutions were discussed in [40; pp. 338-344]. Related work on iterative methods for linear integral and/or operator equations include Bakušinskii [4], Bialy [9], Diaz and Metcalf [11], Fridman [12], [13], Groetsch [17], Kammerer and Nashed [21], [23], Krianev [30], Lardy [33], Landweber [31], Shaw [64], Strand [67], and the survey by Patterson [53]. Explicit convergence rates are given in [23], [40].

6.5. Methods of additive decompositions: Splitting methods

Splitting methods for the equation (6.1) when A is a nonsingular matrix involve a decomposition of A in the form $A = B - C$, where B is nonsingular. Iterative methods based on additive decomposition are then of the form:

$$(6.6)\quad \begin{cases} x_{n+1} = B^{-1}Cx_n + B^{-1}y = (I - B^{-1}A)x_n + B^{-1}y \\ = (I - B^{-1}A)^{n+1}x_0 + [(I - B^{-1}A)^n + \dots + (I - B^{-1}A) + I]B^{-1}y \ . \end{cases}$$

The formula (6.6) can also be used if A is a square singular matrix. In this case, if $y \in \mathcal{R}(A)$, then $AA^\dagger y = y$ and we can rewrite (6.6), using $A^\dagger$, in the form

$$(6.7)\qquad x_{n+1} = A^\dagger y + (I - B^{-1}A)^{n+1}(x_0 - A^\dagger y) \ .$$

Keller [27] investigated convergence of this method for singular but compatible systems. For the case $y \notin \mathcal{R}(A)$, one can either apply the splitting method to the (compatible) normal equations $A^*Ax = A^*y$ for least-squares solutions, or proceed to a direct splitting. Some results in the first direction are given by Korganoff and Pavel [29], where derived practical procedures are given, and the difficulties associated with direct splitting are described. Direct splitting methods which circumvent the intermediary use of A^*A or AA^* and their powers are developed under some technical conditions in Berman and Plemmons [6] and [7]. Some splitting methods for linear operator equations are given in Zlobec [73], Kammerer and Plemmons [24]. The concept of a regular splitting of a nonsingular matrix is extended to rectangular linear systems and operator equations by replacing A^{-1} by $A^{\dagger}$ and by considering mappings that leave a cone invariant and using the concept of cone monotonicity. Stringent technical conditions are associated with the definition of a proper splitting.

The splitting $A = M - N$ of a rectangular matrix A is called proper if the range and null spaces of A and M are equal. With this idea the usual splitting of a nonsingular matrix can be extended to the general case of a linear system $Ax = b$. The iterative method

$$x_{n+1} = M^{\dagger} N x_n + M^{\dagger} b ,$$

where $A = M - N$ is a proper splitting, converges to $A^{\dagger}b$ if and only if the spectral radius of $M^{\dagger}N$ is less than one. Several extensions have been given by weakening the conditions of the splitting.

REFERENCES

[1] S. N. Afriat, Orthogonal and oblique projectors and the characteristics of pairs of vector spaces, Proc. Cambridge Philos. Soc., 53 (1957), 800-816.

[2] P. M. Anselone, Collectively Compact Operator Approximation Theory, Prentice-Hall, Englewood Cliffs, N. J., 1971.

[3] P. M. Anselone and R. H. Moore, Approximate solutions of integral and operator equations, J. Math. Anal. Appl., 9 (1964), 264-277.

[4] A. B. Bakušinskii, A general method of constructing regularizing algorithms for a linear incorrect equation in Hilbert space, USSR Computational Math. and Math. Phys., 7 (1967), 279-287.

[5] A. Ben-Israel, On error bounds for the generalized inverse, SIAM J. Numer. Anal., 3 (1966), 585-592.

[6] A. Berman and R. J. Plemmons, Monotonicity and the generalized inverse, SIAM J. Appl. Math., 22 (1972), 155-161.

[7] A. Berman and R. J. Plemmons, Cones and iterative methods for best least-squares solutions of linear systems, SIAM J. Numer. Anal., 11 (1974), 145-154.

[8] F. J. Beutler and W. L. Root, The operator pseudo-inverse in control and systems identification, this Volume.

[9] H. Bialy, Iterative Behandlung linearer Functionalgleichungen, Arch. Rational Mech. Anal., 4 (1959), 166-176.

[10] R. E. Cline, Representations for the generalized inverse of sums of matrices, SIAM J. Numer. Anal., 2 (1965), 99-114.

[11] J. B. Diaz and F. T. Metcalf, On iterative procedures for equations of the first kind, $Ax = y$ and Picard's criterion for the existence of a solution, Math. Comp., 24 (1970), 923-935.

[12] V. M. Fridman, A method of successive approximation for Fredholm integral equations of the first kind, Uspehi Mat. Nauk, 11 (1956), 233-234.

[13] V. M. Fridman, On the convergence of the method of steepest descent type, Uspehi Mat. Nauk, 17 (1962), 201-204.

[14] M. B. Gagua, On the approximate solution of linear equations, Trudy Vycisl. Centra Akad. Nauk Gruzin. SSSR, 3 (1963), 27-47.

[15] C. W. Groetsch, Steepest descent and least squares solvability, Canad. Math. Bull., 17 (1974), 275-276.

[16] C. W. Groetsch, Representations of the generalized inverse, J. Math. Anal. Appl., 49 (1975), 154-157.

[17] C. W. Groetsch, Some aspects of Mann's iterative method for approximating fixed points, Conference on Computing Fixed Points and Applications, Clemson University, June, 1974.

[18] R. J. Hanson and C. L. Lawson, Extensions and applications of the Householder algorithm for solving linear least squares problems, Math. Comp., 23 (1969), 787-812.

[19] S. Hildebrandt and E. Wienholtz, Constructive proofs of representation theorems in separable Hilbert spaces, Comm. Pure Appl. Math., 17 (1964), 369-373.

[20] W. J. Kammerer and M. Z. Nashed, Steepest descent for singular linear operators with nonclosed range, Applicable Anal., 1 (1971), 143-159.

[21] W. J. Kammerer and M. Z. Nashed, A generalization of a matrix iterative method of G. Cimmino to best approximate solution of linear integral equations of the first kind, Atti. Accad. Naz. Lincei Lincei Rend. Cl. Sci. Fis. Mat. Natur., Ser. VIII, 51 (1971), 20-25.

[22] W. J. Kammerer and M. Z. Nashed, On the convergence of the conjugate gradient method for singular linear operator equations, SIAM J. Numer. Anal., 9 (1972), 165-171.

[23] W. J. Kammerer and M. Z. Nashed, Iterative methods for best approximate solutions of linear integral equations of the first and second kinds, J. Math. Anal. Appl., 40 (1972), 547-573.

[24] W. J. Kammerer and R. J. Plemmons, Direct iterative methods for least-squares solutions to singular operator equations, J. Math. Anal. Appl., 49 (1975), 312-526.

[25] L. V. Kantorovich and G. P. Akilov, Functional Analysis in Normed Spaces, Pergamon Press, Oxford, England, 1964.

[26] T. Kato, Perturbation Theory for Linear Operators, Springer-Verlag, Berlin-Heidelberg-New York, 1966.

[27] H. B. Keller, On the solution of singular and semidefinite systems by iteration, SIAM J. Numer. Math., 2 (1965), 281-290.

[28] J. J. Koliha, Power convergence and pseudoinverses of operators between Banach spaces, J. Math. Anal. Appl., 48 (1974), 446-449.

[29] A. Korganoff and M. Pavel, Iterative methods for pseudo-inverses computation: methods of additive decomposition, 1973.

[30] A. V. Krianev, Iterative method of solution of incorrect problems, Zh. Vychisl. Mat.i Mat. Fiz., 14 (1974), 25-35.

[31] L. Landweber, An iteration formula for Fredholm integral equations of the first kind, Amer. J. Math., 73 (1951), 615-624.

[32] L. J. Lardy, Some iterative methods for linear operator equations with applications to generalized inverses, Tech. Report 73-65, Dept. of Math., University of Maryland, College Park, November, 1973.

[33] L. J. Lardy, A series representation for the generalized inverse of a closed linear operator, Tech. Report 74-18, Dept. of Math., University of Maryland, College Park, April, 1974.

[34] C. L. Lawson and R. J. Hanson, Solving Least Squares Problems, Prentice-Hall, Englewood Cliffs, N. J., 1974.

[35] S. F. McCormick and G. H. Rodrigue, A uniform approach to gradient methods for linear operator equations, J. Math. Anal. Appl., 49 (1975), 275-285.

[36] C. D. Meyer, Jr., Generalized inversion of modified matrices, SIAM J. Appl. Math., 24 (1973), 315-323.

[37] R. H. Moore and M. Z. Nashed, Approximation to generalized inverses of linear operators, SIAM J. Appl. Math., 27 (1974), 1-16.

[38] R. H. Moore and M. Z. Nashed, Approximation-projection methods for generalized inverses in Banach spaces and their finite dimensional implementation, in preparation.

[39] M. Z. Nashed, Steepest descent for singular linear operator equations, SIAM J. Numer. Anal., 7 (1970), 358-362.

[40] M. Z. Nashed, Generalized inverses, normal solvability, and iteration for singular operator equations, pp. 311-359 in Nonlinear Functional Analysis and Applications (L. B. Rall, editor), Academic Press, New York, 1971.

[41] M. Z. Nashed, On application of generalized splines and generalized inverses in regularization and projection methods, Proceedings of ACM 1973, Annual National Conference of the Association for Computing Machinery, pp. 415-419.

[42] M. Z. Nashed, On moment-discretization and least-squares solutions of integral equations of the first kind, Tech. Summary Rept. #1371, Mathematics Research Center, the University of Wisconsin, 1974; J. Math. Anal. Appl., to appear.

[43] M. Z. Nashed, Regularization and numerical analysis of ill-posed operator equation, Bull. Amer. Math. Soc., to appear.

[44] M. Z. Nashed, Aspects of generalized inverses in analysis and regularization, this Volume.

[45] M. Z. Nashed, The spectral theory as a tool in linear operator equations, to appear.

[46] M. Z. Nashed, Measurability of generalized inverses of random linear operators in Banach spaces and random generalized matrices, Notices Amer. Math. Soc., 22(1975), 31.

[47] M. Z. Nashed and H. Salehi, Measurability of generalized inverses of random linear operators, SIAM J. Appl. Math., 25(1973), 681-692.

[48] M. Z. Nashed and G. F. Votruba, A unified approach to generalized inverses of linear operators: I. Algebraic, topological and projectional properties, Bull. Amer. Math. Soc., 80(1974), 825-830.

[49] M. Z. Nashed and G. F. Votruba, A unified operator theory of generalized inverses, this Volume.

[50] M. Z. Nashed and Grace Wahba, Rates of convergence of linear integral and operator equations, Math. Comp., 28(1974), 69-80.

[51] B. Noble, Computations of generalized inverses, and other matters, this Volume.

[52] A. M. Ostrowski, General existence criteria for the inverse of an operator, Amer. Math. Monthly, 74 (1967), 826-827.

[53] W. M. Patterson, 3rd, Iterative Methods for the Solution a Linear Operator Equation in Hilbert Space - A Survey, Lecture Notes in Math., Vol. 394, Springer-Verlag, Berlin-New York, 1974.

[54] R. Penrose, A generalized inverse for matrices, Proc. Cambridge Philos. Soc., 51(1955), 406-413.

[55] W. V. Petryshyn, Direct and iterative methods for the solution of linear operator equations in Hilbert spaces, Trans. Amer. Math. Soc., 105(1962), 136-175.

[56] W. V. Petryshyn, On generalized inverses and the uniform convergence of $(I-\beta K)^n$ with applications, J. Math. Anal. Appl., 18(1967), 417-439.

[57] W. V. Petryshyn, On projectional-solvability and the Fredholm alternative for equations involving linear A-proper operators, Arch. Rational Mech. Anal., 30 (1968), 270-284.

[58] W. V. Petryshyn, On the approximation-solvability of equations involving A-proper and pseudo A-proper mappings, Bull. Amer. Math. Soc., 81(1975), to appear.

[59] R. J. Plemmons, Monotonicity and iterative approximations involving rectangular matrices, Math. Comp., 26 (1972), 853-858.

[60] N. I. Polsky, On the convergence of certain approximate methods in analysis, Ukrainian Math. J., 7 (1955), 56-70.

[61] P. M. Prenter, A collocation method for the numerical solution of integral equations, SIAM J. Numer. Anal., 10 (1973), 570-581.

[62] L. Rakovshchick, Approximate solutions of equations with normally resolvable operators, U.S.S.R. Computational Math. and Math. Phy., 6 (1966), 3-11.

[63] C. B. Shaw, Jr., Improvement of the resolution of an instrument by numerical solution of an integral equation, J. Math. Anal. Appl., 37 (1972), 83-112.

[64] C. B. Shaw, Jr., Best accessible estimation: convergence properties and limiting form of direct and reduced versions, J. Math. Anal. Appl., 44 (1973), 531-552.

[65] D. W. Showalter and A. Ben-Israel, Representation and computation of the generalized inverse of a bounded linear operator between Hilbert spaces, Atti Accad. Naz. Lincei Rend. Cl. Sci. Fis. Mat. Natur., Ser. VIII, 48 (1970), 120-130.

[66] G. W. Stewart, On the continuity of the generalized inverse, SIAM J. Appl. Math., 17 (1969), 33-45.

[67] O. N. Strand, Theory and methods related to the singular-function expansion and Landweber's iteration for integral equation of the first kind, SIAM J. Numer. Anal., 11 (1974), 798-925.

[68] F. Stummel, Diskrete Knovergenz linearer Operation. I., Math. Ann., 190(1970), 45-92, II. Math. Z., 120 (1971), 231-264.

[69] K. Tanabe, Characterization of linear stationary iterative processes for solving a singular system of linear equations, Numer. Math., 22(1974), 349-359.

[70] K. Tanabe, A projection method for solving a singular equations and its applications, Numer. Math., 17(1971), 203-214.

[71] P.-Å. Wedin, Perturbation bounds in connection with singular value decomposition, BIT, 12(1972), 99-111.

[72] P. -Å. Wedin, Perturbation theory for pseudo-inverses, BIT, 13(1973), 217-232.

[73] S. Zlobec, On computing the best least squares solutions in Hilbert spaces, Dept. of Math., McGill University, Montreal, 1972.

School of Mathematics
Georgia Institute of Technology
Atlanta, Georgia 30332

The Operator Pseudoinverse in Control and Systems Identification

Frederick J. Beutler
William L. Root

Table of Contents

PART 0

INTRODUCTION

Extension of the concept of the pseudoinverse of a linear mapping in finite dimensional linear space to a linear operator in Hilbert and even Banach space has been of interest for some years [B-3][B-4][D-1][N-1][P-3]. Our concern in this paper is with the (Moore-Penrose) pseudoinverse of a linear transformation between Hilbert spaces. Part of our work deals with basic theory, which we then use for applications to systems identification and the quadratic regulator problem.

Since the finite dimensional (matrix) pseudoinverse can easily be interpreted geometrically in terms of certain orthogonal projections on Euclidean spaces, generalization to Hilbert spaces is completely natural; indeed, if one limits consideration to bounded operators with closed range, much of the theory of matrix pseudoinverses generalizes directly to Hilbert space [D-1][B-3]. However, widening the class of transformations to encompass densely defined closed linear operators introduces substantive complications; domains of definition must be carefully treated, and account taken of the fact that neither the sum nor the product of densely defined closed operators necessarily enjoys the same property. Nevertheless, a good part of the theory carries over in a satisfying fashion.

There are genuine difficulties, however, when an operator has non-closed range, for then the pseudoinverse is unbounded and only densely defined; moreover, many of the usual pseudoinverse characterizations become invalid. In case the range of the operator is not closed, the pseudoinverse usually fails to provide an answer to real problems (e.g., in statistical estimation), since one cannot tolerate a "solution" which is only sometimes meaningful. To circumvent this impediment, we consider (1) shrinking the image Hilbert space and endowing it with a new topology which insures that the range of the operator is closed, and (2) replacing the original operator or its pseudoinverse by a better behaved approximation. Either procedure can be applied as desired, but since some change is necessarily engendered thereby in the figure of merit which is optimized, the changed formulation may no longer be legitimate for its intended application. Roughly speaking, change in the Hilbert space or modifying the original operator is related to proper mathematical modelling of a physical situation, while approximations to the pseudoinverse are more closely connected to pseudoinverse theory as such.

There is good reason to analyze pseudoinverses of arbitrary densely defined linear closed operators, rather than just bounded operators and/or those whose ranges are closed. Our examples (and doubtless many others) will indicate this. On the other hand, it seems evident that we cannot easily drop the requirements that an operator be closed and densely defined.

Part I of this work is devoted to a self-contained exposition of the basic theory of the pseudoinverse for densely defined linear closed operators with arbitrary range. It is self-contained in the sense that no results are borrowed from the literature on pseudoinverses, finite-dimensional or otherwise. Much of the material covered in Part I is available elsewhere in some form, but we believe some of our results for unbounded operators represent extensions. The theory we develop includes the case of operators that do not have closed range, but any questions regarding approximations and change in topologies are deferred until later. Appendix A consists of a collection of lemmas ("exercises for the reader") relevant to unbounded operators; they are intended to facilitate the reading of Part I and of the later sections.

In Part II, a Gauss-Markov theorem on statistical estimation is proved under the hypothesis that both the quantity to be estimated and the observations are elements of Hilbert spaces. This theorem is an improved version of a theorem stated earlier in [R-5], where the proof was not given; the proof given here suffices for the previously stated result. Our theorem applies only to non-singular covariance operators, and reduces to the classical Gauss-Markov theorem when the spaces are finite dimensional. To obtain our result (in Hilbert spaces) it becomes necessary to treat the inversion of an unbounded non-closed operator, which carries us beyond the subject matter of Part I. Consequently, an additional hypothesis and some further argument are required.

Appendix B (needed for Part II) is a brief development of the more elementary aspects of the theory of Hilbert-space-valued random variables. The use of Bochner integrals enables us to establish the needed facts very quickly, thus sparing the reader the chore of studying the literature on the subject.

Two applications of engineering interest are discussed; the first is a problem in unknown system identification (Part III), while the second (Part IV) is an optimal control problem known to workers in the field as the quadratic regulator problem. In Part III, Volterra-Frechet integral polynomials are taken to represent a class of unknown systems in input-output form, and the Gauss-Markov theorem of Part II is applied to the estimation of their kernels. There is also in this section some discussion of the applicability of the pseudoinverse to identification problems in general.

In one classical form of the quadratic regulator problem, it is required to find the minimum energy input which will move a system from some initial state to the origin at a designated time. Part IV offers a reformulation which generalizes this problem to admit a greater variety of linear constraints, possibly including some which are incompatible and/or unattainable by the system. The solution always exists as a pseudoinverse (even if the system is described by an unbounded operator), and reduces to the classical result if the system is capable of meeting the constraint. Another related quadratic regulator problem, usually called the free endpoint problem, is also treated in Part IV. Here, a quadratic loss function involving both input and output is to be minimized by appropriate choice of input. We show this loss function to be conveniently represented if we change the topology on the image Hilbert space of the input-output operator; moreover, the new topology guarantees the range of this operator to be closed. The solution to a generalization of the free endpoint quadratic regulator problem can then be directly obtained by applying the pseudoinverse. In short, the pseudoinverse is a unifying influence on extensions of some quadratic regulator problems, and as such provides an insight not generally obtained through the more classical approaches.

In both Parts III and IV we encounter as a recurrent theme the necessity of inverting, in some acceptable sense, operators with nonclosed range. The technique of changing the topology on a reduced image space is employed in both places, but not in identical fashion. In particular, Part III describes a general approach which may also be useful in dealing with problems other than those considered here.

Approximations to the pseudoinverse constitutes the subject matter of Part V. Some known matrix approximations are examined in the context of unbounded operators whose range may not be closed. Properties of these approximations, and of other techniques suggested for operators of non-closed range, are investigated. Systematic use is made of the polar decomposition of operators, and of the functional calculus applicable to the resulting positive operators expressed as a spectral representation. We believe the material of this section to be largely original, especially in their coverage of unbounded operators. Part V also contains some results on the limits to be expected from any choice of approximations.

It may be useful to the reader to mention that the entire paper is dependent on Part I, but that Parts IV and V are independent of one another, and of Parts II and III.

PART I

PSEUDOINVERSE OPERATORS IN HILBERT SPACE

The pseudoinverse of a matrix has a rich literature, and has become sufficiently well recognized to constitute the subject of two recent books [A-2][R-1]. Much of the underlying theory is phrased in matrix-theoretic concepts, even though some of the principal optimization applications are more clearly motivated by the "best approximation" property, which the pseudoinverse matrix possesses with respect to Euclidean norms. In particular, it has been observed that application of the pseudo-inverse matrix solves certain optimal control (quadratic regulator [K-1]) and minimum mean square estimation [A-1] problems.

The emphasis on norm minimization suggests a function analytic rather than an algebraic approach to the pseudoinverse. Indeed, it seems natural to attempt to extend pseudoinverses to a Hilbert space context [D-1], since Hilbert space itself is no more than a generalization of finite dimensional linear vector space with Euclidean norm. The necessary extension is in fact easily accomplished for bounded linear operators of closed range [D-1][B-3]. This class of course includes all bounded linear operators of finite dimensional range, and a fortiori, operators on finite dimensional spaces. Thus, the Hilbert space pseudoinverse theory is not only a legitimate extension of the matrix theory, but also represents an approach eminently suitable to optimization questions.

If we abandon the assumption that an operator is necessarily bounded and/or equipped with closed range, new complexities are encountered. These must be faced squarely if application to control or estimation theory are to be contemplated in any degree of generality. One may recall, for instance, that the differentiation operator in an L_2 space is unbounded [R-2], so that the input-output relation of a dynamical system described by differential equations is represented by an operator whose range is not a closed set. On the other hand, we shall see that the Gauss-Markov estimation problem in separable Hilbert space is equivalently formulated in terms of unbounded operators.

In order to avoid a complete chaos of pathological behaviors, we shall suppose that we seek the pseudoinverse of a linear operator $A: H_1 \to H_2$, where A is a closed operator ([R-2], Section 115) densely defined on the Hilbert space H_1, with Hilbert space H_2 as range space; such an operator will be called DDC (densely defined closed). Roughly speaking, the pseudoinverse A^+ of A will solve the following minimization problem: given $z \in H_2$, A^+ delivers $x_0 = A^+z$, where $x_0 \in H_1$ minimizes the norm of $z - Ax$ over $x \in H_1$, and where x_0 is the element of least norm accomplishing the minimization. These notions will be made precise, and we shall also attempt to determine the conditions under which A^+ exists, what its properties are, and how it may be characterized.

We begin by defining the symbols we shall need for our analysis. $D(A)$ is the domain of A, and $R(A)$ its range; an overline denotes closure, so $\overline{R(A)}$ is the least subspace containing the range of A. By a subspace we shall consistently mean a closed linear manifold in the appropriate Hilbert space. Thus, the null space $N(A)$ of A is a subspace because A is DDC. An orthogonal complement is indicated by the standard symbol $\perp$, as for example $N(A)^{\perp}$ is the complement of the null space of A. We use P to stand for an orthogonal projection, and particularly P_R for the projection on $\overline{R(A)}$ and P_M for the projection on $N(A)^{\perp}$. In addition, standard symbols are employed for norm, adjoint, inverse and the like.

For the minimization problem, it is convenient to define some special symbols which are used recurrently. Thus, let

$$\delta_z = \inf_{x \in D(A)} \|z - Ax\|, \qquad z \in H_2 \tag{1.1}$$

$$K_z = \{x: Ax = P_R z\} \qquad \text{(preimage of } P_R z \text{ under } A) \tag{1.2}$$

$$F_A = \{z: P_R z \in R(A)\} . \tag{1.3}$$

Finally, the subscript r indicates the restriction of any operator on H_1 to the subspace $N(A)^{\perp}$, or the restriction to $\overline{R(A)}$ of an operator with domain space H_2. Specifically, $A_r: N(A)^{\perp} \rightarrow \overline{R(A)}$ and $A_r^*: \overline{R(A)} \rightarrow N(A)^{\perp}$. Properties of restrictions of operators and their combinations are summarized in Appendix A to the extent that they are required in what follows.

As we have noted, our applications of the pseudoinverse rest on its "best approximation" property. In intuitive terms, the pseudoinverse, when applied to $z \in H_2$, must give us the best approximate solution $x_0 = A^+ z$ to the functional equation

$$z = Ax \tag{1.4}$$

We use BAS as the acronym for "best approximate solution," the term being more precisely described by

Definition 1.1: $x_0 \in H_1$ is a BAS of (1.4) if

$$\|z - Ax_0\| = \delta_z \tag{1.5}$$

and

$$\|x_0\| < \|x\| \tag{1.6}$$

for any other x which also attains the infimum (1.6).

Remark: Uniqueness of the BAS constitutes part of its definition, being an immediate consequence of the strict inequality (1.6). The definition does not assert the existence of the BAS; indeed, there is always a sequence $\{x_n\} \in D(A)$ approaching the infimum (1.1), but without any element attaining it as required by Definition 1.1. However, the question of existence of the BAS is quickly settled by

Theorem 1.1: A BAS exists iff $z \in F_A$ [cf. (1.3)]; whenever a BAS x_0 exists, it is unique and satisfies

$$Ax_0 = P_R z \tag{1.7}$$

and

$$x_0 \in N(A)^{\perp} . \qquad (1.8)$$

Conversely, assume x_0 satisfies (1.7) and (1.8). Then x_0 is the BAS, and is the only element of H_1 satisfying these two equations.

Proof: We know $\inf_{y \in \overline{R(A)}} \|z-y\| = \delta_z$, the infimum being uniquely attained by $z_1 = P_R z$. It follows that (1.5) is met iff x is such that $Ax = P_R z$. But x of this type exists only if $z \in F_A$, in which case K_z [see (1.2)] is non-empty, and (1.5) and (1.7) are both equivalent to $x \in K_z$.

Suppose now x', $x'' \in K_z$. Then $Ax' = Ax''$, or $(x' - x'') \in N(A)$. Hence any $x \in K_z$ has the orthogonal decomposition

$$x = x_0 + x_1 \qquad (1.9)$$

where $x_0 \in N(A)^{\perp}$ is the same for every $x \in K_z$, and $x_1 \in N(A)$. To verify $x_0 \in K_z$, we use Lemma A.1 to argue $x_0 \in D(A)$, and obtain $Ax = Ax_0$ from (1.9) and $x_1 \in N(A)$. Now apply the Pythagorean theorem to (1.9); the strict inequality (1.6) follows unless $x_1 = \theta$ and consequently $x = x_0$. Thus x_0 is a BAS satisfying (1.8).

For the converse, note (1.7) means $z \in F_A$, whence a BAS x_0' exists and satisfies (1.7) and (1.8). But $(x_0 - x_0') \in N(A)$ from (1.7), whereas $(x_0 - x_0') \in N(A)^{\perp}$ by (1.8). Therefore $x_0 = x_0'$, or x_0 is seen to be the BAS as claimed. The same argument shows that at most one vector can satisfy (1.7) and (1.8). |||

Theorem 1.1 is readily specialized to operators whose ranges are closed sets, viz.

Corollary 1.1: If $R(A)$ is a closed set, the BAS always exists.

Proof: For any $z \in H_2$, $P_R z \in \overline{R(A)} = R(A)$, so $F_A = H_2$. |||

We have seen that every $z \in F_A$ has associated with it a unique BAS $x_0 \in H_1$, thereby suggesting an operator which transform elements of H_2 to elements of H_1. More formally:

<u>Definition 1.2:</u> An operator $A^+: H_2 \to H_1$ is called a <u>pseudoinverse operator</u> (henceforth abbreviated PI) if $D(A^+) = F_A$, and if, for each $z \in F_A$

$$x_0 = A^+ z \tag{1.10}$$

is the BAS.

How can we recognize an operator as a PI? The following criterion is often helpful.

<u>Lemma 1.1:</u> The linear operator $A^+: H_2 \to H_1$ is a PI iff $D(A^+) = F_A$ and

$$N(A^+) = R(A)^{\perp} \tag{1.11}$$

$$\overline{R(A^+)} = N(A)^{\perp} \tag{1.12}$$

and

$$AA^+ y = y \quad \text{for all} \quad y \in R(A) . \tag{1.13}$$

Proof: (Sufficiency) For $z \in F_A$, again call $P_R z = z_1$. Then $z_1 \in R(A)$, and (1.13) asserts $A^+ z_1 \in K_z$, i.e. $Ax_0 = P_R z$ with $x_0 = A^+ z_1$. Second, this $x_0 \in N(A)^{\perp}$ by (1.12), so we conclude x_0 is the BAS [compare (1.8) in Theorem 1.1]. To complete the proof, we need only show $A^+ z = A^+ z_1$. To this end, we decompose $z = z_1 + z_2$, where z_1 is as before and $z_2 \in N(A^+)$. Now $z \in D(A^+) = F_A$, and $A^+ z_2 = \theta$, whence $A^+ z = A^+ z_1 = x_0$ as required by (1.10).

(Necessity) Clearly, $R(A^+) \subset N(A)^{\perp}$, for otherwise (1.8) fails. To prove $R(A^+)$ dense in $N(A)^{\perp}$, observe $D(A) \cap N(A)^{\perp}$ to be dense in $N(A)^{\perp}$ (cf. Lemma A.4), and each $x \in [D(A) \cap N(A)^{\perp}]$ is a BAS for $y = Ax$ (Theorem 1.1). Since then $x = A^+ y$, $[D(A) \cap N(A)^{\perp}] \subset R(A^+)$. Thus (1.12) is necessary.

If (1.13) is not satisfied, $A^+z \notin K_z$ for some $z \in R(A) \subset F_A$, so (1.7) does not hold. Respecting (1.11), consider that the BAS for $z \perp R(A)$ is perforce the null vector from (1.7) and (1.8), requiring that $N(A^+) \supset R(A)^\perp$ if A^+ is to be a PI. On the other hand, (1.13) demands $N(A^+) \subset R(A)^\perp$, which then leads to the validity of (1.11). |||

Remark: The conditions (1.11) and (1.13) of Lemma 1.1 already insure that $R(A)^\perp$ and $R(A)$ are in the domain of A^+, wherefrom $F_A \subset D(A^+)$. To see this, observe that A^+ is linear by hypothesis, and that

$$F_A = R(A) \oplus R(A)^\perp \tag{1.14}$$

is then in $D(A^+)$. We note F_A is dense in H_2, which implies A^+ to be densely defined because $F_A \subset D(A^+)$. However, if $D(A^+)$ is larger than F_A we cannot claim A^+ to be DDC, whereas (as we shall prove below) A^+ must be a closed operator if $D(A^+) = F_A$.

There are many alternative forms for a matrix PI [R-1], but these fail to carry over to unbounded operators in direct fashion. The source of difficulty is that combinations of unbounded operators need not be DDC, and may in fact be defined only on the null vector; hence, manipulations of such operators is anything but routine. Nevertheless, we can demonstrate some of the properties of the PI by an explicit construction, which (inter alia) exhibits the PI as a linear DDC operator.

Theorem 1.2: The PI exists as the uniquely defined linear DDC operator

$$A^+ = A_r^{-1} P_R . \tag{1.15}$$

Proof: Since a unique BAS corresponds to each $z \in F_A$, the PI is properly and uniquely described on F_A. It then suffices to prove that the right side of (1.15) constitutes a linear DDC operator which meets the conditions set forth in Lemma 1.1.

In the first place, Lemma A. 6 asserts A_r^{-1} to be defined and DDC. Then, since P_R is bounded, $A_r^{-1} P_R$ is likewise a closed operator. $A_r^{-1} P_R$ is also clearly linear. To show $A_r^{-1} P_R$ densely defined, we note its domain includes both its null space $R(A)^{\perp}$ and the domain of A_r^{-1}, namely $R(A)$ [see Lemma A. 2]. By linearity, we therefore have $D(A_r^{-1} P_R) \supset F_A$, the latter having the form (1. 14); since F_A is dense in H_2, so is $D(A_r^{-1} P_R)$.

We have proved that A^+ [as given by (1. 15)] is linear DDC, with $D(A^+) \supset F_A$. To show $D(A^+) = F_A$, consider $z \notin F_A$. $P_R z \notin R(A) = D(A_r^{-1})$, which means $z \notin D(A_r^{-1} P_R)$.

Finally, we must verify (1. 11), (1. 12) and (1. 13) of Lemma 1. 1. From (1. 15), $N(A^+) \supset R(A)^{\perp}$, but also $A_r^{-1} y \neq \theta$ for non-null $y \in R(A)$; in fact, $A_r^{-1} : \overline{R(A)} \to N(A)^{\perp}$ has domain $R(A)$ and possesses an inverse A_r. Consequently, the A^+ of (1. 15) satisfies (1. 11).

Next, we see from Lemma A. 4 that $\overline{R(A_r^{-1})} = N(A)^{\perp}$. Moreover, $R(A^+) = R(A_r^{-1})$, since each range depends only on preimages in $R(A)$, and for any $y \in R(A)$, $A^+ y = A_r^{-1} y$. Thus, (1. 12) has been proven.

For any $y \in R(A)$, we now obtain

$$AA^+ y = AA_r^{-1} y = A_r A_r^{-1} y = y \tag{1. 16}$$

where the middle equality results from the definition of A_r as the restriction of A to $N(A)^{\perp}$, which is precisely the range space of A_r^{-1}. Therefore, (1. 13) is validated also, and the proof of the theorem is complete. |||

<u>Corollary 1. 2</u>: A^+ is bounded iff $R(A)$ is closed.

Proof: A^+ is bounded iff A_r^{-1} is bounded, and the latter is equivalent to $R(A)$ closed by Lemma A. 8. |||

The matrix pseudoinverse is often defined not by its best approximation property (Definition 1. 2), but rather as the (unique) matrix satisfying the identities [G-2]

$$AA^+A = A, \quad A^+AA^+ = A^+, \quad AA^+ = P_R, \quad A^+A = P_M. \tag{1. 17}$$

These same identities – suitably modified for DDC operators – are relevant to the PI as per Definition 1.2. Indeed, the modified identities (1.17) follow from our definition. Conversely, a DDC operator for which a weaker form of (1.17) is valid must be a PI. We now proceed to state precisely and prove these relationships.

Theorem 1.3: Let A^+ be the PI for the DDC operator A. Then

$$AA^+A = A \tag{1.18}$$

$$A^+AA^+ = A^+ \tag{1.19}$$

$$AA^+ \subset P_R, \quad \text{with} \quad D(AA^+) = F_A \tag{1.20}$$

and

$$A^+A \subset P_M, \quad \text{with} \quad D(A^+A) = D(A) \ . \tag{1.21}$$

Proof: (1.20) is a direct consequence of (1.11) and (1.13) in Lemma 1.1. In view of (1.11), both sides of (1.19) yield the null vector when applied to $R(A)^{\perp}$, so we need consider only $y \in R(A)$, and demonstrate $A^+AA^+y = A^+y$ for (1.19). But $AA^+y = y$ by (1.13), and so (1.19) follows.

To prove (1.21), decompose $x \in D(A)$ as

$$x = x_1 + x_2, \quad x_1 \in N(A)^{\perp}, \quad x_2 \in N(A) \ . \tag{1.22}$$

Then $x_1 \in D(A)$ by Lemma A.1, and

$$A^+Ax = A^+Ax_1 = A_r^{-1} P_R A_r x_1 = A_r^{-1} A_r x_1 = x_1 \tag{1.23}$$

which shows A^+A acts like P_M for every $x \in D(A)$. With this result it is easy to prove (1.18), since from (1.23) and (1.22) above

$$AA^+Ax = Ax_1 = Ax \tag{1.24}$$

for all $x \in D(A)$. |||

Although A^+A and AA^+ are DD according to (1.21) and (1.20), one cannot assume these to be closed operators. In fact, A^+A (respectively AA^+) closed corresponds precisely to the boundedness of A (respectively A^+), as is seen from

Corollary 1.3: Let A^+ be the PI of the DDC operator A. Then A is bounded iff A^+A is a closed operator; A^+ is bounded [or equivalently, $R(A)$ is a closed set] iff AA^+ is a closed operator.

Proof: Because of the duality evident in Theorem 1.3, we need only consider the first assertion. By (1.21), A^+A is bounded and DD, so A^+A is closed iff it is defined on all H_1 as a bounded operator. Of course, A (a DDC operator now defined on all H_1) is then bounded also. On the other hand, A bounded and A^+ DDC implies A^+A to be a closed operator.

We could state a direct converse to Theorem 1.3, but it is noteworthy that less restrictive requirements can be substituted for (1.20) and (1.21). More specifically, we have

Theorem 1.4: Suppose $A^+: H_2 \to H_1$ is a linear closed operator with $D(A^+) \subset F_A$. Let this operator satisfy (1.18) and (1.19), and assume further

$$A^+A \quad \text{is a symmetric operator} \tag{1.25}$$

and

$$AA^+ \quad \text{is a symmetric operator.} \tag{1.26}$$

Then A^+ is the PI of the (linear DDC) operator A.

Proof: We again turn to the sufficiency conditions of Lemma 1.1, starting with (1.13). If $y \in R(A)$, we may take $x \in D(A)$ such that $Ax = y$, whence (1.18) becomes $AA^+y = y$ for each $y \in R(A)$. This disposes of (1.13).

Our next task is to prove (1.11), or $N(A^+) = R(A)^\perp$. Actually, $N(AA^+) = R(A)^\perp$ suffices, because $N(A^+) = N(AA^+)$ by the chain

$$N(A^+) \subset N(AA^+) \subset N(A^+AA^+) = N(A^+) \tag{1.27}$$

The usual argument on ranges and null spaces produces

$$R(AA^+)^{\perp} = N([AA^+]^*) \subset N(AA^+) \ , \tag{1.28}$$

the right hand inclusion following from (1.26). However, the containment in (1.28) is even an equality, because $D(AA^+)$ is dense in $D([AA^+]^*)$, so that $N(AA^+)$ is not only dense in $N([AA^+]^*)$, but is also closed by (1.27). We then have

$$R(AA^+)^{\perp} = N(AA^+) \ . \tag{1.29}$$

Finally, $R(A) = R(AA^+)$ from

$$R(A) \supset R(AA^+) \supset R(AA^+A) = R(A) \ . \tag{1.30}$$

The combination of these equalities therefore yields $N(A^+) = R(A)^{\perp}$, as was to be shown.

The remaining equality (1.12) of Lemma 1.1 can be written as $R(A^+)^{\perp} = N(A)$, which is like (1.11) except that the symbols A and A^+ are interchanged. But the hypotheses on A and A^+ are symmetrical, so that the desired identity can be demonstrated in the same fashion as $R(A)^{\perp} = N(A^+)$.

We must still show A^+ to have the correct domain. The domain clearly includes $R(A)$, since $AA^+y = y$ for all $y \in R(A)$. Also, the null space of A^+ is $R(A)^{\perp}$, so by the linearity of A^+

$$D(A^+) \supset R(A) \oplus R(A)^{\perp} = F_A \ ; \tag{1.31}$$

in view of the original hypothesis $D(A^+) \subset F_A$, we then have $D(A^+) = F_A$. |||

Remark: In the literature on the matrix PI (1.25) and (1.26) are replaced by the stronger hypotheses that A^+A and AA^+ are self-adjoint. Such assumptions are inappropriate here, since (by Corollary 1.3), they would limit the applicability of Theorem 1.4 to bounded operators A and A^+. Moreover, even the (apparently) weaker assumptions A^+A and AA^+ closed symmetric already imply $A^+A = P_M$, $AA^+ = P_R$, with both A and A^+ bounded.

Many identities have been developed for the matrix pseudoinverse [R-1][A-2]. Most of these equalities are retained in a weaker version when applied to unbounded operators. The two presented below, however, are generalized without change in their statement, although the proofs necessarily become sophisticated beyond mere modifications of the usual matrix arguments.

<u>Theorem 1.5</u>: $(A^+)^+ = A$. (1.32)

Proof: We construct $(A^+)^+$ by the method of Theorem 1.2. In the first place, $(A^+)_r = A_r^{-1}$ from (1.15). Secondly, $R(A^+) = N(A)^{\perp}$ by (1.12). But then $(A^+)_r^{-1} = A_r$ and the projection on the closure of the range of A^+ is P_M, so that the application of (1.15) to A^+ yields

$$(A^+)^+ = A_r P_M = AP_M = A. \qquad (1.33)$$

The right hand equality is here justified by Lemma A.1 and the orthogonal decomposition (A.1). |||

<u>Theorem 1.6</u>: $(A^+)^* = (A^*)^+$. (1.34)

Proof: The construction of Theorem 1.2 exhibits the PI of A^* to be

$$(A^*)^+ = (A^*_r)^{-1} P_M ; \qquad (1.35)$$

by Lemma 1.1, this PI has null space $N[(A^+)^*] = R(A^+)^{\perp} = N(A)$, with the right hand equality being obtained from (1.12). Since both operators in (1.34) have the same null space, it suffices to compare their restrictions $(A^+)^*{}_r$ and $[(A^*)^+]_r$ to $N(A)^{\perp}$. Using successively Lemma A.9 and the representation of Theorem 1.2, we obtain

$$(A^+)^*{}_r = (A^+)_r{}^* = (A_r^{-1})^* . \qquad (1.36)$$

But also

$$[(A^*)^+]_r = (A^*{}_r)^{-1} = (A_r^*)^{-1} = (A_r^{-1})^* \quad ; \qquad (1.37)$$

here the equalities follow respectively from (1. 35), Lemma A. 9 and the interchange of inversion and the adjoint operation (see [R-2], Section 117). |||

It is also characteristic of the matrix pseudoinverse that it may be expressed in various forms involving combinations of other matrices related to A [R-1][A-2]. For example, among the alternative formulations of the matrix pseudoinverse we have $A^+ = A^*(AA^*)_r{}^{-1}$ and $A^+ = (A^*A)_r{}^{-1}A^*$. Since such forms also occur in applications to infinite dimensional problems (e. g., Gauss-Markov estimation), it would be desirable if they were also valid in the more general case of DDC operators. But again, operators A which are unbounded and/or have non-closed range lead to complications which must be taken into account. To this end, we need

Definition 1. 3: $A^x: H_2 \to H_1$ is a restricted pseudoinverse operator (acronym: RPI) for the linear DDC operator A if $A^x \subset A^+$.

Remark: A^x need be neither densely defined nor bounded to be a RPI.

Theorem 1. 7: $A' = A^*(AA^*)_r^{-1} P_R \qquad (1.38)$

is a RPI with domain

$$D(A') = R(AA^*) \oplus R(A)^{\perp} . \qquad (1.39)$$

Proof: By Lemma A. 14, $(AA^*)_r$ is invertible, and $(AA^*)_r^{-1} = (A_r^*)^{-1} A_r^{-1}$. Thus (1. 38) makes sense and

$$A' = A^*[(A_r^*)^{-1} A_r^{-1}]P_R . \qquad (1.40)$$

Now $A^*(A_r^*)^{-1} = A_r^*(A_r^*)^{-1}$ is a restriction of the identity. In other words, (1.30) implies

$$A' \subset A_r^{-1} P_R = A^+ \ . \tag{1.41}$$

Let us determine $D(A')$. It is clear A' is linear and well defined on $R(A)^{\perp}$. In $\overline{R(A)}$, A' is at best defined on $R(A)$ because of (1.41) and the extent of $D(A^+)$; we therefore limit consideration to $y \in R(A)$. Now $D[(AA^*)_r^{-1}] = R[(AA^*)_r] = R(AA^*)$, the equality on the right following from Lemmas A.2 and A.10 and Corollary A.2. Hence $y \in R(A)$ is in the domain of $(AA^*)_r^{-1}$ iff $y \in R(AA^*)$. Further, such a y leads to

$$(AA^*)_r^{-1} y \in R[(AA^*)_r^{-1}] = D[(AA^*)_r] =$$

$$D(A_r A_r^*) \subset D(A_r^*) \subset D(A^*) \ . \tag{1.42}$$

Here we have again used Lemma A.12 for the equality on the right. Since $(AA^*)_r^{-1} y$ automatically falls in $D(A^*)$ as shown by (1.42), and since $R(A) \supset R(AA^*)$, $D(A')$ is correctly given by (1.39) and A' is an RPI. |||

Corollary 1.4: $A' = A^+$ iff $R(AA^*) = R(A)$; in particular, A' is the PI if $R(A)$ is closed (or equivalently, any of the conditions of Lemmas A.8 or A.15 are met).

Proof: From (1.39) and (1.14), $D(A') = D(A^+)$ iff

$$R(AA^*) = R(A) \ . \quad ||| \tag{1.43}$$

Should $R(A)$ be closed (which is true if any of the conditions of Lemmas A.8 or A.15 are satisfied), Corollary A.4 states (1.43) to be valid.

Remark: A' is densely defined in any event because $\overline{R(AA^*)} = \overline{R(A)}$ as indicated by Corollary A.3. Also, (1.41) verifies the existence of a closed extension for A'; it is plausible that A^+ is actually the minimal closed extension of A', although neither proof nor counterexample has been found to elucidate the question.

Analogous to A', but differing in its domain, is

$$A'' = (A^*A)^{-1}A^* . \tag{1.44}$$

If A is a matrix, or even if A is bounded and $R(A)$ closed, A' and A'' are both bounded and coincide with A^+, and there is little advantage of one with respect to the other. But when A is an unbounded operator of closed range, we can guarantee only $A' = A^+$ (see Corollary 1.4 above) and conclude nothing further on A'', whereas A bounded (with arbitrary range) gives rise to $A'' = A^+$ without any simplification of A'.

<u>Theorem 1.8</u>: A'' is an RPI with domain

$$D(A'') = [D(A^*) \cap R(A)] \oplus R(A)^{\perp} . \tag{1.45}$$

Proof: The proof is much like that of Theorem 1.7, so we shall omit some of the details. Since $\overline{R(A)} = N(A^*)^{\perp}$ we write

$$A'' = (A^*A)_r^{-1} A_r^* P_R \tag{1.46}$$

$$= A_r^{-1}[(A_r^*)^{-1} A_r^*]P_R . \tag{1.47}$$

The term $[(A_r^*)^{-1} A_r^*]$ is a restriction of the identity on $\overline{R(A)}$ to $D(A_r^*)$, so that

$$A'' \subset A_r^{-1} P_R = A^+ . \tag{1.48}$$

The domain of A'' evidently contains $R(A)^{\perp}$. On $\overline{R(A)}$, this domain is limited to vectors in $R(A)$, and [by (1.46) even to $y \in [R(A) \cap D(A^*)]$. For such y, $A_r^* y \in R(A_r^*) = D[(A_r^*)^{-1}]$, and $[(A_r^*)^{-1} A_r^*]y = y$. But $y \in R(A) = D(A_r^{-1})$, whence $A_r^{-1}[(A_r^*)^{-1} A_r^*]y$ is defined also. Thus $y \in [R(A) \cap D(A)]$ implies $y \in D(A'')$, and $D(A'')$ is described by (1.45) by the linearity of A''. |||

Corollary 1.5: $A'' = A^+$ iff $R(A) \subset D(A^*)$; in particular, A'' is the PI if A is bounded (or equivalently, if any of A^*, AA^* or A^*A are bounded).

Proof: If $R(A) \subset D(A^*)$, the $D(A'')$ of (1.45) coincides with $D(A^+)$. The second assertion of the corollary is then immediate, since any of the hypotheses imply $D(A^*) = H_2$. |||

Remark: Attention is called to a limited duality between A' and A''. If any of A_r^{-1}, $(A_r^*)^{-1}$, $(AA^*)_r^{-1}$ or $(A^*A)_r^{-1}$ are bounded, A' is the PI A^+, and is thus a bounded operator defined on all H_2. On the other hand, if any of A, A^*, AA^* or A^*A [i.e., A_r, A_r^*, $(AA^*)_r$ or $(A^*A)_r$] are bounded, A'' is the same as the PI A^+, and is thus a DDC operator on H_2.

PART II

A GAUSS-MARKOV THEOREM FOR HILBERT SPACE

A natural extension of the classical theory of linear unbiased minimum-variance estimators (LUMV estimators) is to the case that both the vector of unknown parameters and the vector of observations are infinite-dimensional in the sense that both are elements of separable Hilbert spaces. Although one may well argue that in any actual data processing scheme only a finite set of parameters will be estimated and only a finite set of data will be used, a Hilbert space version of the theory is of interest as providing characterization of limit cases. Indeed, there are many estimation problems which are inherently infinite-dimensional, and for which reduction to finite sets of observations and parameters represents an approximation. The subject of identification of unknown systems to be discussed briefly in Part III provides examples of such problems. In this Part we state and prove A Gauss-Markov theorem for Hilbert space, but we leave comments about its application to Part III.

Necessary and sufficient conditions for the existence of an LUMV estimator in Hilbert space, and a characterization of the estimator in terms of reproducing-kernel Hilbert space concepts, is given in [P-2]. Our objective here is somewhat different; we wish to exhibit reasonable sufficient conditions for the existence of an LUMV estimator, that also guarantee an extension of the standard classical formula for the estimator. Let H_1 and H_2 be separable Hilbert spaces. Consider the linear model

$$z = Bx + w \qquad (2.1)$$

where $x \in H_1$, where B is a continuous linear transformation from all of H_1 into H_2, and where w is an H_2-valued random variable. Hilbert-space-valued random variables are discussed below and in Appendix B, where definitions are given for the mathematical expectation of an H-valued random variable w and for the covariance operator of w, as well as conditions for the existence of the covariance. We assume the expected value of w, Ew, exists and equals zero and that a covariance operator, K, for w exists. The element x represents the vector of unknown parameters to be estimated, and z represents the vector of observations. A linear estimator C is defined to be a continuous linear transformation from all of H_2 into H_1; $\hat{x} = Cz$ is then an estimate of x. Since z is an H_2-valued random variable, Cz is an H_1-valued random variable (see Appendix B). We call $E\|Cz-E(Cz)\|^2 = E\|Cw\|^2$ the variance of the estimator C; it will be seen below that if K has finite trace the variance is always finite. We say C is an unbiased estimator for x if E[Cz] exists and equals x. If C is unbiased, then the variance equals the mean-squared error, $E\|Cz-x\|^2$. C^0 is LUMV if it is linear, unbiased and of finite variance, and $E\|C^0z-x\|^2 \leq E\|Cz-x\|^2$ for all linear unbiased estimators C. In the classical case H_1 and H_2 are finite-dimensional Euclidean spaces. Then, if K is strictly positive definite, and if $N(B) = \theta$, an LUMV estimator always exists and is given, e.g., by the formula†

$$C = (B^*K^{-1}B)_r^{-1}B^*K^{-1} \quad . \tag{2.2}$$

It is this formula we extend to the case H_1 and H_2 are Hilbert spaces. If one puts $A = K^{-\frac{1}{2}}B$, then (2.2) can be written

$$\begin{aligned}\hat{x} &= (A^*A)_r^{-1}A^*(K^{-\frac{1}{2}}z)\\ &= A^+(K^{-\frac{1}{2}}z) \quad .\end{aligned} \tag{2.3}$$

The standard proof of (2.2) is carried out by first proving that C is in fact the PI_1 when $K = I$, then transforming z in the model (2.1) by $K^{-\frac{1}{2}}$, that is by re-norming the

† The notation is consistent with that of Part I.

observation space H_2 . Essentially, this† is what we do below, although it is not quite literally what we do because of certain technical considerations.

Before getting to the theorem, some further elucidation of the conditions already imposed and their implications is warranted. The discussion to follow also points out the reasonableness, and sometimes the necessity, of certain additional conditions. First, there are various ways of establishing the existence of a quantity w that can reasonably be called a Hilbert-space-valued random variable, that satisfies $E(w) = \theta$, and that has associated with it a bounded, self-adjoint, non-negative-definite operator K such that

$$(Ky, z) = E[(y, w)\overline{(z, w)}] \quad . \tag{2.4}$$

K is called the covariance operator (since the matrix form of (2.4) characterizes the
sional random vector). An outline of a simple way of defining w is relegated to Appendix B, since the material involved is largely foreign to the rest of the paper. It will be observed that the construction given in Appendix B automatically implies that K is nuclear (i.e., is compact with finite trace). This condition is also reasonable as a requirement imposed on the estimation problem; indeed, if $E\|w\|^2 < \infty$ and the probability measure is countably additive, K must be nuclear (see Appendix B), and $\operatorname{Tr} K = E\|w\|^2$. The condition that K be strictly definite is necessary for the forthcoming formula to be meaningful; see the footnote below.

The definition of an LUMV estimator is lifted from the classical one for the finite-dimensional case. The now additional restriction that the estimator be continuous as well as linear is almost necessary for the estimation problem to

---

† It would be interesting to extend a Gauss-Markov theorem to Hilbert space with the condition that K be nonsingular removed, but we have not done this. The classical method and formula we use do not appear to be suitable. It would appear more promising , for example, to try to extend the method and formula in [A-2].

make sense. In fact, the estimator must be everywhere-defined or it is useless; for technical reasons we want it to be a closed operator. Hence it must be bounded. The condition that B be continuous is certainly a technical convenience. As far as the modelling of real problems is concerned, it seems natural to require B to be continuous.

Now if the unknown element x in (2.1) has a non-zero component in the nullspace of B, that component in no way affects the observation z, and it cannot possibly be estimated. Hence we can only estimate the orthogonal projection of x on $N(B)^{\perp}$. In a notation consistent with that used in Part 1, we let P_M be the projection on $N(B)^{\perp}$, B_r be the restriction of B to $N(B)^{\perp}$ and put $x_1 = P_M x$. Henceforth we are concerned only with estimates of x_1.

Since w and z are H_2-valued random variables, Cw and Cz are H_1-valued random variables for any (bounded linear) estimator C (see Appendix B). Further, the mathematical expectations and covariance operators exist, and we have

$$\begin{aligned} E[Cz] &= E[CBx + Cw] = CBx \\ &= CBx_1 . \end{aligned} \tag{2.5}$$

The unbiasedness condition (for x_1) then becomes $CBx_1 = x_1$, or

$$CB_r x_1 = x_1 . \tag{2.6}$$

Thus, the restriction of C to R(B) is necessarily B_r^{-1}, which is another way of stating the unbiasedness condition. From this observation we can make the simple but important inference that B must have closed range. In fact,

$$\begin{aligned} \|C\| = \sup_{\|y\|=1} \|Cy\| &\geq \sup_{\substack{\|y\|=1 \\ y\in R(B)}} \|Cy\| \\ &= \sup_{\substack{\|y\|=1 \\ y\in R(B)}} \|B_r^{-1} y\| = \|B_r^{-1}\| . \end{aligned}$$

Thus C cannot be bounded unless B_r^{-1} is bounded, but this is equivalent to B having closed range (Lemma A. 8). The condition that B have closed range becomes an essential hypothesis.

The variance of a finite variance linear unbiased estmator C for x_1 is

$$E\|Cz - x_1\|^2 = E\|CBx + Cw - x_1\|^2$$
$$= E\|Cw\|^2 . \qquad (2.7)$$

Let $\{\phi_n\}$, $n = 1, 2, \ldots$, be a c. o. n. s. for H_1, chosen so that a subsequence of the ϕ_n's exactly spans $N(B)^{\perp}$. Since K, the covariance of w (see Appendix B), is required to be a nuclear operator, we have for C unbiased,

$$E\|Cz - x_1\|^2 = E\|Cw\|^2$$
$$= E[\sum^{\infty} |(w, C^* \phi_n)|^2] = \sum^{\infty} E(C^* \phi_n, w)\overline{(C^* \phi_n, w)}$$
$$= \sum^{\infty} (K C^* \phi_n, C^* \phi_n) = \sum^{\infty} (CKC^* \phi_n, \phi_n)$$
$$= Tr(CKC^*) . \qquad (2.8)$$

In fact, CKC^* is nuclear since K is nuclear and C is bounded, and so $Tr(CKC^*)$ exists (cf. [G-1], Chapter 1). We see that the conditions on K and C guarantee that the estimator is of finite variance. We summarize these comments and make an obvious addition in

Lemma 2.1: Let the B of (2.1) be a DDC linear operator and let w be an H_2-valued random variable with a nuclear covariance operator K and with $Ew = 0$. Then a necessary condition that a linear unbiased estimator C for x_1 exist is that B have closed range. If B has closed range a finite variance linear unbiased estimator does exist, in particular

$C = B^+$ is one. For these assertions to hold K does not have to be strictly definite and B does not have to be bounded.

Proof: The first assertion has been proved above. By Corollary 1.2, B^+ is a bounded operator. It meets the condition for unbiasedness, (2.6), and the error variance is $\mathrm{Tr}(B^+KB^{+*})$, which is finite as has been pointed out above. |||

Remark: By Corollaries 1.4 and 1.5 it follows that B^+ can be expressed in the form B', and if B is bounded, in the form B''. B^+ provides an estimator whose range is $N(B)^\perp$. However, it is fairly obvious that if $\overline{R(B)} \neq H_2$ and $N(B) \neq \theta$ finite variance unbiased estimators C exist such that $R(C) \cap N(B) \neq \theta$.

We proceed now to a consideration of LUMV estimation. The key step in getting the LUMV estimator is solving a minimization problem – a slightly complicated variant of the basic minimization problem that led to the pseudoinverse. We treat this problem first out of context. Let H_1 and H_2 be Hilbert spaces (not necessarily separable). Let K be a bounded self-adjoint, strictly positive-definite linear operator on H_2. Let B be a bounded linear operator with $D(B) = H_1$ and closed range $R(B) \subset H_2$. Let ϕ be an arbitrary element of $N(B)^\perp$. We call the following minimization problem, problem I:

$$\begin{aligned} &\text{find } c \in H_2 \text{ to minimize } (Kc, c) \\ &\text{subject to } B^* c = \phi \end{aligned} \tag{I}$$

It is convenient to change the form of this problem. Since K is self-adjoint and strictly positive-definite, it has a self-adjoint strictly positive-definite square root $K^{\frac{1}{2}}$. Put $\xi = K^{\frac{1}{2}}c$, from which we have $c = K^{-\frac{1}{2}}\xi$. Then problem II is defined:

$$\begin{aligned} &\text{find } \phi \in H_2 \text{ to minimize } \|\xi\|^2 \\ &\text{subject to } B^* K^{-\frac{1}{2}} \xi = \phi \end{aligned} \tag{II}$$

I and II are exactly equivalent in the following sense; if a ξ_0 exists that solves II, then $K^{-\frac{1}{2}}\xi_0$ is defined and $c_0 = K^{-\frac{1}{2}}\xi_0$ solves I; if a c_0 exists that solves I, then $\xi_0 = K^{-\frac{1}{2}}c_0$ is such that $B^*K^{-\frac{1}{2}}\xi_0$ is defined, and ξ_0 solves II. Consequently, we can restrict our attention to II. Except for one thing we could apply the results on BAS directly to problem II; the difficulty is that the (in general unbounded) operator $B^*K^{-\frac{1}{2}}$ need not be closed, although it is densely defined. We can introduce hypotheses to guarantee that it is closable, but then just replacing it with its closure is not good enough, because a minimizing element ξ_0 in the domain of the closure would not need to belong to $D(B^*K^{-\frac{1}{2}})$. The difficulty is circumvented by working with the adjoint.

<u>Theorem 2.1</u>: Problem II has a unique solution ξ_0 if

(1) $K^{-\frac{1}{2}}B$ is densely defined on H_1

(2) $B_r^{-1}K^{\frac{1}{2}}P_RK^{-\frac{1}{2}}$ is a bounded operator, where P_R is the projection on $\overline{R(K^{-\frac{1}{2}}B)}$ (This operator is well-defined as will be shown).

The solution ξ_0 is given by $\xi_0 = K^{\frac{1}{2}}S^*\phi$, where S is the bounded continuous extension of $B_r^{-1}K^{\frac{1}{2}}P_rK^{-\frac{1}{2}}$ to all of H_2 (it will be seen that $B_r^{-1}K^{\frac{1}{2}}P_rK^{-\frac{1}{2}}$ is densely defined and such an extension exists). Further, S has closed range; $R(S) = N(B)^{\perp}$.

Proof: $K^{-\frac{1}{2}}B$ is closed since $K^{-\frac{1}{2}}$ is closed and B is bounded; it is densely defined by hypothesis (1). Thus, if we put $A = K^{-\frac{1}{2}}B$, A is DDC and admits a PI, A^+. By Theorem 1.2, $A^+ = A_r^{-1}P_R$, where P_R is the projection on $\overline{R(A)}$ and A_r is the natural restriction of A to $N(A)^{\perp}$. We note certain preliminary facts about the transformation A: first,

$$N(A) = N(B), \tag{2.9}$$

or, equivalently,

$$\overline{R(A^*)} = N(A)^{\perp} = N(B)^{\perp} ; \qquad (2.10)$$

second,

$$A_r = K^{-\frac{1}{2}} B_r \qquad (2.11)$$

where B_r is as before the restriction of B to $N(B)^{\perp}$; third, A_r^{-1} is bounded, or, equivalently, A has closed range. The equality (2.9) is valid because $N(A) = N(K^{-\frac{1}{2}}B)$ and the only element $K^{-\frac{1}{2}}$ carries into the null element is the null element. Then (2.11) follows immediately from (2.10). To show that A_r^{-1} is bounded one can verify directly from (2.11) that $A_r^{-1} = B_r^{-1}K^{\frac{1}{2}}$, and note that B_r^{-1} is bounded.

To show that $B_r^{-1}K^{\frac{1}{2}}P_R K^{-\frac{1}{2}}$ is well-defined we must show that $D(B_r^{-1}) \supset R(K^{\frac{1}{2}}P_R K^{-\frac{1}{2}})$. In fact, $\overline{R(K^{\frac{1}{2}}P_R K^{-\frac{1}{2}})} \subset \overline{R(K^{\frac{1}{2}}P_R)} = \overline{R(K^{\frac{1}{2}}K^{-\frac{1}{2}}B)} = \overline{R(B)} = R(B) = R(B_r)$. The one perhaps non-obvious step in this chain is the first equality, which one can verify straightforwardly using the fact that $K^{\frac{1}{2}}$ is bounded.

Now consider the transformation $A^+K^{-\frac{1}{2}} = A_r^{-1}P_R K^{-\frac{1}{2}} = B_r^{-1}K^{\frac{1}{2}}P_R K^{-\frac{1}{2}}$. Since A^+ is closed and bounded, $D(A^+K^{-\frac{1}{2}}) = D(K^{-\frac{1}{2}})$, which is dense in H_2. By hypothesis (2) $A^+K^{-\frac{1}{2}}$ is bounded. Consequently, $A^+K^{-\frac{1}{2}}$ can be extended by continuity to a closed bounded linear transformation S defined on all of H_2.

We show that SB is the orthogonal projection P_M on $N(B)^{\perp}$. Let $x \in D(K^{-\frac{1}{2}}B)$ and put $x = x_1 + x_2$, $x_1 \in N(B)^{\perp}$, $x_2 \in N(B)$. Then,

$$SBx = A^+K^{-\frac{1}{2}}B(x_1 + x_2) = A^+A(x_1 + x_2) = x_1$$

by Lemma A.1 and Theorem 1.3. Since SB is continuous and

everywhere defined and agrees with P_M on a dense subset of H_1, $SB = P_M$. But then SB is self-adjoint, and so $B^* S^* = P_M$.

We can now verify that $\xi_0 = K^{\frac{1}{2}} S^* \phi$ solves problem II. First, since $\phi \in N(B)^{\perp}$

$$B^* K^{-\frac{1}{2}} \xi_0 = B^* K^{-\frac{1}{2}} K^{\frac{1}{2}} S^* \phi = B^* S^* \phi = \phi$$

so the linear constraint is satisfied. Now, A^* is closed and $A^* \supset B^* K^{-\frac{1}{2}}$, since for all $x \in D(A)$ and $y \in D(K^{-\frac{1}{2}})$,

$$(Ax, y) = (K^{-\frac{1}{2}} Bx, y) = (x, B^* K^{-\frac{1}{2}} y) \quad .$$

Consequently, any ξ satisfying $B^* K^{-\frac{1}{2}} \xi = \phi$ satisfies $A^* \xi = \phi$, and $A^*(\xi - \xi_0) = \theta$. Thus, $\xi - \xi_0 \in N(A^*)$. It will follow that $\|\xi_0\|^2$ is a minimum if $\xi_0 \in N(A^*)^{\perp}$, for then

$$\|\xi\|^2 = \|\xi - \xi_0\|^2 + \|\xi_0\|^2 \geq \|\xi_0\|^2 \quad .$$

But ξ_0 does belong to $N(A^*)^{\perp}$. For $SK^{\frac{1}{2}} \supset (A^+ K^{-\frac{1}{2}}) K^{\frac{1}{2}} = A^+$, and since A^+ is everywhere defined, $SK^{\frac{1}{2}} = A^+$. Then $K^{\frac{1}{2}} S^* = (A^+)^*$, so $\xi_0 \in \overline{R((A^+)^*)}$. But by Theorem 1.6 and Lemma 1.1, $\overline{R((A^+)^*)} = \overline{R((A^*)^+)} = N(A^*)^{\perp}$.

To establish uniqueness, let us note first that $\phi \in R(A^*)$, and hence by Theorem 1.1 the problem of minimizing $\|\xi\|^2$ subject to $A^* \xi = \phi$ has a unique solution ξ'. ξ' is characterized as that one element which satisfies $A^* \xi' = \phi$ and $\xi' \in N(A^*)^{\perp}$. Since ξ_0 has been shown to satisfy both these conditions, $\xi' = \xi_0$. Then, <u>a fortiori</u>, ξ_0 is a unique solution to problem II.

Finally, we observe that $\underline{R(S) \supset R(SB)} = R(P_M) = N(B)^{\perp}$, while on the other hand, $\overline{R(S)} = R(A^+ K^{-\frac{1}{2}}) \subset \overline{R(A^+)} = N(A)^{\perp} = N(B)^{\perp}$. Hence $R(S) = N(B)^{\perp}$. |||

<u>Corollary 2.1</u>: The conclusion of Theorem 2.1 holds if hypothesis (2) is replaced by any of the following:

(2') $A^{+}K^{-\frac{1}{2}}$ is bounded;

(2") $A^{*}(AA^{*})_r^{-1}P_R K^{-\frac{1}{2}}$ is bounded;

(2"') $(A^{*}A)_r^{-1}A^{*}K^{-\frac{1}{2}}$ is bounded.

Correspondingly, S may be defined as the continuous extension of any of the operators defined in (2'), (2") or (2"').

Proof: From the proof of Theorem 2.1 we have $A^{+} = B_r^{-1}K^{\frac{1}{2}}P_R$, and from Corollary 1.4 $A^{+} = A^{*}(AA^{*})_r^{-1}P_R$, so the assertions involving (2') and (2") follow. For (2"'), we note that $(A^{*}A)_r^{-1}A^{*} \subset A^{+}$ by Theorem 1.8.

Furthermore,

$$(A^{*}A)_r^{-1}A^{*}K^{-\frac{1}{2}} \supset (A^{*}A)_r^{-1}B^{*}K^{-\frac{1}{2}}K^{-\frac{1}{2}} = (A^{*}A)_r^{-1}B^{*}K^{-1} .$$

But $R(B^{*}K^{-1}) \subset R(A^{*})$, and by Corollary A.4 $D[(A^{*}A)_r^{-1}] = R(A^{*}A) = R(A^{*})$, so $D[(A^{*}A_r)^{-1}A^{*}K^{-\frac{1}{2}}] = D(K^{-\frac{1}{2}})$, which is dense in H_2. Thus the continuous extension of the operator of (2"') is the same S as before. |||

Theorem 2.1, or equivalently Corollary 2.1, can now be applied "coordinate-wise" to give a Gauss-Markov theorem. For this, we need again that the Hilbert spaces be separable. We retain the notation of Theorem 2.1.

<u>Theorem 2.2</u>[1]: In the linear model given by (2.1) let H_1 and H_2 be separable Hilbert spaces, and let w be an H_2-valued random variable with mean zero and strictly positive definite, nuclear covariance operator K. Assume:

(1) B is a <u>bounded</u> linear operator with $D(B) = H_1$, $R(B) \subset H_2$, and $R(B) = \overline{R(B)}$.

(2) $K^{-\frac{1}{2}}B$ is densely defined on H_1.

---

[1] See also [R-5].

(3) $A^+K^{-\frac{1}{2}}$ is a bounded operator (or equivalently the operators defined by (2), (2") or (2"') of Theorem 2.1 and Corollary 2.1 are bounded).

Then $C = S$ is an LUMV estimator for x_1, where $x_1 = P_M x$. The solution $C = S$ is unique.

Proof: For C to be an LUMV estimator of x_1 it must be bounded, satisfy $CBx_1 = x_1$ and minimize $\sum^{\infty}(CKC^* \phi_i, \phi_i)$ over the class of all C satisfying the other conditions. From Theorem 2.1, S is bounded and $SB = P_M$, so $S = C$ satisfies the first two conditions.

Now the condition $CBx_1 = x_1$ implies $CB = P_M$. Hence $CB = B^* C^*$, and the condition $CBx_1 = x_1$ can be rewritten as $B^* C^* x_1 = x_1$. Recall that the c.o.n.s. $\{\phi_i\}$ was chosen so that a subsequence exactly spans $N(B)^{\perp}$. Put $c_i = C^* \phi_i$, $i = 1, 2, \ldots$. Then the unbiasedness condition becomes

$$B^* c_i = \phi_i \quad \text{for those} \quad \phi_i \in N(B)^{\perp} \tag{2.12}$$

and the expression to be minimized is

$$\sum_{i=1}^{\infty} (K c_i, c_i) \ . \tag{2.13}$$

If we can find bounded C so that the corresponding c_i satisfy (2.12), so that all c_i corresponding to $\phi_i \in N(B)$ are zero, and so that each non-zero term of the sum in (2.13) is minimized, then C is LUMV, provided the sum in (2.13) is finite. However, the finiteness of the sum is automatic since C is bounded (see (2.8) and the succeeding comments). But c_i minimizes (Kc_i, c_i) subject to $B^* c_i = \phi_i$ if $c_i = K^{-\frac{1}{2}}(K^{\frac{1}{2}} S^* \phi_i) = S^* \phi_i$ by Theorem 2.1 and the equivalence of problems I and II. Furthermore, $S^* \phi_i = \theta$ for all $\phi_i \in N(B)$ because $N(S^*) = R(S)^{\perp} = N(B)$. Thus $C^* = S^*$ provides a minimizing set of c_i's, and $C = S$ is LUMV. The uniqueness follows from Theorem 2.1. |||

Corollary 2.1: If the subspace $R(B)$ is invariant under K, then conditions (2) and (3) of Theorem 2.2 are automatically satisfied, and B^+ is the unique LUMV estimator.

Proof: Since K is self-adjoint, $R(B)^\perp$ as well as $R(B)$ is invariant under K, that is $K[R(B)] \subset R(B)$ and $K[R(B)^\perp] \subset R(B)^\perp$. Then since K is nonsingular on H_2, its restriction K_B to $R(B)$ is a 1:1 mapping of $R(B)$ into $R(B)$. K_B retains the properties of self-adjointness, strict positivity and nuclearity; K_B^{-1} is defined, and $D(K_B^{-1})$ is dense in $R(B)$. By Lemma A.8, B_r is bounded from below, hence $D(K^{-1}B_r)$ is dense in $N(B)^\perp$, whence by Lemma A.1, $D(K^{-1}B)$ is dense in H_1. This implies condition (2) of Theorem 2.2.

We now consider $B_r^{-1}K^{\frac{1}{2}}P_R K^{-\frac{1}{2}}$ where P_R, it will be recalled, is the projection on $\overline{R(K^{-\frac{1}{2}}B)}$. Let P_B be the projection on $R(B)$. It is immediate from the hypothesis that P_B commutes with K, from which it follows that P_B commutes with $K^{-\frac{1}{2}}$ (c.f. [R-2], p. 143). Then, since B has closed range, $\overline{R(K^{-\frac{1}{2}}B)} = \overline{R(K^{-\frac{1}{2}}P_B)} = \overline{R(P_B K^{-\frac{1}{2}})} = \overline{R(B)}$. Thus $P_R = P_B$, and $B_r^{-1}K^{\frac{1}{2}}P_R K^{-\frac{1}{2}} = B_r^{-1}K^{\frac{1}{2}}P_B K^{-\frac{1}{2}} = B_r^{-1}P_B K^{\frac{1}{2}}K^{-\frac{1}{2}} = B_r^{-1}K^{\frac{1}{2}}K^{-\frac{1}{2}}$. This operator is bounded, so condition (3) is satisfied, and its continuous extension is B^+. |||

Remarks: (1) Although in the case of Corollary 2.1 the LUMV estimator does not depend on K, the error variance, as given by (2.8), does. The result of Corollary 2.1 is to be expected, of course, from the classical case. Intuitively, Corollary 2.1 is saying that if the noise in $R(B)^\perp$ is uncorrelated with the noise in $R(B)$, one may as well first project the observations on $R(B)$.

(2) The hypotheses of Theorem 2.2 are presumably somewhat awkward to verify in many instances. An example where they are satisfied, involving conditions similar to but somewhat less restrictive than the commutativity condition of Corollary 2.1, can be constructed as follows. The verifications are routine and will not be given. Let B be

written in the form DU where D is a self-adjoint operator on H_2 and U is partially isometric from H_1 into H_2. Suppose D has discrete spectrum, so $D\psi_i = \lambda_i \psi_i$, where $\{\psi_i\}$ is a c.o.n.s. for R(B) and $\lambda_i > 0$. Let $\{\eta_i\}$ be a c.o.n.s. for $R(B)^\perp$. Let $K = K_1P_1 + K_2P_2 + K_3P_3$, where: K_1 is a nonsingular covariance operator on $M_1 = V\{\psi_1, \ldots, \psi_N; \eta_1, \ldots, \eta_M\}$[1], K_2 is a nonsingular covariance operator on $M_2 = V\{\psi_{N+1}, \eta_{N+2}, \ldots\}$, K_3 is a nonsingular covariance operator on $M_3 = V\{\eta_{M+1}, \eta_{M+2}, \ldots\}$, and P_i is the orthogonal projection on M_i. Then conditions (2) and (3) of Theorem 2.2 are satisfied.

(3) The condition that B have closed range is, as we have seen, essential, but it is often not satisfied in the kinds of applications one would like to consider. In some instances, one can replace the "natural" H_2, which is typically ℓ_2 or L_2, by a smaller Hilbert space H_2' which still contains all elements $y \in R(B) \subset H_2$, but in which the range of B becomes a closed set. This cannot be done satisfactorily unless the noise is small enough in an appropriate sense. We shall describe this procedure in connection with the example of Part III.

---

1 V{...} denotes the subspace spanned by the vectors.

PART III

AN APPLICATION TO SYSTEM IDENTIFICATION

To identify an unknown input-output system means to determine a suitable mathematical model for the unknown system from incomplete prior knowledge of the system, using data obtained by measurement of outputs and either measurement or prior knowledge of corresponding inputs. A model is suitable if (i) it will reproduce the behavior of the system well enough, according to some set criterion, when the system is stimulated by any one of the class of inputs of interest, and (ii) it is in a useful form. Both the criterion of fit and the usefulness of the model will depend to some degree on what the identification is to be used for, e.g., to permit control of the system, to allow transmission of information through the system, to yield predictions of future behavior, etc. There is usually no reason why there should be only one acceptable model. A definition as general as the one above encompasses a tremendous variety of problems; any two of them may have very little in common with each other, and acceptable solutions may involve disparate mathematical methods. However, at bottom, identification problems are inversion problems, as will be pointed out specifically below, so it is not surprising that certain examples involve generalized inverses. Often the measurements are noisy enough that the basic inversion required becomes a problem of statistical estimation.

To fix ideas, let u denote an input to a system, and y the corresponding output. We suppose $u \in \mathcal{U}$, the class of inputs of interest, and $y \in \mathcal{Y}$, any fixed class of outputs containing all y corresponding to $u \in \mathcal{U}$. The sets $\mathcal{U}$ and $\mathcal{Y}$ will be assigned mathematical structures as seem appropriate (of course, in modelling a problem, these structures are not unique). We assume there is a functional relationship from u to y, that is, for each input $u \in \mathcal{U}$ there is one corresponding output y[1]; then we have

$$y = F(u), \quad u \in \mathcal{U} \quad . \tag{3.1}$$

The function F characterizes the system in question completely, of course, as far as input and output data are concerned (as long as we think of the system as a "black box"), and we shall refer to the mapping F as the "system." This simple terminology carries with it the implication that what one is perhpas accustomed to calling one dynamical system, but with different initial states, here becomes a collection of systems.

Now suppose the system F is unknown and we want to identify it. First, either from prior knowledge, or purely as a working hypothesis, we postulate a class of systems to which (we hope) the unknown system must belong. In different language, we postulate a set $\mathcal{N}$ of mappings from $\mathcal{U}$ into $\mathcal{Y}$ which we assume contains the unknown F. Then we carry out if possible a set of experiments to yield data to fix F closely enough within the class $\mathcal{N}$. Temporarily we can deal with the mapping F as an abstract entity, but eventually it must be represented concretely. The final representation of the estimate of F is the identification. Often an unknown dynamical system is represented in terms of a differential equation, the parameters of which are to be determined by the identification. This really amounts to representing F^{-1}. Another type of representation is directly in terms of integral operators, and this is the kind of representation we work with here.

Suppose now that $\mathcal{U}, \mathcal{Y}$ and $\mathcal{N}$ are given as sets and that in addition $\mathcal{Y}$ is a linear space. We do not yet need to require any mathematical structure for $\mathcal{U}$. There is then a linear structure imposed on $\mathcal{N}$ in the ordinary way; i.e., one defines αF, α a scalar, and $F_1 + F_2$, $F, F_1, F_2 \in \mathcal{N}$, by the equations

[1] There has to be some relation between inputs and outputs or identification makes no sense. There are more general situations of course, e.g., an input u might determine a probability distribution on outputs.

$$[\alpha F](u) = \alpha F(u) \tag{3.2}$$

$$[F_1 + F_2](u) = F_1(u) + F_2(u) . \tag{3.3}$$

Let $\mathcal{L}$ be the linear space of mappings from $\mathcal{U}$ into $\mathcal{Y}$ generated by $\mathcal{N}$. Consider first the problem of noise-free identification of F in which the outputs $y = F(u)$, $u \in \mathcal{U}$, can be known exactly. We interchange the roles of u and F and regard u as determining a mapping from $\mathcal{N}$, and even from $\mathcal{L}$, into $\mathcal{Y}$. So we can write

$$y = U(F) = F(u), \quad F \in \mathcal{L} \tag{3.4}$$

for each $u \in \mathcal{U}$, where U is the mapping corresponding to the input u. The problem of finding F is now the problem of inverting U. U is always a linear mapping. In fact:

$$U(\alpha F_1 + \beta F_2) = [\alpha F_1 + \beta F_2](u) = \alpha F_1(u) + \beta F_2(u)$$
$$= \alpha\, U(F_1) + \beta\, U(F_2) .$$

Thus, in a basic sense, noise-free identification is always a linear problem. If the output of the system can be observed only in the presence of noise w, then the model (3.1) is replaced by

$$z = F(u) + w \tag{3.5}$$

and (3.4) is replaced by

$$z = U(F) + w . \tag{3.6}$$

Equation (3.6) is an abstract version of the usual linear model for statistical estimation.

This simple observation that identification is basically a linear problem, under fairly general conditions, is often not appreciated. It indicates the potential applicability of generalized inverses to identification, quite apart from the particular example to follow. We note, obviously, that the linearity can be lost by using representations of the unknown systems which do not preserve it, and such representations are often used, sometimes for good reasons.

In the generic example to follow, a situation is to be considered in which $\mathcal{Y}$ is a Hilbert space, so that w and z are Hilbert-space-valued random variables. $\mathcal{H}$, the space of systems, will also be a Hilbert space, and U, which is necessarily linear, will be a bounded operator, sometimes with closed range. The model used in the example has been chosen so that we can illustrate the application of theorems in Parts I and II, and hopefully provide some insight into certain aspects of identification theory. However, the model is not typically used in practice, partly because of the computational complexity it introduces. A number of comments need to be made to place what we are doing in better perspective, but these are postponed till after the example.

We model the class of systems to which the unknown system is to belong with a class of Volterra-Frechet polynomials. In particular, consider transformations of the form

$$y(t) = [H(u)](t) = \sum_{n=1}^{N} \int_0^{\tau} \cdots \int_0^{\tau} k_n(\nu_1, \ldots, \nu_n) u(t-\nu_1) \ldots u(t-\nu_n) d\nu_1 \ldots d\nu_n, \quad t \in T \tag{3.7}$$

where T is a finite interval in R^1, τ is a positive number or $+\infty$, N is a fixed positive integer, and the functions u, $k_1, \ldots, k_N, y$ are real or complex-valued. Let $u \in L_2(R^1)$ and $k_n \in L_2([0, \tau]^n)$, $n = 1, \ldots, N$; that is, let k_n satisfy

$$\int_0^{\tau} \cdots \int_0^{\tau} |k_n(\nu_1, \ldots, \nu_n)|^2 \, d\nu_1, \ldots, d\nu_n < \infty \; . \tag{3.8}$$

Clearly, if one permutes the arguments ν_i of the k_n's the integrals in (3.7) are unchanged. Consequently, one can symmetrize each k_n, i.e., replace $k_n(\nu_1, \ldots, \nu_n)$ by $\frac{1}{n!} \sum_\pi k_n(\nu_{\pi(1)}, \ldots, \nu_{\pi(n)})$ where the sum is over all permutations π of n integers, and we suppose this is done. The symmetric kernels of n arguments form a closed linear subspace of $L_2([0, \tau]^n)$ which we denote $\mathfrak{L}_n$.

The transformation defined by (3.7) represents a system which, in systems engineering terminology, is causal with finite memory. Clearly, by changing the interval of integration, the causality condition can be removed. If $T = [0, \tau]$ and if $u(t) = 0$ for $t < 0$, then the integrals can be rewritten with the variable upper limit t without changing the transformation.

It is convenient to introduce the notations: F_n for the integral operator with kernel k_n, and y_n for the nth integral in (3.7). Then (3.7) can be written

$$y = \sum_{n=1}^{N} F_n(u) = \sum_{n=1}^{N} y_n \tag{3.9}$$

We identify the operators F_n with their symmetric kernels k_n, and define $\|F_n\|$ to be the L_2 norm of k_n. Thus, the F_n form a Hilbert space that is isometrically isomorphic to $\mathfrak{L}_n$ under the correspondence $F_n \leftrightarrow k_n$, and to cut down the verbiage we say simply that $F_n \in \mathfrak{L}_n$. We define a norm for F by $\|F\|^2 = \sum_{n=1}^{N} \|F_n\|^2 = \sum_{n=1}^{N} \|k_n\|^2$, and, with again an abuse of notation, regard F as an element of $\mathfrak{L}_1 \oplus \cdots \oplus \mathfrak{L}_N$, which we denote by $\mathfrak{L}$.

An application of the Schwarz inequality shows that

$$|y_n(t)|^2 \leq \|k_n\|^2 \|u\|^{2n} , \tag{3.10}$$

and hence that $y \in L_2(T)$, with

$$\|y_n\|^2 \leq m\|k_n\|^2\|u\|^{2n} \tag{3.11}$$

where m is the length of the interval T. Thus F is a mapping from $L_2(R^1)$ into $L_2(T)$, and it is bounded on bounded sets. It can be shown without much difficulty that if $y_n = F_n(u)$, $y'_n = F_n(u')$ then

$$\|y_n - y'_n\|^2 \leq C(n)\|u-u'\|^2[\max(\|u\|, \|u'\|)]^{2(n-1)} \tag{3.12}$$

(c.f., [R-4]), where $C(n)$ is a constant for fixed n. From (3.12) it follows that H is a continuous mapping from $L_2(R^1)$ into $L_2(T)$. We do not need this fact in what follows, so we do not bother to prove (3.12).

We are now ready to discuss identification of systems modeled by (3.7) in accordance with the ideas leading to (3.6). Let $\mathcal{U}$, the class of systems to which the unknown system is to belong, be $\mathfrak{L}$. Reinterpret (3.7) as $y = U(F)$, where $F = \{k_1, \ldots, k_N\} \in \mathfrak{L}$, and U is the operator determined by the right side of (3.7) with fixed u. U is a linear operator from $\mathfrak{L}$ into $L_2(T)$. It follows easily from (3.11) that U is a bounded linear operator. Actually, U is compact. To see this, let us again introduce notation for each term in the sum in (3.7). Let U_n be n^{th} integral operator in the sum; U_n has kernel $[u(t-\nu_1)\ldots(u(t-\nu_n)]$. We have $y = \sum_{n=1}^{N} U_n(k_n) = \sum_{n=1}^{N} U_n(F_n)$ with an obvious abuse of notation. Further, if we regard U_n as the operator on $\mathfrak{L}$ which carries F_n into y_n and all F_j, $j \neq n$, into zero, we have

$$y = \sum_{n=1}^{N} U_n(F) . \tag{3.13}$$

Now each U_n is Hilbert-Schmidt, in fact

$$\int_T \int_{-\infty}^{\infty} \int_{-\infty}^{\infty} |u(t-\nu_1)\ldots u(t-\nu_n)|^2 d\nu_1 \ldots d\nu_n dt \leq m\|u\|^{2n} . \tag{3.14}$$

Thus, each U_n is compact, and from (3.13) it follows that U is compact.

Suppose that observations of the output are made in the presence of additive noise w, where w is to be represented as an $L_2(T)$-valued random variable with mean zero and nuclear covariance operator K (see Appendix B). The model is then of the form of (3.6), which we repeat,

$$z = U(F) + w , \tag{3.6}$$

and we wish to estimate $F \in \mathcal{H} = \mathcal{L}$. Consider first (non-statistical) least-squares estimation of F –for this, of course, we disregard the statistical properties of w; w is just an error. U is DDC, so by Theorem 1.1 a BAS exists iff $z \in F_U$, the set of z's for which the projection of z on $\overline{R(U)}$ belongs to $R(U)$. The BAS is given by U^+z, where $D(U^+) = F_U$. Since U is bounded, $U^+ = U'' = (U^*U)_r^{-1}U^*$ by Corollary 1.5. These results, though they describe the situation that obtains, are of little practical interest if $R(U)$ is not closed, because an unbounded estimator that is not always defined is of little value. Unfortunately, $R(U)$ is not closed unless U is a degenerate operator, since U is compact. If we consider the estimation of F by linear unbiased estimators from a statistical point of view, the same difficulty obviously arises. In particular, Theorem 2.2 does not guarantee an LUMV estimator unless (aside from other hypotheses) $R(U)$ is closed. Fortunately, this difficulty can be circumvented mathematically by replacing the observation space $L_2(T)$ with a smaller Hilbert space in which the range of U is closed; we do this below. This replacement changes the model of the problem in such fashion that, speaking loosely, some output information is lost. This loss of information may or may not be of consequence, depending on the noise, as we shall see.

We digress from the example temporarily to construct the new Hilbert space that replaces the observation space H_2 . This can be done in more generality than is needed for the example.

Lemma 3.1: Let T be a DDC operator on a Hilbert space H. Then the linear space $D(T)$ is itself a (complete) Hilbert space with the inner product

$$(x, y)'_T = (x, y) + (Tx, Ty), \qquad x, y \in D(T) \quad . \tag{3.15}$$

If T is bounded from below, $D(T)$ is also a Hilbert space with the inner product

$$(x, y)_T = (Tx, Ty), \qquad x, y \in D(T) \quad . \tag{3.16}$$

Proof: Since T is closed, the graph of T, $G(T)$, is a Hilbert space with the norm $\|\{x, Tx\}\|^2 = \|x\|^2 + \|Tx\|^2$ ([D-2], p. 1186). That is, (3.15) defines an inner product for $G(T)$, so $D(T)$ is a Hilbert space. If T is bounded from below there is $\alpha > 0$ such that $\|Tx\| \geq \alpha\|x\|$, $x \in D(T)$. Then

$$\|x\|_T^2 = \|Tx\|^2 \leq \|x\|^2 + \|Tx\|^2 \leq (\alpha^{-1} + 1)\|x\|_T^2 ,$$

so the norms given by (3.15) and (3.16) are topologically equivalent. |||

Theorem 3.1: Let B be a DDC linear transformation from the Hilbert space H_1 into the Hilbert space H_2. Then there is a Hilbert space $\tilde{H}_2$ formed on the linear set $F_B \subset H_2$, but with a different norm, such that

(1) $R(B) \subset \tilde{H}_2$ and $R(B)$ is closed in $\tilde{H}_2$.

(2) The orthogonal complement of $R(B)$ in H_2 is the same set as the orthogonal complement of $R(B)$ in $\tilde{H}_2$, and furthermore the H_2-norm of $y \in R(B)^{\perp}$ is the same as the $\tilde{H}_2$-norm.

(3) $\tilde{B}$, the mapping B reinterpreted as a mapping from H_1 into $\tilde{H}_2$, has closed range, and is bounded iff B is bounded.

Proof: B^*, being DDC, has a polar decomposition, $B^* = JT$, where $T = (BB^*)^{\frac{1}{2}}$ is self-adjoint and J is partially isometric ([D-2], XII,7.7). Then $B = B^{**} = T^*J^* = TJ^*$ since J is bounded ([D-2], XII,1.6), and J^* is also partially isometric ([D-2], XII, 7.6) with initial domain $\overline{R(B^*)} = N(B)^{\perp}$ and final domain $\overline{R(B)}$.

To avoid some confusion of notation later we put $M = \overline{R(B)}$ and $N = R(B)^{\perp}$, so that $H_2 = M \oplus N$. Now $\overline{R(T)} = \overline{R(B)} = M$, and the restriction of T to M, T_r, has a self-adjoint inverse T_r^{-1} by Lemma A.14. We note further that $D(T_r^{-1}) = R(T) = R(B)$. In fact, from $B = TJ^*$ it follows that $R(T) \supset R(B)$. To show that $R(B) \supset R(T)$, suppose $y \in R(T)$, then $y = Tz$ for some $z \in N(T)^{\perp} = \overline{R(T)} = \overline{R(B)}$. But $z \in \overline{R(B)}$ is given by $z = J^*x$ for some $x \in N(B)^{\perp}$, since J^* restricted to $N(B)^{\perp}$ is a unitary mapping onto $\overline{R(B)}$. Thus $y = TJ^*x = Bx$, so $y \in R(B)$.

We can now apply Lemma 3.1 to $R(B) = D(T_r^{-1})$, with T_r^{-1} replacing T, and call the resulting Hilbert space $\tilde{M}$. Then we define $\tilde{H}_2 = \tilde{M} \oplus N$.

The assertion (1) is now immediate since $\tilde{M}$ is itself a Hilbert space. (2) is obvious from the definition of $\tilde{H}_2$. $\tilde{R}(B) = \overline{\tilde{R}(B)}$ is just a restatement of (1) since $R(B)$ and $\tilde{R}(B)$ are the same set. We have left to show that $\tilde{B}$ is bounded iff B is bounded. Let P_1 and P_2 be the projections on $\tilde{M}$ and N respectively. For $x \in H_1$, let x_1 be the projection of x on $N(B)^{\perp}$ and x_2 the projection on $N(B)$. Now, using the norm induced by (3.15) for $\tilde{M}$, and denoting norms in $\tilde{H}_2$ by $\|\cdot\|_{\sim}$, we have for $x \in D(B)$

$$\|\tilde{B}x\|_{\sim}^2 = \|Bx\|^2 + \|T_r^{-1}Bx\|^2$$

$$= \|Bx\|^2 + \|T_r^{-1}TJ^*x_1\|^2$$

$$= \|Bx\|^2 + \|x_1\|^2$$

which proves the assertion. |||

Remark 1. If B is bounded the norm established by (3.16) can be used for $\tilde{M}$, and then $\tilde{B}$ is partially isometric.

Remark 2. If B is compact, $T^2 = BB^*$ is compact as well as self-adjoint and positive, so that $T^2 e_n = \lambda_n e_n$, $n = 1, 2, \ldots$, where the e_n are orthonormal and span $\overline{R(T)} = \overline{R(B)}$, and where $\lambda_n > 0$ and converge monotonically to zero. The norm induced by (3.16) is applicable in this case, and we have for $y \in \tilde{H}_2$,

$$\|y\|_{\sim}^2 = \sum \lambda_n^{-1} |(y, e_n)|^2 + \|P_2 y\|^2 \qquad (3.17)$$

where the norm in the last term is the norm of H_2.

Let us return to the identification problem for which the model is given by (3.6), or by (3.4) in the noise-free case, and for which $\mathcal{U}$ is the class of Volterra-Frechet polynomials which has been introduced. The operator U now plays the role of the operator B in the preceding theorem; otherwise we shall use the notation of that theorem and of Remark 2. For the noise-free case, since $R(U) \subset \tilde{H}_2$ (as a set), we can clearly replace the observations space H_2 and the equation $y = U(F)$ with the space $\tilde{H}_2$ and the equation $\tilde{y} = \tilde{U}(F)$, where $\tilde{H}_2$ and $\tilde{U}$ are as given by Theorem 3.1 with $B = U$. When there is additive observation noise, however, so that the model is given by (3.6), this replacement may mean taking unjustified liberties with the model. For if $L_2(T)$ is the "actual" observation space, then for some values of $w(\omega)$, z will not lie in $\tilde{H}_2$; changing the mathematical model would in this case actually mean changing the original problem. The only case where no harm is done in changing to $\tilde{H}_2$ is when $w(\omega)$ belongs to the set $\tilde{H}_2$ with probability one.

A condition which guarantees this is given in the theorem to follow. Furthermore, the condition will prove to be necessary in a sense to be specified. Since $w(\omega)$ is an H_2-valued random variable, in the sense defined in Appendix B^*, $\{(w, e_n)\}$, with the e_n chosen as in Remark 2, is a sequence of scalar-valued random variables. We suppose w has mean zero and nuclear covariance operator K; then $E[(w, e_n)] = 0$, $E|(w, e_n)|^2 = (Ke_n, e_n) = \text{variance } [(w, e_n)]$. From Remark 2 it follows that

$$\|w(\omega)\|_{\sim}^2 = \sum_{n=1}^{\infty} \lambda_n^{-1} |(w(\omega), e_n)|^2 + \|P_2 w(\omega)\|^2 \tag{3.18}$$

for all ω for which the right side is finite.

Theorem 3.2: If

$$\sum_{n=1}^{\infty} \lambda_n^{-1}(Ke_n, e_n) < \infty \tag{3.19}$$

then $w(\omega) \in \tilde{H}_2$ (as a subset of H_2) with probability one. If $\tilde{w}(\omega) \in \tilde{H}_2$ is defined to equal $w(\omega)$ for all ω for which $\|w(\omega)\|_{\sim}$ is finite and zero otherwise, then $\tilde{w}(\omega)$ is an $\tilde{H}_2$-valued random variable with mean zero and with a nuclear covariance operator $\tilde{K}$.

Conversely, if $w(\omega) \in \tilde{H}_2$ with probability one and $\tilde{w}$ is defined as above, then the existence of a nuclear covariance operator $\tilde{K}$ for $\tilde{w}$ implies that (3.19) is satisfied.

Proof: From (3.18) it follows that

$$E\|w(\omega)\|_{\sim}^2 = \sum_{n=1}^{\infty} \lambda_n^{-1}(Ke_n, e_n) + E\|P_2 w(\omega)\|^2 . \tag{3.20}$$

Since the second term of the right-hand side of (3.20) is finite by the properties of w, the condition (3.19) guarantees the finiteness of (3.20) which in turn implies that $w(\omega) \in \tilde{H}_2$ with probability one.

With the notation used in Theorem 3.1, $\tilde{H}_2 = \tilde{M} \oplus N$. We establish a c.o.n.s. for $\tilde{H}_2$ by separately specifying c.o.n. systems for $\tilde{M}$ and N. The c.o.n.s. for N can be taken arbitrarily: we denote it by $\{f_m\}$. For $\tilde{M}$ it can immediately be verified, using the fact from Remark 2 that

$T_r^{-2} e_n = \lambda_n^{-1} e_n$, that $\{\tilde{e}_n\}$ is a c.o.n.s. in $\tilde{M}$ if $\tilde{e}_n = \lambda_n^{\frac{1}{2}} e_n$. Then, with $\tilde{w}$ defined as in the statement of the theorem, we have with probability one,

$$(\tilde{w}, f_n)_\sim = (w, f_n) \tag{3.21}$$

and

$$\begin{aligned}(\tilde{w}, \tilde{e}_n)_\sim &= (T_r^{-1} w, T_r^{-1} \tilde{e}_n) \\ &= (w, T_r^{-2}(\lambda_n^{\frac{1}{2}} e_n)) \\ &= \lambda_n^{-\frac{1}{2}} (w, e_n) \ .\end{aligned} \tag{3.22}$$

From (3.21) and (3.22) and the fact that w is weakly measurable it follows that $(\tilde{w}, f_n)_\sim$ and $(\tilde{w}, \tilde{e}_n)_\sim$ are measurable scalar-valued functions. Then a limit argument implies that $\tilde{w}$ is weakly measurable, and hence is an $\tilde{H}_2$-valued random variable (see Appendix B).

It now follows from the fact that $E\|\tilde{w}\|_\sim^2 = E\|w\|_\sim^2 < \infty$, which has already been established, that $\tilde{w}$ has a bounded (and hence self-adjoint) covariance operator $\tilde{K}$. The proof is the same as the corresponding proof in Appendix B. To show that $\tilde{K}$ is nuclear we show that $\tilde{K}^{\frac{1}{2}}$ is Hilbert-Schmidt and use the fact ([G-1], p. 39) that the square of a Hilbert-Schmidt operator is nuclear. In fact

$$\begin{aligned}\sum_{n=1}^{\infty} \|\tilde{K}^{\frac{1}{2}} \tilde{e}_n\|_\sim^2 = \sum_{n=1}^{\infty} (\tilde{K} \tilde{e}_n, \tilde{e}_n)_\sim &= \sum_{n=1}^{\infty} E|(\tilde{e}_n, \tilde{w})_\sim|^2 \\ &= \sum_{n=1}^{\infty} \lambda_n^{-1} E|(e_n, w)|^2 \\ &= \sum_{n=1}^{\infty} \lambda_n^{-1} (Ke_n, e_n) < \infty\end{aligned} \tag{3.23}$$

and,

$$\sum_{n=1}^{\infty} \| \tilde{K}^{\frac{1}{2}} f_n \|_{\sim}^{2} = \sum_{n=1}^{\infty} (\tilde{K} f_n, f_n)_{\sim}$$

$$= \sum_{n=1}^{\infty} (K f_n, f_n) < \infty \ , \qquad (3.24)$$

where we have used (3.21) and (3.22).

The finiteness of the left sides of (3.23) and (3.24) demonstrates that $\tilde{K}^{\frac{1}{2}}$ is Hilbert-Schmidt ([G-1], p. 34). The existence of the mean of $\tilde{w}$ follows as in Appendix B. Equations (3.21) and (3.22) together with the fact that $Ew = \theta$ imply that $E\tilde{w} = 0$.

To prove the converse we suppose $\tilde{K}$ exists and is nuclear. Then $\tilde{K}$ has finite trace, so a fortiori the sum $\sum_{n=1}^{\infty} (\tilde{K} \tilde{e}_n, \tilde{e}_n)_{\sim}$ must be finite. But this sum is the same as $\sum_{n=1}^{\infty} \| \tilde{K}^{\frac{1}{2}} \tilde{e}_n \|_{\sim}^{2}$, and the conclusion follows by (3.23). |||

Thus, in the identification example, if the conclusion (3.19) is satisfied, we can replace H_2, U and w by $\tilde{H}_2$, $\tilde{U}$ and $\tilde{w}$, respectively, and then apply Theorem 2.2 to the new model with the assurance that the new model is just as faithful a representation of the real-life problem as was the original one. Theorem 2.2 will apply, of course, only if the covariance operator of the noise is such that conditions (2) and (3) of that theorem hold. These conditions do not conflict, at least in general, with (3.19), as consideration of Corollary 2.1 and the remark (2) following it will show.

As indicated earlier there are a number of comments to be made, both regarding the example as such, and its place in identification theory.

Remarks (1). The condition (3.19) can be interpreted as specifying that a "noise-to-signal-ratio" is finite, or more properly that the sum of noise-to-signal ratios in each orthogonal component is finite. (Ke_n, e_n) is the variance of the noise component (w, e_n), and λ_n is the square of the "signal" component along e_n.

(2) There is no difficulty in extending the system model of (3.7) to the finite-dimensional vector case where u and y are vector-valued (c. f. [R-3]). The same results follow.

(3) The restriction that F be time-invariant can be relaxed in the following way with no essential changes resulting. Let the kernels k_n be of the form:

$$k_n(t, \nu_1, \ldots, \nu_n) = \sum_{i=1}^{M} \alpha_i(t)\, k_n^{(i)}(\nu_1, \ldots, \nu_n). \quad (3.25)$$

With M finite and suitable restrictions on $\alpha_i(t)$, the function spaces are still Hilbert spaces, and U is still compact (c. f. [R-4]). Certain classes of time-varying systems can be modeled this way.

(4) The estimate one obtains is, of course, only for F_1, the projection of F on $N(U)^\perp$. How "large" $N(U)^\perp$ is depends of course on U. If τ is finite, one can choose a sequence of inputs $u_1, u_2, \ldots,$ of duration $m - \tau$, say (recall that m is the length of T); then, allowing τ units of time between inputs (dead-time), make repeated measurements. If p measurements are made, the output space becomes the p-fold product of $L_2(T)$, the input space becomes the p-fold product of $L_2(T-\tau)$, and U is suitably determined by $u_1, \ldots, u_p$. $\mathfrak{X}$, however, remains effectively the same. Clearly, $N(U)^\perp$ is a monotone nondecreasing sequence of subspaces with increasing p. One would expect to do a better job of identifying F as more measurements are made, if the u_i are suitably chosen (see Remark (5)). But making the P repeated measurements is just one special way of making one measurement where the input has duration $pm - \tau$ and the output has duration pm.

(5) From the point of view of identification theory as reasonably constrained by practical considerations, we feel a better approach than what has been done here is the following. Assume a priori that $\mathfrak{X}$ is a compact subset of $\mathfrak{L}$ and

that $\mathcal{U}$ is a compact subset of $L_2(R^1)$. Then, given $\epsilon > 0$, the model (3.4) can be replaced by a finite set of linear equations with finitely many unknown parameters to be found with the property that the solutions for the parameters in the finite set of equations will determine the true system in $\mathcal{U}$ uniformly to within the specified $\epsilon > 0$. With the addition of observation noise w, the problem becomes that of estimating parameters in a finite-dimensional version of (3.6). Then the classical theory of pseudoinverses and of LUMV estimation applies (c.f. [R-3], for example. One can at least argue that the compactness assumptions required are reasonable for most problems.

When identification is studied from the point of view of finite-dimensional approximations, the question raised in the previous remark about the number of measurements necessary (or the required duration of one measurement) is easier to phrase satisfactorily, and is often answerable in principle at least (c.f. [R-3]).

(6) In fairness to the reader unfamiliar with the area, it should be pointed out that the application of Volterra polynomials to problems in system identification has been considered for some time (c.f. [B-2]) but as has been indicated, has not been widely used in practice. A much more common approach is to estimate parameters directly in a state-variable model, using techniques from the Kalman-Bucy recursive filtering theory.

PART IV

ON THE QUADRATIC REGULATOR PROBLEM†

In optimal control theory, no problem has received more attention or attained a greater degree of maturity than the one indicated by the title of this section (see [I-1] for detailed summaries and bibliographical information). Various versions of the quadratic regulator problem (hereafter designated QRP) have appeared in standard texts [A-3] [B-6], perhaps because minimization of a quadratic loss functional applied to a linear dynamical system with linearly constrained inputs leads to a simple and easily understood solution. Here we shall exhibit the PI as unifying apparently disparate aspects of the QRP, providing generalizations beyond its customary form, and having broad capabilities to solve QRP specified in infinite dimensional spaces and by unbounded operators.

The optimal control literature generally relates the QRP to the notion of a linear dynamical system, as described by‡

$$\dot{x} = C(t)x + B(t)u \qquad x(0) = x^o \ . \tag{4.1}$$

Here $u(\cdot)$ and $x(\cdot)$ –usually referred to as input and output, respectively–are finite dimensional vector functions of time over $[0, T]$, $C(\cdot)$ and $B(\cdot)$ are matrix functions of consistent dimension and assuring that a unique solution $x \in L_2^n$ corresponds to every input $u \in L_2^p$. By L_2^m we mean a Hilbert function space; $v \in L_2^m$ implies that $v(t)$ is an m dimensional vector for each $t \in [0,T]$, with v having measurable components v_i, and being possessed of the L_2^m norm

---

† The authors wish to thank Professor Elmer G. Gilbert for his helpful comments on this section.

‡ The PI was first applied to a QRP for a linear dynamical system in [K-1].

$$|||v||| = \sqrt{\sum_{i=1}^{m} \int_0^T |v_i(t)|^2 dt} = \sqrt{\int_0^T \|v(t)\|^2 dt} . \qquad (4.2)$$

We can now state one form of the QRP, sometimes called the fixed endpoint QRP. It is desired to choose $u \in L_2^p$ which moves the system from $x(0) = x^o$ to a designated $x(T) = x^1$, and among the u satisfying this constraint find the u_0 having minimum L_2^p norm. (See [B-6], p. 137). As usually phrased and solved, this QRP presupposes conditions insuring that any x^1 can actually be attained by the system (4.1) at time T; the solution is then said to exist. But actually, this is an unnecessarily narrow interpretation of the fixed endpoint problem due to the limitation of the techniques sometimes used. Application of the PI places this QRP in a natural setting admitting easy generalizations.

Let us begin our analysis by describing a classical approach to the fixed endpoint QRP, and thus exhibiting the limitations resulting therefrom. Since

$$x(T) = \int_0^T W(T, t)\, B(t)\, u(t) dt + W(T, 0)x^o \qquad (4.3)$$

in terms of the state transition matrix $W(\cdot, \cdot)$, the constraint $x(T) = x^1$ can be expressed for an n dimensional x^1 by the simultaneous linear equations

$$(u, h_i) = a_i , \qquad i = 1, 2, \ldots, n \qquad (4.4)$$

with a_i the $i^{\underline{th}}$ component of the n vector $x^1 - W(T, 0)x^o$, i.e.

$$a_i = [\delta_{1i}\, \delta_{2i} \ldots \delta_{ni}] \{x^1 - W(T, 0)x^o\} ,$$

and $h_i(\cdot)$ the p component column vector function

$$h_i(t) = [W(T, t)B(t)]^* \begin{bmatrix} \delta_1 \\ \vdots \\ \delta_{ni} \end{bmatrix} . \tag{4.5}$$

To complete the classical solution, let M be the subspace of L_2^p spanned by $h_1, h_2, \ldots, h_n$, and write u in terms of the orthogonal decomposition

$$u = \sum_1^n \alpha_i h_i + u_1 \tag{4.6}$$

with $u_1 \in M^{\perp}$. If (4.6) is substituted in (4.4), we see u meets the required constraint $x(T) = x^1$ iff

$$\sum_{j=1}^n \alpha_j (h_j, h_i) = a_i \qquad i = 1, 2, \ldots, n . \tag{4.7}$$

The second term u_1 in (4.6) plays no role in meeting the constraint, and (by the Pythagorean theorem for orthogonal elements in Hilbert space) only serves to increase $|||u|||$; consequently, $|||u|||$ is minimized uniquely by choosing

$$u_1 = \theta . \tag{4.8}$$

Evidently, the system can satisfy the final state constraint $x(T) = x^1$ for arbitrary x^1 iff the matrix with elements $\{(h_i, h_j)\}$ is nonsingular. This matrix is Gramian [A-1], which means that its invertibility is equivalent to the linear independence of the h_i, $i = 1, 2, \ldots, n$ in L_2^p . If we say—following standard terminology in linear systems theory—that the system is controllable over $[0, T]$ if, for every x^o and x^1 there exists an input $u \in L_2^p$ taking the system from $x(0) = x^o$ to $x(T) = x^1$, we have attained the following result: the system (4.1) is controllable over $[0, T]$ iff the n rows of the matrix $W(T, \cdot)\, B(\cdot)$ are linearly independent in L_2^p . This criterion is well known in systems theory [C-1]; our arguments constitute a short and elegent proof of its validity.

We now turn to a PI formulation generalizing the fixed endpoint QRP in several directions. First, the QRP can be solved even if the system is not controllable. In fact, the optimal input u_o to (4.1) minimizes $\|x(T) - x^1\|$ over all $u \in L_2^p$, regardless of controllability; if the system (4.1) should be controllable, this norm is zero and the u_o obtained via the PI model agrees with the optimum resulting from the classical problem statement. Second, we shall be able to consider constraints not only on the endpoint $x(T)$, but also on x and u. These three classes of constraints can all be expressed in terms of linear functionals on u, so that they can be viewed in unified form. It is also noteworthy that the PI yields valid optima even when the constraints are incompatible with one another. Finally, the PI model permits us to consider dynamical systems (4.1) in which $u(t)$ and $x(t)$ are Hilbert-spaced-valued (i.e., infinite dimensional) for each $t \in [0,T]$, although we will continue to be restricted to a finite set of constraints.

To apply the PI to the fixed endpoint QRP, we need to express the input-output system relation by a linear transformation, that is

$$Lu = x \quad , \tag{4.9}$$

where u is any element of a Hilbert space H_1, and $L: H_1 \rightarrow H_2$ is a linear bounded operator defined for every such u. In general, a dynamical system such as (4.1) fails to satisfy (4.9), but we may write (substituting $\tilde{x}$ for x)

$$\tilde{x}(t) = (Lu)(t) + S(t)x^o \quad ; \tag{4.10}$$

then, taking $x = \tilde{x} - Sx^o$, we see (4.9) to be applicable without loss of generality.

The linear constraints are considered next. They represent desired system behavior, but do not preclude the possibility that they cannot be attained; in the latter event, the PI automatically chooses a u_o which causes the constraint values to be approached as closely as possible. As we have mentioned, there are three types of constraints, all of which ultimately lead to constraint equations having the form

$$(u, f_i) = a_i \ .$$

A linear functional on u is already in the stated form. As for a linear functional on x, we may write $G_i(x) = a_i$ which, by the Riesz representation theorem and the definition $f_i = L^* g_i$ gives rise to

$$G_i(x) = (x, g_i) = (Lu, g_i) = (u, L^* g_i) = (u, f_i) = a_i \ ; \quad (4.11)$$

the constraint is then in the proper form. Lastly, choosing $f_i = h_i$ as in (4.5), and defining a_i as in the equation above, (4.5) renders $(u, f_i) = [x(T)]_i$ and $a_i = x_i^1$, whence $(u, f_i) = a_i$, $i = 1, 2, \ldots, n$ is equivalent to $x(T) = x^1$.

Application of the PI involves an operator equation to which the PI gives the BAS. Thus, the constraints expressed as $(u, f_i) = a_i$ must be translated into a more appropriate form for this purpose. Accordingly, let H_3 be a separable Hilbert space with a maximal linearly independent set $\{y_n\}$, and define $A: L_2^p \to H_3$ by

$$Au = \sum (u, f_i) y_i \ . \quad (4.12)$$

Then, with

$$z = \sum a_i y_i \quad (4.13)$$

we obtain

$$Au = z \quad (4.14)$$

as the operator equation embodying all the desired constraints.

If a finite constraint set is set forth, the range of A is a finite dimensional set in H_3 and perforce closed. By Theorem 1.2 and Corollary 1.2, A possesses a PI A^+ which is bounded and delivers a BAS (cf. Definition 1.1)

$$u_o = A^+ z = \sum a_i A^+ y_i \ , \qquad (4.15)$$

which means $u_o \in L_2^p$ is the unique input to the dynamical system satisfying

$$\inf_{u \in L_2^p} \|z - Au\|^2 = \inf_{u \in L_2^p} \{ \sum_{i,j} [(u, f_i) - a_i][\overline{(u, f_j) - a_j}](y_i, y_j) \} = \qquad (4.16)$$

$$\sum_{i,j} [(u_o, f_i) - a_i][(u_o, f_j) - a_j](y_i, y_j) \ ,$$

and having smallest L_2^p norm (i.e., least energy) attaining the infimum (4.16). By choosing suitable families $\{y_n\}$ we are thus able to prove existence and uniqueness of optimal controls for a variety of quadratic loss functions. We remark parenthetically that if there exists a $u \in L_2^p$ exactly satisfying all the constraints $(u, f_i) = a_i$, (4.16) becomes zero, and u_o is the element of smallest norm which meets the constraints. In particular, if the constraints represent $x(T) = x^1$ and the system (4.1) is controllable, the u_o furnished by the PI (4.15) is the same as the optimum obtained as the solution of the classical fixed endpoint QRP.

In applying the PI to the QRP in the manner described, we need not be concerned with the consistency of the constraints; if the constraints are incompatible, the PI mediates among them to produce a compromise BAS. To illustrate this property, we give an example involving a simple constraint. Consider therefore the system (4.4) with $n = 2$, $h_1 = h_2 = h$ and $a_1 \neq a_2$, so that (4.7) is clearly impossible to satisfy. We may, however, apply (4.15) to determine the BAS u_o. The latter naturally depends on the choice of y_1 and y_2, as can be seen from (4.16). If we choose y_1 and y_2 orthonormal (for convenience), we have $N(A)^\perp = V\{h\}$, $R(A) = V\{y_1 + y_2\}$, $Ah = \|\|h\|\|^2 (y_1 + y_2)$, and

$z = a_1y_1 + a_2y_2$. In view of $P_R z = \frac{1}{2}(a_1 + a_2)(y_1 + y_2)$, the description of A^+ by Theorem 1.2 leads to

$$u_o = A^+ z = \tfrac{1}{2}(a_1 + a_2) \frac{h}{|||h|||^2} .$$

Thus far, reference in (4.1) has been to dynamical systems in which $x(t)$ and $u(t)$ are finite dimensional vectors of fixed dimension for each t . Since we have made no use of the finite dimensionality, we may as well generalize (4.1) as follows: $u(t) \in K_1$ for almost every $t \in [0, T]$, K_1 being a separable Hilbert space. Assume $u(\cdot)$ strongly (Bochner) measurable [H-2], and $u \in H(K_1)$, that is

$$|||u|||^2 = \int_0^T \|u(t)\|^2 dt < \infty . \tag{4.17}$$

Further, in (4.1), let $B(\cdot)$ be an essentially uniformly bounded measurable operator valued function which for each t carries K_1 into a new Hilbert space K_2, and let $C(\cdot)$ be locally Bochner integrable, with $C(t)$ an endomorphism on K_2 for almost every t . Then, if (4.1) is interpreted in integral equation form, there is a unique solution $x(\cdot)$ which is uniformly bounded (cf. [M-1]) and hence belongs to $H(K_2)$; indeed, one obtains (4.9) and (4.10) as before, with $L: H(K_1) \to H(K_2)$ and $S(T): K_2 \to K_2$ both bounded operators. The remainder of the analysis is unchanged, provided that there is a finite constraint set, which then implies $R(A)$ in (4.14) to be finite dimensional.

The finite dimensionality of $R(A)$ [with A specified by (4.12)] becomes inconsistent with a constraint of the type $x(T) = x^1$ whenever K_2 is infinite dimensional. At this point we could confine ourselves to a finite constraint set as represented by the projection of x^1 onto a hyperplane of finite dimension. Alternatively, we might attempt to deal more directly with infinite dimensional $x(T)$ (or any infinite constraint set) by choosing $\{f_i\}$ and $\{y_i\}$ in

$$Au = \sum_{1}^{\infty} (u, f_i)y_i$$

so that A [cf. also (4.12)] is DDC with closed range. Only then is the PI A^+ defined for arbitrary z (Corollary 1.2) and hence for every choice of x^1. Moreover, z makes sense only if the sum $\sum a_i y_i$ appearing in (4.13) is convergent in H_3. The requirements on A and z are formidable, and cannot be met in general. For instance, the obvious terminal constraint choice of $\{y_i\}$ as a complete orthonormal set in $H_3 = K_2$ and the a_i as projections of x^1 on $V\{y_i\}$ lead to a convergent sum for z, but in combination with a DDC operator A whose range is not closed.

The limitation of the above PI technique to a finite constraint set is avoided in a somewhat different QRP model, which is termed the free endpoint QRP. As usually described in the optimal control literature, the free endpoint QRP is concerned with moving the system (4.1) from $x(0) = x^0$ toward the origin at time T with minimum expenditure of energy, simultaneously maintaining x as small as possible. More precisely ([B-6], Section 21), one wishes to attain the infimum of the loss function

$$J(u) = \int_0^T [\|u(t)\|^2 + \|Q_2 x(t)\|^2]dt + \|Q_3 x(T)\|^2 \qquad (4.18)$$

where the $\|\cdot\|$ are Euclidean norms on the relevant finite dimensional vector space. Analysis of the free endpoint QRP revolves around the existence and uniqueness of the optimal control [i.e., u attaining the infimum of (4.18)], and the computation of the optimum whenever it does exist.

In what follows, we shall generalize the free endpoint QRP to infinite dimensional spaces, and demonstrate that the optimal input is (uniquely) provided by the PI even for some models involving unbounded operators. Moreover, our loss function

$$I(u) = \|Q_1(u-z_1)\|^2 + \|Q_2(x-z_2)\|^2 + \|Q_3[x(T) - z_3]\|^2 \qquad (4.19)$$

will be more general than the $J(\cdot)$ of (4.18). The z_i in (4.19) may be thought of as target values of the input, output and terminal state sought in the system design; they also reflect the change of variable (4.10), so that (4.19) becomes equivalent to (4.18) if we specify in particular $z_1 = \theta$, $z_2(t) = -S(t)x^o$ and $z_3 = -S(T)x^o$. We should mention, however, that any interpretation of (4.19) in terms of an optimal control problem is merely an intuitive convenience, since we shall make no use of any special properties of the dynamical system (4.1).

For treating the minimization problem incompletely posed by (4.19), we establish the following structure. Let $u \in H_1$ (an arbitrary Hilbert space), and take $L: H_1 \to H_2$, with

$$Lu = x \quad . \tag{4.20}$$

It will be supposed that L–like all other operators mentioned here–is linear and DDC. We shall assume further that H_2 is a Hilbert function space, so that we are writing x for $x(\cdot)$, and that there is a linear DDC operator $S: H_1 \to H_3$ such that

$$x(T) = Su \quad . \tag{4.21}$$

We introduce yet another Hilbert space, $K = H_1 \times H_2 \times H_3$; its norm is the standard

$$\|\{u, v, w\}\|^2 = \|u\|^2 + \|v\|^2 + \|w\|^2 \tag{4.22}$$

The operator $A: H_1 \to K$ is defined by

$$Au = \{u, Lu, Su\} \tag{4.23}$$

and we have

$$D(A) = D(L) \cap D(S) \quad . \tag{4.24}$$

Since L and S are both closed, $G(A)$ is a closed graph in K, and A itself is a closed operator. But $D(A)$ is not necessarily dense in H_1, so A must be assumed densely defined by hypothesis. Lastly, let us define $Q: K \to K$ by the relation

$$Q\{u, v, w\} = \{Q_1 u, Q_2 v, Q_3 w\} \tag{4.25}$$

where each Q_i is not only bounded, but also bounded from below. The latter condition is expressed by the existence of an $\alpha > 0$ such that

$$\|Q_i y_i\| \geq \alpha \|y_i\| \quad \text{for all } y_i \in H_i . \tag{4.26}$$

We can now state

<u>Theorem 4.1</u>: Let L and S be DDC operators as defined above, with $D(L) \cap D(S)$ dense in H_1. Assume that each Q_i is bounded and bounded below. Then, for each $z = \{z_1, z_2, z_3\} \in K$ there exists a unique $u_o \in H_1$ satisfying

$$I(u_o) = \inf_{u \in H_1} [I(u)] , \tag{4.27}$$

with I as in (4.19). This u_o is given by

$$u_o = (QA)^+ Qz . \tag{4.28}$$

Proof: It is clear from (4.25) and the properties of the Q_i that Q is an endomorphism on K, and is bounded below; from this (and that A is DDC) one shows QA to be a closed operator. Furthermore, A is densely defined, so the same is true for QA, and in fact QA is DDC .

The range of QA is closed because [see Lemma A. 8(e)] for all $u \in D(A)$

$$\|QAu\| \geq \|Q_1 u\| \geq \alpha \|u\| , \tag{4.29}$$

and the inequality also shows that QA is one-to-one. In the light of these properties, we can now consider the BAS for the equation

$$(QA)u = Qz \; ; \tag{4.30}$$

since QA is DDC with R(QA) closed, the BAS is furnished for every $(Qz) \in K$ (Corollaries 1.1 and 1.2) by the PI. This means the u_o of (4.28) satisfies

$$\|Qz - (QA)u_o\| = \inf_{u \in H_1} \|Qz - (QA)u\| \; . \tag{4.31}$$

If we compare the norm of $Qz - (QA)u$ with the expression (4.19) for I(u), we see at once they are identical, so that (4.31) is equivalent to (4.27).

To prove u_o is the unique element attaining the infimum of (4.31), note (from the proof of Theorem 1.1) that $(QA)u_o$ must be the projection of Qz on R(QA). But QA is one-to-one, so u_o is the only element of H_1 such that $(QA)u = P_R(Qz)$. |||

Various other combinations of assumptions also yield free endpoint QRP having the same solution (4.28). Most of these are of little interest, and can be reproduced by the reader as needed. However, we may want to consider bounded L and S, especially since these correspond to the dynamical system (4.1) with infinite dimensional vector valued functions as per our earlier discussion. We find that the assumptions on Q_2 and Q_3 can then be relaxed, as is indicated by

Corollary 4.1: If L and S are bounded operators, if each Q_i is bounded and Q_1 is bounded from below, the conclusions of Theorem 4.1 continue to hold.

Proof: QA is now a bounded operator which is everywhere defined. The remainder of the proof of Theorem 4.1 remains unchanged, the same assertions being valid throughout. |||

PART V

PSEUDOINVERSE OPERATOR APPROXIMATIONS

In the literature on the matrix PI, it is sometimes proposed that the PI be obtained by a sequence of approximations involving AA^* or A^*A (see [A-2], Theorem 3.4 and [L-2], p. 167). Such forms exhibit computational simplicity, since they substitute the inversion of a strictly positive (Hermitian) matrix for more complicated combinations of inverses. When approximations of the same type are applied to DDC operators, the objectives are necessarily different. Computational simplicity is no longer at issue; we are now concerned with analytical tractability, particularly for unbounded operators whose range may not be a closed set.

One goal is to circumvent the problems that inevitably arise when $R(A)$ is not closed, and A^+ is unbounded and not defined everywhere; we have already sought solutions in Parts III and IV by changing the topology, but now we also consider approximating sequences of operators. The description of A^+ and approximations thereto in terms of the positive operators AA^* or A^*A not only generalizes matrix approximations of the same type, but also suggests use of the spectral representation and the associated functional calculus. We shall find that the PI can be expressed in terms of positive operators (and the functional calculus) as a consequence of the use of the polar decomposition ([D-2], XII.7.7); indeed, the polar decomposition proves to be an extremely powerful tool in the analysis of the PI, and will be used often in this section. Finally, we remark that the approximations considered below are closely related to the operators A' and A'' defined by (1.38) and (1.44), respectively; these operators are studied extensively in Part I, and find application in Parts II and III.

Since we are dealing with DDC operators, we shall face questions that fail to arise in the context of the matrix PI. In addition to the boundedness and domain properties as in Part I, we shall need to investigate the mode of convergence for each approximation to the PI. In particular, we shall show that certain approximation sequences always converge to the PI strongly, but in norm iff $R(A)$ is a closed set.

In what follows, we shall freely use the notation and results of Part I and Appendix A, although we shall assist the reader with specific references whenever possible. For future reference, we also call the reader's attention to the well known content of

Lemma 5.1: Every DDC operator A has a polar decomposition

$$A = VS \tag{5.1}$$

where $V: H_1 \to H_2$ is a partial isometry from $N(A)^{\perp}$ onto $\overline{R(A)}$, and S is the positive operator $(A^*A)^{\frac{1}{2}}$ with spectral representation

$$S = (A^*A)^{\frac{1}{2}} = \int_0^{\infty} \lambda dG_{\lambda} \ . \tag{5.2}$$

In (5.2), G_o is the projection on $N(A)$. Alternatively, A may be decomposed as

$$A = TU^* \ , \tag{5.3}$$

$U: H_2 \to H_1$ being a partial isometry from $\overline{R(A)}$ onto $N(A)^{\perp}$ [and null on $R(A)^{\perp}$], and $T: H_2 \to H_1$ the positive operator

$$T = (AA^*)^{\frac{1}{2}} = \int_0^{\infty} \lambda dE_{\lambda} \ ; \tag{5.4}$$

here E_o is the projection on $R(A)^{\perp}$.

Proof: The first decomposition is stated and proved in [D-2], XII.7.7, and the spectral representation discussed in XII.2 of the same reference. Now apply this decomposition to A^*, i.e., $A^* = UT$. On taking the adjoint of A^*, we obtain (5.3); the equality is a consequence of U bounded ([R-2], Section 115).

From the equality ([D-2], XII. 2. 6)

$$\|Ty\|^2 = \int_0^\infty \lambda^2 d\|E_\lambda y\|^2 \tag{5.5}$$

and the properties of the resolution of the identity, we see that E_o is the projection on $N(T) = R(T)^\perp = R(AA^*)^\perp = R(A)^\perp$ (compare Corollary A. 3). The proof that G_o is the projection on N(A) is similar. |||

The representations (5.1) and (5.3) lead to a natural way of expressing approximations to the PI when these involve

$$AA^* = T^2 \quad \text{or} \quad A^*A = S^2 \ . \tag{5.6}$$

To conveniently compare the approximations with A^+, we then write the latter in terms of T (or S) also. Furthermore, the formulation of A^+ in terms of the polar decomposition may be of some independent interest. Motivated by these considerations, we state and prove

Theorem 5.1: The PI A^+ of A can be represented as[†]

$$A^+ = U_r T_r^{-1} P_R \tag{5.7}$$

or alternatively

$$A^+ = S_r^{-1} V_r^{-1} P_R \ . \tag{5.8}$$

[†]To avoid possible confusion, we again remind the reader that the restrictions applied to any operator are to the null-space or closure of the range of A. The identities and relationships given in Appendix A apply to U_r, T_r, etc. only because the null and/or range spaces coincide with N(A) and/or $\overline{R(A)}$.

Proof: We apply the characterization of $A^+ = A_r^{-1} P_R$ of Theorem 1.2. From the definition of U as a partial isometry, $N(U^*) = N(A)$, so the results of Appendix A on restrictions of operators are applicable. By Lemma A.9, we may write U_r^* to denote either $(U^*)_r$ or $(U_r)^*$ interchangeably; hence, for $x \in N(A)^\perp$, $A_r x = Ax = TU^* x = TU_r^* x$. Furthermore, U_r^* is a unitary operator on the restricted spaces, with $R(U_r^*) = R(U^*) = \overline{R(A)}$ so that we even have

$$A_r = TU_r^* = T_r U_r^* = T_r U_r^{-1} . \tag{5.9}$$

To compute A_r^{-1} from (5.9)–this inverse exists by Lemma A.6–we first argue that T_r is invertible; this follows because (5.9) implies $A_r U_r = T_r$, with both operators on the left side of the equality having inverses. With the existence of T_r^{-1}, inversion of (5.9) yields $A_r^{-1} = U_r T_r^{-1}$ (c.f., [R-2], Section 114), and this proves (5.7).

The argument on (5.8) is analogous and even easier. $\overline{R(S)} = N(A)^\perp$, which means $A = V_r S$, V_r being a unitary operator from $N(A)^\perp$ to $\overline{R(A)}$. Clearly, $A_r = V_r S_r$ from which $A_r^{-1} = S_r^{-1} V_r^{-1}$. Then A^+ is given by (5.8) as claimed. |||

The restrictions of S and T are easily identified in terms of the spectral representations (5.2) and (5.4). Since the spectral representations will appear again later in this section, we need

<u>Lemma 5.2</u>: Let

$$H_\lambda = G_\lambda - G_o \quad \text{and} \quad F_\lambda = E_\lambda - E_o . \tag{5.10}$$

Then S_r and T_r have the respective spectral representations

$$S_r = \int_0^\infty \lambda \, dH_\lambda \quad \text{and} \quad T_r = \int_0^\infty \lambda \, dF_\lambda \tag{5.11}$$

Proof: The verifications are identical for S_r and T_r, so we need prove the spectral representation only for the first of these. From Theorem 5.1, G_o is the projection on $N(A)$, whence $G_\lambda x = H_\lambda x$ for any $x \in N(A)^\perp$; this means the spectral integrals (5.2) and (5.11) are the same when applied to any such x. It remains to show that H_λ is a resolution of the identity on the Hilbert space $N(A)^\perp$. Evidently, H_λ is a right continuous increasing family of projections with $H_o = \theta$ and $\lim_{\lambda \to \infty} H_\lambda = I - G_o$. The latter is the projection on $N(A)^\perp$, which is the identity in the restricted space. Thus, H_λ satisfies all the conditions required of a resolution of the identity. |||

We are now ready to define the approximations to the PI which constitute the principal objects of our study. For any positive number σ, let

$$A'_\sigma = A^*[\sigma I + (AA^*)]^{-1} . \tag{5.12}$$

In the finite dimensional case ([A-2], Theorem 3.4 and [L-2], p. 167) the limit of A'_σ exists as $\sigma \to 0$, and in fact

$$\lim_{\sigma \to 0} A'_\sigma = A^+ . \tag{5.13}$$

Since one is only required to invert a positive matrix to implement A'_σ, (5.12) appears as an attractive formula for approximating the PI. Some matrix manipulations on (5.12), followed by a passage to the limit, yield

$$\lim_{\sigma \to 0} A'_\sigma = A' , \tag{5.14}$$

where we have previously defined

$$A' = A^*(AA^*)_r^{-1} P_R . \tag{5.15}$$

Clearly, A' is another form of the PI in finite dimensional spaces; this also is obtained as a special case of our Corollary 1.4.

As we shall see, the simple and straightforward results pertaining to finite dimensional spaces continue to hold in the more general context of operators whose range is closed. In the event $R(A)$ is not closed, A' is an RPI as shown by Theorem 1.7, whereas A'_σ behaves quite differently from A', and moreover fails to converge to A^+ in norm as $\sigma \to 0$. The comparison of A^+, A and A'_σ for operators of non-closed range is of particular interest in Part II, where the PI of a Hilbert-Schmidt operator makes its appearance.

The essential facts regarding A'_σ are summarized in

Theorem 5.2: The operator A'_σ given by (5.12) is bounded and defined on all H_2, and the same is true for

$$AA'_\sigma = (AA^*)[\sigma I + (AA^*)]^{-1} . \tag{5.16}$$

The restriction of A'_σ to $D(A^+)$ converges strongly to A^+. The convergence

$$\lim_{\sigma \to 0} A'_\sigma = A^+ \tag{5.17}$$

is uniform (i.e., in norm) iff $R(A)$ is a closed set.

Remark: Since AA'_σ is defined for all $z \in H_2$, one might conjecture that A'_σ accomplishes what A^+ cannot. As we know, the BAS fails to exist and $z \notin D(A^+)$ if $P_R z \notin R(A)$ (see Theorem 1.1), but it is reasonable to expect that A'_σ delivers something close to the optimum as σ becomes small. These possibilities will be explored later in this section.

Proof: It is known that $[\sigma I + (AA^*)]^{-1}$ is bounded and everywhere defined ([R-2], Section 118), and since A^* is DDC, the operator A' is closed. We also have

$$R\{[\sigma I + (AA^*)]^{-1}\} = D[\sigma I + (AA^*)] = D(AA^*) \subset D(A^*) \tag{5.18}$$

so A'_σ is everywhere defined and, being closed, must be bounded according to the closed graph theorem. The same reasoning shows AA'_σ bounded and everywhere defined also.

For a comparison of A^+ and A'_σ it suffices to consider the application of these operators to $z \in R(A)$. In fact, $N(A^+) = R(A)^\perp$ as shown in Lemma 1.1, and we now prove $N(A'_\sigma) \supset R(A)^\perp$. Suppose then $z \in R(A)^\perp$, which implies $[\sigma I + (AA^*)]^{-1} z = \sigma^{-1} z \in R(A)^\perp$, or in other words, $R(A)^\perp$ reduces $[\sigma I + (AA^*)]^{-1}$. But $N(A^*) = R(A)^\perp$, so we obtain $A'_\sigma z = \theta$.

To demonstrate the asserted strong convergence of A'_σ to A^+, we apply $(A'_\sigma - A^+)$ to $z \in R(A)$. Since $z = Ax$ for some $x \in N(A)^\perp$, and since $A^+Ax = x$ for such an element (Theorem 1. 3), we have

$$\|(A'_\sigma - A^+)z\|^2 = \|\{A^*[\sigma I + (AA^*)]^{-1}A\}x - x\|^2 . \quad (5.19)$$

All operations on the right side of (5.19) take place in the restricted Hilbert spaces $N(A)^\perp$ and $\overline{R(A)}$ because $[\sigma I + (AA^*)]^{-1}$ reduces $\overline{R(A)}$ as we have shown. Thus we obtain

$$A_r^*[\sigma I_r + (AA^*)_r]^{-1}A_r = U_rT_r[\sigma I_r + T_r^2]^{-1} T_rU_r^{-1} . \quad (5.20)$$

Now U_r is unitary over the restricted Hilbert spaces, so that (5.19) becomes

$$\|\{T_r[\sigma I_r + T_r^2]^{-1} T_r - I_r\}y\|^2 = \int_0^\infty \left|\frac{\sigma}{\sigma + \lambda^2}\right|^2 d\|F_\lambda y\|^2 \quad (5.21)$$

for $U_r^{-1}x = y \in D(T_r)$; the rules applicable to such calculations may be found in [R-2], Section 128. On putting these last three equations together we find

$$\lim_{\sigma \to 0} \|A'_\sigma z - A^+z\|^2 = \lim_{\sigma \to 0} \int_0^\infty \left|\frac{\sigma}{\sigma + \lambda^2}\right|^2 d\|F_\lambda y\|^2 . \quad (5.22)$$

The measure generated by $\|F_\lambda y\|^2$ is finite, and the integrand in (5.22) is bounded by unity. Consequently, the Lebesgue dominated convergence theorem insures that the right side of (5.22) is zero, thus yielding the strong convergence asserted by the theorem.

Suppose now $R(A)$ is not closed. Then A'_σ remains bounded but A^+ is unbounded (Corollary 1.2), so A'_σ cannot converge in norm to A^+. On the other hand, for $y \in R(A)$ it is always true that

$$\begin{aligned}\|(A'_\sigma - A^+)y\|^2 &= \|U_r[T_r(\sigma I_r + T_r^2)^{-1} - T_r^{-1}]y\|^2 \\ &= \|[T_r(\sigma I_r + T_r^2)^{-1} - T_r^{-1}]y\|^2\end{aligned} \tag{5.23}$$

by the characterization (5.7) of A^+, and the fact that U_r is unitary. The norm of (5.23) can once more be written in terms of the spectral representation, viz.

$$\|(A'_\sigma - A^+)y\|^2 = \int_0^\infty \left|\frac{\lambda}{\sigma+\lambda^2} - \lambda^{-1}\right|^2 d\|F_\lambda y\|^2 = \int_0^\infty \left|\frac{\lambda^{-1}\sigma}{\sigma+\lambda^2}\right|^2 d\|F_\lambda y\|^2 . \tag{5.24}$$

If $R(A)$ is closed, $R[(AA^*)_r]$ is likewise closed (Corollary A.3), whence T_r^2 and <u>a fortiori</u> T_r have bounded inverses. But then ([R-2], Section 128) there exists an $\alpha > 0$ such that $F_\lambda = \theta$ whenever $\lambda < \alpha$. The lower limit of integration in (5.24) can then be changed from zero to α, with the result that the integrand is bounded by $\alpha^{-6}\sigma^2$ over the entire interval of integration. Consequently,

$$\|(A'_\sigma - A^+)y\|^2 \le \alpha^{-6}\sigma^2\|y\|^2 \tag{5.25}$$

and hence A'_σ converges to A^+ in norm as σ tends toward zero. |||

In Section I, we introduced not only $A' = A^*(AA^*)_r^{-1}P_R$, but also another RPI which enjoyed a limited duality with A'. The latter operator was $A'' = (A^*A)_r^{-1}A^*$, whose domain was quite complicated (see Theorem 1.8) in general; nevertheless, A'' proved useful in connection with the applications of Section II. The relation between A' and A'' is mirrored by a corresponding similarity of A'_σ to A''_σ, where

$$A''_\sigma = [\sigma I + (A^*A)]^{-1}A^*, \quad \sigma > 0 \ . \qquad (5.26)$$

The principal properties of A''_σ are stated in

<u>Theorem 5.3:</u> A''_σ is bounded and is possessed of the domain $D(A''_\sigma) = D(A^*)$. The (closed minimal) extension of A''_σ to an operator defined on all H_2 converges strongly (as $\sigma \to 0$) to A^+ on $D(A^+)$, and converges to A^+ in norm iff $R(A)$ is a closed set.†

<u>Remark:</u> If A (and hence A^*) is unbounded, A''_σ cannot be closed. To avoid the ensuing pathologies, we therefore state the convergence parts of the above theorem in terms of the closed extension of A''_σ.

<u>Proof:</u> The domain of A''_σ is indeed $D(A^*)$ because $[\sigma I + (A^*A)]^{-1}$ is bounded and everywhere defined on H_1. To show A''_σ bounded, observe that $(A^*)'_\sigma = A[\sigma I + (A^*A)]^{-1}$ is bounded and everywhere defined, as may be seen by applying the last theorem to A^* in place of A. Then its adjoint $[(A^*)'_\sigma]^*$ is likewise bounded and defined everywhere on H_2. The adjoint satisfies (see [R-2], Section 115c)

$$[(A^*)'_\sigma]^* \supset A''_\sigma \ . \qquad (5.27)$$

† Strong convergence of A''_σ for bounded A is already asserted in [B-1], p. 60.

Thus A''_σ is bounded on $D(A^*)$, and since the latter is dense in H_2, the minimal closed extension of A''_σ is the bounded operator $[(A^*)'_\sigma]^*$.

We call $[(A^*)'_\sigma]^* = A^x_\sigma$. When $R(A)$ is not closed, A^+ is unbounded, and A^x_σ cannot converge to A^+ in norm. On the other hand, A^+ is bounded whenever $R(A)$ is a closed set, and then

$$\|A^x_\sigma - A^+\| = \|(A^x_\sigma - A^+)^*\| = \|(A^*)'_\sigma - (A^+)^*\| = \|(A^*)'_\sigma - (A^*)^+\| \to 0 . \tag{5.28}$$

These relations are justified by: (1) a bounded operator and its adjoint have the same norm ([H-1], Theorem 22.2), (2) $(A^+)^* = (A^*)^+$ by Theorem 1.6, and (3) since $R(A^*)$ is closed also (Corollary A.1) $(A^*)'_\sigma$ converges to $(A^*)^+$ by Theorem 5.2 above.

To complete the proof, we verify the strong convergence of A^x_σ to A^+ on $D(A^+)$. We note at once that $N(A''_\sigma) = N(A^*) = R(A)^\perp = N(A^+)$, so (as in the proof of Theorem 5.2) we confine ourselves to $z \in R(A)$, which we will represent as $z = Ax$, $x \in N(A)^\perp$. Consequently, it is sufficient to show

$$\|A^x_\sigma Ax - A^+Ax\|^2 \to 0 . \tag{5.29}$$

Here again $A^+Ax = x$, and $A^x_\sigma A$ may be replaced by its minimal closed extension $S^2(\sigma I + S^2)^{-1}$. That the latter is in fact the closed minimal extension follows from the definition of A^x_σ and

$$(A^x_\sigma A)^* = A^*(A^*)'_\sigma = S^2(\sigma I + S^2)^{-1} , \tag{5.30}$$

in which we have used (5.16), and adopted once more the notation $S^2 = A^*A$. The minimal extension is obtained from

(5.30) by taking the adjoint of both sides of the equation, and noting that

$$S^2(\sigma I+S^2)^{-1} = \int_0^\infty \frac{\lambda^2}{\sigma+\lambda^2} dG_\lambda \qquad (5.31)$$

wherefrom $S^2(\sigma I + S^2)^{-1}$ is self-adjoint. A repetition of the argument employed in Theorem 5.2 indicates that the restriction of $S^2(\sigma I + S^2)^{-1}$ to $N(A)^\perp$ is simply $S_r^2(\sigma I_r + S_r^2)^{-1}$. The left side of (5.29) is then

$$\|(A_\sigma^x A - A^+A)x\|^2 = \|[S_r^2(\sigma I_r+S_r^2)^{-1} - I_r]x\|^2 = \int_0^\infty \left|\frac{\sigma}{\sigma+\lambda^2}\right|^2 d\|H_\lambda x\|^2 . \qquad (5.32)$$

The desired convergence now follows, because the integral in (5.32) tends toward zero with σ, as is easily shown through use of the dominated convergence theorem. |||

Of course, A'_σ and A''_σ are merely two of the many possible approximations to A^+. Except for the convenience of dealing with positive operators, we might as well have chosen

$$A_\sigma^+ = U_r(T_\sigma)_r^{-1} P_R . \qquad (5.33)$$

with

$$(T_\sigma)_r = \int_0^\infty (\lambda+\sigma) dF_\lambda . \qquad (5.34)$$

When so defined, A_σ^+ is bounded, and converges to A^+ in the same modes as A'_σ (vide Theorem 5.2); proofs of these claims follow the same lines of argument already used in Theorem 5.2. Alternatively, one might have taken the T_σ as

$$T_\sigma = \int_{\lambda\geq\sigma} \lambda dE_\lambda , \qquad (5.35)$$

or perhaps based an approximation on the decomposition (5.8) as a starting point. If we apply (5.34), or use either of the above approximations in the context of the other form of the polar decomposition (5.8), we again obtain strong convergence to A^+ in general, and convergence in norm whenever R(A) is closed. From the practical viewpoint, little is gained by applying the approximations mentioned earlier in this paragraph; no computational advantage is evident. We do, however, assure that a solution exists for every $z \in H_2$, for the above approximations have the effect of closing the range of A.

When R(A) is not closed, the introduction of a new topology can be used to modify the minimization problem (perhaps in undesirable fashion) to insure that R(A) is closed in the new topology. One such technique has been analyzed in detail in Part III, and its validity demonstrated by Theorem 3.1. Another appeared in disguised form as part of Theorem 4.1. Here we describe an approximation similar to the latter, but in more explicit form. Let us take K_σ as the Hilbert space $K_\sigma = H_1 \times H_2$, with norm specified by

$$\|\{x, y\}\|^2 = \sigma \|x\|^2 + \|y\|^2, \qquad \sigma > 0 . \tag{5.36}$$

Given the DDC operator $A: H_1 \to H_2$, we define the new operator $A_1: H_1 \to K_\sigma$ as follows:

$$A_1 x = \{x, Ax\} . \tag{5.37}$$

Because A is closed, A_1 has closed range ([H-2], Definition 2.11.2). If we now let $z_1 = \{\theta, z\}$, there always exists a BAS (Definition 1.1) for the functional equation $A_1 x = z_1$, and the BAS is furnished by the PI of A_1, i.e., $x_o = A_1^+ z_1$. In this instance, the BAS x_o is the unique element attaining the infimum of

$$\|A_1 x - z_1\|^2 = \|\{x, Ax\} - \{\theta, z\}\|^2 = \sigma\|x\|^2 + \|Ax - z\|^2 . \tag{5.38}$$

Formally speaking, one hopes that for small σ the infimum of (5.38) is close to the infimum of $\|Ax-z\|$, and that this is accomplished without excessive growth of $\|x\|$.

Although the above touches on the question, we have been unable to solve the following: given $z \in H_2$ and $\alpha > 0$, is there a (unique?) $x \in H_1$ of smallest norm satisfying $\|Ax-z\| \leq \delta + \alpha$?[†] If the answer is affirmative, how is the optimum x_o related to z? A partial conjecture suggests a construction of such x_o on a "piece-by-piece" basis. In the first place, this problem (in fact, any BAS linear problem) can be reduced without loss of generality to an equivalent formulation in which $x \in N(A)^{\perp}$, $z \in N(A)^{\perp}$, $\delta = 0$, and A is a positive operator on $N(A)^{\perp}$ (we omit the proof). In the new form, the optimization problem can be viewed in terms of the spectral representation of A, whose associated resolution of the identity now divides $N(A)^{\perp}$ into mutually orthogonal subspaces with projections $(E_{\lambda_{j+1}} - E_{\lambda_j})$. Now if we take $\Delta_j x = \lambda_j^{-1}(E_{\lambda_{j+1}} - E_{\lambda_j})z$, we diminish $\|Ax-z\|^2$ by $\|(E_{\lambda_{j+1}} - E_{\lambda_j})z\|^2$ while enlarging $\|x\|^2$ the amount $\lambda_j^{-2}\|(E_{\lambda_{j+1}} - E_{\lambda_j})z\|^2$; these are estimates tending to true values as $|\lambda_{j+1} - \lambda_j| \to 0$. To obtain $\|Ax-z\| \leq \alpha$, we take as many increments $\Delta_j x$ as are needed to bring this norm down to the desired level. For each increment $\Delta_j x$, we choose for

---

[†] The answer to a related problem is better understood: If A is a bounded operator and C a closed convex bounded subset of H_1, there exists $x_o \in C$ satisfying $\inf_{x \in C}\|Ax-z\| = \|Ax_o-z\|$. To prove the assertion, consider $\{x_n\} \in C$ such that $\|Ax_n-z\|$ tends to the infimum. Now C is weakly compact, so a subsequence of $\{x_n\}$ converges to x_o, say. This subsequence has, in turn, another subsequence whose Cesaro sums $\{w_n\}$ converge (in norm) to x_o. Lastly, $\|Aw_n-z\|$ tends to the infimum, and $Aw_n \to Ax_o$ in norm. This result is stated as an exercise for compact operators in [B-1], p. 59.

the index j the one (not already used) corresponding to the largest λ_j, and hence the smallest possible $\Delta_j x$. We remark that the orthogonality and reducing property of the incremental operators on z imply

$$\|Ax-z\| = \|E_{\lambda_{n+1}} z\| \quad \text{and} \quad \|x\|^2 = \sum \lambda_j^{-2} \|(E_{\lambda_{j+1}} - E_{\lambda_j})z\|^2 , \tag{5.39}$$

where $x = \sum \Delta_j x$, and there are a total of n terms with the λ_j arranged in decreasing order. If we formally pass to the limit, we find that

$$x = \int_{\beta}^{\infty} \lambda^{-1} \, dE_{\lambda} z \tag{5.40}$$

in which β is chosen as the largest number consistent with $\|Ax-z\| \leq \alpha$. Since β is dependent on z, (5.40) does not define a linear operator, so we must assume that the solution of the optimization problem (if this is indeed the solution) represents a nonlinear operator.

To gain further insight into the behavior of PI approximations when $P_R z \notin R(A)$, let us study an example involving A'_σ. Let us call

$$x_\sigma = A'_\sigma z, \qquad P_R z \notin R(A) , \tag{5.41}$$

and evaluate $\|Ax_\sigma - z\|$ and $\|x_\sigma\|$. It will become apparent that $\|x_\sigma\| \to \infty$ as $\|Ax-z\| \to \delta$; as we shall show later in Theorem 5.4, this undesirable behavior of x_σ is inevitable. However, by analyzing the specific case (5.41) we will obtain sharper results on the variation of $\|Ax_\sigma - z\|$ and $\|x_\sigma\|$ as a function of σ.

As we saw in the proof of Theorem 5.2, $N(A'_\sigma) = R(A)^{\perp}$, and not only A'_σ but also AA'_σ is defined on all $^\sigma H_2$. Therefore, the first norm becomes

$$\|Ax_\sigma - z\|^2 = \|AA'_\sigma z - P_R z\|^2 + \|(I-P_R)z\|^2$$

$$= \delta_z^2 + \int_0^\infty \left|\frac{\sigma}{\sigma+\lambda^2}\right|^2 d\|F_\lambda P_R z\|^2 , \qquad (5.42)$$

in which we have used the fact that $AA'_\sigma = T^2[\sigma I + T^2]^{-1}$, thus giving rise to a spectral representation and corresponding form for norms. From the right hand integral in (5.42), one infers that $\|Ax_\sigma - z\|$ decreases monotonically with σ, possessing the limit †

$$\lim_{\sigma \to 0} \|Ax_\sigma - z\| = \delta_z . \qquad (5.43)$$

Thus, we can approach the infimum (1.1) as closely as desired by choosing σ sufficiently small. The convergence does not require $P_R z \in R(A)$; in fact, A'_σ can be applied to $z \in H_2$ which leaves the PI A^+ undefined, and retains the desirable property of minimizing $\|Ax-z\|$ even for such z.

But our analysis remains incomplete without consideration of the norm of x_σ. Since $A'_\sigma = U_r T_r(\sigma I + T_r^2)^{-1} P_R$ a direct computation shows

$$\|x_\sigma\|^2 = \int_0^\infty \left|\frac{\lambda}{\sigma+\lambda^2}\right|^2 d\|F_\lambda P_R z\|^2 . \qquad (5.44)$$

Since the integrand is monotonically increasing as σ tends towards zero, the same is true of $\|x_\sigma\|$. In fact, if $P_R z \in R(A)$, $P_R z$ is in the domain of T_r^{-1} [see (5.7) and the argument following], and

$$\|x_\sigma\| \nearrow \|T_r^{-1} P_R z\| . \qquad (5.45)$$

† Equation (5.43) is actually a simple corollary to the strong convergence of A'_σ asserted in Theorem 5.2. Obviously, A''_σ then produces convergence in the sense of (5.43) also.

From (5.7), we recognize the right side of (5.45) as the norm of the BAS x_o. Accordingly, as σ tends towards zero

$$\|Ax_\sigma - z\| \searrow \|Ax_o - z\| \text{ and } \|x_\sigma\| \nearrow \|x_o\| \tag{5.46}$$

for all z such that $P_R z \in R(A)$; the norms are strictly decreasing and increasing, respectively.

When $P_R z \notin R(A)$, the right side of (5.44) tends toward infinity (see [D-2], Theorem XII.2.6) because then $P_R z \notin D(T_r^{-1})$ from (5.9). We may therefore conclude that

$$\|x_\sigma\| \nearrow \infty \text{ for } P_R z \notin R(A) . \tag{5.47}$$

The rate at which this norm tends to infinity can be bounded from above by majorizing the right side of (5.44). Indeed, we always have

$$\|x_\sigma\| < \sigma^{-1} , \tag{5.48}$$

whether or not $P_R z$ belongs to the range of A.

That $\|Ax_\sigma - z\| \to \delta_z$ is accompanied by $\|x_\sigma\| \to \infty$ is not merely a coincidence attributable to a poor choice of operator or sequence $\{x_\sigma\}$. Rather, the comparative behavior of the two norms is immutable, as is indicated by

Theorem 5.4: Suppose $P_R z \notin R(A)$, and $\{x_n\} \in D(A)$ such that

$$\|Ax_n - z\| \to \delta_z . \tag{5.49}$$

Then

$$\|x_n\| \to \infty . \tag{5.50}$$

Proof: By an argument identical with that leading to the left hand equality in (5.42), we can replace (5.49) by

$$\|Ax_n - y\| \to 0 \tag{5.51}$$

if we take $y = P_R z$. Let us then suppose $\{x_n\}$ bounded and reason by contradiction. If $\{x_n\}$ is bounded, it is weakly compact, and we may assume without loss of generality that the entire sequence converges weakly, say

$$x_n \xrightarrow[w]{} x_o \ . \tag{5.52}$$

Then $\{x_n\}$ has another subsequence–which we again denote by $\{x_n\}^n$ for brevity in notation—such that ([R-2], Section 38)

$$w_m = m^{-1} \sum_1^m x_j \tag{5.53}$$

and

$$w_m \longrightarrow x_o \text{ in norm.} \tag{5.54}$$

But also, the triangle inequality yields

$$\|Aw_m - y\| \leq m^{-1} \sum_1^m \|Ax_j - y\| \to 0 \ , \tag{5.55}$$

with the right hand convergence following from (5.51). But A is a closed operator, and $\|Aw_m - y\| \to 0$, together with (5.54), implies $x_o \in D(A)$ and $Ax_o = y$. This contradicts the original hypothesis $y \notin R(A)$. |||

APPENDIX A

SOME PROPERTIES OF OPERATORS IN HILBERT SPACES

In the classical literature on linear unbounded operators in Hilbert spaces [D-2][R-2][A-1], one is generally confined to densely defined closed (DDC) operators; operators which are not DDC fail to yield significant useful results, and are therefore seldom considered. But even the DDC assumption leads to a much more complex structure than is encountered for bounded operators (which can be trivially extended to DDC operators). For instance, if one of A and B is bounded and the other DDC, AB need not be DDC . Thus, applications to pseudoinverses requires some knowledge on the characteristics of certain compositions of DDC operators and their restrictions to subspaces. Consequently, we offer here a series of technical lemmas, to which we shall refer in the body of this work from time to time. Although each lemma is quite easy, and perhaps not interesting in itself, we have not found these results conveniently accessible in the standard literature.

Our notation is consistent throughout the paper. A is a linear DDC operator from a Hilbert space H_1 to another, H_2. The operator A has domain D(A), and is endowed with range R(A) and nullspace N(A); its restriction to $N(A)^{\perp}$ will be called A_r. Projections on N(A) and $N(A)^{\perp}$ are denoted by P_N and P_M, respectively. Analogously, $B: H_2 \to H_1$ will have a restriction B_r to $\overline{R(A)} = N(A^*)^{\perp}$, and the corresponding projection on this substance is P_R.

<u>Lemma A.1:</u> $x \in D(A)$ iff $P_M x \in D(A)$.

Proof: Consider the orthogonal decomposition

$$x = P_M x + x_1 , \tag{A.1}$$

where $x_1 \in N(A) \subset D(A)$. It is clear from the linearity of A that $P_M x \in D(A)$ implies $x \in D(A)$; conversely, if $x \in D(A)$, so is $P_M x = x - x_1$. |||

Lemma A. 2: $R(A_r) = R(A)$.

Proof: Since $R(A_r) \subset R(A)$ is obvious, we need only prove the set inclusion in the opposite direction. To this end, take any $z \in R(A)$, so that there exists an $x \in D(A)$ such that $Ax = z$. The decomposition (A. 1) then yields $A_r(P_M x) = Ax = z$ because $Ax_1 = \theta$. |||

The next lemma applies to any restriction of A, and we take A_r to be such. Of course, our interest here is the specialization of the result below to $K = N(A)^{\perp}$.

Lemma A. 3: Any linear operator $A_r \subset A$ has a closed extension $\overline{A_r}$. If, in particular, $D(A_r) = D(A) \cap K$, where K is a subspace, A_r is a closed operator.

Proof: The graph $G(A)$ of A is a Hilbert space under the usual norm, and $G(A_r) \subset G(A)$. Clearly, A_r and hence $G(A_r)$ can be extended to a linear manifold dense in some subspace of $G(A)$. We may therefore suppose $\overline{G(A_r)}$ already extended in this fashion. Since $\{\theta, y\} \notin \overline{G(A_r)}$ for any $y \neq \theta$, $\overline{A_r}$ can be defined uniquely in terms of $\overline{G(A_r)}$ ([H-2], Section 2.11), and $\overline{A_r}$ is a closed operator because its graph is closed.

Assume now K is a subspace; we show $G(A_r)$ is then complete in $G(A)$. In fact, if $\{\{x_n, A_r x_n\}\}$ is a Cauchy sequence in $G(A_r)$, $\{x_n\} \in D(A_r)$, and since K is closed, $x_n \to x_o \in K$. Moreover, $\{x_n, A_r x_n\} = \{x_n, Ax_n\} \to \{x_o, Ax_o\} \in G(A)$. Thus $x_o \in D(A) \cap K = D(A_r)$, which means $\{x_o, Ax_o\} \in G(A_r)$. This completes the proof. |||

We return to the notation of A_r as the restriction of A to the subspace $N(A)^{\perp}$. By virtue of the lemmas above already proven we see A_r is a closed linear operator from the Hilbert space $N(A)^{\perp} \subset H_1$ into the Hilbert space $\overline{R(A)} \subset H_2$. Its domain $D(A_r)$ consists precisely of the set $D(A_r) = \{u : u = P_M x,\ x \in D(A)\}$, and its range coincides with that of A. That A_r is in fact a DDC operator now follows from

<u>Lemma A.4</u>: $A_r : N(A)^{\perp} \to \overline{R(A)}$ is densely defined.

Proof: Since A is densely defined, there is for each $x \in N(A)^{\perp}$ a sequence $\{x_n\} \in D(A)$ such that $x_n \to x$. But $\|x - P_M x_n\| \leq \|x - x_n\| \to 0$, and $\{P_M x_n\} \in D(A_r)$. |||

It sometimes becomes necessary to deduce the properties of an operator from those of its restriction, or to extend an operator from $N(A)^{\perp}$ to H_1 [$\overline{R(A)}$ to H_2]. Then the extended operator is well behaved is assured by

<u>Lemma A.5</u>: The linear operator $A: H_1 \to H_2$ is DDC if

(a) $N(A)$ is a closed set

and (b) its restriction A_r is DDC on $N(A)^{\perp}$. (A.2)

Proof: Since A is linear, $N(A)$ is a subspace. Let A_q be the restriction of A to $N(A)$, and take $G(A_q)$ to be the corresponding graph, consisting of elements $\{x, \theta\}$ with $x \in N(A)$. Then $G(A_q)$ is a subspace in $H_1 \times H_2$, and is orthogonal to $G(A_r)$. We observe further [cf. Lemma A.1 and the linearity of A] that $G(A) = G(A_r) \oplus G(A_q)$, which exhibits $G(A)$ to be closed. To demonstrate that A is densely defined, first note that $D(A) = D(A_r) \oplus N(A)$. If now $x \in H_1$, there exists $\{x_{n1}\} \in D(A_r)$ such that $x_{n1} \to P_M x$, whence $x_{n1} + P_N x = x_n \to x$. As $\{x_n\} \in D(A)$, this shows $D(A)$ dense in H_1. |||

We next exhibit some results on inverses of operators related to A. If A is invertible with range dense in H_2, there is little content to the theory of the pseudoinverse. We therefore do not suppose A has an inverse, but rather subsume it under the more general results that follow. The first of these is

Lemma A. 6: A_r has a DDC inverse A_r^{-1}.

Proof: The inverse exists if A_r is one-to-one. For $x_1, x_2 \in N(A)^{\perp}$ $A_r x_1 = A_r x_2$ is equivalent to $Ax_1 = Ax_2$, which shows $x_1 - x_2 \in N(A)$. But $x_1 - x_2 \in N(A)^{\perp}$ also, which means $x_1 = x_2$; thus, $A_r: N(A)^{\perp} \to \overline{R(A)}$ is one-to-one. That A_r^{-1} is closed follows from the fact that A_r is closed (cf. Lemma A. 3 and [H-2], Theorem 2.11.15). Finally, $D(A_r^{-1}) = R(A_r)$, and the latter is dense in $\overline{R(A)}$ from Lemma A. 2. |||

Lemma A. 7: A_r^* has a DDC inverse, and

$$(A_r^*)^{-1} = (A_r^{-1})^* . \tag{A. 3}$$

Proof: A_r^* and $(A_r^{-1})^*$ are themselves DDC, whereupon the desired conclusions follow from a well known argument based on $G(A_r)$; see [R-2], Section 117 for details. |||

Lemma A. 8: The following conditions are equivalent:

(a) A_r^{-1} is bounded.

(b) $(A_r^*)^{-1}$ is bounded.

(c) $R(A)$ is a closed set.

(d) $A_r A_r^{-1}$ is a closed operator.

(e) There exists $m > 0$ such that $m\|x\| \le \|A_r x\|$ for all $x \in D(A_r)$.

(f) There exists $k > 0$ such that $k\|y\| \le \|A_r^* y\|$ for all $y \in D(A_r^*)$.

Proof: Since $D(A_r^{-1}) = R(A)$ and A_r^{-1} is a closed operator (cf. Lemma A. 6), condition (c) implies (a) by the closed graph theorem ([R-2], Section 117). Conversely, the domain of a closed bounded operator is a closed set, whence (c) follows from (a) and Lemma A. 2.

To prove (b) from (a), consider (A. 3) and the fact that (a) implies $(A_r^{-1})^*$ bounded. On the other hand, (b) requires $(A_r^{-1})^*$ bounded, and (a) follows.

Respecting (e), observe $D(A_r) = R(A_r^{-1})$, so we may put $x = A_r^{-1}y$, with y uniquely defined and ranging over $D(A_r^{-1})$ as x ranges over $D(A_r)$. This means $\|A_r^{-1}y\| \leq m^{-1}\|y\|$ for all $y \in D(A_r^{-1})$, i. e. $\|A_r^{-1}\| \leq m^{-1}$. For the converse, one proves (a) from (e) by retracing the above steps in reverse order. The equivalence of (b) and (f) is similarly demonstrated.

We turn our attention to the relation between (a) and (d). $A_rA_r^{-1}$: $\overline{R(A)} \to \overline{R(A)}$ is the restriction in this space of the identity operator to its domain $D(A_rA_r^{-1}) = R(A)$. Thus $A_rA_r^{-1}$ is bounded and densely defined. Consequently, $A_rA_r^{-1}$ is a closed operator iff its domain $R(A)$ is a closed set. |||

It is tempting to assert that (b) is equivalent to yet another condition, namely

$$\text{(g)} \quad R(A^*) \text{ is a closed set.} \tag{A. 4}$$

However, the proof given in the first paragraph (of the proof of the lemma) fails because the connection between $(A^*)_r$ and A_r^* is not clear. We recall here that by B_r for $B: H_2 \to H_1$ we mean the restriction of B to $\overline{R(A)}$

<u>Lemma A. 9</u>:

$$(A^*)_r = A_r^* \tag{A. 5}$$

Proof: For any $x \in D(A)$, $y \in D(A^*)$ we have

$$(Ax, y) = (x, A^*y) = (x_1, A^*y) = (x_1, A^*y_1) = (x_1, (A^*)_ry_1) \tag{A. 6}$$

where $x_1 = P_M x$ and $y_1 = P_R y$. At the same time

$$(Ax, y) = (Ax_1, y_1) = (A_r x_1, y_1) . \tag{A. 7}$$

The right sides of (A. 6) and (A. 7) are valid for all $x_1 \in D(A_r)$ from Lemma A. 1. Therefore, $A_r^* \supset (A^*)_r$, the latter being defined at least on $D(A^*) \cap N(A^*)^{\perp}$. To demonstrate equality, consider any $y \in D(A_r^*)$. For such y and all $x_1 \in D(A_r)$

$$(A_r x_1, y) = (x_1, A_r^* y) = (x_1, y^*) , \tag{A. 8}$$

in which $y^* \in \overline{R(A_r^*)}$, and $\overline{R(A_r^*)} = \overline{D(A_r)}$ because [see Lemma A. 7] $(A_r^*)^{-1}$ is densely defined. But any $x \in D(A)$ has the orthogonal decomposition $x = x_1 + x_2$ with $x_1 \in D(A_r)$ and $x_2 \in N(A)$ (cf. again Lemma A. 1). Hence in (A. 8) $A_r x_1 = Ax_1 = Ax$, while $(x_1, y^*) = (x, y^*)$ because y^* is orthogonal to $N(A)$. In other words, $(Ax, y) = (x, y^*)$, or $y \in D(A^*)$. The proof is completed by noting that the inclusion $D(A_r^*) \subset D(A^*)$ just obtained yields

$$D(A_r^*) = D(A_r^*) \cap \overline{R(A)} \subset D(A^*) \cap \overline{R(A)} = D[(A^*)_r] . \quad |||$$

The validity of Lemma A. 9 permits us to use the notation A_r^* to replace both $(A^*)_r$ and A_r^*. This lemma may also be applied to strengthen the results of Lemma A. 8, viz.

<u>Corollary A. 1</u>: Lemma A. 8 remains true if condition (g) of (A. 4) is added; i. e., conditions (a) through (g) are equivalent.

Proof: $R(A^*) = R[(A^*)_r]$, an assertion whose proof we defer for the moment. Since $R(A_r^*)$ is closed or not according as $(A_r^*)^{-1}$ is bounded or unbounded, (b) and (g) hold simultaneously. By Lemma A. 8, (b) is equivalent to the other properties (a) to (f), so the proof is finished. |||

Lemma A.10: (Duality of A_r and A_r^*) Lemmas A.1, A.2, A.4, A.5 and A.8 continue to hold if symbols are consistently interchanged as follows:

$$A \text{ with } A^*,\ A_r \text{ with } A_r^*.\ P_M \text{ with } P_R, \text{ and } H_1 \text{ with } H_2 . \qquad (A.9)$$

Proof: $A^*: H_2 \to H_1$ is DDC, and A_r^* is its restriction to $\overline{R(A)}$, where $\overline{R(A)} = N(A^*)^\perp$. The arguments of Lemmas A.1, A.2, A.4 and A.5 therefore lead to the desired results in terms of the symbol substitution (A.9). No proof is required for Lemma A.8, in view of its inherent symmetry with the addition of Corollary A.1. |||

Our next objects of study are the restrictions of A^*A and AA^*. As we have pointed out earlier, compositions of DDC operators need be neither closed nor densely defined. However, it is known that A^*A and AA^* are not only DDC, but even self-adjoint positive operators ([R-2], Section 119). The restriction $(A^*A)_r: N(A)^\perp \to N(A)^\perp$ is then closed by Lemma A.3, and similarly for $(AA^*)_r: \overline{R(A)} \to \overline{R(A)}$. It is less obvious that $(A^*A)_r$ and $(AA^*)_r$ are densely defined and self-adjoint. We now prove these facts, together with other properties of $(A^*A)_r$ and $(AA^*)_r$ useful elsewhere in this work.

Lemma A.11: $N(A) = N(A^*A)$: $N(A^*) = N(AA^*)$. (A.10)

Proof: We prove only the first set equality. Clearly, $N(A^*A) \supset N(A)$. Consider then $x \in N(A^*A)$; we compute

$$0 = (A^*Ax, x) = \|Ax\|^2 , \qquad (A.11)$$

which demonstrates that $x \in N(A)$. |||

Lemma A.11 above enables us to write $(A^*A)_r: N(A^*A)^\perp \to N(A^*A)^\perp$, showing through use of identical arguments that

Corollary A.2: Lemmas A.1, A.2, A.4 and A.6 remain valid if A is replaced by A^*A and A_r by $(A^*A)_r$.

In particular, $(A^*A)_r$ is densely defined and possesses a DDC inverse.

<u>Corollary A. 3:</u> $\overline{R[(A^*A)_r]} = \overline{R(A^*A)} = \overline{R(A^*)}$; $\overline{R[(AA^*)_r]} = \overline{R(AA^*)}$

$$= \overline{R(A)} \ . \qquad (A.12)$$

Proof: Lemma A. 2 yields $R[(A^*A)_r] = R(A^*A)$, while $R(A^*A)^\perp = R(A^*)^\perp$ follows from the standard result $N(A) = R(A^*)^\perp$ and (A. 10). The second set of equalities is similarly obtained. |||

To verify the self-adjointness of $(A^*A)_r$ [as well as $(AA^*)_r$], and to provide a useful alternative form, we state first

<u>Lemma A.12:</u> $(A^*A)_r = A_r^* A_r$ and $(AA^*)_r = A_r A_r^*$. (A. 13)

Proof: The left side of the first equality satisfies

$(A^*A)_r x = A^*Ax$ for any $x \in D(A^*A) \cap N(A)^\perp$. For such x,

$$A_r^* A_r x = A_r^*(Ax) = A^*Ax \ ; \qquad (A.14)$$

the latter equality is true because $Ax \in \overline{R(A)}$, and A_r^* is the restriction of A^* to $\overline{R(A)}$. We have thus shown $(A^*A)_r \subset A_r^* A_r$. To complete the proof, we demonstrate that $D(A_r^* A_r) \subset D[(A^*A)_r]$. In fact, $A_r^* A_r x$ is defined only if $x \in D(A) \cap N(A)^\perp$ and $Ax \in D(A^*) \cap \overline{R(A)}$. But $Ax \in \overline{R(A)}$ in any case, so the second condition becomes merely $Ax \in D(A^*)$. That is, $x \in D(A^*A) \cap N(A)^\perp = D[(A^*A)_r]$. |||

<u>Remark:</u> The proof makes implicit use of Lemma A. 9, since we deduce the domain of A_r^* as that of $(A^*)_r$, whereas the applications of (A. 13) interpret A_r^* as $A_r{}^*$.

<u>Lemma A. 13:</u> $(A^*A)_r$ and $(AA^*)_r$ are self-adjoint operators on $N(A)^\perp$ and $\overline{R(A)}$, respectively.

Proof: Although it was shown earlier that $(A^*A)_r$ is DDC, this follows again from Lemma A.12 and the fact that A^*A: $N(A)^\perp \to N(A)^\perp$ is DDC. Indeed, $A_r^* A_r$ is self-adjoint on this space ([R-2], Section 119), so the same is true for $(A^*A)_r$ by (A.13). As usual, the assertion regarding $(AA^*)_r$ is analogously proven. |||

Corollary A.2 already claims that $(A^*A)_r$ and $(AA^*)_r$ possess DDC inverses. We now repeat and sharpen this result via

<u>Lemma A.14:</u> $(A^*A)_r: N(A)^\perp \to N(A)^\perp$ and $(AA^*)_r: \overline{R(A)} \to \overline{R(A)}$ have self-adjoint inverses given respectively by

$$(A^*A)_r^{-1} = A_r^{-1} A_r^{*-1} \quad \text{and} \quad (AA^*)_r^{-1} = A^{*-1} A_r^{-1} \quad . \tag{A.15}$$

Proof: We recall that $(A^*A)_r = A_r^* A_r$ by (A.13). Then, since A_r and A_r^* have inverses from Lemmas A.6 and A.7, respectively, $(A^*A)_r$ has the inverse stated in (A.15) (see [R-2], Section 114h). The inverse of a self-adjoint operator is likewise self-adjoint ([R-2], Section 119), and so the self-adjointness of $(A^*A)_r^{-1}$ is a consequence of Lemma A.13. |||

Finally, we give the conditions under which the inverses of $(A^*A)_r$ and $(AA^*)_r$ are bounded, and their ranges closed.

<u>Lemma A.15:</u> Suppose

(a) any one of the ranges of A, A^*, (A^*A) or (AA^*) is closed

or (b) any of the inverses A_r^{-1}, A_r^{*-1}, $(A^*A)_r^{-1}$, or $(AA^*)_r^{-1}$ bounded.

Then all the ranges in (a) are closed, and all inverses (b) bounded.

Proof: By virtue of Lemma A. 2, and its application to adjoint and composite operators via Lemma A.10 and Corollary A. 2, we consider the ranges of the restrictions of the operators appearing in (a). Then the arguments of Lemma A. 8 demonstrate the desired result for A and A^*, and also yield the relation between range and inverse for (A^*A) and (AA^*).

If now A_r^{-1} and A_r^{*-1} are bounded, formula (A. 15) verifies the boundedness of $(A^*A)_r^{-1}$ and $(AA^*)_r^{-1}$. Conversely, suppose $(A^*A)_r^{-1}$ bounded. Then $R(A^*A)$ is a closed set, and we have from (A. 12) and $R(A^*A) \subset R(A^*)$

$$\overline{R(A^*)} = R(A^*A) \subset R(A^*) \subset \overline{R(A^*)} \; ; \qquad \text{(A. 16)}$$

this means $R(A^*) = R(A^*A)$, or $R(A^*)$ is closed. The latter enables us to show that all the other conditions of (a) and (b) are met. Likewise, $R(AA^*)$ closed or $(AA^*)_r^{-1}$ bounded leads to $R(A)$ closed, and hence the truth of each condition (a) and (b). |||

In the course of the above proof, we have additionally deduced

Corollary A. 4: If any of the conditions (a) or (b) of Lemma A. 15 are satisfied,

$$R(A^*A) = R(A^*) \text{ and } R(AA^*) = R(A) , \qquad \text{(A. 17)}$$

each of these ranges being a closed set. |||

It is well known that if A_1 and A_2 are DDC, $(A_1A_2)^* \supset A_2^*A_1^*$, but this sheds little light on the closed extension of the right side. However, if either A_1 or A_2 are bounded, at least partial results can be secured ([R-2], Section 115). Our interest lies in taking A_2 bounded, since this occurs in Gauss-Markov estimation. More specifically we have

Lemma A. 16: Suppose

a. A_1 is densely defined and closable, with closure $\overline{A}_1$.

b. A_2 is bounded and everywhere defined.

c. $\overline{A}_1 A_2$ is densely defined.

Then the closed extension of $A_2^* A_1^*$ is $(\overline{A}_1 A_2)^*$.

Proof: Since A_2^* is bounded and A_1 is densely defined, $A_2^* A_1^*$ is densely defined and hence has an adjoint; in fact, A_2^* being bounded implies (see again [R-2], Section 115)

$$(A_2^* A_1^*)^* = \overline{A}_1 A_2 . \tag{A.18}$$

Taking adjoints once more then yields the conclusion, viz.

$$\overline{A_2^* A_1^*} = (A_2^* A_1^*)^{**} = (\overline{A}_1 A_2)^* . \quad ||| \tag{A.19}$$

Remark: It does not follow that $A_2^* A_1^* = (\overline{A}_1 A_2)^*$, because the left side need not be a closed operator. This can be demonstrated via a counter-example in which A_1 and A_2 are each self-adjoint, and A_2 has finite dimensional range.

APPENDIX B

HILBERT-SPACE-VALUED RANDOM VARIABLES

In discussing statistical estimation in Hilbert space we deal with random quantities taking values in a Hilbert space. There is an extensive literature concerning Hilbert-space-valued random variables, or what is essentially the same thing, probability measures in a Hilbert space (see [P-1]). However, what we need is relatively simple and can be presented briefly in a form that directly fits our requirements.

Let $(\Omega, \mathcal{A}, P)$ be a complete probability space, where $\mathcal{A}$ is a σ-algebra of subsets of Ω and P is a probability measure on $\mathcal{A}$. Let $\{w_i\}$, $i = 1, 2, \ldots,$ be a sequence of (real or complex-valued) random variables on $(\Omega, \mathcal{A})$ such that

$$\sum^{\infty} E|w_i|^2 = \sum^{\infty} \int_{\Omega} |w_i(\omega)|^2 \, dP(\omega) < \infty \ . \tag{B. 1}$$

We note that, by the Fubini theorem, the inequality (B. 1) implies

$$E(\sum^{\infty} |w_i|^2) = \int_{\Omega} \sum^{\infty} |w_i(\omega)|^2 \, dP(\omega)$$

$$= \sum^{\infty} E|w_i|^2 < \infty \ . \tag{B. 2}$$

Now let $\{\phi_n\}$, $n = 1, 2, \ldots,$ be a complete orthonormal system (c. o. n. s.) in H_2 and define $w(\omega)$ by

$$\begin{aligned} w(\omega) &= \sum w_i(\omega)\phi_i, \quad \omega \in \Lambda \\ &= 0, \quad \omega \in \Lambda^c \end{aligned} \tag{B. 3}$$

where $\Lambda \subset \Omega$ is the set of all ω for which $\sum^{\infty} |w_i(\omega)|^2 < \infty$. By (B.2) and the assumption of completeness of P, $\Lambda^c \in \mathcal{A}$, and $P(\Lambda^c) = 0$. Thus, for all ω, $w(\omega)$ is an element of H_2, and for a.e. ω

$$\|w(\omega)\|^2 = \sum_i^{\infty} \|w_i(\omega)\|^2 \quad . \tag{B.4}$$

We can now define $\tilde{w}_n(\omega) = (w(\omega), \phi_n)$ for all ω. The $\tilde{w}_n$ are measurable ($\mathcal{A}$) and are equal to the w_n except on a fixed set of probability zero. It should cause no confusion if we drop the tilde and henceforth simply denote $(w(\omega), \phi_n)$ as w_n. Then (B.4) holds for all ω. Since the (w, ϕ_n) are all measurable it is easy to establish that w is weakly measurable, that is that (w, y) is a measurable scalar-valued function for each $y \in H_2$. Then, since H_2 is separable, $w(\cdot)$ is also strongly measurable (see [H-2] Section 3.5 for both definition and proof). We call w a Hilbert-space-valued random variable.

The condition (B.1) is sufficient to establish both a mathematical expectation of w and a covariance operator. Since, by (B.2) and (B.4),

$$E\|w\|^2 = \int_{\Omega} \|w(\omega)\|^2 \, dP(\omega) < \infty \tag{B.5}$$

then,

$$\int_{\Omega} \|w(\omega)\| \, dP(\omega) < \infty \quad . \tag{B.6}$$

The strong measurability of w and (B.6) imply that $w(\omega)$ is Bochner-integrable $(\Omega, \mathcal{A}, P)$ (see [H-2], Section 3.7). We define

$$Ew = \int_{\Omega} w(\omega) \, dP(\omega) \tag{B.7}$$

where the integral is a Bochner integral. An application of a basic theorem of Hille ([H-2], Theorem 3.7.12) gives

$$(y, Ew) = \int_{\Omega} (y, w) dP .$$

Hence, if the w_i have mean zero it follows that $Ew = \theta$.

Also, we have from (B.4) that

$$E|(y, w)\overline{(z, w)}| \leq \|y\| \cdot \|z\| \; E\|w\|^2 < \infty . \qquad \text{(B. 8)}$$

Since $\psi(y, z) = E\{(y, w)\overline{(z, w)}\}$ is a Hermitian symmetric conjugate bilinear form, which by (B. 8) is bounded, there exists a bounded symmetric operator K defined on all of H_2 such that

$$(Ky, z) = \psi(y, z) \quad . \qquad \text{(B. 9)}$$

K is obviously nonnegative definite; it is called the covariance operator of w . The assumption (B. 1) further implies that K is nuclear, that is that it is compact and has finite trace.† One can verify directly from (B. I) that the bounded operator K is Hilbert-Schmidt, and hence compact, and then, using the compactness, verify that it has finite trace.

---

† We refer the reader to [G-1], pp. 26-47. A nonnegative self-adjoint compact operator T has an o. n. s. of eigenvectors $\{\eta_n\}$ that span $\overline{R(T)}$, and nonnegative eigenvalues λ_n; $T\eta_n = \lambda_n \eta_n$. By definition, T is nuclear if

$$\sum_{n=1}^{\infty} (T\eta_n, \eta_n) = \sum_{n=1}^{\infty} \lambda_n < \infty .$$

If T is nuclear,

$$\sum_{n=1}^{\infty} (T\phi_n, \phi_n) = \sum_{n=1}^{\infty} (T\eta_n, \eta_n) = Tr(T) \quad ,$$

where $\{\phi_n\}$ is any c. o. n.s.

However, a shorter proof is as follows. Since K is self-adjoint and nonnegative, it has a nonnegative square root $K^{\frac{1}{2}}$. Then

$$\sum^{\infty} \| K^{\frac{1}{2}} \phi_n \|^2 = \sum^{\infty} (K\phi_n, \phi_n) = \sum^{\infty} E|(\phi_n, w)|^2$$

$$= \sum^{\infty} E|w_n|^2 < \infty \qquad \text{(B. 10)}$$

The convergence of $\sum \| K^{\frac{1}{2}} \phi_n \|^2$ ensures that $K^{\frac{1}{2}}$ is Hilbert-Schmidt (cf. [G-1], p. 34). Then $K = K^{\frac{1}{2}} \cdot K^{\frac{1}{2}}$ is nuclear (cf. [G-1], p. 39).

Finally, we observe that K is strictly positive definite if and only if $(K\phi_n, \phi_n) > 0$ for all n, that is, if and only if $E|w_n|^2 > 0$ for all n.

The preceding remarks may be summarized as a theorem:

Theorem B.1: Let $(\Omega, \mathcal{A}, P)$ be a complete probability space and $\{w_n\}$, $n = 1, 2, \ldots,$ a sequence of scalar-valued random variables satisfying (B. 1). Let $\{\phi_n\}$ be a c. o. n. s. in the separable Hilbert space H. Then w as defined by (B. 3) is an H-valued random variable, i. e., a strongly measurable H-valued function on $(\Omega, \mathcal{A})$. The expectation of w as defined by the Bochner integral in (B. 7) exists, and a unique bounded self-adjoint, nonnegative-definite covariance operator K exists that satisfies

$$(Ky, z) = E(y, w)\overline{(z, w)}, \quad y, z \in H .$$

Further, K is nuclear.

K is strictly positive definite if and only if $E|w_n|^2 > 0$, for all n, and $Ew = 0$ if and only if $Ew_n = 0$ for all n.

Remark: The nuclearity of K is essential in the following sense. Suppose v is an H-valued random variable satisfying $E\|v\|^2 < \infty$. Then a covariance operator K for v exists and is nuclear. In fact, the existence of K follows exactly as above, and K is bounded, self-adjoint and non-negative definite. Then for any c. o. n. s. $\{\phi_n\}$ the first two equalities in (B. 10) hold. But

$$\sum^{\infty} E|(\phi_n, v)|^2 = E\|v\|^2 ,$$

which is finite. The rest of the argument follows as before.

One standard situation is for the observations to be functions in an L_2 space. This leads us to discuss stochastic processes whose sample functions belong almost surely to L_2. Let $w(t, \omega)$ be a measurable, separable, real or complex-valued stochastic process with t belonging to the parameter set $T \subset R^1$, and with probability space $(\Omega, \mathcal{A}, P)$ (c. f. [D-3], Chapter 2). T can be an interval, or all of R^1, or indeed any measurable subset of R^1 with positive measure. $w(t, \cdot)$ is a random variable on $(\Omega, \mathcal{A}, P)$. As is conventional, we usually suppress the probability variable and write $w(t)$ instead of $w(t, \omega)$. Let $w(t)$ satisfy

$$\int_T E|w(t)|^2 dt < \infty \quad . \tag{B. 11}$$

Then $R(t, s) \overset{d}{=} E\, w(t)\overline{w(s)}$ exists for a. e. t and a. e. s, and

$$\int_T |w(t, \omega)|^2 dt < \infty$$

for all ω in some Λ, $P(\Lambda) = 1$. We put $R(t, s) = 0$ for those values of t, s for which $Ew(t)\overline{w(s)}$ is not defined. We define $w(\omega) \in L_2(T)$ to be the Lebesgue equivalence class of functions $w(\cdot, \omega)$ for each $\omega \in \Lambda$, and to be the zero element otherwise. Then we have

Theorem B. 2: $w(\omega)$ as just defined is an H-valued random variable $(H = L_2(T))$. It has a nuclear covariance operator K given by

$$[Ky](t) = \int_T R(t, s)y(s)ds \tag{B.12}$$

for $y \in L_2(T)$. The expected value of $w(\omega)$ exists and is characterized by

$$(Ew, y) = E(w, y) = \int_T E\,w(t)\,\overline{y(t)}\,dt \tag{B.13}$$

for all $y \in L_2(T)$.

Proof: The assertion that w is an H-valued random variable means that it is strongly measurable. Since $L_2(T)$ is separable this is equivalent to weak measurability. We have for any $y \in L_2(T)$,

$$(w(\omega), y) = \int_T w(t, \omega)\,\overline{y(t)}dt, \quad \omega \in \Lambda \tag{B.14}$$

$$= 0, \quad \omega \in \Lambda^c \ .$$

The integral in (B.14) is a measurable function of ω (c.f. [D-3], Theorem 2.7) hence (w, y) is measurable. Thus w is an $L_2(T)$-valued random variable. From (B.11) we have $E\|w\|^2 < \infty$. Hence, as in the preceding theorem and the remark following, w has a nuclear covariance operator K . Furthermore,

$$(Ky, z) = E[(y, w)\overline{(z, w)}]$$

$$= E\{\int_T y(t)\,\overline{w(t)}dt \int_T z(s)w(s)ds\}$$

$$= \int_T \int_T R(s, t)\,\overline{z(s)}\,y(t)dt \tag{B.15}$$

where the interchange of integrals is justified by the Fubini theorem after the following calculation:

$$\int_T \int_T E|w(s)\overline{w(t)}|\,|y(t)|\,|z(s)|\,dt\,ds$$

$$\leq \int_T \int_T [R(s,s)]^{\frac{1}{2}} [R(t,t)]^{\frac{1}{2}} |y(t)|\,|z(s)|\,dt\,ds$$

$$\leq (\int_T R(s,s)ds)\,\|y\|\,\|z\| < \infty$$

by the assumption (B.11). Since (B.15) holds for all $y, z \in L_2(T)$, (B.12) is proved. The expected value of w is given by the Bochner integral

$$Ew = \int_\Omega w(\omega)\,dP(\omega) \tag{B.16}$$

which exists because

$$\int_\Omega \|w(\omega)\|\,dP(\omega) < \infty\;, \tag{B.17}$$

which in turn follows from the square integrability guaranteed by (B.11). (B.13) follows from (B.16) and a theorem used previously ([H-2], Theorem 3.7.12). |||

Now let C be a bounded linear transformation from all of H_2 into H_1, and let w be an H_2-valued random variable random variable satisfying $E\|w\|^2 < \infty$. Then Cw is an H_1-valued random variable. Clearly $Cw(\omega)$ is defined for all ω. Also Cw is weakly (and hence strongly) measurable since $(Cw, x) = (w, C^*x)$, $x \in H_1$, and w is weakly measurable. Finally,

$$\int_\Omega \|Cw(\omega)\|^2\,dP(\omega) \leq \|C\|^2 \int_\Omega \|w(\omega)\|^2\,dP(\omega) < \infty\;.$$

Thus $E[Cw]$ is defined, and a covariance operator K_2 for Cw is defined. We have, furthermore, again by the Hille theorem,

$$E[Cw] = \int_\Omega Cw(\omega)\, dP(\omega) = C \int_\Omega w(\omega) dP(\omega) = C(Ew)$$

and

$$(K_2 u, x) = E[(u, Cw)\overline{(x, Cw)}]$$
$$= E[(C^* u, w)\overline{(C^* x, w)}] = (KC^* u, C^* x) ,$$

from which $K_2 = CKC^*$.

REFERENCES

[A-1] N. I. Akhiezer & I. M. Glazman, Theory of Linear Operators in Hilbert Space, vol. I and II (tr. by M. Nestell), Frederick Ungar Publishing Co., New York, 1961 and 1963.

[A-2] A. Albert, Regression and the Moore-Penrose Pseudoinverse, Academic Press, New York, 1972.

[A-3] M. Athans & P. L. Falb, Optimal Control, McGraw-Hill, New York, 1966.

[B-1] A. V. Balakrishnan, Introduction to Optimization Theory in a Hilbert Space, Lecture Notes in Operations Research and Mathematical Systems, vol. 42, Springer-Verlag, New York, 1970.

[B-2] A. V. Balakrishnan, "Determination of nonlinear systems from input-output data," Proc. of the Princeton University Conference on Identification Problems in Communication and Control Systems (1963), 31-49.

[B-3] F. J. Beutler, "The operator theory of the pseudo-inverse I. Bounded operators," J. Math. Anal. and Appl., 10 (1965), 451-470.

[B-4] F. J. Beutler, "The operator theory of the pseudo-inverse II. Unbounded operators with arbitrary range," J. Math. Anal. and Appl., 10 (1965), 472-493.

[B-5] R. Bouldin, "The pseudo-inverse of a product, SIAM J. Appl. Math. 24 (1973), 489-495.

[B-6] R. W. Brockett, Finite Dimensional Linear Systems, Wiley, New York, 1970.

[C-1] C. T. Chen, Introduction to Linear System Theory, Holt, Rinehart & Winston, New York, 1970.

[D-1] C. Desoer and B. Whalen, "A note on pseudoinverses," J. SIAM, 11 (1963), 442-447.

[D-2] N. Dunford & J. T. Schwartz, Linear Operators, vol. I and II, Interscience Publishers, New York, 1958 and 1963.

[D-3] J. L. Doob, Stochastic Processes, Wiley, New York, 1952.

[G-1] I. M. Gelfand & N. Ya. Vilenkin, Generalized Functions, vol. 4: Applications of Harmonic Analysis (tr. by A. Feinstein), Academic Press, New York, 1964.

[G-2] T. N. Greville, "The pseudoinverse of a rectangular or singular matrix and its application to the solution of systems of linear equations," J. Soc. Indus. Appl. Math., 1 (1959), 38-43.

[H-1] P. R. Halmos, Introduction to Hilbert Space and the Theory of Spectral Multiplicity, Chelsea, New York, 1951.

[H-2] E. Hille & R. S. Phillips, Functional Analysis and Semi-Groups (rev. ed.), Am. Math. Soc. Colloq. Pub., Am. Math. Soc., Providence, R.I., 1958.

[I-1] IEEE Trans. Automatic Control, Special Issue on Linear-Quadratic-Gaussian Problem, AC-16 (1971).

[K-1] R. Kalman, Y. Ho & K. Narendra, "Controllability of linear dynamical systems," in Contribution to Differential Equations, vol. 1, Wiley, New York, 1962; pp. 189-213.

[L-1] W. S. Loud, "Generalized inverses and generalized Green's functions," SIAM J. Appl. Math., 12 (1970).

[L-2] D. G. Luenberger, Optimization by Vector Space Methods, Wiley, New York, 1969.

[M-1] J. L. Massera & J. J. Schaffer, "Linear differential equations and functional analysis," Ann. of Math., 67 (1958), 517-573.

[M-2] N. Minamide and K. Nakamura, "A restricted pseudoinverse and its application to constrained minima," SIAM J. Appl. Math., 19 (1970), 167-177.

[N-1] M. Z. Nashed, A Retrospective and Prospective Survey of Generalized Inverses of Operators, Mathematics Research Center Report no. 1125, University of Wisconsin, Madison, 1971.

[P-1] K. R. Parthasarathy, Probability Measures in Metric Spaces, Academic Press, New York, 1967.

[P-2] E. Parzen, "An approach to time series analysis," Ann. Math. Stat., 32 (1961), 951-989.

[P-3] W. A. Porter, Modern Foundations of System Engineering, Macmillan, New York, 1966.

[P-4] W. A. Porter & J. P. Williams, "Extensions of the minimum effort control problem," J. Math. Anal. and Appl., 10 (1966), 536-549.

[R-1] C. R. Rao & S. K. Mitra, Generalized Inverse of Matrices and Its Application, Wiley, New York, 1971.

[R-2] F. Riesz & B. Sz.-Nagy, Functional Analysis (tr. by L. Boron), Frederick Ungar Publishing Co., New York, 1955.

[R-3] W. L. Root, On the Modelling of Systems for Identification, Report, University of Michigan, Ann Arbor, 1972; to be published in SIAM J. Control.

[R-4] W. L. Root, "On the structure of a class of system identification problems," Automatica, 7 (1971), 219-231.

[R-5] W. L. Root, "On the modelling and estimation of communication channels," Multivariate Analysis-III, (ed. R. Krishnaiah) Academic Press, New York, 1973.

University of Michigan
Ann Arbor, Michigan

Applications of Generalized Inverses to Programming, Games and Networks

Adi Ben-Israel

1. Introduction

Generalized inverses are applicable to various engineering and management models where linear transformations are used to describe a process, its rate of change, or its cost. Often such applications are straightforward, but even then, using generalized inverses may result in simpler notation and theory or in a more efficient computation. In other cases, a generalized inverse cannot be avoided. Whether used for convenience or out of necessity, a generalized inverse most suitable for the problem at hand has to be chosen from the many available generalized inverses, see, e.g., [10].

The applications discussed here were selected so as to minimize duplication and overlap with other papers in this volume. Hence the omission (from this paper) of applications to control and systems, and the little space (see Theorem 2.2 and Corollaries 2.3, 2.4) devoted here to extremal and least-squares properties of generalized inverses.

Section 2 is an introduction to the Bott-Duffin inverse and its application to electrical network analysis. Further details can be found in [13], [21], [40] and ([10], Sections 2.9, 2.12 and 3.5).

A special class of linear programs, which are explicitly solvable by using $\{1\}$-inverses, is discussed in Section 3. For other applications of generalized inverses to linear programming see, e.g., [15], [19], [20], [38], [39] and [47].

An application of (suitably defined) generalized inverses to the solution of Diophantine linear equations is described in Section 4. These results may prove very useful in integer programming, as indicated by recent work of A. Charnes and F. Granot.

Section 5 gives a characterization of equilibrium points of bimatrix games in terms of certain submatrices (of the payoff matrices) and their Moore-Penrose inverses. Interestingly, the value of a two-person zero-sum game turns out to be

$$\frac{1}{e^T A(I/J)^{\dagger} e}$$

where $A(I/J)$ is a submatrix (generalized saddle point) of the payoff matrix A.

Other applications of generalized inverses to mathematical programming include the gradient projection method ([43], [44], [46] and [32], where the Moore-Penrose inverse is used implicitly), least-squares solution of nonlinear equations ([3], [6], [22] and [48]), quadratic programming (see, e.g., [12], [25] and [36]) chance constrained programming [16], and modular design [18].

The terminology, and notation, of [10] are used here. In particular, we denote by

$R^{m\times n}[C^{m\times n}]$-the $m\times n$ real [complex] matrices

$R_r^{m\times n}[C_r^{m\times n}]$-the same with rank r

and for any matrix A

$A^T[A^*]$-the transpose [conjugate transpose]

$R(A)$-the range of A,

$N(A)$-the null space of A,

$A^{\dagger}$- the Moore-Penrose inverse of A.

Finally, for a given $A\epsilon C^{m\times n}$ and $1\leq i, j, \ldots, k\leq 4$, an $\{i, j, \ldots, k\}$-inverse of A is any solution of equations (i), (j), ..., (k) in the Penrose equations:

(1) $AXA = A$

(2) $XAX = X$

(3) $(AX)^* = AX$

(4) $(XA)^* = XA$.

2. The Bott-Duffin inverse and network analysis.

Consider the system

(2.1) $$Ax + y = b, \quad x \in L, \quad y \in L^{\perp} ,$$

with given $A \in C^{n\times n}$, $b \in C^n$ and a subspace L of C^n. If the matrix $(AP_L + P_{L^{\perp}})$ is nonsingular, then (2.1) is consistent for all $b \in C^n$, and the solution

(2.2) $$x = P_L(AP_L + P_{L^{\perp}})^{-1}b, \quad y = b - Ax$$

is unique. The transformation

(2.3) $$A^{(-1)}_{(L)} \triangleq P_L(AP_L + P_{L^{\perp}})^{-1} ,$$

was introduced and studied by Bott and Duffin [13] (who called it the constrained inverse of A). We call it the Bott-Duffin inverse of A (with respect to L).

In this section, the application of the Bott-Duffin inverse to electrical network analysis is outlined. The discussion is restricted to DC networks.

Let the graph of the network, consisting of m nodes and n branches, be represented by its node-branch incidence matrix M. We denote by

x-the vector of branch voltages

y-th vector of branch currents

and recall that the Kirchhoff laws define two complementary orthogonal subspaces

$N(M)$ - the currents satisfying the Kirchhoff current law

$R(M^T)$ - the voltages satisfying the Kirchhoff voltage law, ([55], see also [10] Section 2.12)

With each branch b_j $(j = 1, \ldots, n)$ we associate a series voltage generator of v_j volts and a parallel current generator of w_j amperes, related to the branch voltage x_j and current y_j by the Ohm law

$$a_j(x_j - v_j) + (y_j - w_j) = 0, \qquad (j = 1, \ldots, n) \; , \tag{2.4}$$

where a_j is the conductance of the branch b_j. Denoting the network conductance matrix by

$$A \triangleq [\operatorname{diag} a_j]$$

we write the network equations

$$Ax + y = Av + w \tag{2.5}$$

$$x \in R(M^T), \; y \in N(M) \tag{2.6}$$

which are (2.1) with $b = Av + w$, $L = R(M^T)$.

In AC networks without mutual coupling, equations (2.4) with complex constants and variables still hold for the branches. The complex a_j is then the admittance of the branch b_j. AC networks with mutual coupling due to transformers, are still represented by (2.5), where the admittance matrix A is symmetric, its off-diagonal elements giving the mutual couplings, see, e.g., [13].

The nonsingularity of $(AP_{R(M^T)} + P_{N(M)})$, which guarantees the existence of the Bott-Duffin inverse $A^{(-1)}_{(R(M^T))}$ and of a unique solution to the network equations (2.5)-(2.6), is established by the following

Lemma 2.1

Let $A \in C^{n \times n}$ and let L be a subspace of C^n such that

$$(Ax, x) \neq 0, \quad \forall\, 0 \neq x \in L .$$

Then $(AP_L + P_{L^\perp})$ is nonsingular.

Proof

If $Ax + y = 0$ for some $x \in L$ and $y \in L^\perp$, then $Ax \in L^\perp$ and therefore $(Ax, x) = 0$. □

The unique solution of (2.5)-(2.6) is therefore

(2.7) $$x^0 = A^{(-1)}_{(R(M^T))}(Av + w)$$

(2.8) $$y^0 = (I - AA^{(-1)}_{(R(M^T))})(Av + w) .$$

The matrix $A^{(-1)}_{(R(M^T))}$, called the <u>transfer matrix</u> of the network, carries a complete description of the electrical properties of the network. Indeed, from (2.7), the (i, j)th entry of $A^{(-1)}_{(R(M^T)}$ is the voltage across the branch b_i as a result of a one ampere current source in b_j, $(i, j = 1, \ldots, n)$.

For a nonsingular matrix A, it follows from the identity

$$(A^{-1}P_{L^\perp} + P_L)^{-1} = (AP_L + P_{L^\perp})^{-1}A$$

that $(A^{-1})^{(-1)}_{(L^\perp)}$ exists if and only if $A^{(-1)}_{(L)}$ exists. Since the conductance matrix A is nonsingular, (2.5) can be rewritten as

(2.9) $$A^{-1}y + x = A^{-1}w + v$$

and the unique solution of (2.9)-(2.6) is

(2.10) $$y^0 = (A^{-1})^{(-1)}_{(N(M))}(A^{-1}w + v)$$

(2.11) $$x^0 = (I - A^{-1}(A^{-1})^{(-1)}_{(N(M))})(A^{-1}w + v) .$$

The matrix $(A^{-1})^{(-1)}_{(N(M))}$ is called the dual transfer its (i, j)th entry giving the current in branch b_i as a result of a one volt source in b_j .

The correspondence between the solutions (2.7)-(2.8) and (2.10)-(2.11) is called electrical duality. A sample result is

$$A^{-1}(A^{-1})^{(-1)}_{(N(M))} + A^{(-1)}_{(R(M^T))} A = I$$

obtained by comparing (2.7) and (2.11). See [13], [21] and [50] for further results on electrical duality.

We recall, for a function $q: C^n \to C$, that a stationary point of q in L is a point $x^0 \in L$ at which the gradient

$$\nabla q(x^0) \in L^{\perp} .$$

The value of q at a stationary point is called a stationary value.

Theorem 2.2 ([13]).

Let $A \in C^{n \times n}$ be Hermitian, and let L be a subspace of C^n such that $A^{(-1)}_{(L)}$ exists. Then for any two vectors v, w in C^n, the quadratic function

(2.12) $$q(x) = \frac{1}{2}(x - v)^* A(x - v) - w^* x$$

has a unique stationary value in L when

(2.13) $$x = A^{(-1)}_{(L)}(Av + w) .$$

Conversely, if the Hermitian matrix A and the subspace L are such that for any two vectors v, w in C^n, the quadratic equation (2.12) has a stationary value in L, then $A^{(-1)}_{(L)}$ exists and the stationary point is unique for any v, w and given by (2.13).

<u>Proof</u>

Differentiating (2.12) we see that the sought stationary point $x \in L$ satisfies

$$\nabla q(x) = A(x-v) - w \in L^{\perp} .$$

Setting $y = -\nabla q(x)$ we conclude that x is a stationary point of q in L if and only if x is a solution of

$$Ax + y = Av + w, \quad x \in L, \quad y \in L^{\perp} . \tag{2.14}$$

Thus the existence of a stationary value of q for any v, w is equivalent to the consistency of (2.14) for any v, w, i.e., to the existence of $A^{(-1)}_{(L)}$, in which case (2.13) is the unique stationary point in L. □

If the matrix A, in Theorem 2.2, is positive definite then $A^{(-1)}_{(L)}$ exists (by Lemma 2.1) and the function q in (2.12) is strictly convex (so that a stationary point of q is its unique minimizer). Theorem 2.2 then gives:

<u>Corollary 2.3</u>

Let $A \in C^{n \times n}$ be Hermitian positive-definite and let L be a subspace of C^n. Then for any v, w in C^n the function

$$q(x) = \frac{1}{2}(x-v)^* A(x-v) - w^* x \tag{2.12}$$

has a unique minimum in L, when

$$x = A^{(-1)}_{(L)}(Av + w) . \tag{2.13}$$ □

Returning to electrical networks, Corollary 2.3 gives the classical variational principle of Kelvin [53] and Maxwell ([35], pp. 903-908), stating that the rate of energy dissipation in the network is minimized.

Corollary 2.4

Let $\{A, M, v, w\}$ represent an electrical network. Then

(a) The vector of branch voltages

$$x^0 = A^{(-1)}_{(R(M^T))} (Av + w) \tag{2.7}$$

is the unique minimizer of

$$q(x) = \frac{1}{2}(x - v)^* A(x-v) - w^* x \tag{2.14}$$

in $R(M^T)$, and the vector y^0 of branch currents is

$$y^0 = -\nabla q(x^0) = -A(x^0 - v) + w \in R(M^T)^{\perp} = N(M) \ . \tag{2.15}$$

(b) The vector y^0 is the unique minimizer of

$$p(y) = \frac{1}{2}(y-w)^* A^{-1}(y-w) - v^* y \tag{2.16}$$

in $N(M)$, and the vector x^0 is

$$x^0 = -\nabla p(y^0) = -A^{-1}(y^0 - w) + v \in N(M)^{\perp} = R(M^T) \ . \tag{2.17}$$

□

Corollary 2.4 shows that the voltage x is uniquely determined by the function (2.12) to be minimized subject to the Kirchhoff voltage law. Kirchhoff's current law and Ohm's law then follow from (2.15).

Dually the current y is uniquely determined by the

function (2.16) and the Kirchhoff current law.

Corollary 2.4 is a special case of the duality theory of convex programming.

For other references on applications of generalized inverses to electrical network theory see [1], [2], [7], [21], [49], [50] and [51].

3. Explicit solutions of interval linear programs.

An interval linear program (ILP) is a linear program of the form

$$\text{(3.1)} \qquad \max \; c^T x$$
$$\text{s.t.}$$
$$\text{(3.2)} \qquad b^- \leqq Ax \leqq b^+$$

where the matrix $A \in R^{m \times n}$ and the vectors $b^- \leqq b^+ \in R^m$, $c \in R^n$ are given. Clearly any linear program with bounded variables can be formulated as an ILP. For example, the "standard" linear program

$$\text{(3.3)} \qquad \max\{c^T x: \; Ax = b, \; x \geqq 0\}$$

can be written as

$$\text{(3.4)} \qquad \max\left\{c^T x: \begin{bmatrix} b \\ 0 \end{bmatrix} \leqq \begin{bmatrix} A \\ I \end{bmatrix} x \leqq \begin{bmatrix} b \\ Me \end{bmatrix}\right\},$$

where e is a vector of ones and M is an upper bound on the components of x.

An ILP{(3.1), (3.2)} is called feasible if

$$\text{(3.5)} \qquad F \triangleq \{x: b^- \leqq Ax \leqq b^+\} \neq \emptyset$$

and bounded if feasible and if

$$\max\{c^T x: \; x \in F\} < \infty \; .$$

Since, by definition, $F = F + N(A)$, it follows that a feasible ILP is bounded if and only if

$$c \perp N(A) \quad . \tag{3.6}$$

Any bounded ILP can be solved explicitly in the following two cases:

(i) $A \in R_m^{m \times n}$, the full row rank case.

(ii) $A \in R_m^{(m+1) \times n}$.

These explicit solutions, obtained in [8] and [42], are presented below. The remaining case, $A \in R_r^{m \times n}$ with $r < m-1$, can be solved iteratively using the explicit solutions of a sequence of appropriately chosen subproblems, see [42] for details.

The full row rank case is solved in the following.

Theorem 3.1 ([8])

Let $A \in R_m^{m \times n}$ and let (3.6) hold. Then the set of optimal solutions of the ILP {(3.1), (3.2)} is

$$A^{(1)} \eta(b^-, b^+, A^{(1)T}c) + N(A) \tag{3.7}$$

where $A^{(1)}$ is any {1}-inverse of A, and for any three vectors u, v, w in R^m the set $\eta(u, v, w)$ consists of the vectors $\eta = [\eta_i]$ given componentwise by

$$\eta(u, v, w) = \left\{\eta = [\eta_i] : \eta_i = \begin{cases} u_i & \text{if } w_i < 0 \\ v_i & \text{if } w_i > 0, \\ \lambda u_i + (1-\lambda)v_i & \text{if } w_i = 0 \\ \quad 0 \leq \lambda \leq 1 \end{cases} \quad i = 1, \ldots, m\right\} . \tag{3.8}$$

Proof.

Let $A \in R_m^{m \times n}$. Then for any {1}-inverse of A, $A^{(1)}$,

(3.9) $$z = Ax$$

(3.10) $$x = A^{(1)}z + N(A)$$

is a one to one correspondence between all vectors $z \in R^m$ and all translations of $N(A)$ in R^n. Substituting (3.9) and (3.10) in the ILP {(3.1), (3.2)}, and using (3.6), one obtains the following equivalent ILP

(3.1') $$\max \; c^T A^{(1)} z$$
s.t.

(3.2') $$b^- \leqq z \leqq b^+$$

whose optimal solution is obviously

(3.11) $$\eta(b^-, b^+, A^{(1)T} c)$$

which, substituted in (3.10), gives (3.7). □

Checking the orthogonality condition (3.6) does not require special computations, since the computation of $A^{(1)}$ gives a basis of $N(A)$ as a by-product, see [8] or [10], Section 1.4.

We turn now to the second case of explicitly solvable ILP's, where the rank of the coefficient matrix is one less than its number of rows. We can write such an ILP as

(3.1) $$\max \quad c^T x$$
s.t.

(3.2) $$b^- \leqq Ax \leqq b^+$$

(3.12) $$b^-_{m+1} \leq h^T Ax \leq b^+_{m+1}$$

where $A \in R_m^{m \times n}$ and $0 \neq h \in R^m$. Assuming the boundedness condition (3.6), one gets, by substituting (3.9) and (3.10) in the ILP {(3.1), (3.2), (3.12)}, the equivalent ILP

(3.1') $$\max \; c^T A^{(1)} z \quad \text{s.t.}$$

(3.2') $$b^- \leqq z \leqq b^+$$

(3.12') $$b^-_{m+1} \le h^T z \le b^+_{m+1} \; .$$

The set of optimal solutions of the subproblem {(3.1'), (3.2')} is given by (3.11). If there is a $z \in \eta(b^-, b^+, A^{(1)T}c)$ which satisfies (3.12), then z is an optimal solution of the ILP {(3.1'), (3.2'), (3.12')}, and substituting z in (3.10) gives an optimal solution of the ILP {(3.1), (3.2), (3.12)}.

By the definition (3.8), the set $\eta(b^-, b^+, A^{(1)T}c)$ consists of a single vector unless, for some $1 \le i \le m$,

$$(A^{(1)T}c)_i = 0 \text{ and } b_i^- \neq b_i^+ \; .$$

Even then, determining whether (3.12') is satisfied by some $z \in \eta(b^-, b^+, A^{(1)T}c)$ is easy, since $\{h^T z: z \in \eta(b^-, b^+, A^{(1)T}c)\}$ is an interval with endpoints

(3.13) $$h^T z^- = \sum_{I \in I(-)} h_i b_i^- + \sum_{i \in I(+)} h_i b_i^+ + \sum_{\substack{i \in I(0) \\ h_i < 0}} h_i b_i^+ + \sum_{\substack{i \in I(0) \\ h_i \ge 0}} h_i b_1^-$$

and

(3.14) $$h^T z^+ = \sum_{i \in I(-)} h_i b_i^- + \sum_{i \in I(+)} h_i b_i^+ + \sum_{\substack{i \in I(0) \\ h_i < 0}} h_i b_i^- + \sum_{\substack{i \in I(0) \\ h_i \ge 0}} h_i b_i^+ ,$$

where

(3.15) $$\left.\begin{matrix} I(-) \\ I(+) \\ I(0) \end{matrix}\right\} \triangleq \{i: i = 1, \ldots, m, \; (A^{(1)T}c)_i \begin{cases} < \\ > \\ = \end{cases} 0 \} \; .$$

If

$$[b^-_{m+1}, b^+_{m+1}] \cap [h^T z^-, h^T z^+] = \phi \tag{3.16}$$

i.e. if (3.12') is violated by every $z \in \eta(b^-, b^+, A^{(1)T}c)$, we can assume, without loss of generality, that the upper bound b^+_{m+1} is exceeded, i.e.,

$$h^T z^- > b^+_{m+1} .$$

Let the vector $z^- \in \eta(b^-, b^+, A^{(1)T}c)$ be given by

$$z_i^- = \begin{cases} b_i^- & \text{if } i \in I(-) \text{ or } [i \in I(0) \text{ and } h_i \geq 0] \\ b_i^+ & \text{if } i \in I(+) \text{ or } [i \in I(0) \text{ and } h_i < 0] \end{cases} \tag{3.17}$$

and denote

$$\Delta \triangleq b^+_{m+1} - h^T z^- < 0 . \tag{3.18}$$

From (3.17) it is clear that the only components z_i^- which can be changed in order to move (3.12') toward feasibility (while maintaining feasibility in (3.2')) are those indexed by

$$Q \triangleq \{i : 1 \leq i \leq m,\ h_i \neq 0,\ \frac{(A^{(1)T}c)i}{h_i} > 0\} \ . . \tag{3.19}$$

Since

$$\gamma_i = \frac{(A^{(1)T}c)i}{h_i} \tag{3.20}$$

is the marginal cost of moving toward feasibility in (3.12') by changing z_i^- for $i \in Q$, we reorder the (say q in number) indices in Q by

(3.21) $$Q = \{k_1, k_2, \ldots, k_q\}$$

where

(3.22) $$\gamma_{k_1} \le \gamma_{k_2} \le \ldots \le \gamma_{k_q} .$$

We can now obtain an optimal solution of the ILP {(3.1'), (3.2'), (3.12')}, or determine that the problem is infeasible, by changing the components

$$\{z_i^- : \quad i = k_1, k_2, \ldots, k_q\}$$

one at a time, as much as possible without violating (3.2'), until (3.12') is satisfied. If (3.12') is not satisfied after all these changes, then the ILP is infeasible.

We summarize these observations in:

Theorem 3.2 ([42])

Let the ILP {(3.1), (3.2), (3.12)} be given as above. Then

(a) If some $z \in \eta(b^-, b^+, A^{(1)T})$ satisfies (3.12') then it is an optimal solution of the ILP {(3.1'), (3.2'), (3.12')}, and an optimal x is obtained by (3.10).

(b) If (3.16) holds, let $\Delta < 0$ and $Q = \{k_1, k_2, \ldots, k_q\}$ be as above. Then the ILP is infeasible if and only if

(3.23) $$\sum_{i \in Q} \delta_i h_i > \Delta$$

where

(3.24) $$\delta_i = \begin{cases} b_i^- - z_i^- & \text{if } h_i > 0 \\ b_i^+ - z_i^- & \text{if } h_i < 0 \end{cases}$$

for $i \in Q$.

(c) If (3.16) holds and (3.23) does not hold, let $\{\delta_i : i \in Q\}$ be as above and let

$$p \triangleq \min\{i : 1 \le i \le q, \sum_{j=1}^{i} \delta_{k_j} h_{k_j} \le \Delta\} \tag{3.25}$$

$$\theta \triangleq \frac{\Delta - \sum_{j=1}^{p-1} \delta_{k_j} h_{k_j}}{h_{k_p}} . \tag{3.26}$$

Then an optimal solution of the ILP {(3.1'), (3.2'), (3.12')} is

$$z_i = \begin{cases} z_i^-, & i \notin Q \text{ or } i \in \{k_{p+1}, \ldots, k_q\} \\ z_i^- + \delta_i, & i \in \{k_1, \ldots k_{p-1}\} \\ z_i^- + \theta, & i = k_p . \end{cases} \tag{3.27}$$

□

For further details, extensions of the above results, and applications of ILP see [9], [41], [42], [57] and [58].

4. Integer and mixed integer solutions of linear equations.

Let K denote the ring of integers {0, ±1, ±2, ...} . In this section we study the problems

$$AXB = C, \quad X \in K^{n \times p} \tag{4.1}$$

$$Ax = b , \quad x \in K^n \tag{4.2}$$

$$Ax + Cy = b, \quad x \in K^n \tag{4.3}$$

where A, B, C are integral matrices of appropriate orders and $b \in K^m$ .1/

1/ In principle, no generality is gained by assuming the matrices A, B, C to be rational.

Let $A \in K^{m\times n}$ be given. A matrix $X \in R^{n\times m}$ is called a left [right] integral $\{i, j, \ldots, k\}$-inverse of A if X is an $\{i, j, \ldots, k\}$-inverse of A and if $XA \in K^{n\times n}$ [$AX \in K^{m\times m}$]. The set of all left [right] integral $\{i, j, \ldots, k\}$-inverses of A is denoted by $A^L\{i, j, \ldots, k\}$ [$A^R\{i, j, \ldots, k\}$].

A $\{1, 2\}$-inverse of $A \in K^{m\times n}$ which is both left and right integral, was constructed in [26] as follows.

Lemma 4.1 ([26])

Let $A \in K^{m\times n}$ and let

(4.4) $$PAQ = S$$

be the Smith normal form of A (see, e.g., [34], pp. 42-44 or [10], Section 2.11) so that $P, P^{-1} \in K^{m\times m}$; $Q, Q^{-1} \in K^{n\times n}$ and S is diagonal. Then the matrix

(4.5) $$QS^{\dagger}P$$

is in $A^L\{1, 2\} \cap A^R\{1, 2\}$.

Proof.

Obvious □

The following two results are proved exactly as their non-integral counterparts in [37].

Theorem 4.2 ([17])

Let $A \in K^{m\times n}$, $B \in K^{p\times q}$, $C \in K^{m\times q}$ be given. Then the equation

(4.1) $$AXB = C, \quad X \in K^{n\times p}$$

is consistent if and only if for any $A^{(\ell)} \in A^L\{1\}$, $B^{(r)} \in B^R\{1\}$

(4.6) $$A^{(\ell)}CB^{(r)} \in K^{n\times p}$$

and

(4. 7) $$AA^{(\ell)} CB^{(r)}B = C \ ,$$

in which case the general solution of (4. 1) is

(4. 8) $$X = A^{(\ell)}CB^{(r)} + Y - A^{(\ell)}AYBB^{(r)}, \quad Y \in K^{n \times p} \ . \qquad \square$$

<u>Corollary 4. 3</u> ([17])

Let $A \in K^{m \times n}$, $b \in K^m$. Then the equation

(4. 2) $$Ax = b, \quad x \in K^n$$

is consistent if and only if for any $A^{(\ell)} \in A^L\{1\}$,

(4. 9) $$A^{(\ell)}b \in K^n$$

and

(4. 10) $$AA^{(\ell)}b = b \ ,$$

in which case the general solution of (4. 2) is

(4. 11) $$x = A^{(\ell)}b + (I - A^{(\ell)}A)y, \quad y \in K^n \ . \qquad \square$$

In order to solve the mixed integer equation (4. 3), we rewrite it as

$$Cy = b - Ax \quad x \in K^n$$

which, for any $x \in K^n$, is consistent in y if and only if

(4. 12) $$CC^{(1)}(b - Ax) = b - Ax$$

where $C^{(1)}$ is any {1}-inverse of C . Taking $C^{(1)} \in C^R\{1\}$, it follows from (4. 12) that the mixed integer equation (4. 3) is equivalent to the equation

(4.13) $$(I - CC^{(1)})Ax = (I - CC^{(1)})b, \quad x \in K^n$$

which is of the type (4.2), solved, in Corollary 4.3.

For further details and results see [26], [17], [14] and ([10], Section 2.11).

5. Equilibrium points of bimatrix games.

Let $M \triangleq \{1, 2, \ldots, m\}$, $N \triangleq \{1, 2, \ldots, n\}$, and let two $m \times n$ real matrices $A = [a_{ij}]$ and $B = [b_{ij}]$ be given. The bimatrix game $\{A, B\}$ is the two person game where, if player I chooses $i \in M$ and player II chooses $j \in N$, the payoffs are a_{ij} to player I and b_{ij} to player II.

We denote the set of probability vectors in R^n by P_n

$$P_n \triangleq \{x \in R^n_+ : e^T x = 1\}$$

where e denotes the vector of ones, of appropriate dimension.

A point $\{x, y\} \in R^m \times R^n$ is called an equilibrium point of the bimatrix game $\{A, B\}$ if

(5.1) $$x \in P_m ,$$

(5.1') $$y \in P_n ,$$

(5.2) $$B^T x \leq (x^T By)e ,$$

(5.2') $$Ay \leq (x^T Ay)e .$$

Some references on equilibrium points of bimatrix games are [29], [33], [31] and [30].

In this section, which is based on [11], equilibrium points of a bimatrix game $\{A, B\}$ are characterized in terms of certain submatrices of A and B and their Moore-Penrose inverses.

Since the equilibrium points are unchanged if the same constant is added to every entry of A and B, we can assume, without loss of generality, that A and B are positive matrices.

For any subsets $I \subset M$, $J \subset N$ we denote their complements by

$$\bar{I} = \{i \in M: i \notin I\}$$

$$\bar{J} = \{j \in N: j \notin J\} .$$

For any $A = [a_{ij}] \in R^{m \times n}$, $A(I/J)$ denotes the submatrix $[a_{ij}]$, $i \in I$, $j \in J$. Similarly, for any $x = [x_i] \in R^n$, $x(I)$ denotes the subvector $[x_i]$, $i \in I$.

Theorem 5.1 ([11])

Let A, B be positive $m \times n$ matrices and let $x \in R^m_+$, $y \in R^n_+$. Then $\{x, y\}$ is an equilibrium point of $\{A, B\}$ if and only if there are sets $I \subset M$, $J \subset N$ for which the following eight conditions are satisfied.

(a) $$e = B(I/J)^T u \quad \text{for some } u \geqq 0 ,$$

(a') $$e = A(I/J)v \quad \text{for some } v \geqq 0 ,$$

(b) $$x(\bar{I}) = 0 ,$$

(b') $$y(\bar{J}) = 0 ,$$

(c) $$x(I) = B(I/J)^{T\dagger} e \left[\frac{1 - e^T w}{e^T B(I/J)^{\dagger} e} \right] + w, \quad \text{for some } w \in N(B(I/J)^T) ,$$

(c') $$y(J) = A(I/J)^{\dagger} e \left[\frac{1 - e^T z}{e^T A(I/J)^{\dagger} e} \right] + z, \quad \text{for some } z \in N(A(I/J)) ,$$

(d) $$B(I/N)^T x(I) \leq \left[\frac{1 - e^T w}{e^T B(I/J)^{\dagger} e}\right] e ,$$

(d') $$A(M/J) y(J) \leq \left[\frac{1 - e^T z}{e^T A(I/J)^{\dagger} e}\right] e .$$

Proof

If. Let $x \in R_+^m$ and $y \in R_+^n$ satisfy the eight conditions (a)-(d').

First we show that $\{x, y\}$ satisfies (5.1), (5.1'), for which it suffices to prove

$$e^T x = 1 = e^T y .$$

Now $e^T w \neq 1$ (where the vector w is defined by (c)), for if $e^T w = 1$ then (d) implies that $x(I) = 0$ (since B is positive), so by (c) $w = 0$ contradicting $e^T w = 1$. Therefore

(5.3) $$e^T x = e^T x(I), \text{ by (b)}$$
$$= 1, \quad \text{by (c)} .$$

Similarly we prove that $e^T y = 1$.

To prove (5.2') we calculate

(5.4) $$x^T A y = x(I)^T A(I/J)\, y(J), \quad \text{by (b), (b')} ,$$

$$= x(I)^T A(I/J)\, A(I/J)^{\dagger} e \left[\frac{1 - e^T z}{e^T A(I/J)^{\dagger} e}\right] + x(I)^T A(I/J) z, \quad \text{by (c')} ,$$

$$= x(I)^T e \left[\frac{1 - e^T z}{e^T A(I/J)^{\dagger} e}\right] , \quad \text{since } z \in N(A(I/J)) \text{ and } e \in R(A(I/J)) \text{ by (a')} ,$$

$$= \frac{1 - e^T z}{e^T A(I/J)^{\dagger} e} , \quad \text{by (5.3)} .$$

The inequality (5.2') now follows from (d'), (b') and (5.4). The inequality (5.2) is similarly proved.

Only if. Let $\{x, y\}$ be an equilibrium point and define

$$I \triangleq \{i \in M: \sum_{j=1}^{n} a_{ij} y_j = x^T Ay\} \tag{5.5}$$

$$J \triangleq \{j \in N: \sum_{i=1}^{m} b_{ij} x_i = x^T By\} \ . \tag{5.5'}$$

Both I, J are nonempty since $I = \phi$ implies that $Ay < (x^T Ay)e$, which by (5.3) implies $x^T Ay < x^T Ay$.

From the definition (5.1)-(5.2') of an equilibrium point $\{x, y\}$ it follows that $\{x, y\}$ satisfies the "complementary slackness" conditions

$$y^T [B^T x - (x^T By)e] = 0 \tag{5.6}$$

$$x^T [Ay - (x^T Ay)e] = 0 \ , \tag{5.6'}$$

which, together with (5.2), (5.2'), (5.5) and (5.5'), imply that $\{x, y\}$ satisfies (b) and (b').

Using (5.5) and (b') we get

$$A(I/J)y(J) = (x^T Ay) e \tag{5.7}$$

which proves (a') since $y(J) \geq 0$, $x^T Ay > 0$. (a) is similarly proved.

The general solution of (5.7) is

$$y(J) = A(I/J)^{\dagger} e(x^T Ay) + z, \quad z \in N(A(I/J)) \ , \tag{5.8}$$

which gives

$$\begin{aligned} 1 &= e^T y \\ &= e^T y(J), \qquad \text{by (b')} \ , \\ &= (e^T A(I/J)^{\dagger} e)(x^T Ay) + e^T z \ . \end{aligned} \tag{5.9}$$

We can assume, without loss of generality, that[1]

$$e^T A(I|J)^{\dagger} e \neq 0 \quad , \tag{5.10}$$

so that (5.9) gives

$$x^T Ay = \frac{1 - e^T z}{e^T A(I|J)^{\dagger} e} \tag{5.11}$$

which, substituted in (5.8) and (5.2') proves (c') and (d'). (c) and (d) are similarly proved. □

An important special case of bimatrix games are two-person zero-sum games, in which case

$$B = -A \tag{5.12}$$

and the inequalities (5.2) and (5.2)' reduce to

$$A^T x \geqq (x^T Ay) e \tag{5.13}$$

$$Ay \leqq (x^T Ay) e \tag{5.13'}$$

which, in turn, are equivalent to the "saddle point" inequality:

$$u^T Ay \leqq x^T Ay \leqq x^T Av \tag{5.14}$$

for all $u \in P_m$, $v \in P_n$.

Theorem 5.1 holds for the case (5.12) if we drop the assumption that B is a positive matrix, and if we replace (a) by

---

1/ The example

$$A(I|J) = \begin{bmatrix} 3 & 1 \\ 4 & 2 \end{bmatrix}, \quad A(I|J)^{\dagger} = \frac{1}{2}\begin{bmatrix} 2 & -1 \\ -4 & 3 \end{bmatrix}$$

shows that (5.10) does not follow from the positivity of $A(I|J)$. However we can assume (5.10) since a positive constant may be added to all elements of A and B without changing the equilibrium points.

$e \in R(B(I|J)^T)$. The conditions (c), (c'), (d) and (d') are simplified in the case (5.12) since then

(5.15) $\qquad e^T w = 0, \quad \text{by } e \in R(A(I|J)), \ w \in N(A(I|J)^T)$,

and

(5.16) $\qquad e^T z = 0, \quad \text{by } e \in R(A(I|J)^T), \ z \in N(A(I|J))$.

Assuming (5.10) again, we infer from (5.11) and (5.16) that the value of the 2-person 0-sum game $\{A\}$ is

(5.17) $$\min_{y \in P_n} \max_{x \in P_m} \{x^T A y\} = \frac{1}{e^T A(I|J)^\dagger e} = \max_{p \in P_m} \min_{y \in P_n} \{x^T A y\}$$

which is an interesting property of the Moore-Penrose inverse. For further details see [11] and [4].

REFERENCES

[1] W. N. Anderson, Jr., Shorted operators, SIAM J. Appl. Math. 20 (1971), 520-525.

[2] W. N. Anderson, Jr. and R. J. Duffin, Series and parallel addition of matrices, J. Math. Anal. Appl. 26 (1969), 576-594.

[3] A. Ben-Israel, On iterative methods for solving nonlinear least squares problems over convex sets, Israel J. Math. 5 (1967), 211-224.

[4] A. Ben-Israel, On optimal solutions of 2-person 0-sum games, Atti Accad. Naz. Lincei Rend. Cl. Sci. Fis. Mat. Natur. Ser. VIII XLIV (1968), 274-278.

[5] A. Ben-Israel, Linear equations and inequalities on finite dimensional, real or complex, vector spaces: A unified theory, J. Math. Anal. Appl. **27** (1969), 367-389.

[6] A. Ben-Israel, On Newton's method in nonlinear programming, pp. 339-352 in Proc. Princeton Sympos. Math. Prog. (H. W. Kuhn, Editor), Princeton University Press, Princeton, N. J., 1970, vi + 620 pp.

[7] A. Ben-Israel and A. Charnes, Generalized inverses and the Bott-Duffin network analysis, J. Math. Anal. Appl. **7** (1963), 428-435. Erratum, ibid **18** (1967), 393.

[8] A. Ben-Israel and A. Charnes, An explicit solution of a special class of linear programming problems, Operations Research **16** (1968), 1167-1175.

[9] A. Ben-Israel, A. Charnes, A. P. Hurter, Jr. and P. D. Robers, On the explicit solution of a special class of linear economic models, Operations Research **18** (1970), 462-470.

[10] A. Ben-Israel and T. N. E. Greville, Generalized Inverses: Theory and Applications, Pure and Applied Mathematics, A. Wiley-Interscience Series of Texts, Monographs and Tracts, John Wiley and Sons, New York, 1974.

[11] A. Ben-Israel and M. J. L. Kirby, A characterization of equilibrium points of bimatrix games, Atti Accad. Naz. Lincei Rend. Cl. Sci. Fis. Mat. Natur. Ser. VIII, **XLVI** (1969), 196-201.

[12] J. C. G. Boot, Quadratic Programming, North-Holland Publishing Co., Amsterdam, 1964.

[13] R. Bott and R. J. Duffin, On the algebra of networks, Trans. Amer. Math. Soc., **74** (1953), 99-109.

[14] V. J. Bowman and C-A Burdet, On the general solution to systems of mixed-integer linear equations, SIAM J. Appl. Math. 26 (1974), 120-125.

[15] A. Charnes and W. W. Cooper, Structural sensitivity analysis in linear programming and an exact product form left inverse, Naval Res. Logist. Quart. 15 (1968), 517-522.

[16] A. Charnes, W. W. Cooper and G. L. Thompson, Constrained generalized medians and hypermedians as deterministic equivalents for two-stage linear programs under uncertainty, Management Sci. 12 (1965), 83-112.

[17] A. Charnes and F. Granot, Existence and representation of Diophantine and mixed Drophantine solutions to linear equations and inequalities, Center for Cybernetic Studies, The University of Texas, Austin, January 1973.

[18] A. Charnes and M. J. L. Kirby, Modular design, generalized inverses and convex programming, Operations Research 13 (1965), 836-847.

[19] R. E. Cline, Representations for the Generalized Inverse of Matrices with Applications in Linear Programming, doctoral dissertation, Purdue University, Layfayette, Ind., 1963.

[20] R. E. Cline and L. D. Pyle, The generalized inverse in linear programming -- an intersection projection method and the solution of a class of structured linear programming problems, SIAM J. Appl. Math. 24 (1973), 338-351.

[21] R. J. Duffin, Network models, pp. 65-91 in Mathematical Aspects of Electrical Network Analysis, SIAM-AMS Proc. Vol. III (H. S. Wilf and F. Harary, Editors) Amer. Math. Soc., Providence, R.I., 1971, 206 pp.

[22] R. Fletcher, Generalized inverses for nonlinear equations and optimization, pp. 75-85 in Numerical Methods for Nonlinear Algebraic Equations (P. Rabinowitz, Editor) Gordon and Breach, London, 1970, xi + 199 pp.

[23] M. Gerstenhaber, Theory of convex polyhedral cones, Chapter 18 in Activity Analysis of Production and Allocation (T. C. Koopmans, Editor), J. Wiley, New York, 1951.

[24] T. N. E. Greville, Some applications of the pseudoinverse of a matrix, SIAM Review **2** (1960), 15-22.

[25] D. Hearn and W. D. Randolph, Dual approaches to quadratically constrained quadratic programs, presented at the VIII International Symposium on Mathematical Programming, Stanford University, August 1973.

[26] M. F. Hurt and C. Waid, A generalized inverse which gives all the integral solutions to a system of linear equations, SIAM J. Appl. Math. **19** (1970), 547-550.

[27] M. J. L. Kirby, Generalized Inverses and Chance Constrained Programming, doctoral dissertation in Applied Mathematics, Northwestern University, Evanston, Ill., June 1965.

[28] V. Klee, Some characterizations of convex polyhedra, Acta Mathematica **102** (1959), 79-107.

[29] H. W. Kuhn, An algorithm for equilibrium points in bimatrix games, Proc. Nat. Acad. Sci. USA **47** (1961), 1657-1662.

[30] C. E. Lemke, Bimatrix equilibrium points and Mathematical programming, Management Sci. **11** (1965), 681-689.

[31] C. E. Lemke and J. T. Howson, Jr., Equilibrium points of bimatrix games, J. Soc. Indust. Appl. Math. 12 (1964), 413-423.

[32] O. L. Mangasarian, Equivalence in nonlinear programming, Naval Res. Logist. Quart. 10 (1963), 299-306.

[33] O. L. Mangasarian, Equilibrium points of bimatrix games, J. Soc. Indust. Appl. Math. 12 (1964), 778-780.

[34] M. Marcus and H. Minc, A Survey of Matrix Theory and Matrix Inequalities, Allyn and Bacon, Boston, Mass., 1964, xvi + 180 pp.

[35] J. C. Maxwell, Treatise of Electricity and Magnetism, 3rd Edition, Oxford University Press, 1892.

[36] D. L. Nelson, T. O. Lewis and T. L. Boullion, A quadratic programming technique using matrix pseudoinverses, Indust. Math. 21 (1971), 1-21.

[37] R. Penrose, A generalized inverse for matrices, Proc. Cambridge Philos. Soc. 51 (1955), 406-413.

[38] L. D. Pyle, The Generalized Inverse in Linear Programming, doctoral dissertation, Purdue University, Lafayette, Ind., 1960.

[39] L. D. Pyle, The generalized inverse in linear programming. Basic structure, SIAM J. Appl. Math. 22 (1972), 335-355.

[40] C. R. Rao and S. K. Mitra, Theory and application of constrained inverse of matrices, SIAM J. Appl. Math. 24 (1973), 473-488.

[41] P. D. Robers and A. Ben-Israel, An interval programming algorithm for discrete linear L_1 approximation problems, J. Approximation Th. 2 (1969), 323-336.

[42] P. D. Robers and A. Ben-Israel, A suboptimization method for interval linear programming: A new method for linear programming, Linear Algebra and Its Appl. 3 (1970), 383-405.

[43] J. B. Rosen, The gradient projection method for nonlinear programming. Part I: Linear constraints. J. Soc. Indust. Appl. Math. 8 (1960), 181-217.

[44] J. B. Rosen, The gradient projection method for nonlinear programming. Part II: Nonlinear constraints, J. Soc. Indust. Appl. Math. 9 (1961), 514-532.

[45] J. B. Rosen, Minimum and basic solutions to singular linear systems, J. Soc. Indust. Appl. Math. 12 (1964), 156-162.

[46] J. B. Rosen, Gradient projection as a least squares solution of Kuhn-Tucker conditions, Computer Sciences Department, The University of Wisconsin, Madison, Wis., July 1965.

[47] J. B. Rosen, Chebyshev solution of large linear systems, J. Comput. Syst. Sci. 1 (1967), 29-43.

[48] H. D. Scolnik, On the solution of non-linear least squares problems, pp. 1258-1265 in Information Processing 71, North-Holland Publishing Company, 1972.

[49] G. E. Sharpe and G. P. H. Styan, Circuit duality and the general network inverse, IEEE Trans. Circuit Th. 12 (1965), 22-27.

[50] G. E. Sharpe and G. P. H. Styan, A note on the general network inverse, IEEE Trans. Circuit Th. 12 (1965), 632-633.

[51] G. E. Sharpe and G. P. H. Styan, A note on equifactor matrices, Proc. IEEE 55 (1967), 1226-1227.

[52] W. Thomson (Lord Kelvin), Cambridge and Dublin Math. J. (1848), 84-87.

[53] H. Uzawa, A theorem on convex polyhedral cones, Chapter 2 in Studies in Linear and Non-linear programming (K. J. Arrow, L. Hurwicz and H. Uzawa, Editors), Stanford University Press, Stanford, Calif., 1958.

[54] H. Weyl, Reparticion de couriente en una red conductora, Revista Matematica Hispano-Americana 5 (1923), 153-164.

[55] H. Weyl, The elementary theory of convex polyhedra, pp. 3-18 in Contributions to the Theory of Games, Vol. I (H. W. Kuhn and A. W. Tucker, Editors), Ann. Math. Studies No. 24, Princeton University Press, Princeton, N. J., 1950.

[56] S. Zlobec and A. Ben-Israel, On explicit solutions of interval linear programs, Israel J. Math. 8 (1970), 12-22.

[57] S. Zlobec and A. Ben-Israel, Explicit solutions of interval linear programs, Operations Research 21 (1973), 390-393.

Department of Applied Mathematics, Technion-Israel Institute of Technology and Department of Industrial Engineering and Management Sciences, Northwestern University. Research sponsored by the United States Army under Contract No. DA-31-124-ARO-D-462.

Statistical Applications of the Pseudo Inverse

Arthur Albert

0. Introduction

Generalized and pseudo inverses find their primary statistical applications in linear models. Many formulas, derivations and explanations become simpler and more unified when these concepts are employed.

Speaking as a mathematical statistician, it is my personal opinion that the Moore-Penrose Pseudo Inverse makes conceptual-life simpler than other generalized inverses. Traditional wisdom claims that the opposite condition obtains in the computational domain. I must say, I'm not totally convinced on that score.

In this presentation we will briefly survey a selected (but not exhaustive) list of statistical topics upon which the Moore-Penrose Pseudo Inverse can be fruitfully borne to bear. For the purposes of this survey, an effort has been made to pick up where Charles Rhode left off in his 1968 article. We therefore steer clear of such important issues as classical least squares (nonsingular covariances), recursive estimation, Henderson's method for estimating variance components, and the stationary probability vector for ergodic Markov chains. Extensive bibliographies can be found in Lewis and Boullion (1968) and Mitra and Rao (1971a).

In this paper we will deal with the following topics:

I. Constrained Least Squares

II. Maximum Likelihood Estimation and the Singular Normal Distribution

III. The Gauss Markov Theorem (Singular Covariance)
IV. When is a Naive L. S. E. a B. L. U. E.?
V. The Distribution of Quadratic Forms in Normal Random Variables
VI. Sums of Squares
VII. Conditional Expectations and Covariances for Normal Random Variables

In what follows, we have tried to conform to this notational convention:

Sets are upper case script
Matrices are upper case Latin
Vectors are lower case Latin
Scalars are lower case Greek
Random variables are boldface lower case Greek
Random vectors are boldface lower case Latin.

I. Constrained Least Squares

Let

$$\mathcal{S}(F;y) = \{u: u \text{ minimizes } \|y-Fu\|^2\} , \tag{1.1}$$

where F is a given matrix and y is a given vector.

$$u \in \mathcal{S}(F;y) \Leftrightarrow u = F^+ y + (I - F^+F)w \text{ for some } w . \tag{1.2}$$

In words, $\mathcal{S}(F;y)$ is the set of vectors, u, which can be represented in the form $u = F^+y + (I - F^+F)w$ where w is free to vary. $u_0 = F^+y$ is, by the way, the unique vector of minimum norm in $\mathcal{S}(F;y)$.

If $y \in \mathcal{R}(F)$, (the range of F) it can be expressed in the form $y = Fu$ for some u. In this case the equation

$$Fu = y \tag{1.3}$$

has at least one solution in u. All solutions to (1.3) necessarily minimize $\|y-Fu\|^2$. Consequently if we let

$$\mathfrak{J}(F;y) = \{u: Fu = y\}$$

then $\mathfrak{J} = \emptyset$ unless $y \in \mathfrak{R}(F)$ in which case $\mathfrak{J}(F;y) = \mathfrak{S}(F;y)$.

Now consider the problem of minimizing

(1.4) $$\|z - Hx\|^2$$

subject to

(1.5) $$Fx = y:$$

If $\mathfrak{J}(F;y)$ is empty, the problem has no solution. If $\mathfrak{J}(F;y)$ is not empty ($\Leftrightarrow y \in \mathfrak{R}(F) \Leftrightarrow FF^+y = y$), this problem is the same as minimizing (1.4) over the set of x's of the form

(1.6) $x(w) = F^+y + (I - F^+F)w$ where w is free to vary.

Thus

$$\min_{x \in \mathfrak{J}(F;y)} \|z - Hx\|^2 = \min_w \|z - Hx(w)\|^2$$

$$= \min_w \|z - H[F^+y + (I - F^+F)w]\|^2$$

$$= \min_w \|\bar{z} - \bar{H}w\|^2$$

where

(1.7) $$\bar{z} = z - HF^+y$$

and

(1.8) $$\bar{H} = H(I - F^+F).$$

Any and all w of the form

(1.9) $$w(v) = \bar{H}^{+}\bar{z} + (I - \bar{H}^{+}\bar{H})v$$

$$\text{minimize} \quad \|\bar{z} - \bar{H}w\|^2 .$$

Thus, any and all x's of the form

(1.10) $$\begin{aligned} x[w(v)] &= F^{+}y + (I - F^{+}F)\, w(v) \\ &= F^{+}y + (I - F^{+}F)\,[\bar{H}^{+}\bar{z} + (I - \bar{H}^{+}\bar{H})v] \end{aligned}$$

minimize (1.4) over $\mathfrak{J}(F;y)$. The minimum norm solution obtains when $v = 0$:

Of all vectors which minimize $\|z - Hx\|^2$ over $\mathfrak{J}(F;y)$,

(1.11) $$x_0 = \bar{H}^{+}\bar{z}^{+}(I - \bar{H}^{+}H)F^{+}y$$

is the one having minimum norm.

Comment: These results are proved and amplified upon in Chapter III of Albert (1972).

II. The Singular Normal Density and Maximum Likelihood Estimates

If $\underset{\sim}{z} \sim N(Hx, V^2)$, where V is possibly singular, the density for $\underset{\sim}{z}$ is

(2.1) $$f(z;Hx) = \begin{cases} (2\pi)^{-rk(V)/2}\, d^{-1} \exp -1/2(z-Hx)'\,(V^2)^{+}(z-Hx) \\ \\ 0 \qquad \text{otherwise} \end{cases}$$

where d is the product of V's nonzero eigenvalues.

The maximum likelihood estimator (hereafter "m. l. e.") for x is obtained by minimizing

(2.2) $$(z - Hx)'\,(V^2)^{+}\,(z - Hx)$$

over all x such that $z - Hx \in \mathfrak{R}(V)$.

(The likelihood is zero for other x's .)

$$\{x: z - Hx \in \mathcal{R}(V)\} = \{x: (I - VV^{+})(z - Hx) = 0\}$$

$$= \{x: \tilde{H}x = \tilde{z}\} = \mathcal{J}(\tilde{H};\tilde{z})$$

where

(2.3) $\tilde{H} = (I - VV^{+})H = [I - V^2(V^2)^{+}]H$

and

(2.4) $\tilde{z} = (I - VV^{+})z = [I - V^2(V^2)^{+}]z$.

Thus, the m.l.e., $\hat{x}$, minimizes

(2.5) $(z - Hx)'(V^2)^{+}(z - Hx) = \|V^{+}(z - Hx)\|^2 = \|z^{*} - H^{*}x\|^2$

subject to

(2.6) $x \in \mathcal{J}(\tilde{H};\tilde{z})$, (where $H^{*} = V^{+}H$ and $z^{*} = V^{+}z$) .

The m.l.e. for x in the singular normal case is thus obtainable as a constrained least squares estimator (hereafter "l.s.e.").

Comment: An extensive treatment of the singular normal distribution is to be found in Chapter X of Rao and Mitra (1971a).

III. Best Linear Unbiased Estimates; The Gauss-Markov Theorem.

Suppose

(3.1) $$\underset{\sim}{z} = Hx + \underset{\sim}{v}$$

where H is a known matrix, $\underset{\sim}{v}$ is a zero mean random vector with

(3.2) $$\mathrm{Cov}(\underset{\sim}{v}, \underset{\sim}{v}) = V^2$$

and x is an unknown vector parameter.

g'x is said to be an estimable parametric function (e. p. f.) if it admits an unbiased estimator that is linear in the data. The last is true if and only if

$$(\exists l)\ (\mathcal{E} l'\underset{\sim}{z} = g'x \text{ for all } x) \iff$$

$$(\exists l)\ (\forall x)\ (l'Hx = g'x) \iff (\exists l)\ (H'l = g)$$

$$\iff g \in \mathcal{R}(H') \iff H^+Hg = g \ .$$

Thus, g'x is an e. p. f. $\iff H^+Hg = g \iff \mathfrak{J}(H';g) \neq \emptyset$.

The variance of $l'\underset{\sim}{z}$ is $l'V^2l = \|Vl\|^2$. The best linear unbiased estimator for an e. p. f. g'x is $\hat{l}'\underset{\sim}{z}$ where $\hat{l}$ minimizes $\|Vl\|^2$ over $\mathfrak{J}(H';g)$. Although $\hat{l}$ is not necessarily unique, it turns out that $\hat{l}'\underset{\sim}{z}$ is unique (w. p. 1) and that $\hat{l}'\underset{\sim}{z} = g'\hat{\underset{\sim}{x}}$ w. p. 1 where

(3.3) $$\hat{\underset{\sim}{x}} = H^+\underset{\sim}{z} - H^+V^2Q(QV^2Q)^+\underset{\sim}{z}$$

$$Q = I - HH^+ \ .$$

In the case where V (or V^2) is nonsingular, (3.3) reduces to

(3.4) $$\hat{\underset{\sim}{x}} = (H'V^{-2}H)^+ (H'V^{-2}\underset{\sim}{z}) \ .$$

The Gauss-Markov Theorem may therefore be extended to claim that every e. p. f., g'x, has a unique best linear unbiased estimator (hereafter "b. l. u. e."), namely $\hat{\underset{\sim}{x}}$, where $\hat{\underset{\sim}{x}}$, is given by (3.3).

The present work is abstracted from Albert (1973a), which in turn was motivated by earlier work of Zyskind and Martin (1969). The main result of that paper stated that any estimable function has an essentially unique b. l. u. e. of the form $g'\underset{\sim}{x}^*$ where $\underset{\sim}{x}^*$ is any solution to the normal equation

(3.5) $$H'(V^2)^* H x = H'(V^2)^* \underset{\sim}{z}$$

and $(V^2)^*$ is any matrix chosen from a certain subclass of g-inverses for V^2. A rather complicated characterization of the allowable class of g-inverses was given there.

More recently, Rao and Mitra (1971b) have shown that $g'\hat{\underset{\sim}{x}} = g'\tilde{\underset{\sim}{x}}(\lambda)$ for all $\lambda > 0$, where $\tilde{\underset{\sim}{x}}(\lambda)$ is of the form

(3.6) $$\tilde{\underset{\sim}{x}}(\lambda) = [H'(V^2 + \lambda^2 HH')^- H]^- H'(V^2 + \lambda^2 HH')^- \underset{\sim}{z} .$$

Here $[\ ]^-$ is any g-inverse for the matrix in square brackets. (A g-inverse for A is any solution, in X, to the equation $AXA = A$.)

IV. When is a Naive L. S. E. a B. L. U. E.?

The naive l. s. e. for x in the model

(4.1) $$\underset{\sim}{z} = H\underset{\sim}{x} + \underset{\sim}{v}$$

$$\text{Cov}(\underset{\sim}{v}, \underset{\sim}{v}) = V^2$$

is

(4.2) $$\hat{\underset{\sim}{x}}_N = H^+ \underset{\sim}{z} .$$

The B. L. U. E. for x is

(4.3) $$\hat{\underset{\sim}{x}}_B = H^+ \underset{\sim}{z} - H^+ V^2 Q(QV^2Q)^+ \underset{\sim}{z} .$$

$$\hat{\underset{\sim}{x}}_N = \hat{\underset{\sim}{x}}_B \quad \text{w. p. 1}$$

if and only if $\mathcal{E}\underset{\sim}{y}\underset{\sim}{y}' = 0$, where

(4.4) $$\underset{\sim}{y} = H^+ V^2 Q(QV^2Q)^+ \underset{\sim}{z} .$$

$$\mathcal{E}\underset{\sim}{y}\underset{\sim}{y}' = AA' \tag{4.5}$$

where

$$A = H^+V^2Q(QV^2Q)^+V \ . \tag{4.6}$$

Thus, $\hat{\underset{\sim}{x}}_N = \hat{\underset{\sim}{x}}_B$ if and only if $A = 0$.

But

$$A = H^+V^2Q(QV^2Q)^+V = 0$$

$$\Longleftrightarrow (H^+V)\,(VQ)\,[(VQ)'\,(VQ)]^+\,(VQ)' = 0$$

$$\Longleftrightarrow (H^+V)\,(VQ)\,(VQ)^+ = 0$$

$$\Longleftrightarrow (H^+V)\,(VQ) = 0$$

$$\Longleftrightarrow (HH^+)\,V^2Q = 0$$

$$\Longleftrightarrow QV^2(HH^+) = 0$$

$$\Longleftrightarrow QV^2H = 0$$

$$\Longleftrightarrow \mathcal{R}(V^2H) \subseteq \mathcal{R}(H) \ .$$

Theorem: $\hat{\underset{\sim}{x}}_N = \hat{\underset{\sim}{x}}_B$ (w. p. 1) $\Longleftrightarrow \mathcal{R}(V^2H) \subseteq \mathcal{R}(H)$.

Comment: These results seem to date back to work by Rao (1965). Later contributions were made by Kruskal (1968) and Mitra and Rao (1969). An exhaustive treatment of current aspects of this problem can be found in Chapter VIII of Rao and Mitra (1971a).

V. The Distribution Theory for Quadratic Forms in Normal R. V. 's.

If $\underset{\sim}{z} \sim N(\mu, R)$ and if A is symmetric, $\underset{\sim}{z}'A\underset{\sim}{z}$ has a noncentral chisquare distribution if and only if

$$(RA)^2R = (RA)R \tag{5.1}$$

$$RA\mu \in \mathcal{R}[RAR] \tag{5.2}$$

$$(A\mu)' R(A\mu) = \mu' A\mu \quad . \tag{5.3}$$

In this case, $\underset{\sim}{z}' A\underset{\sim}{z} \sim \chi^2(k, \delta)$ where $k = \text{trace}(AR)$ degrees of freedom and the nonentrality parameter is $\delta = \mu' ARARA\mu$.

As a very special case, if $A = R^+$, (5.1)-(5.3) hold and we see that $\underset{\sim}{z}' R^+ \underset{\sim}{z}$ has a chisquare distribution with $k = \text{trace}(R^+R) = \text{rank}(R)$df and noncentrality parameter $\delta = \mu'(R^+R)^2R^+\mu = \mu' R^+\mu$.

In cases where R is nonsingular, (5.2) always holds since

$$RA\mu = RAR(R^{-1}\mu) \quad .$$

(5.1) and (5.3) will hold if and only if

$$ARA = A \iff (AR)^2 = (AR) \, .$$

Thus, when R is nonsingular, $\underset{\sim}{z}' A\underset{\sim}{z}$ will have a noncentral chisquare distribution if and only if AR is idempotent. In this case, $z' Az \sim \chi^2(k, \delta)$ where $k = \text{tr}(AR)$ and $\delta = \mu' ARA\mu$.

Even more generally,

$\underset{\sim}{z}' A\underset{\sim}{z} + 2b'\underset{\sim}{z} + c$ has a noncentral chisquare distribution if and only if

$$(RA)^2R = RAR \tag{5.4}$$

$$R(A\mu + b) \in \mathcal{R}[RAR] \tag{5.5}$$

$$(A\mu + b)' R(A\mu + b) = \mu' A\mu + 2b'\mu + c \quad . \tag{5.6}$$

In this case, $\underset{\sim}{z}' A\underset{\sim}{z} \sim \chi^2(k, \delta)$ where $k = \text{tr}(AR)$ and $\delta = (A\mu + b)' RAR(A\mu + b)$.

If $Q_1 = z'Az$ and $Q_2 = z'Bz$ then Q_1 and Q_2 are independent if and only if

(5.7) $$RARBR = 0$$

(5.8) $$RARB\mu = RBRA\mu = 0$$

(5.9) $$\mu'ARB\mu = 0 \quad .$$

When R is nonsingular or if A and B are n.n.d., Q_1 and Q_2 are independent if and only if

(5.10) $$ARB = 0 \quad .$$

Comment: When R is nonsingular, these results simplify. See Chapter II of Searle (1971). The general case is dealt with in Chapter IX of Rao and Mitra (1971a).

VI. Sums of Squares.

Suppose

$$z = \sum_{i=1}^{m} \xi_i u_i + \text{residual}$$

where the vectors $u_1, u_2, \ldots, u_m$ are orthonormal and the residual is assumed to be small. If the ξ's are not known, they can be estimated by

$$\hat{\xi}_i = (z'u_i) \quad .$$

Indeed

$$\|z\|^2 = \sum_{i=1}^{m} \hat{\xi}_i^2 + \|z - \sum_{i=1}^{m} \hat{\xi}_i u_i\|^2$$

and one measure of the "contribution of u_i to the magnitude of z" is the size of $\hat{\xi}_i^2$.

In more complicated linear models, it is often assumed that

$$(6.1) \qquad \underset{\sim}{z} = Hx + \underset{\sim}{v}$$

where $\underset{\sim}{v} \sim N(0, \sigma_0^2 I)$,

H is some specified matrix and x is unknown. In an effort to deduce information about the structural relationship between $\underset{\sim}{z}$ and H (how strongly is $\underset{\sim}{z}$ affected by each of the columns of H ?) it is common for the statistician to pick a judiciously chosen set of mutually orthogonal projections, $P_1 P_2, \ldots, P_k$, which sum to the identity and to examine the magnitude of $\|P_j \underset{\sim}{z}\|^2$. The intuitive idea behind this is that

$$\|\underset{\sim}{z}\|^2 = \sum_{j=1}^{k} \|P_j \underset{\sim}{z}\|^2$$

and that $\|P_j \underset{\sim}{z}\|^2$ "explains" how much of $\underset{\sim}{z}$ lies in the linear manifold of vectors upon which P_j projects.

The linear model (6.1), exists in two varieties. In the first, x is viewed as an unknown vector parameter. In this case, it turns out that the terms $\|P_1 \underset{\sim}{z}\|^2$, $\|P_2 \underset{\sim}{z}\|^2, \ldots,$ $\|P_k \underset{\sim}{z}\|^2$ are mutually independent and for each j, $\|P_j \underset{\sim}{z}\|^2 / \sigma_0^2 \sim \chi^2(k_j, \delta_j)$ (which reads "$\|P_j \underset{\sim}{z}\|^2 / \sigma_0^2$ has a noncentral chisquare distribution with k_j degrees of freedom and noncentrality parameter δ_j) where $k_j = \text{rank}(P_j)$ and $\delta_j = \|P_j Hx\|^2 / \sigma_0^2$. This result is an easy consequence of the results in section V of this review, but it depends heavily on the assumed covariance structure for $\underset{\sim}{z}$ (namely $\text{Cov}(\underset{\sim}{z}, \underset{\sim}{z}) = \sigma_0^2 I$.)

The second version of the model (6.1), assumes that x and H can be partitioned:

$$(6.2) \qquad x = \begin{bmatrix} x_0 \\ \cdots \\ \underset{\approx}{x}_1 \\ \cdots \\ \vdots \\ \cdots \\ \underset{\approx}{x}_r \end{bmatrix}, \quad H = (H_0 \vdots H_1 \vdots \cdots \vdots H_r)$$

so that

$$(6.3) \qquad Hx = H_0 x_0 + \sum_{j=1}^{r} H_j \underset{\approx}{x}_j$$

where x_0 is viewed as an unknown vector parameter, but each of the other $\underset{\approx}{x}_i$ is viewed as an unobservable vector random variable. Indeed, it is assume that

$$\begin{bmatrix} \underset{\sim}{V} \\ \underset{\approx}{x}_1 \\ \vdots \\ \underset{\approx}{x}_r \end{bmatrix}$$

has a zero mean normal distribution with covariance structure

$$\operatorname{cov}(\underset{\approx}{x}_j, \underset{\sim}{V}) = 0 \quad \text{and} \quad \operatorname{cov}(\underset{\approx}{x}_i, \underset{\approx}{x}_j) = \begin{cases} \sigma_i^2 I & \text{if } i = j \\ 0 & \text{if } i \neq j \end{cases}.$$

This induces a new covariance structure on $\underset{\approx}{z}$, namely

$$\operatorname{cov}(\underset{\approx}{z}, \underset{\approx}{z}) = W = \sigma_0^2 I + \sum_{j=1}^{r} \sigma_j^2 H_j H_j' .$$

The distribution theory for the $\| P_j \underset{\sim}{z} \|^2$ is still of interest but, because of the change in $\underset{\sim}{z}$'s covariance (from $\sigma_0^2 I$ to W), is no longer simple. Under certain special circumstances (which fortunately seem to occur in many cases of practical importance) W has properties which allow a relatively simple distribution theory for the $\| P_j \underset{\sim}{z} \|^2$:

Theorem 6.1: Assume $\sigma_0^2 > 0$: The terms $\| P_1 \underset{\sim}{z} \|^2, \ldots, \| P_k \underset{\sim}{z} \|^2$ are mutually independent and each $\| P_j \underset{\sim}{z} \|^2$ has the same distribution as some scalar multiple of a noncentral chi-square r.v. regardless of the values of $\sigma_0^2, \sigma_1^2, \ldots, \sigma_r^2$ if and only if there are scalars, λ_{ij}, such that

(6.4) $$H_j H_j' P_i = \lambda_{ij} P_i \qquad (i = 1, 2, \ldots, k; \; j = 1, 2, \ldots, r) .$$

In this case, the distribution of $\| P_i \underset{\sim}{z} \|^2 / \gamma_i$ is $\chi^2(k_i, \delta_i)$ where

(6.5) $$\left. \begin{aligned} \gamma_i &= \sigma_0^2 + \sum_{j=1}^{r} \lambda_{ij} \sigma_j^2 \\ k_i &= \text{rank}\,(P_i) \\ \delta_i &= \| P_i H_0 x_0 \|^2 / \gamma_i \end{aligned} \right\} \qquad i = 1, 2, \ldots, k ,$$

and the mean square value for $P_i \underset{\sim}{z}$ is

(6.6) $$\mathcal{E} \| P_i \underset{\sim}{z} \|^2 = \gamma_i (\delta_i + k_i) .$$

Since $\| \underset{\sim}{z} \|^2 = \sum_{i=1}^{r} \| P_i \underset{\sim}{z} \|^2$, this theorem tells us when a given partition of the sum of squares has the chisquare-independence properties.

The following closely related result states conditions for the existence of some such sum of squares decomposition.

Theorem 6.2: There exists a resolution of the identity † $(P_1, P_2, \ldots, P_k)$ such that for all nonnegative values of $\sigma_0^2, \ldots, \sigma_r^2$ the r.v.'s $\|P_1 \underset{\sim}{z}\|^2$, $\|P_2 \underset{\sim}{z}\|^2, \ldots, \|P_k \underset{\sim}{z}\|^2$ are mutually independent and have distributions that are the same as scalar multiples of noncentral chisquare r.v.'s if and only if $H_i H_i'$ and $H_j H_j'$ commute for $i, j = 1, 2, \ldots, r$.

If $Hx = \sum_{i=0}^{r} H_i x_i$, the "natural" resolution of the identity is $(P_0, P_1, \ldots, P_{r+1})$, where

$$
(6.7) \qquad
\begin{aligned}
P_0 &= H_0 H_0^+ \\
P_1 &= \overline{H}_1 \overline{H}_1^+ ; \quad \overline{H}_1 = (I - P_0) H_1 \\
P_2 &= \overline{H}_2 \overline{H}_1^+ ; \quad \overline{H}_2 = (I - P_0 - P_1) H_2 \\
&\vdots \\
P_r &= \overline{H}_r \overline{H}_r^+ ; \quad \overline{H}_r = (I - \sum_{j=0}^{r} P_j) H_r \\
P_{r+1} &= (I - \sum_{j=0}^{r} P_j) \ .
\end{aligned}
$$

For each j, the column vectors of $\overline{H}_j$ are obtained by projecting the corresponding column vectors of H_j, onto the orthogonal complement of the linear manifold spanned by the column vectors of $H_0, H_1, \ldots, H_{j-1}$. Furthermore, P_j is the projection on the linear manifold spanned by the column vectors of $\overline{H}_j$. (This procedure is exactly a Gramm-Schmidt Orthogonalization if each of the matrices H_i happen

† Which means that the projections are mutually orthogonal and sum to I .

to consist of exactly one column.)

Example: The two way ANOVA with interaction:

Suppose

$$\underset{\sim}{\zeta}_{ijk} = \mu + \alpha_i + \beta_j + \varepsilon_{ij} + \nu_{ijk} \quad (i=1,\ldots,I;\ j=1,\ldots,J;\ k=1,\ldots,K) \ . \tag{6.8}$$

We can write this model in the form (6.1)-(6.3) if we take

$$H = \begin{bmatrix} 1 \\ 1 \\ \vdots \\ 1 \end{bmatrix} \quad \text{an } N \times 1 \text{ matrix where } N = IJK \tag{6.9}$$

and

$$x_0 = [\mu] \ ,$$

$$H_1 = (h_{11} \vdots h_{12} \vdots \cdots \vdots h_{1I}) \quad \text{and} \quad x_1 = \begin{bmatrix} \alpha_1 \\ \alpha_2 \\ \vdots \\ \alpha_I \end{bmatrix} \tag{6.10}$$

where the s-th component of h_{1i} is

$$h_{1i}(s) = \begin{cases} 1 & \text{if } i = 1 + [s/KJ] \\ 0 & \text{otherwise} \end{cases} \quad i=1,\ldots,I;\ s=1,\ldots,N \ . \tag{6.11}$$

(Here, $[\xi]$ is the integer part of ξ.)

(6.12)
$$H_2 = (h_{21} \vdots h_{22} \vdots \cdots \vdots h_{2J}), \quad x = \begin{bmatrix} \beta_1 \\ \beta_2 \\ \vdots \\ \beta_J \end{bmatrix}$$

where

(6.13)
$$h_{2j}(s) = \begin{cases} 1 & \text{if } j = 1 + [s/N] - J[s/KJ] \\ 0 & \text{otherwise} \end{cases} \qquad \begin{array}{l} j = 1, 2, \ldots, J \\ s = 1, \ldots, N \end{array} .$$

(6.14)
$$H_3 = (h_{31} \vdots h_{32} \vdots \cdots \vdots h_{3N}), \quad x_3 = \begin{bmatrix} \varepsilon_{11} \\ \vdots \\ \varepsilon_{1J} \\ \varepsilon_{21} \\ \vdots \\ \varepsilon_{I1} \\ \vdots \\ \varepsilon_{IJ} \end{bmatrix}$$

where

(6.15)
$$h_{3k}(s) = \begin{cases} 1 & \text{if } k = 1 + [s/K] \\ 0 & \text{otherwise} \end{cases} \qquad \begin{array}{l} k = 1, 2, \ldots, N; \\ s = 1, 2, \ldots, N \end{array} .$$

Defining

$$P_0 = H_0 H_0^+$$
$$P_j = \bar{H}_j \bar{H}_j^+ \qquad j = 1, 2, 3$$

and

$$P_4 = I - P_0 - P_1 - P_2 - P_3$$

where $\bar{H}_j = (I - P_0 - P_1 \cdots - P_{j-1})H_j$,

the s. s. decomposition

$$\|\underset{\sim}{z}\|^2 = \sum_{i=0}^{4} \|P_j \underset{\sim}{z}\|^2$$

takes the form

$$(6.16)\quad \sum_{ijk} \underset{\sim}{\zeta}_{ijk}^2 = \sum_{ijk} \underset{\sim}{\zeta}_{...}^2 + \sum_{ijk} (\underset{\sim}{\zeta}_{i..} - \underset{\sim}{\zeta}_{...})^2$$

$$+ \sum_{ijk} (\underset{\sim}{\zeta}_{.j.} - \underset{\sim}{\zeta}_{...})^2 + \sum_{ijk} (\underset{\sim}{\zeta}_{ij.} - \underset{\sim}{\zeta}_{i..} - \underset{\sim}{\zeta}_{.j.} + \underset{\sim}{\zeta}_{...})^2$$

$$+ \sum_{ijk} (\underset{\sim}{\zeta}_{ijk} - \underset{\sim}{\zeta}_{ij.})^2 .$$

In the fixed effects model where x_0, x_1, x_2 and x_3 are viewed as unknown deterministic parameters and the $\underset{\sim}{\nu}_{ijk}$ are i. i. d. $N(0, \sigma_0^2)$, the terms on the right side of (6. 16) are mutually independent. When divided by σ_0^2, they have noncentral chisquare distributions with respective noncentrality parameters:

$$\delta_0 = N\mu^2/\sigma_0^2, \quad \delta_1 = JK\sum_{i=1}^{I}(\alpha_i - \alpha_. + \varepsilon_{i.} - \varepsilon_{..})^2 ,$$

$$\delta_2 = IK\sum_{j=1}^{J}(\beta_j - \beta_. + \varepsilon_{.j} - \varepsilon_{..})^2$$

$$\delta_3 = K\sum_{i=1}^{I}\sum_{j=1}^{J}(\varepsilon_{ij} - \varepsilon_{i.} - \varepsilon_{.j} + \varepsilon_{..})^2$$

$$\delta_4 = 0 .$$

It turns out that

(6.17) $$H_j H_j' P_i = \lambda_{ij} P_i \qquad i = 0, 1, 2, 3, 4; \quad j = 1, 2, 3$$

where

(6.18)
$$\lambda_{01} = \lambda_{11} = KJ$$
$$\lambda_{02} = \lambda_{22} = KI$$
$$\lambda_{03} = \lambda_{13} = \lambda_{23} = \lambda_{33} = K$$
$$\lambda_{ij} = 0 \qquad \text{otherwise.}$$

Therefore, if a mixed model is assumed where the α's, β's and ε's are treated as r.v.'s, Theorem 6.1 guarantees that the terms on the right side of (6.16) remain independent and $\|P_j \underset{\sim}{z}\|^2/\gamma_j$ has a noncentral chisquare distribution with noncentrality parameter $\|P_j H_0 x_0\|^2/\gamma_j$ $(j = 0, 1, 2, 3, 4)$ where

(6.19)
$$\|P_0 H_0 x_0\|^2 = N\mu^2$$
$$\|P_j H_0 x_0\|^2 = 0 \qquad i = 1, 2, 3, 4$$

and

(6.20)
$$
\begin{aligned}
\gamma_0 &= \sigma_0^2 + KJ\sigma_1^2 + KI\sigma_2^2 + K\sigma_3^2 \\
\gamma_1 &= \sigma_0^2 + KJ\sigma_1^2 \qquad\qquad\; + K\sigma_3^2 \\
\gamma_2 &= \sigma_0^2 \qquad\qquad\;\; + KI\sigma_2^2 + K\sigma_3^2 \\
\gamma_3 &= \sigma_0^2 \qquad\qquad\qquad\qquad\;\; + K\sigma_3^2 \\
\gamma_4 &= \sigma_0^2 \;.
\end{aligned}
$$

The mean squares for each term on the right side of (6.16) are:

$$
\begin{aligned}
\mathcal{E}\|P_0\underset{\sim}{z}\|^2 &= N\mu^2 + \gamma_0 \\
\mathcal{E}\|P_1\underset{\sim}{z}\|^2 &= (I-1)\gamma_1 \\
\mathcal{E}\|P_2\underset{\sim}{z}\|^2 &= (J-1)\gamma_2 \\
\mathcal{E}\|P_3\underset{\sim}{z}\|^2 &= [IJ+1 - (I+J)]\gamma_3 \\
\mathcal{E}\|P_4\underset{\sim}{z}\|^2 &= IJ(K-1)\gamma_4 \;.
\end{aligned}
$$

Comment: See Chapter VI of Albert (1972) for an extensive treatment of the sum of squares decomposition for the fixed effects model. The results for the mixed model are new and are to be found in Albert (1973b).

VII. Conditional Expectations and Covariances

(7.1)
$$
S = \begin{array}{c} \\ r \\ s \end{array}\!\!\begin{array}{c} \begin{array}{cc} \;\; r \;\;& \;\; s \;\; \end{array} \\ \left(\begin{array}{c|c} S_{11} & S_{12} \\ \hline S_{12}' & S_{22} \end{array}\right) \end{array}
$$

is a covariance matrix if and only if

(7.2) $S_{11} \geq 0, \quad S_{22} - S'_{12} S^{+}_{11} S_{12} \geq 0$ and $S_{11} S^{+}_{11} S_{12} = S_{12}$.

(Albert (1972), Ch. IX).

If

$$\underset{\sim}{x} = \begin{matrix} r \\ s \end{matrix}\begin{pmatrix} \underset{\sim}{x}_1 \\ --- \\ \underset{\sim}{x}_2 \end{pmatrix}$$

has a multivariate normal distribution with mean zero and covariance given by (7.1), then $\underset{\sim}{x}$ has the same distribution as

$$\underset{\sim}{x}^{*} = \begin{matrix} r \\ s \end{matrix}\begin{pmatrix} \underset{\sim}{x}^{*}_1 \\ --- \\ \underset{\sim}{x}^{*}_2 \end{pmatrix}$$

where

(7.3) $\underset{\sim}{x}^{*}_1 = S^{\frac{1}{2}}_{11} \underset{\sim}{w}_1, \quad x^{*}_2 = S'_{12} S^{+}_{11} \underset{\sim}{x}^{*}_1 + (S_{22} - S'_{12} S^{+}_{11} S_{12})^{\frac{1}{2}} \underset{\sim}{w}_2$

and

$$\begin{matrix} r \\ s \end{matrix}\begin{pmatrix} \underset{\sim}{w}_1 \\ --- \\ \underset{\sim}{w}_2 \end{pmatrix} \sim N(0, I) \ .$$

Since $\underset{\sim}{x}^{*}_1$ is independent of $\underset{\sim}{w}_2$,

(7.4) $m_{2|1}(x) = \mathcal{E}(\underset{\sim}{x}_2 | \underset{\sim}{x}_1 = x) = \mathcal{E}(\underset{\sim}{x}^{*}_2 | \underset{\sim}{x}^{*}_1 = x) = S'_{12} S^{+}_{11} x$

provided $x \in \mathcal{R}[S_{11}]$.

(Since $\underset{\approx}{x}_1, \underset{\approx}{x}_1^* \sim N(0, S_{11})$, it must be that $\underset{\approx}{x}_1, \underset{\approx}{x}_1^* \in \mathcal{R}[S_{11}]$ w. p. 1.)

Furthermore

$$(7.5)\quad S_{22|1}(x) = \mathcal{E}\{[\underset{\approx}{x}_2 - m_{2|1}(x)][\underset{\approx}{x}_2 - m_{2|1}(x)]' \,|\, \underset{\approx}{x}_1 = x\}$$

$$= \mathcal{E}\{[\underset{\approx}{x}_2^* - m_{2|1}(x)][\underset{\approx}{x}_2^* - m_{2|1}(x)]' \,|\, \underset{\approx}{x}_1^* = x\}$$

$$= \mathcal{E}[(S_{22} - S_{12}' S_{11}^+ S_{12})^{\frac{1}{2}} \underset{\approx}{w}_2][(S_{22} - S_{12}' S_{11}^+ S_{12})^{\frac{1}{2}} \underset{\approx}{w}_2]'$$

$$= S_{22} - S_{12}' S_{11}^+ S_{12}$$

since $\underset{\approx}{x}_1^*$ is independent of $\underset{\approx}{w}_2$.

(7.4) and (7.5) furnish streamlined derivations for the conditional expectation and conditional covariance formulas for zero mean normal vectors. The general case is an easy extension:

For if

$$\underset{\approx}{x} = \begin{matrix} r \\ s \end{matrix}\begin{pmatrix} \underset{\approx}{x}_1 \\ --- \\ \underset{\approx}{x}_2 \end{pmatrix} \sim N(m, S)$$

where

$$m = \begin{matrix} r \\ s \end{matrix}\begin{pmatrix} m_1 \\ --- \\ m_2 \end{pmatrix}$$

then

$$\underset{\approx}{x} - m = \begin{pmatrix} \underset{\approx}{x}_1 - m_1 \\ ------ \\ \underset{\approx}{x}_2 - m_2 \end{pmatrix} \sim N(0, S)$$

so

$$\mathcal{E}(\underset{\approx}{x}_2 - m_2 | \underset{\approx}{x}_1 - m_1 = x - m_1) = S_{12}' S_{11} (x - m_1) .$$

Thus

$$\mathcal{E}(\underset{\approx}{x}_2 - m_2 | \underset{\approx}{x}_1 = x) = S_{12}' S_{11}^{+} (x - m_1)$$

and hence

(7.6) $$\mathcal{E}(\underset{\approx}{x}_2 | \underset{\approx}{x}_1 = x) = m_2 + S_{12}' S_{11}^{+} (x - m_1) .$$

Covariances are not affected by addition of deterministic quantities, so

(7.7) $$\begin{aligned} \operatorname{cov}(\underset{\approx}{x}_2, \underset{\approx}{x}_2 | \underset{\approx}{x}_1 = x) &= \operatorname{cov}(\underset{\approx}{x}_2 - m_2, \underset{\approx}{x}_2 - m_2 | \underset{\approx}{x}_1 - m_1 = x - m_1) \\ &= S_{22} - S_{12}' S_{11}^{+} S_{12} . \end{aligned}$$

Comment: An early application of pseudo-inverses to conditional expectations can be found in a paper by Marsaglia (1965).

Formulas (7.6) and (7.7) can be viewed as rules for updating the prior distribution of $\underset{\approx}{x}_2$ in the light of the datum $\underset{\approx}{x}_1$, provided $\underset{\approx}{x}_2$ and $\underset{\approx}{x}_1$ have a joint normal distribution. This point of view has been exploited most successfully by Kalman (1960). A development of the Kalman filtering equations along the present lines can be found in Chapter IX of Albert (1972).

REFERENCES

Albert, Arthur (1972) Regression and the Moore-Penrose Pseudoinverse, Academic Press.

Albert, Arthur (1973a) "The Gauss Markov Theorem for Regression Models with Possibly Singular Covariances." S. I. A. M. J. Appl. Math. 24 pp. 182-187.

Albert, Arthur (1973b) "Orthogonal Designs and the Mixed Model ANOVA." Boston Univ. Math. Dept. TR 73-6. (Submitted for publication.)

Kalman, R. E. (1960) "A New Approach to Linear Filtering and Prediction Problems". J. Basic. Engrg. 82 pp. 35-45.

Kruskal, W. (1968) "When are Gauss-Markov and Least Squares Estimates Identical? A Coordinate Free Approach". Ann. Math. Stat. 39 pp. 70-75.

Lewis, T. O., Odell, P. L. and Boullion, T. L. (1968) "A Bibliography on Generalized Matrix Inverses" in Proceedings of the Symposium on Theory and Applications of Generalized Inverses of Matrices, held at Texas Technical College.

Marsaglia, G. (1965) "Conditional Means and Covariances of Normal Variables with Singular Covariance Matrices," J. A. S. A. 59 pp. 1205-1204.

Mitra, S. K. and Rao, C. R. (1969) "Conditions for Optimality and Validity of Simple Least Squares Theory", Ann. Math. Stat. 40 pp. 1617-1624.

Rao, C. R. and Mitra, S. K. (1971a) Generalized Inverse of Matrices and Its Applications. John Wiley.

Rao, C. R. and Mitra, S. K. (1971b) "Further Contributions to the Theory of Generalized Inverse of Matrices and Its Applications." Sankhya Ser. A 33 pp. 289-300.

Rao, C. R. (1965) "Least Squares Theory Using an Estimated Dispersion Matrix and Its Application to the Measurement of Signals." Proc. of Fifth Berkeley Symp. on Math. Stat. 1 pp. 355-372. Univ. of Calif. Press. (Presented 1965; In print 1967).

Rhode, C. A. (1968) "Special Applications of the Theory of Generalized Matrix Inversion to Statistics" in Proceedings of the Symposium on Theory and Applications of Generalized Inverses of Matrices, held at Texas Technical College.

Searle, S. R. (1971) Linear Models. John Wiley.

Zyskind, G. and Martin, F. B. (1969) "On Best Linear Estimation and the Gauss-Markov Theorem in Linear Models with Arbitrary Covariance Structure." S. I. A. M. J. Appl. Math. 17 pp. 1190-1202.

Department of Mathematics
Boston University
Boston, Mass.

Estimation and Aggregation in Econometrics: An Application of the Theory of Generalized Inverses

John S. Chipman

CONTENTS

"Briefly, and in its most concrete form, the object of statistical methods is the reduction of data. A quantity of data, which usually by its mere bulk is incapable of entering the mind, is to be replaced by relatively few quantities which shall adequately represent the whole, or which, in other words, shall contain as much as possible, ideally the whole, of the relevant information contained in the original data."

R. A. Fisher

Introduction

The reduction of data, no less central an objective in econometrics than in statistics generally, is still more far-reaching in its application. One is interested not only in estimating parameters in models, but also in estimating the models themselves. Not only is one concerned with estimating relationships among variables, but one is concerned perhaps even to a greater degree with the problem of how to choose, measure, and combine these variables to begin with. It is not that economic theory is lacking in appropriate hypotheses, but rather that the complexity of these hypotheses is very great in comparison with the degrees of freedom exhibited by the necessarily non-experimental data.

Consequently, only a small fraction, possibly none at all, of the parameters of the "true model" we believe in are estimable, or "identifiable"; in general, only a small number of functions (e.g., linear combinations) of these parameters are estimable, a number at most equal to the degrees of freedom provided by the data. In linear systems this means that observation matrices corresponding to these "true" models will generally have ranks less than the number of independent variables of the system, or, at least, indistinguishably less, taking account of the inevitable rounding errors and errors of measurement. One is thus led to the construction of simplified aggregative models which can be regarded as representations, or estimates, of the "true" models we really believe in, and which have fewer, or at least not more, degrees of freedom than are provided by the data.

A natural tool of analysis to use in dealing with these problems is provided by the concept of a generalized inverse X^- of a matrix X, defined as any matrix such that $XX^-X=X$ (cf. Rao [86]). Such a concept arose naturally in the treatment by Koopmans [58, p. 75] of the problem of multicollinearity (i.e., of the failure of an observation matrix to have full rank equal to the number of independent variables)--a problem first dealt with in the econometric literature by Frisch [38, 39]. Koopmans defined the "partial inverse" of a symmetric non-negative definite matrix M as "that matrix which, by the orthogonal transformation which brings M into the form

$$\begin{bmatrix} 0 & 0 & \cdots & 0 \\ 0 & \mu_2 & \cdots & 0 \\ \cdots & \cdots & \cdots & \cdots \\ 0 & 0 & \cdots & \mu_k \end{bmatrix}, \text{ itself assumes the form } \begin{bmatrix} 0 & 0 & \cdots & 0 \\ 0 & 1/\mu_2 & \cdots & 0 \\ \cdots & \cdots & \cdots & \cdots \\ 0 & 0 & \cdots & 1/\mu_k \end{bmatrix}."$$

This is equivalent to a characterization introduced by Rao [86] which can serve to define the Moore-Penrose generalized inverse of any matrix X (cf. Chipman and Rao [19, pp. 5-6]). The latter inverse is defined by Penrose [80] as the unique matrix A (which always exists) satisfying the four properties

$$(\dagger)\quad \begin{array}{ll} \text{(i)}\ XAX = X\ ; & \text{(ii)}\ AXA = A\ ; \\ \text{(iii)}\ XA = (XA)'\ ; & \text{(iv)}\ AX = (AX)' \end{array}$$

(where a prime denotes transposition), and denoted $X^{\dagger}$. The Moore-Penrose inverse does not, however, turn out to be the most appropriate concept in the present study; rather, we shall be concerned with matrices A satisfying one or more of the properties

$$(\ddagger)\quad \begin{array}{ll} \text{(i)}\ XAX = X\ ; & \text{(ii)}\ AXA = A\ ; \\ \text{(iii)}\ XAV = (XAV)'\ ; & \text{(iv)}\ AXU = (AXU)'\ , \end{array}$$

where U and V are symmetric non-negative definite matrices, including at least the first (cf. Chipman [15, 16]). If U and V are both positive definite, the existence and uniqueness of a matrix A satisfying ($\ddagger$)--which will be denoted $X^{\ddagger}$--follows trivially from the existence and uniqueness of $X^{\dagger}$. It is shown in Theorem 1.2 that if U (but not necessarily V) is positive definite, there exists a matrix $A = X^{\ddagger}$ satisfying ($\ddagger$), and $X^{\ddagger}V$ is unique. (Generalized inverses of type ($\ddagger$) have also been discussed by Mitra [73].)

We shall generally be concerned with matrices A satisfying only a subset of properties ($\ddagger$) including always property (i). Since Rao's notation X^{-} will be used to denote any matrix with the latter property, in order to economize on notation (as well as to avoid excessive use of subscripts) we shall use the notation $X^{\ddagger}$ also to denote a matrix A satisfying only properties (i) and (iii) of ($\ddagger$); when necessary to avoid ambiguity, the symbol $X_V^{\ddagger}$ will be used. Likewise, the symbol $X^{\ddagger}$ (or $X_U^{\ddagger}$) will sometimes be used to denote a matrix A satisfying properties (i) and (iv) of ($\ddagger$). The symbol X^{+} will be used for a number of special purposes. In Theorem 1.6, X^{+} will denote any matrix A satisfying properties (i) and (iii) of (†), while in Lemma 1.3, V^+ will denote a generalized inverse of V satisfying some additional properties relating to the matrix X. In Part 2 the symbol G^+ will be used to denote simply a particular distinguished generalized inverse of G which will be fixed during the course of the discussion. The term reflexive generalized inverse of X will designate a matrix A satisfying properties (i) and (ii) of (†).

A further generalization of the concept defined by (‡) turns out to be desirable, particularly in Part 2. If U and V are given symmetric non-negative definite matrices, we shall denote by $X^{\#}$ (or sometimes by $X^{\natural}$) any matrix A satisfying some or all of the properties

$$(\#)\quad \begin{array}{l}\text{(i) } XAXU = XU\ ;\quad \text{(ii) } AXAV = AV\ ;\\ \text{(iii) } XAV = (XAV)'\text{; (iv) } AXU = (AXU)'\ ,\end{array}$$

including at least property (i); generally, $X^{\#}$ will denote a matrix A satisfying properties (i) and (iv) of (#). Such a matrix will be termed a generalized quasi-inverse of X. The existence of a generalized quasi-inverse of X satisfying properties (i) and (iv) of (#) is shown in Lemma 1.9. The transpose $X^{\#'}$ of a generalized quasi-inverse of X will (in section 2.3) be termed a generalized quasi-inverse of X'. Matrices A satisfying (#) are described by Rao and Mitra [89, p. 527]. The importance of generalized quasi-inverses derives from the fact that they behave like generalized inverses except on a set of measure zero, i.e., with probability 1 (U and V being interpreted as--possibly singular--covariance matrices). They play an essential role in the analysis of best approximate aggregation and disaggregation in Part 2.

Part 1 of this study deals with the linear regression model (1.1.1) in which X is the $n \times k$ matrix of n observations on k independent variables, and V is (up to a proportionality factor) the $n \times n$ sample covariance matrix of the residual errors in the regression. "Multicollinearity" refers to circumstances in which X has rank less than k (or, because of rounding errors or errors of measurement or for any other reasons, is closely approximated by a matrix with rank less than k); approximate multicollinearity is very common, in fact is generally to be expected except in cases where variables have been either omitted or aggregated expressly to avoid it. Singularity of V could result from certain types of sampling covariance patterns, and sometimes is inherent in the economic structure of the problem; an example of the latter type has been given by Theil [107, p. 275]. The $k \times k$ matrix U is interpreted (section 1.5) as the prior variance of the $k \times 1$ vector β of regression coefficients, the latter being considered as a random variable in the Bayesian sense, with a subjective probability distribution; it enters into the definition of the bias in the estimation of β (formulas (1.4.1) and (1.4.5)). The four properties (‡) may be given a simple interpretation in terms of the corresponding properties of the estimator

$$(*) \qquad \tilde{\beta} = X^{\ddagger}y + (I - X^{\ddagger}X)\gamma$$

of β, where y is the $n \times 1$ vector of observations on the dependent variable and γ is some $k \times 1$ vector:

1. If and only if $X^{\ddagger}$ satisfies property (i) of (‡), (*) furnishes an unbiased linear (affine) estimator of every estimable function of β (Theorem 1.1).

2. If and only if $X^{\ddagger}$ satisfies properties (i) and (iii) of (‡), (*) furnishes the best linear unbiased estimator of every estimable function of β (Theorem 1.4, Definition 1.1).

3. If and only if $X^{\ddagger}$ satisfies properties (i) and (iii) of (‡) and property (ii) of (#), (a) the estimator (*) has minimum variance in the class of linear (affine) estimators of β which furnish unbiased estimators for every estimable function of β, and (b) if Y is any matrix complementary to X (Definition 1.3) and z is any vector in the column space of Y, and if γ is a solution of the equation $Y\gamma = z$, (*) furnishes the best linear conditionally unbiased estimator of β subject to $Y\beta = z$ (Theorem 1.11, Definition 1.4). Under these circumstances the estimator (*) is equivalent with probability 1 to an estimator (*) in which $X^{\ddagger}$ satisfies

properties (i), (ii), and (iii) of (‡).

4. If and only if $X^{\ddagger}$ satisfies properties (i), (iii), and (iv) of (‡) and property (ii) of (#), where U, w, and w_0 are defined by (1.4.5) and (1.4.2) and satisfy (1.4.14), and $\gamma = ww_0^{-1}$, then (*) is the best linear minimum bias estimator of β (Theorem 1.12, Definition 1.5). Under these circumstances the estimator (*) is equivalent, with probability 1, to one in which $X^{\ddagger}$ satisfies all four properties of (‡).

It is clear from this catalogue that it is the four properties of (‡) rather than of the Moore-Penrose inverse (†) which are the appropriate properties of generalized inverses that should be used in the analysis of the regression model. In fact, a more general analysis than the one being presented here would show that it is the four properties of (#) which constitute the really fundamental ones. This general situation results when U is singular and the condition rank XUX' = rank X is violated, implying that prior linear restrictions are imposed on the regression coefficients. A treatment of linear restrictions from this point of view is reserved for a later occasion; in the meantime one may refer to more traditional approaches which will be found in Chipman and Rao [17, 18], Chipman [15, 16], Goldman and Zelen [40], Pringle and Rayner [83], and Theil [103, 107].

Part 2 of this study deals with the aggregation problem. Most contemporary work on this problem stems from the classic contribution by Theil [101] (see also [105], [106]), who formulated the problem in a precise manner and derived conditions for what we shall call <u>perfect aggregation.</u> Another major step forward in formulating and clarifying the nature of the problem was accomplished in an important study by Malinvaud [68] (see also [67]). Following up suggestions on the part of Malinvaud [68] and Hurwicz [53] that aggregation be considered as a statistical decision problem, W. D. Fisher in a pioneering paper [34] formulated the problem of what we shall call <u>best approximate aggregation</u> (see also [35]). The treatment presented here builds largely on the work of Theil, Malinvaud, and W. D. Fisher, although a number of other writings have also been influential (Hatanaka [47], Fei [27], Rosenblatt [92], and Ijiri [54]); however, our formulation of the problems of best approximate aggregation and disaggregation differs from Fisher's in a number of essential respects.

Briefly, the problem of best approximate aggregation is formulated as follows, for the linear case (section 2.2). Let $\mathcal{X}$ and $\mathcal{Y}$ be subsets of vector spaces of dimension n and m respectively, and let $F : \mathcal{X} \to \mathcal{Y}$ be a linear transformation. The elements x and y of $\mathcal{X}$ and $\mathcal{Y}$ are vectors (whose first components may be taken = 1) of values of independent and dependent variables respectively, and the mapping F is interpreted as representing the true model of the economy. Now we are given linear transformations $G : \mathcal{X} \to \mathcal{X}^*$ and $H : \mathcal{Y} \to \mathcal{Y}^*$ to subspaces of smaller dimension n^* and m^* respectively. The elements x^* and y^* of $\mathcal{X}^*$ and $\mathcal{Y}^*$ are vectors (whose first components may again be taken = 1) of values of artificial aggregative independent and dependent variables; typically, they will be price indexes, quantity indexes, etc., and G and H are the rules for the construction of such indexes. We now seek an aggregative model $F^* : \mathcal{X}^* \to \mathcal{Y}^*$ which is in some sense as faithful as possible a copy of F. To define this criterion precisely, let $\tilde{y}^* = HFx$ be the conditional forecast one would make of y^* if one knew F and x, and $\hat{y}^* = F^*Gx$ the corresponding forecast one would make on the basis of the model F^*. If x is a random variable with moment matrix $M = \mathcal{E}xx'$, then the aggregative model F^* which minimizes the mean square error $\mathcal{E}(\hat{y}^* - \tilde{y}^*)(\hat{y}^* - \tilde{y}^*)'$ is given by (see Theorem 2.1)

$$F^* = HFG^\# + Z^*(I - GG^\#) \quad , \tag{a}$$

where $G^\#$ is any generalized quasi-inverse of G with respect to M satisfying properties (i) and (iv) of (#) (with U replaced by M). If rank GMG' = rank G, $G^\#$ is a generalized inverse of G.

We also consider the problem of <u>best approximate disaggregation</u>. F^* having been chosen as in (a), we seek a linear transformation $\overline{H}: \mathcal{Y}^* \to \mathcal{Y}$ enabling us to forecast the "true" dependent variables $y = Fx$ by means of the formula $\hat{y} = \overline{H}\hat{y}^* = \overline{H}F^*Gx$. The transformation $\overline{H}$ is a "blowing-up" formula enabling one to predict particular prices, or employment levels in particular industries, on the basis of aggregative forecasts of price indexes and employment indexes. Since $\tilde{y} = FG^{\#}Gx$ has moment matrix

$$W = \mathcal{E}\tilde{y}\ \tilde{y}' = FG^{\#}GMG'G^{\#'}F' ,$$

the disaggregation operator $\overline{H}$ which minimizes the mean square error $\mathcal{E}(\hat{y} - y)(\hat{y} - y)'$ is given by $\overline{H} = H^{\#}$, where $H^{\#}$ is a generalized quasi-inverse of H with respect to W, satisfying properties (i) and (iv) of (#) with W in place of U. This is proved in Theorem 2.2.

From this two-fold process, of first choosing $G^{\#}$ so as to minimize aggregation bias and then choosing $H^{\#}$ so as to minimize disaggregation bias, we obtain an approximative model $\hat{F} = H^{\#}HFG^{\#}G$. The final problem is to choose the transformations G and H themselves in an optimal manner. Given a set $\mathcal{S}$ of pairs (G, H), representing reasonable candidates for possible price and quantity index formulas, etc., one then selects G and H so as to minimize

$$\| F - H^{\#}HFG^{\#}G \|_M = \sqrt{\text{trace}(F - H^{\#}HFG^{\#}G)MF'} .$$

The resulting model $\hat{F}$ is our "estimate" of the model F. Statistically, F itself is in general not estimable, but FM always is. If costs of computation, difficulty of manipulation of large and complex systems, etc., could be ignored, then FM would be the information to aim for; the above formulation assumes that the saving in costs and added convenience more than compensate for falling short of this ideal.

PART 1. THEORY OF ESTIMATION

1.1. Best linear unbiased estimation

Our point of departure is the univariate multiple regression model

$$(1.1.1) \qquad y = X\beta + \varepsilon, \quad \mathcal{E}\varepsilon = 0, \quad \operatorname{Var} \varepsilon = \mathcal{E}\varepsilon\varepsilon' = \sigma^2 V ,$$

where y is a random $n \times 1$ vector of observations on a "dependent" variable, X is an $n \times k$ matrix of observations on k "independent" (explanatory) variables, and ε is the $n \times 1$ vector of random disturbances with zero mean and variance matrix $\sigma^2 V$, V being an $n \times n$ symmetric non-negative definite matrix, assumed known. The symbol $\mathcal{E}$ denotes the expectation operator, which is taken to be conditional on X. The estimation problem will be considered to be that of finding a linear (affine) function of y (which will depend on X) which is in some sense a good estimator of the $k \times 1$ vector β of regression coefficients. (Estimation of σ^2 will not be taken up.)

We shall denote a linear (affine) estimator of β by

$$(1.1.2) \qquad \tilde{\beta} = Ay + c .$$

In the development to follow, we shall relate various properties of this estimator to the corresponding properties enjoyed by the respective generalized inverses (or quasi-inverses) A of X.

A linear functional ($1 \times k$ vector) ψ is called (unbiasedly) estimable (following Bose [12]) if an unbiased linear (affine) estimator of $\psi\beta$ exists, i.e., if there exists an affine function $a_0 + ay$ whose expected value is $\psi\beta$, for all β. Clearly, a necessary and sufficient condition for ψ to be unbiasedly estimable is that $\psi = aX$, i.e., that ψ be in the row space of X, and that $a_0 = 0$; the estimator is then given by ay. We now state a simple and well known result (cf. e.g., Chipman [15]):

Theorem 1.1. In order that the estimator (1.1.2) should furnish an unbiased estimator $\psi\tilde{\beta}$ for every estimable function $\psi\beta$, it is necessary and sufficient that A and c satisfy

$$XAX = X; \quad Xc = 0 , \tag{1.1.3}$$

i.e., that $\tilde{\beta} = X^{-}y + (I - X^{-}X)z$ for arbitrary z.

Proof. The result follows from the requirement that

$$\mathcal{E}\,\psi\tilde{\beta} = \mathcal{E}\,aX\tilde{\beta} = aX\mathcal{E}\,\tilde{\beta} = aXAX\beta + aXc = aX\beta = \psi\beta$$

hold identically in β, by the definition of unbiasedness.

Lemma 1.1. Let

$$W = \begin{bmatrix} W_{11} & W_{12} \\ W_{21} & W_{22} \end{bmatrix} \tag{1.1.4}$$

be any non-negative definite symmetric matrix. Then the equation

$$CW_{22} = W_{12} \tag{1.1.5}$$

has a solution. Further, for any such solution, CW_{21} is unique.

Proof. We may write $W = S'S$ where $S = [S_1, S_2]$, whence (1.1.5) becomes $CS_2'S_2 = S_1'S_2$. Since

$$[I - S_2'S_2(S_2'S_2)^{-}]S_2'S_2[I - (S_2'S_2)^{-}S_2'S_2] = 0 ,$$

we have $S_2 = S_2(S_2'S_2)^{-}S_2'S_2$, so that $W_{12} = W_{12}W_{22}^{-}W_{22}$, whence a solution to (1.1.5) is given by $C = S_1'S_2(S_2'S_2)^{-} =$

$W_{12}W_{22}^-$. If C_1 and C_2 are two solutions of (1.1.5) then $(C_1 - C_2)S_2' = 0$, hence $(C_1 - C_2)W_{21} = (C_1 - C_2)S_2'S_1 = 0$.

Q.E.D.

The following basic result is implicit in, and was inspired by, the work of Zyskind [112].

Theorem 1.2. Let X be any $n \times k$ matrix, V any $n \times n$ symmetric non-negative definite matrix, and U any $k \times k$ symmetric positive definite matrix. Then there exists a generalized inverse $X^\ddagger$ of X satisfying the four properties of $(\ddagger)$. A forteriori, there exists an $X^\ddagger$ satisfying properties (i), (iii), and (iv) of $(\ddagger)$ and property (ii) of (#); and for any such $X^\ddagger$, $X^\ddagger V$ is unique.

Proof. Without loss of generality we may choose $U = I_k$, since the theorem then holds for $\bar{X} = XU^{-\frac{1}{2}}$ and $\bar{X}^\ddagger = U^{\frac{1}{2}}X^\ddagger$. Let p be the rank of X, and let B be an $n \times p$ matrix whose columns form a basis for the column space of X; then $X = BC$ for some $p \times k$ matrix C (of rank p). Define $Q_1' = B^\dagger = (B'B)^{-1}B'$, and let Q_2 be such that $[Q_1, Q_2]$ is an $n \times n$ non-singular matrix and $Q_2'Q_1 = 0$. Then

$$Q'X = \begin{bmatrix} Q_1' X \\ Q_2' X \end{bmatrix} = \begin{bmatrix} C \\ 0 \end{bmatrix} ,$$

since $X = Q_1B'BC$. Define

$$W = Q'VQ ,$$

with partitioning as in (1.1.4). Finally let us define

(1.1.6) $$M = C^\dagger[I, -W_{12}W_{22}^\dagger] = X'Q_1(Q_1'XX'Q_1)^{-1}[I, -Q_1'VQ_2(Q_2'VQ_2)^\dagger] .$$

We verify that $MQ'XM = M$, $Q'XMQ'X = Q'X$, and $MQ'X = X'Q_1(Q_1'XX'Q_1)^{-1}Q_1'X$, which is symmetric. Further,

$$(1.1.7)\quad Q'XMW = \begin{bmatrix} W_{11} - W_{12}W_{22}^{\dagger}W_{21} & W_{12} - W_{12}W_{22}^{\dagger}W_{22} \\ 0 & 0 \end{bmatrix} .$$

From Lemma 1.1, (1.1.5) has a solution, whence by Penrose's solvability theorem [80] $W_{12} = W_{12}W_{22}^{\dagger}W_{22}$. Thus the matrix (1.1.7) is symmetric, and

$$(1.1.8)\qquad M = (Q'X)^{\ddagger} ,$$

where $(Q'X)^{\ddagger}$ is a generalized inverse of $Q'X$ satisfying the four properties of (‡), with I for U and W for V; in particular, $Q'X(Q'X)^{\ddagger}W$ is symmetric. We now verify readily that the matrix

$$(1.1.9)\qquad X^{\ddagger} = (Q'X)^{\ddagger}Q'$$

satisfies the four properties of (‡), with $U = I$; in particular, $XX^{\ddagger}V = Q'^{-1}(Q'X)(Q'X)^{\ddagger}WQ^{-1}$, which is symmetric. This proves existence.

Now let A and B be any two matrices both satisfying properties (i), (iii), and (iv) of (‡) and property (ii) of (#), where U is positive definite. We have (underlining expressions that are to be altered at the next step):

$$\underline{AV} = A\underline{XAV} = AVA'\underline{X'} = A\underline{VA'X'}B'X' = AXA\underline{VB'X'} = A\underline{XAX}BV = AXBV$$

and

$$\underline{BV} = \underline{BX}BV = U\underline{X'B'}U^{-1}BV = \underline{UX'A'X'}B'U^{-1}BV = AX\underline{UX'B'U^{-1}}BV =$$

$$AX\underline{BXBV} = AXBV \;.$$

Q. E. D.

For any square matrix C we define $C \succcurlyeq 0$ to mean that C is non-negative definite, i.e., $x'Cx \geqq 0$ for all x. $A \succcurlyeq B$ means that $A - B \succcurlyeq 0$. It is shown in Chipman [15] that the relation $\succcurlyeq$ is a partial ordering as between symmetric matrices. The following result is proved in Chipman [16, p. 120, Lemma 1.2]:

Lemma 1.2. (Generalized Schwarz Inequality). Let X be an $n \times k$ matrix and V an $n \times n$ symmetric non-negative definite matrix, and let $X^{\ddagger}$ be any matrix satisfying properties (i) and (iii) of ($\ddagger$). Then for any $m \times n$ matrix A,

$$AVA' \succcurlyeq AXX^{\ddagger}VX^{\ddagger'}X'A' , \tag{1.1.10}$$

with equality holding if and only if

$$AV = AXX^{\ddagger}V . \tag{1.1.11}$$

We have seen that a necessary and sufficient condition for a linear function $\psi\beta$ to be unbiasedly estimable is that $\psi = aX$ for some a. By Penrose's solvability theorem [80] this is equivalent to the condition that ψ satisfy

$$\psi = \psi X^{\ddagger}X , \tag{1.1.12}$$

where $X^{\ddagger}$ is a generalized inverse of X satisfying properties (i) and (iii) of ($\ddagger$). The general solution is then given by

$$a = \psi X^{\ddagger} + z(I - XX^{\ddagger}) , \tag{1.1.13}$$

where z is arbitrary. In choosing an estimator ay of $\psi\beta$, it remains to determine a in some optimal manner, that is, to find an optimal choice of z in (1.1.13). It follows from Theorem 1.3 below that the optimal z's are those that satisfy $z(I - XX^{\ddagger})V = 0$; in particular, one such optimal choice is $z = 0$.

Definition 1.1. A linear (affine) function $a_0 + ay$ is said to be a best linear unbiased estimator (blue), or Gauss-Markoff estimator, of the estimable function $\psi\beta$ in the model (1.1.1), if it is of minimum variance $\text{Var}\ ay = \mathcal{E}\, a\varepsilon\varepsilon' a' = \sigma^2 aVa'$ in the class of unbiased linear (affine) estimators of $\psi\beta$, i.e., those for which $a_0 = 0$ and $\psi = aX$. That is, $a_0 + ay$ is a blue of $\psi\beta$ if and only if $a_0 = 0$ and a minimizes aVa' subject to $aX = \psi$. An estimator (1.1.2) will be called a Gauss-Markoff estimator of β if it furnishes a blue $\psi\tilde{\beta}$ for every estimable function $\psi\beta$.

Theorem 1.3 (Zyskind [112]). Each of the following equivalent conditions is necessary and sufficient for a linear function ay to be a best linear unbiased estimator of its expectation $\mathcal{E}\, ay = \psi\beta$, where $X^{\ddagger}$ is any generalized inverse of X satisfying properties (i) and (iii) of ($\ddagger$):

(a) $aV = \psi X^{\ddagger} V$;

(b) $a(I - XX^{\ddagger})V = 0$;

(c) $Va' = Xb$ for some b .

Proof. By Lemma 1.2 we have

$$aVa' \geqq aXX^{\ddagger}VX^{\ddagger\prime}X'a' = \psi X^{\ddagger}VX^{\ddagger\prime}\psi' ,$$

with equality holding if and only if

$$aV = aXX^{\ddagger}V = \psi X^{\ddagger}V . \tag{1.1.14}$$

This proves (a) and (b). From property (iii) of $X^{\ddagger}$, the first equation of (1.1.14) is equivalent to

$$Va' = XX^{\ddagger}Va' . \tag{1.1.15}$$

By Penrose's solvability theorem, (1.1.15) is equivalent to (c).

Q.E.D.

An analogous result can now be stated in terms of the properties of the matrix A in formula (1.1.2) for the estimator of β (cf. Chipman [16]).

Theorem 1.4. In order that a linear estimator satisfying (1.1.3) furnish a best linear unbiased estimator $\psi\tilde{\beta}$ of every estimable function $\psi\beta$ in the model (1.1.1), i.e., in order that it be a Gauss-Markoff estimator, it is necessary and sufficient that A satisfy property (iii) of (‡). If $\tilde{\beta}_i = A_i y + c_i$, $i = 1, 2$, are any two such estimators satisfying (1.1.3) and property (iii) of (‡), then for any estimable functional ψ, $\psi\tilde{\beta}_1 = \psi\tilde{\beta}_2$ with probability 1, the equality holding without qualification if V is positive definite.

Proof. Let $X^{\ddagger}$ satisfy properties (i) and (iii) of (‡). If $\psi = aX$ is an estimable functional, the variance of $\psi\tilde{\beta}$ is given by $\sigma^2 aXAVA'X'a'$; thus, in order that $\tilde{\beta}$ should furnish a best linear unbiased estimator $\psi\tilde{\beta}$ for every estimable function $\psi\beta$, it is necessary and sufficient that A be a solution to the problem

$$\text{(1.1.16)} \qquad \text{minimize } XAVA'X' \text{ subject to } XAX = X .$$

By Lemma 1.2 we have, for any A satisfying (1.1.3),

$$XAVA'X' \geq XAXX^{\ddagger}VX^{\ddagger\prime}X'A'X' = XX^{\ddagger}VX^{\ddagger\prime}X' ,$$

with equality holding if and only if

$$\text{(1.1.17)} \qquad XAV = XX^{\ddagger}V .$$

Now if A satisfies (1.1.17) then $XAV = XX^{\ddagger}V = (XX^{\ddagger}V)' = (XAV)'$, so A satisfies property (iii) of (‡); conversely, if A satisfies property (iii) of (‡) then

$$\underline{XAV} = VA'\underline{X'} = \underline{VA'X'}X^{\ddagger\prime}X' = XA\underline{VX^{\ddagger\prime}X'} = \underline{XAX}X^{\ddagger}V = XX^{\ddagger}V .$$

If $\tilde{\beta}_i = A_i y + c_i$ are two estimators, $i = 1, 2$, such that c_1 and c_2 both satisfy (1.1.3) and A_1 and A_2 both satisfy properties (i) and (iii) of (‡), and if $\psi = aX$ is an estimable functional, then $\psi\tilde{\beta}_1 - \psi\tilde{\beta}_2 = aX(A_1 - A_2)y$ hence

$$\mathcal{E}(\psi\tilde{\beta}_1 - \psi\tilde{\beta}_2) = aX(A_1 - A_2)X\beta = 0$$

from property (i) of (‡), and

$$\text{Var}(\psi\tilde{\beta}_1 - \psi\tilde{\beta}_2) = aX(A_1 - A_2)V(A_1 - A_2)'X'a' = 0$$

from property (iii), implying that $\psi\tilde{\beta}_1 - \psi\tilde{\beta}_2 = 0$ with probability 1. If V is positive definite then $XA_1 = XA_2$ from (1.1.17), completing the proof.

Q.E.D.

Owing to the above uniqueness result, if one's interest is limited to obtaining best linear unbiased estimators of estimable functions, it suffices to consider any convenient subclass of matrices A satisfying properties (i) and (iii) of (‡). One such subclass, defined by Theorem 1.5 below, can be given a useful characterization in terms of certain generalized inverses of V, to be denoted V^+, whose existence is established in the following lemma which is due to Zyskind and Martin [113]:

Lemma 1.3 (Zyskind and Martin). Given any $n \times k$ matrix X of rank p and any $n \times n$ symmetric non-negative definite matrix V of rank r, there exists a generalized inverse V^+ of V which satisfies, in addition to property (i) of (†), the condition

(1.1.18) $$VV^{+'}X = X\Gamma \quad \text{for some } \Gamma ,$$

i.e., which is such that the column space of X is invariant under the projection $VV^{+'}$. Moreover, V^+ may be chosen to be symmetric and positive definite, and therefore such as to satisfy the condition

(1.1.19) $$\text{rank } X'V^{+}X = \text{rank } X \quad .$$

If the columns of X are contained in the column space of V, then any generalized inverse V^{-} of V satisfies (1.1.18) and (1.1.19).

Proof. Let $P = [P_1, P_2]$ be an orthogonal matrix diagonalizing V to

(1.1.20) $$P'VP = \begin{bmatrix} P_1' \\ P_2' \end{bmatrix} V[P_1, P_2] = \begin{bmatrix} \Lambda_1 & 0 \\ 0 & 0 \end{bmatrix} = \Lambda \quad ,$$

where P_1 is $n \times r$ and Λ_1 is an $r \times r$ diagonal matrix with positive diagonal elements. Let the p columns of the $n \times p$ matrix $Q = [Q_1, Q_2]$ form an orthonormal basis for the column space of X in such a way that the s columns of Q_1 form a basis for the intersection of the column spaces of X and V. Then for some $n \times s$ matrix C we have, using (1.1.20),

(1.1.21) $$Q_1 = VC = P_1\Lambda_1 P_1'C = P_1 A_1 \quad ,$$

where $A_1 = \Lambda_1 P_1'C$, so that, defining

(1.1.22) $$A = \begin{bmatrix} A_1 \\ 0 \end{bmatrix} = P'Q_1, \qquad B = \begin{bmatrix} B_1 \\ B_2 \end{bmatrix} = \begin{bmatrix} P_1' Q_2 \\ P_2' Q_2 \end{bmatrix} = P'Q_2 \quad ,$$

we have, for some $p \times k$ matrix K of rank p,

(1.1.23) $$X = [Q_1, Q_2]K = P[A, B]K = P \begin{bmatrix} A_1 & B_1 \\ 0 & B_2 \end{bmatrix} \begin{bmatrix} K_1 \\ K_2 \end{bmatrix} .$$

Now we show that the $(n-r)\times(p-s)$ matrix B_2 has full column rank $p-s$. Suppose $B_2x_2 = 0$ for some vector x_2; then since $A_1'B_1 = A'B = A'P'PB = Q_1'Q_2 = 0$ we have

$$\begin{bmatrix} A_1' & 0 \\ B_1' & B_2' \end{bmatrix}\begin{bmatrix} A_1 & B_1 \\ 0 & B_2 \end{bmatrix}\begin{bmatrix} 0 \\ x_2 \end{bmatrix} = \begin{bmatrix} A_1'A_1 & 0 \\ 0 & B_2'B_2 \end{bmatrix}\begin{bmatrix} 0 \\ x_2 \end{bmatrix} = \begin{bmatrix} 0 \\ 0 \end{bmatrix}.$$

Since $[A, B] = P'[Q_1, Q_2]$ has full column rank p, this implies $x_2 = 0$; therefore rank $B_2 = p - s$.

Now it is readily verified from (1.1.20) that

$$V^{-} = P\begin{bmatrix} \Lambda_1^{-1} & M \\ L & N \end{bmatrix}P' \tag{1.1.24}$$

is a generalized inverse of V, where L, M, N are arbitrary. It follows then from (1.1.24), (1.1.23), and (1.1.21) that

$$VV^{-'}X = [Q_1, P_1(B_1 + \Lambda_1 L'B_2)]K . \tag{1.1.25}$$

Since B_2 has full column rank it has a left inverse, say $B_2^{\dagger} = (B_2'B_2)^{-1}B_2'$; choosing $L' = -\Lambda_1^{-1}B_1B_2^{\dagger}$ in (1.1.24) and defining

$$V^{+} = P\begin{bmatrix} \Lambda_1^{-1} & M \\ -B_2^{\dagger'}B_1'\Lambda_1^{-1} & N \end{bmatrix}P' , \tag{1.1.26}$$

(1.1.25) becomes, on account of (1.1.23) and the fact that $K^{\dagger} = K'(KK')^{-1}$ is a right inverse of K,

$$VV^{+'}X = [Q_1, 0]K = XK^{\dagger}\begin{bmatrix} K_1 \\ \\ 0 \end{bmatrix},$$

proving (1.1.18).

Choosing $M = -\Lambda_1^{-1}B_1B_2^{\dagger}$ and $N = I + B_2^{\dagger'}B_1'\Lambda_1^{-1}B_1B_2^{\dagger}$, and defining

$$T = \begin{bmatrix} \Lambda_1^{-\frac{1}{2}} & -\Lambda_1^{-\frac{1}{2}}B_1B_2^{\dagger} \\ \\ 0 & I \end{bmatrix} P' ,$$

we see that $V^+ = T'T$, which is symmetric and positive definite. This implies (1.1.19).

Finally, if $X = VC$ for some C then $VV^-X = VV^-VC = VC = X$, so (1.1.18) is satisfied for $\Gamma = I$. Further, $X'V^-X = C'VV^-VC = C'VC$, hence the rank of $X'V^-X$ is the same as that of $VC = X$, establishing (1.1.19).

Q. E. D.

The following lemma will be used repeatedly in the sequel. Most of it is comprised in Rao and Mitra [89, pp. 22-3] (see also Rao [86]), but since they do not provide a complete proof it seems worthwhile to present one here.

Lemma 1.4. (a) Let B and C be any two matrices for which the product BC is defined.

(i) If rank BC = rank C, the rows of C are contained in the row space of BC, i.e., the equation $ABC = C$ has a solution; it follows that $C(BC)^-BC = C$, i.e., that $(BC)^-B$ is a generalized inverse of C. Consequently,

(1.1.27) $\quad C(BC)^-BC = C$ if and only if rank BC = rank C.

Likewise, if rank BC = rank B, the columns of B are

contained in the column space of BC, i.e., the equation $BCD = B$ has a solution; it follows that $BC(BC)^{-}B = B$, i.e., that $C(BC)^{-}$ is a generalized inverse of B. Consequently,

(1.1.28) $$BC(BC)^{-}B = B \text{ if and only if } \operatorname{rank} BC = \operatorname{rank} B .$$

(ii) Let $\operatorname{rank} B = \operatorname{rank} BC = \operatorname{rank} C$. Then any solution D of the equation $BCD = B$ is a generalized inverse of C, and any solution of the form $D = (BC)^{-}B$ is a reflexive generalized inverse of C. Likewise, any solution A of the equation $ABC = C$ is a generalized inverse of B, and any solution of the form $A = C(BC)^{-}$ is a reflexive generalized inverse of B.

(b) Let W be any matrix such that $\operatorname{rank} C'WC = \operatorname{rank} C$. Then:

(iii) The equations

(1.1.29) $$C(C'WC)^{-}C'WC = C, \quad C'WC(C'WC)^{-}C' = C'$$

hold for any choice of generalized inverse $(C'WC)^{-}$. Thus the equations

(1.1.30) $$LC'WC = WC, \quad C'WCR = C'W$$

are solvable. Any solutions L and R of these respective equations are generalized inverses of C' and C respectively, and any solutions of the form

(1.1.31) $$L = WC(C'WC)^{-}, \; R = (C'WC)^{-}C'W$$

are reflexive generalized inverses of C' and C respectively.

(iv) $C(C'WC)^{-}C'$ is invariant with respect to the choice of generalized inverse $(C'WC)^{-}$.

<u>Proof</u>. (i) Let B and C be of orders $\ell \times m$ and $m \times n$ respectively, and let r be the common rank of C and

BC . Then there exists an $r \times n$ matrix N whose rows form a basis for the row space of C, hence $C = MN$, where M is an $m \times r$ matrix of rank r . Define $A = M(BM)^{\dagger}$; since $BC = BMN$ has rank r, the $\ell \times r$ matrix BM has rank r , hence $(BM)^{\dagger}$ is a left inverse of BM . Thus,

$$ABC = ABMN = MN = C \quad ,$$

i.e., the rows of C are contained in the row space of BC . This implies

$$C(BC)^{-}BC = ABC(BC)^{-}BC = ABC = C \ .$$

Since $C(BC)^{-}BC = C$ obviously implies rank BC = rank C , (1.1.27) follows. The second statement is proved similarly.

(ii) From (i), the equations $ABC = C$ and $BCD = B$ are both solvable, hence for any respective solutions A and D we have $CD = ABCD = AB$, hence $CDC = ABC = C$ and $BAB = BCD = B$. Choosing $A = C(BC)^{-}$ and $D = (BC)^{-}B$, that $ABA = A$ and $DCD = D$ follows at once from the fact that $C(BC)^{-}BC = C$ and $BC(BC)^{-}B = B$, from (i).

(iii) Setting $B = C'W$, the first equation of (1.1.29) follows from (i) . Transposing the equation $C'WC(C'WC)^{-}C'WC = C'WC$ we see that $(C'WC)^{-'}$ is a generalized inverse of $C'W'C$, and since rank $C'W'C$ = rank C' = rank C, analogously to the first equation of (1.1.29) we have $C(C'WC)^{-'}C'W'C = C$. Transposing this, we obtain the second equation of (1.1.29).

Premultiplying the first equation of (1.1.29) by W, and post-multiplying the second by W, we obtain (1.1.30) for L and R given by (1.1.31); thus, the equations (1.1.30) are solvable. Now, since rank C = rank $C'WC \leqq$ rank $C'W \leqq$ rank C, clearly rank $C'W$ = rank $C'WC$ = rank C; choosing $B = C'W$ and $D = R$ in (ii), it follows that any solution R of the second equation of (1.1.30) is a generalized inverse of C . Choosing R as in (1.1.31), that $RCR = R$ follows immediately from the second equation of (1.1.29). The remaining statement is proved similarly, by choosing $B = C'W'$ and $D = L'$ in (ii), and using the first equation of (1.1.29).

(iv) Let $(C'WC)_1^-$ and $(C'WC)_2^-$ be any two generalized inverses of $C'WC$. Then from (1.1.29) we have

$$[C(C'WC)_1^- - C(C'WC)_2^-]C'$$

$$= [C(C'WC)_1^- - C(C'WC)_2^-]C'WC(C'WC)^-C'$$

$$= [C(C'WC)_1^- C'WC - C(C'WC)_2^- C'WC](C'WC)^-C'$$

$$= 0 \; .$$ Q.E.D.

Theorem 1.5. Let V^+ be a generalized inverse of V satisfying condition (1.1.19) of Lemma 1.3. Then:

(a) A solution A exists to the equation

(1.1.32) $$X'V^+XA = X'V^+ \; .$$

(b) Any solution A of (1.1.32) satisfies property (i) of (‡).

(c) Any solution to (1.1.32) of the form

(1.1.33) $$A = (X'V^+X)^-X'V^+$$

satisfies property (ii) of (‡).

(d) If V^+ is symmetric, any solution A of (1.1.32) satisfies

(1.1.34) $$V^+XA = A'X'V^+ \; .$$

(e) If V^+ satisfies condition (1.1.18) of Lemma 1.3, then any solution A of (1.1.32) satisfies property (iii) of (‡).

Proof. Conclusions (a), (b), and (c) follow immediately from Lemma 1.4 (iii). Conclusion (d) follows upon premultiplying both sides of (1.1.32) by A', and transposing. It remains to establish (e).

The general solution of (1.1.32) is

$$A = (X'V^{+}X)^{-}X'V^{+} + [I - (X'V^{+}X)^{-}X'V^{+}X]Z \ , \tag{1.1.35}$$

where Z is arbitrary. By (1.1.19) and Lemma 1.4 (iii) this yields

$$XA = X(X'V^{+}X)^{-}X'V^{+} \ , \tag{1.1.36}$$

from the first equation of (1.1.29). Now by Penrose's solvability theorem, (1.1.18) is equivalent to

$$XX^{-}VV^{+'}X = VV^{+'}X \ , \tag{1.1.37}$$

where X^{-} is any generalized inverse of X. Choosing X^{-} to be any solution A of (1.1.32)--as we may, from (b)--(1.1.37) becomes, on account of (1.1.36),

$$X(X'V^{+}X)^{-}X'V^{+}VV^{+'}X = VV^{+'}X \ .$$

This implies

$$\begin{aligned} XAV &= X(X'V^{+}X)^{-}X'V^{+}V \\ &= X(X'V^{+}X)^{-}X'V^{+}VV^{+'}X(X'V^{+}X)^{-'}X' \\ &= VV^{+'}X(X'V^{+}X)^{-'}X' \\ &= VA'X' \ . \end{aligned}$$

Q.E.D.

Formula (1.1.33) generalizes Aitken's well-known formula for the case in which X and V have full rank (cf. [3]). Unfortunately, however, Lemma 1.3 does not provide an explicit expression for V^{+}. On the other hand, Rao and Mitra [89, pp. 46, 148] give an alternative representation (distinct from (1.1.33)) in the form

(1.1.38) $$X^{\ddagger} = [X'(V + XX')^{-}X]^{-}X'(V+XX')^{-} \quad ,$$

which coincides with (1.1.33) (with $V^{+} = V^{-}$) when the columns of X are contained in the column space of V. It seems that the matrix $(V + XX')^{-}$ in (1.1.38) need not be a generalized inverse of V. Further representations (when X has full rank) have been obtained by Holly [49]; see also Pringle and Rayner [83, pp. 114-117], Theil [107, pp. 273-293], Malinvaud [69, pp. 198-199], and Rao [88]. In a recent paper, Mitra [74] has shown that the Zyskind-Martin inverse V^{+} enjoys optimality properties not shared by $(V + XX')^{-}$.

1.2. Relation between least squares and best linear unbiased estimators.

The so-called Gauss-Markoff theorem states that when $V = I$ in the model (1.1.1), the ordinary least squares estimator ψb of any estimable function $\psi\beta$ is best linear unbiased. Until relatively recently it was not generally appreciated that the hypothesis can be weakened to allow $V \neq I$, provided V and X are related in a certain way.

Definition 1.2. An estimator $b = f(y)$ is called a least squares (or ordinary least squares) estimator of β in the model (1.1.1) if it minimizes $(y - Xb)'(y - Xb)$. For any linear functional ψ, ψb is called a least squares estimator of $\psi\beta$ if b is a least squares estimator of β.

Theorem 1.6. A function $b = f(y)$ is a least squares estimator of β in the model (1.1.1) if and only if b is a solution of the equation

(1.2.1) $$Xb = XX^{+}y \quad ,$$

where X^{+} is any generalized inverse of X satisfying properties (i) and (iii) of (†). A solution to (1.2.1) always exists. Furthermore, if ψ is an estimable functional, the least squares estimator of $\psi\beta$ is $\psi X^{+}y$, which is unique.

Proof. Let X^{+} satisfy properties (i) and (iii) of $(\dagger)$, and write

$$y - Xb = (I - XX^{+})y + X(X^{+}y - b) \quad .$$

Then

$$(y - Xb)'(y - Xb) = y'(I - XX^{+})y + (X^{+}y - b)'X'X(X^{+}y - b) \quad ,$$

which attains a minimum if and only if (1.2.1) holds. That a solution to (1.2.1) exists follows immediately from Penrose's solvability theorem [80]. If ψ is estimable then $\psi = aX$ for some a, hence if b is any solution of (1.2.1),

$$\psi b = aXb = aXX^{+}y = \psi X^{+}y \quad .$$

Uniqueness follows from the fact that if A and B both satisfy conditions (i) and (iii) of $(\dagger)$, then

$$\underline{XA} = A'\underline{X'} = \underline{A'X'}\,\underline{B'X'} = \underline{XAX}B = XB \; .$$

Q. E. D.

We now investigate conditions under which least squares estimators are best linear unbiased estimators of estimable functions in the model (1.1.1). In the case in which X and V are both of full rank, a sufficient condition was first obtained by Anderson [5]; this condition--which turns out to be crucial in the analysis of autocorrelation of the residuals, and underlies the theory of the Durbin-Watson test (cf. [26])--is that the columns of X be linear combinations of a set of k characteristic vectors of V. It was shown subsequently by Magness and McGuire [66] (without reference to Anderson) that the condition is also necessary --again under the assumption that X and V are both of full rank. An analogous necessary and sufficient condition, stated in Theorem 1.7 below, was obtained by Zyskind [112] for the general case; an equivalent necessary and sufficient condition was obtained independently by Rao [87] (see also Watson [109], Kruskal [60], Rao and Mitra [89, p. 155]).

In the light of Theorems 1.4 and 1.6, the question of the conditions under which least squares estimators of estimable functions are best linear unbiased reduces to the question of the conditions under which a generalized inverse X^+ of X satisfying properties (i) and (iii) of (†) also satisfies property (iii) of (‡).

Theorem 1.7 (Zyskind [112]). Each of the following equivalent conditions is necessary and sufficient for the least squares estimator of any estimable function in the model (1.1.1) to be best linear unbiased:

(a) $XX^+V = VXX^+$,

where X^+ is a generalized inverse of X satisfying properties (i) and (iii) of (†);

(b) $VX = X\Gamma$ for some Γ,

i.e., the column space of X is invariant under V;

(c) $X = P_1K$ for some K,

where P_1 is an $n \times p$ matrix, p being the rank of X, and $P = [P_1, P_2]$ is an orthogonal matrix diagonalizing V to

$$P'VP = \Lambda \quad , \tag{1.2.2}$$

i.e., there is a subset of p orthogonal characteristic vectors of V which form a basis for the column space of X.

Proof. From Theorem 1.4, the best linear unbiased estimator of any estimable function $\psi\beta$ is $\psi X^{‡}y$, where $X^{‡}$ satisfies properties (i) and (iii) of (‡). From Theorem 1.6, the least squares estimator of $\psi\beta$ is ψX^+y, where X^+

satisfies properties (i) and (iii) of (†). For the least squares estimator to be best linear unbiased it is therefore necessary and sufficient that X^+ satisfy property (iii) of (‡), i.e., that

$$XX^+V = VX^{+'}X' = VXX^+ ,$$

establishing (a).

We now show that (b) is equivalent to (a). By Penrose's solvability theorem [80], (b) is equivalent to

$$VX = XX^+VX .$$

This implies $VXX^+ = XX^+VXX^+ = XX^+VX^{+'}X'$, hence VXX^+ is symmetric and (a) holds. Conversely if (a) holds then post-multiplying by X we obtain (b) for $\Gamma = X^+VX$.

Finally we show that (c) is equivalent to (b). Let (c) hold. From (1.2.2) we have

$$VP = V[P_1, P_2] = [P_1, P_2]\begin{bmatrix} \Lambda_1 & 0 \\ 0 & \Lambda_2 \end{bmatrix} = [P_1\Lambda_1, P_2\Lambda_2] = P\Lambda .$$

Since K must have rank p,

$$VX = VP_1K = P_1\Lambda_1K = XK^\dagger\Lambda_1K = X\Gamma$$

where $\Gamma = K^\dagger\Lambda_1K$. Thus (b) holds. Conversely let (b) hold, and let $Q = [Q_1, Q_2]$ be an orthogonal matrix such that the p columns of the $n \times p$ matrix Q_1 form a basis for the column space of X, i.e.,

$$X = Q_1L \text{ for some } L .$$

Then

$$Q_2'VQ_1 = Q_2'VXL^\dagger = Q_2'X\Gamma L^\dagger = Q_2'Q_1L\Gamma L^\dagger = 0 ,$$

hence

$$Q'VQ = \begin{bmatrix} Q_1'VQ_1 & 0 \\ 0 & Q_2'VQ_2 \end{bmatrix} .$$

Let R_1 and R_2 be orthogonal matrices diagonalizing $Q_i'VQ_i$ to $R_i'Q_i'VQ_iR_i = \Lambda_i$, $i = 1, 2$, and define

$$R = \begin{bmatrix} R_1 & 0 \\ 0 & R_2 \end{bmatrix}, \quad P = [P_1, P_2] = [Q_1R_1, Q_2R_2] = QR .$$

We verify that P is orthogonal and that (1.2.2) holds. Further,

$$X = Q_1L = P_1R_1'L = P_1K ,$$

where $K = R_1'L$. Thus, (c) holds.

Q. E. D.

Mathematically, Theorem 1.7 provides conditions under which the set of solutions A to equations (i) and (iii) of (†) is a subset of the set of solutions A to equations (i) and (iii) of (‡) . To be explicit, let us define

(1.2.3) $$\mathfrak{X}^{\dagger}_{1,3} = \{A: XAX = X \;\&\; XA = X'A'\}$$

and

(1.2.4) $$\mathfrak{X}^{\ddagger}_{1,3} = \{A: XAX = X \;\&\; XAV = VA'X'\} ;$$

then Theorem 1.7 furnishes conditions under which $\mathfrak{X}^{\dagger}_{1,3} \subseteq \mathfrak{X}^{\ddagger}_{1,3}$. Note, however, that this inclusion must hold, whenever the sets intersect; for if $A_0 \in \mathfrak{X}^{\dagger}_{1,3} \cap \mathfrak{X}^{\ddagger}_{1,3}$ then for any $A_1 \in \mathfrak{X}^{\dagger}$ we have $XA_1 = XA_0$ by Theorem 1.6, hence $XA_1V = XA_0V = VA_0'X' = VA_1'X'$, so $A_1 \in \mathfrak{X}^{\ddagger}_{1,3}$. On the other hand, under the conditions of Theorem 1.7 the best linear unbiased estimators of estimable functions coincide almost surely with their least squares estimators, since if $A_0 \in \mathfrak{X}^{\dagger}_{1,3} \cap \mathfrak{X}^{\ddagger}_{1,3}$ and $A_2 \in \mathfrak{X}^{\ddagger}_{1,3}$ then $XA_2V = XA_0V$ by Theorem 1.4', hence for any estimable functional $\psi = aX$ the best linear unbiased estimator aXA_2y of $\psi\beta$ is equal to the least squares estimator aXA_0y with probability 1.

The following theorem provides conditions under which the particular subset of $\mathfrak{X}^{\ddagger}_{1,3}$ defined by Theorem 1.5(d) (with V replaced by any symmetric non-negative definite matrix H) is actually a subset of $\mathfrak{X}^{\dagger}_{1,3}$.

Theorem 1.8. Let H be any $n \times n$ symmetric non-negative definite matrix, let X be any $n \times k$ matrix, and let H^+ be any symmetric generalized inverse of H satisying

$$\text{rank } X'H^+X = \text{rank } X \quad . \tag{1.2.5}$$

Let a linear estimator $\tilde{\beta}$ of β in the model (1.1.1) be given by (1.1.2), where b satisfies (1.1.3) and A is any solution of

$$X'H^+XA = X'H^+ \quad . \tag{1.2.6}$$

Then a necessary and sufficient condition for $\tilde{\beta}$ to be a least squares estimator of β is that

$$H^+X = X\Gamma \text{ for some } \Gamma \quad . \tag{1.2.7}$$

Proof. By Theorem 1.6, $\tilde{\beta}$ is a least squares estimator of β if and only if XA is symmetric. From Theorem 1.5(d) this implies

$$H^+XA = A'X'H^+ = XAH^+ \quad . \tag{1.2.8}$$

Postmultiplying (1.2.8) by X we obtain (1.2.7), for $\Gamma = AH^{+}X$, by virtue of Theorem 1.5(b). Conversely, by Penrose's solvability theorem (1.2.7) is equivalent to

$$XX^{-}H^{+}X = H^{+}X \tag{1.2.9}$$

for any generalized inverse X^{-} of X. Choosing $X^{-} = (X'X)^{-}X'$ and transposing, (1.2.9) becomes

$$X'H^{+} = X'H^{+}X(X'X)^{-}X' \quad ,$$

whence by (1.2.5) and Lemma 1.4 we have

$$X(X'H^{+}X)^{-}X'H^{+} = X(X'H^{+}X)^{-}X'H^{+}X(X'X)^{-}X' = X(X'X)^{-}X' \ . \tag{1.2.10}$$

It follows from (1.1.36) (with V replaced by H) that XA is symmetric.

Q.E.D.

Illustration 1: Grouped Observations. Economic data customarily come tabulated in group form. For instance, if we wish to examine the relationship between consumer incomes and expenditures, we cannot normally obtain data on the incomes and expenditures of individual households; instead, the range of possible incomes is partitioned into intervals, and data are provided on the number of households in each income group and the average income and expenditure in the group. The theory of grouped observations has been developed by Prais and Aitchison [82], who introduced the concept of a "grouping matrix"

$$G = \begin{bmatrix} g_1 & 0 & \cdots & 0 \\ 0 & g_2 & \cdots & 0 \\ \cdots & \cdots & \cdots & \cdots \\ 0 & 0 & \cdots & g_{n^*} \end{bmatrix} \tag{1.2.11}$$

which is an $n^* \times n$ block diagonal matrix whose ith block g_i is a $1 \times n_i$ row vector of non-negative elements (not all zero), where $\sum_{i=1}^{n^*} n_i = n$. More generally, G is a non-negative matrix with at most one non-zero element in each column. (In Part 2 we shall discuss grouping matrices extensively in connection with the grouping of different variables; here we are concerned with the gouping of different observations on the same variable.) In the following discussion it will be assumed that the components of each g_i are all equal to $1/n_i$. This assumption implies that G has rank n^* and is row stochastic, i.e., has row sums equal to one. It follows also that the matrix

$$H = G^{\dagger}G = G'(GG')^{-1}G \tag{1.2.12}$$

is non-negative, symmetric, idempotent, and doubly stochastic (i.e., has unit row and column sums). H is a block diagonal matrix whose diagonal blocks $H_i = g_i'(g_i g_i')^{-1} g_i$ are idempotent matrices of unit rank whose elements are all equal to $1/n_i$; the rank of H is of course n^*.

When, as will be assumed, the observations on the dependent variable y and the independent variables X in the model (1.1.1) are all grouped in the same fashion, the appropriate model for the grouped data is, in place of (1.1.1) and for the case $V = I$,

$$Gy = GX\beta + G\varepsilon;\ \mathcal{E}G\varepsilon = 0;\ \text{Var}\, G\varepsilon = \sigma^2 GG' . \tag{1.2.13}$$

A linear functional ψ which is estimable with respect to the model (1.1.1) remains estimable with respect to the model (1.2.13) provided the rank of GX is the same as that of X; for in that case, if $(GX)^{\ddagger}$ is a generalized inverse of GX satisfying properties (i) and (iii) of (‡) with V replaced by GG', we have, from (1.2.12) and Lemma 1.4,

$$X(GX)^{\ddagger}GX = X(X'HX)^{-}X'HX = X ,$$

whence by Penrose's solvability theorem there exists an A^* such that $X = A^*GX$, i.e., the rows of X are contained in the row space of GX. Under these conditions, the estimator

(1.2.14) $$\hat{\beta} = (X'HX)^{-}X'Hy$$

furnishes a best linear unbiased estimator $\psi\hat{\beta}$ of any estimable function $\psi\beta$, within the class of linear functions of Gy. On the other hand, the best linear unbiased estimator of $\psi\beta$ within the class of linear functions of y is given by $\psi\tilde{\beta}$, where $\tilde{\beta} = X^{+}y$ is a least squares estimator of β.

We may now ask the question: when does the estimator (1.2.14) furnish a best linear unbiased estimator of every estimable function $\psi\beta$, in the class of linear functions of y? That is, when is the estimator $\psi\hat{\beta}$ calculated from grouped observations the same as the least squares estimator $\psi\tilde{\beta}$ that would be used if ungrouped data were available?

Theorem 1.8 is immediately applicable to this problem. Since H is idempotent, it is its own generalized inverse, and we can take $H^{+} = H$. A necessary and sufficient condition for the estimator $\psi\hat{\beta}$ given by (1.2.14) to be the least squares estimator of $\psi\beta$ whenever ψ is estimable and rank GX = rank X, is

(1.2.15) $$HX = X\Gamma \quad \text{for some } \Gamma .$$

Owing to the symmetry and idempotency of H, this is equivalent to condition (1.1.18) of Lemma 1.3, and thus by Theorem 1.5(e), (1.2.15) is also the condition that $\psi\hat{\beta}$ be the best linear unbiased estimator of $\psi\beta$ in the model (1.1.1) when $V = I$.

Since $H^{+} = H$, (1.2.15) is equivalent to condition (b) of Theorem 1.7, which is in turn equivalent to condition (c) of that theorem, where the columns of P are characteristic vectors of H (in place of V). We can clearly replace (c) by the condition that

(1.2.16) $$X = JK \quad \text{for some } K ,$$

where J is an $n \times p$ matrix (p being the rank of X) whose columns are any set of p mutually orthogonal characteristic vectors of H. Now H, being symmetric and idempotent, has two distinct characteristic roots 1 and 0 with multiplicities $n^* = \text{rank } G$ and $n - n^*$ respectively. The equation $HHu = Hu$ shows that any non-vanishing linear combination Hu of columns of H is a characteristic vector of H corresponding to the unit characteristic root, and likewise the equation $H(I - H)u = 0$ shows that any non-vanishing linear combination $(I - H)u$ of columns of $I - H$ is a characteristic vector of H corresponding to the zero characteristic root. We may let J in (1.2.16) consist of any mutually orthogonal set of these characteristic vectors.

To interpret the meaning of (1.2.16) let us consider the special case of the model (1.1.1) in which

$$X = [\iota, x], \quad V = I \ , \tag{1.2.17}$$

where ι is a column vector of n ones and x a column vector of (unrecorded) observations on a single independent variable, and X has rank 2. Define

$$Z = [\iota, z] = [\iota, x - \iota\iota^{\dagger}x] = X \begin{bmatrix} 1 & -\iota^{\dagger}x \\ 0 & 1 \end{bmatrix} , \tag{1.2.18}$$

where $z = (I - \iota\iota^{\dagger})x$ is the vector whose components are the deviations of the components of x from their sample means $\bar{x} = \iota^{\dagger}x$. Since (1.2.18) is a non-singular transformation between X and Z we may equally well express (1.2.16) in terms of Z in place of X. Since H is stochastic, it has ι as one of its characteristic vectors corresponding to the unit characteristic root; thus we may express our condition in the form

$$[\iota, z] = [\iota, v] \begin{bmatrix} k_{11} & k_{12} \\ k_{21} & k_{22} \end{bmatrix} , \tag{1.2.19}$$

where v is a characteristic vector of H which is orthogonal to ι. Premultiplying (1.2.19) by ι' we see that $k_{11} = 1$

and $k_{12} = 0$, whence we deduce from (1.2.19) that $k_{21} = 0$; thus, in order that (1.2.14) (where the generalized inverse is now a regular inverse) should be the best linear unbiased estimator of β in the model (1.1.1), (1.2.17), it is necessary and sufficient that $z = x - \iota\bar{x}$ be a characteristic vector of H orthogonal to ι.

If we choose z to be a characteristic vector of H corresponding to the zero characteristic root, i.e., $z = (I - H)u \neq 0$, then since the column sums in each block $I_{n_i} - H_i$ of $I - H$ are equal to zero it follows that the mean value of the components of x in each group is equal to the grand mean $\bar{x}$:

$$Gx = G[(I - H)u + \iota\iota^{\dagger}x] = \iota\iota^{\dagger}x = \iota\bar{x} \ .$$

This means that the second column of GX is a multiple $\bar{x}$ of the first, hence GX has rank 1, contradicting the assumption that rank GX = rank $X = 2$.

The only remaining possibility is to choose z to be a characteristic vector of H corresponding to the unit characteristic root, and orthogonal to ι, i.e., $z = Hu \neq 0$, where $\iota'Hu = 0$. But inspection of the matrix H shows that any linear combination of columns of H has the property that the components in each group are equal to one another. We have proved:

In the model (1.1.1), (1.2.17) where rank GX = rank $X = 2$, a necessary and sufficient condition for the estimator $\hat{\beta} = (X'HX)^{-1}X'Hy$ of (1.2.12) to be best linear unbiased in the class of linear functions of y, i.e., to be equal to the least squares estimator of β in (1.1.1), is that the observations on x be equal to one another within each group.

<u>Illustration 2</u>: <u>Two-Stage Least Squares Estimation</u>. The two-stage least squares method in simultaneous equations models was introduced by Theil [103], and independently by Basmann [10], as a method of estimating the $m \times 1$ and $k_1 \times 1$ vectors β and γ of parameters in the equation

$$y = Y\beta + Z_1\gamma + u \ , \tag{1.2.20}$$

where y is an $n \times 1$ vector of observations on the dependent variable in the first equation of a system of $m + 1$ equations, Y is an $n \times m$ matrix of observations on the m remaining endogenous variables in the system, Z_1 is an $n \times k_1$ matrix consisting of the first k_1 columns of the $n \times k$ matrix

$$Z = [Z_1, Z_2] \tag{1.2.21}$$

of observations on the $k = k_1 + k_2$ variables exogenous to the system, and u is an $n \times 1$ random residual such that $\mathcal{E}(u|Z) = 0$. There are by hypothesis m additional equations like (1.2.20) in the system which, in conjunction with (1.2.20) itself, may be expressed in the solved or "partially reduced" form

$$Y = Z\Pi + V \ , \tag{1.2.22}$$

where Π is a $k \times m$ matrix of "reduced form" parameters, and V is an $n \times m$ matrix of random errors (not to be confused with the matrix V of (1.1.1)) such that $\mathcal{E}(V|Z) = 0$. As an example, (1.2.20) might represent the demand for a certain commodity as a function of its price, other prices, and incomes, etc.; the remaining equations (1.2.22) could represent the supply function for this commodity as well as demand and supply functions for other commodities. It is assumed by Theil and Basmann that both Z and the matrix

$$X = [Y, Z_1] \tag{1.2.23}$$

have full rank; however, it has been pointed out by Fisher and Wadycki [36] and by Swamy and Holmes [100] that it is not necessary to assume that Z has full rank (see also Neeleman [117]). The same is true of X .

In the first stage of the two-stage procedure, one obtains a least squares estimator of Π in (1.2.22), namely

$$P = Z^{+}Y \ , \tag{1.2.24}$$

where Z^+ is a generalized inverse of Z satisfying properties (i) and (iii) of (†). Defining

$$\hat{Y} = ZP = ZZ^+Y = Z(Z'Z)^-Z'Y \tag{1.2.25}$$

(the last equality following from the fact that Z^+ is any solution of the normal equations $Z'ZZ^+ = Z'$, and $ZZ^+ = Z(Z'Z)^-Z'$ by Lemma 1.4) and substituting (1.2.25) for Y in (1.2.20) one obtains

$$y = \hat{Y}\beta + Z_1\gamma + \hat{u} \ , \tag{1.2.26}$$

where

$$\hat{u} = u + \hat{V}\beta, \quad \hat{V} = Y - \hat{Y} = (I - ZZ^+)Y = (I - ZZ^+)V \ , \tag{1.2.27}$$

so that $\mathcal{E}(\hat{u}|Z) = 0$. The second stage consists in estimating the parameters β and γ in (1.2.26) by least squares. Defining

$$\hat{X} = [\hat{Y}, \ Z_1], \quad \alpha' = (\beta', \gamma') \ , \tag{1.2.28}$$

the required estimator is

$$\hat{\alpha} = \hat{X}^+ y \ , \tag{1.2.29}$$

where $\hat{X}^+$ is a generalized inverse of $\hat{X}$ satisfying properties (i) and (iii) of (†).

Noting from (1.2.23), (1.2.25), and (1.2.28) that

$$ZZ^+X = ZZ^+[Y, Z] \begin{bmatrix} I_{m+k_1} \\ \\ 0 \end{bmatrix} = [\hat{Y}, Z] \begin{bmatrix} I_{m+k_1} \\ \\ 0 \end{bmatrix} = \hat{X} \ , \tag{1.2.30}$$

and choosing $\hat{X}^+ = (\hat{X}'\hat{X})^-\hat{X}'$, we may express (1.2.29) in the form

$$\hat{\alpha} = [X'Z(Z'Z)^-Z'X]^-X'Z(Z'Z)^-Z'y \ . \tag{1.2.31}$$

Let us now ask the question: when is the two-stage least squares estimator (1.2.31) actually a direct least squares estimator of α in (1.2.20)? Theorem 1.8 is immediately applicable to this problem. We have $H = Z(Z'Z)^{-}Z'$, which is symmetric and idempotent, hence equal to its own generalized inverse. A necessary and sufficient condition is then

$$ZZ^{+}X = Z(Z'Z)^{-}Z'X = X\Gamma \quad \text{for some } \Gamma \,, \tag{1.2.32}$$

and a sufficient condition for (1.2.32) to hold is in turn that the columns of Y be in the column space of Z, since in that case it follows from (1.2.23) that the columns of X are in the column space of Z, i.e., $X = ZC$ for some C, consequently $ZZ^{+}X = ZZ^{+}ZC = ZC = X$. This result was obtained by Fisher and Wadycki [36]; see also Rayner [118]. Two special cases may be noted: (1) rank $Z = n$, which will occur if $n \leqq k$ and Z has full rank (in which case $ZZ^{+} = I$ and $\hat{V} = 0$); (2) $V = 0$ in (1.2.22), i.e., the remaining equations in the simultaneous equations system hold exactly, without error (hence, again $\hat{V} = 0$). The result is of interest mainly on account of the fact--which has often been noticed by empirical workers--that, unexpectedly, ordinary direct least squares estimates are frequently quite close to two-stage and other simultaneous equations estimates. The result extends trivially to other methods of simultaneous equations estimation, as Swamy and Holmes [100] have shown; see also Khazzoom [119].

1.3. Complementary linear restrictions

So far, no use has been made of properties (ii) and (iv) of (‡). According to Theorem 1.4 the Gauss-Markoff criterion of best linear unbiasedness for estimable functions involves precisely properties (i) and (iii) of (‡); satisfaction of the additional two properties of (‡) can therefore only be of relevance to properties of estimators $\psi\tilde{\beta}$ of "non-estimable" functions $\psi\beta$, or more precisely, of linear functions $\psi\beta$ which are not unbiasedly estimable. For those who consider unbiasedness to be a sine qua non, there would be no point in going any further. But from the viewpoint of doing the best that one can with the available data, such an attitude seems unnecessarily rigid. We shall therefore characterize properties of the estimator (1.1.2) when A satisfies the remaining properties of (‡); in this section we consider the additional property (ii).

Theorem 1.9. In order that the estimator (1.1.2) of β in the model (1.1.1) should have minimum variance in the class of estimators (1.1.2) furnishing unbiased estimators for all (unbiasedly) estimable functions $\psi\beta$, it is necessary and sufficient that A and c satisfy (1.1.3) and

$$\text{(1.3.1)} \qquad \text{(ii)}\ AXAV = AV; \qquad \text{(iii)}\ XAV = VA'X' .$$

If V is positive definite, this reduces to the condition that $Xc = 0$ and that A satisfy properties (i), (ii), and (iii), of (‡).

Proof. In view of Theorem 1.1 and Definition 1.1, it suffices to show that A is a solution to the problem

$$\text{(1.3.2)} \qquad \text{minimize } AVA' \text{ subject to } XAX = X$$

if and only if A satisfies (1.3.1). Let $X^{\ddagger}$ satisfy properties (i), (ii), and (iii) of (‡). By Lemma 1.2, A is a solution to (1.3.2) if and only if (1.1.11) holds. The problem therefore reduces to showing that (1.1.11) is equivalent to (1.3.1), given that A satisfies (1.1.3).

Let A satisfy (1.1.3) and (1.1.11). Then property (ii) of (1.3.1) follows from

$$AX\underline{AV} = A\underline{XAX}X^{\ddagger}V = \underline{AXX^{\ddagger}V} = AV ,$$

and property (iii) of (1.3.1) from

$$X\underline{AV} = \underline{XAX}X^{\ddagger}V = \underline{XX^{\ddagger}V} = VX^{\ddagger'}\underline{X'} = VX^{\ddagger'}X'A'X' = VA'X' .$$

Conversely, let A satisfy (1.1.3) and (1.3.1). Then (1.1.11) follows from

$$\underline{AV} = A\underline{XAV} = AVA'\underline{X'} = A\underline{VA'X'}X^{\ddagger'}X' = \underline{AXAV}X^{\ddagger'}X' = A\underline{VX^{\ddagger'}X'} = AXX^{\ddagger}V.$$

The final statement of the theorem is immediate.

Q. E. D.

Another interpretation will now be given to properties (i), (ii), and (iii) of (‡) in terms of implicit complementary linear restrictions on β in the model (1.1.1). The following development is based on Chipman [15].

Definition 1.3. An $m \times k$ matrix Y will be said to be complementary to the $n \times k$ matrix X if the following two conditions are satisfied:

1. $\operatorname{rank} X + \operatorname{rank} Y = p + q = k$.

2. $uX + vY = 0$ implies $uX = vY = 0$.

Y will be said to be polar to X if condition 2 is replaced by the stronger condition

2'. $Y'X = 0$.

If Y is complementary to X and z is in the column space of Y, the equation

$$Y\beta = z \tag{1.3.3}$$

will be called a set of complementary linear restrictions for the model (1.1.1).

The following result is evident, but a proof is included for completeness.

Lemma 1.5. If Y is complementary to X, the augmented matrix $Z' = [X', Y']$ has rank k, hence $Z'Z = X'X + Y'Y$ is non-singular.

Proof. Let the $p \times k$ matrix X_1, consisting of p rows of X, constitute a basis for the row space of X, and let the $q \times k$ matrix Y_1, consisting of q rows of Y, constitute a basis for the row space of Y. Let the $1 \times p$ and

$1 \times q$ vectors u_1 and v_1 consist of the corresponding components of the $1 \times n$ and $1 \times m$ vectors u and v, and let the remaining components of u and v be zeros. Then from condition 2 of Definition 1.3, $uX + vY = 0$ implies $uX = u_1X_1 = 0$ and $vY = v_1Y_1 = 0$. Since both X_1 and Y_1 have full row rank, this implies $u_1 = 0$ and $v_1 = 0$. Therefore the hypothesis $u_1X_1 + v_1Y_1 = 0$ implies $(u_1, v_1) = 0$, whence the matrix $[X_1', Y_1']$, and therefore also the matrix $[X', Y']$, has rank k. Since Z and $Z'Z$ have the same rank (see, e.g., Afriat [2]), $X'X + Y'Y$ is non-singular.

Q. E. D.

Lemma 1.6. Let A be a generalized inverse of X satisfying properties (i) and (ii) of (‡) and property (1.1.34) of Theorem 1.5, where V^+ is a symmetric non-negative definite generalized inverse of V satisfying conditions (1.1.18) and (1.1.19) of Lemma 1.3. Then there exists a matrix Y, complementary to X, such that

$$A = (X'V^+X + Y'Y)^{-1}X'V^+ . \tag{1.3.4}$$

Conversely, if Y is complementary to X and A is defined by (1.3.4) then A satisfies properties (i) and (ii) of (‡) and (1.1.34).

Proof. Replacing X by $(V^+)^{\frac{1}{2}}X$ in Lemma 1.5 it is clear that the matrix $X'V^+X + Y'Y$ is non-singular, for any Y complementary to X. Since A satisfies properties (i) and (ii) of (‡), it has rank p (the rank of X). Let Y be any matrix polar to A', i.e., such that

$$YA = 0 \tag{1.3.5}$$

and rank $Y = q = k - p$. Then for any (u, v) such that $uX + vY = 0$,

$$uXA = uXA + vYA = 0 ,$$

hence by property (i) of (‡), $uX = 0$ and thus $vY = 0$;

therefore Y is a complementary to X. From property (i) of (‡), (1.1.34), and (1.3.5), we have

$$(X'V^{+}X + Y'Y)A = X'A'X'V^{+} = X'V^{+} , \tag{1.3.6}$$

from which (1.3.4) immediately follows.

To prove the converse, we find exactly as in Chipman [15, Lemma 1.1] that if Y is complementary to X and V^{+} is symmetric non-negative definite and satisfies (1.1.19) then

$$X(X'V^{+}X + Y'Y)^{-1}Y' = 0 . \tag{1.3.7}$$

(In the usual case in which Y is of order $q \times k$, this follows readily upon defining X_1 as in the proof of Lemma 1.5 and defining $[A_1, B]' = [X_1', Y']^{-1}$, where A_1 and B are of orders $k \times p$ and $k \times q$ respectively; then $YB = I_q$ and $X = PX_1$ for some $n \times p$ matrix P, hence $XB = 0$ and $(X'V^{+}X + Y'Y)B = Y'$, yielding (1.3.7).) From (1.3.7) it follows that if A is defined by (1.3.4),

$$\begin{aligned} AX &= (X'V^{+}X + Y'Y)^{-1}(X'V^{+}X + Y'Y - Y'Y) \\ &= I - (X'V^{+}X + Y'Y)^{-1}Y'Y . \end{aligned} \tag{1.3.8}$$

From (1.3.4), (1.3.7), and (1.3.8) it now follows that A satisfies properties (i) and (ii) of (‡); property (1.1.34) is immediate.

Q.E.D.

Note that from (1.3.6), (1.3.4), and (1.3.5) (or (1.3.7)) we have

$$X'V^{+}X(X'V^{+}X + Y'Y)^{-1}X'V^{+}X = X'V^{+}X , \tag{1.3.9}$$

so that $(X'V^{+}X + Y'Y)^{-1}$ is a generalized inverse of $X'V^{+}X$. This observation, in conjunction with Lemma 1.6, enables us to add the following information to Theorem 1.5:

<u>Theorem 1.10</u>. Let V^{+} be a symmetric non-negative

definite generalized inverse of V satisfying properties (1.1.18) and (1.1.19) of Lemma 1.3. Then a solution A of (1.1.32) satisfies properties (i), (ii), and (iii) of (‡) if and only if

$$A = (X'V^+X)^- X'V^+ , \tag{1.3.10}$$

for some generalized inverse $(X'V^+X)^-$ of $X'V^+X$, which may always be chosen to be $(X'V^+X + Y'Y)^{-1}$, for some matrix Y complementary to X.

Note that the statement of Theorem 1.10 contains the unfortunate qualification that A is a solution of (1.1.32). It does not appear to be known whether the representation (1.3.10) holds for any matrix A satisfying properties (i), (ii), (iii) of (‡), although when V is positive definite it is well known--and it follows directly from Lemma 1.6--that this is indeed the case (cf. Chipman [16, p. 122], Pringle and Rayner [83, p. 26]). The result would follow if it could be shown that any matrix A satisfying properties (i) and (iii) of (‡) is a solution of (1.1.32) for some (symmetric non-negative definite) generalized inverse V^+ of V satisfying conditions (1.1.18) and (1.1.19) of Lemma 1.3. Zyskind and Martin [113, p. 1194] claim that this is so, but the result does not appear to follow from their analysis; the difficulty is that in the proof of Lemma 1.3 the matrix $VV^{+'}$ projects the columns of Q_2 into the origin, which is only one way of projecting them into the intersection of the column spaces of X and V. (On this point see also the discussion in Pringle and Rayner [83, p. 116].)

Defining

$$B = (X'V^+X + Y'Y)^{-1} Y' , \tag{1.3.11}$$

we can obtain a relation analogous to (1.3.8) for BY, from which we deduce that B is a generalized inverse of Y satisfying properties (i), (ii), and (iii) of (†). Henceforth we shall therefore denote

$$X^{‡} = (X'V^+X + Y'Y)^{-1} X'V^+, \quad Y^{‡} = (X'V^+X + Y'Y)^{-1} Y' . \tag{1.3.12}$$

Definition 1.4. An estimator (1.1.2) of β in the model (1.1.1) is said to be conditionally unbiased subject to the complementary linear restrictions (1.3.3), where Y is complementary to X and z is in the column space of Y, if $\mathcal{E}\tilde{\beta} = \beta$ for all β satisfying (1.3.3). An estimator (1.1.2) is called a best linear conditionally unbiased estimator of β if it is conditionally unbiased and of minimum variance in the class of conditionally unbiased estimators of β subject to (1.3.3).

Lemma 1.7. In order that the estimator (1.1.2) be conditionally unbiased subject to the complementary linear restrictions (1.3.3) it is necessary and sufficient that A and c satisfy the equation

$$\mathcal{E}\tilde{\beta} = AX\beta + c = \beta + Y^{\ddagger}(z - Y\beta) \tag{1.3.13}$$

identically in β, or equivalently, that

$$c = Y^{\ddagger}z \quad \text{and} \quad AX = I - Y^{\ddagger}Y \ . \tag{1.3.14}$$

Proof. Since the equation (1.3.3) is solvable, by assumption, and $Y^{\ddagger}$ is a generalized inverse of Y, we have $YY^{\ddagger}z = z$ by Penrose's solvability theorem, and the general solution of (1.3.3) is

$$\beta = Y^{\ddagger}z + (I - Y^{\ddagger}Y)\gamma \ , \tag{1.3.15}$$

where γ is arbitrary. Since we require

$$\mathcal{E}\tilde{\beta} = AX\beta + c = \beta \quad \text{for all } \beta \text{ satisfying } Y\beta = z \ , \tag{1.3.16}$$

substituting (1.3.15) in (1.3.16) and making use of (1.3.7) we obtain

$$AX\gamma + c = \gamma + Y^{\ddagger}(z - Y\gamma) \quad \text{for all } \gamma \ , \tag{1.3.17}$$

which is the same as the second functional equation of (1.3.13). Setting $\gamma = 0$ we obtain the first equation of

(1.3.14), and substituting this back in (1.3.17) we obtain the second equation of (1.3.14).

Q.E.D.

It is interesting to note that the generalized inverse $Y^{\ddagger}$ plays the role of a Lagrangean multiplier in (1.3.13).

The following theorem provides a link between the criterion of best linear conditional unbiasedness subject to complementary linear restrictions and the criterion of Theorem 1.9.

Theorem 1.11. In order that an estimator (1.1.2) be a best linear conditionally unbiased estimator of β in the model (1.1.1), subject to the complementary linear restrictions (1.3.3), that is to say, in order that it be of minimum variance in the class of estimators (1.1.2) satisfying (1.3.14), it is necessary and sufficient that A satisfy the additional conditions (1.3.1). Any such estimator is equivalent, with probability 1 (and without qualification if V is positive definite) to the unique estimator

(1.3.18) $$\hat{\beta} = X^{\ddagger}y + Y^{\ddagger}z = (X'V^{+}X + Y'Y)^{-1}(X'V^{+}y + Y'z) ,$$

where $X^{\ddagger}$ and $Y^{\ddagger}$ are defined by (1.3.12) and are such that $X^{\ddagger}$ satisfies properties (i), (ii), and (iii) of ($\ddagger$) and $Y^{\ddagger}$ satisfies properties (i), (ii), and (iii) of ($\dagger$).

Proof. In view of Lemma 1.7, the problem is to minimize $\widetilde{\text{Var}\,\beta} = \sigma^2 AVA'$ subject to (1.3.14). Since clearly $X^{\ddagger}X + Y^{\ddagger}Y = I$ from (1.3.12), the problem is to characterize the set of solutions A to the problem

(1.3.19) $$\text{minimize } AVA' \text{ subject to } AX = X^{\ddagger}X .$$

From Lemma 1.6, $X^{\ddagger}$ satisfies properties (i) and (ii) of ($\ddagger$). Since from (1.3.7) we have

$$X'V^{+}XX^{\ddagger} = (X'V^{+}X + Y'Y - Y'Y)(X'V^{+}X + Y'Y)^{-1}X'V^{+} = X'V^{+} ,$$

$X^{\ddagger}$ is a solution of (1.1.32). Since V^{+} satisfies conditions (1.1.18) and (1.1.19) of Lemma 1.3 (by assumption), it follows from Theorem 1.5(e) that $X^{\ddagger}$ satisfies property (iii) of ($\ddagger$). Applying Lemma 1.2 to (1.3.19) it follows immediately that

$$AVA' \geqslant X^{\ddagger}VX^{\ddagger\prime} \tag{1.3.20}$$

for all A satisfying

$$AX = X^{\ddagger}X \ , \tag{1.3.21}$$

equality holding in (1.3.20) if and only if

$$AV = X^{\ddagger}V \ . \tag{1.3.22}$$

Condition (1.3.22) is necessary and sufficient for A to be a solution of (1.3.19).

It remains to show that if A satisfies (1.3.21), then it satisfies (1.3.22) if and only if it satisfies (1.3.1). If A satisfies (1.3.21) and (1.3.22) then

$$AV = X^{\ddagger}V = X^{\ddagger}XX^{\ddagger}V = AXAV \ ,$$

establishing property (ii) of (1.3.1), and

$$XAV = XX^{\ddagger}V = (XX^{\ddagger}V)' = (XAV)' \ ,$$

establishing property (iii). Conversely, if A satisfies (1.3.21) and (1.3.1) then (1.3.22) follows from (1.3.21) and the last equation in the proof of Theorem 1.9.

Finally, if the estimators $\tilde{\beta}$ and $\hat{\beta}$ are given by (1.1.2) and (1.3.18) respectively, where c satisfies (1.3.14) and A satisfies (1.3.21) and (1.3.22), then $\tilde{\beta} - \hat{\beta} = (A - X^{\ddagger})y$ hence $\mathcal{E}(\tilde{\beta} - \hat{\beta}) = 0$ from (1.3.21) and $\mathrm{Var}(\tilde{\beta} - \hat{\beta}) = 0$ from (1.3.22).

Q.E.D.

As an extremely simple example we may consider the model

$$(1.3.23)\quad \eta_t = e^{\beta_1+\beta_2 t}\,\xi_{t3}^{\beta_3}\,\xi_{t4}^{\beta_4}\,\xi_{t5}^{\beta_5}\,e^{\varepsilon_t} \qquad (t = 1, 2, \ldots, n) \;,$$

where η_t denotes the demand for butter at time t, ξ_{t3} and ξ_{t4} are respectively the price of butter and price of margarine at time t, ξ_{t5} is income at time t, and ε_t is a random error such that $\mathcal{E}(\varepsilon_t | \xi_{sj}) = 0$ for $s, t = 1, 2, \ldots, n$ and $j = 3, 4, 5$. We may assume that there is multicollinearity present which takes the form

$$(1.3.24)\quad \xi_{t4} = \lambda\xi_{t3} \quad (\lambda > 0), \quad \xi_{t5} = \mu e^{\rho t} (\mu, \rho > 0)$$

$$(t = 1, 2, \ldots, n) \;,$$

i.e., the price of butter and price of margarine remain proportional to one another, and income grows at an exponential rate. Denoting

$$(1.3.25)\quad y_t = \log \eta_t, \; x_{t1} = 1, \; x_{t2} = t, \; x_{tj} = \log \xi_{tj} \; (t = 3, 4, 5) \;,$$

we may assume that the $n \times 5$ matrix $X = [x_{tj}]$ has rank 3. Denoting

$$(1.3.26)\qquad Z = \begin{bmatrix} \log\lambda & 0 & 1 & -1 & 0 \\ \log\mu & \rho & 0 & 0 & -1 \end{bmatrix} \;,$$

we see that $XZ' = 0$, hence Z is polar to X. The estimable functionals ψ are those which are orthogonal to Z'. It is clear that no β_j is estimable unless $\lambda = \mu = 1$, and even then only β_1 is estimable. However, $\beta_3 + \beta_4$ and $\beta_2 + \rho\beta_5$ are estimable.

Now let us suppose that an investigator decides to drop the independent variables x_{t4} (logarithm of the price of margarine) and x_{t5} (logarithm of income) from the regression. This is tantamount to imposing the linear restriction $Y\beta = 0$, where

$$Y = \begin{bmatrix} 0 & 0 & 0 & 1 & 0 \\ 0 & 0 & 0 & 0 & 1 \end{bmatrix}. \tag{1.3.27}$$

Since (1.3.25) implies that the fourth and fifth columns of X are linearly dependent on the first three, Y is clearly complementary to X. Assuming that $\operatorname{Var} \varepsilon = \sigma^2 I$, the estimator

$$\hat{\beta} = (X'X + Y'Y)^{-1} X'y$$

will furnish the best linear unbiased estimator $\widehat{\psi\beta}$ of any estimable function $\psi\beta$. For instance, $\hat{\beta}_3$ will be the best linear unbiased estimator of $\beta_3 + \beta_4$. A similar result holds if (1.3.27) is replaced by

$$Y = \begin{bmatrix} 0 & 0 & -w & 1 & 0 \\ 0 & 0 & 0 & 0 & 1 \end{bmatrix}. \tag{1.3.28}$$

where $w > 0$; the restriction $Y\beta = 0$ then implies the non-estimable constraint $\beta_2 = w\beta_1$.

It is sometimes desirable to have a quick way of checking whether a matrix Y is complementary to X, given that a specified matrix Z is polar to X. If X is $n \times k$ of rank p, and Z is $q \times k$ of rank $q = k - p$, then a $q \times k$ matrix Y is complementary to X if and only if $\operatorname{rank} YZ' = q$. For, let $\operatorname{rank} YZ' = q$ and let u, v be such that $uX + vY = 0$; then

$$vYZ' = uXZ' + vYZ' = (uX + vY)Z' = 0$$

whence $v = 0$. Therefore $vY = 0$, hence Y is complementary to X. Conversely, let Y be complementary to X; then, premultiplying both sides of the equation

$$\begin{bmatrix} X \\ Y \end{bmatrix} Z' = \begin{bmatrix} 0 \\ YZ' \end{bmatrix}$$

by $(X'X + Y'Y)^{-1}[X', Y']$, we obtain

$$Z' = (X'X + Y'Y)^{-1}Y'YZ' ,$$

whence rank $YZ' \geqq$ rank $Z' = q$. Since YZ' is a $q \times q$ matrix it follows that rank $YZ' = q$.

In the above example we verify that with Y as in (1.3.28) and Z as in (1.3.26), det $YZ' = w + 1$, hence Y is complementary to X so long as $w \neq -1$.

1.4. Best linear minimum bias estimation

It remains now to provide an interpretation of property (iv) of (‡). The following criterion, which is based on a characterization due to Penrose [81], was introduced in Chipman [15]. See also Rao and Mitra [89, p. 138], Pringle and Rayner [83, p. 107], Schönfeld [94], Neeleman [117], and Rao [88].

Definition 1.5. The estimator (1.1.2) is said to be a minimum bias estimator of β in the model (1.1.1) if A and c are such as to minimize the bias matrix

$$[I - AX, c]\begin{bmatrix} W & w \\ w' & w_0 \end{bmatrix}\begin{bmatrix} I - X'A' \\ c' \end{bmatrix} , \tag{1.4.1}$$

where

$$\begin{bmatrix} W & w \\ w' & w_0 \end{bmatrix} = \begin{bmatrix} H'H & H'h \\ h'H & h'h \end{bmatrix} = \begin{bmatrix} H' \\ h' \end{bmatrix} [H, h] \tag{1.4.2}$$

is some given symmetric non-negative definite matrix, H being of order $(k + 1) \times k$ and h of order $(k + 1) \times 1$. An estimator (1.1.2) is called a best linear minimum bias estimator of β in the model (1.1.1) if it has minimum variance $\sigma^2 AVA'$ in the class of linear estimators with minimum bias.

Lemma 1.8. In order that A and c should minimize (1.4.1) it is necessary and sufficient that c satisfy, for arbitrary d,

$$c = -(I - AX)ww_0^{\dagger} + d(I - w_0 w_0^{\dagger}) \tag{1.4.3}$$

(which is equal to $-(I - AX)ww_0^{-1}$ if $w_0 > 0$, and is otherwise arbitrary), and that A minimize

$$(I - AX)\,U(I - AX)' \;, \tag{1.4.4}$$

where

$$U = W - w_0^{\dagger}ww' = [I, -ww_0^{\dagger}]\begin{bmatrix} W & w \\ w' & w_0 \end{bmatrix}\begin{bmatrix} I \\ -w_0^{\dagger}w' \end{bmatrix}. \tag{1.4.5}$$

Proof. Since necessarily $h(h'h)^{\dagger}h'h = h$, we find from (1.4.1), (1.4.2), and (1.4.5) that

$$[I - AX, c]\begin{bmatrix} W & w \\ w' & w_0 \end{bmatrix}\begin{bmatrix} I - X'A' \\ c' \end{bmatrix} = (I - AX)U(I - AX)' +$$

$$[(I - AX)H'h + ch'h](h'h)^{\dagger}[h'H(I - AX)' + h'hc'] \;.$$

This reaches a minimum with respect to c if and only if

$$ch'h = -(I - AX)H'h \tag{1.4.6}$$

--an equation which is always solvable (see Lemma 1.1). The general solution of (1.4.6) is (recalling (1.4.2)) given by (1.4.3), where d is arbitrary. It remains then to minimize (1.4.4) with respect to A .

Q. E. D.

Lemma 1.9. Let X be any $n \times k$ matrix, and let U be any $k \times k$ symmetric non-negative definite matrix. Then a $k \times n$ matrix A minimizes (1.4.4) if and only if it satisfies

$$(1.4.7) \qquad \text{(i)}\ XAXU = XU; \qquad \text{(iv)}\ AXU = UX'A' ,$$

or equivalently, if and only if it satisfies

$$(1.4.8) \qquad AXUX' = UX' .$$

A solution A to (1.4.8) always exists.

If rank XUX' = rank X, then condition (i) of (1.4.7) may be replaced by condition (i) of (‡), and in this case any solution of (1.4.8) of the form

$$(1.4.9) \qquad A = UX'(XUX')^{-}$$

also satisfies property (ii) of (‡). If $A = X^{\#}$ satisfies (1.4.7), the minimum bias is given by

$$(1.4.10) \qquad \inf_{A}(I - AX)\,U(I - AX)' = (I - X^{\#}X)U .$$

Proof. First we verify the equivalence of (1.4.7) and (1.4.8). From (1.4.7) we have $AXUX' = UX'A'X' = UX'$. Conversely, from (1.4.8) we have $UX'A' = AXUX'A' = AXU$ and thus $XAXU = XUX'A' = XU$. Now, from

$$[I - XUX'(XUX')^{-}]XUX'[I - XUX'(XUX')^{-}]' = 0$$

it follows that $[I - XUX'(XUX')^{-}]XU = 0$, which is Penrose's necessary and sufficient condition for the solvability of (1.4.8). Thus, (1.4.8) always has a solution, and the general solution is given by

$$(1.4.11) \qquad A = UX'(XUX')^{-} + Z[I - XUX'(XUX')^{-}]$$

Now let $X^{\#}$ be a solution of (1.4.8); then for any conformable matrix A we have

$$(1.4.12)\qquad \begin{aligned}&(I - AX)\,U(I - AX)'\\ &= (I - X^{\#}X)\,U(I - X^{\#}X)' + (X^{\#} - A)XUX'(X^{\#} - A)' \;.\end{aligned}$$

The second term on the right is non-negative definite, and vanishes if and only if

$$(1.4.13)\qquad (X^{\#} - A)XU = 0 \;.$$

If A satisfies (1.4.13) then $AXUX' = X^{\#}XUX' = UX'$; conversely, if A satisfies (1.4.8) then $(X^{\#} - A)XUX'(X^{\#} - A)' = 0$, which is equivalent to (1.4.13). Therefore A minimizes (1.4.12) if and only if (1.4.8) holds.

If rank XUX' = rank X then $XAX = X$ from (1.4.10) and Lemma 1.4. Choosing the particular set of solutions (1.4.9) of (1.4.11) we have $AXA = A$, again from Lemma 1.4. Expression (1.4.10) is immediate.

Q. E. D.

From the last two lemmas and Theorem 1.9 we have immediately the following result:

Theorem 1.12. A necessary and sufficient condition for the linear function (1.1.2) to be a minimum bias estimator of β in the model (1.1.1) is that A be a generalized quasi-inverse of X satisfying properties (i) and (iv) of (#) and that c satisfy (1.4.3), where U, w, and w_0 are defined by (1.4.5) and (1.4.2). If U and w_0 satisfy

$$(1.4.14)\qquad \text{rank } XUX' = \text{rank } X, \quad w_0 > 0$$

--a condition which is always fulfilled if the matrix (1.4.2) is positive definite--this reduces to the condition that A be a generalized inverse of X satisfying properties (i) and (iv) of (‡) and that

$$(1.4.15)\qquad c = -(I - AX)ww_0^{-1} \;.$$

If (1.4.14) holds, then a necessary and sufficient condition for (1.1.2) to be a best linear minimum bias estimator of β is that c satisfy (1.4.15) and that A satisfy properties (i), (iii), and (iv) of (‡) and property (ii) of (#). When V is positive definite this reduces to the condition that A satisfy all four properties of (‡).

If the matrix (1.4.2)--and hence the matrix U of (1.4.5)--is positive definite, any best linear minimum bias estimator of β is equivalent, with probability 1, to the estimator

$$\hat{\beta} = X^{\ddagger}y - (I - X^{\ddagger}X)ww_0^{-1} , \tag{1.4.16}$$

where $X^{\ddagger}$ satisfies all four properties of (‡).

Proof. An immediate consequence of Lemmas 1.8 and 1.9, Theorem 1.9, and Theorem 1.2.

Q.E.D.

In conclusion we demonstrate the equivalence of best linear minimum bias estimation and best linear conditionally unbiased estimation, when the matrix (1.4.2) is positive definite. The latter assumption is essential for such equivalence, and is a natural one under the circumstances. The requirement that $w_0 > 0$ is natural enough, since otherwise c would be arbitrary (see Lemma 1.8). Likewise, singularity of U would be tantamount to the imposition of prior linear restrictions on β--a topic we do not take up here. (On this subject see Chipman and Rao [18], Chipman [15], Goldman and Zelen [40], Pringle and Rayner [83].)

Theorem 1.13. Let (1.3.3) be a given set of complementary linear restrictions for the model (1.1.1); then there exists a matrix (1.4.2) in terms of which any best linear conditionally unbiased estimator of β subject to (1.3.3) is, with probability 1 (and without qualification if V is positive definite) a best linear minimum bias estimator of β. Conversely, given any positive definite matrix (1.4.2), there exists a set of complementary linear restrictions (1.3.3) for the model (1.1.1) such that any best linear minimum bias estimator of β is, with probability 1 (and without qualification if V is positive definite) a best linear conditionally unbiased estimator of β subject to (1.3.3).

Proof. Let the linear estimator (1.1.2) be a best linear conditionally unbiased estimator of β subject to (1.3.3). By Theorem 1.11 it is equivalent, with probability 1, to the estimator (1.3.19), where $X^{\ddagger}$ (as defined by (1.3.12)) satisfies properties (i), (ii), and (iii) of ($\ddagger$). Let the matrix (1.4.2) be defined by

$$(1.4.17)\quad U = (X'V^{+}X + Y'Y)^{-1},\quad w_0 = 1,\quad w = -Y^{\ddagger}z,\quad W = U + ww' \quad .$$

Then $X^{\ddagger}$ also satisfies property (iv) of ($\ddagger$), and (1.3.19) is equivalent to (1.4.16) on account of (1.3.7). Therefore (1.3.19) is a best linear minimum bias estimator of β when (1.4.2) is defined by (1.4.17).

Conversely, let (1.1.2) be a best linear minimum bias estimator of β, where the matrix (1.4.2) is given and positive definite. Without loss of generality let it be normalized so that $w_0 = 1$. By Theorem 1.12, (1.1.2) is then equivalent, with probability 1, to the estimator (1.4.16), where $X^{\ddagger}$ satisfies all four properties of ($\ddagger$). Since U is positive definite, U^{-1} has a smallest positive characteristic root $\lambda > 0$; likewise, $X'V^{+}X$ has a largest characteristic root $\mu \geqq 0$. Let $0 < \delta < \lambda/\mu$; then for any $k \times 1$ vector x,

$$\frac{x'X'V^{+}Xx}{x'\delta^{-1}U^{-1}x} \leqq \max_u \frac{u'X'V^{+}Xu}{u'u} \Big/ \min_u \frac{u'\delta^{-1}U^{-1}u}{u'u} = \frac{\delta\mu}{\lambda} < 1$$

(cf., e.g., Bellman [11, p. 110]), hence $\delta^{-1}U^{-1} - X'V^{+}X$ is positive definite. Defining Y and z by

$$(1.4.18)\qquad Y'Y = \delta^{-1}U^{-1} - X'V^{+}X,\quad z = -Yw \quad ,$$

(1.4.16) becomes equivalent to (1.3.19), hence (1.4.16) is a best linear conditionally unbiased estimator of β subject to (1.3.3), when Y and z are given by (1.4.18).

Q.E.D.

1.5. Bayesian estimation

In the Bayesian formulation, the $k \times 1$ vector β in (1.1.1) is considered to be a random variable distributed independently of ε. The regression equation (1.1.1) then defines the joint distribution of β and y. The unconditional probability distribution of β is known as the prior distribution, and its conditional distribution given y as the posterior distribution. Under certain conditions, the mean of the latter distribution, known as the "posterior mean", may be considered to be an acceptable "estimator" of β in the sense of being a function of the random variable y which best approximates the random variable β.

Since Bayesian analysis of the regression model is a problem in multivariate analysis, we start by considering the more general problem in terms of the joint distribution of two random vectors x_1 and x_2 in place of β and y; thus, we consider the conditional distribution of x_1 given x_2, and the problem of selecting a function of x_2 that best approximates x_1. This more general treatment will also serve as the foundation for the theory of best approximate aggregation to be developed in Part 2.

Let x_1 and x_2 be two random vectors of orders $n_1 \times 1$ and $n_2 \times 1$ respectively, with (finite) means and variances

$$\mathcal{E}\begin{bmatrix} x_1 \\ x_2 \end{bmatrix} = \begin{bmatrix} \mu_1 \\ \mu_2 \end{bmatrix}, \quad \mathrm{Var}\begin{bmatrix} x_1 \\ x_2 \end{bmatrix} = \begin{bmatrix} \Sigma_{11} & \Sigma_{12} \\ \Sigma_{21} & \Sigma_{22} \end{bmatrix}. \tag{1.5.1}$$

Definition 1.6. If x_1 and x_2 are two jointly distributed random vectors satisfying (1.5.1), the minimum mean square linear (affine) regression of x_1 on x_2 (cf.

Cramér [21, p. 272]), also called the wide sense conditional expectation of x_1 given x_2, denoted $\hat{\mathcal{E}}(x_1 | x_2)$ (cf. Doob [24, p. 77]), is any linear (affine) function

$$\tilde{x}_1 = Ax_2 + c \tag{1.5.2}$$

which minimizes the mean square error

$$\mathcal{E}(\tilde{x}_1 - x_1)(\tilde{x}_1 - x_1)' = \mathcal{E}(Ax_2 + c - x_1)(Ax_2 + c - x_1)' \tag{1.5.3}$$

in the class of linear functions (1.5.2). Such a function $\hat{\mathcal{E}}(x_1 | x_2)$ is also called the best linear (affine) predictor of x_1 given x_2.

Theorem 1.14. If x_1 and x_2 have a joint distribution satisfying (1.5.1), the minimum mean square linear regression of x_1 on x_2 is given by

$$\hat{x}_1 = \hat{\mathcal{E}}(x_1 | x_2) = \hat{A}x_2 + \mu_1 - \hat{A}\mu_2 = \mu_1 + \hat{A}(x_2 - \mu_2) \tag{1.5.4}$$

where $\hat{A}$ is any solution of the equation

$$A\Sigma_{22} = \Sigma_{12}, \tag{1.5.5}$$

which always exists. The corresponding minimum mean square error is

$$\mathcal{E}(\hat{x}_1 - x_1)(\hat{x}_1 - x_1)' = \Sigma_{11} - \hat{A}\Sigma_{21} = \Sigma_{11} - \Sigma_{12}\Sigma_{22}^{-}\Sigma_{21}, \tag{1.5.6}$$

which is unique. The random variable (1.5.4) is unique in the sense that if $\hat{A}_1$ and $\hat{A}_2$ are any two solutions of (1.5.5). the corresponding random variables (1.5.4) are equal to one another with probability 1; hence,

$$\hat{\mathcal{E}}(x_1 | x_2) = \mu_1 + \Sigma_{12}\Sigma_{22}^{-}(x_2 - \mu_2) \text{ with probability } 1. \tag{1.5.7}$$

Proof. Let a linear function be given by (1.5.2). Then we have

$$\mathcal{E}(\tilde{x}_1 - x_1)(\tilde{x}_1 - x_1)' = A\Sigma_{22}A' - A\Sigma_{21} - \Sigma_{12}A' + \Sigma_{11} +$$

$$(A\mu_2 - \mu_1 + c)(A\mu_2 - \mu_1 + c)' ,$$

which is minimized with respect to c when $c = \mu_1 - A\mu_2$, yielding

$$\mathcal{E}(\tilde{x}_1 - x_1)(\tilde{x}_1 - x_1)' = \Sigma_{11} - A\Sigma_{21} - \Sigma_{12}A' + A\Sigma_{22}A' . \tag{1.5.8}$$

In view of (1.5.4) it remains to show that A must be a solution of (1.5.5). Let $\hat{A}$ be such a solution, which must exist by virtue of Lemma 1.1; then (1.5.8) is equivalent to

$$\mathcal{E}(\tilde{x}_1 - x_1)(\tilde{x}_1 - x_1)' = \Sigma_{11} - \hat{A}\Sigma_{21} + (A - \hat{A})\Sigma_{22}(A - \hat{A})'$$

which is minimized with respect to A when $A\Sigma_{22} = \hat{A}\Sigma_{22} = \Sigma_{12}$.

Uniqueness of (1.5.6) follows from the antisymmetry of the relation $\succcurlyeq$, or directly from Lemma 1.1. Invariance of (1.5.4) follows from the fact that if $\hat{A}_1$ and $\hat{A}_2$ are any two solutions of (1.5.5), the difference between the corresponding two expressions (1.5.4) has zero mean and variance, hence they are equal with probability 1. Choosing the particular solution $\hat{A} = \Sigma_{12}\Sigma_{22}^{-}$, we obtain (1.5.7).

Q.E.D.

The above result will be found in Foster [37] and Chipman [15, 16]. The following result, generalizing the well known development to be found in Anderson [6, p. 23], will be found in Marsaglia [70], Harris and Helvig [46], Chipman [16], and Pringle and Rayner [83, p. 70].

Theorem 1.15. Let x_1 and x_2 be jointly normally distributed random vectors satisfying (1.5.1). Then (1.5.4) and (1.5.6) give the conditional mean $\mathcal{E}(x_1 | x_2)$ and conditional variance $\mathrm{Var}(x_1|x_2)$ respectively, for any solution A of (1.5.5).

Proof. Defining the linear transformation

$$\begin{bmatrix} y_1 \\ y_2 \end{bmatrix} = \begin{bmatrix} I & -A \\ 0 & I \end{bmatrix} \begin{bmatrix} x_1 \\ x_2 \end{bmatrix} , \tag{1.5.9}$$

y_1 and y_2 are jointly normally distributed, hence independent if and only if

$$0 = \mathcal{E}(y_1 - \mathcal{E}y_1)(y_2 - \mathcal{E}y_2)' = \Sigma_{12} - A\Sigma_{22} ,$$

which is equivalent to (1.5.5). Let A satisfy (1.5.5); then, since $x_2 = y_2$, and y_1 and y_2 are independent,

$$\mathcal{E}(y_1|x_2) = \mathcal{E}(y_1|y_2) = \mathcal{E}y_1 = \mu_1 - A\mu_2 .$$

Since $x_1 = y_1 + Ax_2$ from (1.5.9), we therefore have

$$\mathcal{E}(x_1|x_2) = \mathcal{E}(y_1|x_2) + Ax_2 = \mu_1 + A(x_2 - \mu_2) \tag{1.5.10}$$

as in (1.5.4). From (1.5.9) and (1.5.5) we have

$$\mathrm{Var}\begin{bmatrix} y_1 \\ y_2 \end{bmatrix} = \begin{bmatrix} I & -A \\ 0 & I \end{bmatrix} \begin{bmatrix} \Sigma_{11} & \Sigma_{12} \\ \Sigma_{21} & \Sigma_{22} \end{bmatrix} \begin{bmatrix} I & 0 \\ -A' & I \end{bmatrix} = \begin{bmatrix} \Sigma_{11} - A\Sigma_{21} & 0 \\ 0 & \Sigma_{22} \end{bmatrix} . \tag{1.5.11}$$

Since $\mathrm{Var}(x_1|x_2) = \mathrm{Var}(y_1|x_2) = \mathrm{Var}(y_1|y_2) = \mathrm{Var}\, y_1$, (1.5.11) gives the conditional variance

$$\mathrm{Var}\,(x_1|x_2) = \Sigma_{11} - A\Sigma_{21} . \tag{1.5.12}$$

Q.E.D.

In some applications it is natural to impose the restriction that the function (1. 5. 2) is homogeneous. Corresponding to Definition 1. 6 we then have

Definition 1. 7. If x_1 and x_2 are two jointly distributed vector random variables with moment matrix

(1. 5. 13)
$$\mathcal{E}\begin{bmatrix} x_1 \\ x_2 \end{bmatrix}\begin{bmatrix} x_1 \\ x_2 \end{bmatrix}' = \begin{bmatrix} M_{11} & M_{12} \\ M_{21} & M_{22} \end{bmatrix} ,$$

the best homogeneous linear predictor of x_1 given x_2, denoted $\hat{\rho}(x_1|x_2)$, is any homogeneous linear function Bx_2 which minimizes the mean square error

(1. 5. 14)
$$\mathcal{E}\,(Bx_2 - x_1)(Bx_2 - x_1)'$$

in the class of homogeneous linear functions Bx_2 .

The analogue of Theorem 1. 14 is now the following.

Theorem 1. 16. Let x_1 and x_2 be jointly distributed vector random variables with moment matrix given by (1.5.13). The best homogeneous linear predictor of x_1 given x_2 is then

(1. 5. 15)
$$\hat{\rho}(x_1|x_2) = \hat{B}x_2 ,$$

where $\hat{B}$ is any solution of the equation

(1. 5. 16)
$$BM_{22} = M_{12} ,$$

which always exists. For any such $\hat{B}$, (1. 5. 15) is unique with probability 1. The corresponding (unique) minimum mean square error is

(1. 5. 17)
$$\mathcal{E}(\hat{B}x_2 - x_1)(\hat{B}x_2 - x_1)' = M_{11} - M_{12}M_{22}^{-}M_{21} .$$

Proof. If $\hat{B}$ is a solution of (1.5.16) then for any conformable B we have

$$\mathcal{E}(Bx_2 - x_1)(Bx_2 - x_1)' = M_{11} - \hat{B}M_{21} + (B - \hat{B})M_{22}(B - \hat{B})' ,$$

which reaches a minimum if and only if $BM_{22} = \hat{B}M_{22} = M_{12}$, yielding (1.5.16) and (1.5.17). If $\hat{B}_1$ and $\hat{B}_2$ are any two solutions of (1.5.16) then defining $z = \hat{B}_1x_2 - \hat{B}_2x_2$ we have $\mathcal{E}zz' = 0$, whence $z = 0$ with probability 1 .

Q.E.D.

It is often convenient to treat affine functions as homogeneous ones by the device of introducing an extra dummy variable that takes on the single value 1 . Such a procedure will be found useful in section 2.2 below. Accordingly, if we define

$$(1.5.18) \qquad y_1 = \begin{bmatrix} 1 \\ x_1 \end{bmatrix}, \quad y_2 = \begin{bmatrix} 1 \\ x_2 \end{bmatrix}, \quad B = \begin{bmatrix} 1 & 0 \\ c & A \end{bmatrix},$$

then (1.5.2) may be written in the form $\tilde{y}_1 = By_2$. Combining (1.5.18), (1.5.13), and (1.5.1) so that

$$\begin{bmatrix} M_{11} & M_{12} \\ M_{21} & M_{22} \end{bmatrix} = \begin{bmatrix} \mu_1\mu_1' + \Sigma_{11} & \mu_1\mu_2' + \Sigma_{12} \\ \mu_2\mu_1' + \Sigma_{21} & \mu_2\mu_2' + \Sigma_{22} \end{bmatrix},$$

we conclude at once from Theorem 1.16 that the best homogeneous linear predictor of y_1 given y_2 is

$$(1.5.19) \qquad \hat{\rho}(y_1|y_2) = \hat{B}y_2 = \begin{bmatrix} 1 & 0 \\ \hat{c} & \hat{A} \end{bmatrix}\begin{bmatrix} 1 \\ x_2 \end{bmatrix} = \begin{bmatrix} 1 \\ \hat{c} + \hat{A}x_2 \end{bmatrix},$$

where $\hat{B}$ is any solution of the equation

$$(1.5.20)\quad \begin{bmatrix} 1 & 0 \\ c & A \end{bmatrix} \begin{bmatrix} 1 & \mu_2' \\ \mu_2 & \mu_2\mu_2' + \Sigma_{22} \end{bmatrix} = \begin{bmatrix} 1 & \mu_2' \\ \mu_1 & \mu_1\mu_2' + \Sigma_{12} \end{bmatrix} .$$

Taking account of the fact that $\Sigma_{12}\Sigma_{22}^{-}\Sigma_{22} = \Sigma_{12}$, from Lemma 1.1 (or Lemma 1.4), we verify that one such solution is

$$(1.5.21)\quad \hat{B} = \begin{bmatrix} 1 & 0 \\ \mu_1 - \Sigma_{12}\Sigma_{22}^{-}\mu_2 & \Sigma_{12}\Sigma_{22}^{-} \end{bmatrix} .$$

Substituting (1.5.21) in (1.5.19) we obtain

$$(1.5.22)\quad \hat{\rho}(y_1|y_2) = \begin{bmatrix} 1 \\ \mu_1 + \Sigma_{12}\Sigma_{22}^{-}(x_2 - \mu_2) \end{bmatrix} .$$

On the other hand, applying Theorem 1.14 to (1.5.18) and (1.5.1) we obtain

$$(1.5.23)\quad \hat{\mathcal{E}}(y_1|y_2) = \begin{bmatrix} 1 \\ \mu_1 \end{bmatrix} + \begin{bmatrix} 0 & 0 \\ 0 & \Sigma_{12} \end{bmatrix} \begin{bmatrix} 0 & 0 \\ 0 & \Sigma_{22} \end{bmatrix}^{-} \begin{bmatrix} 0 \\ x_2 - \mu_2 \end{bmatrix}$$

$$= \begin{bmatrix} 1 \\ \mu_1 + \Sigma_{12}\Sigma_{22}^{-}(x_2 - \mu_2) \end{bmatrix} .$$

It follows then from (1.5.22), (1.5.23), and (1.5.7) that whenever (1.5.18), (1.5.13), and (1.5.1) hold, we have

$$(1.5.24)\quad \hat{\rho}(y_1|y_2) = \hat{\mathcal{E}}(y_1|y_2) = \begin{bmatrix} 1 \\ \hat{\mathcal{E}}(x_1|x_2) \end{bmatrix} \text{ with probability } 1 .$$

This formula will be found useful in section 2.2.

The usefulness of the concept of wide sense conditional expectation stems in part from the fact that it reduces to ordinary or "strict sense" conditional expectation when the random variables are normally distributed; this follows directly from Theorems 1.14 and 1.15, which imply that if x_1 and x_2 are jointly normally distributed then $\mathcal{E}(x_1|x_2) = \hat{\mathcal{E}}(x_1|x_2)$. A second reason is that the operation $\hat{\mathcal{E}}(\cdot\,|\,\cdot)$ shares many of the properties of combination enjoyed by $\mathcal{E}(\cdot\,|\,\cdot)$. Thus, it is immediate from (1.5.7) that

$$\mathcal{E}\{\hat{\mathcal{E}}(x_1|x_2)\} = \mathcal{E}x_1 , \tag{1.5.25}$$

which is a property shared by (in fact, is the defining property of) the strict sense conditional expectation $\mathcal{E}(x_1|x_2)$. Likewise, if L_0 and L_1 are $m \times 1$ and $m \times n_1$ matrices, and $L = [L_0, L_1]$, then

$$\hat{\mathcal{E}}(L_0 + L_1x_1|x_2) = L_0 + L_1\hat{\mathcal{E}}(x_1|x_2) \text{ with probability 1} \tag{1.5.26}$$

(and $\hat{\rho}(Ly_1|y_2) = L\hat{\rho}(y_1|y_2)$); and if x_1^1 and x_1^2 are two $n_1 \times 1$ random vectors jointly distributed with x_2,

$$\hat{\mathcal{E}}(x_1^1 + x_1^2|x_2) = \hat{\mathcal{E}}(x_1^1|x_2) + \hat{\mathcal{E}}(x_1^2|x_2) \text{ with probability 1}. \tag{1.5.27}$$

These properties are also shared by the ordinary (strict sense) conditional expectation. On the other hand, the following property of conditional expectations does not extend to wide sense conditional expectations (where $n_1 = n_2$):

$$\mathcal{E}|x_1| < \infty,\ \mathcal{E}|x_2'x_1| < \infty \text{ imply } \mathcal{E}(x_2'x_1|x_2) = x_2'\mathcal{E}(x_1|x_2) \quad \text{a.s.} \tag{1.5.28}$$

(cf. Doob [24, p. 22]), where $\mathcal{E}|x_1| < \infty$ means that each component of x_1 has finite expectation. As a counterexample,† let the joint density of x_1 and x_2 (when $n_1 = n_2 = 1$)

† I am indebted to John Geweke for working out this example.

be given by the following table:

	$x_1 = 1$	$x_1 = 3$	$\mathcal{E}x_1 = 2$
$x_2 = 0$	.1	.1	.2
$x_2 = 1$	.1	.2	.3
$x_2 = 2$	.3	.2	.5
$\mathcal{E}x_2 = 1.3$	.5	.5	$\mathcal{E}x_2x_1 = 2.5$

Then we verify that the relevant functions are given by the following table:

	$x_2 = 0$	$x_2 = 1$	$x_2 = 2$
$\hat{\mathcal{E}}(x_2x_1 \mid x_2)$	16/61	121/61	226/61
$x_2\hat{\mathcal{E}}(x_1 \mid x_2)$	0	125/61	230/61
$\mathcal{E}(x_2x_1 \mid x_2)$	0	7/3	18/5
$x_2\mathcal{E}(x_1 \mid x_2)$	0	7/3	18/5

This shows that (1.5.28) does not hold if $\mathcal{E}$ is replaced by $\hat{\mathcal{E}}$. The situation is depicted in Figure 1, where for aid in reading the diagram smooth curves have been drawn through the relevant points.

A third and perhaps overriding reason for the usefulness of the concept of wide sense conditional expectation is that it furnishes the best linear (affine) approximation to the true conditional expectation. We have seen that if the underlying distribution is multivariate normal, then the two concepts actually coincide, i.e., the conditional expectation minimizes the mean square error in the class of linear (affine) functions of the conditioning variable. However, if the assumptions of normality and linearity are both dropped, and some very mild regularity assumptions are substituted,

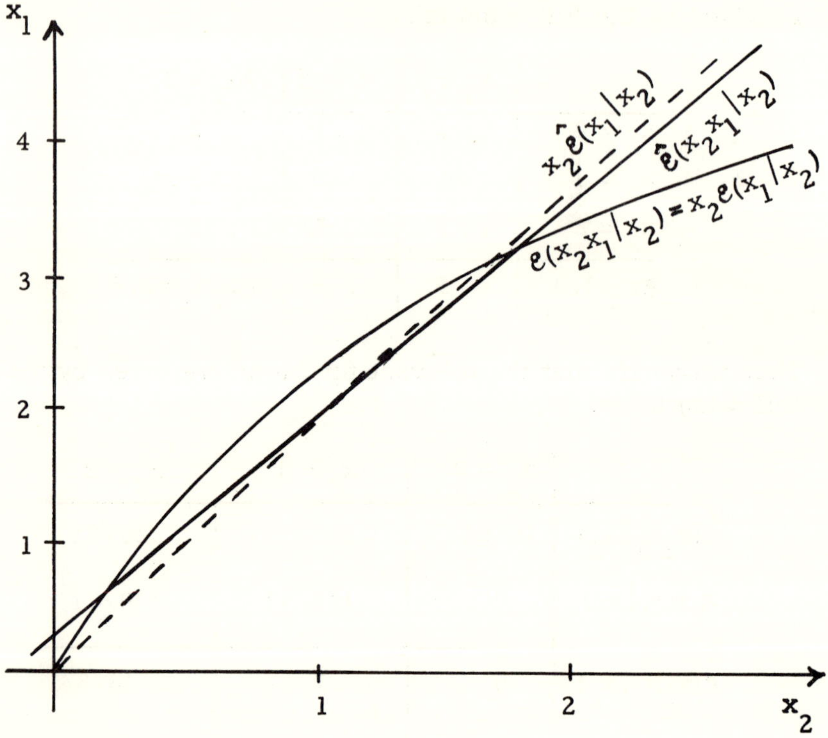

Figure 1

then we may conclude that the true (strict sense) conditional expectation minimizes the mean square error in the class of all (regular) functions of the conditioning variable. This is the substance of the following fundamental result (cf. Cramér [21, pp. 77-8], Doob [24, pp. 271-2]), which includes the previous one as a special case.

Theorem 1.17. Let x_1 and x_2 have a joint distribution satisfying (1.5.1). Then among all Borel measurable† functions

† Cf. Doob [24, p. 600].

(1. 5. 29) $$\tilde{x}_1 = f(x_2)$$

such that

(1. 5. 30) $$\mathcal{E}\tilde{x}_1\tilde{x}_1' < \infty$$

(i. e., such that $mI - \mathcal{E}\tilde{x}_1\tilde{x}_1'$ is positive definite for sufficiently large m), that which minimizes the mean square error

(1. 5. 31) $$\mathcal{E}(\tilde{x}_1 - x_1)(\tilde{x}_1 - x_1)' = \mathcal{E}(f(x_2) - x_1)(f(x_2) - x_1)'$$

is given by

(1. 5. 32) $$\hat{x}_1 = \mathcal{E}(x_1 | x_2) \quad .$$

Proof. The mean square error (1. 5. 31) may be decomposed as

$$\mathcal{E}(x_1 - \tilde{x}_1)(x_1 - \tilde{x}_1)' = \mathcal{E}[(x_1 - \hat{x}_1) + (\hat{x}_1 - \tilde{x}_1)][(x_1 - \hat{x}_1) + (\hat{x}_1 - \tilde{x}_1)]'$$

(1. 5. 33) $$= \mathcal{E}(x_1 - \hat{x}_1)(x_1 - \hat{x}_1)' + \mathcal{E}(\hat{x}_1 - \tilde{x}_1)(\hat{x}_1 - \tilde{x}_1)' -$$
$$\mathcal{E}(\hat{x}_1 - \tilde{x}_1)(x_1 - \hat{x}_1)' - \mathcal{E}(x_1 - \hat{x}_1)(\hat{x}_1 - \tilde{x}_1)' \; ,$$

where $\hat{x}_1$ is defined by (1. 5. 32). Denoting by x_{1j} the jth component of the vector x_1, we have clearly $\mathcal{E}x_{1i}^2 < \infty$ and $\mathcal{E}\tilde{x}_{1j}^2 < \infty$ from (1. 5. 1) and (1 . 5. 30) respectively, hence

$$\mathcal{E}|x_{1i}\tilde{x}_{1j}| \leqq \sqrt{\mathcal{E}x_{1i}^2}\,\sqrt{\mathcal{E}\tilde{x}_{1j}^2} < \infty$$

by the Schwarz inequality. Thus we have

(1. 5. 34) $$\mathcal{E}|x_1| < \infty, \quad \mathcal{E}|x_1\tilde{x}_1'| < \infty .$$

By a theorem of Doob [24, p. 22], inequalities (1. 5. 34) and the fact that $\tilde{x}_1 = f(x_2)$ is a Borel measurable function of x_2 imply

(1. 5. 35) $$\mathcal{E}(x_1\tilde{x}_1'|x_2) = \mathcal{E}(x_1|x_2)\tilde{x}_1' \text{ with probability 1 } .$$

Taking expectations of both sides of (1. 5. 35) we obtain

(1. 5. 36) $$\mathcal{E}\{[x_1 - \mathcal{E}(x_1|x_2)]\tilde{x}_1'\} = 0 .$$

Since $\mathcal{E}|x_1x_1'| < \infty$ from (1. 5. 1), Lemma 1. 10 below implies that (1. 5. 32) satisfies $\mathcal{E}|\hat{x}_1\hat{x}_1'| < \infty$, and thus by the same argument as used above we conclude that $\mathcal{E}|x_1\hat{x}_1'| < \infty$ and therefore

(1. 5. 37) $$\mathcal{E}\{[x_1 - \mathcal{E}(x_1|x_2)]\hat{x}_1'\} = 0 .$$

From (1. 5. 36), (1. 5. 37), and (1. 5. 32) it follows that the last term in the development on the right side of (1. 5. 33) vanishes:

(1. 5. 38) $$\mathcal{E}(x_1 - \hat{x}_1)(\hat{x}_1 - \tilde{x}_1)' = \mathcal{E}(x_1 - \hat{x}_1)\hat{x}_1' - \mathcal{E}(x_1 - \hat{x}_1)\tilde{x}_1' = 0 .$$

By transposition, the second-to-last term in (1. 5. 33) also vanishes, hence the mean square error becomes

(1. 5. 39) $$\mathcal{E}(x_1 - \tilde{x}_1)(x_1 - \tilde{x}_1)' = \mathcal{E}(x_1 - \hat{x}_1)(x_1 - \hat{x}_1)' + \mathcal{E}(\hat{x}_1 - \tilde{x}_1)(\hat{x}_1 - \tilde{x}_1)' .$$

This reaches its minimum with respect to $\tilde{x}_1$ when and only when the second term on the right vanishes, which occurs if and only if $\tilde{x}_1 = \hat{x}_1$ with probability 1.

Q. E. D.

The above result required the use of the following technical lemma.

Lemma 1.10. If x_1 and x_2 have a joint distribution with finite moment matrix (1. 5.13), then the conditional expectation (1. 5. 32) satisfies

$$\mathcal{E}\hat{x}_1\hat{x}_1' \leqslant \mathcal{E}x_1x_1' \tag{1. 5. 40}$$

(i. e., the matrix $\mathcal{E}x_1x_1' - \mathcal{E}\hat{x}_1\hat{x}_1'$ is non-negative definite). Thus, $\mathcal{E}\hat{x}_1\hat{x}_1' < \infty$.

Proof. By Jensen's inequality for conditional expectations (cf. Doob [24, p. 33]), if $F(x_1)$ is any symmetric, non-negative definite, matrix-valued convex function of the vector x_1, then

$$F(\mathcal{E}\{x_1|x_2\}) \leqslant \mathcal{E}\{F(x_1)|x_2\} \text{ with probability 1 .} \tag{1. 5. 41}$$

(Convexity of F means that if x_1^0 and x_1^1 are two values of its argument, and $0 < t < 1$, then

$$F[(1 - t)x_1^0 + tx_1^1] \leqslant (1 - t)F(x_1^0) + tF(x_1^1).\ ) \tag{1. 5. 42}$$

We verify that the function

$$F(x_1) = x_1x_1' \tag{1. 5. 43}$$

satisfies (1. 5. 42) and the remaining stipulated conditions, hence applying (1. 5. 41) to (1. 5. 43) we obtain

$$\hat{x}_1\hat{x}_1' \leqslant \mathcal{E}\{x_1x_1'|x_2\} \text{ with probability } 1 \quad . \tag{1.5.44}$$

Taking expectations of both sides of (1.5.44), we obtain (1.5.40).

Q.E.D.

We may now proceed to apply the above results to the regression model (cf. Chipman [15, 16]). Let β and ε in the regression equation (1.1.1) be independently distributed random vectors with means and variances given by

$$\mathcal{E}\begin{bmatrix}\beta\\ \varepsilon\end{bmatrix} = \begin{bmatrix}\bar{\beta}\\ 0\end{bmatrix}, \quad \mathrm{Var}\begin{bmatrix}\beta\\ \varepsilon\end{bmatrix} = \begin{bmatrix}U & 0\\ 0 & V\end{bmatrix} , \tag{1.5.45}$$

where $\bar{\beta}$ is the <u>prior mean</u> of β, and U its <u>prior variance</u>. From (1.1.1) we have

$$\begin{bmatrix}\beta\\ y\end{bmatrix} = \begin{bmatrix}I_k & 0\\ X & I_n\end{bmatrix}\begin{bmatrix}\beta\\ \varepsilon\end{bmatrix} ,$$

whence we obtain the counterpart of (1.5.1):

$$\mathcal{E}\begin{bmatrix}\beta\\ y\end{bmatrix} = \begin{bmatrix}\bar{\beta}\\ X\bar{\beta}\end{bmatrix}, \quad \mathrm{Var}\begin{bmatrix}\beta\\ y\end{bmatrix} = \begin{bmatrix}U & UX'\\ XU & XUX' + V\end{bmatrix} . \tag{1.5.46}$$

Adopting the criterion of Definition 1.6 as our criterion for good estimation, we seek a linear (affine) function

$$\tilde{\beta} = Ay + c \tag{1.5.47}$$

which minimizes the mean square error (or "Bayes risk")

$$\mathcal{E}(\tilde{\beta} - \beta)(\tilde{\beta} - \beta)' \quad . \tag{1.5.48}$$

From Theorem 1.14 our desired estimator is

$$\hat{\beta} = \hat{\mathcal{E}}(\beta \mid y) = \bar{\beta} + UX'(XUX' + V)^{-} (y - X\bar{\beta}) \quad . \tag{1.5.49}$$

This may be defined as the <u>wide sense posterior mean</u> of β given y.

With (1.5.46) in place of (1.5.1), the condition $c = \mu_1 - A\mu_2$ of Theorem 1.14 becomes $c = (I - AX)\bar{\beta}$, and (1.5.8) yields the decomposition

$$\mathcal{E}(\tilde{\beta} - \beta)(\tilde{\beta} - \beta)' = (I - AX)U(I - AX)' + AVA' \quad . \tag{1.5.50}$$

Now if U is positive definite (which means that no prior linear restrictions are imposed on β), then as $U \to \infty$ (in the sense that the smallest characteristic root of U tends to infinity), the criterion of minimization of mean square error (1.5.50), given by Definition 1.6, approaches the criterion of best linear minimum bias estimation, given by Definition 1.5. This was pointed out in Chipman [15]. In the special case in which X and V are also of full rank, this yields the justification for Gauss-Markoff estimation (or least squares estimation when the conditions of Theorem 1.7 hold) originally provided by Jeffreys [55] (see also W. D. Fisher [32], Radner [85], Chipman [15]), and which can be obtained immediately from the identity

$$X'V^{-1}(XUX' + V) = (X'V^{-1}X + U^{-1})UX' \quad ,$$

yielding

$$\hat{\mathcal{E}}(\beta \mid y) = (X'V^{-1}X + U^{-1})^{-1}X'V^{-1}y + [I - (X'V^{-1}X + U^{-1})^{-1}X'V^{-1}X]\bar{\beta} \, ,$$

and thus

$$\lim_{U \to \infty} \hat{\mathcal{E}}(\beta \mid y) = (X'V^{-1}X)^{-1}X'V^{-1}y \quad .$$

PART 2. THEORY OF AGGREGATION

2.1. The nature of the aggregation problem

The aggregation problem has been formulated in the following manner by Malinvaud [68]. Let there be a detailed model of the economy, represented as a mapping $f: \mathcal{X} \to \mathcal{Y}$ from a set $\mathcal{X}$ of explanatory or "exogenous" variables x to a set $\mathcal{Y}$ of explained or "endogenous" variables y. Typically, $\mathcal{X}$ and $\mathcal{Y}$ may be regarded as topological spaces and f as a continuous function; in practical applications, $\mathcal{X}$ and $\mathcal{Y}$ are generally vector spaces and f is frequently a linear transformation. The components of a vector $x \in \mathcal{X}$ are typically entities such as prices, incomes, resource supplies, and extra-economic variables such as indicators of climatic conditions, etc.; the components of a vector $y \in \mathcal{Y}$ are typically entities such as consumption, investment, and output.

At best, the detailed model can only be regarded as an approximate copy of the real world; even so, it is invariably much too complicated to be applied directly to empirical observations. The appropriate dimensionality of the spaces $\mathcal{X}$ and $\mathcal{Y}$, if the detailed model is to come at all close to being a true copy of reality, would run into the hundreds of thousands, or perhaps even hundreds of millions, corresponding to the immense variety of goods people produce and consume, the numerous business establishments, large and small, that are in operation, and the sizable populations in different parts of the globe, all with heterogeneous tastes. In practice, therefore, one has to deal with aggregates: price indices instead of individual prices, total income instead of individual incomes, etc. The aggregation procedure consists in mapping the spaces $\mathcal{X}$ and $\mathcal{Y}$ into spaces $\mathcal{X}^*$ and $\mathcal{Y}^*$ of smaller dimension; these mappings may be denoted $g: \mathcal{X} \to \mathcal{X}^*$ and $h: \mathcal{Y} \to \mathcal{Y}^*$ respectively. The aggregation problem is the problem of finding a mapping $f^*: \mathcal{X}^* \to \mathcal{Y}^*$ which is in some sense compatible with the mappings f, g, and h. Such a mapping is considered as defining a simplified model, or consolidated or aggregative model, to be used in applications.

By "compatibility" of the mappings f, g, h, and f^* we could mean that they must be logically consistent (in a sense shortly to be made precise), or we could mean simply that they should agree in some approximate sense. The first state of affairs will be described as one of perfect aggregation, and the second one of approximate aggregation. Perfect aggregation can never be expected to be possible; nevertheless, the conditions for perfect aggregation are worth studying because they help us understand the kinds of restrictions on the mappings or their domains that must be approximately fulfilled in order for a consolidated model to be a good approximation of a detailed one. We shall therefore start by considering this case.

As described in extenso in Theil [101] (although not in this abstract manner) there are two distinct ways of approaching the problem of perfect aggregation. One is to assume that the underlying observation set $\mathcal{X}$ is unrestricted and to look for restrictions on the structure of the mappings f, g, h, that must be imposed in order for perfect aggregation to be possible. Another is to assume that the mappings f, g, h are unrestricted and can be chosen arbitrarily, and to look for constraints that must be imposed on the structure of the observation set $\mathcal{X}$, i.e., to determine a subset $\mathcal{X}_0$ of $\mathcal{X}$ such that consistency is achieved provided that observable values of the explanatory variables are confined to this subset and thus f and g are restricted to $\mathcal{X}_0$. We shall examine these two approaches in turn, then consider mixed cases in which the mappings and the observation set are subject to compensating restrictions. Thereafter we shall consider the problem of approximate aggregation.

A. Perfect aggregation: restricted structure, unrestricted domain. It is very helpful to consider the problem with the aid of the kinds of diagrams used extensively in algebraic topology, and which were first introduced into the study of the aggregation problem by Malinvaud [67, 68] (see Figure 2).

The problem of perfect aggregation consists in determining whether, or under what conditions, this diagram commutes, i.e., the mappings f, g, h, f^* satisfy

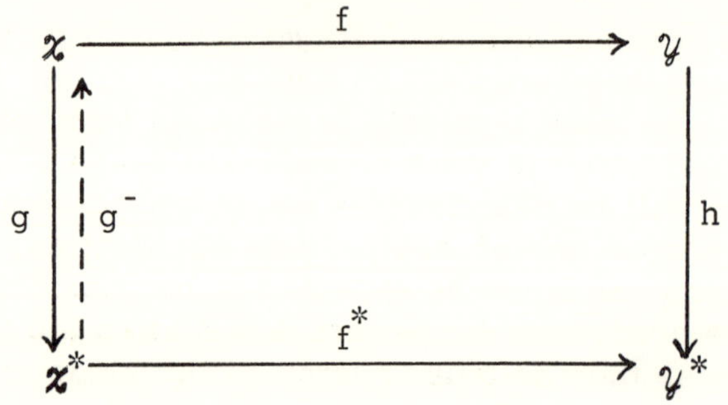

Figure 2

$$f^* g = hf \quad , \tag{2.1.1}$$

where, for instance, hf denotes the composition $h \circ f$ of the two functions f and h .

The problem as just posed is too general to be very interesting. After stating that the four functions must satisfy (2.1.1), there is little more to be said. This situation can be remedied by adding more structure to the problem, but of course there are a number of alternative ways in which this can be--and has been--done. I shall discuss three such ways.

(1) One can specify f to have certain types of properties given by economic theory, and require that f^* should have analogous properties; one then asks when there exist mappings g and h satisfying (2.1.1). This is the general approach introduced by Klein [56, 57]; it has not proved too fruitful, however, leading either to negative or to rather special results (cf. Malinvaud [68]).

(2) One can require the mappings g and h to satisfy certain special properties, and then try to interpret the meaning of (2.1.1). This approach has achieved prominence in production theory (cf. Solow [97], F. M. Fisher [28, 29, 31]). Thus, we may consider $\mathfrak{X}$ to be the non-negative orthant of n-dimensional Euclidean space, E^n_+, consisting of

vectors of inputs into a production process giving rise to a single output, say, so that $\mathcal{Y} = E^1_+$. Let the n inputs be partitioned into n^* sectors corresponding to various types of labor, various types of capital, etc., and let g be a formula for constructing an aggregate labor index, an aggregate capital index, etc., from the respective data on the corresponding detailed inputs. Then we may represent $\mathcal{X}$ as a cartesian product $\mathcal{X} = \mathcal{X}_1 \times \mathcal{X}_2 \times \ldots \times \mathcal{X}_{n^*}$, where $\mathcal{X}_i = E^{n_i}_+$ and $\sum_{i=1}^{n^*} n_i = n$, and g as a vector $g = (g_1, g_2, \ldots, g_{n^*})$, where $g_i\colon E^{n_i}_+ \to E^1_+$ and $\mathcal{X}^* = E^{n^*}_+$. Such a mapping $g\colon \mathcal{X} \to \mathcal{X}^*$ will be called a grouping mapping or grouping function. Letting $\mathcal{Y}^* = \mathcal{Y} = E^1_+$ and letting h be the identity mapping, the condition (2.1.1) simply states that for all $x = (x^1, x^2, \ldots, x^{n^*}) \in \mathcal{X}$, where $x^i \in \mathcal{X}_i$, we have

$$f(x) = f^*(g_1(x^1), g_2(x^2), \ldots, g_{n^*}(x^{n^*})) \; ; \tag{2.1.2}$$

that is to say, when g is a grouping function and h the identity mapping, (2.1.1) states that f is a separable function.

Much of the discussion in the literature is concerned with finding necessary and sufficient conditions for separability. A well known condition due to Leontief [62], which requires assuming that the mappings are twice continuously differentiable, is

$$\begin{vmatrix} \partial f/\partial x_i^r & \partial^2 f/\partial x_i^r \partial x_k^s \\ \partial f/\partial x_j^r & \partial^2 f/\partial x_j^r \partial x_k^s \end{vmatrix} = 0 \quad \text{for } s \neq r \; . \tag{2.1.3}$$

An analogous condition dispensing with the requirement of differentiability has been obtained by Stigum [99] and the question of the approximate fulfillment of the separability condition has been discussed by F. M. Fisher [30]. Further characterizations have been studied in depth by Debreu [22], Gorman [42], and Sertel [95].

(3) One can take the mappings g and h as given and ask what properties a mapping f must have in order for the equation (2.1.1) to have a continuous solution, f^*. This approach is in the spirit of early investigations by Dresch [25], May [71, 72], Pu [84], and Nataf [77], and corresponds closely to that of the fundamental work by Theil [101]; it will be the basic approach adopted here.

Suppose there exists a mapping $g^- : \mathfrak{X}^* \to \mathfrak{X}$ such that

$$gg^-g = g \tag{2.1.4}$$

(see Figure 2). Such a mapping will be called a generalized inverse of the mapping g, under composition. So long as such a g^- exists, we may apply Penrose's solvability condition [80, Theorem 2], which is valid for any associative algebra. Thus, for any g^- satisfying (2.1.4), if a solution f^* to (2.1.1) exists then clearly

$$hf = hfg^-g \quad . \tag{2.1.5}$$

Conversely, if (2.1.5) holds then

$$f^* = hfg^- \tag{2.1.6}$$

is a particular solution of (2.1.1). Thus, (2.1.5) furnishes a necessary and sufficient condition for the solvability of (2.1.1).

It is easy to see (by a simple argument using the axiom of choice) that any function g has a generalized inverse. However, assuming that f, g, and h are continuous, a condition such as (2.1.5) cannot be expected to be very useful unless g^- is continuous. In the case of a linear transformation g between vector spaces we of course know that g has a generalized inverse which is itself a linear transformation; but it is certainly not the case that any continuous function g has a continuous generalized inverse g^-. In the case of grouping mappings, however, under some simple and fairly plausible conditions a continuous generalized inverse exists which is also reflexive, i.e., satisfies $g^-gg^- = g^-$. For each $i = 1, 2, \ldots, n^*$ let there be an n_i-tuple

$(a_{i1}, a_{i2}, \ldots, a_{in_i})$ such that either $g_i(x^i) = c_i$ for all $x^i \in \mathcal{X}_i$ or else the function ψ_i defined by

$$\psi_i(\lambda) = g_i(\lambda a_{i1}, \lambda a_{i2}, \ldots, \lambda a_{in_i})$$

is a strictly monotone function of λ. Define $\psi_i^-(x_i^*) = c_i^\dagger$ if g_i is constant and $\psi_i^-(x_i^*) = \psi_i^{-1}(x_i^*)$ if ψ_i is monotone. Let

$$\text{(2.1.7)} \quad \mathcal{X} = \mathcal{X}_1 \times \mathcal{X}_2 \times \ldots \times \mathcal{X}_{n^*}, \quad \mathcal{X}^* = \mathcal{X}_1^* \times \mathcal{X}_2^* \times \ldots \times \mathcal{X}_{n^*}^*,$$

where $\mathcal{X}_i = E_+^{n_i}$, $\mathcal{X}_i^* = E_+^1$, and $\sum_{i=1}^{n^*} n_i = n$. Since g is a grouping mapping, it is defined by $g = (g_1, g_2, \ldots, g_{n^*})$ where $g_i: \mathcal{X}_i \to \mathcal{X}_i^*$. We shall define the generalized inverse mapping g^- by $g^- = (g_1^-, g_2^-, \ldots, g_{n^*}^-)$, where g_i^-: $\mathcal{X}_i^* \to \mathcal{X}_i$, $i = 1, 2, \ldots, n^*$ and each g_i^- is defined by

$$\text{(2.1.8)} \quad g_i^-(x_i^*) = (a_{i1}\, \psi_i^-(x_i^*), a_{i2}\psi_i^-(x_i^*), \ldots, a_{in_i}\, \psi_i^-(x_i^*)) .$$

If i is such that g_i is constant then clearly $g_i(g_i^-(g_i(x^i)))$ $= g_i(x^i) = c_i$ and $g_i^-(g_i(g_i^-(x_i^*))) = g_i^-(x_i^*) = (a_{i1}\, c_i^\dagger, a_{i2}\, c_i^\dagger,$ $\ldots, a_{in_i}\, c_i^\dagger)$; if i is such that ψ_i is monotone then

$$g_i(g_i^-(x_i^*)) = g_i(a_{i1}\psi_i^{-1}(x_i^*), a_{i2}\,\psi_i^{-1}(x_i^*), \ldots, a_{in_i}\,\psi_i^{-1}(x_i^*))$$

$$= \psi_i(\psi_i^{-1}(x_i^*)) = x_i^* ,$$

i.e., g_i^- is a right inverse, hence a reflexive generalized inverse, of g_i. From the definitions of g and g^- it follows immediately that g^- is a reflexive generalized inverse of g, and it is continuous.

The above sufficient condition may be expected to hold in a wide variety of cases. For instance, in particular it will hold if each g_i is an increasing function of each of its n_i arguments.

Formulas (2.1.5) and (2.1.6) may actually be used to determine the class of admissible functions f and to solve for f^*. As a simple illustration, let $\mathcal{X} = E_+^n$ consist of vectors x whose components x_i are the incomes of n individuals, and let $\mathcal{Y} = E_+^n$ consist of vectors y whose components y_i are the consumption expenditures of the same individuals. Let $f = (f_1, f_2, \ldots, f_n)$ be the function relating each individual's consumption expenditures to his own income, i.e., $y_i = f_i(x_i)$, $i = 1, 2, \ldots, n$. Define the mappings g and h by

$$x^* = g(x) = \sum_{i=1}^{n} x_i \, , \quad y^* = h(y) = \sum_{i=1}^{n} y_i \tag{2.1.9}$$

respectively. We verify that the mapping g^- defined by

$$g^-(x^*) = (x^*/n, x^*/n, \ldots, x^*/n) \tag{2.1.10}$$

satisfies (2.1.4), and that the functional equation (2.1.5) becomes

$$\sum_{j=1}^{n} f_j(x_j) = \sum_{j=1}^{n} f_j\left(\sum_{i=1}^{n} x_i/n\right) . \tag{2.1.11}$$

This is a generalization of Jensen's functional equation (cf. Aczél [1, p. 43]), and may be solved as follows on the assumption that it holds for all non-negative x_j's and that at least one f_j is continuous at some point. For each $k = 1, 2, \ldots, n$, (2.1.11) implies

$$f_k(x_k) + \sum_{j \neq k} f_j(0) = \sum_{j=1}^{n} f_j(x_k/n) \quad .$$

Combined with (2.1.11) this yields (taking $x_k = \sum_{i=1}^{n} x_i$)

$$\text{(2.1.12)} \quad \sum_{j=1}^{n} f_j(x_j) = f_k(\sum_{i=1}^{n} x_i) + \sum_{j \neq k} f_j(0) \quad (k = 1, 2, \ldots, n) \; .$$

Defining

$$\text{(2.1.13)} \qquad a_j = f_j(0), \qquad \varphi_j(x_j) = f_j(x_j) - a_j \; ,$$

(2.1.12) becomes

$$\text{(2.1.14)} \quad \sum_{j=1}^{n} \varphi_j(x_j) = \varphi_k(\sum_{i=1}^{n} x_i) \qquad \text{for } k = 1, 2, \ldots, n \; ,$$

which implies that the functions φ_k are the same, i.e.,

$$\text{(2.1.15)} \qquad \varphi_k(x) = \varphi(x) \qquad \text{for } k = 1, 2, \ldots, n \; .$$

Thus, (2.1.14) becomes

$$\text{(2.1.16)} \qquad \sum_{j=1}^{n} \varphi(x_j) = \varphi(\sum_{i=1}^{n} x_i) \; .$$

This is known as Cauchy's functional equation (cf. Aczél [1, p. 31]); provided it is assumed that φ is continuous at a point (which it must be since some f_j is), it has the general solution

$$\text{(2.1.17)} \qquad \varphi(x) = cx \; .$$

Substituting (2.1.17) and (2.1.15) in (2.1.13) we obtain

$$f_j(x_j) = a_j + cx_j \qquad (j = 1, 2, \ldots, n) \; . \tag{2.1.18}$$

Substituting (2.1.18), (2.1.9), and (2.1.10) in (2.1.6) we obtain for the aggregate function f^*

$$f^*(x^*) = \sum_{i=1}^{n} (a_i + c\frac{x^*}{n}) = \sum_{i=1}^{n} a_i + cx^* = a + cx^* \; . \tag{2.1.19}$$

This is the famous Keynesian "consumption function", and the required hypothesis (2.1.18) underlying its use has been known for a long time (cf. Hansen [45, p. 231]). Generalizations of (2.1.18) for the multi-commodity case have been obtained by Gorman [41] and Nataf [78,79], using differentiability assumptions. The strong requirement of linearity in (2.1.18) (and in general), resulting from the linearity of g and h (as in (2.1.9)), was found by Nataf [77] using differentiability assumptions; see also Green [43].

Returning to the solution (2.1.6) of (2.1.1) we may observe that even though g^- need not be unique, and thus f^* need not be invariant with respect to choice of g^-, nevertheless the restriction of f^* to its observable domain $g(\mathcal{X}) \subseteq \mathcal{X}^*$ is of course unique and invariant with respect to choice of g^-, on account of (2.1.5). If g is surjective, i.e., if $g(\mathcal{X}) = \mathcal{X}^*$, then f^* is actually invariant with respect to the choice of g^- since in that case g^- is a <u>right inverse</u> of g, i.e.,

$$gg^- = 1_{\mathcal{X}^*} \; , \tag{2.1.20}$$

where $1_{\mathcal{X}^*}$ is the identity mapping on $\mathcal{X}^*$. Thus, if g^+ is any other generalized inverse of g, since it must also satisfy (2.1.5) and (2.1.20) we have upon postmultiplying (2.1.1) by g^-,

(2.1.21) $$f^* = hfg^- = hfg^+gg^- = hfg^+ \ .$$

B. Perfect aggregation: unrestricted structure, restricted domain. Let us now suppose that the mapping $f: \mathcal{X} \to \mathcal{Y}$ is unrestricted, but that the underlying data x are restricted to some proper subset $\mathcal{X}_0 \subset \mathcal{X}$ of its domain. A typical example of this would be a situation in which prices of different commodities move up and down together, or in which the incomes of certain individuals move up and down together. Let us further assume that the mapping $g: \mathcal{X} \to \mathcal{X}^*$ has been chosen in such a way that, when restricted to $\mathcal{X}_0$, it is one-to-one (injective); denote this restricted mapping by $g_0 = g \mid \mathcal{X}_0$, where $g_0: \mathcal{X}_0 \to \mathcal{X}^*$. Then g_0 has a left inverse $g_0^-: \mathcal{X}^* \to \mathcal{X}_0$, satisfying

(2.1.22) $$g_0^- \circ g_0 = 1_{\mathcal{X}_0} \ ,$$

where $1_{\mathcal{X}_0}$ is the identity mapping on $\mathcal{X}_0$. (Note that g_0^- need not be continuous, even though g_0 is; conditions for g_0^- to be continuous will be taken up presently.) Defining the inclusion mapping $i: \mathcal{X}_0 \subseteq \mathcal{X}$, we shall define the functions $\bar{g}: \mathcal{X}^* \to \mathcal{X}$ and $p: \mathcal{X} \to \mathcal{X}$ by

(2.1.23) $$\bar{g} = i \circ g_0^-, \quad p = \bar{g} \circ g \ .$$

We note also from the definitions of g_0 and i that

(2.1.24) $$g_0 = g \circ i \ .$$

From (2.1.23), (2.1.24), and (2.1.22) it follows that $p \circ i = \bar{g} \circ g \circ i = i \circ g_0^- \circ g_0 = i$, hence

(2.1.25) $$\bar{g} \circ g \circ \bar{g} = \bar{g} \ ,$$

i.e., g is a generalized inverse of $\bar{g}$.

For any $x_0 \in \mathcal{X}_0$ we have $p(x_0) = \bar{g}(g(x_0)) = i(g_0^-(g(i(x_0)))) = g_0^-(g_0(x_0)) = x_0$, hence $x_0 \in p(\mathcal{X})$. On the other hand, if $x' = p(x)$ for some $x \in \mathcal{X}$ then $x' = \bar{g}(g(x)) = i(g_0^-(g(x))) = g_0^-(g(x)) \in \mathcal{X}_0$. Thus,

$$\mathcal{X}_0 = p(\mathcal{X}) \; (= \bar{g} \circ g(\mathcal{X})) \quad . \tag{2.1.26}$$

We define the function $p_0 \colon \mathcal{X} \to \mathcal{X}_0$ by $p_0(x) = p(x)$ for $x \in \mathcal{X}$ (see Figure 3).

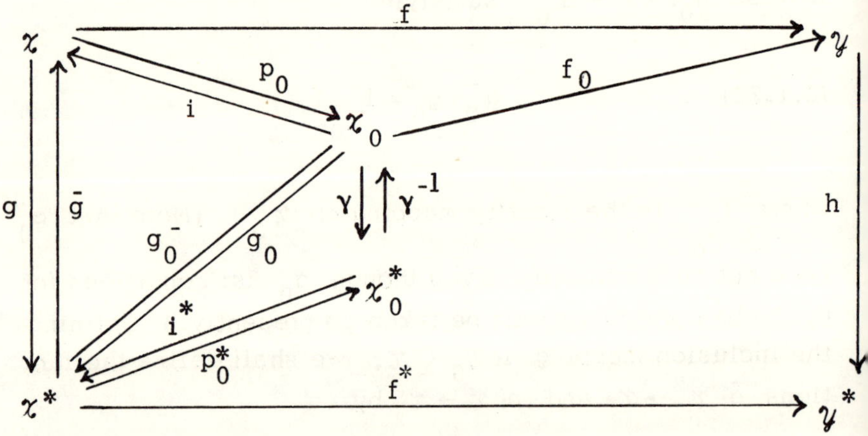

Figure 3

Given that the underlying data are restricted to $\mathcal{X}_0$, perfect aggregation requires that $f^*(g(x)) = h(f(x))$ for all $x \in \mathcal{X}_0 = p(\mathcal{X})$, i.e., that $f^* \circ g \circ p = h \circ f \circ p$, or equivalently,

$$f^* g\bar{g}g = hf\bar{g}g \quad . \tag{2.1.27}$$

Penrose's solvability condition for (2.1.27) is $hf\bar{g}g(g\bar{g}g)^{-}g\bar{g}g = hf\bar{g}g$, which is certainly satisfied in view of (2.1.25) and the fact that $\bar{g}$ is a generalized inverse of $g\bar{g}g$. A solution of (2.1.27) is

$$f^{*} = hf\bar{g} \ . \tag{2.1.28}$$

This solution has the same form as (2.1.6). Unlike the previous case, however, $\bar{g}$ need not be a generalized inverse of g, and the solution (2.1.28) is not such that $f^{*}|g(\mathcal{X})$ is invariant with respect to the choice of $\bar{g}$.

As observed above, even if g is continuous $\bar{g}$ need not be, in which case f^{*} may not be continuous. It is clearly desirable to find conditions under which $\bar{g}$ is continuous when f, g, and h are, in which case f^{*} will be also. The condition we shall impose is that the restriction of g to $\mathcal{X}_0$ induces a homeomorphism between $\mathcal{X}_0$ and a retract of $\mathcal{X}^{*}$. Specifically, let us define

$$\mathcal{X}_0^{*} = g(\mathcal{X}_0) \ (= g_0(\mathcal{X}_0)) \ . \tag{2.1.29}$$

Let $\gamma: \mathcal{X}_0 \to \mathcal{X}_0^{*}$ be the mapping defined by $\gamma(x) = g_0(x)$ $(= g(x))$ for all $x \in \mathcal{X}_0$, and let i^{*} be the inclusion mapping $i^{*}: \mathcal{X}_0^{*} \subseteq \mathcal{X}^{*}$. Then γ satisfies

$$g_0 = i^{*} \circ \gamma \tag{2.1.30}$$

(see Figure 3). We shall assume that γ is a homeomorphism, i.e., that γ is a bijection and $\gamma^{-1}: \mathcal{X}_0^{*} \to \mathcal{X}_0$ is continuous; and that $\mathcal{X}_0^{*}$ is a retract of $\mathcal{X}^{*}$, i.e., that there exists a continuous function $p_0^{*} = \mathcal{X}^{*} \supset \mathcal{X}_0^{*}$ (called a retraction of $\mathcal{X}^{*}$ onto $\mathcal{X}_0^{*}$ ----- cf. Hu [52]) with the property that

$$p_0^{*} \circ i^{*} = 1_{\mathcal{X}_0^{*}} \ , \tag{2.1.31}$$

where $1_{\mathcal{X}_0^*}$ is the identity mapping on $\mathcal{X}_0^*$. The composed function

$$g_0^- = \gamma^{-1} \circ p_0^* \tag{2.1.32}$$

is then a continuous extension of γ^{-1} over $\mathcal{X}^*$, and is thus a continuous left inverse of g_0. Consequently $\bar{g}$, as defined by (2.1.23), is continuous; furthermore, we have

$$\mathcal{X}_0 = \bar{g}(\mathcal{X}^*) \quad (= g_0^-(\mathcal{X}^*)) . \tag{2.1.33}$$

Let $p^*: \mathcal{X}^* \to \mathcal{X}^*$ be defined by $p^*(x^*) = p_0^*(x^*)$ for $x^* \in \mathcal{X}^*$. Since p_0^* is a retraction, p^* is idempotent, i.e., satisfies $p^* \circ p^* = p^*$ (cf. Hu [52, p. 17]). The converse is also true; hence, since $p: \mathcal{X} \to \mathcal{X}$, as defined by (2.1.23), is idempotent as a consequence of (2.1.25), and is continuous (since $\bar{g}$ is), it follows that $p_0: \mathcal{X} \supseteq \mathcal{X}_0$ is a retraction of $\mathcal{X}$ onto $\mathcal{X}_0$, i.e.,

$$p_0 \circ i = 1_{\mathcal{X}_0} , \tag{2.1.34}$$

where $1_{\mathcal{X}_0}$ is the identity mapping on $\mathcal{X}_0$; hence $\mathcal{X}_0$ is a retract of $\mathcal{X}$. Note that from the definitions (and from (2.1.30) and (2.1.32)) we have

$$p = i \circ p_0, \qquad p^* = i^* \circ p_0^* = g \circ \bar{g} . \tag{2.1.35}$$

A special case of interest is that in which $g(\mathcal{X}) = \mathcal{X}_0^*$. Owing to (2.1.29) and (2.1.26) this implies that $g = g_0 \circ p_0$. From (2.1.23), (2.1.34), and (2.1.22) we then have

(2.1.36) $$g \circ \bar{g} \circ g = (g_0 \circ p_0) \circ (i \circ g_0^-) \circ (g_0 \circ p_0)$$

$$= g_0 \circ (p_0 \circ i) \circ (g_0^- \circ g_0) \circ p_0 = g_0 \circ p_0 = g$$

i.e., $\bar{g}$ is a generalized inverse of g. Conversely, (2.1.36) implies $g(\mathcal{X}) = g(p(\mathcal{X})) = g(\mathcal{X}_0) = \mathcal{X}_0^*$, hence the condition $g(\mathcal{X}) = \mathcal{X}_0^*$ is necessary and sufficient for $g \circ \bar{g} \circ g = g$.

A still more special case of interest is that in which $\mathcal{X}^* = \mathcal{X}_0^*$; then $\mathcal{X}_0^*$ is trivially a retract of $\mathcal{X}^*$. In this case $i^* = p_0^* = 1_{\mathcal{X}_0^*}$ and consequently from (2.1.23), (2.1.24), (2.1.30), and (2.1.32) we have

(2.1.37) $$g \circ \bar{g} = g_0 \circ g_0^- = i^* \circ p_0^* = 1_{\mathcal{X}_0^*},$$

i.e., $\bar{g}$ is a right inverse of g. Conversely, (2.1.37) implies $\mathcal{X}^* = g(\bar{g}(\mathcal{X}^*)) = g(\mathcal{X}_0) = \mathcal{X}_0^*$ from (2.1.33) and (2.1.29), hence the condition $\mathcal{X}^* = \mathcal{X}_0^*$ is necessary and sufficient for $g \circ \bar{g} = 1_{\mathcal{X}_0^*}$. In practice, this result can always be ensured by suitable choice of g, e.g., by replacing $g: \mathcal{X} \to \mathcal{X}^*$ by $g' = \mathcal{X} \to \mathcal{X}_0^*$, where $g' = p_0^* \circ g$; then we may define $\bar{g}': \mathcal{X}_0^* \to \mathcal{X}$ by $\bar{g}' = i \circ \gamma^{-1}$, and from (2.1.24), (2.1.30), and (2.1.31) we have $g' \circ \bar{g}' = 1_{\mathcal{X}_0^*}$. This situation may be regarded as fairly typical, since if $\mathcal{X}_0^*$ has smaller dimension than $\mathcal{X}^*$ this means that the selection of $g: \mathcal{X} \to \mathcal{X}^*$ has failed to take full advantage of the possibilities of aggregation.

As a simple example we may pursue the previous illustration and suppose that the incomes of the n individuals remain proportional to one another. Then $\mathcal{X}_0$ is a ray in E^n_+, and may be represented parametrically by $\mathcal{X}_0 = \{x \in E^n_+ : x_i = \lambda_i t,\ t \geqq 0\}$, where $\lambda_i \geqq 0$ and $\sum_{i=1}^n \lambda_i = 1$. From the definition (2.1.9) of g we have $\mathcal{X}^* = E^1_+$, hence we may take $\mathcal{X}^*_0 = \mathcal{X}^*$, and $t = x^*$. The mapping $g: \mathcal{X} \to \mathcal{X}^*$ may be represented in matrix form as $x^* = Gx$, where $G = (1, 1, \ldots, 1)$, and of course the same operator represents $\gamma: \mathcal{X}_0 \to \mathcal{X}^*_0$ when its domain is restricted to $\mathcal{X}_0$. The unique inverse mapping $\gamma^{-1}: \mathcal{X}^*_0 \to \mathcal{X}_0$ is given by $\gamma^{-1}(x^*) = (\lambda_1 x^*, \lambda_2 x^*, \ldots, \lambda_n x^*)$, and $\bar{g}: \mathcal{X}^* \to \mathcal{X}$ is defined as the function that coincides with γ^{-1} but has $\mathcal{X}$ as range, i.e.,

$$\bar{g}(x^*) = (\lambda_1 x^*, \lambda_2 x^*, \ldots, \lambda_n x^*) \ . \tag{2.1.38}$$

We may represent $\bar{g}$ in matrix form by the transformation $x = \bar{G}x^*$, where $\bar{G}$ is the transpose of $(\lambda_1, \lambda_2, \ldots, \lambda_n)$; $\bar{G}$ also represents γ^{-1} when its range is considered as $\mathcal{X}_0$. We have $G\bar{G} = 1$ (the identity mapping on $\mathcal{X}^*_0$); thus $\bar{G}$ is a right inverse, hence a reflexive generalized inverse, of G. The idempotent matrix $P = \bar{G}G$ represents the projection p, and when its range is considered as $\mathcal{X}_0$ it represents the retraction p_0; furthermore it represents the identity mapping on $\mathcal{X}_0$ when restricted to $\mathcal{X}_0$ (its invariant subspace) as regards both domain and range. (This brings out the fact that generalized inverses may be considered as representations, in terms of a convenient coordinate system, of true inverses of linear transformations between subspaces of linear spaces; cf. Nashed [115, pp. 315ff.], Nashed and Votruba [116], and Kruskal [61]. They thus partake of the advantages of the "coordinate-free approach" to linear transformations without sacrificing the natural coordinate system in which, after all, empirical data are necessarily expressed.) Returning to our substantive example, we have from (2.1.28), (2.1.9), and (2.1.38) the "aggregate consumption function"

$$f^*(x^*) = \sum_{i=1}^{n} f_i(\lambda_i x^*) \ , \tag{2.1.39}$$

which may be contrasted with the special form given by (2.1.19).

C. Perfect aggregation: partially restricted structure, partially restricted domain. In general, while we can often expect the data to be restricted to a proper subset $\mathcal{X}_0$ of $\mathcal{X}$, it is generally too much to expect--unless the mapping $g: \mathcal{X} \to \mathcal{X}^*$ is especially chosen for the purpose--that the restriction of g to $\mathcal{X}_0$ will be one-to-one. However, it may be possible to factor g into $g'' \circ g'$, where $g': \mathcal{X} \to \mathcal{X}'$ and $g'': \mathcal{X}' \to \mathcal{X}^*$, in such a way that the restriction of g' to $\mathcal{X}_0$ is one-to-one. In this case the problem breaks down into a combination of the two preceding problems. The situation is depicted in Figure 4, where it is assumed that h is decomposed into a similar factorization $h = h'' \circ h'$ (which can always be done, since one can always choose one of the factors to be an identity mapping).

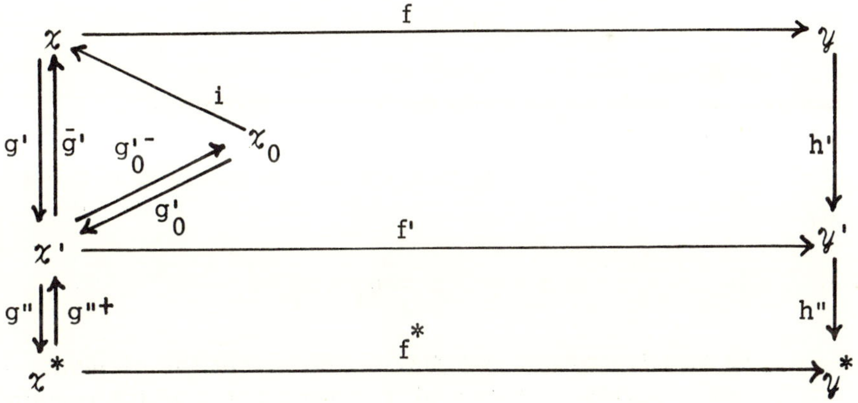

Figure 4

For definiteness let $\mathcal{X}$ and $\mathcal{X}^*$ have the form (2.1.7), and define

$$(2.1.40)\quad \mathcal{X}_I = \mathop{\times}_{i=1}^{r} \mathcal{X}_i,\ \mathcal{X}_{II} = \mathop{\times}_{i=r+1}^{n} \mathcal{X}_i,\ \mathcal{X}_I^* = \mathop{\times}_{i=1}^{r} \mathcal{X}_i^*,\ \mathcal{X}_{II}^* = \mathop{\times}_{i=r+1}^{n^*} \mathcal{X}_i^* .$$

Then

$$(2.1.41)\qquad \mathcal{X} = \mathcal{X}_I \times \mathcal{X}_{II},\qquad \mathcal{X}^* = \mathcal{X}_I^* \times \mathcal{X}_{II}^*,$$

and we shall define

$$(2.1.42)\qquad \mathcal{X}' = \mathcal{X}_I^* \times \mathcal{X}_{II},\qquad \mathcal{X}_0 = \mathcal{X}_I^0 \times \mathcal{X}_{II},$$

where $\mathcal{X}_I^0 \subset \mathcal{X}_I$. Given two mappings $g_I: \mathcal{X}_I \to \mathcal{X}_I^*$ and $g_{II}: \mathcal{X}_{II} \to \mathcal{X}_{II}^*$, we define the mappings $g': \mathcal{X} \to \mathcal{X}'$ and $g'': \mathcal{X}' \to \mathcal{X}^*$ by

$$(2.1.43)\qquad g' = g_I \times 1_{\mathcal{X}_{II}},\qquad g'' = 1_{\mathcal{X}_I^*} \times g_{II},$$

where $1_{\mathcal{X}_{II}}$ and $1_{\mathcal{X}_I^*}$ denote the identity mappings on $\mathcal{X}_{II}$ and $\mathcal{X}_I^*$ respectively. We then define the mapping $g: \mathcal{X} \to \mathcal{X}^*$ by

$$(2.1.44)\qquad g = g'' \circ g' = g_I \times g_{II} .$$

As an illustration, we may suppose that the elements of $\mathcal{X}_I$ are r-tuples of prices of r commodities which remain proportionate to one another, and are therefore confined to the one-dimensional ray $\mathcal{X}_I^0 \subset \mathcal{X}_I$; then $g_I(x_I)$ may be considered to be a price index formed from these r prices.

We now assume: (1) that the restriction of g' to $\mathcal{X}_0$ induces a homeomorphism from $\mathcal{X}_0$ to a retract of $\mathcal{X}'$, or equivalently, that the restriction of g_I to $\mathcal{X}_I^0$ induces a homeomorphism from $\mathcal{X}_I^0$ to a retract of $\mathcal{X}_I^*$; and (2) that the image of $\mathcal{X}_0$ under g' is the same as that of $\mathcal{X}$, i.e., $g'(\mathcal{X}_0) = g'(\mathcal{X})$, or equivalently, $g_I(\mathcal{X}_I^0) = g_I(\mathcal{X}_I)$. It follows from the discussion of Case B that g' has a continuous, reflexive, generalized inverse $\bar{g}': \mathcal{X}' \to \mathcal{X}$. Specifically, we define $g_0' = g' | \mathcal{X}_0$, and $g_I^0 = g_I | \mathcal{X}_I^0$, so that

$$(2.1.45) \qquad g_0 = g' \circ i = (g_I \times 1_{\mathcal{X}_{II}}) \circ (i_I \times 1_{\mathcal{X}_{II}})$$

$$= (g_I \circ i_I) \times 1_{\mathcal{X}_{II}} = g_I^0 \times 1_{\mathcal{X}_{II}}$$

where $i_I: \mathcal{X}_I^0 \subset \mathcal{X}_I$. Then we define a continuous left inverse of g_0' by $g_0'^- = g_I^{0-} \times 1_{\mathcal{X}_{II}}$, where $g_I^{0-}: \mathcal{X}_I^* \to \mathcal{X}_I^0$ is a continuous left inverse of g_I^0. The mappings $\bar{g}': \mathcal{X}' \to \mathcal{X}$ and $\bar{g}_I: \mathcal{X}_I^* \to \mathcal{X}_I$ are then defined by

$$(2.1.46) \quad \bar{g}' = i \circ g_0'^- = (i_I \times 1_{\mathcal{X}_{II}}) \circ (g_I^{0-} \times 1_{\mathcal{X}_{II}})$$

$$= (i_I \circ g_I^{0-}) \times 1_{\mathcal{X}_{II}} = \bar{g}_I \times 1_{\mathcal{X}_{II}} .$$

In the first stage we seek a mapping $f': \mathcal{X}' \to \mathcal{Y}'$ such that $f'(g'(x)) = h'(f(x))$ for all $x \in \mathcal{X}_0$. As before (see (2.1.26), (2.1.23), and (2.1.33)) we have

$$(2.1.47) \qquad \mathcal{X}_0 = \bar{g}'g'(\mathcal{X}) = \bar{g}_I g_I(\mathcal{X}_I) \times \mathcal{X}_{II} ,$$

hence we require f' to satisfy the commutativity condition $f'g'\bar{g}'g' = h'f\bar{g}'g'$. This condition always holds, since $\bar{g}'$ satisfies $\bar{g}'g'\bar{g}' = \bar{g}'$, and thus we may choose

$$f' = h'f\bar{g}' \quad . \tag{2.1.48}$$

Proceeding to the second stage, this f' is now required to satisfy $f^*g' = h''f'$, which means that f must satisfy

$$f^*g'' = hf\bar{g}' \quad . \tag{2.1.49}$$

Let us assume that g_{II} possesses a continuous, reflexive, generalized inverse as defined, say, by (2.1.8). Denoting it by $g^+_{II}: \mathcal{X}^*_{II} \to \mathcal{X}_{II}$, we define $g''^+: \mathcal{X}^* \to \mathcal{X}'$ by

$$g''^+ = 1_{\mathcal{X}^*_{II}} \times g^+_{II} \quad . \tag{2.1.50}$$

Then from (2.1.46), (2.1.50), and (2.1.43) we see that

$$\bar{g}'g''^+g''g' = \bar{g}_I g_I \times g^+_{II}\, g_{II} \quad . \tag{2.1.51}$$

Defining the mapping $g^-: \mathcal{X}^* \to \mathcal{X}$ by

$$g^+ = \bar{g}' \circ g''^+ = (\bar{g}_I \times 1_{\mathcal{X}_{II}}) \circ (1_{\mathcal{X}^*_I} \times g^+_{II}) = \bar{g}_I \times g^+_{II} \quad , \tag{2.1.52}$$

we see immediately from (2.1.51) and (2.1.52) that g^+ is a reflexive generalized inverse of g, and it is continuous. The Penrose solvability condition for (2.1.49) may therefore by written

$$hf\bar{g}' = hf\bar{g}'g''^+g'' = hfg^+g'' \quad , \tag{2.1.53}$$

and if this is fulfilled a solution of (2.1.49) is

(2.1.54) $$f^* = hfg^+ .$$

The solvability condition (2.1.53) may be stated in an equivalent form. Postmultiplying by g' it yields

(2.1.55) $$hf\bar{g}'g' = hfg^+g .$$

Conversely, observing from (2.1.52), (2.1.44), (2.1.50), (2.1.46), and (2.1.43) that

(2.1.56) $$g^+g\bar{g}' = \bar{g}_I \times g_{II}^+ g_{II} = g^+g'' ,$$

we obtain (2.1.53) back again upon postmultiplying (2.1.55) by $\bar{g}'$.

A simpler condition equivalent to (2.1.53) and (2.1.55) holds in the special case in which h is linear and f is additive, i.e., in which

(2.1.57) $$h(y_I + y_{II}) = h(y_I) + h(y_{II}) \quad \text{for } y_I, y_{II} \in \mathcal{Y}$$

and

(2.1.58) $$f(x_I, x_{II}) = f_I(x_I) + f_{II}(x_{II}) \quad \text{for } x_I \in \mathcal{X}_I , \quad x_{II} \in \mathcal{X}_{II} .$$

Since (2.1.55) can be written in the form

(2.1.59) $$hf \circ (\bar{g}_I g_I \times 1_{\mathcal{X}_{II}}) = hf \circ (\bar{g}_I g_I \times g_{II}^+ g_{II})$$

on account of (2.1.43), (2.1.44), (2.1.46), and (2.1.52), together with (2.1.57) and (2.1.58), (2.1.59) yields

$$hf_I\bar{g}_I g_I(x_I) + hf_{II}(x_{II}) = hf\bar{g}_I g_I(x_I) + hf_{II}g_{II}^+ g_{II}(x_{II}) \ ,$$

or equivalently,

$$hf_{II}(x_{II}) = hf_{II}g_{II}^+ g_{II}(x_{II}) \ . \tag{2.1.60}$$

Defining the cartesian products

$$\mathcal{X}_A = \mathcal{X}_I \times \mathcal{X}_{II}^*, \quad \mathcal{X}_A^0 = \mathcal{X}_I^0 \times \mathcal{X}_{II}^* \tag{2.1.61}$$

and the mappings $g_A: \mathcal{X} \to \mathcal{X}_A$, $g_B: \mathcal{X}_A \to \mathcal{X}^*$ by

$$g_A = 1_{\mathcal{X}_I} \times g_{II}, \quad g_B = g_I \times 1_{\mathcal{X}_{II}^*} \ , \tag{2.1.62}$$

as well as the generalized inverse mappings

$$g_A^+ = 1_{\mathcal{X}_I} \times g_{II}^+, \quad \bar{g}_B = \bar{g}_I \times 1_{\mathcal{X}_{II}^*} \ , \tag{2.1.63}$$

so that

$$g = g_B \circ g_A \quad \text{and} \quad g^+ = g_A^+ \bar{g}_B \ , \tag{2.1.64}$$

we see that when (2.1.57) and (2.1.58) hold we may write (2.1.60) in the form

$$hf = hfg_A^+ g_A = hf \circ (1_{\mathcal{X}_I} \times g_{II}^+ g_{II}) \ . \tag{2.1.65}$$

In any event, whether or not conditions (2.1.57) and (2.1.58) hold, (2.1.65) always implies (2.1.55), since

$$(2.1.66)\qquad hfg_A^+ g_A \bar{g}' g' = hf \circ (1_{\mathcal{X}_I} \times g_{II}^+ g_{II}) \circ (\bar{g}_I g_I \times 1_{\mathcal{X}_{II}})$$
$$= hfg^+ g \ .$$

A detailed analysis of the reverse factorization (2.1.64) is contained in section 2.2. (See Figure 5.)

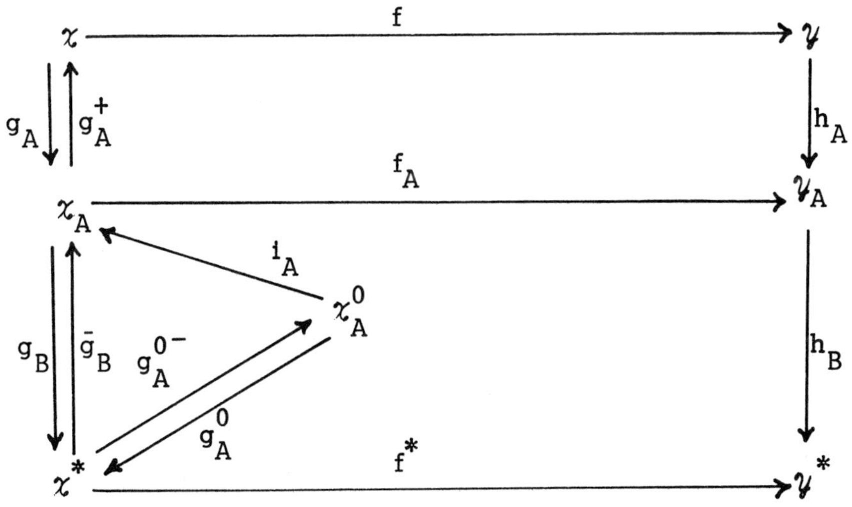

Figure 5

D. Best approximate aggregation. The conditions for perfect aggregation considered up to now cannot be expected to hold exactly. There may nevertheless be compelling practical reasons for wishing to deal with an approximative aggregate model. Accordingly, we may wish to choose a mapping $f^*: \mathcal{X}^* \to \mathcal{Y}^*$ which will fulfil the conditions for perfect aggregation as closely as possible. One way to accomplish this is to find a suitable indicator of the distance $d(f^*g, hf)$ between the two functions f^*g and hf, and to choose f^* so as to minimize this distance; such an indicator should register zero distance whenever any of the conditions for perfect aggregation is fulfilled.

This may be accomplished by introducing a stochastic structure into the formulation. Let x now be considered to be a random variable with finite first and second moments, and let f, g, and h, be such that the joint distribution of $x^* = g(x)$ and $y^* = hf(x)$ has finite first and second moments. Our (matrix) indicator of the distance between f^*g and hf, or <u>aggregation bias</u>, may be defined as the mean square error

$$d(f^*g, hf) = \mathcal{E}[f^*(x^*) - y^*][f^*(x^*) - y^*]' \tag{2.1.67}$$
$$= \mathcal{E}[f^*g(x) - hf(x)][f^*g(x) - hf(x)]' \; .$$

The criterion (2.1.67) can be given a natural interpretation. Following W. D. Fisher [34], we may suppose that there are two investigators, each of whom has a different role to play. The first investigator, we may assume, knows g and h and has sufficient information concerning past observations of $x \in \mathcal{X}$ and $y \in \mathcal{Y}$ to be able to acquire knowledge of (to "estimate") f and the distribution of x, and thus of the joint distribution of x and y. The second investigator, on the other hand, has access only to observations on $x^* \in \mathcal{X}^*$, and he wishes to use these for the purpose of forecasting y^*. It is the job of the first investigator, therefore, to furnish the second investigator with the function f^*, or at least with a means of acquiring (estimating) it on the basis of past observations on x^* and y^*.

From Theorem 1.17 it follows immediately that (2.1.67) achieves a minimum over the set of Borel measurable functions f^* such that $f^*(x^*)$ has finite first and second moments, if and only if

$$f^*(x^*) = \mathcal{E}(y^* | x^*) = \mathcal{E}\{hf(x) | g(x) = x^*\} \; . \tag{2.1.68}$$

Not suprisingly, therefore, what the second investigator needs to know is the conditional expectation of y^* given

x^* (or more generally, the conditional distribution), and this conditional expectation is precisely the function f^*. As is manifest from (2.1.68), f^* depends on f, g, and h, since these functions, together with the distribution of x, determine the joint distribution of x^* and y^*; but in general the dependence is not of any simple kind. However, if the composed function hf is <u>affine</u>, i.e., equal to the sum of a constant function and a homogeneous linear function, then (2.1.68) becomes

$$f^*(x^*) = \mathcal{E}\{hf(x)\,|\,x^*\} = hf(\mathcal{E}\{x\,|\,x^*\}) \quad . \tag{2.1.69}$$

Defining the function $g^\natural\colon \mathcal{X}^* \to \mathcal{X}$ by

$$g^\natural(x^*) = \mathcal{E}(x\,|\,x^*) \quad , \tag{2.1.70}$$

(2.1.69) then yields

$$f^* = hfg^\natural \quad , \tag{2.1.71}$$

which has the same form as (2.1.6), (2.1.28), and (2.1.54). If g in turn is affine, then in circumstances to be described in section 2.2 $g^\natural$ will be a generalized inverse, and in some cases a reflexive generalized inverse, of g.

If the second investigator is content to have an affine approximation $\hat{f}^*$ of f^*, then in accordance with Theorem 1.14 he can, from knowledge of the first and second moments of the joint distribution of x^* and y^*, replace (2.1.68) by the wide sense conditional expectation

$$\hat{f}^*(x^*) = \hat{\mathcal{E}}(y^*\,|\,x^*) \quad . \tag{2.1.72}$$

This furnishes the function which minimizes (2.1.67) in the class of all affine functions of x^*. For the case in which the functions f, g, h, and f^* are all affine, a detailed analysis is presented in section 2.2.

It is easy to verify that the matrix (2.1.67) must vanish if any of the conditions for perfect aggregation holds. This is immediate for the case (2.1.1). If $x \in \mathcal{X}_0$ then from (2.1.26) the distribution of x is concentrated in the subspace for which $x = \bar{g}(g(x))$, hence applying the expectation operator to this subspace and using (2.1.27), (2.1.67) again vanishes. A similar analysis holds for the mixed cases. Thus, when conditions for perfect aggregation hold, an f^* always exists for which (2.1.67) vanishes, and such an f^* necessarily coincides with the conditional mean (2.1.68), by virtue of Theorem 1.17.

Even though the function f^* defined by (2.1.68) cannot in general be expressed in the form (2.1.71), where $g^\natural$ is defined by (2.1.70), it is reasonable to ask whether there might exist some other function $\bar{g}: \mathcal{X}^* \to \mathcal{X}$ satisfying an analogous equation

(2.1.73) $$hf\bar{g} = f^* .$$

This question turns out to be of particular importance for the disaggregation problem, to be discussed presently. The Penrose necessary and sufficient condition for the solvability of (2.1.73) is

(2.1.74) $$hf(hf)^- f^* = f^* .$$

A sufficient condition is that $f^*(\mathcal{X}^*) \subseteq hf(\mathcal{X})$; for, denoting $\mathcal{Y}^{*\prime} = hf(\mathcal{X})$ and defining $i': \mathcal{Y}^{*\prime} \subseteq \mathcal{Y}^*$, and defining the mappings $\psi: \mathcal{X} \to \mathcal{Y}^{*\prime}$ and $\psi^*: \mathcal{X}^* \to \mathcal{Y}^{*\prime}$ by $\psi(x) = hf(x)$ and $\psi^*(x^*) = f^*(x^*)$ respectively, so that

(2.1.75) $$hf = i' \circ \psi, \quad f^* = i' \circ \psi^* ,$$

since ψ is surjective by definition, it has a right inverse $\psi^-: \mathcal{Y}^{*\prime} \to \mathcal{X}$, and we may define $\bar{g}: \mathcal{X}^* \to \mathcal{X}$ by

(2.1.76) $$\bar{g} = \psi^- \psi^*$$

(see Figure 6). Then from (2.1.75) and (2.1.76) we have

$$hf\bar{g} = i'\psi\psi^{-}\psi^{*} = i'\psi^{*} = f^{*} .$$

Since i' has a left inverse $p'\colon \mathcal{Y}^{*} \supseteq \mathcal{Y}^{*'}$ (which is a retraction if it is continuous), we may take $(hf)^{-} = \psi^{-}p'$, and (2.1.74) is verified directly.

Unfortunately, the function $\bar{g}$ of (2.1.76) need not be continuous, since ψ^{-} need not be. The problem of finding conditions under which there exists a continuous function $\bar{g}$ such that $\psi\bar{g} = \psi^{*}$ is known as the "lifting problem" (cf. Spanier [98, pp. 65-6, 74]), and such a $\bar{g}$ is called a "lifting" of ψ^{*}. I shall not attempt to go into such conditions here.

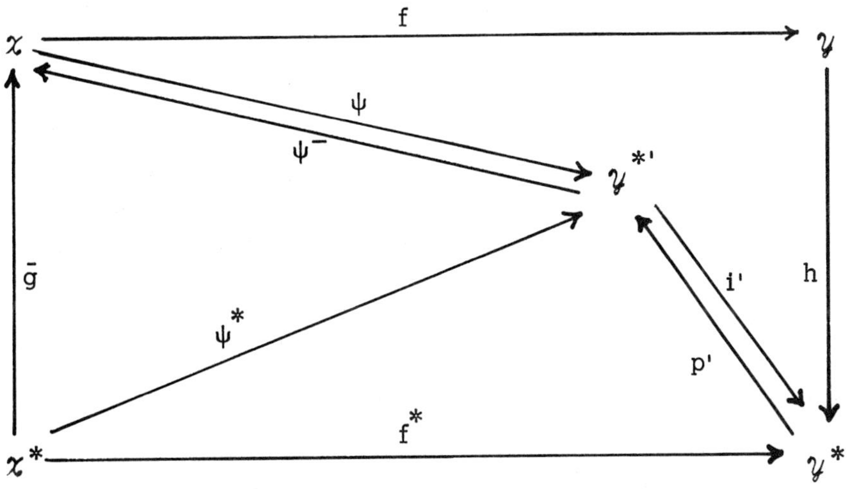

Figure 6

E. Perfect disaggregation; the "blowing-up problem". Even though it may be considered desirable--as well as sufficient for many descriptive purposes--to work with an aggregative model $f^* = \mathcal{X}^* \to \mathcal{Y}^*$ rather than the detailed one $f: \mathcal{X} \to \mathcal{Y}$, one may still wish for certain policy purposes to be able to forecast the elements of $\mathcal{Y}$ rather than simply those of $\mathcal{Y}^*$. For example, while the aggregative model may allow one to forecast the general price level or the general level of unemployment, one may still wish to have some rule, i.e., a function $h^-: \mathcal{Y}^* \to \mathcal{Y}$, which will furnish a best possible prediction of the levels of individual commodity prices or of unemployment in specific industries or regions. This is the "blowing-up problem" or disaggregation problem.

Let us suppose that an aggregative mapping $f^*: \mathcal{X}^* \to \mathcal{Y}^*$ has already been chosen, and that it is expressible in the form $f^* = hf\bar{g}$ for some mapping $\bar{g}: \mathcal{X}^* \to \mathcal{X}$. We have seen (formulas (2.1.6), (2.1.28), (2.1.54)) that this is always possible when conditions for perfect aggregation hold, and is possible under certain conditions just described when aggregation is imperfect and f^* is given by (2.1.68). In such circumstances, if an investigator had at his disposal the data x^* (but not x) and knowledge of f, a possible forecast he could make of y consistent with his aggregative model f^* would be $f(\bar{g}(x^*))$. If, however, he was either ignorant of or unwilling to work with f, he might want to find a "degrouping function" $\bar{h}$ yielding a forecast $\bar{h}(f^*(x^*))$ as close as possible to $f(\bar{g}(x^*))$. We say that conditions for perfect relative disaggregation hold if these two forecasts always coincide, i.e., if an $\bar{h}$ can be found such that

$$\bar{h}f^* = f\bar{g}, \quad \text{where} \quad f^* = hf\bar{g} \; . \tag{2.1.77}$$

The Penrose solvability condition for (2.1.77) is

$$f\bar{g}(hf\bar{g})^-hf\bar{g} = f\bar{g} \; , \tag{2.1.78}$$

and if it holds a solution of (2.1.77) is

(2.1.79) $$\bar{h} = f\bar{g}(hf\bar{g})^{-} = f\bar{g}f^{*-} .$$

Actually, a somewhat milder criterion for perfect relative disaggregation may be substituted for (2.1.77). Since only a subset $g(\mathcal{X})$ of $\mathcal{X}$ will actually be observed, we need only require that $\bar{h}f^{*}g(x) = f\bar{g}g(x)$ for all $x \in \mathcal{X}$, or

(2.1.80) $$\bar{h}hf\bar{g}g = f\bar{g}g .$$

The corresponding solvability condition is of course

(2.1.81) $$f\bar{g}g(hf\bar{g}g)^{-}hf\bar{g}g = f\bar{g}g ,$$

with the formal solution

(2.1.82) $$\bar{h} = f\bar{g}g(hf\bar{g}g)^{-} .$$

Clearly (2.1.77) implies (2.1.80), and if $\bar{g}$ satisfies $\bar{g}g\bar{g} = \bar{g}$ as in (2.1.25) then the two conditions are equivalent. This will be the case if conditions for perfect aggregation hold that are of type B, or if they are of type A and $\bar{g} = g^{-}$ is chosen to be a reflexive generalized inverse of g. Thus, in many cases there would be no loss of generality in considering (2.1.77) as our criterion for perfect relative disaggregation in place of the less stringent condition (2.1.80); however, it is just as easy to deal with the latter, hence we shall do so and settle on (2.1.80) as the pertinent criterion. It should be noted that in considering the problem of perfect relative disaggregation there is no need to assume that the aggregation is itself perfect; we need only assume in what follows that f^{*} is some given mapping of the form $f^{*} = hf\bar{g}$, where $\bar{g}$ is continuous.

To Malinvaud [68, p. 112] is due the insight that the possibility of perfect relative disaggregation hinges on the dimensionality of the image space $f\bar{g}g(\mathcal{X})$ (or the image space $f\bar{g}(\mathcal{X}^{*})$ in case the criterion (2.1.77) is used) not exceeding that of $\mathcal{Y}^{*}$. Let us denote this image space by

(2.1.83) $$\mathcal{Y}_0 = f\bar{g}g(\mathcal{X}) .$$

Our <u>sufficient condition for perfect relative disaggregation</u> will be that <u>the restriction of</u> h <u>to</u> $\mathcal{Y}_0$ <u>induces a homeomorphism from</u> $\mathcal{Y}_0$ <u>to a retract of</u> $\mathcal{Y}^*$. Specifically, let us define

(2.1.84) $$\mathcal{Y}_0^* = h(\mathcal{Y}_0) = hf\bar{g}g(\mathcal{X}) .$$

Let $\varphi: \mathcal{X} \to \mathcal{Y}_0$ be the mapping that agrees with $f\bar{g}g$ on $\mathcal{X}$, i.e., $\varphi(x) = f\bar{g}g(x)$ for $x \in \mathcal{X}$, and let j be the inclusion mapping $j: \mathcal{Y}_0 \subseteq \mathcal{Y}$. Then

(2.1.85) $$f \circ \bar{g} \circ g = j \circ \varphi .$$

Further, let $h_0: \mathcal{Y}_0 \to \mathcal{Y}^*$ and $\eta: \mathcal{Y}_0 \to \mathcal{Y}_0^*$ be defined as the mappings that agree with h, i.e., $h_0(y) = \eta(y) = h(y)$ for all $y \in \mathcal{Y}_0$, and let j^* be the inclusion mapping $j^*: \mathcal{Y}_0^* \subseteq \mathcal{Y}^*$, so that

(2.1.86) $$h_0 = h \circ j = j^* \circ \eta .$$

Our postulate is that the inverse mapping $\eta^{-1}: \mathcal{Y}_0^* \to \mathcal{Y}_0$ exists and is continuous, and that j^* has a continuous left inverse $q_0^*: \mathcal{Y}^* \to \mathcal{Y}_0^*$ (so that $\mathcal{Y}_0^*$ is a retract of $\mathcal{Y}^*$), i.e.,

(2.1.87) $$q_0^* \circ j^* = 1_{\mathcal{Y}_0^*} ,$$

where $1_{\mathcal{Y}_0^*}$ is the identity mapping on $\mathcal{Y}_0^*$. (See Figure 7.)

We define

(2.1.88) $$h_0^- = \eta^{-1} \circ q_0^* ; \ \bar{h} = j \circ h_0^- ; \ q_0 = h_0^- \circ h .$$

Finally, we define the mappings $q: \mathcal{Y} \to \mathcal{Y}$ and $q^*: \mathcal{Y}^* \to \mathcal{Y}^*$ by

(2.1.89) $$q = j \circ q_0 = \bar{h} \circ h ; \quad q^* = j^* \circ q_0^* = h \circ \bar{h} .$$

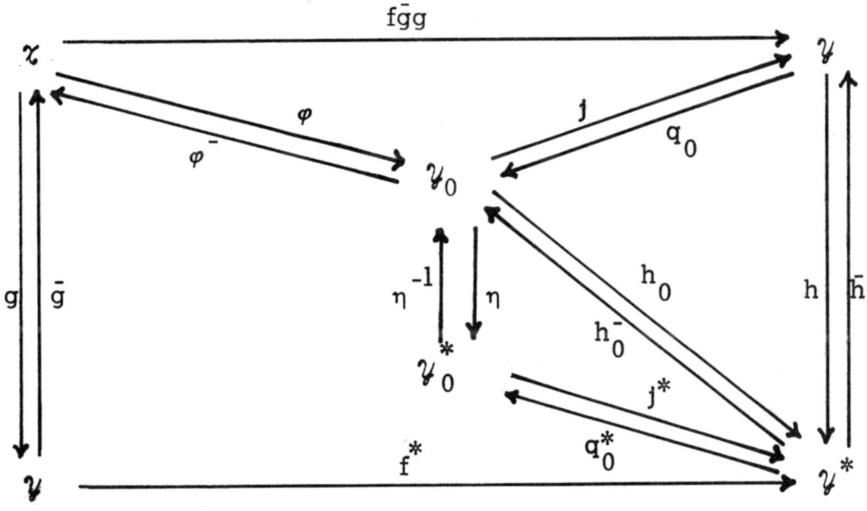

Figure 7

From the above definitions we obtain immediately

(2.1.90) $$q_0 \circ j = h_0^- \circ h \circ j = h_0^- \circ h_0 = \eta^{-1} \circ q_0^* \circ j^* \circ \eta$$
$$= 1_{\mathcal{Y}_0} ,$$

so that q_0 is a retraction of $\mathcal{Y}$ onto $\mathcal{Y}_0$ (hence $\mathcal{Y}_0$ is a retract of $\mathcal{Y}$) and h_0^- is a continuous left inverse of h_0. From (2.1.88), (2.1.86), and (2.1.90) we then obtain

(2.1.91) $$\bar{h} \circ h \circ j = j ,$$

and together with (2.1.88) this implies that

(2.1.92) $$\bar{h} \circ h \circ \bar{h} = \bar{h} ,$$

i.e., that h is a generalized inverse of $\bar{h}$. (Thus, the mappings q and q^* of (2.1.89) are idempotent.) From (2.1.85) and (2.1.91) it now follows that

$$\bar{h}hf\bar{g}g = \bar{h}hj\varphi = j\varphi = f\bar{g}g \ , \tag{2.1.93}$$

i.e., $\bar{h}$, as defined by (2.1.88), satisfies (2.1.80).

Let us now confirm that the solution $\bar{h}$ to (2.1.80) just obtained may be expressed in the form (2.1.82). First we may verify that the solvability condition (2.1.81) holds. Since φ is surjective by definition, it possesses a right inverse $\varphi^{-}: \mathcal{Y}_0 \to \mathcal{X}$, i.e.,

$$\varphi \circ \varphi^{-} = 1_{\mathcal{Y}_0} \ . \tag{2.1.94}$$

(Note that φ^{-} need not, and in general will not, be continuous, as the example $\mathcal{X} = E^1$, $\varphi(x) = x^3 - x^2$ shows.) We then verify with the help of (2.1.85), (2.1.86), (2.1.90), and (2.1.94) that one choice for a generalized inverse of $hf\bar{g}g$ is

$$(hf\bar{g}g)^{-} = (hj\varphi)^{-} = \varphi^{-}h_0^{-} \ . \tag{2.1.95}$$

Substituting (2.1.95) and (2.1.85) into the left side of (2.1.81) and using (2.1.90), (2.1.91), and (2.1.94), we obtain the expression on the right. Formula (2.1.82) results from

$$f\bar{g}g(hf\bar{g}g)^{-} = j\varphi(hj\varphi)^{-} = j\varphi\varphi^{-}h_0^{-} = jh_0^{-} = \bar{h} \ . \tag{2.1.96}$$

Note that $\bar{h}$ is continuous, since h_0^{-} is, even though φ^{-} and hence $(hf\bar{g}g)^{-}$ need not (and in general will not) be.

F. <u>Best approximate disaggregation</u>. If the condition (2.1.80) for perfect relative disaggregation does not hold exactly for any $\bar{h}$, we may introduce an indicator of the discrepancy analogous to (2.1.67). Supposing x to be a random variable with finite first and second moments, let us assume that the random variables defined by

$$(2.1.97) \qquad \tilde{y} = f\bar{g}g(x), \quad \hat{y}^* = f^*g(x) = hf\bar{g}g(x)$$

have a joint distribution with finite first and second moments. Our indicator of <u>relative disaggregation bias</u> is then

$$(2.1.98) \qquad d(\bar{h}f^*g, f\bar{g}g) = \mathcal{E}[\bar{h}(\hat{y}^*) - \tilde{y}][h(\hat{y}^*) - \tilde{y}]'$$

$$= \mathcal{E}[\bar{h}f^*g(x) - f\bar{g}g(x)][\bar{h}f^*g(x) - f\bar{g}g(x)]'.$$

From Theorem 1.17 it immediately follows that a necessary and sufficient condition for (2.1.98) to achieve a minimum over the set of Borel measurable functions $\bar{h}$ such that $\bar{h}(\hat{y}^*)$ has finite first and second moments is that $\bar{h}$ be given by

$$(2.1.99) \qquad \bar{h}(\hat{y}^*) = \mathcal{E}(\tilde{y} \mid \hat{y}^*) = \mathcal{E}\{f\bar{g}g(x) \mid f^*g(x) = \hat{y}^*\} .$$

In dealing with the problem of best approximate disaggregation, there is no reason to limit oneself to obtaining a good approximation to $\tilde{y} = f\bar{g}g(x)$ rather than to $y = f(x)$ itself, even though one could rarely expect to obtain a perfect approximation in the latter case. The appropriate criterion is then given by an indicator of <u>absolute disaggregation bias</u>, defined as

$$(2.1.100) \qquad d(\bar{h}f^*g, f) = \mathcal{E}[\bar{h}(\hat{y}^*) - y][\bar{h}(\hat{y}^*) - y]'$$

$$= \mathcal{E}[\bar{h}f^*g(x) - f(x)][\bar{h}f^*g(x) - f(x)]' .$$

In accordance with Theorem 1.17, a necessary and sufficient condition for (2.1.99) to achieve a minimum over the set of Borel measurable functions $\bar{h}$ for which $\bar{h}(\hat{y}^*)$ has finite first and second moments is that $\bar{h}$ be given by

$$(2.1.101) \qquad \bar{h}(\hat{y}^*) = \mathcal{E}(y \mid \hat{y}^*) = \mathcal{E}\{f(x) \mid f^*g(x) = \hat{y}^*\} .$$

It is shown in section 2.2 below that when f, g, and h are affine and $\bar{g}$ is given by $g^\natural$ in (2.1.70), the two disaggregation functions (2.1.99) and (2.1.101) coincide.

2.2. Linear aggregation and disaggregation.

Let $\mathcal{X}, \mathcal{Y}, \mathcal{X}^*, \mathcal{Y}^*$ be vector spaces of dimensions n, m, n^*, m^* respectively, and let f, g, h, f^* be linear transformations. The latter may be written in matrix form

$$y = Fx, \quad x^* = Gx, \quad y^* = Hy, \quad y^* = F^* x^* . \tag{2.2.1}$$

It is in the nature of the case that $n^* < n$ and $m^* \leqq m$.

Typically, G will be a grouping matrix, defined as a matrix (of order $n^* \times n$) with non-negative elements and at most one-zero element in each column; likewise with H . (The transpose of a grouping matrix will also be called a grouping matrix.) For an appropriate numbering of the explanatory variables, G may then be written in the form of a block diagonal matrix

$$G = \begin{bmatrix} g_1 & 0 & \cdots & 0 \\ 0 & g_2 & \cdots & 0 \\ \cdots & \cdots & \cdots & \cdots \\ 0 & 0 & \cdots & g_{n^*} \end{bmatrix} \tag{2.2.2}$$

where each g_i is a row vector with n_i non-negative components, and $\sum_{i=1}^{n^*} n_i = n$. For example, if the first n_1 components of x are prices, it would be natural for the components of g_1 to be non-negative weights summing to 1 , forming a price index x_1^*; if the next n_2 components of x are individual incomes, the appropriate g_2 would be a row vector of ones, so that x_2^* would be aggregate income; and so on. In general, since linear systems are typically not homogeneous, a component of x^* and one or more components of x will be the dummy variable 1 .

A generalized inverse of G will be denoted G^-, and defined as any matrix satisfying

$$GG^-G = G . \tag{2.2.3}$$

Such a matrix always exists, and in the case of grouping matrices the block diagonal matrix whose diagonal blocks are column vectors g_i^-, where $g_i g_i^- g_i = g_i$, is a generalized inverse of G.

Note that as long as $g_i \neq 0$ for $i = 1, 2, \ldots, n^*$, a grouping matrix G must have the full rank n^*, hence

$$GG^- = I \tag{2.2.4}$$

i.e., G^- is a right inverse of G. This can generally be expected to be the case in applications, since it would be pointless to carry a row of zeros in G when instead the row could be removed and the dimensionality of χ^* correspondingly reduced. However, since most of the analysis that follows does not depend on G being a grouping matrix, we shall consider the general case in which G can have any rank $\leqq n^*$, although special attention will be paid to the case (2.2.4).

We may now consider the alternative approaches discussed in the preceding section.

A. Perfect aggregation: restricted structure, unrestricted domain. The consistency requirement corresponding to (2.1.1) is of course

$$F^*G = HF . \tag{2.2.5}$$

By Theorem 2 of Penrose [80], a solution F^* of (2.2.5) exists if and only if

$$HF = HFG^-G , \tag{2.2.6}$$

where G^- is any generalized inverse satisfying (2.2.3), in which case the general solution of (2.2.5) is

$$F^* = HFG^- + Z^*(I - GG^-) , \tag{2.2.7}$$

where Z^* is arbitrary. We are free to choose $Z^* = 0$, so a subclass of solutions of (2.2.5) is of course

$$F^* = HFG^- \quad . \tag{2.2.8}$$

If G has rank n^*, so that (2.2.4) holds, then it is clear from (2.2.7) that (2.2.8) is the general solution of (2.2.5). Moreover, <u>the solution</u> (2.2.8) is <u>unique, independently of the choice of</u> G^-. This follows from the fact that (2.2.6) must also hold for any other generalized inverse of G, e.g. the Moore-Penrose inverse $G^\dagger$, hence from (2.2.6) and (2.2.4), (2.2.8) becomes

$$F^* = HFG^- = HFG^\dagger GG^- = HFG^\dagger \ , \tag{2.2.9}$$

which is unique.

The Penrose solvability condition (2.2.6) is expressed as a redundant set of homogeneous bilinear restrictions $HF(I - G^-G) = 0$ on the elements of F. It is sometimes convenient in order to interpret the meaning of these restrictions, or for purposes of hypothesis testing, to replace $I - G^-G$ by a basis R for its column space. Equivalently, let a <u>restriction matrix</u> R be defined as any $n \times (n-q)$ matrix of rank $n-q$, where $q \leq n^*$ is the rank of G, such that $GR = 0$ (i.e., R' is polar to G). Then (2.2.6) is equivalent to the condition

$$HFR = 0 \quad . \tag{2.2.10}$$

As an illustration, let

$$G = \begin{bmatrix} 1 & 1 & 0 & 0 & 0 \\ 0 & 0 & 1 & 1 & 1 \end{bmatrix} , \qquad H = \begin{bmatrix} 1 & 1 & 0 & 0 \\ 0 & 0 & 1 & 1 \end{bmatrix} .$$

Then we have, say,

$$HFR = \begin{bmatrix} 1 & 1 & 0 & 0 \\ 0 & 0 & 1 & 1 \end{bmatrix} \left[\begin{array}{cc|ccc} f_{11} & f_{12} & f_{13} & f_{14} & f_{15} \\ f_{21} & f_{22} & f_{23} & f_{24} & f_{25} \\ \hline f_{31} & f_{32} & f_{33} & f_{34} & f_{35} \\ f_{41} & f_{42} & f_{43} & f_{44} & f_{45} \end{array}\right] \begin{bmatrix} 1 & 0 & 0 \\ -1 & 0 & 0 \\ 0 & 0 & 1 \\ 0 & -1 & 0 \\ 0 & 0 & -1 \end{bmatrix}$$

$$= \begin{bmatrix} f_{11}+f_{21}-(f_{12}+f_{22}) & f_{13}+f_{23}-(f_{14}+f_{24}) & f_{13}+f_{23}-(f_{15}+f_{25}) \\ f_{31}+f_{41}-(f_{32}+f_{42}) & f_{33}+f_{43}-(f_{34}+f_{44}) & f_{33}+f_{43}-(f_{35}+f_{45}) \end{bmatrix}.$$

Thus, (2.2.10) states that each block of F must have equal column sums. The same conclusion can of course also be read off from (2.2.5), which states that the rows of HF are contained in the row space of G :

$$\left[\begin{array}{cc|ccc} f_{11}+f_{21} & f_{12}+f_{22} & f_{13}+f_{23} & f_{14}+f_{24} & f_{15}+f_{25} \\ \hline f_{31}+f_{41} & f_{32}+f_{42} & f_{33}+f_{43} & f_{34}+f_{44} & f_{35}+f_{45} \end{array}\right]$$

$$= \begin{bmatrix} f^*_{11} & f^*_{12} \\ f^*_{21} & f^*_{22} \end{bmatrix} \begin{bmatrix} 1 & 1 & 0 & 0 & 0 \\ 0 & 0 & 1 & 1 & 1 \end{bmatrix} = \left[\begin{array}{cc|ccc} f^*_{11} & f^*_{11} & f^*_{12} & f^*_{12} & f^*_{12} \\ \hline f^*_{21} & f^*_{21} & f^*_{22} & f^*_{22} & f^*_{22} \end{array}\right].$$

The elements of F^* are thus the common column sums of the corresponding blocks of F .

B. Perfect aggregation: unrestricted structure, restricted domain. Let the underlying data be assumed to be restricted to a subspace $\mathcal{X}_0$ of $\mathcal{X}$ having the property that G induces a one-to-one correspondence between $\mathcal{X}_0$ and a linear subspace $\mathcal{X}_0^*$ of $\mathcal{X}^*$. Then

(2.2.11) $$G\mathcal{X}_0 = \mathcal{X}_0^* = P^*\mathcal{X}^*, \quad \text{where} \quad P^{*2} = P^* ,$$

and there exists an $n \times n^*$ matrix $\bar{\Gamma}$ such that

(2.2.12) $$\bar{\Gamma}\mathcal{X}_0^* = \mathcal{X}_0 .$$

Thus, $\bar{\Gamma}G\mathcal{X}_0 = \mathcal{X}_0$ and $G\bar{\Gamma}\mathcal{X}_0^* = \mathcal{X}_0^*$, hence the matrices G and $\bar{\Gamma}$ represent the mappings $\gamma: \mathcal{X}_0 \to \mathcal{X}_0^*$ and $\gamma^{-1}: \mathcal{X}_0^* \to \mathcal{X}_0$ when confined to the respective domains and ranges of the latter. Defining

(2.2.13) $$\bar{G} = \bar{\Gamma} P^* ,$$

we have $\bar{G}P^* = \bar{G}$ from the idempotency of P^*, hence

(2.2.14) $$\mathcal{X}_0 = \bar{\Gamma}\mathcal{X}_0^* = \bar{\Gamma}P^*\mathcal{X}^* = \bar{G}\mathcal{X}^*$$

and thus

(2.2.15) $$\bar{G}G\mathcal{X}_0 = \bar{G}\mathcal{X}_0^* = \bar{G}P^*\mathcal{X}^* = \bar{G}\mathcal{X}^* = \mathcal{X}_0 .$$

Consequently, $\bar{G}G\bar{G}\mathcal{X}^* = \bar{G}G\mathcal{X}_0 = \mathcal{X}_0 = \bar{G}\mathcal{X}^*$, that is,

(2.2.16) $$\bar{G}G\bar{G} = \bar{G} .$$

From (2.2.14) and (2.2.11) we have $G\bar{G}\mathcal{X}^* = G\mathcal{X}_0 = P^*\mathcal{X}^*$, hence

(2.2.17) $$P^* = G\bar{G} .$$

Defining $P = \bar{G}G$, we have $P^2 = P$ from (2.2.16); further, from (2.2.15) we have $\mathcal{X}_0 \subseteq \bar{G}G\mathcal{X}$, and from (2.2.14) we have $\bar{G}G\mathcal{X} \subseteq \bar{G}\mathcal{X}^* = \mathcal{X}_0$, hence

(2.2.18) $$\mathcal{X}_0 = \bar{G}G\mathcal{X} = P\mathcal{X}, \quad \text{where} \quad P^2 = P .$$

Thus, the subspace $\mathcal{X}_0 \subset \mathcal{X}$ is characterized by

(2.2.19) $$\mathcal{X}_0 = \{x \in \mathcal{X}:\ x = \bar{G}Gx\} ,$$

where $\bar{G}$ satisfies (2.2.16).

Our criterion for perfect aggregation is now

$$F^* Gx = HFx \text{ for all } x \in \mathcal{X}_0 .$$

In view of (2.2.19) this is equivalent to

(2.2.20) $$F^* G\bar{G}G = HF\bar{G}G ,$$

as in (2.1.27). Since (2.2.16) implies that $\bar{G}$ is a generalized inverse of $G\bar{G}G$, Penrose's solvability condition for (2.2.20) is necessarily satisfied and the general solution is

(2.2.21) $$F^* = HF\bar{G} + Z^*(I - G\bar{G}) .$$

Now if $G\mathcal{X} = \mathcal{X}_0^*(= G\mathcal{X}_0)$ then from (2.2.15) and (2.2.11) we have $G\bar{G}G\mathcal{X} = G\bar{G}\mathcal{X}_0^* = G\mathcal{X}_0 = \mathcal{X}_0^* = G\mathcal{X}$, i.e.,

(2.2.22) $$G\bar{G}G = G .$$

Conversely, if (2.2.22) holds then from (2.2.18) and (2.2.11) we have $G\mathcal{X} = G\bar{G}G\mathcal{X} = G\mathcal{X}_0 = \mathcal{X}_0^*$. Thus, the condition $G\mathcal{X} = \mathcal{X}_0^*$ is necessary and sufficient for $\bar{G}$ to be a generalized inverse of G . In many applications this is a reasonable assumption to make, since if $G\mathcal{X}$ has higher dimensionality than $G\mathcal{X}_0$ then G, in effect, preserves a needless amount of information about $\mathcal{X}$.

If $\mathcal{X}^* = \mathcal{X}_0^*$ then from (2.2.11) we have $P^* = I_{n^*}$; the converse is obviously also true. Owing to (2.2.17) this means that the condition $\mathcal{X}^* = G\mathcal{X}_0(= \mathcal{X}_0^*)$ is necessary and sufficient for $\bar{G}$ to be a right inverse of G . When this is the case, (2.2.21) reduces to the unique solution

$$F^* = H F \bar{G} . \tag{2.2.23}$$

Unlike the solution (2.2.8), this one is not invariant with respect to the choice of right inverse $\bar{G}$, but depends on the particular $\bar{G}$ entering into the characterization (2.2.19) of $\mathcal{X}_0$.

A case of particular interest is that in which G is a grouping matrix and $\bar{G}$ has the form

$$G_D^{\ddagger} = DG'(GDG')^{\dagger} , \tag{2.2.24}$$

where D is a block diagonal matrix with blocking conformable with that of G, and is such that (a) GDG' has the same rank as G, (b) the $n_i \times n_i$ diagonal blocks D_i of D are symmetric non-negative definite matrices, and (c) the elements of the matrices D_i are all non-negative. It was shown in Lemma 1.4 that condition (a) implies that $G_D^{\ddagger}$ is a reflexive generalized inverse of G, i.e., it satisfies (2.2.16) and (2.2.22). In fact, it satisfies the four properties

$$\text{(i) } GG_D^{\ddagger}G = G; \quad \text{(ii) } G_D^{\ddagger}GG_D^{\ddagger} = G_D^{\ddagger}; \tag{2.2.25}$$

$$\text{(iii) } GG_D^{\ddagger} = G_D^{\ddagger\prime}G' ; \quad \text{(iv) } G_D^{\ddagger}GD = DG'G_D^{\ddagger\prime} .$$

Condition (b) implies that GDG', and hence $(GDG')^{\dagger}$, is a diagonal matrix with non-negative diagonal elements, and together with condition (c) this implies that $G_D^{\ddagger}$ is a (column-wise) grouping matrix. Its form may be most quickly apprehended by consideration of the following example in which D is chosen to be a diagonal matrix $D = \text{diag}(d_1, d_2, \ldots, d_n)$:

$$(2.2.26) \qquad G = \begin{bmatrix} 1 & 1 & 0 & 0 & 0 \\ 0 & 0 & 1 & 0 & 1 \end{bmatrix}, \quad G_D^{\ddagger} = \begin{bmatrix} \frac{d_1}{d_1+d_2} & 0 \\ \frac{d_2}{d_1+d_2} & 0 \\ 0 & \frac{d_3}{d_3+d_5} \\ 0 & 0 \\ 0 & \frac{d_5}{d_3+d_5} \end{bmatrix} .$$

The effect of the restriction $x = G_D^{\ddagger}Gx$ is therefore to constrain the variables x_i to be proportionate to one another within groups.

C. Perfect aggregation: partially restricted structure, partially restricted domain. We shall now confine our attention to the case in which G is a grouping matrix. We start with the observation that a simple factorization holds for grouping matrices. With G as in (2.2.2), where each g_i is a $1 \times n_i$ vector, let

$$(2.2.27) \quad r+s = n^*, \quad \sum_{i=1}^{r} n_i = n', \quad \sum_{i=r+1}^{n^*} n_i = n'', \quad \sum_{i=1}^{n^*} n_i = n' + n'' = n \quad .$$

Then we have a factorization

(2.2.28)

$$\begin{bmatrix} g_1 & & & & & & \\ & g_2 & & & & & \\ & & \ddots & & & & \\ & & & g_r & & & \\ & & & & g_{r+1} & & \\ & & & & & \ddots & \\ & & & & & & g_{n^*} \end{bmatrix} = \begin{bmatrix} 1 & & & & & & \\ & 1 & & & & & \\ & & \ddots & & & & \\ & & & 1 & & & \\ & & & & g_{r+1} & & \\ & & & & & \ddots & \\ & & & & & & g_{n^*} \end{bmatrix} \begin{bmatrix} g_1 & & & & & \\ & g_2 & & & & \\ & & \ddots & & & \\ & & & g_r & & \\ & & & & I & \\ & & & & & \ddots \\ & & & & & & I \end{bmatrix},$$

which can be written as

$$(2.2.29)\quad G = \begin{bmatrix} G_{11} & 0 \\ 0 & G_{22} \end{bmatrix} \begin{bmatrix} I_r & 0 \\ 0 & G_{22} \end{bmatrix} \begin{bmatrix} G_{11} & 0 \\ 0 & I_{n''} \end{bmatrix} = G_2 G_1 ,$$

where G_{11} is of order $r \times n'$, G_{22} of order $s \times n''$, and thus G_2 is of order $n^* \times (r+n'')$ and G_1 is of order $(r+n'') \times n$. Defining D as in (2.2.24) and partitioning it conformably to the partition of G, as

$$(2.2.30)\qquad D = \begin{bmatrix} D_{11} & 0 \\ 0 & D_{22} \end{bmatrix},$$

where D_{11} and D_{22} are diagonal matrices of orders n' and n'' respectively, we shall define

$$(2.2.31)\qquad G_1^{\ddagger} = \begin{bmatrix} G_{11}^{\ddagger} & 0 \\ 0 & I_{n''} \end{bmatrix} = \begin{bmatrix} D_{11}G_{11}'(G_{11}D_{11}G_{11}')^{\dagger} & 0 \\ 0 & I_{n''} \end{bmatrix},$$

where it is assumed that rank $G_{11}D_{11}G_{11}'$ = rank G_{11}, so that $G_1^{\ddagger}$ satisfies the four conditions analogous to (2.2.25), with D replaced by D_{11}.

We shall now assume that the underlying data are confined to the set $\mathcal{X}_0 = G_1^{\ddagger} G_1 \mathcal{X}$. We can assume that there is a factorization of H,

(2.2.32) $$H = H_2 H_1 ,$$

corresponding to that of G (we can always choose $H_1 = I$). In accordance with the analysis of the preceding section (see Figure 3) we break the problem up into two stages. In stage 1 we seek a transformation $F_1: \mathcal{X}_1 \to \mathcal{Y}_1$ (where $G_1: \mathcal{X} \to \mathcal{X}_1$ and $H_1: \mathcal{Y} \to \mathcal{Y}_1$) satisfying $F_1 G_1 x = H_1 F x$ for all $x \in \mathcal{X}_0$, or equivalently, as in (2.2.20) (recalling that $G_1^{\ddagger}$ is a generalized inverse of G_1),

(2.2.33) $$F_1 G_1 = H_1 F G_1^{\ddagger} G_1 .$$

A solution F_1 of (2.2.33) necessarily exists, and since $G_1^{\ddagger}$ satisfies property (ii) of (2.2.25) the general solution of (2.2.33) is

(2.2.34) $$F_1 = H_1 F G_1^{\ddagger} + Z_1(I - G_1 G_1^{\ddagger}) .$$

In stage 2 we seek a transformation $F^*: \mathcal{X}^* \to \mathcal{Y}^*$ (where $G_2: \mathcal{X}_1 \to \mathcal{X}^*$ and $H_2: \mathcal{Y}_1 \to \mathcal{Y}^*$) satisfying $F^* G_2 x = H_2 F_1 x$ for all $x \in \mathcal{X}_1$, i.e., satisfying

(2.2.35) $$F^* G_2 = H_2 F_1 .$$

The Penrose solvability condition for (2.2.35) is

(2.2.36) $$H_2 F_1 = H_2 F_1 G_2^- G_2 .$$

Postmultiplying (2.2.35) by G_1 and using (2.2.33), (2.2.32), and (2.2.29), we obtain

$$F^*G = HFG_1^{\ddagger}G_1 \quad . \tag{2.2.37}$$

Now let G_2^+ be defined by

$$G_2^+ = \begin{bmatrix} I_r & 0 \\ 0 & G_{22}^- \end{bmatrix} , \tag{2.2.38}$$

where G_{22}^- is any generalized inverse of G_{22}. From (2.2.31) and (2.2.38) we verify that

$$G^+ = G_1^{\ddagger}G_2^+ \tag{2.2.39}$$

is a generalized inverse of G. Applying Penrose's solvability condition to (2.2.37) with G^+ chosen as in (2.2.39) and (2.2.38), we obtain

$$HFG_1^{\ddagger}G_1 = HFG_1^{\ddagger}G_1G^+G = HFG^+G \quad . \tag{2.2.40}$$

Thus, whenever (2.2.37) is solvable it can be written in the form

$$F^*G = HFG^+G \quad , \tag{2.2.41}$$

which has the general solution

$$F^* = HFG^+GG^- + Z^*(I - GG^-) \quad . \tag{2.2.42}$$

A particular solution of (2.2.41) is clearly

$$F^* = HFG^+ = HFG_1^{\ddagger}G_2^+ \quad . \tag{2.2.43}$$

If G has full rank, so that G^- is a right inverse of G, then (2.2.43) is the unique solution of (2.2.41), independently of the choice of G_{22}^- in (2.2.38); for, from (2.2.36) (with G_2^- replaced by G_2^+) we have, making use of (2.2.32) and (2.2.34),

$$\text{(2.2.44)} \quad HFG_1^{\dagger}G_2^{+} = H_2F_1G_2^{+} = H_2F_1G_2^{+}G_2G_2^{-} = H_2F_1G_2^{-} = HFG_1^{\dagger}G_2^{-},$$

where G_2^- is any generalized inverse (hence right inverse) of G_2.

Consider now in place of (2.2.28) the reverse factorization (2.2.45)

$$\begin{bmatrix} g_1 & & & & & & \\ & g_2 & & & & & \\ & & \ddots & & & & \\ & & & g_r & & & \\ & & & & g_{r+1} & & \\ & & & & & \ddots & \\ & & & & & & g_{n^*} \end{bmatrix} = \begin{bmatrix} g_1 & & & & & & \\ & g_2 & & & & & \\ & & \ddots & & & & \\ & & & g_r & & & \\ & & & & 1 & & \\ & & & & & \ddots & \\ & & & & & & 1 \end{bmatrix} \begin{bmatrix} I & & & & & & \\ & I & & & & & \\ & & \ddots & & & & \\ & & & I & & & \\ & & & & g_{r+1} & & \\ & & & & & \ddots & \\ & & & & & & g_{n^*} \end{bmatrix}$$

which may be written in the notation of (2.2.27) and (2.2.29) as

$$\text{(2.2.46)} \quad G = \begin{bmatrix} G_{11} & 0 \\ 0 & G_{22} \end{bmatrix} = \begin{bmatrix} G_{11} & 0 \\ 0 & I_s \end{bmatrix} \begin{bmatrix} I_{n'} & 0 \\ 0 & G_{22} \end{bmatrix} = G_BG_A ,$$

where G_B is of order $n^* \times (n' + s)$ and G_A is of order $(n' + s) \times n$. Defining the generalized inverses

$$(2.2.47) \qquad G_A^+ = \begin{bmatrix} I_{n'} & 0 \\ 0 & G_{22}^- \end{bmatrix}, \quad G_B^\ddagger = \begin{bmatrix} G_{11}^\ddagger & 0 \\ 0 & I_s \end{bmatrix},$$

we verify from (2.2.29), (2.2.31), and (2.2.39) that the constraints given by (2.2.40) that must be imposed on HF for the solvability of (2.2.37) may be written in the form

$$(2.2.48) \quad HF(G_1^\ddagger G_1 - G^+G) = HF(I_n - G_A^+G_A) = HF\begin{bmatrix} 0 & 0 \\ 0 & I-G_{22}^-G_{22} \end{bmatrix} = 0 .$$

Let R_A be a matrix of order $n \times (n'' - s)$, where $n'' - s = n - (n'+s)$ is the rank of $I_n - G_A^+G_A$, whose columns form a basis for the column space of $I_n - G_A^+G_A$. Then the restrictions (2.2.48) may be expressed in the equivalent form

$$(2.2.49) \qquad HFR_A = 0 .$$

As an example, let

$$G = \begin{bmatrix} 1 & 1 & 0 & 0 & 0 \\ 0 & 0 & 1 & 1 & 1 \end{bmatrix} = \begin{bmatrix} 1 & 0 & 0 & 0 \\ 0 & 1 & 1 & 1 \end{bmatrix} \begin{bmatrix} 1 & 1 & 0 & 0 & 0 \\ 0 & 0 & 1 & 0 & 0 \\ 0 & 0 & 0 & 1 & 0 \\ 0 & 0 & 0 & 0 & 1 \end{bmatrix} = G_2G_1$$

and

$$D = \begin{bmatrix} 3 & 0 & 0 & 0 & 0 \\ 0 & 2 & 0 & 0 & 0 \\ 0 & 0 & 1 & 0 & 0 \\ 0 & 0 & 0 & 1 & 0 \\ 0 & 0 & 0 & 0 & 1 \end{bmatrix}, \qquad G_1^\ddagger = \begin{bmatrix} 3/5 & 0 & 0 & 0 \\ 2/5 & 0 & 0 & 0 \\ 0 & 1 & 0 & 0 \\ 0 & 0 & 1 & 0 \\ 0 & 0 & 0 & 1 \end{bmatrix}.$$

The set $\mathcal{X}_0$ is the set of x for which $x_1 = (3/5)(x_1 + x_2)$ and $x_2 = (2/5)(x_1 + x_2)$. The reverse factorization is

$$G = \begin{bmatrix} 1 & 1 & 0 & 0 & 0 \\ 0 & 0 & 1 & 1 & 1 \end{bmatrix} = \begin{bmatrix} 1 & 1 & 0 \\ 0 & 0 & 1 \end{bmatrix} \begin{bmatrix} 1 & 0 & 0 & 0 & 0 \\ 0 & 1 & 0 & 0 & 0 \\ 0 & 0 & 1 & 1 & 1 \end{bmatrix} = G_B G_A \ .$$

Choosing $m^* = m = 1$ and $H = 1$, the constraint (2.2.49) becomes

$$FR_A = (f_1, f_2, f_3, f_4, f_5) \begin{bmatrix} 0 & 0 \\ 0 & 0 \\ 1 & 1 \\ -1 & 0 \\ 0 & -1 \end{bmatrix} = (0, 0) \ ,$$

i.e., $f_3 = f_4 = f_5$.

It is of some interest to confirm that the preceding analysis can also be carried through in terms of the reverse factorization (2.2.45). Let us suppose that there is a corresponding factorization $H = H_B H_A$. At stage A we start with the transformations $G_A\colon \mathcal{X} \to \mathcal{X}_A$ and $H_A\colon \mathcal{Y} \to \mathcal{Y}_A$, and seek a transformation $F_A\colon \mathcal{X}_A \to \mathcal{Y}_A$ such that $H_B F_A G_A x = HFx$ for all $x \in G_1^{\ddagger} G_1 \mathcal{X} = \mathcal{X}_0$, i.e., such that

$$H_B F_A G_A G_1^{\ddagger} G_1 = HFG_1^{\ddagger} G_1 \ . \tag{2.2.50}$$

From (2.2.46), (2.2.31), and (2.2.29) we see that

$$(2.2.51) \qquad (I-G_A)G_1^{\ddagger}G_1 = G_1^{\ddagger}G_1$$

hence (2.2.50) is actually equivalent to the apparently stronger condition

$$(2.2.52) \qquad H_B F_A G_A = HF .$$

(Here, the linearity of the transformations is crucial; compare (2.1.65).) The Penrose solvability condition for (2.2.52) is

$$(2.2.53) \qquad HF = HFG_A^+ G_A$$

as in (2.1.65), and this is the same as (2.2.48). The general solution of (2.2.52) is

$$(2.2.54) \qquad F_A = H_B^- HFG_A^+ + Z_A - H_B^- H_B Z_A G_A G_A^+ .$$

At stage B we proceed with transformations $G_B: \mathcal{X}_A \to \mathcal{X}^*$ and $H_B: \mathcal{Y}_A \to \mathcal{Y}^*$. However, since $G_B G_A = G_2 G_1 \triangleq G$ we must restrict the partially aggregated data to a subset $\mathcal{X}_A^0 = G_B^{\ddagger} G_B \mathcal{X}_A \subset \mathcal{X}_A$, hence we only require F^* to satisfy $F^* G_B x_A = H_B F_A x_A$ for $x_A \in \mathcal{X}_A^0$, or (making use of (2.2.54) and of the fact that $G_A^+ G_B^{\ddagger} = G_1^{\ddagger} G_2^+$ from (2.2.47), (2.2.31), (2.2.38), and (2.2.39)),

$$(2.2.55) \quad F^* G_B = H_B F_A G_B^{\ddagger} G_B = HFG^+ G_B + H_B Z_A (I - G_A G_A^+) G_B^{\ddagger} G_B .$$

Such an F^* always exists, given that (2.2.52) holds. Post-multiplying (2.2.55) by G_A and observing that the matrices

$G_A G_A^+$ and $G_B^+ G_B$ commute, we obtain (2.2.41) as before, with general solution (2.2.42). The situation is depicted in Figure 4.

D. Best approximate aggregation. The conditions for perfect aggregation that we have just set forth, whether of type A, B, or C, cannot of course be expected to hold exactly. At best they will hold approximately, and the problem is then to find for given G and H a transformation F^* which provides in some sense a best approximation to the (in general) inconsistent equation (2.2.5). The following formulation has been influenced by the seminal work of W. D. Fisher [34], though it differs from it in a significant respect in that Fisher takes as given a mapping $\bar{H}: \mathcal{Y}^* \to \mathcal{Y}$ rather than a mapping $H: \mathcal{Y} \to \mathcal{Y}^*$, and his main concern is to find a simplified mapping $\hat{F} = \bar{H}F^*G: \mathcal{X} \to \mathcal{Y}$ which best approximates F (i.e., a mapping $F^*: \mathcal{X}^* \to \mathcal{Y}^*$ such that $\bar{H}F^*G$ best approximates F). In the present formulation I shall instead take G and H as given, and break the problem up into two stages: the aggregation problem, of finding a mapping $F^*: \mathcal{X}^* \to \mathcal{Y}^*$ such that F^*G best approximates HF, and which is expressible in the form $HF\bar{G}$ for some $\bar{G}: \mathcal{X}^* \to \mathcal{X}$; and the disaggregation problem, of finding a "blowing-up transformation" $\bar{H}: \mathcal{Y}^* \to \mathcal{Y}$ such that $\bar{H}F^*$ best approximates $F\bar{G}$ (or more generally, such that $\bar{H}F^*G$ best approximates $F\bar{G}G$). The disaggregation problem is taken up in subsections E and F below, and Fisher's formulation will be discussed in subsection G.

Let the model be reformulated in stochastic terms, so that the first equation of (2.2.1) is replaced by the specification

$$(2.2.56) \qquad y = Fx + e, \quad \mathcal{E}(e|x) = 0, \quad \mathcal{E}(ee'|x) = \Sigma ,$$

where x and the error term e are both random variables, x having mean and variance

$$(2.2.57) \qquad \mathcal{E}x = \mu, \quad \text{Var } x = \mathcal{E}(x-\mu)(x-\mu)' = \Theta .$$

The moment matrix of x is defined as

$$M = \mathcal{E}xx' = \mu\mu' + \Theta \quad . \tag{2.2.58}$$

We shall assume that F, Σ, and M, in addition to G and H, are known. (The estimation problem is taken up in section 2.3.)

An investigator is concerned with making conditional forecasts of $y^* = Hy$ given $x^* = Gx$. Let us assume that he does so on the basis of the aggregative model

$$y^{**} = F^*x^* + e^*, \quad \mathcal{E}^*(e^*|x^*) = 0, \quad \mathcal{E}^*(e^*e^{*'}|x^*) = \Sigma^* \quad . \tag{2.2.59}$$

(The symbol $\mathcal{E}^*$ stands for the expectation operator associated with the model (2.2.59); it need not and in general does not coincide with the operator $\mathcal{E}$ associated with the "true" model (2.2.56). On the other hand we shall assume that the distribution of x^* in (2.2.59) is determined by (2.2.57), i.e., that $\mathcal{E}x^* = G\mu$ and $\text{Var } x^* = G\Theta G'$.) The investigator's conditional forecast of y^* will then be defined by

$$\hat{y}^* = \mathcal{E}^*(y^{**}|x^*) = F^*x^* = F^*Gx \quad . \tag{2.2.60}$$

On the other hand, the "true" y^* is determined from (2.2.56) by

$$y^* = HFx + He \tag{2.2.61}$$

so that an investigator having both the model (2.2.56) and data on x at his disposal would forecast y^* by

$$\tilde{y}^* = H\mathcal{E}(y|x) = HFx \quad . \tag{2.2.62}$$

Our indicator of aggregation bias is then

(2.2.63) $$\mathcal{E}(\hat{y}^* - \tilde{y}^*)(\hat{y}^* - \tilde{y}^*)' =$$

$$\mathcal{E}(F^* x^* - \tilde{y}^*)(F^* x^* - \tilde{y}^*)' =$$

$$(F^* G - HF)M(F^* G - HF)' \quad .$$

The <u>mean square error of forecast</u> is defined by

(2.2.64) $$\mathcal{E}(\hat{y}^* - y^*)(\hat{y}^* - y^*)' = \mathcal{E}(F^* x^* - y^*)(F^* x^* - y^*)' \quad ,$$

and this is equal to

(2.2.65) $$\mathcal{E}\{\mathcal{E}\{(\hat{y}^* - y^*)(\hat{y}^* - y^*)' | x\}\} = (F^* G - HF) M(F^* G - HF)' + H\Sigma H' \quad .$$

Since (2.2.65) and (2.2.63) differ only by a constant, minimization of forecast error is of course equivalent to minimization of aggregation bias.

From (2.2.1), (2.2.61), (2.2.62), and (2.2.58) we verify that

(2.2.66)
$$\mathcal{E}\begin{bmatrix} \tilde{y}^* \\ x^* \end{bmatrix}\begin{bmatrix} \tilde{y}^* \\ x^* \end{bmatrix}' = \begin{bmatrix} HFMF'H' & HFMG' \\ GMF'H' & GMG' \end{bmatrix},$$

$$\mathcal{E}\begin{bmatrix} y^* \\ x^* \end{bmatrix}\begin{bmatrix} y^* \\ x^* \end{bmatrix}' = \begin{bmatrix} HFMF'H' + H\Sigma H' & HFMG' \\ GMF'H' & GMG' \end{bmatrix},$$

and thus it follows from Theorem 1.16 that (2.2.63) and (2.2.65) both reach a minimum if and only if F^* is a solution of the equation

(2.2.67) $$F^* GMG' = HFMG' \quad .$$

In accordance with the notation of section 1.5, therefore, for any F^* satisfying (2.2.67) we have, with (2.2.60),

$$\mathcal{E}^*(y^{**}|x^*) = F^*x^* = \hat{\rho}(\tilde{y}^*|x^*) = \hat{\rho}(y^*|x^*) \quad , \tag{2.2.68}$$

that is, (2.2.60) gives the best homogeneous linear predictor of y^* as a function of x^*.

The following theorem provides an elaboration of the foregoing result.

Theorem 2.1. The matrix (2.2.63) of aggregation bias attains a minimum with respect to F^* if and only if

$$F^* = HFG^{\#} + Z^*(I - GG^{\#}) \quad , \tag{2.2.69}$$

where Z^* is arbitrary and $G^{\#}$ is any generalized quasi-inverse of G satisfying

$$\text{(i)}\ GG^{\#}GM = GM, \quad \text{(iv)}\ G^{\#}GM = MG'G^{\#'} \quad , \tag{2.2.70}$$

or equivalently,

$$G^{\#}GMG' = MG' \quad . \tag{2.2.71}$$

A solution $G^{\#}$ to (2.2.71) always exists. The conditional forecast (2.2.68), regarded as a random variable, is invariant with respect to the choice of $G^{\#}$ among solutions of (2.2.71) so long as F^* is given by (2.2.69), and the (unique) minimum aggregation bias is given by

$$\inf_{F^*} (F^*G - HF)M(F^*G - HF)' = HF(I - G^{\#}G)MF'H' \quad . \tag{2.2.72}$$

If rank GMG' = rank G then $G^{\#}$ is a generalized inverse of G. In particular this will hold if the rows of G are contained in the row space of M.

Proof. The equivalence of (2.2.70) and (2.2.71), and the fact that a solution $G^{\#}$ to (2.2.71) exists, were shown in Lemma 1.9. Now let $G^{\#}$ satisfy (2.2.70), and write

$$F^{*}G - HF = (F^{*} - HFG^{\#})G - HF(I - G^{\#}G) \quad .$$

From (2.2.71) it then follows that

$$(2.2.73) \qquad (F^{*}G - HF)M(F^{*}G - HF)' =$$

$$HF(I - G^{\#}G)M(I - G^{\#}G)'F'H' + (F^{*} - HFG^{\#})GMG'(F^{*} - HFG^{\#})' \; .$$

The second term on the right is non-negative definite, and vanishes if and only if

$$(2.2.74) \qquad (F^{*} - HFG^{\#})GMG' = 0 \quad ,$$

i.e., if and only if, on account of (2.2.71), F^{*} satisfies (2.2.67). From Penrose's theorem the solvability of (2.2.71) is equivalent to the condition $MG'(GMG')^{-}GMG' = MG'$ (which follows directly from Lemma 1.4(i), since rank GMG' = rank MG'), and this in turn implies the solvability of (2.2.74); thus, F^{*} minimizes (2.2.73) if and only if (2.2.67) holds.

The general solution of (2.2.67) is

$$(2.2.75) \qquad F^{*} = HFMG'(GMG')^{-} + Z^{*}[I - GMG'(GMG')^{-}] \; ,$$

which has the form (2.2.69) for the particular solution of (2.2.71) given by

$$(2.2.76) \qquad G^{\#} = MG'(GMG')^{-} \; .$$

Conversely, if F^{*} is given by (2.2.69) for any $G^{\#}$ satisfying (2.2.70), it is clear from (2.2.71) that F^{*} satisfies (2.2.74).

Invariance follows from the fact that if $(G^{\#})_1$ and $(G^{\#})_2$ both satisfy (2.2.70) then

$$(2.2.77)\quad \underline{(G^{\#})_1 GM} = \underline{(G^{\#})_1 GMG'}(G^{\#})_2' = \underline{MG'(G^{\#})_1' G'}(G^{\#})_2'$$
$$= \underline{MG'(G^{\#})_2'}$$
$$= (G^{\#})_2 GM \ .$$

Thus, if $\hat{y}_i^* = \hat{F}_i^* x^* = [HF(G^{\#})_i + Z_i^*(I - G(G^{\#})_i)]Gx$, $i = 1, 2$, are two conditional forecasts of y^*, we have

$$\mathcal{E}(\hat{y}_1^* - \hat{y}_2^*)(\hat{y}_1^* - \hat{y}_2^*)' =$$

$$[HF(G^{\#})_1 - HF(G^{\#})_2]GMG'[HF(G^{\#})_1 - HF(G^{\#})_2]' = 0 \ ,$$

whence $\hat{y}_1^* = \hat{y}_2^*$ with probability 1.

The expression (2.2.72) follows at once from (2.2.73), (2.2.74), and (2.2.70).

If rank GMG' = rank G then from the general solution

$$G^{\#} = MG'(GMG')^- + Z^*[I - GMG'(GMG')^-]$$

of (2.2.71) we have $GG^{\#}G = G$ by Lemma 1.4. If the rows of G are contained in the row space of M then $G = KM^{\frac{1}{2}}$ for some K and some square root $M^{\frac{1}{2}}$ of M, since M and $M^{\frac{1}{2}}$ have the same row space. From condition (i) of (2.2.70) we then have

$$0 = (I - GG^{\#})GM^{\frac{1}{2}} = (I - GG^{\#})KM \ ,$$

which implies $0 = (I - GG^{\#})KM^{\frac{1}{2}} = (I - GG^{\#})G$.

Q. E. D.

Remark: It is not hard to see that any solution in the class (2.2.69) may be represented as $F^* = HFG^\#$ for some (in general, different) $G^\#$ satisfying (2.2.71).

Treatment in terms of affine transformations. In order to relate the foregoing results to those of section 2.1 it will be helpful to provide an explicit indication of the correspondence between homogeneous and affine transformations. Typically the functional relationships arising in practice are affine rather than homogeneous, although by convention we can always translate one into the other. Thus we may suppose that the column vectors x and x^* have 1 as their first component, and denote

$$(2.2.78)\quad x = \begin{bmatrix} 1 \\ x_1 \end{bmatrix}, \quad G = \begin{bmatrix} 1 & 0 \\ 0 & G_1 \end{bmatrix}, x^* = \begin{bmatrix} 1 \\ x_1^* \end{bmatrix} = \begin{bmatrix} 1 \\ G_1 x_1 \end{bmatrix}.$$

Likewise, we may partition F and F^* as

$$(2.2.79)\qquad F = [F_0, F_1], \quad F^* = [F_0^*, F_1^*]$$

respectively, where F_0 and F_0^* are column matrices. Denoting $\mathcal{X} = \{1\} \times \mathcal{X}_1$ and $\mathcal{X}^* = \{1\} \times \mathcal{X}_1^*$, we may think of $\mathcal{X}_1$ and $\mathcal{X}_1^*$ as corresponding to the spaces $\mathcal{X}$ and $\mathcal{X}^*$ of section 2.1. Defining $\mathcal{E}x_1 = \mu_1$ and $\mathcal{E}(x_1 - \mu_1)(x_1 - \mu_1)' = \Theta_1$ we have

$$(2.2.80)\qquad \mu = \begin{bmatrix} 1 \\ \mu_1 \end{bmatrix}, \quad \Theta = \begin{bmatrix} 0 & 0 \\ 0 & \Theta_1 \end{bmatrix},$$

so that (2.2.58) becomes

(2.2.81)
$$M = \begin{bmatrix} 1 & \mu_1' \\ \mu_1 & \mu_1\mu_1' + \Theta_1 \end{bmatrix}.$$

We shall define the affine functions $f: \mathcal{X}_1 \to \mathcal{Y}$ and $f^*: \mathcal{X}_1^* \to \mathcal{Y}^*$ by

(2.2.82)
$$f(x_1) = F_0 + F_1 x_1, \quad f^*(x_1^*) = F_0^* + F_1^* x_1^* .$$

We shall also define $g: \mathcal{X}_1 \to \mathcal{X}_1^*$ by

(2.2.83)
$$g(x_1) = G_1 x_1$$

where G_1 is defined by (2.2.78).

From (2.2.78) and (2.2.81) we may verify immediately (taking account of the fact that $\Theta_1 G_1'(G_1\Theta_1 G_1')^- G_1\Theta_1 G_1' = \Theta_1 G_1'$ from Lemma 1.4, since rank $G_1\Theta_1 G_1' =$ rank $\Theta_1 G_1'$) that one choice for a solution $G^\#$ of (2.2.71) is

(2.2.84)
$$G^\# = \begin{bmatrix} 1 & 0 \\ (I - G_1^\natural G_1)\mu_1 & G_1^\natural \end{bmatrix}, \quad \text{where } G_1^\natural = \Theta_1 G_1'(G_1\Theta_1 G_1')^- .$$

Since the joint distribution of x_1 and x_1^* has mean and variance

(2.2.85)
$$\mathcal{E}\begin{bmatrix} x_1 \\ x_1^* \end{bmatrix} = \begin{bmatrix} \mu_1 \\ G_1\mu_1 \end{bmatrix}, \quad \text{Var}\begin{bmatrix} x_1 \\ x_1^* \end{bmatrix} = \begin{bmatrix} \Theta_1 & \Theta_1 G_1' \\ G_1\Theta_1 & G_1\Theta_1 G_1' \end{bmatrix},$$

it follows immediately from Theorem 1.14 and formula (1.5.24) that

$$(2.2.86)\quad G^{\#}x^{*} = G^{\#}\begin{bmatrix} 1 \\ x_1^{*} \end{bmatrix} = \begin{bmatrix} 1 \\ \hat{\mathcal{E}}(x_1 | x_1^{*}) \end{bmatrix} = \hat{\mathcal{E}}(x | x^{*}) = \hat{\rho}(x | x^{*}) \quad .$$

We may therefore define the affine function $g^{\natural}: \mathcal{X}_1^{*} \to \mathcal{X}_1$ by

$$(2.2.87)\qquad g^{\natural}(x_1^{*}) = \hat{\mathcal{E}}(x_1 | x_1^{*}) = (I - G_1^{\natural}G_1)\mu_1 + G_1^{\natural} x_1^{*} \quad .$$

Similarly, the joint distribution of y^{*} (defined by (2.2.61)) and x_1^{*} has mean and variance

$$(2.2.88)\qquad \mathcal{E}\begin{bmatrix} y^{*} \\ x_1^{*} \end{bmatrix} = \begin{bmatrix} HF_0 + HF_1\mu_1 \\ G_1\mu_1 \end{bmatrix} ,$$

$$\operatorname{Var}\begin{bmatrix} y^{*} \\ x_1^{*} \end{bmatrix} = \begin{bmatrix} HF_1\Theta_1F_1'H' + H\Sigma H' & HF_1\Theta_1G_1' \\ G_1\Theta_1F_1'H' & G_1\Theta_1G_1' \end{bmatrix} ,$$

hence if we choose $F^{*} = HFG^{\#}$, where $G^{\#}$ is given by (2.2.84), and define f^{*} by (2.2.82), it follows from Theorem 1.14 that the wide sense conditional expectation of y^{*} given x_1^{*} is

$$(2.2.89)\qquad \hat{\mathcal{E}}(y^{*} | x_1^{*}) = HF_0 + HF_1\mu_1 + HF_1G_1^{\natural}(x_1^{*} - G_1\mu_1) = f^{*}(x_1^{*})$$

$$= \mathcal{E}^{*}(y^{**} | x_1^{*}) \quad .$$

Writing $h(y) = Hy$, it follows from (2.2.82) and (2.2.87) that (2.2.89) may also be written as

(2.2.90) $$\hat{\mathcal{E}}(y^* | x_1^*) = h \circ f \circ g^{\natural}(x_1^*) \ ,$$

i.e., $f^* = hfg^{\natural}$ as in (2.1.71). It is easy to see that (2.2.89) also gives the expression for $\hat{\mathcal{E}}(\tilde{y}^* | x_1^*)$. If the distribution of x is normal then it follows from Theorem 1.15 that the wide sense conditional expectation is equal to the true conditional expectation, and thus the conditional expectation $\mathcal{E}^*(y^{**} | x_1^*)$ derived from the aggregative model (2.2.59) actually coincides with the true conditional expectation $\mathcal{E}(y^* | x_1^*)$ derived from the detailed model (2.2.56); in any event--whether or not x is normal--(2.2.89) states that the conditional expectation $\mathcal{E}^*(y^{**} | x^*)$ derived from the aggregative model (2.2.59) always coincides with the wide sense conditional expectation $\hat{\mathcal{E}}(y^* | x_1^*)$ derived from the detailed model (2.2.56).

According to Theorem 2.1, in order for $G^{\#}$ to be a generalized inverse of G it is sufficient that rank G = rank GMG'. Now from (2.2.78) and (2.2.84) we verify that $GG^{\#}G = G$ if and only if $G_1 G_1^{\natural} G_1 = G_1$, and a necessary and sufficient condition for this is in turn that rank G_1 = rank $G_1 \Theta_1 G_1'$. Finally we see from (2.2.83) and (2.2.87) that $g^{\natural}$ is a generalized inverse (under composition) of g if and only if $G_1^{\natural}$ is a generalized inverse of G_1.

<u>Relation between best approximate and perfect aggregation</u>. For the criterion of minimization of aggregation bias to be consistent with the criteria for perfect aggregation, we should of course find that when the latter are fulfilled the bias matrix (2.2.63) vanishes. Let us verify that this is indeed the case.

<u>Case A</u>. In terms of the stochastic formulation (2.2.56), (2.2.57), we may interpret the assumption that the domain of F is unrestricted to mean that Θ, and hence M, is positive definite. In that case it is clear that the commutativity condition (2.2.5) is necessary and sufficient for the bias matrix (2.2.63) to be equal to zero. Alternatively, direct substitution of the general solution (2.2.7) in (2.2.63) yields

(2.2.91) $(F^*G-HF)M(F^*G-HF)' = HF(I-G^-G)M(I-G^-G)'F'H'$

which vanishes for any generalized inverse G^- of G, on account of the solvability condition (2.2.6). Since the positive definiteness of M implies that rank G = rank GMG', it follows from Theorem 2.1 that $G^\#$ is such a generalized inverse.

Case B. If F is unrestricted but its domain is restricted by the condition

(2.2.92) $(I-\bar{G}G)x = 0$ with probability 1

(see (2.2.18)), where $\bar{G}$ is any matrix satisfying (2.2.16), then from (2.2.58) we have

$$\mathcal{E}(I-\bar{G}G)xx'(I-\bar{G}G)' = (I-\bar{G}G)M(I-\bar{G}G)' = 0 \ ,$$

which implies

(2.2.93) $(I-\bar{G}G)M = 0 \ .$

This implies in turn that $\bar{G}$ satisfies (2.2.71), and thus qualifies as a choice of $G^\#$ in Theorem 2.1. The minimum aggregation bias is then zero from (2.2.72). Alternatively, direct substitution of (2.2.21) in (2.2.63) yields

(2.2.94) $(F^*G-HF)M(F^*G-HF)' = HF(I-\bar{G}G)M(I-\bar{G}G)'F'H' \ ,$

which vanishes on account of (2.2.93).

Case C. In this case we may assume that the domain of F is restricted by

(2.2.95) $(I-G_1^\ddagger G_1)x = 0$ with probability 1 ,

where G_1 and $G_1^\ddagger$ are defined by (2.2.29) and (2.2.31), and F itself is constrained by (2.2.40) (or (2.2.48)), i.e.,

$$HF(G_1^{\ddagger}G_1 - G^{+}G) = 0 \quad , \tag{2.2.96}$$

where G and G^{+} are defined by (2.2.29) and (2.2.39) respectively. From (2.2.95) and (2.2.58) we obtain

$$(I - G_1^{\ddagger}G_1)M = 0 \quad . \tag{2.2.97}$$

Together, (2.2.96) and (2.2.97) yield

$$HF(I - G^{+}G)M = HF(I - G_1^{\ddagger}G_1)M = 0 \quad . \tag{2.2.98}$$

Now, substitution of (2.2.42) in (2.2.63) yields

$$(F^{*}G - HF)M(F^{*}G - HF)' = HF(I - G^{+}G)M(I - G^{+}G)'F'H' \ ,$$

which vanishes on account of (2.2.98). From (2.2.98) we also verify directly that F^{*}, as given by (2.2.42), satisfies (2.2.67).

D'. <u>Optimal selection of grouping matrices.</u> Up to now the transformations G and H have been taken as given. However, the most challenging problem of aggregation theory is that of appropriate selection of these transformations. We now outline one formulation of this problem when the sole objective is that of the optimal choice of an aggregative model F^{*} for purposes of making aggregative forecasts. Subsequently, in subsection F', we shall take up a more general formulation in which the objective of making disaggregative forecasts is also taken into account.

Let there be specified a certain set $\mathscr{S}$ of pairs (G, H) of grouping matrices, say of given orders $n^{*} \times n$ and $m^{*} \times m$. For each such pair one can determine an F^{*} according to (2.2.69) so as to minimize the aggregation bias

(2.2.63), and compute the corresponding minimum bias (2.2.72). One would then like to select a pair $(G, H) \in \mathcal{S}$ which minimizes the minimum aggregation bias. However, the relation $\succcurlyeq$ defines only a partial ordering of symmetric non-negative definite matrices of a given order (in this case m^*) hence, even if H is of order $m^* \times m$ for all $(G, H) \in \mathcal{S}$ (so that the difference of two matrices of the form (2.2.72) is defined), a minimum matrix (2.2.72) over the set $\mathcal{S}$ need not necessarily exist. (Example: let $m^* = m = n = 2$, $n^* = 1$, and let $\mathcal{S}$ consist of the two pairs $((1, 0), I_2)$ and $((0, 1), I_2)$.) It may therefore be appropriate to replace the matrix indicator (2.2.63) by a numerical one; this will obviously be required when dealing with pairs (G, H) which differ in respect to the order of the second matrix, H. Given a numerical indicator of aggregation bias, we may formulate the problem as that of choosing $(G, H) \in \mathcal{S}$ so as to minimize the minimum aggregation bias; alternatively, one might specify sets $\mathcal{S}_{n^*, m^*}$ of pairs of grouping matrices of given orders, then specify some maximum admissible level of aggregation bias, and determine the critical dimensions n^*, m^* which satisfy this restriction.

For any $m^* \times n$ matrix A we may define its M-norm by

$$\|A\|_M = \sqrt{\text{trace } AMA'} \quad . \tag{2.2.99}$$

A logical choice for a numerical indicator of aggregation bias is then the M-norm of $F^*G - HF$; from (2.2.72) the minimum aggregation bias is then measured by

$$\inf_{F^*} \|F^*G - HF\|_M = \sqrt{\text{trace } F'H'HF(I - G^{\#}G)M} \quad . \tag{2.2.100}$$

The problem of minimizing (2.2.100) over the set $\mathcal{S}_{n^*, m^*}$ of all pairs of grouping matrices (G, H) of orders $n^* \times n$ and $m^* \times m$ respectively, is a quadratic programming problem. If one restricts the set still further by, say, requiring the elements of G and H to be either zeros or ones, the problem becomes one of the integer programming type; the latter type of problem has been analyzed in considerable detail by W. D. Fisher [34, 35].

An alternative numerical indicator of aggregation bias closely related to (2.2.100) has been suggested by Ijiri [54] (for the case $M = I$). Let the "aggregation coefficient" ρ (by analogy to the correlation coefficient) be defined by

$$\rho^2 = 1 - \|F^*G - HF\|_M^2 / \|HF\|_M^2 . \tag{2.2.101}$$

Then from (2.2.72) and (2.2.70) we find readily that

$$\sup_{F^*} \rho = \|HFG^{\#}G\|_M / \|HF\|_M . \tag{2.2.102}$$

A possible alternative criterion would then be to choose (G, H) so as to maximize $\sup \rho$ over $\mathcal{G}$. In the special case $m = m^* = 1$ and $n^* = 2$ in which x, x^*, and G have the form (2.2.78) and G_1 is a row matrix with one unit element and the remaining components equal to zero, this problem reduces to that of the optimal selection of one out of a preassigned set of explanatory variables--which is a problem originally formulated by Hotelling [51] (see also Chow [20]).

E. Perfect disaggregation. The disaggregation problem will be posed in the following manner. Let us suppose that an aggregative model $F^*: \mathcal{X}^* \to \mathcal{Y}^*$ has already been chosen, and that is has the form

$$F^* = HF\bar{G} , \tag{2.2.103}$$

where $\bar{G}$ is some $n \times n^*$ matrix. (If any of the conditions for perfect or best approximate aggregation holds, this amounts to choosing particular solutions from (2.2.7), (2.2.21), (2.2.42), or (2.2.69) respectively; we shall verify below that this involves no loss of generality.) An investigator wishing to forecast $y \in \mathcal{Y}$ on the basis of the model F^*, and having access only to observations $x^* \epsilon \mathcal{X}^*$, will wish to have at his disposal a "degrouping transformation" $\bar{H}: \mathcal{Y}^* \rightarrow \mathcal{Y}$, enabling him to "blow up" his aggregative forecast

$$\hat{y}^* = F^* x^* \tag{2.2.104}$$

into a detailed forecast

$$\hat{y} = \bar{H}\hat{y}^* = \bar{H}F^* x^* . \tag{2.2.105}$$

Had he, on the other hand, had knowledge of F (though still not of x), he would have been able to forecast y instead by

$$\tilde{y} = F\bar{G}x^* . \tag{2.2.106}$$

This may be considered as an implicit forecast on his part of the variable y, in the sense that when aggregated it reproduces his forecast $\hat{y}^* = H\tilde{y}$ of y^*. In case the data are restricted by (2.2.14), so that each x satisfies $x = \bar{G}x^*$ for some $x^* \epsilon \mathcal{X}^*$, then (2.2.106) gives the true forecast $\tilde{y} = Fx = y$. On the other hand, if condition (2.2.6) for perfect aggregation holds, and if we take $\bar{G} = G^-$, then even though $\hat{y}^* = HFG^- x^*$ is invariant with respect to the choice of G^- for all $x^* \epsilon G\mathcal{X}$, the same is not true of $\tilde{y} = FG^- x^*$. Thus, (2.2.106) depends on the particular choice of $\bar{G}$; the question of the proper choice of $\bar{G}$ is therefore part of the disaggregation problem, and will be taken up in subsection F .

Given the choice of $\bar{G}$, we shall say that conditions for <u>perfect relative disaggregation</u> hold when the two forecasts $\hat{y}$ and $\tilde{y}$ of (2.2.105) and (2.2.106) coincide for all $x^* \epsilon G\mathcal{X}$, i.e., when $\bar{H}$ satisfies $\bar{H}F^*Gx = F\bar{G}Gx$ for all $x \epsilon \mathcal{X}$. Owing to (2.2.103) this is equivalent to the condition that there exists a solution $\bar{H}$ to the equation

$$\bar{H}HF\bar{G}G = F\bar{G}G . \tag{2.2.107}$$

This condition falls short of providing ultimate or absolute perfection in that it ensures only that $\hat{y}$ be equal to the hypothetical value $\tilde{y}$ rather than to the "true" value $y = Fx$; thus, its fulfillment allows one merely to do the best that could be hoped for given that information may already have been irretrievably lost through the process of aggregation. In case $\hat{y}$ is equal to the true value $y = Fx$ for all $x \in \mathcal{X}$, we shall say that conditions for perfect absolute disaggregation hold; these are equivalent to the existence of a solution $\overline{H}$ to the equation

$$\text{(2.2.108)} \qquad \overline{H}HF\overline{G}G = F \; .$$

In case the domain of F is restricted to $\overline{G}\mathcal{X}^*$, (2.2.107) corresponds to perfect absolute aggregation. Another case in which this is so is that in which the domain of F is unrestricted but F is assumed to satisfy $F = \Phi G$ for some matrix Φ ; condition (2.2.6) for perfect aggregation would then be replaced by the stronger condition

$$\text{(2.2.109)} \qquad F = FG^{-}G \; ,$$

hence with $\overline{G} = G^{-}$ the two conditions (2.2.107) and (2.2.108) are obviously the same. In some applications (2.2.109) may be a reasonable assumption to make; e.g., in the example considered in subsection A, y_i might represent an individual's expenditure on class i of commodities, x_1 and x_2 his income from two different sources (e.g., salary and dividends), and x_3, x_4, and x_5 three different categories of his wealth (say, cash, securities, and real estate); the restriction $FR = 0$ replacing (2.2.10) would then state that the income and wealth effects would not depend on the source of income or the category of wealth.

Condition (2.2.109) is rarely fulfilled in practice; if it held, there would in effect be no aggregation problem. We shall therefore concentrate attention on condition (2.2.107) rather than (2.2.108); however, when we come in subsection F to consider the problem of best approximate disaggregation, it will be the discrepancy between the two sides of

(2.2.108) that will have to be regarded as providing the fundamental concept underlying the criterion of disaggregation bias rather than the corresponding discrepancy in (2.2.107).

Before proceeding to an analysis of (2.2.107) we may note that the assumed homogeneity of the transformation $\bar{H}$ entails no substantive restriction provided the symbols are suitably interpreted. Thus, we may suppose that the vectors y and y^*, like x and x^* in (2.2.78), have the dummy variable 1 as first component, so that

$$(2.2.110) \qquad y = \begin{bmatrix} 1 \\ y_1 \end{bmatrix}, \quad y^* = \begin{bmatrix} 1 \\ y_1^* \end{bmatrix}.$$

If x, x^*, and G are given by (2.2.78), we may redefine the domains and ranges of the relevant mappings so that we have $f\colon \mathcal{X}_1 \to \mathcal{Y}_1$, $h\colon \mathcal{Y}_1 \to \mathcal{Y}_1^*$, and $f^*\colon \mathcal{X}_1^* \to \mathcal{Y}_1^*$, where $\mathcal{Y} = \{1\} \times \mathcal{Y}_1$ and $\mathcal{Y}^* = \{1\} \times \mathcal{Y}_1^*$. If we require these functions to be affine, the matrices F, H, and $\bar{G}$ must have the forms

$$(2.2.111) \quad F = \begin{bmatrix} 1 & 0 \\ F_0 & F_1 \end{bmatrix}, \quad H = \begin{bmatrix} 1 & 0 \\ 0 & H_1 \end{bmatrix}, \quad \bar{G} = \begin{bmatrix} 1 & 0 \\ \bar{G}_0 & \bar{G}_1 \end{bmatrix}.$$

The required degrouping matrix $\bar{H}$ will then be of the form

$$(2.2.112) \qquad \bar{H} = \begin{bmatrix} 1 & 0 \\ \bar{H}_0 & \bar{H}_1 \end{bmatrix}.$$

In formulas (2.2.88)-(2.2.90) above it will then only be necessary to replace the symbols H and y^* by H_1 and y_1^* respectively. This formulation includes the case in which f is homogeneous, since we can set $F_0 = 0$ in (2.2.111); while this does not require f^* to be homogeneous (unless $\bar{G}_0 = 0$), it is always possible to force f^* to be homogeneous if desired.

Penrose's necessary and sufficient condition for the existence of a solution $\bar{H}$ to (2.2.107) is

$$F\bar{G}G(HF\bar{G}G)^{-}HF\bar{G}G = F\bar{G}G \ , \tag{2.2.113}$$

as in (2.1.81). But this in turn is equivalent to

$$\text{rank } HF\bar{G}G = \text{rank } F\bar{G}G \ , \tag{2.2.114}$$

by Lemma 1.4. We shall take (2.2.114) to be our necessary and sufficient condition for the possibility of perfect relative disaggregation. If it holds, (2.2.107) has the general solution

$$\bar{H} = F\bar{G}G(HF\bar{G}G)^{-} + Z[I - HF\bar{G}G(HF\bar{G}G)^{-}] \ . \tag{2.2.115}$$

Defining $\mathcal{Y}_0 = F\bar{G}G\mathcal{X}$ and $\mathcal{Y}_0^* = HF\bar{G}G\mathcal{X}$ as in (2.1.83) and (2.1.84) condition (2.2.114) ensures that the transformation H induces a one-to-one correspondence between $\mathcal{Y}_0$ and $\mathcal{Y}_0^*$, and its inverse is given precisely by $\bar{H}$ when restricted to $\mathcal{Y}_0^*$; this is just what is stated by (2.2.107). Thus, (2.2.114) corresponds precisely to the condition specified in section 2.1 that the restriction of h to $\mathcal{Y}_0$ should induce a homeomorphism from $\mathcal{Y}_0$ to a retract $\mathcal{Y}_0^*$ of $\mathcal{Y}_0$. What (2.2.114) states is that in the case of linear transformations a sufficient (as well, of course, as necessary) condition for this is that $\mathcal{Y}_0$ and $\mathcal{Y}_0^*$ have the same dimension.

It is immediate from (2.2.115) and Lemma 1.4 that if

$$\text{rank } HF\bar{G}G = \text{rank } H \ , \tag{2.2.116}$$

then $\bar{H}$ is a generalized inverse of H.

In case the condition

$$\text{rank } HF\bar{G}G = \text{rank } HF\bar{G} \tag{2.2.117}$$

holds in addition to (2.2.114), then $G(H\bar{F}\bar{G}G)^- = (H\bar{F}\bar{G})^-$ by Lemma 1.4, hence (2.2.115) may be written

$$\bar{H} = F\bar{G}(HF\bar{G})^- + Z[I - HF\bar{G}(HF\bar{G})^-] \quad . \tag{2.2.118}$$

This will be the case, in particular, when G is a generalized inverse of $\bar{G}$, so that (2.2.107) becomes equivalent to the condition

$$\bar{H}HF\bar{G} = F\bar{G} \tag{2.2.119}$$

as in (2.1.77), i.e., $\hat{y} = \bar{H}\bar{F}^* x^* = F\bar{G}x^* = \tilde{y}$ for all $x^* \in \mathcal{X}^*$, and (2.2.114) reduces to

$$\text{rank } HF\bar{G} = \text{rank } F\bar{G} \quad . \tag{2.2.120}$$

The meaning of conditions (2.2.114) and (2.2.116) can be brought out by considering the case in which G and H have full rank and $\bar{G}$ is a generalized inverse of G. In that case, for given G and H (2.2.114) constitutes a restriction on F over and above any restriction that may hold of the type (2.2.36) or (2.2.6) (or even (2.2.109)). That is, even if such restrictions hold it is still possible in the absence of restriction (2.2.114) to ensure that

$$\text{rank } HFG^-G = \min(m^*, n^*), \ \text{rank } FG^-G = \min(m, n^*) \ , \tag{2.2.121}$$

in the sense that these equalities can be made to hold by means of a slight perturbation in the elements of F, if necessary. Given that $n > n^*$ and $m \geqq m^*$, if the elements of F are free to vary subject to restrictions such as (2.2.6) then (2.2.114) is essentially equivalent to the condition

$$m^* \geqq \min(m, n^*) \ , \tag{2.2.122}$$

that is, either $n^* \geqq m = m^*$ or $m > m^* \geqq n^*$. In words: _If no specific a priori conditions on_ F _are imposed other than conditions for perfect aggregation, an essential necessary and sufficient condition for perfect relative disaggregation is_ that the number of dependent variables in the aggregative model be either equal to the number of dependent variables in the detailed model, or at least as great as the number of independent variables in the aggregative model. This criterion is due to Malinvaud [68, p. 112].

In the same sense, we may say that (2.2.116) is essentially equivalent to the condition

$$n^* \geqq m^* \quad . \tag{2.2.123}$$

Taken together, conditions (2.2.114) and (2.2.116) are therefore essentially equivalent to

$$m^* = \min(m, n^*) \quad , \tag{2.2.124}$$

that is, either $n^* \geqq m = m^*$ or $m > m^* = n^*$.

While (2.2.122) provides a convenient rule of thumb that may be regarded as covering most cases that arise in practice (analogously to the role of the "order condition" for identifiability in simultaneous equations models--cf. Koopmans, Rubin, and Leipnik [59]), there do exist cases in which meaningful restrictions can be specified that guarantee fulfillment of the rank condition (2.2.114) even though the order condition (2.2.122) does not hold. We now provide such an example.

Let there be two individuals whose consumption expenditures are y_1 and y_2 respectively, whose incomes are x_1 and x_2 respectively, and whose wealth holdings are x_3 and x_4 respectively; let the aggregative model specify aggregate consumption $y_1^* = y_1 + y_2$ as an affine function of aggregate income $x_1^* = x_1 + x_2$ and aggregate wealth $x_2^* = x_3 + x_4$. In terms of the notation (2.2.78), (2.2.110), and (2.2.111) we have

$$G = \begin{bmatrix} 1 & 0 & 0 & 0 & 0 \\ 0 & 1 & 1 & 0 & 0 \\ 0 & 0 & 0 & 1 & 1 \end{bmatrix}, \quad H = \begin{bmatrix} 1 & 0 & 0 \\ 0 & 1 & 1 \end{bmatrix}, \quad F = \begin{bmatrix} 1 & 0 & 0 & 0 & 0 \\ f_{10} & f_{11} & f_{12} & f_{13} & f_{14} \\ f_{20} & f_{21} & f_{22} & f_{23} & f_{24} \end{bmatrix} .$$

Choosing $\bar{G} = G^{\dagger}$, the relevant rank condition is (2.2.120). We verify that

$$FG^{\dagger} = \begin{bmatrix} 1 & 0 & 0 \\ f_{10} & \frac{1}{2}(f_{11} + f_{12}) & \frac{1}{2}(f_{13} + f_{14}) \\ f_{20} & \frac{1}{2}(f_{21} + f_{22}) & \frac{1}{2}(f_{23} + f_{24}) \end{bmatrix}$$

and

$$HFG^{\dagger} = \begin{bmatrix} 1 & 0 & 0 \\ f_{10}+f_{20} & \frac{1}{2}(f_{11}+f_{21}+f_{12}+f_{22}) & \frac{1}{2}(f_{13}+f_{23}+f_{14}+f_{24}) \end{bmatrix} .$$

Perfect relative disaggregation therefore requires

(d) $$\begin{vmatrix} f_{11} + f_{12} & f_{13} + f_{14} \\ f_{21} + f_{22} & f_{23} + f_{24} \end{vmatrix} = 0 .$$

On the other hand, perfect aggregation requires, in accordance with (2.2.6),

(a) $$f_{11} + f_{21} = f_{12} + f_{22}, \quad f_{13} + f_{23} = f_{14} + f_{24} .$$

Conditions (d) and (a) are certainly independent.

Now it is reasonable to assume that each person's consumption depends only on his own income and wealth, so that $f_{12} = f_{14} = 0$ and $f_{21} = f_{23} = 0$. If this additional assumption is made the above conditions cease to be independent since (a) obviously then implies (d). If we suppose (a) to hold then under this assumption we find readily that (2.2.118) yields

$$\bar{H} = FG^{\dagger}(HFG^{\dagger})^{\dagger} = \begin{bmatrix} 1 & 0 \\ f_{10} - f_{20} & \frac{1}{2} \\ f_{20} - f_{10} & \frac{1}{2} \end{bmatrix} .$$

The disaggregation rule, or "blowing-up" formula, thus becomes

$$\hat{y}_1 = f_{10} - f_{20} + \frac{1}{2} y_1^*$$

$$\hat{y}_2 = f_{20} - f_{10} + \frac{1}{2} y_1^* .$$

In the above example we have $\min(m, n^*) = \min(3, 3) = 3 > 2 = m^*$, hence (2.2.122) is violated; (2.2.114) nevertheless holds so long as $f_{12} = f_{14} = f_{21} = f_{23} = 0$ and $f_{11}/f_{13} = f_{22}/f_{24}$, so that (d) holds. Condition (2.2.116) also holds (as does (2.2.123)), hence $\bar{H}$ is a generalized inverse (in fact, a right inverse) of $\bar{H}$. The following is an example of a case in which (2.2.124) holds. Let

$$G = H = \begin{bmatrix} 1 & 0 & 0 \\ 0 & 1 & 1 \end{bmatrix}, \quad F = \begin{bmatrix} 1 & 0 & 0 \\ f_{10} & f_{11} & f_{12} \\ f_{20} & f_{21} & f_{22} \end{bmatrix} .$$

We find readily that

$$\bar{H} = FG^{\dagger}(HFG^{\dagger})^{\dagger} = \begin{bmatrix} 1 & 0 \\ \dfrac{f_{10}(f_{21}+f_{22}) - f_{20}(f_{11}+f_{12})}{f_{11}+f_{12}+f_{21}+f_{22}} & \dfrac{f_{11}+f_{12}}{f_{11}+f_{12}+f_{21}+f_{22}} \\ \dfrac{f_{20}(f_{11}+f_{12}) - f_{10}(f_{21}+f_{22})}{f_{11}+f_{12}+f_{21}+f_{22}} & \dfrac{f_{21}+f_{22}}{f_{11}+f_{12}+f_{21}+f_{22}} \end{bmatrix}.$$

As before, $\bar{H}$ is a right inverse of H.

The example used to illustrate Case A is another one in which perfect relative disaggregation is possible (whether or not the condition (2.2.10) for perfect aggregation holds), and as we shall see in section 2.4 there is an important class of models--the Leontief models--in which (2.2.124) holds and conditions (2.2.114) and (2.2.116) are both satisfied.

If $F^* = HF\bar{G}$ has rank m^* then F^{*-} is a right inverse of F^*, hence if (2.2.120) holds the solution (2.2.118) becomes $\bar{H} = F\bar{G}F^{*-}$, which is invariant with respect to the choice of F^{*-}. If also $m > m^* = n^*$ then of course F^{*-1} is the only possible choice.

Let us now verify that when $\bar{G}$ is chosen so as to satisfy any of the conditions for perfect or best approximate aggregation, there is no loss of generality in choosing a particular solution for F^* of the form (2.2.103). If (2.2.6) holds and $\bar{G} = G^-$, it is obvious from (2.2.7) that the condition $\bar{H}F^*G = FG^-G$ for perfect relative disaggregation is equivalent to (2.2.107). If the domain of F is restricted to $\mathcal{X}_0$ as defined by (2.2.19), where $\bar{G}$ satisfies (2.2.16), perfect relative disaggregation requires that $\bar{H}F^*G\bar{G}G = F\bar{G}G\bar{G}G = F\bar{G}G$, and it is clear from (2.2.21) that this is again equivalent to (2.2.107). A similar analysis holds for the mixed case (2.2.42), where $\bar{G} = G^+$.

Turning to the stochastic formulation (2.2.56)-(2.2.58) we shall take as our criterion of almost perfect relative disaggregation the condition that the random variables $\hat{y}$ and $\tilde{y}$ of (2.2.105) and (2.2.106) be equal to one another with probability 1. By Tchebycheff's inequality a necessary and sufficient condition for this is that the mean square error

$$\mathcal{E}(\hat{y} - \tilde{y})(\hat{y} - \tilde{y})' = \mathcal{E}(\bar{H}\hat{y}^{*} - \tilde{y})(\bar{H}\hat{y}^{*} - \tilde{y})'$$

$$= (\bar{H}F^{*} - F\bar{G})GMG'(\bar{H}F^{*} - F\bar{G})' \qquad (2.2.125)$$

should vanish, and given (2.2.103) this is equivalent to the condition

$$\bar{H}HF\bar{G}GM = F\bar{G}GM \ . \qquad (2.2.126)$$

If, instead, F^{*} is given by (2.2.69) where $G^{\#}$ satisfies (2.2.70) then it is clear from property (i) of (2.2.70) that the vanishing of (2.2.125) is again equivalent to (2.2.126), where $\bar{G} = G^{\#}$. This shows that there is no loss of generality in choosing $Z^{*} = 0$ in (2.2.69).

Applying the Penrose solvability condition to (2.2.126) we obtain from Lemma 1.4

$$\text{rank } HF\bar{G}GM = \text{rank } F\bar{G}GM \ , \qquad (2.2.127)$$

which is our <u>necessary and sufficient condition for almost perfect relative disaggregation.</u>

If we choose $\bar{G} = G^{\#}$ in (2.2.126) and postmultiply both sides by $G'(GMG')^{-}$ then by (2.2.71) we obtain

$$\bar{H}HFG^{\#} = FG^{\#} \qquad (2.2.128)$$

for the particular choice of $G^{\#}$ given by (2.2.76). This coincides with (2.2.119) for $\bar{G} = G^{\#} = MG'(GMG')^{-}$.

<u>Link between stochastic and deterministic formulations.</u> It is interesting to note in comparing (2.2.127) and (2.2.114) that the stochastic formulation has the effect of

replacing the domain $\mathcal{X}$ of F and G by $M\mathcal{X}$, since (2.2.127) states that $\bar{H}HF\bar{G}Gx = F\bar{G}Gx$ for all $x \in M\mathcal{X}$. If Θ is singular, the distribution of x will be concentrated in a proper affine subspace of $\mathcal{X}$ whose dimension is the rank of Θ; in general this affine subspace, or hyperplane, does not coincide with $M\mathcal{X}$ but is a subset of it; however, there are two interesting cases in which it does coincide with it, and a detailed analysis of these cases may be found helpful in gaining a concrete picture of the situation.

Let S be an $n \times n$ matrix such that $SS' = \Theta$, and let N be any matrix polar to Θ, so that $NS = 0$ and rank N + rank S = n. Since $\mathcal{E}N(x - \mu) = 0$ and $\mathcal{E}N(x - \mu)(x - \mu)'N' = N\Theta N' = 0$ from (2.2.57), it follows that $N(x - \mu) = 0$ with probability 1, i.e., that $x - \mu$ is almost surely in the column null space of N, which is the same as the column space of S (see for instance Thrall and Tornheim [108, p.80]); thus, $x - \mu = Sz$ for some $z \in \mathcal{X}$, with probability 1. The distribution of x is therefore concentrated in the affine subspace (hyperplane)

$$\mathcal{X}_0 = \{x \in \mathcal{X}: x = \mu + Sz \text{ for some } z \in \mathcal{X}\} . \tag{2.2.129}$$

Letting z be a random variable with $\mathcal{E}z = 0$ and $\mathcal{E}zz' = S'(SS')^{-}S$, we may represent the random variable x as $x = \mu + Sz$ with probability 1.

As our first case let $\mathcal{X} = \{1\} \times \{\mathcal{X}_1\}$, so that x, μ, and Θ have the representation (2.2.78), (2.2.80) and define

$$S = \begin{bmatrix} 0 & 0 \\ 0 & S_1 \end{bmatrix}, \quad L = \begin{bmatrix} 1 & 0 \\ \mu_1 & S_1 \end{bmatrix}, \tag{2.2.130}$$

where S_1 is such that $S_1S_1' = \Theta_1$. Then any $x \in \mathcal{X}_0$ has the representation

$$(2.2.131)\quad \begin{bmatrix} 1 \\ x_1 \end{bmatrix} = \begin{bmatrix} 1 \\ \mu_1 \end{bmatrix} + \begin{bmatrix} 0 & 0 \\ 0 & S_1 \end{bmatrix} \begin{bmatrix} 1 \\ z_1 \end{bmatrix} = \begin{bmatrix} 1 \\ \mu_1 + S_1 z_1 \end{bmatrix} = \begin{bmatrix} 1 & 0 \\ \mu_1 & S_1 \end{bmatrix} \begin{bmatrix} 1 \\ z_1 \end{bmatrix},$$

i.e., $x = \mu + Sz = Lz$. From (2.2.130) and (2.2.81) we have $LL' = M$, and since $LL'(LL')^- L = L$ it follows that $x = Lz = Mv$, where $v = (LL')^- Lz$. Thus, the distribution of x is concentrated in $M\mathcal{X}$.

The second case is that in which the transformations are constrained to be homogeneous in the substantive variables, and the linear restriction defined by (2.2.129) is also homogeneous, i.e., $\mathcal{X}_0$ is a linear subspace of $\mathcal{X}$. This means that $\mathcal{X}_0$ contains the origin, hence $\mu = -\Theta u$ for some u; therefore $N\mu = 0$, and thus $Nx = 0$ for almost all x. Defining $L = [\mu, S]$ we have $LL' = \mu\mu' + \Theta = M$ from (2.2.58). L has the same rank as S, since μ is in the column space of S, and thus N is polar to M; consequently, x is almost surely in the column space of M, that is, the distribution of x is concentrated in the linear subspace $\mathcal{X}_0 = M\mathcal{X}$.

F. Best approximate disaggregation. We have taken as our criterion of almost perfect relative disaggregation bias the vanishing of the matrix (2.2.125); accordingly, this matrix will be defined as the matrix of relative disaggregation bias. As was remarked earlier, the reason for concentrating on the discrepancy between $\hat{y}$ and $\tilde{y} = F\bar{G}Gx$ rather than the discrepancy between $\hat{y}$ and $y = Fx$ is that we can expect the equality $\hat{y} = \tilde{y}$ to hold in some practical cases, but we can rarely expect to have $\hat{y} = y$. However, it is the latter discrepancy that is obviously the fundamental one in analyzing disaggregation bias. Accordingly, we define the matrix of absolute disaggregation bias as the mean square error

$$\begin{aligned} \mathcal{E}(\hat{y} - y)(\hat{y} - y)' &= \mathcal{E}(H\hat{y}^* - y)(H\hat{y}^* - y)' \\ (2.2.132)\qquad &= (\bar{H}F^*G - F)M(\bar{H}F^*G - F)' . \end{aligned}$$

Our criterion for <u>best approximate absolute disaggregation</u> is then defined by the condition that $\bar{H}$ be chosen so as to minimize (2.2.132).

With F^* given by (2.2.103) and $\hat{y}^*$ by (2.2.104) we verify from (2.2.56) and (2.2.58) that

$$\mathcal{E}\begin{bmatrix} y \\ \hat{y}^* \end{bmatrix}\begin{bmatrix} y \\ \hat{y}^* \end{bmatrix}' = \begin{bmatrix} FMF' & FMG'\bar{G}'F'H' \\ HF\bar{G}GMF' & HF\bar{G}GMG'\bar{G}'F'H' \end{bmatrix}. \tag{2.2.133}$$

It follows from Theorem 1.16 that (2.2.132) achieves a minimum with respect to $\bar{H}$ if and only if $\bar{H}$ is a solution of the equation

$$\bar{H}HF\bar{G}GMG'\bar{G}'F'H' = FMG'\bar{G}'F'H' \ . \tag{2.2.134}$$

For any such $\bar{H}$ we have

$$\hat{y} = \bar{H}\hat{y}^* = \hat{\rho}(y\,|\,\hat{y}^*) \ , \tag{2.2.135}$$

i.e., $\bar{H}\hat{y}^*$ is the best homogeneous linear predictor of y given $\hat{y}^*$.

We may of course equally well consider the criterion of <u>best approximate relative disaggregation</u>, defined by the condition that $\bar{H}$ be chosen so as to minimize (2.2.125). With $\tilde{y}$ given by (2.2.106) we verify as above that

$$\mathcal{E}\begin{bmatrix} \tilde{y} \\ \hat{y}^* \end{bmatrix}\begin{bmatrix} \tilde{y} \\ \hat{y}^* \end{bmatrix}' = \begin{bmatrix} F\bar{G}GMG'\bar{G}'F' & F\bar{G}GMG'\bar{G}'F'H' \\ HF\bar{G}GMG'\bar{G}'F' & HF\bar{G}GMG'\bar{G}'F'H' \end{bmatrix}, \tag{2.2.136}$$

and the minimizing $\bar{H}$ is any solution of the equation

$$\bar{H}HF\bar{G}GMG'\bar{G}'F'H' = F\bar{G}GMG'\bar{G}'F'H' \ . \tag{2.2.137}$$

We shall denote such a solution $\bar{H}$ by $H^{\#}$ since, defining

$$W = F\bar{G}GMG'\bar{G}'F' \quad , \tag{2.2.138}$$

we see that $H^{\#}$ is any solution of

$$H^{\#}HWH' = WH' \tag{2.2.139}$$

and is thus a generalized quasi-inverse of H satisfying

$$\text{(i)} \quad HH^{\#}HW = HW, \quad \text{(iv)} \quad H^{\#}HW = WH'H^{\#\prime} \quad . \tag{2.2.140}$$

The best homogeneous linear predictor of $\tilde{y} = F\bar{G}^{*}x^{*}$ given $\hat{y}^{*} = HF\bar{G}x^{*}$ is then given by the random variable

$$\hat{\tilde{y}} = H^{\#}\hat{y}^{*} = \hat{\rho}(\tilde{y}\,|\,\hat{y}^{*}) \quad . \tag{2.2.141}$$

In the above development it has been assumed that F^{*} is given by (2.2.103), where $\bar{G}$ is arbitrary. Consider, however, the case in which F^{*} is given by (2.2.69). Then it is clear that the above formulas (2.2.133)-(2.2.141) go through without change, with $\bar{G}$ replaced by $G^{\#}$. With $\bar{G} = G^{\#}$, however, it follows from (2.2.71) that equations (2.2.134) and (2.2.137) are identical. Thus, <u>when F^{*} is chosen in accordance with the criterion of best approximate aggregation, the criteria of best approximate absolute and relative disaggregation turn out to be the same</u>. That is, we then have

$$\hat{y} = \hat{\rho}(y\,|\,\hat{y}^{*}) = \bar{H}\hat{y}^{*} = H^{\#}\hat{y}^{*} = \hat{\rho}(\tilde{y}\,|\,\hat{y}^{*}) = \hat{\tilde{y}} \quad . \tag{2.2.142}$$

This leads us to our basic theorem on best approximate disaggregation.

<u>Theorem 2.2.</u> The matrix (2.2.132) of absolute disaggregation bias attains a minimum with respect to $\bar{H}$ if and only if

$$\bar{H} = F(F^*G)^{\#} + Z[I - F^*G(F^*G)^{\#}] \ , \tag{2.2.143}$$

where $(F^*G)^{\#}$ is any generalized quasi-inverse of F^*G with respect to M, satisfying properties (i) and (iv) of (2.2.70) with G replaced by F^*G. The minimum absolute disaggregation bias is

$$\inf_{\bar{H}} (\bar{H}F^*G - F)M(\bar{H}F^*G - F)' = F[I - (F^*G)^{\#}F^*G]MF' \ . \tag{2.2.144}$$

The matrix (2.2.125) of relative disaggregation bias, defined with respect to the model (2.2.103), attains a minimum with respect to $\bar{H}$ if and only if $\bar{H} = H^{\#}$, where $H^{\#}$ is any generalized quasi-inverse of H with respect to W, satisfying (2.2.140), where W is given by (2.2.138). The general solution of (2.2.139) gives the explicit expression

$$H^{\#} = F\bar{G}G(HF\bar{G}G)^{\#} + Z[I - HF\bar{G}G(HF\bar{G}G)^{\#}] \ , \tag{2.2.145}$$

where $(HF\bar{G}G)^{\#} = (F^*G)^{\#}$ is as defined above. An alternative expression is

$$H^{\#} = F\bar{G}(HF\bar{G})^{\#} + Z[I - HF\bar{G}(HF\bar{G})^{\#}] \ , \tag{2.2.146}$$

where $(HF\bar{G})^{\#}$ is any generalized quasi-inverse of $HF\bar{G}$ with respect to GMG' (i.e., satisfying properties (i) and (iv) of (2.2.70) with G replaced by $HF\bar{G}$ and M replaced by GMG'). The minimum relative disaggregation bias is equal to

$$\begin{aligned} \inf_{\bar{H}} (I - \bar{H}H)W(I - \bar{H}H)' &= (I - H^{\#}H)W \\ &= F\bar{G}G[I - (HF\bar{G}G)^{\#}HF\bar{G}G]MG'\bar{G}'F' \ , \end{aligned} \tag{2.2.147}$$

which can also be expressed as

(2.2.148) $$(I - H^{\#}H)W = F\bar{G}[I - (HF\bar{G})^{\#}HF\bar{G}]GMG'\bar{G}'F' .$$

If F^{*} is given by (2.2.69), so that $\bar{G}$ in (2.2.125) is replaced by $G^{\#}$, the expressions (2.2.145) and (2.2.146) remain valid, with $\bar{G} = G^{\#}$; further,

(2.2.149) $$(F^{*}G)^{\#} = (HFG^{\#}G)^{\#} = G^{\#}G(HFG^{\#}G)^{\#} = G^{\#}G(F^{*}G)^{\#} ,$$

so that the solutions (2.2.143) and (2.2.145) are the same, i.e., the absolute and relative disaggregation bias are simultaneously minimized if and only if $\bar{H} = H^{\#}$, where

(2.2.150) $$H^{\#} = F(HFG^{\#}G)^{\#} + Z[I - HFG^{\#}G(HFG^{\#}G)^{\#}] .$$

Under these same circumstances we have

(2.2.151) $$\begin{aligned}&(H^{\#}F^{*}G - F)M(H^{\#}F^{*}G - F)' \\ &\geqslant (H^{\#}F^{*} - FG^{\#})GMG'(H^{\#}F^{*} - FG^{\#})' ,\end{aligned}$$

i.e., the minimum absolute disaggregation bias exceeds or equals the minimum relative disaggregation bias; and equality holds in (2.2.151) if and only if F satisfies

(2.2.152) $$FM = FG^{\#}GM .$$

The minimum relative disaggregation bias is equal to zero, i.e., relative disaggregation is almost perfect, if and only if

(2.2.153) $$\text{rank } HFG^{\#}GM = \text{rank } FG^{\#}GM .$$

If rank HWH' = rank H, i.e., if

(2.2.154) $$\text{rank } HFG^{\#}GM = \text{rank } H ,$$

then $H^{\#}$ is a generalized inverse of H. If

$$\text{rank } HFG^{\#}GM = \text{rank } HFG^{\#}G \tag{2.2.155}$$

then $(HFG^{\#}G)^{\#}$ is a generalized inverse of $HFG^{\#}G$, and if also rank $GMG' = $ rank G then $(HFG^{\#}G)^{\#}$ is a generalized inverse of $F^{*}G$.

Proof. The proof of the first part of the theorem is virtually identical with that of Theorem 2.1, with an obvious change of notation. Equalities (2.2.149) follow from (2.2.71) and the explicit general solutions of the equations analogous to (2.2.71) for the respective generalized quasi-inverses.

To establish (2.2.151) and (2.2.152), let us define

$$K = [I - (HFG^{\#}G)^{\#}HFG^{\#}G]M[I - (HFG^{\#}G)^{\#}HFG^{\#}G]' . \tag{2.2.156}$$

We verify that

$$(I - G^{\#}G)K = (I - G^{\#}G)M , \tag{2.2.157}$$

so that $G^{\#}$ is a generalized quasi-inverse of G not only with respect to M but also with respect to K in place of M in (2.2.70). It follows that

$$K - G^{\#}GKG'G^{\#'} = (I - G^{\#}G)K(I - G^{\#}G)' \tag{2.2.158}$$

and consequently, just as in Lemma 1.2,

$$FKF' \geqslant FG^{\#}GKG'G^{\#'}F' , \tag{2.2.159}$$

with equality holding if and only if

$$F(I - G^{\#}G)K = F(I - G^{\#}G)M = 0 . \tag{2.2.160}$$

The remaining statements of the theorem are straightforward consequences of Lemma 1.4. Q. E. D.

Condition (2.2.152) is the stochastic counterpart of (2.2.109), and as was pointed out with respect to the latter it can rarely be expected to hold. Consequently the inequality (2.2.151) can generally be expected to be strict, i.e., we can generally expect the minimum absolute disaggregation bias to exceed the minimum relative disaggregation bias whenever the aggregative model F^* has been chosen in accordance with the criterion of minimization of aggregation bias. Since, under these circumstances, the criteria of minimization of absolute and relative disaggregation bias lead to the same results, and since zero relative disaggregation bias can be expected to be achieved in a broad variety of applications, the criterion of minimization of relative disaggregation bias is a reasonable one to adopt provided the aggregative model has already been chosen so as to minimize aggregation bias.

We are thus led to the following stepwise, or lexicographic, criterion: choose the aggregative model $F^*: \mathcal{X}^* \to \mathcal{Y}^*$ and the associated degrouping transformation $G^\#: \mathcal{X}^* \to \mathcal{X}$ in such a way as to minimize aggregation bias, and, given this choice, choose the degrouping transformation $H^\#: \mathcal{Y}^* \to \mathcal{Y}$ so as to minimize absolute (or relative) disaggregation bias.

As a further justification of our criterion, it is extremely interesting to observe that whenever the conditions for almost perfect relative disaggregation hold and the degrouping transformation is given by (2.2.145), the transformation $G^\#$ is actually optimal from the standpoint of the criterion of minimization of absolute disaggregation bias. For, under these circumstances the absolute disaggregation bias (2.2.132) is equal to

$$\mathcal{E}(\tilde{y} - y)(\tilde{y} - y)' = F(I - \bar{G}G)M(I - \bar{G}G)'F' \tag{2.2.161}$$

$$= F(I - G^\#G)M(I - G^\#G)'F' + F(G^\# - \bar{G})GMG'(G^\# - \bar{G})'F' ,$$

and owing to (2.2.77) this is minimized if and only if $\bar{G}$ satisfies (2.2.70). This result has a particularly interesting application in the special case in which the conditions (2.2.6) for perfect aggregation hold. In that case, for any generalized inverse G^- of G, the solution (2.2.7) is optimal from the standpoint of minimization of aggregation bias (which is of course zero), and has the property that $y^* = F^* x^* = HFG^- x^*$ for all $x^* \epsilon G\mathcal{X}$, independently of the choice of generalized inverse G^-. On the other hand, $\tilde{y} = FG^- x^*$ will not be invariant (for all $x^* \epsilon G\mathcal{X}$) with respect to the choice of G^- (unless the much stronger condition (2.2.109) holds--which we shall rule out), even though every such choice of G^- must yield a zero relative disaggregation bias. If $\bar{H}$ is given by (2.2.145) (as opposed to (2.2.143)), different choices of generalized inverse G^- will in general yield different (and non-zero) values for the absolute disaggregation bias, and only those which satisfy property (iv) of (2.2.70) will render this absolute disaggregation bias a minimum. An important application of this result arises in the case of affine transformations; to this we now turn.

Treatment in terms of affine transformations. The above results may now be applied to the affine representation (2.2.78), (2.2.80), (2.2.110), and (2.2.111). Let F^* be given by (2.2.69) and $H^\#$ by (2.2.150), so that the random variables of (2.2.104)-(2.2.106) are equal to

$$(2.2.162) \quad \hat{y}^* = HFG^\# Gx, \quad \hat{y} = H^\# \hat{y}^* = H^\# HFG^\# Gx, \quad \tilde{y} = FG^\# Gx$$

with probability 1. Letting $G^\#$ and $G_1^\natural$ be given by (2.2.84) and observing from (2.2.80) that $G^\# G\mu = \mu$, we shall define the vectors η and η_1 by

$$(2.2.163) \qquad \eta = \mathcal{E}\tilde{y} = FG^\# G\mu = \begin{bmatrix} 1 \\ F_0 + F_1\mu_1 \end{bmatrix} = \begin{bmatrix} 1 \\ \eta_1 \end{bmatrix},$$

and the matrices Υ and Υ_1 by

(2.2.164)

$$\Upsilon = \text{Var}\, y = FG^\# G\Theta G' G^{\#\prime} G' F' = \begin{bmatrix} 0 & \\ 0 & F_1 G_1^\natural G_1 \Theta G_1' G_1^{\natural\prime} F_1' \end{bmatrix} = \begin{bmatrix} 0 & 0 \\ 0 & \Upsilon_1 \end{bmatrix}.$$

We verify readily that these expressions are related to the expression (2.2.138) for W (with $\bar{G} = G^{\#}$) by

$$W = \mathcal{E}\tilde{y}\tilde{y}' = \begin{bmatrix} 1 & \eta_1' \\ \eta_1 & \eta_1\eta_1' + \Upsilon_1 \end{bmatrix} = \eta\eta' + \Upsilon \ . \tag{2.2.165}$$

One solution of (2.2.139) is then verified to be

$$H^{\#} = \begin{bmatrix} 1 & 0 \\ (I - H_1^{\natural}H_1)\eta_1 & H_1^{\natural} \end{bmatrix}, \text{ where } H_1^{\natural} = \Upsilon_1 H_1'(H_1\Upsilon_1 H_1')^{-} \ . \tag{2.2.166}$$

Defining the function $h\colon \mathcal{Y}_1 \to \mathcal{Y}_1^*$ by

$$h(y_1) = H_1 y_1 \ , \tag{2.2.167}$$

the optimal degrouping function $h^{\natural}\colon \mathcal{Y}_1^* \to \mathcal{Y}_1$ is equal to

$$\hat{y}_1 = h^{\natural}(\hat{y}_1^*) = \hat{\mathcal{E}}(y_1 | \hat{y}_1^*) = (I - H_1^{\natural}H_1)\eta_1 + H_1^{\natural}\hat{y}_1^* \tag{2.2.168}$$

with probability 1, and since $\bar{G} = G^{\#}$ this is the same as $\hat{\mathcal{E}}(\tilde{y}_1 | \hat{y}_1^*)$.

Explicit expressions for $\hat{y}_1^*$ and $\tilde{y}_1$ are readily obtained. Replacing H and y^* in (2.2.89) by H_1 and y_1^* respectively, we obtain for the function $f\colon \mathcal{X}_1^* \to \mathcal{Y}_1^*$,

$$\hat{y}_1^* = f^*(x_1^*) = \hat{\mathcal{E}}(y_1^* | x_1^*) = H_1\eta_1 + H_1F_1G_1^{\natural}(x_1^* - G_1\mu_1) \ . \tag{2.2.169}$$

Likewise, (2.2.162), (2.2.111), (2.2.84), and (2.2.78) yield

(2.2.170) $\tilde{y}_1 = f \circ g^{\natural}(x_1^*) = \hat{\mathcal{E}}(y_1 | x_1^*) = \eta_1 + F_1 G_1^{\natural}(x_1^* - G_1\mu_1)$.

Of course, $\hat{y}_1^* = H_1\tilde{y}_1 = h(\tilde{y}_1) = h \circ f \circ g^{\natural}(x_1^*)$.

It follows from (2.2.147), (2.2.165), and (2.2.166) that a necessary and sufficient condition for almost perfect relative disaggregation is that

$$(2.2.171) \qquad (I - H^{\#}H)W = \begin{bmatrix} 0 & 0 \\ 0 & (I - H_1^{\natural}H_1)\Upsilon_1 \end{bmatrix} = \begin{bmatrix} 0 & 0 \\ 0 & 0 \end{bmatrix} .$$

Noting from (2.2.166) and Lemma 1.4 that $H_1'(H_1\Upsilon_1H_1')^- = (H_1\Upsilon_1)^-$ we see that (2.2.171) is equivalent to the condition

$$\Upsilon_1 = H_1^{\natural}H_1\Upsilon_1 = \Upsilon_1(H_1\Upsilon_1)^-H_1\Upsilon_1 ,$$

which by Lemma 1.4 is in turn equivalent to the condition that rank $H_1\Upsilon_1$ = rank Υ_1, or again,

$$(2.2.172) \qquad \text{rank } H_1F_1G_1^{\natural}G_1\Theta_1 = \text{rank } F_1G_1^{\natural}G_1\Theta_1 .$$

Now a question of particular interest that arises, notably in connection with Leontief models to be discussed in section 2.4, concerns the homogeneity of the various functions. In Leontief models the function $f: \mathcal{X}_1 \to \mathcal{Y}_1$ is homogeneous, i.e., $F_0 = 0$ in (2.2.111). It is the practice, also, to require $f^*: \mathcal{X}_1^* \to \mathcal{Y}_1^*$ to be homogeneous. We may well ask, however, whether or under what conditions the optimal functions f^*, $g^{\natural}$, and $h^{\natural}$ will be homogeneous, assuming that f, g, and h are. We now turn to this question.

First, let us consider conditions under which the optimal f^* is homogeneous. Owing to (2.2.163) we can write (2.2.169) in the form

$$(2.2.173)\qquad f^*(x_1) = H_1F_0 + H_1F_1(I - G_1^\natural G_1)\mu_1 + H_1F_1G_1^\natural x_1^* .$$

If $F_0 = 0$, a necessary and sufficient condition for homogeneity of f^* is then $H_1F_1(I - G_1^\natural G_1)\mu_1 = 0$. From (2.2.72), (2.2.111), (2.2.84), (2.2.78), and (2.2.81), a necessary and sufficient condition for almost perfect aggregation is

$$(2.2.174)\qquad HF(I - G^\# G)M = \begin{bmatrix} 0 & 0 \\ 0 & H_1F_1(I - G^\natural G_1)\Theta_1 \end{bmatrix} = \begin{bmatrix} 0 & 0 \\ 0 & 0 \end{bmatrix} .$$

The following three conditions are therefore sufficient for f^* to be homogeneous: (1) that f itself be homogeneous, i.e., $F_0 = 0$; (2) that condition (2.2.174) for almost perfect aggregation hold; and (3) that μ_1 be in the column space of Θ_1. This third condition is automatically satisfied if Θ is nonsingular; if Θ is singular it means that the implied almost sure linear restriction on x_1 must be homogeneous.

Next, let us consider the function $g^\natural$. From (2.2.84) and (2.2.78) this is given by

$$(2.2.175)\qquad g^\natural(x_1^*) = (I - G_1^\natural G_1)\mu_1 + G_1^\natural x_1^* .$$

A sufficient condition for $g^\natural$ to be homogeneous is then that

$$(2.2.176)\qquad (I - G_1^\natural G_1)\Theta_1 = 0 \ \text{and}\ \mu_1 = \Theta_1\zeta_1 .$$

In terms of the representation (2.2.131), where $S_1S_1' = \Theta$, (2.2.176) implies that

$$(2.2.177)\qquad x_1 = G_1^\natural G_1x_1 = G_1^\natural x_1^*$$

with probability 1. Condition (2.2.177) implies directly that $g^\natural$ is homogeneous.

Another condition considerably weaker than (2.2.176) turns out to play a significant role in the analysis of input-output models. Writing the necessary and sufficient condition for homogeneity of $g^{\natural}$ as

$$\mu_1 = G_1^{\natural} G_1 \mu_1 \quad , \tag{2.2.178}$$

it states that μ_1 belongs to the column space of $G_1^{\natural} G_1$. A sufficient condition for this is that μ_1 belong to the column space of $\Theta_1 G_1'$, i.e., that

$$\mu_1 = \Theta_1 G_1' \zeta_1^* \tag{2.2.179}$$

for some $(n^* - 1) \times 1$ vector ζ_1^*, since $\Theta_1 G_1' = G_1^{\natural} G_1 \Theta_1 G_1'$ by the defining property of generalized quasi-inverses. It is clear that (2.2.176) implies (2.2.179), with $\zeta_1^* = G_1^{\natural\prime} \zeta_1$. Further, if the second condition of (2.2.176) is assumed to hold--stating that μ_1 belongs to the column space of Θ (which of course is necessarily the case if Θ is non-singular)--then (2.2.179) is actually implied by (2.2.178), i.e., it becomes necessary as well as sufficient for the homogeneity of $g^{\natural}$. We shall thus take (2.2.179) as our basic sufficient condition; in the special case in which Θ_1 is a diagonal matrix and G_1 has exactly one unit element in each column, it states simply that within each group of detailed explanatory variables aggregated into a single explanatory variable, the variances of these detailed explanatory variables are proportional to their means.

Finally, let us consider the function $h^{\natural}$. From (2.2.163) and (2.2.168), if $F_0 = 0$ then a necessary and sufficient condition for homogeneity of $h^{\natural}$ is $(I - H_1^{\natural} H_1) F_1 \mu_1 = 0$. It follows that the following three conditions are sufficient for $h^{\natural}$ to be homogeneous: (1) that f itself be homogeneous; (2) that condition (2.2.171) for almost perfect relative disaggregation hold; and (3) that μ_1 belong to the column space of $\Theta_1 G_1'$, i.e., that (2.2.179) hold; for, under these conditions we have, recalling the definition of Υ_1 in (2.2.164),

$$(I - H_1^{\natural} H_1)\eta_1 = (I - H_1^{\natural} H_1)F_1G_1^{\natural}G_1\Theta_1G_1'\zeta_1^* = 0 \ .$$

It is clear from (2.2.173) that these three conditions are also sufficient for the homogeneity of f^* .

The role of condition (2.2.179) may be brought out by observing that the following four conditions are not sufficient for the homogeneity of $h^{\natural}$: (1) homogeneity of f; (2) condition (2.2.174) for almost perfect aggregation; (3) condition (2.2.171) for almost perfect relative disaggregation; and (4) the second equality of (2.2.176). Under these conditions we have

$$H_1^{\natural} H_1\eta_1 = H_1^{\natural}H_1F_1\mu_1 = H_1^{\natural}H_1F_1G_1^{\natural}G_1\Theta_1\zeta_1 = F_1G_1^{\natural}G_1\mu_1 \ ,$$

and this is equal to η_1 --hence (2.2.168) is homogeneous--if and only if

(2.2.180) $$F_1(I - G_1^{\natural}G_1)\mu_1 = 0 \ .$$

A sufficient condition for (2.2.180) is thus (2.2.178) (or equivalently, (2.2.179)); another is $F_1 = F_1G_1^{\natural}G_1$, which corresponds to (2.2.109) and is much stronger than the condition $H_1F_1 = H_1F_1G_1^{\natural}G_1$ for perfect aggregation corresponding to (2.2.6). To summarize, then: if f, g, and h are homogeneous, condition (2.2.179) is sufficient for the homogeneity of f^* and $g^{\natural}$, and together with (2.2.171) is sufficient for the homogeneity of $h^{\natural}$.

As a simple illustration, let x_1 and x_2 denote the incomes of individuals 1 and 2, and y_1 and y_2 their consumption expenditures. (We now denote by x_i and y_i the ith component of the 2×1 vectors denoted by x_1 and y_1 in (2.2.78) and (2.2.110), and similarly for μ_i in relation to (2.2.80).) Let F, G, and H be given by

$$F = \begin{bmatrix} 1 & 0 & 0 \\ f_{10} & f_{11} & f_{12} \\ f_{20} & f_{21} & f_{22} \end{bmatrix}, \quad G = H = \begin{bmatrix} 1 & 0 & 0 \\ 0 & 1 & 1 \end{bmatrix}$$

and μ, Θ, and M by

$$\mu = \begin{bmatrix} 1 \\ \mu_1 \\ \mu_2 \end{bmatrix}, \quad \Theta = \begin{bmatrix} 0 & 0 & 0 \\ 0 & \theta_1 & 0 \\ 0 & 0 & \theta_2 \end{bmatrix}, \quad M = \begin{bmatrix} 1 & \mu_1 & \mu_2 \\ \mu_1 & \mu_1^2 + \theta_1 & \mu_1\mu_2 \\ \mu_2 & \mu_2\mu_1 & \mu_2^2 + \theta_2 \end{bmatrix}.$$

We find that, with $\theta_1 + \theta_2 > 0$,

$$G^{\#} = \begin{bmatrix} 1 & 0 \\ \theta_2'\mu_1 - \theta_1'\mu_2 & \theta_1' \\ -\theta_2'\mu_1 + \theta_1'\mu_2 & \theta_2' \end{bmatrix}, \quad G^{\dagger} = \begin{bmatrix} 1 & 0 \\ 0 & \frac{1}{2} \\ 0 & \frac{1}{2} \end{bmatrix},$$

where $\theta_i' = \theta_i/(\theta_1 + \theta_2)$. Thus, $g^{\natural}$ is homogeneous if and only if $\mu_1/\theta_1 = \mu_2/\theta_2$, i.e., the variances are proportional to the means. (If $\theta_i = 0$ then $\mu_i = 0$.) We find for the optimal aggregative model

$$HFG^{\#} = \begin{bmatrix} 1 & 0 \\ f_{10}+f_{20}+(\theta_2'\mu_1-\theta_1'\mu_2)(f_{11}+f_{21}-f_{12}-f_{22}) & \theta_1'(f_{11}+f_{21})+\theta_2'(f_{12}+f_{22}) \end{bmatrix}.$$

If f is homogeneous, so that $f_{10} = f_{20} = 0$, the function $f^* = hfg^{\natural}$ will be homogeneous if either (1) the variances are proportional to the means, or (2) the condition $f_{11} + f_{21} = f_{12} + f_{22}$ for perfect aggregation holds. In the latter event, HFG^- is invariant with respect to the choice of G^-, and we verify in particular that $HFG^{\#} = HFG^{\dagger}$, where

$$HFG^{\dagger} = \begin{bmatrix} 1 & 0 \\ f_{10} + f_{20} & \frac{1}{2}(f_{11} + f_{21} + f_{12} + f_{22}) \end{bmatrix} .$$

Now let $f_{10} = f_{20} = 0$, and let us assume that $f_{11} + f_{21} > 0$ and $f_{12} + f_{22} > 0$. Defining

$$\lambda_i = \frac{\theta_1' f_{i1} + \theta_2' f_{i2}}{\theta_1'(f_{11}+f_{21}) + \theta_2'(f_{12}+f_{22})} \qquad (i = 1, 2) ,$$

we find from (2.2.143) that the optimal $\overline{H}$ is

$$\overline{H} = \begin{bmatrix} 1 & 0 \\ (\theta_2'\mu_1 - \theta_1'\mu_2)[\lambda_2(f_{11} - f_{12}) - \lambda_1(f_{21} - f_{22})] & \lambda_1 \\ -(\theta_2'\mu_1 - \theta_1'\mu_2)[\lambda_2(f_{11} - f_{12}) - \lambda_1(f_{21} - f_{22})] & \lambda_2 \end{bmatrix} .$$

The condition $\theta_2\mu_1 = \theta_1\mu_2$ is again sufficient for homogeneity.

If the conditions for perfect aggregation hold the above matrix simplifies to

$$\overline{H} = \begin{bmatrix} 1 & 0 \\ (\theta_2'\mu_1 - \theta_1'\mu_2)(f_{11} - f_{12}) & \lambda_1 \\ -(\theta_2'\mu_1 - \theta_1'\mu_2)(f_{11} - f_{12}) & \lambda_2 \end{bmatrix} ,$$

and an alternative sufficient condition for homogeneity is $f_{11} = f_{12}$ (implying $f_{21} = f_{22}$). This condition corresponds to (2.2.109); in the present illustration the interpretation would be that an individual's consumption expenditures would be affected just as much by an increase in the other individual's income as by an equal increase in his own. This would obviously never be the case in practice; on the contrary, it would be more reasonable to assume $f_{12} = f_{21} = 0$, i.e., that an individual's expenditure depends only on his own income. Under this hypothesis, the condition $\theta_2\mu_1 = \theta_1\mu_2$ is necessary for the homogeneity of h^*.

When the condition $f_{11} + f_{21} = f_{12} + f_{22}$ for perfect aggregation holds, the second of the above expressions for $\bar{H}$ can of course equally well be computed by substituting $F^* = HFG^\dagger$ in (2.2.143). In accordance with Theorem 2.2 it can also be obtained by making the substitution $\bar{G} = G^\#$ in (2.2.145) or (2.2.146); however, the alternative substitution $\bar{G} = G^\dagger$ would in this case lead to an erroneous result, namely (when $f_{10} = f_{20} = 0$)

$$H^\# = \begin{bmatrix} 1 & 0 \\ 0 & (f_{11} + f_{12})/(f_{11} + f_{21} + f_{12} + f_{22}) \\ 0 & (f_{21} + f_{22})/(f_{11} + f_{21} + f_{12} + f_{22}) \end{bmatrix} .$$

F'. Optimal selection of grouping matrices. Theorem 2.2 makes it possible to extend the analysis of subsection D' so as to allow for disaggregation bias as well as aggregation bias. The generalized quasi-inverse $G^\#$ has been chosen so as to minimize aggregation bias, and, given this choice, the generalized quasi-inverse $H^\#$ has been chosen so as to minimize relative and therefore absolute disaggregation bias. A natural indicator of the overall bias is then the minimum absolute disaggregation bias (2.2.144) which, in view of (2.2.71), decomposes as follows:

$$(2.2.181)\qquad (F - H^{\#}HFG^{\#}G)M(F - H^{\#}HFG^{\#}G)' = F(I - G^{\#}G)M(I - G^{\#}G)'F' + (I - H^{\#}H)W(I - H^{\#}H)' \quad .$$

(Here, $H^{\#}$ is given by (2.2.150), and W by (2.2.138) with $\bar{G} = G^{\#}$.) The second term on the right side of (2.2.181) is the minimum relative disaggregation bias, whose vanishing defines the condition for almost perfect relative disaggregation, a necessary and sufficient condition for which is (2.2.153). If the first term also vanishes then, as already remarked, we may say that there is no aggregation problem.

Given a set $\mathcal{S}$ of pairs (G, H) of grouping matrices, for each such pair we choose an optimal $G^{\#}$ in accordance with the criterion of minimization of aggregation bias, and by Theorem 2.1 such a $G^{\#}$ is given by (2.2.70); given such a $G^{\#}$, we choose an optimal $H^{\#}$ in accordance with the criterion of minimization of (absolute or relative) disaggregation bias, and such an $H^{\#}$ is given by (2.2.150). This defines a pair $(G^{\#}, H^{\#})$ as a function of the pair (G, H) . For each such pair $(G^{\#}, H^{\#})$ we compute the M-norm of $F - H^{\#}HFG^{\#}G$, as defined by (2.2.99):

$$(2.2.182)\qquad \| F - H^{\#}HFG^{\#}G \|_M = \sqrt{\| F \|_M^2 - \| H^{\#}HFG^{\#}G \|_M^2} = \sqrt{\text{trace } (F - H^{\#}HFG^{\#}G)MF'} \quad .$$

When (2.2.153) holds this reduces to

$$(2.2.183)\qquad \sqrt{\| F \|_M^2 - \| FG^{\#}G \|_M^2} = \sqrt{\text{trace } F(I - G^{\#}G)MF'} \quad .$$

The norm (2.2.182) provides a numerical measure of minimum bias as a function of (G, H) . The final step that is required is to find a pair $(G, H) \in \mathcal{G}$ which minimizes this minimum bias.

An alternative numerical criterion is given by a generalization of Ijiri's [54] aggregation coefficient:

$$\rho = \sqrt{1 - \|F - H^{\#}HFG^{\#}G\|_M^2 / \|F_M\|^2} \tag{2.2.184}$$

$$= \|H^{\#}HFG^{\#}G\|_M / \|F\|_M .$$

This is a function of G and H, and the objective now would be to maximize ρ over $\mathcal{G}$. When the conditions for almost perfect relative disaggregation hold, (2.2.184) reduces to $\|FG^{\#}G\|_M / \|F\|_M$.

G. Best approximate simplification: W. D. Fisher's formulation. In the formulation of W. D. Fisher [34, 35] the proper object of concern is to find a simplified mapping $\hat{F}: \mathcal{X} \to \mathcal{Y}$ that approximates the "true" detailed mapping $F: \mathcal{X} \to \mathcal{Y}$. Aggregation and disaggregation play a role in this process only to the extent that $\hat{F}$ is postulated to be of the form $\hat{F} = \bar{H}F^*G$, where $G: \mathcal{X} \to \mathcal{X}^*$ and $\bar{H}: \mathcal{Y}^* \to \mathcal{Y}$ are given a priori, and $F^*: \mathcal{X}^* \to \mathcal{Y}^*$ is to be chosen optimally. Thus, we are given an aggregation operator on the explanatory or independent variables, but a disaggregation operator on the explained or dependent variables. The significance of this asymmetry will be discussed below.

Given the model (2.2.56)-(2.2.58) and the transformations G and $\bar{H}$, we seek a forecasting rule of the form

$$\hat{y} = \hat{F}x = \bar{H}F^*Gx \tag{2.2.185}$$

that minimizes expected loss, where the loss function is given by

$$L = (\hat{y} - y)'C(\hat{y} - y) , \tag{2.2.186}$$

C being a given symmetric non-negative definite matrix

which weighs the subjective costs of prediction errors.

We find readily that, for given F, the expected loss conditional on values of the independent variables is

(2.2.187) $$\mathcal{E}(L|x) = x'(\hat{F} - F)'C(\hat{F} - F)x + \text{trace } C\Sigma \quad ,$$

and thus the unconditional expected loss--again, for given F--is

(2.2.188) $$\mathcal{E}L = \mathcal{E}\{\mathcal{E}(L|x)\} = \text{trace}(\hat{F} - F)'C(\hat{F} - F)M + \text{trace } C\Sigma \ .$$

Since trace $C\Sigma$ is constant, our problem is:

(2.2.189) $$\text{Minimize trace } (\bar{H}F^*G - F)'C(\bar{H}F^*G - F)M \text{ with respect to } F^* \ .$$

This problem was solved by W. D. Fisher, initially [34, p. 756, Theorem 4.1] under the assumption that M and C, as well as G and $\bar{H}$, have full rank, and subsequently [35, p. 32, Theorem 2] on the assumption that G and $\bar{H}$, but not necessarily M and C, have full rank. The following version completes the generalization.

Theorem 2.3 (W. D. Fisher). The solution of (2.2.189) is given by

(2.2.190) $$F^* = \bar{H}^\# F G^\# + Z^* - \bar{H}^\# \bar{H} Z^* G G^\# \quad ,$$

where Z^* is arbitrary and $G^\#$ and $\bar{H}^\#$ are generalized quasi-inverses of G and $\bar{H}$ satisfying (2.2.70) and

(2.2.191) $$\text{(i) } C\bar{H}\bar{H}^\#\bar{H} = C\bar{H}; \quad \text{(iv) } C\bar{H}\bar{H}^\# = \bar{H}^{\#\prime}\bar{H}'C$$

respectively, or equivalently, satisfying the respective conditions

(2.2.192) $$G^\# G M G' = M G', \quad \bar{H}'C\bar{H}\bar{H}^\# = \bar{H}'C \ .$$

The minimum unconditional expected loss is equal to, apart from the constant term trace $C\Sigma$,

$$\inf_{F^*} \operatorname{trace}(\bar{H}F^*G - F)'C(\bar{H}F^*G - F)M$$

(2.2.193) $$= \operatorname{trace} C(F - \bar{H}\bar{H}^{\#}FG^{\#}G)MF' .$$

If rank GMG' = rank G and rank $\bar{H}'C\bar{H}$ = rank $\bar{H}$, the matrix $\hat{F} = \bar{H}F^*G$ which minimizes (2.2.187) is given by

(2.2.194) $$\hat{F} = \bar{H}\bar{H}^{\#}FG^{\#}G = \bar{H}(\bar{H}'C\bar{H})^{-}\bar{H}'CFMG'(GMG')^{-}G ,$$

which is invariant with respect to the choices of $G^{\#}$ and $\bar{H}^{\#}$.

Proof. From the decomposition

$$\bar{H}F^*G - F = \bar{H}(F^* - \bar{H}^{\#}FG^{\#})G - (F - \bar{H}\bar{H}^{\#}FG^{\#}G)$$

we find from (2.2.192) and successive use of the commutativity of the trace operation (trace AB = trace BA) that

(2.2.195) $$\operatorname{trace}(\bar{H}F^*G - F)'C(\bar{H}F^*G - F)M =$$
$$\operatorname{trace}(F - \bar{H}\bar{H}^{\#}FG^{\#}G)'C(F - \bar{H}\bar{H}^{\#}FG^{\#}G)M$$
$$+ \operatorname{trace}(F^* - \bar{H}^{\#}FG^{\#})'\bar{H}'C\bar{H}(F^* - \bar{H}^{\#}FG^{\#})GMG' .$$

The second term on the right is non-negative, and vanishes if and only if

(2.2.196) $$\bar{H}'C\bar{H}F^*GMG' = \bar{H}'CFMG'$$

on account of (2.2.192). We verify from (2.2.192) that (2.2.190) satisfies (2.2.196), hence (2.2.196) has a solution and the solution set includes (2.2.190). Conversely, the general solution of (2.2.196) is (2.2.190) for the particular choices of $G^{\#}$ and $\bar{H}^{\#}$ given by

$$G^{\#} = MG'(GMG')^{-}, \quad \bar{H}^{\#} = (\bar{H}'C\bar{H})^{-}\bar{H}'C \tag{2.2.197}$$

respectively. This proves that (2.2.190) defines the set of solutions to the problem (2.2.189).

The minimizing unconditional expected loss (2.2.188) is given (apart from the constant term trace $C\Sigma$) by the first term on the right in (2.2.195). By repeated use of the commutativity of the trace operation and of properties (2.2.70) and (2.2.191) we see that this may be expressed in the form (2.2.193). The final statement follows readily from Lemma 1.4.

Q.E.D.

As remarked above, Fisher's formulation involves an asymmetry in the treatment of independent and dependent variables, an aggregation operator being given in the former case but a disaggregation operator in the latter. As a result, the final aggregation operator $\bar{H}^{\#}$ obtained from Theorem 2.3, as is clear from (2.2.197), need not be a grouping matrix unless $\bar{H}$ itself has the formal properties of a grouping matrix (which is certainly reasonable) and C is block diagonal, with blocking conformable to that of $\bar{H}$ (which is not so reasonable); in fact $\bar{H}^{\#}$ need not even be a non-negative matrix. In such circumstances the aggregative model would lose its intuitive appeal, which is its raison d'être.

The alternative formulation that we have developed here has the advantage of taking H as the starting point rather than $\bar{H}$. On the other hand it has what might appear to be a drawback: instead of being cast in terms of the minimization of a single real-valued function such as (2.2.188), it involves the minimization (in the lexicographic sense) of a two-dimensional function, of which the first component is an indicator of aggregation bias and the second an indicator of disaggregation bias. It might be asked whether it would not be more appropriate to formulate the problem in terms of the minimization of

$$\text{trace}(F' - G'\bar{G}'F'H'\bar{H}')C(F - \bar{H}HF\bar{G}G)M \tag{2.2.198}$$

with respect to $\bar{G}$ and $\bar{H}$, for given G and H. This entails, so long as no non-negativity constraints are imposed on the elements of $\bar{G}$ and $\bar{H}$, the necessary conditions

$$F'H'\bar{H}'C(F - \bar{H}HF\bar{G}G)MG' = 0 , \tag{2.2.199a}$$

$$C(F - \bar{H}HF\bar{G}G)MG'\bar{G}'F'H' = 0 . \tag{2.2.199b}$$

Quite apart from the fact that these equations are difficult to solve, the solution is in general indeterminate as well as counter-intuitive, as examples will show. The reason is not far to seek: minimization of (2.2.198) alone implies that the aggregative model is not taken seriously in its own right, but used only as a means to obtain an optimal simplified model $\hat{F}: \mathcal{X} \to \mathcal{Y}$. If the latter objective is the only one, the aggregative model plays no essential role; the objective would therefore be better achieved by seeking an optimal $\hat{F}$ of reduced rank.

2.3. Consolidation of multivariate multiple regression models

2.3.1. Gauss-Markoff estimation relative to the consolidated model. We consider the multivariate multiple regression model

$$Y = XB + E, \quad \mathcal{E}E = 0 , \tag{2.3.1a}$$

where Y is an $n \times m$ matrix of n random observations on m jointly dependent or "endogenous" variables, X is an $n \times k$ matrix of the corresponding observations on k "independent" or "exogenous" variables (assumed to be "fixed variates"), B is a $k \times m$ matrix of unknown regression coefficients, and E is an $n \times m$ matrix of random errors with zero means. The independent variables may, alternatively, be considered to be random variables, in which case the expectation operator $\mathcal{E}$ represents the conditional expectation given X . Denoting by "row E" the row vector of consecutive rows of E, the description of the model is completed by specifying that

$$\mathcal{E}(\text{row } E)'(\text{row } E) = V \otimes \Sigma \quad , \tag{2.3.1b}$$

where V is an $n \times n$ symmetric non-negative definite matrix (the "sample covariance matrix") and Σ is an $m \times m$ symmetric non-negative definite matrix (the "contemporaneous covariance matrix"); the symbol $\otimes$ denotes the Kronecker product, i.e., $V \otimes \Sigma$ is the $nm \times nm$ matrix $[v_{st}\Sigma]$.

We introduce the transformations

$$X^* = XG, \quad Y^* = YH \quad , \tag{2.3.2}$$

where G and H are (column-wise) grouping matrices of orders $k \times k^*$ and $m \times m^*$ respectively, with $k^* < k$ and $m^* \leqq m$; these are defined as non-negative matrices with at most one positive element in each row.

A consolidated multivariate multiple regression model will be defined as a model

$$Y^* = X^* B^* + E^*, \quad \mathcal{E} E^* = 0 \quad , \tag{2.3.3a}$$

$$\mathcal{E}^*(\text{row } E^*)'(\text{row } E^*) = V \otimes \Sigma^* \quad , \tag{2.3.3b}$$

(where $\mathcal{E}^*$ is the corresponding expectation operator) which is in some sense consistent, either perfectly or only approximately, with the model (2.3.1) and the transformations (2.3.2).

This formulation can readily be reconciled with that of the preceding section, as follows. From the properties of Kronecker multiplication we have, for any conformable matrices A, B, C,

$$\text{row } ABC = \text{row } B \cdot (A' \otimes C); \ \text{col } ABC = (C' \otimes A)\text{col } B \ , \tag{2.3.4}$$

where "col B" denotes the column vector of consecutive columns of B . Thus, (2.3.1a) becomes

$$\text{(2.3.5)} \quad \text{row } Y = \text{row } X \cdot (I_n \otimes B) + \text{row } E; \quad \mathcal{E} \text{ row } E = 0 \ .$$

We may now take $\mathcal{X}$ to be the nk-dimensional space of observation matrices X, and $\mathcal{Y}$ the nm-dimensional space of expected values $\mathcal{E}Y$ of the dependent variables; the mapping f of section 2.1 is then defined by $\mathcal{E}$ row Y = row X $\cdot$ $(I \otimes B)$, where $I \otimes B$ corresponds to the transpose of the transformation F of section 2.2. Likewise, (2.3.2) becomes

$$\text{(2.3.6)} \quad \text{row } X^* = \text{row } X \cdot (I_n \otimes G), \quad \text{row } Y^* = \text{row } Y \cdot (I_n \otimes H) \ ,$$

and of course a similar formula obtains for (2.3.3a):

$$\text{(2.3.7)} \quad \text{row } Y^* = \text{row } X^* \cdot (I_n \otimes B^*) + \text{row } E^*; \ \mathcal{E}^* \text{ row } E^* = 0 \ .$$

These define linear transformations to and between the nk^*- and nm^*- dimensional spaces $\mathcal{X}^*$ and $\mathcal{Y}^*$ respectively.

The present problem differs from that of the preceding section in that the aggregation problem is now combined with the estimation problem--that of estimating B and B^* (or bilinear functions of them) in the models (2.3.1) and (2.3.3) respectively. (Estimation of Σ and Σ^* will not be taken up.) First we consider the estimation problem.

Definition 2.1. A bilinear function $\psi B \varphi$, where ψ is a $1 \times k$ vector and φ an $m \times 1$ vector, is called (unbiasedly) estimable with respect to the model (2.3.1) if there exists an $m \times n$ matrix L such that

$$\text{(2.3.8)} \qquad \mathcal{E}(\text{row } L)(\text{col } Y) = \psi B \varphi \quad \text{for all } B \ .$$

A linear function ψB is called (unbiasedly) estimable with respect to the model (2.3.1) if $\psi B \varphi$ is estimable for all φ .

Lemma 2.1. If $\varphi \neq 0$, the bilinear function $\psi B \varphi$ is estimable with respect to the model (2.3.1) if and only if ψ is in the row space of X. Thus, ψB is estimable with respect to (2.3.1) as long as $\psi B\varphi$ is estimable for some $\varphi \neq 0$.

Proof. Let L be any $m \times n$ matrix; from (2.3.1) and (2.3.4) we have

$$\mathcal{E}(\text{row } L)(\text{col } Y) = \text{row } L \cdot \text{col } XB$$
$$= \text{row } L \cdot (I_m \otimes X)\text{col } B$$
$$= \text{row } LX \cdot \text{col } B \ ,$$

and likewise,

$$\psi B\varphi = \text{col } \psi B\varphi = (\varphi' \otimes \psi)\text{col } B \ .$$

Thus, by Definition 2.1, $\psi B\varphi$ is estimable if and only if

$$\text{row } LX = \varphi' \otimes \psi = \text{row } \varphi\psi \ ,$$

i.e., if and only if $LX = \varphi\psi$. Since $\varphi \neq 0$ this implies $\psi = \varphi^{\dagger} LX \equiv aX$.

Q.E.D.

From (2.3.1) and (2.3.4) we have

$$\text{col } Y = (I_m \otimes X)\text{col } B + \text{col } E \ ;$$

(2.3.9)

$$\mathcal{E} \text{ col } E = 0; \qquad \mathcal{E}(\text{col } E)(\text{col } E)' = \Sigma \otimes V \ ,$$

which expresses the multivariate multiple regression model as a univariate one. A similar expression of course holds for (2.3.3). We can now extend Theorem 1.4 to this model and find a linear (affine) function of col Y which furnishes a best linear unbiased estimator for every estimable function col $\psi B = (I_m \otimes \psi)\text{col } B$.

Theorem 2.4. A linear (affine) function

$$\text{col}\ \tilde{B} = (I_m \otimes A)\text{col}\ Y + \text{col}\ C \quad , \tag{2.3.10}$$

where A and C are $k \times n$ and $k \times m$ matrices respectively, furnishes a best linear unbiased estimator $\text{col}\ \psi\tilde{B} = (I_m \otimes \psi)\text{col}\ \tilde{B}$ of every estimable function $\text{col}\ \psi B = (I_m \otimes \psi)\text{col}\ B$ in the model (2.3.9) if and only if A and C satisfy

$$XAX = X, \qquad XAV = VA'X', \qquad XC = 0 \ . \tag{2.3.11}$$

If $\text{col}\ \tilde{B}_1$ and $\text{col}\ \tilde{B}_2$ are any two affine functions of col Y which furnish best linear unbiased estimators of an estimable function $\text{col}\ \psi B$, then $\psi\tilde{B}_1 = \psi\tilde{B}_2$ with probability 1, the equality holding without qualification if V and Σ are positive definite.

Proof. From Lemma 2.1, ψB is estimable if and only if $\psi = aX$ for some a, so the problem reduces to finding a best linear unbiased estimator of

$$\text{col}\ XB = (I_m \otimes X)\text{col}\ B \tag{2.3.12}$$

in the model (2.3.9). Let

$$\text{col}\ \tilde{B} = J \cdot \text{col}\ Y + \text{col}\ C \tag{2.3.13}$$

be a linear (affine) estimator of col B, where J is an $mk \times mn$ matrix. By Theorems 1.1 and 1.4, in order that (2.3.13) furnish a best linear unbiased estimator $\text{col}\ X\tilde{B}$ of (2.3.12) in the model (2.3.9), it is necessary and sufficient that $\text{col}\ XC = (I_m \otimes X)\text{col}\ C = 0$ (hence $XC = 0$) and that J be a generalized inverse of $I_m \otimes X$ satisfying properties (i) and (iii) of (‡), with V replaced by $\Sigma \otimes V$. It is easily verified that one such generalized inverse is $J = I_m \otimes X^{\ddagger}$, where $X^{\ddagger}$ satisfies properties (i) and (iii) of (‡). This proves the necessity and sufficiency of (2.3.11). The remainder of the proof proceeds exactly as in the proof of Theorem 1.4. Q.E.D.

It follows from Theorem 2.4 that any estimator of B which is Gauss-Markoff in the class of affine functions of col Y is equivalent, with probability 1, to an estimator which can be expressed as an affine function of Y, namely

$$\tilde{B} = AY + C \quad , \tag{2.3.14}$$

where A and C satisfy (2.3.11). Such an estimator is, of course, necessarily Gauss-Markoff in the class of affine functions of Y, and in fact, in the class of bi-affine functions AYK + C of Y. If $X^{\ddagger}$ is any generalized inverse of X satisfying properties (i) and (iii) of (‡), expressing the general solution of the equation XC = 0 in terms of this $X^{\ddagger}$ we may write (2.3.14) in the form

$$\tilde{B} = X^{\ddagger}Y + (I - X^{\ddagger}X)Z \quad , \tag{2.3.15}$$

where Z is arbitrary.

In the context of the aggregation problem, Theorem 2.4 must be interpreted with due reservations. The Gauss-Markoff criterion does not take account of any a priori information an investigator might have concerning B--including knowledge of the fulfillment of certain sets of bilinear restrictions on B that might permit perfect aggregation. For example, the independent variables in the model (2.3.1) will typically include variables such as the incomes of different individuals, and likewise the dependent variables will include the levels of their expenditures. It would be natural to assume that one person's income does not affect another person's expenditures. In general, when restrictions of this kind are taken into account, B and even bilinear functions of B cannot be estimated in an efficient manner independently of Σ (cf. Zellner [110]). These remarks apply with still greater force when (2.3.1) is interpreted as the reduced form of a simultaneous equations model (cf. W. D. Fisher [34], Theil [104]). On the other hand we shall see that a bilinear restriction on B of the form $RBS = 0$, where R and S are given matrices, does permit efficient estimation of certain bilinear functions QBS of B without the intervention of the nuisance parameter Σ (although still not of B itself, unless Σ is a scalar matrix); in fact, this is precisely the situation that arises when conditions for perfect aggregation are fulfilled.

Our procedure will be to assume that an investigator employs an estimator $\tilde{B}^*$ of B^* which is Gauss-Markoff relative to the simplified model (2.3.3), that is, an estimator which would be Gauss-Markoff if the specification (2.3.3) were correct. We shall then examine the properties of this estimator when considered as an estimator of certain bilinear functions of B, on the assumption that the specification (2.3.1) is the correct one. Accordingly, the investigator's estimator will be equal, with probability 1, to a function of Y^* given by

$$\tilde{B}^* = X^{*\ddagger}Y^* + (I - X^{*\ddagger}X^*)Z^* = (XG)^{\ddagger}YH + [I - (XG)^{\ddagger}XG]Z^* , \tag{2.3.16}$$

where $X^{*\ddagger} = (XG)^{\ddagger}$ is any generalized inverse of X^* satisfying properties (i) and (iii) of (‡), and Z^* is arbitrary. Note that this estimator is not an affine, but only a bi-affine, function of Y.

From Theorem 1.4, if $\tilde{B}_1^*$ and $\tilde{B}_2^*$ are any two Gauss-Markoff estimators of B^* for the model (2.3.3), then for any function $\psi^*\tilde{B}^*$ which is estimable with respect to this model, $\psi^*\tilde{B}_1^* = \psi^*\tilde{B}_2^*$ with probability 1. Let us suppose that our investigator is content to use any such estimator, and does not require it to satisfy any further properties; then we may choose any particular representation of $X^{*\ddagger}$. We shall choose the Zyskind-Martin representation of Theorem 1.5, where V^+ is any symmetric non-negative definite generalized inverse of V satisfying conditions (1.1.18) and (1.1.19) of Lemma 1.3 (with X replaced by X^*); for such a V^+, a representation $X^{*\ddagger}$ of the desired kind is defined as any solution A^* of the generalized normal equations

$$X^{*'}V^+X^*A^* = X^{*'}V^+ . \tag{2.3.17}$$

Substituting such a solution in (2.3.16) we get for some arbitrary (and in general, different) Z^* the representation

$$(2.3.18)\quad \tilde{B}^* = (X^{*'}V^+X^*)^- X^{*'}V^+Y^* + [I - (X^{*'}V^+X^*)^- X^{*'}V^+X^*]Z^* \ .$$

Now let us define for some $\nu > 0$ the matrices

$$(2.3.19)\quad M = \nu X'V^+X, \quad G^\# = (G'MG)^- G'M = (X^{*'}V^+X^*)^- X^{*'}V^+X \ .$$

The matrix $G^\#$ corresponds to what Theil [101] has called the "auxiliary regression of the micro-variables on the macro-variables". We verify that it is a generalized quasi-inverse of G satisfying

$$(2.3.20)\qquad (i')\ \ MGG^\#G = MG, \ (iv')\ \ MGG^\# = G^{\#'}G'M$$

(compare (2.2.70)). Property (i') is verified as in Lemma 1.9 by noting that

$$[I - G'MG(G'MG)^-]G'MG[I - (G'MG)^-G'MG] = 0 \ ,$$

and this implies $MG[I - (G'MG)^-G'MG] = 0$ since M is symmetric and non-negative definite. Property (iv') is immediate, and is based on the fact that $X^{*\ddagger}$ satisfies property (1.1.34) of Theorem 1.5.

Since X and $X'V^+X$ have the same row space, by (1.1.19) and Lemma 1.4, so do XG and $X'V^+XG$, hence rank V^+XG = rank XG. Accordingly, if rank XG = rank G then rank $G'MG$ = rank G, and by Lemma 1.4 it follows that $G^\#$ is actually a generalized inverse of G. Under these conditions the matrix $G^\#G$, which is clearly idempotent from the definition (2.3.19), has the same rank as G. If it is now further assumed that G has full rank k^*, so that rank XG = rank $G = k^*$, then $G^\#G = I$, i.e., $G^\#$ is a left inverse of G. These additional assumptions will frequently be invoked in the following development, but only when explicitly mentioned.

The estimator (2.3.18) is Gauss-Markoff relative to

the specification (2.3.3); however, this specification may be incorrect. We shall be interested in analyzing its properties from the standpoint of the model (2.3.1), which we shall consider to be the correct one. To start with, let us compute its expected value relative to the specification (2.3.1); from (2.3.2) and (2.3.19) this is

$$\mathcal{E}\tilde{B}^* = G^{\#}BH + (I - G^{\#}G)Z^* \quad . \tag{2.3.21}$$

If we employ the Zyskind-Martin representations of $X^{\ddagger}$ and $X^{*\ddagger}$, we can obtain a direct relation between the estimators $\tilde{B}$ and $\tilde{B}^*$. Letting $X^{\ddagger}$ be any solution A of the generalized normal equations (1.1.32), as given by (1.1.35), we have from (2.3.19), (2.3.2), (1.1.19), and Lemma 1.4,

$$G^{\#}X^{\ddagger} = X^{*\ddagger} \tag{2.3.22}$$

for the particular solution $X^{*\ddagger} = (X^{*\prime}V^{+}X^{*})^{-}X^{*\prime}V^{+}$ of (2.3.17). Likewise we have $G^{\#}X^{\ddagger}X = G^{\#}$, so that for any Zyskind-Martin representation of $X^{\ddagger}$ in (2.3.15) we have, from (2.3.22) and (2.3.2),

$$G^{\#}\tilde{B}H = X^{*\ddagger}Y^* = \tilde{B}^* \tag{2.3.23}$$

for the particular solution of (2.3.18) with $Z^* = 0$. Taking expectations of both sides of (2.3.23) we obtain a particular solution of (2.3.21).

We are now ready to turn to the aggregation problem.

2.3.2. The aggregation problem. As before, we take up in succession the various cases of perfect aggregation, then best approximate aggregation, perfect disaggregation, and best approximate disaggregation.

A. Perfect aggregation: restricted structure, unrestricted observations. Taking expectations $\mathcal{E}$ and $\mathcal{E}^*$ in

(2.3.1a) and (2.3.3a) respectively, in order for the resulting expressions to be consistent with one another and with the transformations (2.3.2), for all $X \in \chi$, it is clearly necessary and sufficient that

$$GB^{*} = BH \quad , \tag{2.3.24}$$

which is the condition corresponding to condition (2.2.5) of the preceding section. The Penrose solvability condition for (2.3.24) is

$$(I - GG^{-})BH = 0 \quad , \tag{2.3.25}$$

where G^{-} is any generalized inverse of G, and when it is satisfied the general solution of (2.3.24) is

$$B^{*} = G^{-}BH + (I - G^{-}G)Z^{*} \quad . \tag{2.3.26}$$

Thus, when rank XG = rank G, so that $G^{\#}$ in (2.3.19) is a generalized inverse of G, if condition (2.3.25) for perfect aggregation holds then the expected value (2.3.21) (computed with reference to the detailed model (2.3.1)) of the estimator $\tilde{B}^{*}$ of (2.3.18) --which is the Gauss-Markoff estimator of B^{*} relative to the simplified model (2.3.3)--coincides with the general solution (2.3.26) of (2.3.24).

Now, let rank XG = rank G and let D be a block diagonal matrix with non-negative elements and with blocking conformable to that of G, such that rank $G'DG$ = rank G. We define

$$G^{\ddagger}_{M} = (G'MG)^{-}G'M = G^{\#}; \quad G^{\ddagger}_{D} = (G'DG)^{\dagger}G'D \quad . \tag{2.3.27}$$

Then both $G^{\ddagger}_{M}$ and $G^{\ddagger}_{D}$ are (reflexive) generalized inverses of G, and $G^{\ddagger}_{D}$ is a (row-wise) grouping matrix. Under these conditions, if ψ^{*} is any $1 \times k^{*}$ linear functional which is estimable relative to the consolidated model (2.3.3), i.e., is such that $\psi^{*} = a^{*}X^{*}$ for some a^{*}, and if condition (2.3.25) for perfect aggregation holds, then $GG^{\ddagger}_{M}BH = BH = GG^{\ddagger}_{D}BH$ hence

(2.3.28) $$\mathcal{E}\psi^{*}\tilde{B}^{*} = a^{*}XGG^{\ddagger}_{M}BH = a^{*}XGG^{\ddagger}_{D}BH = \psi^{*}G^{\ddagger}_{D}BH = \psi^{*}B^{*} .$$

In the special case in which G has full rank k^{*}, $G^{\ddagger}_{M}$ and $G^{\ddagger}_{D}$ are both left inverses of G and (2.3.28) can be strengthened to

(2.3.29) $$\mathcal{E}\tilde{B}^{*} = G^{\ddagger}_{M}BH = G^{\ddagger}_{M}GG^{\ddagger}_{D}BH = G^{\ddagger}_{D}BH = B^{*} .$$

This has an interesting interpretation. The first equation of (2.3.29) defines a rather complicated relation between $\mathcal{E}\tilde{B}^{*}$ and B which is not easy to describe, since $G^{\ddagger}_{M}$ is not in general a grouping matrix; however, the subsequent two equations tell us that we can replace the first equation by $\mathcal{E}\tilde{B}^{*} = G^{\ddagger}_{D}BH$, which has a ready interpretation: in Theil's [101] terminology it states that if the conditions (2.3.25) for perfect aggregation hold, the expected values of the estimated macro-parameters are functions only of the corresponding micro-parameters, and not of the non-corresponding micro-parameters. An analogous interpretation of (2.3.28) holds in the general case, with respect to estimable functions of the macro-parameters.

Since we are now dealing with the polar case of unrestricted observations, we shall assume that X and Σ both have full rank. Let us also assume that $\operatorname{rank} XG = \operatorname{rank} G = k^{*}$ and $\operatorname{rank} \Sigma H = \operatorname{rank} H = m^{*}$. We now show that under these conditions, if G^{-} is any left inverse of G, and $\tilde{B}^{*}$ is given by (2.3.18), then $\operatorname{col} \tilde{B}^{*}$ is the best linear conditionally unbiased estimator of $\operatorname{col}(G^{-}BH)$ (with respect to the model (2.3.9)) subject to the bilinear restrictions (2.3.25).

Following Chipman and Rao [18], let G^{c} be a $k \times (k - k^{*})$ matrix such that $[G, G^{c}]$ has full rank, and define the $r \times k$ matrix R, where $r = k - k^{*}$, by

(2.3.30) $$\begin{bmatrix} R^{c} \\ R \end{bmatrix} = [G,\ G^{c}]^{-1} .$$

Then R^C is a left inverse of G and G^C a right inverse of R, so we may denote $R^C = G^-$ and $G^C = R^-$. Accordingly, $GG^- + R^-R = I$, and (2.3.25) is equivalent to $R^-RBH = 0$ which in turn is equivalent to

$$RBH = 0 \quad . \tag{2.3.31}$$

We wish to estimate G^-BH subject to (2.3.31). Let us first find the best linear conditionally unbiased estimator of B in the model (2.3.9) subject to the bilinear restriction (2.3.31). This is equivalent to finding the best linear conditionally unbiased estimator of col B in (2.3.9) subject to the linear restriction

$$(H' \otimes R)\text{col } B = 0 \tag{2.3.32}$$

(see (2.3.4)). Defining

$$H^{\ddagger} = (H'\Sigma H)^{-1}H'\Sigma, \quad R^{\ddagger} = M^{-1}R'(RM^{-1}R')^{-1} \quad , \tag{2.3.33}$$

we verify that $(H' \otimes R)^{\ddagger} = H^{\ddagger'} \otimes R^{\ddagger}$ is a right inverse of $H' \otimes R$. The required estimator is given by (cf. Chipman and Rao [18])

$$\text{col } \hat{B} = [I_{mk} - (H' \otimes R)^{\ddagger} (H' \otimes R)] \text{ col } \tilde{B}$$

$$= [(I_m \otimes X^{\ddagger}) - (H^{\ddagger'} H' \otimes R^{\ddagger}RX^{\ddagger})]\text{col } Y \quad ,$$

where $\tilde{B}$ is given by (2.3.15), yielding

$$\hat{B} = X^{\ddagger}Y - R^{\ddagger}RX^{\ddagger}YHH^{\ddagger} \quad , \tag{2.3.34}$$

which depends on Σ (unless Σ is a scalar matrix) and is a bi-affine function of Y. The corresponding estimator of G^-BH is

(2.3.35) $$G^{-}\hat{B}H = G^{-}(I - R^{\ddagger}R)X^{\ddagger}YH \quad ,$$

which is independent of Σ .

To show that (2.3.35) is the same as the estimator $\tilde{B}^{*}$ of (2.3.18) we proceed as follows (cf. Chipman and Rao [18]). Since $RG = 0$ from (2.3.30), we have (writing $G^{\ddagger} = G^{\ddagger}_{M}$ for simplicity in (2.3.27)) $G^{\ddagger}R^{\ddagger} = 0$ from (2.3.27) and (2.3.33), hence $R^{\ddagger}RGG^{\ddagger} = 0$ and $GG^{\ddagger}R^{\ddagger}R = 0$. Since $GG^{\ddagger}$ and $R^{\ddagger}R$ are idempotent of ranks k^{*} and $r = k - k^{*}$ respectively, it follows (by Chipman and Rao [19]) that their sum is idempotent of rank k, i.e., $GG^{\ddagger} + R^{\ddagger}R = I_k$. Thus, (2.3.35) becomes

(2.3.36) $$G^{-}\hat{B}H = G^{-}GG^{\ddagger}X^{\ddagger}YH = G^{\ddagger}X^{\ddagger}YH = X^{*\ddagger}Y^{*} \quad .$$

An extremely simple example will serve to illustrate as well as to delimit the significance of the above result. Let X be partitioned as $X = [\iota, x_1, x_2, \ldots, x_m]$, where ι is a column of n ones and x_i is a column vector whose tth component is the level of individual i's income in year t . Likewise, let Y be partitioned as $Y = [y_1, y_2, \ldots, y_m]$, where y_j is a column vector whose tth component is individual j's consumption expenditures in year t . Here, $k = m + 1$. Let G and H be defined by

$$G = \begin{bmatrix} 1 & 0 \\ 0 & \iota \end{bmatrix} , \qquad H = \iota \quad ,$$

where ι now denotes a column of m ones. Then we have $R = [0, \iota, I]$, where 0 and ι are columns of $m - 1$ zeros and ones respectively and I is the identity matrix of order $m - 1$. Denote $B = [\beta_{ij}]$, where $i = 0, 1, \ldots, m$ and $j = 1, 2, \ldots, m$. Then (2.3.1a) reads

$$y_{tj} = \beta_{0j} + \sum_{i=1}^{m} x_{ti}\beta_{ij} + \varepsilon_{tj} \quad (t = 1, 2, \ldots, n;\ j = 1, 2, \ldots, m) \quad .$$

The condition (2.3.31) states that

$$\sum_{j=1}^{m} \beta_{ij} = \sum_{j=1}^{m} \beta_{1j} \qquad (i = 2, 3, \ldots, m) , \tag{2.3.37}$$

i.e., that the increase in aggregate consumption resulting from an increase in the ith person's income is the same as the increase in aggregate consumption resulting from an equal increase in the first person's income. It would of course be natural to assume that $\beta_{ij} = 0$ for $i \neq j$ and $i \geqq 1$, i.e., that a person's consumption expenditure depends only on his own income, in which case (2.3.37) would state that $\beta_{ii} = \beta_{11}$, i.e., that all individuals have the same marginal propensity to consume out of their own incomes. But the estimator (2.3.36) only takes account of the less stringent condition (2.3.37), and is therefore not fully efficient if the stronger condition holds (unless Σ is a scalar matrix); an efficient method has been developed by Zellner [110].

B. Perfect aggregation: unrestricted structure, restricted observations. Let us suppose that multicollinearity is present, in such a form that the columns of X are constrained to lie in the subspace defined by

$$X = XG\bar{G} \quad , \tag{2.3.38}$$

where $\bar{G}$ is a matrix such that $\bar{G}G\bar{G} = \bar{G}$, as in (2.2.16). Perfect aggregation then requires that $XGB^* = XBH$ for all X satisfying (2.3.38), or

$$G\bar{G}GB^* = G\bar{G}BH \, . \tag{2.3.39}$$

Since $\bar{G}$ is a generalized inverse of $G\bar{G}G$, condition (2.3.39) is necessarily satisfied by virtue of Penrose's criterion $(G\bar{G}G)\bar{G}(G\bar{G}BH) = G\bar{G}BH$. The general solution of (2.3.39) is then

(2.3.40) $$B^* = \bar{G}BH + (I - \bar{G}G)Z^* .$$

Now, let $\psi^* = a^*X^*$ be any linear functional which is estimable with respect to the consolidated model (2.3.3). Since, as remarked above, rank V^+XG = rank XG, it follows from (2.3.19) and Lemma 1.4 that

(2.3.41) $$XGG^{\#}G = XG .$$

Moreover, it follows from (2.3.19) and (2.3.38) (and (2.3.32)) that

(2.3.42) $$G^{\#} = (X^{*'}V^+X^*)^-X^{*'}V^+XG\bar{G} = G^{\#}G\bar{G} .$$

Together, (2.3.41) and (2.3.42) imply that

(2.3.43) $$XGG^{\#} = XGG^{\#}G\bar{G} = XG\bar{G} .$$

It follows then from (2.3.21) and (2.3.40) that

(2.3.44) $$\mathcal{E}\psi^*\tilde{B}^* = a^*XGG^{\#}BH = a^*XG\bar{G}BH = \psi^*\bar{G}BH = \psi^*B^* ,$$

i.e., $\psi^*\tilde{B}^*$ is an unbiased estimator of ψ^*B^* for any solution B^* of (2.3.39). Likewise, if $\psi = aX$ is any linear functional which is estimable with respect to the detailed model (2.3.1), $\psi G\tilde{B}^*$ is an unbiased estimator of $\psi G\bar{G}BH$.

We can go further. Let us take the Zyskind-Martin representations of $X^{\ddagger}$ and $X^{*\ddagger}$ in (2.3.15) and (2.3.18) (with $Z^* = 0$); then it is immediate from (2.3.23) and (2.3.43) that for any ψ^* which is estimable relative to the consolidated model,

(2.3.45) $$\psi^*\tilde{B}^* = a^*XGG^{\#}\tilde{B}H = a^*XG\bar{G}\tilde{B}H = \psi^*\bar{G}\tilde{B}H .$$

From Theorem 2.4, $\psi^*\tilde{B}^*$ is the best linear unbiased estimator of $\psi^*\bar{G}BH$. Likewise, for any ψ which is estimable relative to the detailed model, $\psi G\tilde{B}^* = \psi G\bar{G}\tilde{B}H$ is the best bilinear unbiased estimator of $\psi GGBH$.

It is of interest to consider the special case in which $G^{\#} = G^{\ddagger}_M$ and $\bar{G} = G^{\ddagger}_D$ as in (2.3.27), where D is a conformable block diagonal matrix and rank XG = rank $G'DG$ = rank G, so that $G^{\ddagger}_M$ and $G^{\ddagger}_D$ are generalized inverses and $G^{\ddagger}_D$ is a grouping matrix. The constraint (2.3.38) then becomes

(2.3.46) $$X = XGG_D^{\ddagger} = X^* G_D^{\ddagger} .$$

In Theil's [101] terminology, (2.3.46) states that the micro-variables are proportional to the corresponding macro-variables. Since $G_M^{\ddagger}$ is a generalized inverse, (2.3.42) yields

(2.3.47) $$GG_M^{\ddagger} = GG_D^{\ddagger} .$$

If it is further assumed that rank $G = k^*$, then $G_M^{\ddagger}$ and $G_D^{\ddagger}$ are both left inverses of G and (2.3.47) becomes

(2.3.48) $$G_M^{\ddagger} = G_D^{\ddagger} .$$

From (2.3.23) it is then immediate that

(2.3.49) $$\tilde{B}^* = G_D^{\ddagger} \tilde{B} H .$$

Thus, col $\tilde{B}^*$ is the best linear unbiased estimator of col $G_D^{\ddagger} BH$ in the model (2.3.9).

We may note also that R, as defined by (2.3.30), is complementary to X when rank XG = rank $G = k^*$, since then $uX + vR = 0$ implies $uXG + vRG = 0$, and this in turn implies (on account of (2.3.46)) that $uX = uXGG_D^{\ddagger} = 0$. It follows by Theorem 1.10 that a representation of the Gauss-Markoff estimator (2.3.15) is given by

(2.3.50) $$\tilde{B} = X^{\ddagger} Y = (X'V^+ X + R'R)^{-1} X'V^+ Y .$$

C. Perfect aggregation: partially restricted structure, partially restricted observations. It is hardly necessary to take up this general case. There will be a restriction matrix R some of whose rows are estimable (as in Case A) and some of whose rows are not (as in Case B), and the techniques of Chipman [15] and Goldman and Zelen [40] may be applied in straightforward fashion to obtain analogous results.

D. Best approximate aggregation. We may put the question: can the estimator (2.3.18), which is Gauss-Markoff relative to the specification (2.3.3)(the consolidated model), be rationalized in terms of the criterion of best approximate aggregation? A glance at the form (2.3.23) of this estimator and of the properties (2.3.20) satisfied by $G^{\#}$ reveals in the light of Theorem 2.1 that such a rationalization can be obtained if the matrix M in (2.3.19) can be regarded as a suitable estimate of the "true" contemporaneous moment matrix of the independent variables in the detailed model. For this purpose we shall take ν in (2.3.19) to be given by

(2.3.51) $$\nu = (\iota' V^{+} \iota)^{-1} ,$$

where ι denotes the column of n ones. (We ignore the trivial case $\iota' V^{+} \iota = 0$ which can arise only when $V = 0$.) In the special case in which the hypotheses of Theorem 1.7 are fulfilled, so that we may take $V^{+} = I$, we have $\nu = 1/n$ and $n^{-1} X'X$ is the usual sample moment matrix. However, in the general case, when X is random and $\mathcal{E}(E|X) = 0$ in (2.3.1a), since E and X are then uncorrelated it would not make sense to use $\nu X' V^{+} X$ as an estimate of the true moment matrix unless V entered into the distribution of X in the same way that it enters into the conditional distribution of E given X. We shall therefore formulate the problem in the following manner. First of all it will be assumed that the first column of X is a column of ones, so that we have the partitions

(2.3.52) $$X = [\iota, X_1], \quad G = \begin{bmatrix} 1 & 0 \\ 0 & G_1 \end{bmatrix}, \quad B = \begin{bmatrix} B_0 \\ B_1 \end{bmatrix} .$$

(This involves no loss of generality, since we can always specify $B_0 = 0$.) X is now considered to be a random variable, with mean

(2.3.53) $$\mathcal{E}X = \iota\mu = \iota(1, \mu_1) = (\iota, \iota\mu_1) = (\iota, \mathcal{E}X_1) ,$$

where μ and μ_1 are $1 \times k$ and $1 \times (k-1)$ row vectors respectively, and with variance given by

(2.3.54) $\operatorname{Var}(\operatorname{col} X) = \mathcal{E}[\operatorname{col} X - \operatorname{col}(\iota\mu)][\operatorname{col} X - \operatorname{col}(\iota\mu)]' = \Theta \otimes V$,

where

(2.3.55) $$\Theta = \begin{bmatrix} 0 & 0 \\ 0 & \Theta_1 \end{bmatrix},$$

Θ and Θ_1 being of orders k and $k-1$ respectively. Thus we have

(2.3.56) $\operatorname{Var}(\operatorname{col} X_1) = \mathcal{E}[\operatorname{col} X_1 - \operatorname{col}(\iota\mu_1)][\operatorname{col} X_1 - \operatorname{col}(\iota\mu_1)]' = \Theta_1 \otimes V.$

Finally, we define the random variable $\tilde{Y}^*$ by

(2.3.57) $$\tilde{Y}^* = XBH = \iota B_0 H + X_1 B_1 H ,$$

and the random variables X^* and X_1^* by

(2.3.58) $$X^* = XG = [\iota, X_1 G_1] = [\iota, X_1^*] .$$

In accordance with Theorem 1.14 and formula (2.3.4) we have

(2.3.59) $$\hat{\mathcal{E}}(\operatorname{col} \tilde{Y}^* | \operatorname{col} X_1^*) = \operatorname{col}[\iota(B_0 + \mu_1 B_1)H] +$$

$$(H'B_1'\Theta_1 G_1 \otimes V)(G_1'\Theta_1 G_1 \otimes V)^- [\operatorname{col} X_1^* - \operatorname{col}(\iota\mu_1 G_1)] .$$

Taking $V^- = V^+$ and defining

(2.3.60) $$\eta = B_0 + \mu_1 B_1, \qquad G_1^{\natural} = (G_1'\Theta_1 G_1)^{-} G_1'\Theta_1 ,$$

the right side of (2.3.59) becomes

(2.3.61) $$\mathrm{col}(\iota\eta H) + (H'B_1'G_1^{\natural\,'} \otimes VV^{+})[\mathrm{col}\, X_1^{*} - \mathrm{col}(\iota\mu_1 G_1)] .$$

Applying the inverse of the "col" operation of (2.3.4) to (2.3.61) and denoting the result by $\hat{Y}^{*} = \mathcal{E}(\tilde{Y}^{*}|X_1^{*})$, we obtain

(2.3.62) $$\hat{Y}^{*} = \hat{\mathcal{E}}(\tilde{Y}^{*}|X_1^{*}) = \iota\eta H + VV^{+}(X_1^{*} - \iota\mu_1 G_1)G_1^{\natural}B_1 H .$$

Unfortunately, (2.3.62) expresses $\hat{Y}^{*}$ not as an affine but as a bi-affine function of X_1^{*}. At this point we are therefore forced to introduce the assumption that the columns of X are contained in the column space of V. From Lemma 1.3 this implies that $VV^{+}X = X$ (and, in fact, that V^{+} can be chosen to be any generalized inverse of V); then (2.3.62) reduces to

(2.3.63) $$\hat{Y}^{*} = \hat{\mathcal{E}}(\tilde{Y}^{*}|X_1^{*}) = [\iota, X_1^{*}]\begin{bmatrix} 1 & \mu_1(I - G_1 G_1^{\natural}) \\ 0 & G_1^{\natural} \end{bmatrix}\begin{bmatrix} B_0 H \\ B_1 H \end{bmatrix} = X^{*}G^{\#}BH ,$$

where $G^{\#}$ is a generalized quasi-inverse of G with respect to the matrix M defined by

(2.3.64) $$M = \mu'\mu + \Theta ,$$

i.e., satisfying (2.3.20) for this choice of M. Replacing this M by its estimate (2.3.19) we obtain the required rationalization of the estimator $\tilde{B}^{*}$ of (2.3.23), provided the above condition on the columns of X is fulfilled.

Let us now provide a heuristic justification of

(2.3.65) $$\tilde{M} = (\iota' V^{+} \iota)^{-1} X' V^{+} X$$

as an estimator of M as defined by (2.3.64). From (2.3.53) and (2.3.54) we have the regression model

(2.3.66) $$X = \iota\mu + U, \quad \mathcal{E}U = 0, \quad \mathcal{E}(\text{row } U)'(\text{row } U) = V \otimes \Theta$$

analogous to (2.3.1), and by Theorem 2.4 the Gauss-Markoff estimator of μ is

(2.3.67) $$\tilde{\mu} = \iota^{\ddagger} X \quad \text{where} \quad \iota^{\ddagger} = (\iota' V^{+} \iota)^{-1} \iota' V^{+} .$$

The problem of finding an appropriate estimator of Θ is a more difficult one. The analogous problem in the model (2.3.1) is that of finding an estimator for Σ. A candidate for such an estimator is

(2.3.68) $$\tilde{\Sigma} = \nu Y'(I - XX^{\ddagger})' V^{+} (I - XX^{\ddagger}) Y = \nu [Y' V^{+} (I - XX^{\ddagger}) Y] .$$

It has been shown by Koopmans, Rubin, and Leipnik [59, p. 119] that if $V = I$ and the residuals E in the model (2.3.1) are normally distributed, then the maximum likelihood estimator of Σ when X has full rank is given by (2.3.68) with $\nu = 1/n$. No doubt this result can be generalized. Under the same circumstances but without assuming normality, Zellner [110] has shown that (2.3.68) with $\nu = 1/(n-k)$ provides a consistent estimator of Σ. Evidently a fairly wide range of choices for ν is compatible with consistent estimation. We may conjecture that one such choice is $\nu = 1/(\text{trace } X'V^{+}X)$. The analogous estimator of Θ in the model (2.3.66) would then be

(2.3.69) $$\tilde{\Theta} = (\iota' V^{+} \iota)^{-1} \bar{\bar{X}}' V^{+} \bar{\bar{X}}, \quad \text{where} \quad \bar{\bar{X}} = (I - \iota\iota^{\ddagger}) X .$$

The matrix $\bar{\bar{X}}$ has as elements the deviations of the elements of X from their (generalized) sample means. Putting together (2.3.67), (2.3.69), and (2.3.65) we verify that

(2.3.70) $$\tilde{M} = \tilde{\mu}'\tilde{\mu} + \tilde{\Theta}$$

and that $\tilde{\Theta}$ has the form (2.3.55), where $\tilde{\Theta}_1 = (\iota' V^+ \iota)^{-1} \bar{\bar{X}}_1' V^+ \bar{\bar{X}}_1$ and $\bar{\bar{X}}_1 = (I - \iota\iota^{\ddagger})X_1$.

Our justification for regarding (2.3.65) as an appropriate estimator of (2.3.64) is then based on three basic conditions: (1) the assumption that X has the same autocovariance structure as E, i.e.,

(2.3.71)
$$\mathcal{E}(\varepsilon_{si}\varepsilon_{tj}|X) = v_{st}\sigma_{ij};$$
$$\mathcal{E}(x_{si} - \mu_i)(x_{tj} - \mu_j) = v_{st}\theta_{ij} \quad ;$$

(2) the assumption that the columns of X are contained in the column space of V; and (3) the conjecture that $\tilde{\Theta}$ in (2.3.69) is a consistent estimator of Θ . On this basis we can justify the estimator (2.3.18) for the consolidated model (2.3.3) in terms of the criterion of minimization of aggregation bias.

The first of the above conditions is, unfortunately, rather stringent and unnatural. There seems to be no particular reason for supposing it to hold. If one defines $\tilde{Y}^*$ by (2.3.57) and $\hat{Y}^*$ by

(2.3.72) $$\hat{Y}^* = XGB^* \quad ,$$

where X is fixed, one might think of adopting the alternative criterion

(2.3.73)
$$\text{Minimize } (\hat{Y}^* - \tilde{Y}^*)'V^+(\hat{Y}^* - \tilde{Y}^*)$$
$$= (GB^* - BH)'X'V^+X(GB^* - BH) \quad .$$

From (2.3.19) and Theorem 2.1 it is immediate that this would lead to the solution (2.3.23). In the discussion below of best approximate disaggregation, however, we shall adhere to the interpretation (2.3.63), with M in (2.3.64) replaced by its estimator $\tilde{M}$ of (2.3.65).

E. Perfect disaggregation. In multivariate multiple regression models the disaggregation problem naturally arises in connection with forecasting. One is given a $1\times k^*$ row vector $\hat{x}^* = (1, x_1^*)$ of hypothetical or possibly even known values of the aggregative independent variables, and from this one wishes to obtain a forecast $\hat{y}$ of the values of the detailed dependent variables. This is done in two steps. First one obtains a forecast $\hat{y}^*$ of the values of the aggregative dependent variables. The second step is to "blow up" this forecast by means of the affine function

$$\hat{y} = \hat{y}^* \bar{H} + \bar{H}_0 \tag{2.3.74}$$

where $\hat{y}$ and $\hat{y}^*$ are $1 \times m$ and $1 \times m^*$ row vectors respectively, and the $m^* \times m$ matrix $\bar{H}$ and the $1 \times m$ vector $\bar{H}_0$ are to be determined. For the first step to be possible it is of course necessary (and sufficient) that $\hat{x}^*$ be estimable with respect to the consolidated model, i.e., that

$$(1, \hat{x}_1^*) = \hat{x}^* = a^* X^* = a^*[\iota, X_1^*] = [a^*\iota, a^* X_1^*] \tag{2.3.75}$$

for some a^* satisfying

$$a^*\iota = \sum_{t=1}^{n} a_t^* = 1 \ . \tag{2.3.76}$$

With $\tilde{B}^*$ given by (2.3.23) and $G^\#$ by (2.3.19) and (2.3.51), the aggregative forecast then becomes

$$\hat{y}^* = \hat{x}^* \tilde{B}^* = a^* X^* \tilde{B}^* = a^* X G G^\# \tilde{B} H \ . \tag{2.3.77}$$

Combined with (2.3.74) this yields the detailed forecast

(2.3.78) $$\hat{y} = a^* XGG^{\#}\tilde{B} H\bar{H} + \bar{H}_0 \quad .$$

On the other hand, if one had available an estimate of B itself, one could instead employ the detailed forecast

(2.3.79) $$\tilde{y} = \hat{x}^* G^{\#}\tilde{B} = a^* XGG^{\#}\tilde{B} \quad .$$

We say that conditions for perfect relative disaggregation hold if $\mathcal{E}(\hat{y} \mid X) = \mathcal{E}(\tilde{y} \mid X)$ for all admissible X. The taking of these conditional expectations in (2.3.78) and (2.3.79) involves simply the replacement of $\tilde{B}$ by B. From (2.3.58), (2.3.63) and (2.3.76) we then have

(2.3.80) $$\mathcal{E}(\hat{y} \mid X) = [B_0 + \mu_1(I - G_1G_1^{\natural})B_1]H\bar{H} + \bar{H}_0 + a^* X_1G_1G_1^{\natural}B_1H\bar{H}$$

and

(2.3.81) $$\mathcal{E}(\tilde{y} \mid X) = B_0 + \mu_1(I - G_1G_1^{\natural})B_1 + a^* X_1G_1G_1^{\natural}B_1 \quad .$$

In order that $\mathcal{E}(\hat{y} \mid X) = \mathcal{E}(\tilde{y} \mid X)$ for all X_1 and all a^* it is therefore necessary and sufficient that

(2.3.82) $$G_1G_1^{\natural}B_1H\bar{H} = G_1G_1^{\natural}B_1$$

and, with η defined by (2.3.60),

(2.3.83) $$\bar{H}_0 = \eta(I - H\bar{H}) \quad .$$

The Penrose solvability condition for (2.3.82) is

(2.3.84) $$G_1G_1^{\natural}B_1H(G_1G_1^{\natural}B_1H)^{-}G_1G_1^{\natural}B_1 = G_1G_1^{\natural}B_1 \quad ,$$

which is equivalent in view of Lemma 1.4 to the rank condition

(2.3.85) $$\operatorname{rank} G_1 G_1^{\natural} B_1 H = \operatorname{rank} G_1 G_1^{\natural} B_1 \ .$$

If we are free to perturb the elements of B_1 slightly, then in the absence of specific a priori restrictions (2.3.85) is essentially equivalent to the order condition

(2.3.86) $$m^* \geqq \min(m, k^* - 1) \ ,$$

which is the counterpart of (2.2.124). Since one of the independent "variables" is the constant term which does not actually vary, $k^* - 1$ is actually the number of substantive independent variables in the consolidated model. Thus, (2.3.86) states that either the number of dependent (endogenous) variables in the consolidated model is equal to that in the detailed model (which for practical purposes means that the detailed dependent variables are not aggregated), or else it is at least as great as the number of substantive independent (exogenous) variables in the consolidated model, i.e., the number of independent variables exclusive of the constant term. If (2.3.85) holds, the solution of (2.3.82) is given by

(2.3.87) $$\bar{H} = (G_1 G_1^{\natural} B_1 H)^- G_1 G_1^{\natural} B_1 + [I - (G_1 G_1^{\natural} B_1 H)^- G_1 G_1^{\natural} B_1 H] Z \ .$$

In order to be able to use the disaggregation formula (2.3.74) one would thus need information on η and $G_1 G_1^{\natural} B_1$. This of course is not usually available, but, after all, one cannot expect to get something from nothing.

F. <u>Best approximate disaggregation</u>. Redefining symbols slightly, let us denote in place of (2.3.77),

(2.3.88) $$\hat{y}^* = x^* B^* = x G G^{\#} B H \ ,$$

where B is the "true" matrix of regression coefficients and the $1 \times k^*$ and $1 \times k$ vectors x^* and x are random variables. Likewise, let us denote

(2.3.89) $$y = xB + e$$

where the $1 \times m$ vector e is such that $\mathcal{E}(e|x) = 0$ and $\mathcal{E}(e'e|x) = \Sigma$. Then it is just a matter of notation to see from (2.2.168) that

(2.3.90) $$\hat{y} \equiv \mathcal{E}(y|\hat{y}^*) = \eta(I - HH^{\natural}) + \hat{y}^* H^{\natural} \equiv h^{\natural}(\hat{y}^*)$$

where

(2.3.91) $$H^{\natural} = (H'\Upsilon_1 H)^{-} H'\Upsilon_1, \quad \Upsilon_1 = B_1' G_1^{\natural\prime} G_1' \Theta_1 G_1 G_1 B_1 \quad .$$

In view of (2.3.83), formula (2.3.90) differs from (2.3.74) only in the replacement of $\bar{H}$ by $H^{\natural}$. To apply it one would replace B_1 by its Gauss-Markoff estimator and likewise replace η and Θ_1 by

(2.3.92) $$\tilde{\eta} = \iota^{\ddagger} X X^{\ddagger} Y, \quad \tilde{\Theta}_1 = (\iota' V^+ \iota)^{-1} X_1' V^+ (I - \iota\iota^{\ddagger}) X_1 \quad .$$

Of course, this again requires the use of detailed rather than aggregated observations.

2.3.3. Alternative approaches: random coefficient models. Zellner [111] has considered a formulation in which the regression coefficient matrix B in (2.3.1a) is random, with a prior mean satisfying (2.3.25). If this prior mean $\bar{B}$ is substituted for B in the matrix (2.3.73) of aggregation bias, then this bias matrix of course vanishes. In terms of unconditional expectations (as opposed to conditional expectations given B) the estimator (2.3.16) then yields unbiased estimators for all functionals that are estimable with respect to the consolidated model. Using the results of Theorem 1.7 above (in fact, the equivalent results of Rao [87]), Akkina [4] has extended this result to show that under certain conditions the estimator $\underset{\sim}{B}^*$ of (2.3.16) is Gauss-Markoff in the class of bi-affine functions of Y^* (with respect to the operator $\mathcal{E}$ of the detailed model). (Akkina's results are limited to the univariate case $m = 1$, but are readily extended.)

In terms of the present formulation we may say that if the condition (2.3.25) for perfect aggregation holds approximately, and if the matrix G is selected in an optimal manner according to the procedure outlined in section 2.2.D', the resulting aggregation bias (in terms of conditional distributions given B) will be minimal.

2.4. The Consolidation Problem in Leontief Models

2.4.1. The Leontief model. Leontief's input-output model [63] is based on the system of equations

$$(2.4.1)\quad x_i = \sum_{j=1}^{n} u_{ij} + y_i = \sum_{j=1}^{n} a_{ij}x_j + y_i \quad (i = 1, 2, \ldots, n) ,$$

where x_i is the total[†] output of commodity i, u_{ij} is the amount of commodity i used as input into the production of commodity j, y_i is the net output of commodity i available for consumption or investment, and $a_{ij} = u_{ij}/x_j$ is the "input-output coefficient"--assumed fixed--equal to the amount of commodity i needed as input per unit of output of commodity j. The variables u_{ij} and x_j, and hence the input-output coefficients a_{ij}, are intrinsically nonnegative; the net outputs y_i can be negative on account of imports or temporary inventory decumulation. The units of measurement of all commodities are taken to be their values in a particular base year--the year in which the input-output

† It may also be termed gross output. In some treatments (cf. Leontief [63], McManus [65]) a distinction is made according as to whether the u_{ii} are suppressed and conventionally assumed = 0; we do not adopt this convention here and therefore do not make the distinction

coefficients are calculated--, i.e., units of a dollar's worth in that year; the quantities are therefore comensurate and can be added.

The system (2.4.1) is supplemented by a system of inequalities

$$\sum_{j=1}^{n} v_{ij} = \sum_{j=1}^{n} b_{ij} x_j \leqq \ell_i \qquad (i = 1, 2, \ldots, m) \ , \tag{2.4.2}$$

where v_{ij} stands for the amount of the ith primary resource or factor service (labor, land, natural resource, or--in a static context--capital) used up in the production of commodity j; $b_{ij} = v_{ij}/x_j$ is the factor-output coefficient for factor i in industry j (again assumed fixed), and ℓ_i is the fixed amount of factor i available for use by all industries.

Equating the unit value of output to the costs of all inputs per unit of output (profits and interest being reckoned for this purpose among the costs of capital services), we have $\sum_{i=1}^{n} a_{ij} + \sum_{i=1}^{m} b_{ij} = 1$ for $j = 1, 2, \ldots, n$, whence of course $\sum_{i=1}^{n} a_{ij} \leqq 1$ since the v_{ij} are also intrinsically non-negative. Writing (2.4.1) in matrix form

$$y = (I - A)x \ , \tag{2.4.3}$$

where x and y are $n \times 1$ column vectors and $A = [a_{ij}]$ is the $n \times n$ input-output matrix, this matrix has non-negative elements and column sums less than or equal to unity. If, further, at least one of these column sums is strictly less than unity and at the same time A is indecomposable, or alternatively, if all the column sums of A are less than unity, it follows by a well-known theorem of Frobenius (cf. Solow [96], Debreu and Herstein [23]) that the "Leontief matrix" I - A has an inverse which can be represented by the convergent power series

(2.4.4) $$(I - A)^{-1} = I + A + A^2 + \dots \; .$$

The principal object of input-output analysis is to study the inverse relation

(2.4.5) $$x = (I - A)^{-1}y \; .$$

Thus if, during a period of mobilization, some components of the "bill of goods" y, e.g., consumer goods, must be reduced in order to allow for expansion in other components, e.g., munitions, it is necessary to examine (2.4.5) in order to ascertain the required changes in total outputs x; from this one calculates the resource requirements from (2.4.2) and checks whether they satisfy the indicated feasibility constraints.

If commodities were classified according to homogeneous physical characteristics, the order of the system (2.4.5) would have to be extremely large--in the neighborhood of hundreds of thousands. It is obviously desirable to consolidate the system to one of more manageable size--say in the order of 50 to 100 industrial groupings, or even less. This problem has been extensively studied by a number of authors; among the most significant contributions may be mentioned those of Hatanaka [47], Malinvaud [67], Fei [27], Rosenblatt [91, 92], Theil [101, 102, 106], W. D. Fisher [33], and Ara [7]. For background, special topics, and extensions, one may add the studies of Holzman [50], Balderston and Whitin [8], McManus [64, 65], Hatanaka [48], Charnes and Cooper [13], Rosenblatt [93], Ijiri [54], Morimoto [75, 76], Basevi [9], and Kossov [114]. The present formulation owes a great deal to that of Malinvaud [67].

2.4.2. Empirical construction of input-output matrices. We introduce the matrices (cf. Ara [7])

(2.4.6) $$U = [u_{ij}], \quad X = \text{diag}(x_i) \qquad (i, j = 1, 2, \dots, n) \; .$$

Let $\bar{U} = [\bar{u}_{ij}]$ and $\bar{X} = \text{diag}(\bar{x}_i)$ denote the particular values of these matrices observed in the base year--the year used for calculating the input-output coefficients a_{ij}. These estimated input-output coefficients are defined by

(2.4.7)
$$a_{ij} = \begin{cases} \bar{u}_{ij}/\bar{x}_j & \text{if} \quad \bar{x}_j > 0 \\ 0 & \text{if} \quad \bar{x}_j = 0 \end{cases} .$$

Accordingly, we have

(2.4.8)
$$A = \bar{U}\bar{X}^{\dagger} .$$

Let G be an $n^* \times n$ (row-wise) grouping matrix $(n^* < n)$, i.e., a non-negative matrix with at most one non-zero element in each column. (It will be assumed below that any such non-zero element must be equal to unity.) We define

(2.4.9)
$$\bar{x}^* = G\bar{x}$$

and

(2.4.10)
$$\bar{U}^* = G\bar{U}G', \quad \bar{X}^* = G\bar{X}G' .$$

Defining the consolidated input-out coefficients a^*_{ij} analogously to (2.4.7) we have from (2.4.10)

(2.4.11)
$$A^* = \bar{U}^*\bar{X}^{*\dagger} = G\bar{U}G'(G\bar{X}G')^{\dagger} .$$

Now we shall assume that <u>the rows of G are contained in the row space of $\bar{X}$</u>. This means that any commodity which is not produced in the base year will be excluded from the aggregation; i.e., if $\bar{x}_j = 0$, the jth column of G will be assumed to be vanishing. (This condition ensures that even if $\bar{u}_{ij} > 0$ when $\bar{x}_j > 0$ --contrary to the hypothesis

(2.4.1)--the aggregate flow matrix $\bar{U}^*$ will attach a zero weight to such a $\bar{u}_{ij}$.) Given this assumption, we have $G = C\bar{X}^{\dagger}$ for some C (since $\bar{X}$ and $\bar{X}^{\dagger}$ have the same row space), hence from (2.4.8) we obtain (cf. Penrose [80]), for arbitrary Z ,

$$\bar{U}G' = [A\bar{X} + Z(I - \bar{X}^{\dagger}\bar{X})]\bar{X}^{\dagger}C = A\bar{X}G' \quad . \tag{2.4.12}$$

Substituting (2.4.12) in (2.4.11) we obtain

$$A^* = GA\bar{X}G'(G\bar{X}G')^{\dagger} = GAG^{\ddagger}_{\bar{X}} \ , \tag{2.4.13}$$

where by definition

$$G^{\ddagger}_{\bar{X}} = \bar{X}G'(G\bar{X}G')^{\dagger} \quad . \tag{2.4.14}$$

By virtue of Theorem 2.1, this is a generalized inverse of G satisfying

$$\begin{aligned} &\text{(i)}\ \ GG^{\ddagger}_{\bar{X}}G = G; \quad \text{(ii)}\ \ G^{\ddagger}_{\bar{X}}GG^{\ddagger}_{\bar{X}} = G^{\ddagger}_{\bar{X}}; \\ &\text{(iii)}\ \ G^{\ddagger}_{\bar{X}}G = G'G^{\ddagger\prime}_{\bar{X}}; \quad \text{(iv)}\ \ G^{\ddagger}_{\bar{X}}G\bar{X} = \bar{X}G'G^{\ddagger\prime}_{\bar{X}} \ . \end{aligned} \tag{2.4.15}$$

Note that the matrix A^* of (2.4.13) depends on $\bar{x}$ (the particular vector of outputs observed in the base year). To indicate this dependence explicitly we shall henceforth denote

$$A^*_{\bar{x}} = GAG^{\ddagger}_{\bar{X}} \quad . \tag{2.4.16}$$

We now make the further assumption that <u>the non-zero elements of</u> G <u>are all ones</u>, and that <u>every commodity with positive output in the base year is included in the aggregation.</u>

That is, the aggregation consists in summing the base-year values of commodity flows for all commodities that are actually produced during the base year, so that if $\bar{x}_j > 0$ the jth column sum of G is equal to 1. In conjunction with the assumption made above that the rows of G are contained in the row space of $\bar{X}$, this implies that an industry is aggregated if and only if it has positive output in the base year; in particular, a commodity flow from industry i to industry j (measured in terms of its value in the base year) is included (with a unit weight) if and only if both industry i and industry j produced positive amounts during the base year. (Such flows could of course take place on account of inventory changes.) Formally, we may state our assumption as follows. Let ι and ι^* denote the row vectors of n and n^* ones respectively, and let "sign $\bar{x}$" denote the vector whose ith component is sign $\bar{x}_i$; from the definition (2.4.6) of $\bar{X}$ we then have

$$\bar{X}\iota' = \bar{X}(\text{sign}\,\bar{x}) = \bar{x} \quad . \tag{2.4.17}$$

Our further assumption may be stated as

$$\iota^* G = \text{sign}\,\bar{x}' \quad . \tag{2.4.18}$$

From (2.4.17) and (2.4.18) it follows that

$$\bar{X}G'\iota^{*\prime} = \bar{x} \quad , \tag{2.4.19}$$

hence $\bar{x}$ belongs to the column space of $\bar{X}G'$. We now verify that on the basis of these assumptions (and the definition (2.4.9)),

$$G^{\ddagger}_{\bar{X}}\,\bar{x}^* = G^{\ddagger}_{\bar{X}}\,G\bar{x} = \bar{x} \quad , \tag{2.4.20}$$

i.e., $\bar{x}$ belongs to the subspace $G^{\ddagger}_{\bar{X}}G\mathfrak{X}$, hence $G^{\ddagger}_{\bar{X}}$ behaves like an inverse as between $\bar{x}^*$ and $\bar{x}$ (cf. Malinvaud [67]); this follows from (2.4.19) and (2.4.15):

$$G_X^{\ddagger}G\bar{x} = G_X^{\ddagger}G\bar{X}G'\iota^{*'} = XG'G_X^{\ddagger'}G'\iota^{*'} = \bar{X}G'\iota^{*'} = \bar{x} \ .$$

As an example, suppose there are five industries which we wish to consolidate into two groups consisting of the first two and the last three, and suppose that $\bar{x}_4 = 0$. From (2.4.10) $\bar{U}^*$ becomes

$$\begin{bmatrix} 1 & 1 & 0 & 0 & 0 \\ 0 & 0 & 1 & 0 & 1 \end{bmatrix} \left[\begin{array}{cc:ccc} \bar{u}_{11} & \bar{u}_{12} & \bar{u}_{13} & \bar{u}_{14} & \bar{u}_{15} \\ \bar{u}_{21} & \bar{u}_{22} & \bar{u}_{23} & \bar{u}_{24} & \bar{u}_{25} \\ \hdashline \bar{u}_{31} & \bar{u}_{32} & \bar{u}_{33} & \bar{u}_{34} & \bar{u}_{35} \\ \bar{u}_{41} & \bar{u}_{42} & \bar{u}_{43} & \bar{u}_{44} & \bar{u}_{45} \\ \bar{u}_{51} & \bar{u}_{52} & \bar{u}_{53} & \bar{u}_{54} & \bar{u}_{55} \end{array}\right] \begin{bmatrix} 1 & 0 \\ 1 & 0 \\ 0 & 1 \\ 0 & 0 \\ 0 & 1 \end{bmatrix}$$

$$= \begin{bmatrix} \bar{u}_{11} + \bar{u}_{21} + \bar{u}_{12} + \bar{u}_{22} & \bar{u}_{13} + \bar{u}_{23} + \bar{u}_{15} + \bar{u}_{25} \\ \bar{u}_{31} + \bar{u}_{51} + \bar{u}_{32} + \bar{u}_{52} & \bar{u}_{33} + \bar{u}_{53} + \bar{u}_{35} + \bar{u}_{55} \end{bmatrix}$$

$$= \begin{bmatrix} \bar{u}_{11}^* & \bar{u}_{12}^* \\ \bar{u}_{21}^* & \bar{u}_{22}^* \end{bmatrix} .$$

Thus, each element of $\bar{U}^*$ is equal to the sum of the elements in the corresponding block of $\bar{U}$ --except for elements in deleted rows and columns (the fourth row and column in this case). Likewise, (2.4.16) becomes

$$\begin{bmatrix} 1 & 1 & 0 & 0 & 0 \\ 0 & 0 & 1 & 0 & 1 \end{bmatrix} \left[\begin{array}{cc|ccc} a_{11} & a_{12} & a_{13} & a_{14} & a_{15} \\ a_{21} & a_{22} & a_{23} & a_{24} & a_{25} \\ \hline a_{31} & a_{32} & a_{33} & a_{34} & a_{35} \\ a_{41} & a_{42} & a_{43} & a_{44} & a_{45} \\ a_{51} & a_{52} & a_{53} & a_{54} & a_{55} \end{array}\right] \begin{bmatrix} \bar{x}_1/(\bar{x}_1+\bar{x}_2) & 0 \\ \bar{x}_2/(\bar{x}_1+\bar{x}_2) & 0 \\ 0 & \bar{x}_3/(\bar{x}_3+\bar{x}_5) \\ 0 & 0 \\ 0 & \bar{x}_5/(\bar{x}_3+\bar{x}_5) \end{bmatrix}$$

$$= \begin{bmatrix} \frac{\bar{x}_1}{\bar{x}_1+\bar{x}_2}(a_{11}+a_{21})+\frac{\bar{x}_2}{\bar{x}_1+\bar{x}_2}(a_{12}+a_{22}), & \frac{\bar{x}_3}{\bar{x}_3+\bar{x}_5}(a_{13}+a_{23})+\frac{\bar{x}_5}{\bar{x}_3+\bar{x}_5}(a_{15}+a_{25}) \\ \frac{\bar{x}_1}{\bar{x}_1+\bar{x}_2}(a_{31}+a_{51})+\frac{\bar{x}_2}{\bar{x}_1+\bar{x}_2}(a_{32}+a_{52}), & \frac{\bar{x}_3}{\bar{x}_3+\bar{x}_5}(a_{33}+a_{53})+\frac{\bar{x}_5}{\bar{x}_3+\bar{x}_5}(a_{35}+a_{55}) \end{bmatrix}$$

$$= \begin{bmatrix} a_{11}^* & a_{12}^* \\ a_{21}^* & a_{22}^* \end{bmatrix} .$$

Each element of $A^*_{\bar{x}}$ is equal to a weighted average of the column sums in the corresponding block of A (except for deleted columns--the fourth in this case), the weights being equal to the relative importance (in terms of output) of the commodity within each group.

In case it should happen that the outputs within a particular group of industries were all equal to zero in the base year, the corresponding row and column of $A^*_{\bar{x}}$, as well as the corresponding component of $\bar{x}^*$, would vanish. It is clear that to carry these zeros would be no more than a notational encumbrance; hence without loss of generality we can assume that G contains no vanishing rows and thus has full rank n^*. We shall adopt this convention henceforth, and thus it will follow that

$$GG_{\bar{X}}^{\ddagger} = I \quad , \tag{2.4.21}$$

i.e., that $G_{\bar{X}}^{\ddagger}$ is a right inverse of G. Since (2.4.14) and property (i) of (2.4.15) imply that rank GXG' = rank G, it follows from (2.4.21) that (2.4.14) can henceforth be written $G_{\bar{X}}^{\ddagger} = \bar{X}G'(G\bar{X}G')^{-1}$.

Now defining

$$\bar{y} = \bar{x} - \bar{U}\iota' - (I - A)\bar{x}, \quad \bar{y}^* = G\bar{y} \quad , \tag{2.4.22}$$

we have from (2.4.20), (2.4.21), and (2.4.16),

$$\bar{y}^* = G(I - A)G_{\bar{X}}^{\ddagger}\bar{x}^* = (I - A_{\bar{X}}^*)\bar{x}^* \quad . \tag{2.4.23}$$

Thus, the aggregate entities $\bar{x}^*$, $\bar{y}^*$, $A_{\bar{X}}^*$ satisfy the equation analogous to (2.4.22) for $\bar{x}$, $\bar{y}$, A. However, this is no more than a form of accounting consistency--a minimal requirement of acceptability--and unfortunately does not carry us very far, because the validity of (2.4.23) does not in general extend beyond the particular $\bar{x}^*$ and $\bar{y}^*$ computed in the base year; that is, (2.4.23) does not permit one to extrapolate the relation to values

$$x^* = Gx, \quad y^* = Gy \tag{2.4.24}$$

computed in subsequent years. The consolidation problem in input-output analysis is precisely the problem of finding conditions under which such extrapolation is justifiable.

2.4.3. Aggregation criteria. As before, we may first

consider criteria for perfect aggregation, and then take up criteria for best approximate aggregation and criteria for disaggregation.

A. Perfect aggregation: similarity of technical coefficients. Given the structural relation (2.4.3) and the grouping transformations (2.4.24), in order that there exist a consolidated input-output matrix A^* such that

$$y^* = (I - A^*)x^* \tag{2.4.25}$$

always holds, i.e., such that $G(I - A)x = (I - A^*)Gx$ for all $x \in \mathfrak{X} = E_n^+$ (the non-negative orthant of n-dimensional Euclidean space), it is necessary and sufficient that

$$A^*G = GA \ . \tag{2.4.26}$$

This is the well-known "Hatanaka condition" (cf. Hatanaka [47]), also obtained independently by Malinvaud [67, p.198]. By Penrose's theorem [80] a necessary and sufficient condition for the solvability of (2.4.26) is

$$GA(I - G^-G) = 0 \ , \tag{2.4.27}$$

where G^- is any generalized inverse of G, and the general solution is--taking account of the fact that G has full rank, so that G^- must be a right inverse of G --

$$A^* = GAG^- \ . \tag{2.4.28}$$

This is invariant with respect to the choice of G^-, hence in particular, from (2.4.21), (2.4.27), and (2.4.16),

$$A^* = GAG^-GG_{\frac{}{X}}^{\ddagger} = GAG_{\frac{}{X}}^{\ddagger} = A_{\bar{x}}^* \ . \tag{2.4.29}$$

While (2.4.14) provides the natural choice, it would of course be equally legitimate, when (2.4.27) holds, to choose the Moore-Penrose inverse $G^{\dagger} = G'(GG')^{-1}$, as is done by Ijiri [54]; however, we shall see that in other cases this is no longer an appropriate choice. The invariance of (2.4.28) with respect to choice of G^{-} of the form $G_D^{\dagger} = DG'(GDG')^{-1}$, where D is a non-negative diagonal matrix such that GDG' has full rank, was shown by Rosenblatt [92], who used it to justify the choice $D = \overline{X}$. Rosenblatt also obtained the condition (2.4.27), in terms of $G^{\dagger} = G'(GG')^{-1}$.

We may express (2.4.27) in the more convenient form

(2.4.30) $$GAR = 0 \quad ,$$

where R is an $n \times (n - n^{*})$ matrix of rank $n - n^{*}$ and orthogonal to G, i.e., such that rank $[G', R] = n$ and $GR = 0$, so that the columns of R form a basis for the column space of $I - G^{-}G$. As an example we may consider

$$GAR = \begin{bmatrix} 1 & 1 & 0 & 0 & 0 \\ 0 & 0 & 1 & 1 & 1 \end{bmatrix} \left[\begin{array}{cc|ccc} a_{11} & a_{12} & a_{13} & a_{14} & a_{15} \\ a_{21} & a_{22} & a_{23} & a_{24} & a_{25} \\ \hline a_{31} & a_{32} & a_{33} & a_{34} & a_{35} \\ a_{41} & a_{42} & a_{43} & a_{44} & a_{45} \\ a_{51} & a_{52} & a_{53} & a_{54} & a_{55} \end{array}\right] \begin{bmatrix} 1 & 0 & 0 \\ -1 & 0 & 0 \\ 0 & 1 & 1 \\ 0 & -1 & 0 \\ 0 & 0 & -1 \end{bmatrix}$$

$$= \begin{bmatrix} a_{11}+a_{21}-(a_{12}+a_{22}) & a_{13}+a_{23}-(a_{14}+a_{24}) & a_{13}+a_{23}-(a_{15}+a_{25}) \\ a_{31}+a_{41}+a_{51}-(a_{32}+a_{42}+a_{52}) & a_{33}+a_{43}+a_{53}-(a_{34}+a_{44}+a_{54}) & a_{33}+a_{43}+a_{53}-(a_{35}+a_{45}+a_{55}) \end{bmatrix}$$

The condition (2.4.30) states--and this is of course true in general--that in each submatrix of A which is to be consolidated into a scalar, the column sums must be equal to one another; furthermore, this common column sum is precisely the aggregate input-output coefficient, so that in the above example we have

$$A^* = \begin{bmatrix} a_{11}^* & a_{12}^* \\ a_{21}^* & a_{22}^* \end{bmatrix} = \begin{bmatrix} a_{11} + a_{21} & a_{13} + a_{23} \\ a_{31} + a_{41} + a_{51} & a_{33} + a_{43} + a_{53} \end{bmatrix}.$$

These are the necessary and sufficient conditions for perfect aggregation when the output mix x is unrestricted; they were obtained in this form by McManus [64], Ara [7], and Charnes and Cooper [13].

It is interesting to note that the above conditions are less stringent than the conditions that would be required if the sectors were to be consolidated one at a time. As an illustration, similar to one used by Malinvaud [67], suppose that in the above example we initially sought conditions permitting consolidation of industries 1 and 2 while leaving the remaining three industries unconsolidated. The required condition (2.4.30) would then become

$$GAR = \begin{bmatrix} 1 & 1 & 0 & 0 & 0 \\ 0 & 0 & 1 & 0 & 0 \\ 0 & 0 & 0 & 1 & 0 \\ 0 & 0 & 0 & 0 & 1 \end{bmatrix} \left[\begin{array}{cc|c|c|c} a_{11} & a_{12} & a_{13} & a_{14} & a_{15} \\ a_{21} & a_{22} & a_{23} & a_{24} & a_{25} \\ \hline a_{31} & a_{32} & a_{33} & a_{34} & a_{35} \\ \hline a_{41} & a_{42} & a_{43} & a_{44} & a_{45} \\ \hline a_{51} & a_{52} & a_{53} & a_{54} & a_{55} \end{array}\right] \begin{bmatrix} 1 \\ -1 \\ 0 \\ 0 \\ 0 \end{bmatrix}$$

$$= \begin{bmatrix} a_{11} + a_{21} - (a_{12} + a_{22}) \\ a_{31} - a_{32} \\ a_{41} - a_{42} \\ a_{51} - a_{52} \end{bmatrix} = \begin{bmatrix} 0 \\ 0 \\ 0 \\ 0 \end{bmatrix} .$$

This requires that industries 1 and 2 expend exactly the same amounts on their inputs of each of the commodities 3, 4, 5 per dollar of output, i.e., $a_{i1} = a_{i2}$ for $i = 3, 4, 5$, whereas previously it was required only that they expend the same amounts on the three inputs jointly, i.e., $a_{31} + a_{41} + a_{51} = a_{32} + a_{42} + a_{52}$. While no restrictions are now placed on the technical coefficients of industries 3, 4, and 5, more stringent ones are placed on those of the consolidated industries 1 and 2.

As a simple special case of the above necessary and sufficient condition, if $G = (1, 1, \ldots, 1)$ we require the column sums of A to be equal to one another. Then A can be consolidated into a single 1×1 matrix A^* equal to this common column sum, and we obtain a simple aggregative model of Keynesian type (cf. Chipman [14, Theorem 1], McManus [64]).

It was shown by Ara [7] that if A is indecomposable and satisfies (2.4.26), then A and A^* have the same maximal positive (Frobenius) characteristic root, hence the consolidated Leontief matrix $I - A^*$ is invertible if and only if the unconsolidated one is. For any generalized inverse G^- of G (which is a right inverse since G is assumed to have full rank) it follows readily from (2.4.27) by a trivial induction argument that

$$GA^tG^- = (GAG^-)^t \qquad (t = 0, 1, 2, \ldots) \tag{2.4.31}$$

(cf. Fei [27]), hence from the power series expansion (2.4.4) we obtain

(2.4.32) $$G(I - A)^{-1}G^{-} = (I - GAG^{-})^{-1} ,$$

and this of course is invariant with respect to the choice of right inverse G^{-}. Thus, condition (2.4.27) guarantees perfect aggregation for the inverse system given by (2.4.5), (2.4.24), and

(2.4.33) $$x^{*} = (I - A^{*})^{-1}y^{*} .$$

B. Perfect aggregation: stability of the product mix. If the input-output matrix A is to be unrestricted we must find restrictions on the set of observable outputs x. As before, we shall content ourselves with sufficient conditions. Let D be an $n \times n$ diagonal matrix with non-negative diagonal elements, such that $\text{rank } GDG' = \text{rank } G = n^{*}$, i.e., such that GDG' is non-singular, and define (cf. Rosenblatt [92])

(2.4.34) $$G_D^{\ddagger} = DG'(GDG')^{-1} .$$

Let the observed outputs be required to satisfy

(2.4.35) $$x = G_D^{\ddagger}Gx = G_D^{\ddagger}x^{*} .$$

This states that outputs remain proportional to one another within groups (cf. Malinvaud [67]). Perfect aggregation requires that $A^{*}Gx = GAx$ for all x satisfying (2.4.35), or

(2.4.36) $$A^{*}G = GAG_D^{\ddagger}G ,$$

an equation that is always solvable and whose general solution, by virtue of (2.4.21) and Penrose's theorem, may be written

(2.4.37) $$A^* = GAG_D^\ddagger GG_{\bar{X}}^\ddagger + Z^*(I - GG_{\bar{X}}^\ddagger) = GAG_D^\ddagger \ .$$

We now show (cf. Malinvaud [67]) that given our assumptions (2.4.18) and (2.4.21) concerning G and the fact that it is a grouping matrix, (2.4.35) implies that

(2.4.38) $$G_D^\ddagger = G_{\bar{X}}^\ddagger \ ,$$

hence $A^* = A^*_{\bar{x}}$. To see this, observe first that (2.4.35) must hold in particular for $x = \bar{x}$ and $x^* = \bar{x}^*$; combined with (2.4.20) this implies that

(2.4.39) $$G_D^\ddagger \bar{x}^* = \bar{x} = G_{\bar{X}}^\ddagger \bar{x}^* \ .$$

Now from (2.4.9), (2.4.18), the grouping structure (2.2.2) of G, and the fact that the elements of G and the components of $\bar{x}$ are non-negative, it follows that $\bar{x}^*$ cannot have any zero component unless the corresponding row of G vanishes; but this is ruled out by the assumption that G has full rank. Thus, the components of $\bar{x}^*$ in (2.4.39) are all positive; and since $G_D^\ddagger$ and $G_{\bar{X}}^\ddagger$ are (column-wise) grouping matrices, with at most one positive element in each row which must occur in the same column as between the two matrices, (2.4.39) implies (2.4.38). Thus, <u>(2.4.16) provides the unique solution to (2.4.36) when (2.4.35) holds.</u>

Note that $A^* = GAG^\dagger = GAG'(GG')^{-1}$ would <u>not</u> be a correct solution to (2.4.36) unless the rows of GD were in the row space of G (see Theorem 1.7(b) of section 1.2), which means that the diagonal elements of D would have to be equal to one another within groups, i.e., that the shares of the industries in each group would have to be equal rather than simply proportional to one another. This would be an artificial requirement.

Turning to the inverse system (2.4.5), it is easy to verify that if A is indecomposable, so is $A^* = GAG_{\bar{X}}^\ddagger$. This follows from the fact that each element of A^* is a weighted average of column sums of A; thus, if A^* is (after permutation of rows and columns, if necessary) block triangular, to the zero block there must correspond a zero block in A. Likewise, if the column sums of A are all ≤ 1, so are those of A^*, and the column sums of A^* cannot all be $= 1$ unless those of A are. Accordingly, if A is indecomposable, with column sums ≤ 1 and at least one column sum < 1, the same is true of A^*. Consequently the inverse matrix $(I - A^*)^{-1}$ exists. The inverse system (2.4.33) may therefore be used provided it can be assumed that y satisfies the constraint

$$y = (I - A)G_D^{\dagger} G(I - A)^{-1} y \quad . \tag{2.4.40}$$

It must be admitted that this assumption is rather artificial.

C. Perfect aggregation: mixed cases. Again, it will not be necessary to consider this case, which the interested reader can work out for himself.

D. Best approximate aggregation. We shall limit ourselves to considering the question: can the habitually employed aggregative model

$$y^* = (I - A^*_{\bar{x}})x^* \quad , \tag{2.4.41}$$

where x^* and y^* are defined by (2.4.24) and $A^*_{\bar{x}}$ is constructed according to the conventional formula (2.4.16), be rationalized in terms of the criterion of minimization of aggregation bias?

To answer this question it is necessary either to find a suitable matrix M which can be interpreted as a moment matrix for x, in terms of which (2.4.41) gives the formula for $\hat{\rho}(y^* | x^*)$; or, if homogeneity is not to be imposed a priori, to find a vector μ and a matrix Θ which can be interpreted as the mean and variance respectively of x, in terms of which (2.4.41) gives the expression for $\hat{\mathcal{E}}(y^*|x^*)$. From a casual inspection of (2.4.16) and (2.4.15) one might be tempted by a comparison with (2.2.69) and (2.2.70) to make the identification $M = \bar{X}$, where $\bar{X}$ is the diagonal matrix defined by (2.4.6) whose diagonal elements are the observed outputs $\bar{x}_i$ in the base year. Unfortunately, however, it is clear from (2.2.58) that the moment matrix M cannot be diagonal unless $\mu = 0$, whereas it would be natural to make the identification $\mu = \bar{x}$. This method of rationalizing (2.4.41), while formally possible, is certainly not convincing.

Fortunately, the results of section 2.2.F make it possible to adopt an alternative and more plausible rationalization. We shall make the identification†

$$(2.4.42) \qquad \mu = \bar{x}, \qquad \Theta = \lambda\bar{X} \qquad (\lambda > 0) .$$

From (2.2.169) and (2.2.173) we have

$$(2.4.43) \qquad \mathcal{E}(y^*|x^*) = G(I - A)(I - G \;\; G)\mu + G(I - A)G \;\; x^* ,$$

where $G^\natural$ is a generalized quasi-inverse of G with respect to Θ. From (2.4.15), obviously $G^\ddagger_{\bar{X}}$ qualifies as a choice of $G^\natural$. It follows immediately from (2.4.20) (which is the counterpart of (2.2.178)) that the constant term in (2.4.43) vanishes; this results from the fact that according to the specification (2.4.42), the variances of the x_i's are proportional to their means, which was the interpretation given to condition (2.2.179), whose counterpart is (2.4.19). Since $G^\ddagger_{\bar{X}}$ is a right inverse of G, from (2.4.21), formulas (2.4.43) and (2.4.41) are identical.

In applications, it is not so much the system (2.4.41) that is employed, as the inverse system

---

† The assumption of mean-variance proportionality was specified by W. D. Fisher [33], but with respect to the y_i's rather then the x_i's .

(2.4.44) $$x^* = (I - A^*_{\bar{x}})^{-1} y^* .$$

Let us verify that this may be interpreted as giving the expression for $\hat{\mathcal{E}}(x^* | y^*)$. From Theorem 1.14 we have, with the above identifications,

(2.4.45) $$\hat{\mathcal{E}}(x^* | y^*) = [G - (GFG^{\ddagger}_{\bar{X}})^{\natural} GF]\bar{x} + (GFG^{\ddagger}_{\bar{X}})^{\natural} y^* ,$$

where $F = I - A$ and $(GFG^{\ddagger}_{\bar{X}})^{\natural}$ is a generalized quasi-inverse of $GFG^{\ddagger}_{\bar{X}}$ with respect to $G\bar{X}G'$. Since both $GFG^{\ddagger}_{\bar{X}} = I - A^*_{\bar{x}}$ and $G\bar{X}G'$ have full rank, $(GFG^{\ddagger}_{\bar{X}})^{\natural} = (I - A^*_{\bar{x}})^{-1}$, and substituting this and (2.4.20) in (2.4.45) we see at once that the constant term vanishes and the expression reduces to that of (2.4.44).

E. Perfect disaggregation. The order condition (2.2.122) for perfect relative disaggregation holds--in fact, the stronger condition (2.2.124) holds--since $m = n > m^* = n^*$. Since $G^{\ddagger}_{\bar{X}}$ is a reflexive generalized inverse of G, by condition (ii) of (2.4.15), the rank condition (2.2.120) applies. It states that

$$\text{rank } G(I - A)G^{\ddagger}_{\bar{X}} = \text{rank } (I - A)G^{\ddagger}_{\bar{X}} ,$$

and this certainly holds since $G(I - A)G^{\ddagger}_{\bar{X}} = I - A^*_{\bar{x}}$ has been shown to have the full rank n^* . Thus the equation analogous to (2.2.119) has a unique solution which we may denote

(2.4.46) $$G^{\ddagger}_{W} = (I - A)G^{\ddagger}_{\bar{X}}(I - GAG^{\ddagger}_{\bar{X}})^{-1} ,$$

since it is a right inverse of G satisfying

(2.4.47) $$GG_W^{\ddagger} = I, \qquad G_W^{\ddagger}GW = WG'G_W^{\ddagger\prime}$$

where

(2.4.48) $$W = (I - A)\bar{X}G'(G\bar{X}G')^{-1}G\bar{X}(I - A)' \quad .$$

The matrix (2.4.46) furnishes the blowing-up formula

(2.4.49) $$\hat{y} = G_W^{\ddagger}y^{*} = (I - A)G_{\bar{X}}^{\ddagger}(I - A_{\bar{x}}^{*})^{-1}\, y^{*}$$

enabling one to forecast detailed net outputs from aggregate net outputs. Unfortunately it is not very useful since, given that

$$AG_{\bar{X}}^{\ddagger} = \bar{U}G'(G\bar{X}G')^{-1}$$

from (2.4.12), (2.4.14), and (2.4.21), it requires knowledge of data on $\bar{U}G'$, i.e., on the detailed inputs used up by each industrial aggregate. As it happens, however, this does not really matter, since one is usually interested only in making detailed forecasts of gross output rather than net output. The obvious formula to use for this purpose would be

(2.4.50) $$\hat{x} = G_{\bar{X}}^{\ddagger}(I - A_{\bar{x}}^{*})^{-1}\, y^{*} \quad .$$

Let us verify that (2.4.50) is in fact the optimal disaggregation formula.

F. Best approximate disaggregation. The matrix $G_{\bar{X}}^{\ddagger}$ of (2.4.14) corresponds to the matrix $G^{\#}$ of Theorem 2.1, and the matrix $G_W^{\ddagger}$ of (2.4.46) corresponds to the optimal matrix $\bar{H}$ of Theorem 2.2. It remains to verify that (2.4.50)

gives the expression for $\hat{\mathcal{E}}(x|y^*)$. In terms of the general notation of section 2.2 we have

$$\hat{\mathcal{E}}(x|y^*) = [I - G^{\natural}(HFG^{\natural})^{\natural} HFG^{\natural} G]\mu + G^{\natural}(HFG^{\natural})^{\natural} y^* , \tag{2.4.51}$$

where $G^{\natural}$ is a generalized quasi-inverse of G with respect to Θ, and $(HFG^{\natural})^{\natural}$ is a generalized quasi-inverse of $HFG^{\natural}$ with respect to $G\Theta G'$. Making the identifications $H = G$, $G^{\natural} = G^{\ddagger}_{\bar{X}}$, $F = I - A$, and (2.4.42), and noting as above that $(GFG^{\ddagger}_{\bar{X}})^{\natural} = (I - A^*_{\bar{x}})^{-1}$, we see that (2.4.51) becomes

$$\hat{\mathcal{E}}(x|y^*) = (I - G^{\ddagger}_{\bar{X}} G)\bar{x} + G^{\ddagger}_{\bar{X}}(I - A^*_{\bar{x}})^{-1} y^* , \tag{2.4.52}$$

and this is equivalent to (2.4.50) on account of (2.4.20).

In order to forecast x, therefore, one first forecasts x^* by means of the formula $x^* = (I - A^*_{\bar{x}})^{-1}y^* = (I - GAG^{\ddagger}_{\bar{X}})y^*$, i.e., by operating the inverse of the aggregative Leontief matrix on the aggregative bill of goods. Next, one blows up x^* by the formula $\hat{x} = G^{\ddagger}_{\bar{X}} x^*$. Taking G as in the earlier illustration we have, as an example, just as in (2.2.26),

$$G = \begin{bmatrix} 1 & 1 & 0 & 0 & 0 \\ 0 & 0 & 1 & 0 & 1 \end{bmatrix}, \qquad G^{\ddagger}_{\bar{X}} = \begin{bmatrix} \frac{\bar{x}_1}{\bar{x}_1+\bar{x}_2} & 0 \\ \frac{\bar{x}_2}{\bar{x}_1+\bar{x}_2} & 0 \\ 0 & \frac{\bar{x}_3}{\bar{x}_3+\bar{x}_5} \\ 0 & 0 \\ 0 & \frac{\bar{x}_5}{\bar{x}_3+\bar{x}_5} \end{bmatrix}$$

Thus, to forecast gross output in an individual industry, one multiplies the aggregate gross output of the group to which it belongs by the relative share of this industry's output in the group's output that obtained in the base year. Nothing could be simpler! While the result may not be startling, it is at least reassuring to know that such a simple rule of thumb can be provided with a sound logical basis.

REFERENCES

[1] Aczél, J., Lectures on Functional Equations and Their Applications. New York: Academic Press, 1966.

[2] Afriat, S. N., "Orthogonal and Oblique Projectors and the Characteristics of Pairs of Vector Spaces," Proceedings of the Cambridge Philosophical Society, 53 (Part 4, 1957), 800-816.

[3] Aitken, A. C., "On Least Squares and Linear Combinations of Observations," Proceedings of the Royal Society of Edinburgh, 55 (1935), 42-48.

[4] Akkina, K. R., "Application of Random Coefficient Regression Models to the Aggregation Problem," Econometrica, 42 (March 1974), 369-375.

[5] Anderson, T. W., "On the Theory of Testing Serial Correlation," Skandinavisk Aktuarietidskrift, 31 (1948), 88-116.

[6] Anderson, T. W., Introduction to Multivariate Statistical Analysis. New York: John Wiley & Sons, Inc., 1958.

[7] Ara, Kenjiro, "The Aggregation Problem in Input-Output Analysis," Econometrica, 27 (April 1959), 257-262.

[8] Balderston, J. B., and T. M. Whitin, "Aggregation in the Input-Output Model," in Economic Activity Analysis (edited by Oskar Morgenstern), New York: John Wiley & Sons, Inc., 1954, pp. 79-128.

[9] Basevi, Giorgio, "Aggregation Problems in the Measurement of Effective Protection," in Effective Tariff Protection (edited by Herbert G. Grubel and Harry G. Johnson), Geneva: General Agreement on Tariffs and Trade, 1971, pp. 115-134.

[10] Basmann, R. L., "A Generalized Classical Method of Linear Estimation of Coefficients in a Structural Equation," Econometrica, 25 (January 1957), 77-83.

[11] Bellman, Richard, Introduction to Matrix Analysis. New York: McGraw-Hill Book Company, Inc., 1960.

[12] Bose, R. C., "The Fundamental Theorem of Linear Estimation" (abstract), Proceedings of the Thirty-First Indian Science Congress, 4 (1944), Part III, 2-3.

[13] Charnes, A., and W. W. Cooper, Management Models and Industrial Applications of Linear Programming, Volume I. New York: John Wiley & Sons, Inc., 1961.

[14] Chipman, John S., "The Multi-Sector Multiplier," Econometrica, 18 (October 1950), 355-374.

[15] Chipman, John S., "On Least Squares with Insufficient Observations," Journal of the American Statistical Association, 59 (December 1964), 1078-1111.

[16] Chipman, John S., "Specification Problems in Regression Analysis," in Proceedings of the Symposium on Theory and Application of Generalized Inverses of Matrices (edited by Thomas L. Boullion and Patrick L. Odell), Lubbock, Texas: The Texas Tech Press, 1968, pp. 114-176.

[17] Chipman, John S., and M. M. Rao, "On the Use of Idempotent Matrices in the Treatment of Linear Restrictions in Regression Analysis" (abstract), Biometrics, 14 (June 1958), 290-291.

[18] Chipman, John S., and M. M. Rao, "The Treatment of Linear Restrictions in Regression Analysis," Econometrica, 32 (January-April 1964), 198-209.

[19] Chipman, John S., and M. M. Rao, "Projections, Generalized Inverses, and Quadratic Forms," Journal of Mathematical Analysis and Applications, 9 (August 1964), 1-11.

[20] Chow, Gregory, "The Selection of Variates for Use in Prediction: a Generalization of Hotelling's Solution," mimeographed.

[21] Cramér, Harald, Mathematical Methods of Statistics. Princeton, New Jersey: Princeton University Press, 1946.

[22] Debreu, Gerard, "Topological Methods in Cardinal Utility Theory," in Mathematical Methods in the Social Sciences, 1959 (edited by Kenneth J. Arrow, Samuel Karlin, and Patrick Suppes), Stanford, California: Stanford University Press, 1960, pp. 16-26.

[23] Debreu, Gerard, and I. N. Herstein, "Nonnegative Square Matrices," Econometrica, 21 (October 1953), 597-607.

[24] Doob, J. L., Stochastic Processes. New York: John Wiley & Sons, Inc., 1953.

[25] Dresch, Francis W., "Index Numbers and the General Economic Equilibrium," Bulletin of the American Mathematical Society, 44 (February 1938), 134-141.

[26] Durbin, J., and G. S. Watson, "Testing for Serial Correlation in Least Squares Regression I," Biometrika, 37 (1950), 409-428.

[27] Fei, John Ching-Han, "A Fundamental Theorem for the Aggregation Problem of Input-Output Analysis," Econometrica, 24 (October 1956), 400-412.

[28] Fisher, Franklin M., "Embodied Technical Change and the Existence of an Aggregate Capital Stock," Review of Economic Studies, 32 (October 1965), 263-288.

[29] Fisher, Franklin M., "Embodied Technology and the Existence of Labour and Output Aggregates," Review of Economic Studies, 35 (October 1968), 391-412.

[30] Fisher, Franklin M., "Approximate Aggregation and the Leontief Conditions," Econometrica, 37 (July 1969), 457-469.

[31] Fisher, Franklin M., "The Existence of Aggregate Production Functions," Econometrica, 37 (October 1969), 553-577.

[32] Fisher, Walter D., "On a Pooling Problem from the Statistical Decision Viewpoint," Econometrica, 21 (October 1953), 567-585.

[33] Fisher, Walter D., "Criteria for Aggregation in Input-Output Analysis," Review of Economics and Statistics, 40 (August 1958), 250-260.

[34] Fisher, Walter D., "Optimal Aggregation in Multi-Equation Prediction Models," Econometrica, 30 (October 1962), 774-769.

[35] Fisher, Walter D., Clustering and Aggregation in Economics. Baltimore: The Johns Hopkins Press, 1969.

[36] Fisher, Walter D., and Walter J. Wadycki, "Estimating a Structural Equation in a Large System," Econometrica, 39 (May 1971), 461-465. "Reply," ibid., 40 (September 1972), 909.

[37] Foster, Manus, "An Application of the Wiener-Kolmogoroff Smoothing Theory to Matrix Inversion," Journal of the Society for Industrial and Applied Mathematics, 3 (September 1961), 387-392.

[38] Frisch, Ragnar, "Correlation and Scatter in Statistical Variables," Nordic Statistical Journal, 1 (1929), 36-102.

[39] Frisch, Ragnar, Statistical Confluence Analysis by Means of Complete Regression Systems. Oslo: Universitetets Økonomiske Institutt, 1934.

[40] Goldman, A. J., and M. Zelen, "Weak Generalized Inverses and Minimum Variance Linear Unbiased Estimation," Journal of Research of the National Bureau of Standards [B], 68B (October-December 1964), 151-172.

[41] Gorman, W. M., "Community Preference Fields," Econometrica, 21 (January 1953), 63-80.

[42] Gorman, W. M., "The Structure of Utility Functions," Review of Economic Studies, 35 (October 1968), 367-390.

[43] Green, H. A. John, Aggregation in Economic Analysis: An Introductory Survey. Princeton, New Jersey: Princeton University Press, 1964.

[44] Gupta, K. L., Aggregation in Economics. Rotterdam: Rotterdam University Press, 1969.

[45] Hansen, Alvin H., Fiscal Policy and Business Cycles. New York: W. W. Norton & Company, Inc., 1941.

[46] Harris, W. A., Jr., and T. N. Helvig, "Marginal and Conditional Distributions of Singular Distributions," Publications of Research Institute for Mathematical Sciences, Kyoto University [A], 1 (1966), 199-204.

[47] Hatanaka, M., "Note on Consolidation within a Leontief System," Econometrica, 20 (April 1952), 301-303.

[48] Hatanaka, Michio, The Workability of Input-Output Analysis. Ludwigshafen am Rhein: Fachverlag für Wirtschaftstheorie und Ökonometrie, 1960.

[49] Holly, A., "Sur l'estimation des paramètres d'un modèle linéaire lorsque la matrice des variances-co-variances des résidus est singulière," Laboratoire Georges Darmois, Université Paris IX - Dauphine, January 1973 (mimeographed).

[50] Holzman, Mathilda, "Problems of Classification and Aggregation," in Studies in the Structure of the American Economy (edited by Wassily Leontief), New York: Oxford University Press, 1953, pp. 326-359.

[51] Hotelling, Harold, "The Selection of Variates for Use in Prediction with Some Comments on the General Problem of Nuisance Parameters," Annals of Mathematical Statistics, 11 (September 1940), 271-283.

[52] Hu, Sze-Tsen, Theory of Retracts. Detroit: Wayne State University Press, 1965.

[53] Hurwicz, Leonid, "Aggregation in Macroeconomic Models" (abstract), Econometrica, 20 (July 1952), 489-490.

[54] Ijiri, Yuji, "The Linear Aggregation Coefficient as the Dual of the Linear Correlation Coefficient," Econometrica, 36 (April 1968), 252-259.

[55] Jeffreys, Harold, Theory of Probability, 2nd edition. Oxford: at the Clarendon Press, 1948.

[56] Klein, Lawrence R., "Macroeconomics and the Theory of Rational Behavior," Econometrica, 14 (April 1946), 93-108.

[57] Klein, Lawrence R., "Remarks on the Theory of Aggregation," Econometrica, 14 (October 1946), 303-312.

[58] Koopmans, T., Linear Regression Analysis of Economic Time Series. Haarlem: De Erven F. Bohn N.V., 1937.

[59] Koopmans, T.C., H. Rubin, and R.B. Leipnik, "Measuring the Equation Systems of Dynamic Economics," in Statistical Inference in Dynamic Economic Models (edited by Tjalling C. Koopmans), New York: John Wiley & Sons, Inc., 1950, pp. 53-237.

[60] Kruskal, William, "When are Gauss-Markov and Least Squares Estimators Identical? A Coordinate-Free Approach," Annals of Mathematical Statistics, 39 (February 1968), 70-75.

[61] Kruskal, William, "The Geometry of Generalized Inverses," Journal of the Royal Statistical Society [B], forthcoming.

[62] Leontief, Wassily, "A Note on the Interrelation of Subsets of Independent Variables of a Continuous Function with Continuous First Derivatives," Bulletin of the American Mathematical Society, 53 (April 1947), 343-350.

[63] Leontief, Wassily W., The Structure of American Economy, 1919-1929, 2nd edition. New York: Oxford University Press, 1951.

[64] McManus, M., "General Consistent Aggregation in Leontief Models," Yorkshire Bulletin of Economic and Social Research, 8 (June 1956), 28-48.

[65] McManus, M., "On Hatanaka's Note on Consolidation," Econometrica, 24 (October 1956), 482-487.

[66] Magness, T. A., and J. B. McGuire, "Comparison of Least Squares and Minimum Variance Estimates of Regression Parameters," Annals of Mathematical Statistics, 33 (June 1962), 462-470.

[67] Malinvaud, Edmond, "Aggregation Problems in Input-Output Models," in The Structural Interdependence of the Economy, Proceedings of an International Conference on Input-Output Analysis, Varenna, 27 June-10 July 1954 (edited by Tibor Barna), New York: John Wiley & Sons, Inc., n.d., pp. 187-202.

[68] Malinvaud, E., "L'agrégation dans les modèles économiques," Cahiers du Séminaire d'Econométrie, No. 4 (1956), 69-146.

[69] Malinvaud, E., Statistical Methods of Econometrics, second revised edition. New York: American Elsevier Publishing Company, Inc., 1970.

[70] Marsaglia, George, "Conditional Means and Covariances of Normal Variables with Singular Covariance Matrix," Journal of the American Statistical Association, 59 (December 1964), 1203-1204.

[71] May, Kenneth, "The Aggregation Problem for a One-Industry Model," Econometrica, 14 (October 1946), 285-298.

[72] May, Kenneth, "Technological Change and Aggregation," Econometrica, 15 (January 1947), 51-63.

[73] Mitra, Sujit Kumar, "On a Generalized Inverse of a Matrix and Applications," Sankhyā: The Indian Journal of Statistics, [A], 30 (Part 1, 1968), 107-114.

[74] Mitra, Sujit Kumar, "Unified Least Squares Approach to Linear Estimation in a General Gauss-Markov Model," SIAM Journal of Applied Mathematics, 25 (December 1973), 671-680.

[75] Morimoto, Yoshinori, "On Aggregation Problems in Input-Output Analysis," Review of Economic Studies, 37 (January 1970), 119-126.

[76] Morimoto, Yoshinori, "A Note on Weighted Aggregation in Input-Output Analysis," International Economic Review, 12 (February 1971), 138-143.

[77] Nataf, André, "Sur la possibilité de construction de certains macromodèles," Econometrica, 16 (July 1948), 232-244.

[78] Nataf, André, "Sur des questions d'agrégation en économétrie," Publications de l'Institut de Statistique de l'Université de Paris, 2 (Fasc. 4, 1953), 5-61.

[79] Nataf, André, "Possibilité d'agrégation dans le cadre de la théorie des choix," Metroeconomica, 5 (April 1953), 22-30.

[80] Penrose, R., "A Generalized Inverse for Matrices," Proceedings of the Cambridge Philosophical Society, 51 (1955), 406-413.

[81] Penrose, R., "On Best Approximate Solutions of Linear Matrix Equations," Proceedings of the Cambridge Philosophical Society, 52 (1956), 17-19.

[82] Prais, S. J., and J. Aitchison, "The Grouping of Observations in Regression Analysis," Review of the International Statistical Institute, 22(1954), 1-22.

[83] Pringle, R. M., and A. A. Rayner, Generalized Inverse Matrices, with Applications to Statistics. London: Charles Griffin & Company Limited, 1971.

[84] Pu, Shou Shan, "A Note on Macroeconomics," Econometrica, 14 (October 1946), 299-302.

[85] Radner, R., "Minimax Estimation for Linear Regressions," Annals of Mathematical Statistics, 29 (December 1958), 1244-1250.

[86] Rao, C. Radhakrishna, "Generalized Inverse for Matrices and its Applications to Statistics," in Festschrift for J. Neyman: Research Papers in Statistics (edited by F. N. David), New York: John Wiley & Sons, 1966, pp. 263-279.

[87] Rao, C. Radhakrishna, "A Note on a Previous Lemma in the Theory of Least Squares and Some Further Results," Sankhyā: The Indian Journal of Statistics [A], 30 (Part 3, 1968), 259-266.

[88] Rao, C. Radhakrishna, "Unified Theory of Linear Estimation," Sankhyā: The Indian Journal of Statistics [A], 33 (Part 4, 1971), 371-394.

[89] Rao, C. Radhakrishna, and Sujit Kumar Mitra, Generalized Inverse of Matrices and its Applications. New York: John Wiley & Sons, Inc., 1971.

[90] Rao, K. K., "A Simplified Proof of Gauss-Markov Theorem when the Regression Matrix is of Less than Full Rank," American Mathematical Monthly, 73 (April 1966), 394-395.

[91] Rosenblatt, David, "On Aggregation and Consolidation in Linear Systems," Technical Report C, Department of Statistics, American University, Washington, D. C., August 1956 (mimeographed).

[92] Rosenblatt, David, "Aggregation in Matrix Models of Resource Flows," The American Statistician, 19 (June 1965), 36-39.

[93] Rosenblatt, David, "Aggregation in Matrix Models of Resource Flows II. Boolean Relation Matrix Methods," The American Statistician, 21 (June 1967), 32-37.

[94] Schönfeld, Peter, "Best Linear Minimum Bias Estimation in Linear Regression," Econometrica, 39 (May 1971), 531-544.

[95] Sertel, Murat R., "A Four-Flagged Lemma," Review of Economic Studies, 39 (October 1972), 487-490.

[96] Solow, Robert, "On the Structure of Linear Models," Econometrica, 22 (January 1952), 29-46.

[97] Solow, Robert, "The Production Function and the Theory of Capital," Review of Economic Studies, 23 (No. 2, 1955-56), 101-108.

[98] Spanier, Edwin H., Algebraic Topology. New York: McGraw-Hill Book Company, 1966.

[99] Stigum, Bernt P., "On Certain Problems of Aggregation," International Economic Review, 8 (October 1967), 349-367.

[100] Swamy, P. A. V. B., and James Holmes, "The Use of Undersized Samples in the Estimation of Simultaneous Equation Systems," Econometrica, 39 (May 1971), 455-459.

[101] Theil, H., Linear Aggregation of Economic Relations. Amsterdam: North-Holland Publishing Company, 1954.

[102] Theil, H., "Linear Aggregation in Input-Output Analysis," Econometrica, 25 (January 1957), 111-122.

[103] Theil, H., Economic Forecasts and Policy. Amsterdam: North-Holland Publishing Company, 1958. 2nd edition, 1961.

[104] Theil, H., "The Aggregation Implications of Identifiable Structural Macrorelations," Econometrica, 27 (January 1959), 14-29.

[105] Theil, H., "Alternative Approaches to the Aggregation Problem," in Logic, Methodology, and Philosophy of Science (edited by Ernest Nagel, Patrick Suppes, and Alfred Tarski), Stanford, California: Stanford University Press, 1962, pp. 507-527.

[106] Theil, H., Economics and Information Theory. Amsterdam: North-Holland Publishing Company, and Chicago: Rand McNally & Company, 1967.

[107] Theil, H., Principles of Econometrics. New York: John Wiley & Sons, Inc., 1971.

[108] Thrall, Robert M., and Leonard Tornheim, Vector Spaces and Matrices. New York: John Wiley & Sons, Inc., 1957.

[109] Watson, Geoffrey S., "Linear Least Squares Regression," Annals of Mathematical Statistics, 38 (December 1967), 1679-1699.

[110] Zellner, Arnold, "An Efficient Method of Estimating Seemingly Unrelated Regressions and Tests for Aggregation Bias," Journal of the American Statistical Association, 57 (June 1962), 348-368.

[111] Zellner, Arnold, "On the Aggregation Problem: A New Approach to a Troublesome Problem," in Economic Models, Estimation, and Risk Programming: Essays in Honor of Gerhard Tintner (edited by K. A. Fox, J. K. Sengupta, and G. V. L. Narasimham), Berlin-Heidelberg-New York: Springer-Verlag, 1969, pp. 365-374.

[112] Zyskind, George, "On Canonical Forms, Non-negative Covariance Matrices and Best and Simple Least Squares Linear Estimators in Linear Models," Annals of Mathematical Statistics, 38 (August 1967), 1092-1109.

[113] Zyskind, George, and Frank B. Martin, "On Best Linear Estimation and a General Gauss-Markov Theorem in Linear Models with Arbitrary Nonnegative Covariance Structure," SIAM Journal of Applied Mathematics, 17 (November 1969), 1190-1202.

[114] Kossov, V., "The Theory of Aggregation in Input-Output Models," in Contributions to Input-Output Analysis (edited by A. P. Carter and A. Bródy), Amsterdam: North-Holland Publishing Company, 1972, pp. 241-248.

[115] Nashed, M. Z., "Generalized Inverses, Normal Solvability, and Iteration for Singular Operator Equations," in Nonlinear Functional Analysis and Applications (edited by Louis B. Rall), New York: Academic Press, 1971, pp. 311-359.

[116] Nashed, M. Z., and G. F. Votruba, "A Unified Approach to Generalized Inverses of Linear Operators," Bulletin of the American Mathematical Society, 80 (September 1974), 825-835.

[117] Neeleman, D., Multicollinearity in Linear Economic Models. Tilburg, The Netherlands: Tilberg University Press, 1973.

[118] Rayner, A. C., "A Comment on Estimating a Structural Equation in a Large System," Econometrica, 40 (September 1972), 907.

[119] Khazzoom, J. Daniel, "A Note on the Use of g-Inverse of Matrices for Generalizing the k-Class and 3SLS Estimators in a Large System," unpublished ms., 1974.

Department of Economics
University of Minnesota
Minneapolis, Mn. 55455

Annotated Bibliography on Generalized Inverses and Applications

M.Z. Nashed
L.B. Rall

"You must always invert."

G. G. J. Jacobi

Well, Mr. Jacobi, here it is: all the generalized inversion of two generations of invertors who, knowingly or unknowingly, subscribed to (and extended) your dictum. Please forgive us if we have over-inverted, or if we have not always inverted in the natural and sensible way. (Some of us have inverted with labor and pain by using hints from a dean or a tenure and promotion committee that "you better invert more, or else you would be inverted.")

Introduction

This bibliography is intended to include all sources on the theory, computation and applications of generalized inverses of matrices and linear operators. An effort was put to make the bibliography exhaustive. We have included a few older papers which were not cited in earlier bibliographies on the subject and included all related references up to 1975. In addition we have included a number of selected references on topics which have some interface with generalized inverses, e.g., ill-posed problems and regularization, semi-Fredholm operators, nonlinear least squares problems, nonlinear

Sponsored by the United States Army under Contract No. DAAG29-75-C-0024.

alternative problems and bifurication theory, etc. References on these topics are not complete; the selection of these references is intended to reflect (implicit or explicit) connections with generalized inverses.

Sources (articles, books, theses, etc.) are listed as follows:

Ref. no. in [ ], author, year in ( ), title of paper, journal, pages.

We usually use the standard abbreviations of names of journals as given in the Index to Mathematical Reviews.

Each reference has been assigned one of three classifications: Primary (P), Secondary (S), or Tertiary (T). The letter designating this classification appears below the number of the reference. For example,

[1776] Nashed, M. Z. (1976): Generalized inverses of moral
P principles, I. Case study 1969-75, Arch. Irrational Visibility and Divisibility, 1, to appear.

Three supplements have been added to the main bibliography. Supplements 2 and 3 contain some recent references and more tertiary references, as well as a few older references which we overlooked. Supplement 1 includes additional Ph. D. dissertations.

Multiple authors are listed separately followed by the name of the first-listed joint author and year of publication, repeated as many times as the number of joint papers with that date. Cross references to multiple authors in supplements 2 and 3 are local; they do not include (nor are they included in) the main bibliography.

References in the Annotations to the sources listed in the bibliography are made by a combination of the name of the author (or editor) and either the number of the reference in [ ] or the year of publication inserted in ( ); the latter when the date of publication is to be stressed for historic reasons. In some cases, particularly when the same reference appears frequently, we cite only the number of the reference.

A detailed assessment of the vast literature of generalized inverses based on a limited knowledge and space, is likely to result in some unfairness or controversy. We apologize to all those whose work we may have inadvertibly overlooked in writing these annotations. Lists of omission and/or corrections would be gratefully received.

We thank the many authors on generalized inverses whose contributions, after all, have made this project possible. Special thanks are due to Mrs. Doris Whitmore for typing with care several versions and modifications of this bibliography.

Annotations

The annotations are classified into five major divisions:

I. General

II. Theory of Generalized Inverse Matrices and Operators

III. Generalized Inverses in Analysis

IV. Numerical Analysis, Approximation Theory, and Computational Methods

V. Applications

Each division comprises several parts and each part is subdivided into several sections. Cross references within the Annotations are made by citing the part followed by the appropriate section, e.g. F6 . We give in some cases some explanation and/or historical notes for additional topics not treated explicitly in this volume.

I. General

A. Books, Collection of Articles, Conferences

The earliest book which devotes a substantial portion to the theory of generalized inverses seems to be by Korganoff and Pavel-Parvu (1967). An extensive bibliography as of 1966 was also included therein. A symposium on theory and applications of generalized inverses of matrices was held at Texas Tech University in March 1968 and its proceedings were edited by Boullion and Odell [187]. A bibliography on generalized matrix inverses, containing about 340 references, was compiled in 1968 by Lewis, Boullion and Odell [749] and published in the proceedings [187]. Several books on generalized inverses of matrices and applications have appeared recently: Boullion and Odell (1971), Pringle and Rayner (1971), Rao and Mitra (1971), Albert (1972), Bjerhammar (1973), Ben-Israel and Greville (1974). Other recent books which treat linear least squares problems and associated aspects of generalized inverses include Noble (1969), Lawson and Hanson (1972) and Stewart (1972). The book by Rust and Burrus (1972) treats some aspects of generalized inverses in numerical solutions of ill-conditioned linear equations and related mathematical programming techniques. The books by Graybybill (1969) and C. R. Rao (1972) contain expositions on generalized inverses. All these books deal with generalized inverses of finite matrices (or linear transformations on finite dimensional spaces), except that Ben-Israel and Greville (1974) devote a chapter to generalized inverses of linear operators in Hilbert spaces. Generalized inverses of linear operators in Hilbert and other spaces are developed in an expository paper by Nashed (1971) which include an extensive bibliography on the operator theory of generalized inverses as of 1970. Some notes on this topic from [917] have been revised and incorporated in several sections of the papers by Nashed and by Nashed and Votruba and in the annotations in this volume.

The earliest (sophomore-level) textbook on linear algebra which discusses generalized inverses of matrices seems to be that of Ficken (1967) (paper edition 1963); we also mention the appendix on generalized inverses in Zadeh and Desoer (1963). Other books which contain either a chapter or some brief material on generalized inverses include: Barnett [88], Beltrami [105], Blum [167], Holmes [593], Horst [595], Lancaster [715], Lewis and Odell [752], Luenberger [780], Reid [1129] and Searle [1207]. Generalized inverses are also briefly mentioned or used (explicitly or implicitly) in the context of some applications in Aoki [55], Bellman [101], [1494], Charnes et al. [236], Daniel [294], Drygas [338], Hale [517], Householder [597], Kalman, Falb and Arbib [638], Kaplansky [651], Lanczos [718], Laurent [732], Marchuk and Kuznecov [798], Ortega and Rheinboldt [974], Ostrowski [982], Payne [991], Porter [1042], Reid [1130], Searle [1207], Young and Calvert [1402], Bharucha-Reid [1502], Bart [1484], Boot [1509], Caradus et. al [1520], Dahlquist and Björck [1528], Franklin [1549], Gelb [1552], Skornyakov [1245], Pease [1000] and others.

Early occurrence of generalized inverse concepts (inner or partial inverse, etc.) in books and lecture notes may be found in Baer (1952), Bose (1959), Friedrichs (1953), von Neumann (1950), (1960), Hamburger (1951) and others.

B. Bibliographies and Articles or Books with Extensive Bibliographies: Lewis, Boullion and Odell [749], Boullion and Odell [190], Rao and Mitra [1113]; for operators, see Nashed [915].

C. Theses on Generalized Inverses and Related Topics

1. Ph. D. theses in mathematical sciences (mathematics, applied mathematics, statistics, computer science and operations research)

Amburgey [29], Anderson, P. O. [1425], Anderson, W. N., Jr. [42], Bailes [1426], Barham [1427], Berman [1428], Bhimasankaram [143], Bodin [1464], Borowsky [1429], Boullion [186], Brown [1430], Burris [1431], Businger [1465], Chen [1432], Chitwood [254], Cline [263], Cohoon [1433], Crawford [1434], Dalphin [291], Deutsch [1435], Erdelsky [368], Funderlic [1436], Gately [439], Glassman [1437], Hall [519], Hallum [1438], Harms [1439], Hilgers [583], Ize [1440], Jennings [1467], Johnson [1441], Jones [1442], Joshi [1443], Kaufman [662], Kirby [674], Kublanovskaja [708], Lawson, Linda [736], Lay [737], Lent [744], Lewis, T. O. [747], Lill [1468], Mackichen [1444], Martin [807], Meicler [829], Meyer [837], Morrel [1445], Morris [902], Myller [910], Narain [1446], Nelson [942], O'Neill [1447], Patterson [1448], Perry [1449], Poole [1450], Price [1451], Pringle [1469], Pye [1452], Pyle [1085], Rodrigue [1453], Rohde [1149], Ryan [1470], Scolnik [1203], Sheffield [1222], Shen I-ming [1454], Shurbet [1455], Stallings [1456], Stewart [1457], Stein, J. [1458], Stein, R. A. [1267], Strand [1283], Teitz [1460], Trapp [1325], Tseng [1327], Vanderschel [1461], Van Loan [1462], Votruba [1346], Wall [1351], Weber [1463], Westfall [1372], Worm [1392], Zettl [1408], Zlobec [1413].

2. Ph. D. theses in engineering and related disciplines: Al-Alaoui [13], Anderson, B. M. [38], Kara [652], Manherz [794], Nayak [941], Smith [1247], Cohen [1466], Robers [1758] and others.

3. M. S. theses: Abdennur [4], Anderson, C. L. [39], Dekerlegand [307], Barton [93], Heath [568] and others.

Many of the preceding dissertations treat theory and applications of generalized inverses of matrices or operators. However, we have included in the list several theses which do not deal directly with generalized inverses, but are of tertiary interest to this subject. In addition, there are several earlier dissertations with primary or secondary classification in generalized inverses which are not listed, since their contents were published in papers included in this Bibliography.

D. Historical Aspects of Generalized Inverses

For historical notes and comments and for references to early occurrences of generalized inverses in various algebraic, analytic and statistical contexts, see the books [125], [190], [693] and [1113] and the papers [119], [1128], [915], [533], [917], and [1096].

II. Theory of Generalized Inverse Matrices and Operators

E. Generalized Inverses of Matrices

E1. Most of the theory of generalized inverses of matrices up to 1972 is adequately covered in several monographs and books. See Rao and Mitra (1971), Ben-Israel and Greville (1973) for a comprehensive coverage, and Korganoff and Pavel-Parvu (1967), Boullion and Odell (1971), Pringle and Rayner (1971), Albert (1971) and Bierhammar (1973) for selected topics. We confine the annotations in this section to a few topics in the theory of generalized inverse matrices.

E2. Topics in linear algebra and matrix theory for mathematical background for generalized inverses:

Linear manifolds, range and null spaces, intersection of manifolds. Idempotents and algebraic projectors, algebraic direct sums, complementary subspaces.

Inner products, Euclidean norms, ellipsoidal norms, orthogonal complements, orthogonal projections, the projection theorem, Gram-Schmidt orthonormalization. Norms, vector and matrix norms, unitarily equivalent norms, consistent norms, multiplicative norm. Convergence of matrices, condition number and ill-conditioning.

Canonical forms of matrices: Hermite canonical form (echelon form), triangular reduction, Smith normal form, Householder transformation, full rank factorization and others.

Spectral decomposition, polar decomposition of a matrix, singular values and vectors of a rectangular matrix, Hermitian and normal matrices, diagonalization and bidiagonalization, unitarily similar matrices, unitarily equivalent matrices, matrix functions.

Convex sets and metric projection, strictly convex, uniformly convex norm, generalized Cauchy inequality.

Linear equations and transformations, isometry, partial isometry, normal isometry.

References: [17], [125], [1113], [1272], [101], [1550], [169], [715], [800], [961], [1635], [76], [167], [525], [597], [452], [244], [865].

E3. Algebraic and structural properties; classification and characterization of various subclasses of 1-inverses (g-inverses, inner inverses); duality among subclasses and relations between generalized inverses and projectors; diversity of generalized inverses, existence, properties and construction of 1-inverses, (1, 2)-inverses, (1, 3)-inverses, (1, 2, 3)-inverses, etc.: Rohde [1149], [1150], Hearon and Evans [566] Rao [1104], [1102], Deutsch [317], Nashed and Votruba [934], Sibuya [1237], [1238], Sheffield [1222], Kruskal [700], Goldman and Zelen [458], Morris and Odell [903], Rao and Mitra [1112], Urquhart [1335], [1336], Tewarson [1303], Turbin [1333], and others.

See also the books [125], [190], [1054], [1113] for details and historical accounts, and F2 .

E4. Transformations of generalized inverses under change of projectors; explicit general representations of various generalized inverses in terms of 1-inverses (inner inverses).

Let $\mathcal{V}$ and $\mathcal{W}$ be (real or complex) vector spaces and let L be a linear transformation from $\mathcal{V}$ into $\mathcal{W}$. Given $P^2 = P: \mathcal{V} \to \mathcal{N}(L)$ and $Q^2 = Q: \mathcal{W} \to \mathcal{R}(L)$, both linear and onto, there is a (unique) generalized inverse $M = L^{\dagger}_{P,Q}$ relative to P and Q which is the unique solution of the equations $LX = Q$, $XL = I - P$

and $XLX = X$. (a) If we change P and Q to some other (algebraic) projectors P' and Q', where $\mathcal{R}(P') = \mathcal{N}(L)$ and $\mathcal{R}(Q') = \mathcal{R}(L)$, then

$$L^{\dagger}_{P', Q'} = (2I - ML - P')\, L^{\dagger}_{P, Q}(I - ML + Q') = (I - P')L^{\dagger}_{P, Q}\, Q' \ .$$

Any other (algebraic) generalized inverse is of this form for some linear idempotent maps P' and Q' with the preceding properties. (b) Given two generalized inverses $L^{\dagger}$ and $L^{\#}$, they induce four linear idempotents $Q := LL^{\dagger}$, $P := I - L^{\dagger}L$, $Q' := LL^{\#}$, and $P' := I - L^{\#}L$, which are related as follows: $P' = P + A$, $Q' = Q + B$, where $\mathcal{R}(A) \subset \mathcal{N}(L) \subset \mathcal{N}(A)$ and $\mathcal{R}(B) \subset \mathcal{R}(L) \subset \mathcal{R}(B)$. Equivalently, $P' = P + L^{\dagger} - L^{\#})L$ and $Q' = Q + L(L^{\#} - L^{\dagger})$. See Nashed and Votruba [934]. Various other transformation formulae for inner and generalized inverses and connections with projectors are given in [934], [125], [1113], Deutsch [317], Milne [857]; see also F2.

E5. Ranks of generalized inverses. Results on ranks of 1-inverses and other generalized inverses are given for instance in [1149], [125], [838], [1113], [396]. For any 1-inverse A^- of A, rank $A^- \geq$ rank A. A 1-inverse B of A is also a 2-inverse (outer inverse) iff rank A = rank B. Thus 1-2, 1-2-3, 1-2-4, and the Moore-Penrose inverse all preserve the rank of the original matrix. We always have trace AA^- = trace A^-A = rank A. For the construction of a 1-inverse having any prescribed rank between the rank of A and $\min(m, n)$, and fixed rank solutions of linear equations, see [396], [125], [1113], [874]; similarly for a 2-inverse. For ranks of partitioned matrices, see, for example, [847] and annotations under E11.

E6. The Moore-Penrose inverse. Geometric and analytic properties; $AA^{\dagger}$ and $A^{\dagger}A$ are orthogonal projectors, various characterizations of $A^{\dagger}$: Moore [892], Penrose [1002], Greville [492], Bjerhammar [150], [151], Albert [17], Nashed [915], Rao and Mitra [1113], Rado [1091], Mitra and Bhimasankaram [879], Mitra and Rao [1751] (for seminorms).

Miscellaneous expansions and explicit formulae for $A^{\dagger}$: [125], [1113], [693], [1054], [190], [17], Afriat [7] Zlobec [1412], Turbin [1333] and many others; see also N6, N8, O2 and O3.

Greville notes that C. C. MacDuffee was the first to

point out (in private communications) that an explicit formula for the Moore inverse $A^\dagger$ can be given by using a full-rank factorization of an arbitrary $m \times n$ matrix A of rank r in the form $A = FG$, where F and G are each of rank r. Then

$$A^\dagger = G^*(F^*AG^*)^{-1}F^* = G^*(GG^*)^{-1}(F^*F)^{-1}F^*$$

$$= A^{(1,4)} A A^{(1,3)},$$ where $A^{(1,4)}$ is any (1, 4)-inverse, etc.

E7. Minimal properties of generalized inverses. Generalized inverses and solvability or least squares solvability of linear algebraic equations (see also E6, 8, 9, 10).

For a unified and more general approach to minimal, extremal and best approximate solutions of linear equations in normed spaces, see Nashed and Votruba [934; Sec. 3], where the concepts of left-orthogonal, right-orthogonal, and orthogonal inner inverses are introduced and related to the types of solutions listed above. As a special case, one obtains the usual minimal properties of the various generalized inverses in inner product geometries, as well as generalized inverses with respect to uniformly convex norms, [934; Sec. 4]. Minimal properties of the usual (1, 3)-, (1, 4)-, and the Moore-Penrose inverse are discussed in [125], [1113], [126].

1-inverses and solvability of linear equations; constructions of 1-inverses by Gaussian elimination: [125; 1.4], [1113], [190], [1054], [693], Rao [1100], Sheffield [1223], [934] and others.

(1, 4)-inverses and minimal-norm solutions of compatible equations; characterizations of (1, 4)-inverses: [125; 3.2], [1113], [934], Rosen [1159] and many others.

(1, 3)-inverses and least-squares solutions of inconsistent linear systems; characterizations of (1, 3)-inverses: [125; 3.3], [1113], [934], Goldman and Zelen [458], Tipton and Milnes [1321], and many others.

Least-squares solutions of minimal norms: See E6. Also, Osborne, E. E. [976], [977], Poole [1033] and others listed before.

E8. Generalized inverses in finite dimensional normed space. Metric generalized inverses. Various aspects of (algebraic) generalized inverses of linear transformations in finite dimensional normed spaces are developed in [934]. Specific types of generalized inverses which are subsumed in this setting include the following: the concept of an algebraic generalized inverse relative to another transformation, generalized inverses with specified range and null spaces, generalized inverses corresponding to oblique projectors, weighted generalized inverse, and the group inverse of a linear transformation. (For the structure of a 1-inverse with row and column spaces belonging to specified row and column spaces, see also [868], [1054], [125], [1113] and others). See also comments in E7.

A concept of metric generalized inverse for a (not necessarily linear) transformation was introduced in [932], and developed in more detail in [934]. This concept subsumes several notions of generalized inverses with respect to various uniformly convex norms, the p-q inverse, extremal generalized inverses in Euclidean spaces, etc. For p-q generalized inverse and best approximate solutions with respect to uniformly convex norms, see Boullion and Odell [190; 5.2], Meicler [829], Newman and Odell [956], Newman, Odell and Meicler [957] and Johnson [1441]. Erdelsky [368] considered also generalized inverses in finite dimensional spaces based on essentially strictly convex norms and associated projectors. (See also [125; 3.4] for a detailed development of some of the results of [368]). These generalized inverses are again metric generalized inverses and some of the results can be subsumed in the setting of [934].

E9. Some classes and special types of generalized inverses

9-1. Weighted least-squares problems and weighted generalized inverses. The concept of a weighted generalized inverse (defined in terms of positive definite matrices U and V) is due to Chipman [249] (for the case of nonsingular weights). This concept is equivalent to the Moore-Penrose inverse in the geometry of weighted or "ellipsoidal" inner products induced by the matrices U and V. See, e.g., [934; 4.5]. Weighted least-squares problems with singular (non-negative definite) weights U and V have been considered by several authors, e.g., Goldman and Zelen [458], Zyskind [1422], Watson [1361],

Kruskal [699] and Rao [1103], [1105]. The concept of weighted generalized inverse was further studied by Meicler [829], Ward, Boullion and Lewis [1357] for the case of singular weights, and by others. The relation between generalized inverses on $\mathcal{L}_2$-spaces and certain reproducing kernel spaces also involves use of singular weights; see Nashed and Wahba [939]. For expositions on weighted least-squares problems and generalized inverses, see also [125; 3.3], [190], [915], [1113].

9-2. Milne [857] defines a unique "oblique" generalized inverse of a matrix A in terms of the isomorphic images of the null spaces of A and A^*. This is easily seen to be equivalent to the definition of a generalized inverse induced by nonorthogonal complementary subspaces. (Generalized inverses in finite dimensional spaces induced by nonorthogonal decompositions $\mathbb{R}^n = \mathcal{N}(A) \oplus \mathcal{M}$, $\mathbb{R}^m = \mathcal{R}(A) \oplus \mathcal{S}$ have been studied by several authors including Langenhop [723], Nashed [915], Nashed and Votruba [934] and Robinson [1144]. The sets of oblique and weighted generalized inverses are identical (see Ward, Boullion and Lewis [1358]. Furthermore, every (1, 2)-inverse of a matrix is in fact a Moore-Penrose inverse relative to new inner products (see Nashed [924]).

9-3. Constrained equations and constrained least-squares problems. Constrained generalized inverses. See Minamide and Nakamura [862]; [1103], [17], [915], [125]; see also F15.

The earliest appearance of a concept of a constrained inverse seems to be in the paper of Bott and Duffin [180], which predates Penrose's paper [1002]. Various concepts of restricted or constrained generalized inverses have appeared in the literature often motivated by specific applications. A unifying concept of a constrained generalized inverse was given recently by Rao and Mitra [1116] which brings together various generalized inverses and subclasses of 1-inverses studied in the literature under a common classification scheme. Applications of this concept are indicated in obtaining restricted solutions of consistent linear equations and E-approximate (or projectional) solutions of inconsistent linear equations. A unifying theory for constrained generalized inverses of linear operators with emphasis on topological and proximinal extending and adapting the main results of Nashed and Votruba [934] can also be developed; however the tools and the main results are no longer primarily algebraic.

9-4. The Drazin inverse, the group inverse, spectral generalized inverses: See E13.

9-5. For other classes of generalized inverses see, e.g., Mitra [868], [869], Rao and Mitra [1113] and related cited references.

E10. Generalized inverse of a product (reverse order law): The reverse order law for the inverse of matrix product, i.e., $(AB)^{-1} = B^{-1}A^{-1}$ is generally not transferable to generalized inverse, i.e. in general $(AB)^\dagger \neq B^\dagger A^\dagger$. For example, take $A = (1, 0)$, $B = \binom{1}{1}$, then $(AB)^\dagger = (1)$, while $B^\dagger A^\dagger = (\frac{1}{2}, \frac{1}{2})\binom{1}{0} = (\frac{1}{2})$. A general representation for $(AB)^\dagger$ was developed by Cline [264]: $(AB)^\dagger = B_1^\dagger A_1^\dagger$, where $B_1 = A^\dagger AB$ and $A_1 = AB_1B_1^\dagger$. Sibuya [1238] gave an extension of Cline's formula. Greville [495] showed that a necessary and sufficient condition for $(AB)^\dagger = B^\dagger A^\dagger$ is that $\mathcal{R}(BB^*A^*) \subset \mathcal{R}(A^*)$ and $\mathcal{R}(A^*AB) \subset \mathcal{R}(B)$. In particular if A is $m \times r$, B is $r \times n$ and rank A = rank B $= r$, then $(AB)^\dagger = B^\dagger A^\dagger$. Some cases were also given by Penrose [1002], in particular, the reverse order law holds for $(A^\dagger A)^\dagger$, $(A^*A)^\dagger$, and for the case when either $A^*A = I$ or $B^*B = I$. For an exposition of some results on $(AB)^\dagger$ see [190], [125], [17], [693].

For other papers dealing with the reverse order law, see Erdélyi [370], Bouldin [184], [185], Basket and Katz [95], Pearl [997], Arghiriade [58] {a necessary and sufficient condition (which is implicit in [58]) for $(AB)^\dagger = B^\dagger A^\dagger$ is that A^*ABB^* be range-Hermitian (cf. E13-4)}, Dragomir and Dragomir [1536], Shinozaki and Sibuya [1226], Barwick and Gilbert [94], Hartwig [550], Hartwig and Katz [1742], and Ballentine [1723]. See also F13 for reverse order law for operators.

E11. Generalized inverses of partitioned and bordered matrices; generalized inverse of a sum (cf. [17], [125], [190], [1054; ch. 3], [1113]). These sometimes involve tedious manipulations and complicated identities which have been found to be useful for various results. Greville [493] considered the partition $A_{m+1} = (A_m \mid a_{m+1})$ and provided a nice recursive procedure for computing the generalized inverse a column at a time, which has application to stepwise regression. Some representation for generalized inverses of certain matrix sums were obtained by Cline [266], [267], [268] and applied to computing the generalized inverse of a partitioned matrix; the formula for $(AA^* + BB^*)^\dagger$ is of particular interest. Cline [265] also gave formulae to obtain $A^\dagger$ from $(A:B)$, where A and B are block matrices. These formulae are mathematically elegant but not easily computable. Mitra and Bhimasankaram [880] (see also [1113]) gave formulae for computing several types of 1-inverses (inner or g-inverses) of $(A:a)$, where A is $m \times n$ and a is $m \times 1$, from corresponding 1-inverses of A, and reversely. They also gave formulae for computing various g-inverses of $A + BDC^*$ from those of A when D is nonsingular. Rohde [1149; pp. 47-50] considered generalized inverses of blocks of positive semidefinite matrices. Other papers which contain interesting results on generalized inverses of partitioned and bordered matrices are: Blattner [165], Reid [1125], Hearon [563], Germain-Bonne [446], Cline [265], [270], Burns, Carlson, Haynsworth and Markham [215], Carlson, Haynsworth and Markham [231], Rohde [1150], Pringle and Rayner [1053], [1054], Bhimasankaram [142], [143], Hartwig [550], [548], Ostrowski [980], Meyer [842], [844], [845], [847], Mihalyffy [852], Harwood, Lovass-Nagy and Powers [554], Lovass-Nagy and Powers [771], [772], Vuličević [1347], Hung and Markham [601], [1587]. See also F15 for generalized inverses of block matrices of operators.

Schur complements and generalized inverses: [231], [215], Carlson [230], Anderson et al. [47], Cottle [281], Merris [836].

E12. Spectral theory of rectangular matrices. For a detailed development, see [125; ch. 6]. Topics include:

(i) Singular values and singular vectors expansions, see I1.

(ii) The Autonne-Eckart-Young theorem [72], [358], which extends to rectangular matrices the classical theory for normal matrices by replacing eigenvalues and orthogonal projectors by singular values and partial isometries, respectively. The generalized spectral theory for rectangular matrices is essentially due to Penrose [1002], Lanczos [717] and Hestenes [572], [573],

[574], [575]; see also Rao [1102], Rao and Mitra [1113]. The theory of relatively self-adjoint operators developed by Hestenes, phrased in the terminology of elementary operators and partial isometries (and inspired by the study of partial differential equations), is intimately connected and leads to a generalized spectral theory for linear operators.

(iii) Partial isometries: A partial isometry is a linear operator whose restriction to $\eta(A)^{\perp}$ is an isometry, i. e. $\|Ax\| = \|x\|$ for all $x \in \eta(A)^{\perp}$. Properties, characterizations of partial isometries and/or connections with spectral theory, polar factorization and other matters were studied by von Neumann [952], Murray and von Neumann [909], Halmos [526], Halmos and McLaughlin [527], Halmos and Wallen [528], Hestenes [573], [574], Hearon [561], Erdélyi [369], [373], [374], [375], [376], Erdélyi and Miller [379], Poole and Boullion [1036], Fishel [395], Vogt [1345]. For normal partial isometries, see [373], [528], [561].

(iv) Generalized resolvents of a rectangular matrix: Lancaster [716], Wimmer and Ziebur [1390]; see also [125], [556].

(v) Matrix functions: Gantmacher [435], Rinehart [1137], Frame [409], Dunford and Schwartz [350; vol. 1, pp. 556-565], and [716], Langenhop [725], Merris [835].

(vi) Ternary powers and ternary algebras and generalized inverses: Hestenes [575], [577], [578].

Gauge functions, unitarily equivalent matrix norms and inequalities for singular values: von Neumann [951], Mirsky [865], Wedin [1364], [1365].

(vii) The polar decomposition theorem. Any nonzero $m \times n$ complex matrix A of rank r can be written as $A = BE = EC$, where E is a partial isometry and B, C are Hermitian and positive semidefinite: Autonne [72], Wintner and Murnaghan [1772] for nonsingular matrices; Williamson [1387] and Penrose [1002] for singular and rectangular matrices; Murray and von Neumann [909] for linear operators in Hilbert space; see also Beutler and Root [141] and Hearon [560].

(viii) Closest unitary, orthogonal and Hermitian matrices to a given matrix and related topics: See P9.

E13. Spectral generalized inverses of a square matrix. The Drazin inverse A^d. The group inverse $A^{\#}$. See [125; ch. 4], [1113; ch. 4], [190; ch. 4].

13-1. Various authors (Price, Rohde, Scroggs and Odell, Erdélyi, Englefield, Greville, Odell), have pointed out that for a singular matrix, the Moore-Penrose inverse does not, in general, have the spectral properties possessed by the ordinary inverse. For example $A^{\dagger}$ may have a nonzero eigenvalue that is not the reciprocal of any eigenvalue of the original matrix. Greville [496], [498], [125] discusses certain matrices associated with a given singular square matrix that have some spectral properties of the inverse. He calls two square matrices A and X S-inverses of each other if for every complex λ and every x, x is a λ-vector vector of A of grade p (i.e. $(A - \lambda I)^p x = 0$, $(A - \lambda I)^p x = 0$, $(A - \lambda I)^{p+1} x \neq 0$ iff it is a $\lambda^{\dagger}$-vector of X of grade p.

13.2. A group inverse $A^{\#}$ is a (1, 2)-inverse that commutes with A; equivalently it is a (1, 2)-inverse such that $\mathcal{R}(X) = \mathcal{R}(A)$ and $\mathcal{N}(A) = \mathcal{N}(X)$. There is at most one such inverse. $A^{\#}$ exists $\Longleftrightarrow$ $\mathcal{R}(A)$ and $\mathcal{N}(A)$ are complementary subspaces $\Longleftrightarrow$ rank A = rank A^2, i.e. index $A = 1$. (Positive and negative powers, i.e., A^n, $(A^{\#})^n$, together with the projector $AA^{\#}$ as the unit element consitute an Abelian group). Englefield [367] and Erdélyi [371], [372] called attention to the spectral properties of a group inverse (also called the commuting reciprocal). The group inverse is a special case of the Drazin inverse. If A is of index 1, then $A^{\#}$ is the unique S-inverse of A in the union of all 1-inverses and 2-inverses. If A is diagonalizable, $A^{\#}$ is the only S-inverse of A.

For equivalent condition for the existence of $A^{\#}$, see Cline [270]. For explicit formulae for $A^{\#}$ see [270], [367], [371], [125], [126]. For various aspects of the group inverse, see also [57], [550], [671], [774], [934], [1143], [1239], [1731]. For the role of the group inverse in the theory of finite Markov chains, see Meyer [849].

13-3. The Drazin inverse (see also this volume). For various properties see Drazin [335], Cline [270], Greville [496], and [125]. See also [934] and the references cited therein. The Drazin inverse has the spectral property with respect to nonzero eigenvalues and associated eigenvectors, and other weaker spectral properties (cf. [496], [125]). For various aspects of the theory, computation or applications of the Drazin inverse, see also [125], [190], [223], [498], [499], [1035], [1037], [1038], [1240] and [1244].

13-4. Range-Hermitian matrices (EPr matrices): $\mathcal{R}(A) = \mathcal{R}(A^*)$. These were first studied by Schwerdtfeger [1202] and Pearl [994] (both used the term EPr). The term range-Hermitian was suggested by Greville. For various contributions, see Arghiriade [57], Katz [656], [659], Katz and Pearl [660], Pearl [994], [995], [996], [997], Baskett and Katz [95], Hearon [558], Boros [176], Meyer [843], Ballentine [1723], Hartwig and Katz [1742].

It is easy to show that $A^{\#} = A^{\dagger} \iff A$ is range-Hermitian, i. e., the class of range-Hermitian is characterized as the set of all matrices for which the group inverse and the Moore-Penrose coincide.

13-5. Limits and index of a square matrix: see, e.g., [125], Langenhop [724] and Meyer [848]. See O3-1 .

13-6. Various spectral inverses (S-inverse, S'-inverse, strong and weak spectral inverses), other spectral properties of general or special classes of matrices, eigenvalues of a pseudoinverse matrix: Scroggs and Odell [1205], Rohde [1149], Greville [496], [498], Poole [1450], Poole and Boullion [1036], Wall and Plemmons [1353], Boullion and Odell [189], Ward, Boullion and Lewis [1359], Stallings and Boullion [1260], Sibuya [1240], Ikramov [607].

13-7. Characteristic vectors and values for a rectangular matrix: Milnes [858], Milnes et al. [859], [860], Amburgey [29], Amburgey et al. [30], [32].

E14. Differentiation of generalized inverses. Existence of differentiable generalized inverses, relationships between the derivative of a matrix and that of its generalized inverse;

differentiation of projectors. Pavel-Parvu and Korganoff [989] and Hearon and Evans [567] seem to have been the first to discuss differentiation of generalized inverses; see also Boullion and Odell [190; Sec. 4.3]. Closed formulae for the derivative of a generalized inverse of a matrix function of one or several variables, derivative of a projector and/or other matters in differentiation of pseudoinverses have been considered by Hanson and Lawson [535] (see also [735]), Pérez and Scolnik [1007], Scolnik [1203], [1204], Golub and Pereyra [472], [473], Decell [302], Wedin [1364]. Differentiation of pseudoinverses has been used in nonlinear least squares problems in which some of the variables occur linearly by Golub and Pereyra [472] and Krogh [1608]. For use of derivatives of pseudoinverses in algorithms for constrained minimization problems see also Fletcher [1541], Fletcher and Lill [1544] for the full rank case and Pérez and Scolnik [1007], Scolnik [1203], for the general case. The increment-perturbation identities for operators (see Nashed [924]) can also be used to obtain differentiation results for generalized inverses of linear operators in Banach spaces.

A classic paper on matrix derivatives is Dwyer and MacPhail (1948). For more recent results on matrix differentiation, Neudecker [948]. For various aspects of differentiation in normed and other spaces, see Nashed [916].

E15. Common solutions of linear equations and various systems of matrix equations

15-1. Penrose [1002] proved that a necessary and sufficient condition for the matrix equation $AXB = C$ to have a solution X is that $AA^-CB^-B = C$, where A^- is any inner inverse (1-inverse) of A. Then the general solution is $X = A^-CB^- + Y - A^-AYBB^-$, where Y is arbitrary. As a corollary he obtained the general solution of the consistent equation $Lx = c$ in the form $x = L^-c + (I - P^-P)y$, where y is arbitrary.

15-2. Common solutions of linear equations $Ax = a$ and $Bx = b$ and generalizations. For common solutions of a pair of matrix equations and necessary and sufficient conditions for n matrix equations to have a common solution, see Morris and Odell [904], Shurbet, Lewis and Boullion [1232], Shurbet [1455], Shiozaki and Sibuya [1225]; see also [190; 5.3], [125; 5.4]. For applications of the expressions for the manifold of common solutions to obtain generalized inverses of partitioned matrices see [112], [125; 5.4], [657], [852]; see also related topics in E11.

15-3. $AX + XB = C$: Jameson [610], Boullion and Poole [191]. An algorithm for this system was given by Bartels and Stewart [1725].

15-4. $AX - XB = C$, resultants and generalized inverses: Hartwig [547], [549], Meicler [831].

15-5. $AX = XA = A$: Katz and Pearl [661]. $AX = A$ and $YA = Y$: Meyer [840].

15-6. $AXB = X$: Hartfiel [542].

15-7. $A_1XB_1 = C_1$ and $A_2XB_2 = C_2$: Mitra [876].

15-8. $\sum_{p=1}^{r} f_p(A)\, X\, g_p(B) = C$: Wimmer and Ziebur [1390].

15-9. $A = XYZ$ and $B = ZYX$, and related ones: Brenner and Lim [1732].

E16. <u>Other special topics:</u>

16-1. Series, parallel and hybrid addition of matrices: See these Annotations under Y3.

16-2. Projection on the intersection of two subspaces: See F18(ii).

16-3. Generalized inverse of nonnegative (or positive) semidefinite matrices and/or conditions for positive or nonnegative definiteness in terms of generalized inverses: Rohde [1151], Albert [16], [17], Lewis and Newman [750], Rosenberg [1163], Plemmons and Cline [1028], Healy [1577].

16-4. Generalized inverses of (order) positive matrices and monotonicity: Berman [134], Berman and Plemmons [135], [136], Plemmons [1024], [1026]. For theory of nonnegative matrices, see Seneta [1212].

16-5. Orthogonal and oblique projectors and their relations; characteristics of pairs of vector spaces; characteristic values of products of Hermitian idempotents, products of projectors, and related matters: Afriat [6], [7], Mizel and Rao [887], Hearon [565], Schneider [1194], Greville [500], Nashed and Votruba [931], [934], Shinozaki and Sibuya [1227], deMarr [803], Mitra and Rao [1751], Rao and Mitra [1113], Giles and Kummer [447].

16-6. Rank of the sum of matrices and generalized inverses: Marsaglia [805], Marsaglia and Styan [806], [1626], Meyer [839].

16-7. Generalized inverse versions or extensions of some inequalities and other topics in inequalities. Various classical inequalities: Beckenbach and Bellman [100]; linear inequalities: Berman [1428], Ben-Israel [1498], Ky Fan [1539], Ho and Kashyap [589]; Wielandt inequalities: Mond [889]; spectral inequality of Marcus, Minc and Moyls: Poole and Boullion [1038], inequalities between Hermitian and symmetric forms: Horn [1743]; inequalities among proper values of a linear transformation restricted to subspaces: Perry [1449].

16-8. Schur's complements: See E11.

16-9. Simultaneous diagonalization of semidefinite matrices: See Y4 .

16-10. Generalized inverses of special classes of matrices and forms: Boolean relation matrices: Butler [218], Plemmons [1023], Prasada Rao and Bhaskara Rao [1653]; M-matrices and generalizations: Poole [1651], [1037], Schneider [1675], and references cited in [1037]; Toeplitz matrices: Halmos [526], Cline, Plemmons and Worm [272], Hartwig and Fisher [553], Lovass-Nagy and Powers [778], Pye, Boullion and Atchinson [1082]. (For theory of Toeplitz operators, using Banach algebra techniques, see Douglas [1535]). Petrie matrices and generalized inverses: Arnowitz and Eichinger [64]; Hessenberg matrices: Singh, Poole and Boullion [1244]; circulant matrices, p-potent and r-circulant matrices: Pye,

Boullion and Atchinson [1084], Stallings [1456], Shurbet, Lewis and Boullion [1233], Stallings and Boullion [1262], [1263]; generalized inverses of bilinear forms: Dragomir [332].

16-11. Graphs and nonnegative matrices, graph-theoretic approach to generalized inverses, etc.: Plemmons [1024], Poole [1034]. For a graph theoretic-theoretic approach to matrix inversion by partitioning, see Harary [537]. For graphs in network theory, see, for example, Seshu and Reed [1213]. For generalized inverses of incidence matrices and interval graphs, see Ijiri [606], Boullion and Odell [190; 2.2]; see also Fulkerson and Gross [426].

17. For various other aspects of generalized inverses, see also Arghiriade [57], Arghiriade and Dragomir [61], Ben-Israel and Charnes [119], Bhimasankaram and Rao [144], Bjerhammar [154], [158], Boros and Sturz [177], Brand [199], Ching and Chiu [248], Chipman and Rao [253], Cline and Greville [271], Decell [297], [299], Dokovic [320], Dommanget [322], [323], Ergeváry [359]-[362], Ficken [391], [392], Gabriel [428], [429], [430], Germain-Bonne [446], Giurescu and Gabriel [450], Greville [500], Hartwig [550], [551], [552], Hearon [562], [564], Koop [687], Kruskal [699], Lancaster [715], Langenhop [723], [725], Lovass-Nagy and Powers [778], Malik [787], Meyer [841], [845], Meyer and Painter [850], Mitra [868], [869], Montague and Plemmons [890], Morris [902], Pearl [998], Popa [1039], Presić [1049], Price [1051], Rao [1100], [1102], [1104], Reid [1129], Rivlin [1139], Robinson [1145], Rohde [1151], Ruhe [1166], Searle [1207], [1209], Sibuya [1236], Stojakovic [1278], [1279], [1280], [1281], [1282], Vetter [1342], Villumsen [1343], Wani and Kabe [1356]; also [934].

F. Theory of Generalized Inverses of Linear Operators in Infinite Dimensional Spaces

F1. A unified approach to the operator theory of generalized inverses with a detailed development of algebraic, topological projectional, proximinal-extremal properties, is given by Nashed and Votruba [934] in this volume. Generalized inverses in Banach and topological spaces are treated in detail; most of these results are new. Specializations of this approach yield various earlier definitions of generalized inverses in

Hilbert spaces, which are summarized in [934]. In view of this, the annotations in F2-F9 are confined for the most part to historical remarks, references to earlier contributions and citations of related sections in [934] for new results. See also parts III and IV for other annotations related to generalized inverse operator analysis and approximations.

The earliest work on generalized inverses of densely defined linear operators in Hilbert space seems to be that of Tseng [1327]-[1331] (see also the preface to this volume for historical notes). Several definitions of generalized inverses in Hilbert spaces are discussed in Nashed [915], which also provides an account of various contributions to generalized inverses in functional analysis as of 1970.

F2. Algebraic theory of generalized inverses in infinite dimensional spaces (cf. [934; Sec. 1]): Inner inverses (1-inverses, partial inverses, pseudoinverses): See also Friedrichs [418], Hamburger [530], Sheffield [1222], Deutsch [317] and the annotations E3 for other references.

Outer inverses (2-inverses, semi-inverses, sub-inverses, etc.): [934], [317], [1436], and others (see also E3).

Algebraic aspects of generalized inverses of arbitrary linear transformations: [934], [317], Robinson [1144], Hansen and Robinson [533]; also E3. Structural characterizations of the set of all generalized inverses; transformation of generalized inverses under change of projectors: Nashed and Votruba [934], [931]. Special cases and related aspects are also discussed in Sheffield [1222], Milne [857], Loud [767], Deutsch [317]; see also E4.

F3. The Moore-Penrose (or orthogonal) generalized inverse in Hilbert spaces. This is the most thoroughly studied generalized inverse operator in the literature in view of its least squares property. It is also the most natural extension of the Moore-Penrose inverse from matrices to operators. The case of a bounded linear operator with closed range follows almost verbatim from the matrix case. Some technical modifications are needed to cope with generalized inverses of closed densely-defined (or bounded) linear operators with nonclosed range, and of unbounded operators in general.

3-1. Generalized inverses of bounded linear operators

with closed range, various definitions and characterizations: [915], [934; 5. 3], Desoer and Whalen [316], Beutler [139], Petryshyn [1013], Kurepa [711], Holmes [593], and others.

Extremal (or minimal) properties: [316], [915], [934], [1346].

3-2. Various routes to generalized inverses of bounded linear operators with arbitrary range were taken by Hestenes [574], Beutler [140], Nashed [915], and others. See also [915] and [934] for extensive references. Characterizations and equivalence of various definitions are given in [915], [934]. See also N3. 3.

Extremal properties: Tseng [1330], [1331], Nashed [915], Beutler [139], [140], Erdélyi and Ben-Israel [378] and others (see [915], [934]).

3-3. Generalized inverses of closed densely-defined linear operators with arbitrary range. For various characterizations and extremal properties see Theorem 5. 7 in Nashed and Votruba [934]. A self-contained exposition of some of these aspects is developed in Beutler and Root [141]. Other definitions and properties were given by Tseng [1328], [1329], [1331] (for a summary of Tseng's results, see also [119], [915]), Beutler [140], Hestenes [574], Arghiriade [59], Arghiriade and Dragomir [62], and others.

3-4. Generalized inverses of "arbitrary" linear operators in Hilbert spaces:

Concepts of domain decomposability and carrier: Tseng [1329], Hestenes [574], Arghiriade [59], Nashed and Votruba [931], [934].

Definitions and properties of generalized inverses: Tseng [1328], [1331], Hestenes [574], Arghiriade and Dragomir [61], Arghiriade and Boros [60], Erdélyi [377], Nashed and Votruba [931], [934]. See also remarks 2. 16 and 2. 17 and Sections 5. 3 and 5. 4 in [934] which unify all approaches to generalized inverses of arbitrary linear operators in Hilbert space.

3-5. Generalized inverses in (not necessarily complete) inner product spaces: Arghiriade and Boros [60], Hansen and Robinson [533], and [934; 5. 4D].

F4. Restricted and constrained generalized inverses in Hilbert spaces; constrained best approximation property; generalized inverses and multistage optimization problems: See Nashed [925], [915], [920], Minamide and Nakamura [862], Holmes [593], Porter [1043], Wahba and Nashed [1348], F. J. Hall [519], [520], Rao and Mitra [1116].

F5. Generalized inverses of linear operators in Banach spaces: A general setting was first considered by Nashed and Votruba [931], [932]; details are given in [934]. These papers develop a unifying approach to the operator theory of generalized inverses (projectional, extremal and proximinal properties) when a topology is endowed on the domain and/or the range space of the operator. The approach has a great deal of simplicity and generality, but requires a higher level of abstraction. The main point of departure between this approach and various functional-analytic and/or geometric approaches to generalized inverses of linear operators in Hilbert and Banach space is the following: In this approach the analysis is superimposed on the vector space structure, and generalized inverses in topological spaces are considered by starting from algebraic inner inverses (the existence of the latter is always guaranteed). In contrast, the approaches that have been advanced in the literature treat generalized inverses in analytic settings *ab initio*, ignoring the fact that algebraic generalized inverses always exist regardless of the topology. This usually complicates questions of existence and domain consideration of generalized inverses, and one does not get a clear picture of how the difficulties creep in with the analysis and topology.

Some aspects of pseudoinverses in Banach spaces were considered earlier by Sheffield [1222], Beutler [139], Atkinson [67], Nashed [915] and others, and implicitly considered by Wyler [1394] in the setting of so-called "Green operators". The generalized inverse of a closed linear operator T for which $\mathcal{N}(T)$ and $\mathcal{R}(T)$ have topological complements has been briefly considered by Reid [1128; pp. 19-21], Nashed [915] (the setting also applies to the more general case when $\overline{\mathcal{R}(T)}$ has a topological complement) and others; see [915], [934] for references on various occurrence of generalized inverses in Banach space contexts. We also mention recent occurrences in Koliha [680], [681], Caradus [225], Bouldin [185] of generalized inverses of a bounded linear operator T for which $\mathcal{R}(T)$

[or $\overline{\mathcal{R}(T)}$] and $\eta(T)$ have topological complements; see also Saphar [1184], [1185] for special cases and the literature on Fredholm operators, cited under J.

F6. Extremal and proximinal properties of generalized inverses in Banach spaces and least-squares solutions of minimal norm. These topics are developed in [934], where the concepts of right-orthogonal, left-orthogonal, and orthogonal inner inverses are introduced. The connections between these concepts and the existence of extremal solutions, minimal solutions, and best approximate solutions are demonstrated in the Banach space context. These concepts hinge on the notion of an orthogonally complemented subspace of a normed space.

F7. Metric generalized inverses in Banach spaces. Another approach to generalized inverses in terms of best approximation properties is via the notion of a metric generalized inverse which is used in [934 ; pp. 43-45]. The Moore-Penrose of a closed densely-defined linear operator on a Hilbert space can be viewed also as a metric generalized inverse. However, metric generalized inverses in Banach spaces are quite different from the generalized inverses which are based on the existence of topological complements (F6). Special cases of the metric generalized inverse include the concept of a p-q generalized inverse (Newman and Odell [956]; see also [190]) in the case of finite dimensional spaces, and other definitions used by Holmes [593], Erdelsky [368], and Erdélyi and Ben-Israel [378]; see [934] for details.

F8. Generalized inverses in topological vector spaces (TVS): A general approach was announced in [931] and details were given in [934]. This provides the first treatment of generalized inverses of arbitrary linear operators between two TVS's . Earlier special cases include the following: Generalized inverses for Fredholm operators in TVS were considered by Pietsch [1019] under the title of "σ-transformation" (see also Pietsch [1753], Prössdorf [1658] and the recent paper by Gramsch [1566]). Votruba [1346] defined the generalized inverse of a (continuous) topological homomorphism T in the case when $\mathcal{R}(T)$ and $\eta(T)$ have topological complement. The simple extension to the case when $\overline{\mathcal{R}(T)}$ and $\overline{\eta(T)}$ have topological complements was mentioned in [915].

The concepts of right-topological, left-topological, and topological inner inverses are introduced and developed in [934].

The theory of generalized inverses in topological vector spaces [934] hinges upon the existence of topological complements to certain subspaces associated with an operator. For a survey of complemented subspaces in Banach spaces, see Kadets and Mityagin [628]; the interested reader should certainly examine also the pioneering papers by Murray [908] and Sobczyk [1251]. The theory of extremal and proximinal properties of generalized inverses in Banach spaces [934] is based on the concept of orthogonal complemented subspaces and proximity maps, and uses a notion of orthogonality in Banach spaces due to James. See also the recent papers by Papini [984], [1645] which deal with interesting aspects of orthogonal projections and proximity maps in Banach spaces; for the latter see also the survey paper by Vlasov [1344] and the references cited in [934; p. 46]. See also [1647], [1451], [1591], [1594] and [1463] for related aspects of complementations and/or projectors.

F9. Generalized inverses in "mixed" spaces: Another advantage of the approach in [934] is that it readily provides results on generalized inverses in "mixed" spaces. For example, one may consider a linear operator from a vector space to a Hilbert space, or from a Banach space to a topological vector space, etc. Algebraic, topological, projectional and extremal properties of generalized inverses are immediate by-products of this approach.

F10. Generalized inverses of random linear operators and least squares solutions of random operator equations. Let X, Y be separable Hilbert spaces, and let $(\Omega, \mathcal{B})$ be a measurable space. Let T be a random linear operator from $X \times \Omega$ into Y. Let $T^{\dagger}(\omega)$, for $\omega \in \Omega$, denote the generalized inverse of $T(\omega)$. Questions of measurability of $T^{\dagger}(\omega)$ were investigated by Nashed and Salehi [930], and in particular the following results were established: (i) If T is bounded, then $T^{\dagger}$ is a random operator. (ii) If T is a closed operator with dense domain and if $T^{\dagger}$ is bounded, then $T^{\dagger}$ is a random linear operator under some mild restriction on the domains of $T(\omega)$ and $T^{*}(\omega)$, $\omega \in \Omega$. These results are applied to the measurability of best approximate solutions of random linear operators.

Similar results can be established in Banach spaces. Let $(\Omega, \mathcal{B}, u)$ be a complete probability space, and X, Y be Banach spaces. Let T be an almost surely (a. s.) bounded or densely defined closed linear random linear operator on $\Omega \times X$ into Y. Suppose that a. s. the closure of the range of $T(\omega)$ and the nul space of $T(\omega)$ have topological complements. Results on measurability of the generalized inverse of $T(\omega)$, relative to these complements, are established under mild technical conditions.

For random integral and operator equations, see Bharucha Reid [1502].

F11. Generalized inverses in reproducing kernel Hilbert spaces (RKHS). A Hilbert space $\mathcal{H}$ of real-valued functions on a set S is said to be an RKHS if all the evaluation functionals $f \to f(s)$ for $f \in \mathcal{H}$, $s \in S$ are continuous. A study of generalized inverses of linear operators in RKHS was initiated by Nashed and Wahba [936], [938]. Explicit expressions for generalized inverses and minimal-norm solutions of linear operator equations in RKHS are obtained in several forms. Relationships between generalized inverses in RKHS and $\mathcal{L}_2$-spaces are also established, and the connection the regularization operator of the equation $Af = g$ and $A^\dagger$ in RKHS is demonstrated. In particular, it is shown that the two are the same if $\mathcal{R}(A)$ is closed in an appropriate RKHS. This approach provides a natural and effective setting for regularization and approximation of ill-posed problems when the operator maps one RKHS into another, see [938], [936], [920], [939]; see also L for regularization problems.

F12. Spectral properties of generalized inverses and related topics. See also E12 and E13 for similar topics for matrices. In general there is no tractable relation between the spectrum of a linear operator and the spectrum of its generalized inverse. As in the case of matrices, even the nonzero points of the spectrum $\sigma(A)$ are not in general the reciprocals of the nonzero points of $\sigma(B)$ for some generalized inverse B of A. A generalized inverse which has this latter property is a kind of a spectral generalized inverse. Let A be a bounded linear operator on a Banach space X into X and suppose that each of $\mathcal{N}(A)$ and $\mathcal{R}(A)$ has a topological complement in X (this is equivalent to the existence of a bounded linear operator B such that $ABA = A$ and $BAB = B$; such an A is called a relatively regular operator and such a B is called a bounded generalized inverse of A). Not every bounded linear operator with these properties has a spectral inverse. In a recent paper, Lay [739] investigated conditions under which nonzero points in $\sigma(B)$ are the reciprocals of nonzero points in $\sigma(A)$.

A vector $x \in X$ is a λ-vector for A of grade m if $(\lambda - A)^m x = 0$ and $(\lambda - A)^{m+1} x \neq 0$, where m is a positive integer. Let B be a bounded generalized inverse of A such that $\mathcal{R}(A^p) \subset \mathcal{R}(B)$ for some $p \geqq 1$, and suppose x is a λ-vector of A of grade m, $\lambda \neq 0$. Then x is a λ^{-1}-vector for B of grade m (and, in particular, the reciprocals of the nonzero eigenvalues of A are eigenvalues of B). To deal with the general spectrum of A, the hypotheses have to be strengthened: Let B be a bounded generalized inverse of A such that $A^p B = B A^p$ for some $p \geqq 1$. Then $\{\lambda : \lambda \in \sigma(A), \lambda \neq 0\} \subset \sigma(B)$. Lay also extends the concept of a spectral generalized inverse of a square matrix, introduced by Greville [496], [497] (see also E13) to a bounded linear operator on a Banach space. A bounded generalized inverse B of A is a spectral generalized inverse of A if there exist positive integers p, q, r and s such that $\mathcal{R}(A^p) \subset \mathcal{R}(B^q)$ and $\eta(A^r) \subset \eta(B^s)$. Necessary and sufficient condition for an operator to be a spectral generalized inverse are given. There exist operators which are not generalized inverses but which have the spectral properties of a spectral generalized inverse.

Let A be a relatively regular bounded linear operator. Let f be a univalent function on $\sigma(A)$ with $f(0) = 0$. Then $f(A)$ is relatively regular (Caradus [229]).

A general spectral theory for linear operators was developed by Hestenes [574], [578] (see also annotations E12) in terms of unitary and relatively self-adjoint linear operators. This theory is an extension of the singular values for matrices and reduces to the usual eigenvalue theory when the operator is positive self-adjoint in the usual sense. For connections with the theory of ternary algebras, see Hestenes [575], [577], [578].

For EPr operators, see Campbell and Meyer [222]; also E13-4. For the Drazin inverse of bounded and unbounded operators, see Caradus [228]; also [934] and E13-3.

For partial isometries, normal partial isometries and related topics, see E12. For singular values, see I1. For polar decomposition of an operator, see F16. For other topics see also III.

F13. Generalized inverse of a product (see also E10). Let A and B be bounded linear operators on a (real or complex) Hilbert space $\mathcal{H}$, such that the range of each is a closed subspace of $\mathcal{H}$. The following three conditions are necessary and sufficient for $(AB)^\dagger = B^\dagger A^\dagger$: (i) the range of AB must be closed; (ii) the range of A^* must be invariant under BB^*; (iii) $\mathcal{R}(A^*) \cap \mathcal{N}(B^*)$ must be invariant under A^*A. See Bouldin [184]. The result remains true if conditions (ii) and (iii) were replaced with the following two conditions: (iv) $A^\dagger A$ commutes with BB^*; (v) $BB^\dagger$ commutes with A^*A.

Results on the product of two relatively regular operators in Banach space were given by Atkinson [67], Caradus [227], Koliha [682] and Bouldin [185] and others. The product AB of two relatively regular operators between Banach spaces is relatively regular iff the product QP is relatively regular, where Q is a projector onto a topologoical complement of $\mathcal{N}(A)$ and P is a projector onto $\mathcal{R}(B)$; see [682]. Sufficient conditions for relatively regularity of AB based on the product QP were obtained earlier in [227], while a strengthened form of the main result in [682] was given recently in [185], which also generalizes and refines the main theorem of [183]. For various necessary and sufficient conditions for relative regularity of a product of bounded linear operators in Hilbert spaces, see Bouldin [183], [184], [185]. Results on products of semi-Fredholm operators were obtained by Atkinson [67] and others; see also J.

F14. Operational properties. For generalized inverses of adjoints in topological spaces, see Nashed and Votruba [934; Sec. 6], and in Hilbert spaces, see [934; Sec. 6], Beutler and Root [141; Sections 1 and 5]; [593], [316], [125], [915] and others; see also Wyler [1394], Pietsch [1019] for adjoints of special operators in Banach and topological spaces. Many operational properties and identities which have been developed for generalized inverses of matrices (cf. [17], [190], [125], [1113]) carry over to the context of operators with minor technical changes in statements and/or proofs.

F15. Generalized inverses of block matrices of operators.

Let H_1, H_2 and H_3 be Hilbert spaces. Let $B = \begin{bmatrix} T & C^* \\ C & 0 \end{bmatrix}$ where $T = AA^*$, A and C are bounded linear operators from H_1 to H_2 and H_1 to H_3, respectively. Under the assumption that $A(\mathcal{N}(C))$ is closed and C has a closed range, F. J. Hall [520] shows that B has a closed range and hence a bounded 1-inverse and obtains an expression for it. He also obtains expressions for the operators which are blocks in (1, 3)-inverses and (1, 4)-inverses of B, and proves that these operators are independent of each other. A form of the Moore-Penrose inverse $B^\dagger$ and the restricted generalized inverse and a related constrained minimization problem is indicated. Some results for the case when $\mathcal{R}(C)$ and $A(\mathcal{N}(C))$ are nonclosed are derived; however no expressions for $B^\dagger$ or other inverses of B are obtained for nonclosed ranges.

Several multi-stage constrained minimization problems which lead to generalized inverses of block matrices of operators are discussed by Nashed [920], [925]. For inversions of matrices of operators, see Morrel [1445]. For various aspects of generalized inverses of partitioned and bordered matrices, see E11.

F16. Factorizations of linear operators, polar decomposition, and generalized inverses (see also E12 for matrices):

16-1. Polar decomposition of a bounded linear operator $T = UR$, where U is a partial isometry, see Dunford and Schwartz [350; vol. II, 1245-50], Beutler and Root [141]. Then $T^\dagger = R^\dagger U^*$. See [141], [184], [139], [140], [915] for various aspects of polar factorization and generalized inverses; for matrices, see, e.g., [125] and E12(vii).

16-2. For a bounded linear operator A between Hilbert space with $\mathcal{R}(A)$ closed, we have $A^\dagger = A(AA^*)^\dagger = (A^*A)^\dagger A^*$; for different proofs, see, e.g., [316], [593], [184]. Let X, Y, Z be Hilbert spaces, $B \in \mathcal{L}(Z, Y)$ and $C \in \mathcal{L}(X, Z)$. Assume that B^* and C are onto. Let $A = BC$. Then $A^\dagger = C^\dagger B^\dagger = C^*(CC^*)^{-1}(B^*B)^{-1}B$. (See, e.g., [593]). This extends to Hilbert spaces the expression for the Moore-Penrose inverse, based on full-rank factorization of a matrix (cf. E6). Another interesting formula based on factorization is (see [862], [593]; and [194] for matrices): Let $A \in \mathcal{L}(X, Y)$ with $\mathcal{R}(A)$ closed. If for some Hilbert space Z, there exists an operator $C \in \mathcal{L}(X, Z)$, onto, such that $\mathcal{R}(C^*) = \mathcal{R}(A^*)$, then $A^\dagger = C^*(CA^*AC^*)^{-1}CA^*$. Similarly if there exist $B \in \mathcal{L}(Z, Y)$, B^* onto, such that $\mathcal{R}(B) = \mathcal{R}(A)$, then $A^\dagger = A^*B(B^*AA^*B)^{-1}B^*$. An important result of these representations is to reduce the computation of $A^\dagger$ to the inversion of a matrix when either X or Y is finite dimensional.

16-3. Let M be a positive definite infinite-dimensional matrix function defined on the unit circle. Then $M = AA^*$, where A is a function with Fourier series. This factorization is originally due to Lowdenslager, but his original proof contained an error (see Douglas [328]). Rectification and extensions have been given by Douglas [328], Mandrekar and Salehi [793], Miamee and Salehi [851] and others. Generalized inverses are utilized in some of these factorizations.

F17. <u>Generalized inverses of special classes of linear operators.</u> Continuous linear idempotents; linear functionals: Nashed and Votruba [934; 7.1, 7.2]; $(I - \lambda K)^\dagger$, where K is compact: see I3; $K^\dagger$ and singular-value decomposition, see I1, 2; Fredholm and semi-Fredholm operators, see J1; generalizations of compact operators, Riesz operators, Riesz meromorphic functions, see J4; random operators, see U5; generaliz inverses for distributions: [934; 7-4]; normal operators of finite descent [934; 7.5]; various aspects of normally solvable operators: see J2; see also J5, 6 for various topics in operator theory which have strong interface or applications to aspects of generalized inverses. See also III for generalized inverses in analysis.

F18. <u>Miscellany:</u>

(i) Generalized inverses and solvability of sets of operator equations: the algebraic aspects are essentially the same as for matrices, see E15 and [934].

(ii) A closed formula for the projector $P_{M \cap N}$ on the intersection of two closed subspaces M, N in Hilbert space is given by

$$P_{M \cap N} = \lim_{n \to \infty} P(QP)^n = 2P(P+Q)^{\dagger} Q$$

where P and Q are the projectors on M and N. (See Anderson and Schreiber [48], i.e. $P_{M \cap N}$ is the parallel sum of P and Q; see Y3).

(iii) Topics from functional analysis and operator theory which play an important role in various aspects of the theory of generalized inverses in Hilbert and other spaces include the following: Hilbert, Banach and topological vector spaces; linear operators, bounded operators, closed operators, topological homomorphisms; the projection theorem, the Riesz representation theorem; closed graph theorem, closed range theory, Fredholm alternative theorems, and adjoint operators; convex sets and functions, proximity maps, uniformly convex and strictly convex spaces; various types of operator and norm convergence; positive definite operator, self-adjoint operators, spectral theory for bounded or unbounded operators; see also F8. General references: [1190], [1298], [1401], [1669], [648], [350], [167], [244], [1635], [1042], [780].

G. Generalized Inverses in Various Algebraic Structures

Concepts of generalized inverses make sense in a variety of mathematical structures. For a brief exposition of concepts and results in several structures, see Sec. 8 of the paper by Nashed and Votruba in this volume.

G1. Generalized inverses of matrices over an arbitrary field. The theory of algebraic generalized inverses [934; Sec. 1] specializes to matrices over an arbitrary field. On the other hand, the Moore-Penrose $A^{\dagger}$ (and other inverses which involve an "adjoint" operation) require special consideration over an arbitrary field; the existence depends on the chosen field. Let $\mathfrak{F}$ be a field with an involutory (anti-) automorphism $\lambda \to \bar{\lambda}$, and denote by A^* the matrix $\overline{A'}$; where A' is the transpose of A and $\bar{A}$ is obtained from A by replacing each entry a_{ij} by $\bar{a}_{ji}$. Then $A^{\dagger}$ exists iff A, A^*A and AA^* have the same rank. See Kalman [637], Pearl [999].

G2. Generalized inverses of matrices over general rings. Penrose [1002] considered generalized inverses of matrices over the complex field, and over more general rings (his results for rings were not published). Rado [1091] extended some results on generalized inverses to matrices over any division ring with an involutory anti-automorphism, and established the equivalence of the definition of Moore and Penrose.

G3. Generalized inverses in rings and *-regular rings. Since the set of all $n \times n$ matrices is a ring under multiplication, it is natural to expect that concepts of generalized inverses extend to arbitrary rings. von Neumann [950] introduced the concept of a regular ring, i.e. a ring R in which for every $a \in R$ there exists an element $x \in R$ such that $axa = a$. von Neumann [952], [953] also discussed the relation between regular rings and continuous geometry. Kaplansky calls a ring weakly regular (anti-regular in [934]) if, for each $a \neq 0$, there exists $x \neq 0$ such that $x = xax$ (i.e., every nonzero element has a nonzero outer inverse). He showed that this is equivalent to requiring every right ideal to contain nonzero idemponents. Every regular ring is anti-regular, but not conversely.

There exist infinite-dimensional Banach algebras which are weakly regular. For example, the ring of all continuous real functions on a compact Hausdorff space X is weakly regular iff every open set of X contains an open subset which is both open and closed. In contrast, a Banach algebra is a regular ring iff it is finite dimensional (cf. Kaplansky [649], [650]).

For regular rings, see also Murray and von Neumann [909] McCoy [824], Hartwig [550]. For anti-regular rings, see also [550], [934]. Hansen and Robinson [533] investigated the existence of generalized inverses in the setting of modules over a ring, and related their results to the well-known theory of semi-simple Artinian rings.

A *-regular ring is a regular ring with involution * such that $a^*a = 0$ iff $a = 0$. Penrose (private communication, see Preface) established the existence of a unique element $a^\dagger$ satisfying the four (Penrose) equations for an element a in a *-regular ring. Another proof is due to Kaplansky [650], about the same time that Penrose proved the existence for matrices. Kaplansky based his results on the work of Rickart [1135], [1666] and von Neumann [950], where Moore's original definition is transparent. The existence of inner, outer, the Moore-

Penrose and other generalized inverses for elements of a $*$-regular ring is discussed also by Hartwig [550]. In this lengthy paper, Hartwig also generalizes many of the matrix results in the literature of generalized inverses to $*$-regular rings and provides an excellent exposition of several techniques and applications. Characterizations and computational formulas for various $\{1, 2, 3, 4\}$ inverses are given. The group inverse in $*$-regular rings is also discussed. Necessary and sufficient conditions for $(ab)^\dagger = b^\dagger a^\dagger$ are also given. The thrust of the paper is in the direction of generalized inverses in the ring of 2×2 matrices over a ring, with particular attention being given to computational formulae, bordered matrices, Schur complements, block-rank formulae and EPr elements, etc.

Other references of interest. Regular Banach algebras and singular elements: Kaplansky [649], [651], [1599], [1600], Rickart [1135], [1666], [1667], Brown and McCoy [207], Skornyakov [1245], Naimark [1633], Żelazko [1717]. π-regular rings and Azumaya algebra: Azumaya [74], Rege [1121]. For recent remarks on regular and strongly regular rings, see Raphael [1661]. Special $*$-regular rings: Prijatelj and Vidav [1052]. Some results on invertibility in Banach algebra were recently obtained by Hogan and Langenhop [590]. For the Fredholm elements of a ring, see Barnes [1483].

G4. Generalized inverses in semigroups: Munn [906] obtained necessary and sufficient conditions for an element x of an algebraic semigroup (S, o) to be pseudo-invertible. This condition is simply that some power of x lies in a subgroup of (S, o). Drazin [335] gave necessary and sufficient conditions for x to be pseudo-invertible when (S, o) is a ring. Foulis [408] considered generalized inverses in a special type of semigroups. His work is algebraic in structure and includes as a special case the essence of certain results in the Hilbert space and infinite-dimensional vector space context.

An element $a \in S$ is said to have a Drazin inverse if there exist $x \in S$ such that $a^m = a^{m+1}x$ for some nonnegative integer m, $x = x^2 a$ and $ax = xa$. An element $a \in S$ has at most one Drazin inverse. Necessary and sufficient conditions for the existence of a Drazin inverse and other properties are given in [335]; see also [934]. Since the space $\mathcal{L}(X)$ of all continuous linear operators on a Banach space X into X is a multiplicative semigroup, we can talk about the Drazin inverse of a bounded linear operator. See [934], Caradus [228], and King (a preprint). For references on the Drazin inverse of a matrix, see E-13.3. In another context, Professor Olga Taussky-Todd has informed us that Dade, Taussky and Zassenhauss [290] have shown that an ideal class in an order of a quadratic field has as generalized inverse an element which satisfies the Drazin axioms, but this is not any longer true for number fields of higher degree.

Decell [301] and Decell and Wiginton [306] obtained a characterization of the maximal subgroups of the semigroups of $n \times n$ complex matrices. For aspects of (algebraic) inverse semigroups and generalizations, see Clifford [260], Clifford and Preston [261], Croisot [284], Fitzgerald [397], Munn and Penrose [907], Preston [1050], Bailes [1426], M. R. S. Brown [1430] and the Symposium Proceedings [1657]. For π-regular semigroups, see Losey [766]. For other topics in semigroups, see also Ljapin [758], [759], Magill and Subbiah [784]. For topics related to invertibility in general groups and special spaces, see Liber [753], Arlt [63], Vagner [1337]. For products of EP elements in reflexive semigroups, see Hartwig and Katz [1742].

G5. Generalized inverses of morphisms in categories, generalized inverses of morphisms in balanced concrete categories with concrete factorization; characterizations of regularity: Davis and Robinson [296]. Regularity of composition of two mormorphisms: Noll [963]; also [934].

G6. Generalized inverses in set theory; generalized inverses of nonlinear maps: Rabson [1090], Nashed [915], Nashed and Votruba [934], Davis and Robinson [296]. Operational calculus of linear relations: See, e.g., Arens [56], Brown [209].

III. Generalized Inverses in Analysis

(See also the papers by Hestenes, Nashed, and Rall in this volume)

Notions of generalized or pseudo inverses seem to have first appeared in print in the context of analysis, i.e. in the setting of differential and integral operators rather than in the setting of matrices and algebraic problems (some historical remarks were given earlier by Reid [1128] and Nashed [915]). More specifically, the germ of these ideas may be found in a celebrated 1903 paper of Fredholm, where a particular generalized inverse (called by him "pseudoinverse") of an integral operator of the second kind was given. It may be found also in the work of Hurwitz (1912) on pseudoresolvents, who extended the paper of Fredholm and gave an algebraic characterization and a construction of the classes of all pseudoinverses. Generalized inverses of differential operators were also implicit in the work of Hilbert (1904) on generalized Green's functions, and in the Göttingen theses of Myller (1906), Westfall (1905) and others, dealing with generalized Green's functions for certain classes of compatible differential systems. Roughly speaking, the integral operator whose kernel is a generalized Green's function is a generalized inverse of the differential operator. Other important papers in this direction include papers by Westfall (1909), Bountizky (1909), Harada (1953), and especially Elliott (1928), (1929), W. T. Reid (1931), (1932). The generalized Green's matrix was used specifically by W. T. Reid (1932), Hölder (1935) (and explicitly by D. C. Lewis (1956) in perturbation of periodic solutions). In all of these papers the differential system involved independent boundary conditions in number equal to the order of the systems.

Although there were many precursors to generalized inverses in analysis, the systematic study of various operator-theoretic and approximations aspects of generalized inverses of linear operators was made only in recent years. The thrust of these contributions, including some new results, is contained in this volume.

H. Generalized Inverses of Differential Operators and Generalized Green's Matrices

H1. We refer to Reid [1128] for a vivid historical account of early aspects and precursors of generalized inverses in the context of differential equation. Reid also gives a motivated informal discussion of the interrelations between the concept of a generalized Green's matrix for a differential system and the generalized inverse of a finite matrix. Generalized Green's functions and Green's matrices for differential systems were also studied by Courant and Hilbert [282], Reid [1122], [1123], Smogorshewsky [1250], Dolph and Woodbury [321], Coddington and Levinson [275], Greub and Rheinboldt [489], Lanzos [718; Ch. 5] and others (see [1129], [276]) without reference to generalized inverses. The systematic use of generalized inverses in connection with properties and construction of generalized Green's matrices, which leads to conceptual elegance and structural simplicity, was made by Bradley [197], [198], Reid [1127], [1125], Loud [767], [768], Wyler [1394], and more recently by Chitwood [254], [255], Brown [208], [209], Locker [764] and others.

For use of generalized inverses in various aspects of differential operators and equations, see also Conti [278], Halany and Moro [516], Hestenes [574], [578], Landesman[719], Kallina [631], Wyler [1395], Nashed [919], [925], Locker [762], [764], [761], Lovass-Nagy and Powers [776], [777], [1620], Jones [620], Wong [1773], Zarantonello [1405] and others mentioned in other sections of III.

H2. Generalized Green's matrices and operators; general boundary problems

2-1. An algebraic theory, based on the theory of dual vector spaces, for generalized Green's operators, was developed by Wyler [1394], where necessary and sufficient conditions for the existence of generalized Green's operators are given for the general case, and for operators on Banach space. The set of all generalized Green's operators of a given operator is described and constructed. The results are applied in [1395] to two-point boundary-value problems for systems of ordinary differential equations.

2.2. Bradley [198], [197] investigated the $n \times n$ matrix differential system

$$Y' = AY + Z \quad MY(a) + NY(b) = 0$$

in the case when the homogeneous system is compatible. (Here M and N are constant matrices.) W. T. Reid, in his 1931 paper [1122], discussed such a compatible system and determined a generalized Green's matrix; however, his development did not make use of the Moore-Penrose matrix. Bradley used the Moore-Penrose, which allows for considerable simplification in the construction of a generalized Green's matrix. Bradley gave conditions for the nonhomogeneous system to possess a solution and showed that a generalized Green's matrix for the homogeneous system exists. While this matrix is not unique, a formula is developed which gives the most general generalized Green's matrix in terms of any particular generalized Green's matrix. Furthermore, the concept of a principal generalized Green's matrix is introduced, and it is shown that with respect to certain orthogonality conditions there exists a unique Green's matrix (which is precisely the matrix kernel of the Moore-Penrose inverse of the operator). Explicit and more transparent constructions of the class of all generalized Green's matrices and the analog of the Moore-Penrose inverse are given in two important papers of Loud [767], [768], who also gave the following formula for the unique principal generalized Green's operator (just as in the matrix case): $L^{\dagger} = P_{\mathfrak{R}(L^*)} G P_{\mathfrak{R}(L)}$, where G is any generalized Green's matrix. Loud computed $L^{\dagger}$ explicitly. The concept of a principal generalized Green's matrix was first introduced by Reid [1122], who showed that with respect to certain orthogonality conditions there is a unique generalized Green's matrix. Bradley [198] proved a similar theorem when the number of boundary conditions is different from n. Later, Reid [1127] (see also [1128]) discussed this problem in quite general setting. Reid also used generalized inverses of matrices in the construction of generalized Green's matrices, and in the study of so-called "principal solutions of disconjugate Hamiltonian linear differential systems" [1125]; these topics are also discussed in [1129], [1130].

2-3. Chitwood [254], [255] investigated the $n \times n$ matrix differential system $Y' = AY$ together with boundary conditions of the form $\int_a^b dF(t)\, Y(t) = 0$, where F is an $n \times n$ matrix whose elements are of bounded variation. He showed that if the homogeneous problem is compatible, then the Moore-Penrose inverse of a matrix can be employed to obtain conditions which ensure the existence of a solution to the nonhomogeneous problem. A generalized Green's matrix is constructed and its properties are studied in relation to an adjoint system. A principal generalized Green's matrix (in the sense of W. T. Reid [1127]) is defined and properties analogous to those for the classical case are developed. Of course the classical two-point boundary conditions are subsumed as a special case under the Stieltjes side conditions considered above; thus [255] is an extension of [198]. Earlier, Smogorshewsky [1250] studied an nth order scalar system as well as a system of first order equations under Stieltjes side conditions. The latter system if written in matrix notation would be equivalent to the system studied by Chitwood [255]. For both systems, Smogorshewsky constructed Green's matrices, generalized Green's matrices, and principal matrices, and listed their properties. The adjoint system, however, is not discussed in the Stieltjes case, and the Moore-Penrose is not employed.

Brown [208] also considered Green's functions for linear differential systems under general Stieltjes boundary conditions of the type $\int_0^1 d\nu(t)\, Y(t) = 0$, where ν is $m \times n$ matrix valued measure. The differential system is considered as an operator with domain and range in $\mathfrak{L}^p$. Both the operator and its adjoint are shown to be (normally solvable) Fredholm operators with mutually orthogonal ranges and kernels. Every bounded inner inverse of the operator can be represented on the range by a generalized Green's matrix structurally similar to the Green's matrix. The analogue of the principal generalized Green's matrix of Reid is constructed and shown to determine a generalized inverse via natural projections which includes the Moore-Penrose inverse as a special case. Most of these results parallel those of Chitwood, who considered the case $m = n$. See also [209] which develops some aspects of the theory of "normally solvable" linear relations and constructs a compact partial inverse (generalized Green's matrix).

For related general boundary value problems and a detailed development of the theory when the homogeneous problem is incompatible, see Cole [276; Chap. 6], Reid [1129; Chap. III], Bryan [211], and Tucker [1332], who also dealt with certain aspects of the compatible case. For multi-point problems, see also Locker [763] and Neuberger [946].

2-4. Conti [278] considered the following problem. Let Γ be a real vector space and L a linear operator such that $\mathcal{R}(L) \subset \Gamma$, $\mathcal{N}(L) \neq \{0\}$. Given $y \in \mathcal{R}(L)$, determine the solutions of the first-order linear differential equation: $\dot{x}(t) - A(t)x = f(t)$ such that $Lx = y$. A necessary and sufficient condition in order that such solutions exist is found to be

$$(*) \quad (I_\Gamma - L_U L_U^-)(y - LD_a^+ f) = 0 \ ,$$

where I_Γ is the identity operator on Γ; $L_U = LU(t, s)$, $U(t, s)$ being the evolution operator generated by the matrix $A(t)$; L_U^- is any 1-inverse (inner inverse) of the matrix L_U; and finally $D_a^+: f \to \int_a^t U(t, s) f(s)ds$. Halany and Moro [516] considered a particular operator L of the form $Lx = Mx(a) + Nx(b) + \int_a^b dF(t)\, x(t)$ and showed that condition (*) can be written in a classical form using solutions of a related adjoint problem. This general boundary value problem, involving a Stieltjes integral, is considered in detail, and an application to equations with perturbation term a distribution is given.

2-5. For generalized inverses of random linear operators and measurability of generalized random Green matrices, see Nashed and Salehi [930], and [923].

H3. Applications of generalized inverses to elliptic partial differential equations. Hestenes [574], [578], [577] and Landesman [719] studied aspects of generalized inverses in Hilbert space which are pertinent to the theory of elliptic differential equations. This involves the study of an unbounded operator A from one Hilbert space to another together with adjoint A^*, its generalized inverse or reciprocal $A^\dagger$, and its *-reciprocal $A' = (A^*)^\dagger$. Several characterizations of elliptic operators are given. The $\mathcal{H}_{-k}$ spaces that arise from the so-called negative norms in the study of elliptic partial differential equations are obtained by the use of the *-reciprocal of the operator which maps the function into itself and its first k derivatives. One revealing example, considered by Hestenes [574] (see also [578], [719]) is the interlationships between gradient, divergence and Green's functions, and the generalized inverse.

For the theory of singular quadratic functionals in Hilbert space with applications to partial differential equations, see Hestenes [571], [576], Hestenes and Redheffer [580], Stein [1458].

For Fredholm theory for elliptic partial differential equations, see Walker [1350], [1349]; see also annotations J1.

For a generalization to overdetermined systems of partial differential equations of the notion of diagonal operators, see Mackichan [1444] and recent papers. Although these generalizations are not related to generalized inverses, these involve some deep operator-theoretic results that should be of independent interest in analysis.

H4. Other topics in generalized inverses and differential equations

4-1. Algebraic theory of right invertible operators was considered in a series of papers by Przeworska-Rolewicz (see [1060]-[1076]) with applications to initial value and boundary value problems for linear equations with right invertible operators (see also [1765], [1766]). This work will also appear in a book [1756]. For the existence of a continuous right inverse for linear partial differential operators, see Cohoon [1433].

4-2. The minimum modulus of T is defined by

$$\gamma(T) := \inf \{ \|Tx\| / d(x, \mathcal{N}(T)) : x \notin \mathcal{N}(T) \} ,$$

where $d(x, \mathcal{N}(T))$ is the distance from x to the null space of T. Equivalently $\gamma(T) = \|T^\dagger\|^{-1}$. The concept of the minimum modulus of a closed linear operator in a Banach space was introduced by Kato [654]; this number plays a fundamental role in perturbation theory (cf. [654], [655], [924]). Goldberg and Meir [456] established estimates of $\gamma(T)$ and the existence of corresponding minimizing functions for certain differential operators for finite and infinite intervals. For asymptotic behaviour of the minimum modulus of a Fredholm operator, see Förster and Kaashoek [403].

4-3. Zarantonello [1405] considered the (Moore-Penrose) generalized inverse of a differentiod (pseudodifferential operator) and gave applications to the inversion of the basic differentiation operators ∇, $\nabla\cdot$ and $\nabla \cdot \nabla$. He also developed a Fredholm alternative theory for differentiods having a generalized inverse.

6-4. For use of generalized inverses in analyzing and constructing least squares solutions to two-point boundary value problems, see Locker [764], [762]. See also Mikhlin [854], Mikhlin and Smolitsky [855], Petryshyn [1011] for some classical aspects.

4-4. For applications of generalized inverses to Wiener-Hopf operators and equations: Lent [744], Bjerhammar [156]. For the index theorem for the Wiener-Hopf operators, see Coburn, Douglas and Singer [1526]; see also Palais [1644].

4-5. For a use of generalized inverses in differential equations and the eigenvalue problem $\lambda Tx + Sx$, see Wong [1773]. For Liapunov stability, see Jones [620], Reid [1126]. For an interesting and well-motivated application in dynamics which calls for a specific choice of nonorthogonal projectors in generalized inverses, see Milne [857]. For Riccati differential equations, see Reid [1130], [1124].

4-6. For discretized Dirichlet and Neumann problems in partial differential equations and generalized inverse aspects of some methods for solving singular systems of equations arising from finite-difference approximations, see Dalphin and Lovass-Nagy [292], [293], Dalphin [291], Lovass-Nagy and Powers [1621], Korganoff and Pavel-Parvu [692], [693]; see also Section O for other aspects. For exact and approximate solutions of some rectangular systems of differential equations and initial-value problems, see Lovass-Nagy and Powers [776], [777], [1620].

I. Generalized Inverses of Integral Operators. For historical comments and various results, see also Nashed [915], [925], Rall [1096], Kammerer and Nashed [640].

I1. Singular values and functions were originally introduced for integral operators in two papers of E. Schmidt (1907), where the spectral decomposition of operators of the form K^*K and KK^*, (where K is compact) are given. Using singular values and singular function expansions, Picard (1910) developed an existence criterion for integral equations of the first kind and expressed the (minimal-norm) solution of $Kx = y$ in terms of singular values and functions; see [934] for an exposition and generalization of these results. Other results on singular values were developed by Smithies (1937); see also [1249; Ch. 8]. von Neumann (1937) studied gauge functions of singular values in relation to matrix norms and inequalities. Eckart and Young (1936), (1939), extended the singular-value decomposition and utilized it in certain problems of matrix approximations (see P9). Weyl (1950) and Fan (1949), (1950) used singular values to establish bounds on eigenvalues. The use of truncated singular function expansion as a regularization method has been considered by several authors [534], [1701], [1702], [285], [1339], [1462] and others; see [925]. An algorithm for singular value decomposition was given by Businger and Golub [1517]; see also Golub and Kahan [471] for singular values and generalized inverses, and Businger [1516] for updating singular value decomposition. For other aspects of singular value decomposition in the linear least-squares problems, see Lawson and Hanson [735]. For miscellaneous aspects and other uses of singular values, see Golub [469], Good [479], Hartwig [548], Lawson [734]. A theory of singular values of linear transformation and related topics are developed in detail in the lecture notes by Amir-Moéz [33]; quasi-singular values are discussed by Amir-Moéz [1475]. For some recent results on singular value inequalities, see Thompson [1317], [1318]. Thompson and his collaborators develop also in a series of papers some deep results on singular values of sums and products of linear transformation (cf. [1767] and the references to earlier papers cited therein). Growth estimates for the singular values of $\mathcal{L}_2$-kernels were obtained recently by Cochran [1735], [1527] and Rozenbljum [1759]. Various applications of singular values in theory and approximation of integral equations of the first kind were given recently by Diaz and Metcalf [318], Nashed and Wahba [937], Nashed [920], Kammerer and Nashed [640], Strand [1284], Hilgers [583] and others.

I2. Generalized inverse $K^\dagger$ of a compact operator; Picard's criterion for best approximate solutions of $Kx = y$. These topics are treated in details in [925]. See also [1648], [1249], [1284], [583], [920], [937], Picard's criteria recast in reproducing kernel Hilbert space: [920], [937].

I3. $(I-\lambda K)^\dagger$ and Hurwitz's pseudoresolvent. See Nashed [925; Sec. 2], Rall [1096], Hurwitz (1912). Properties of Fredholm pseudoresolvent and pseudoinverse and comparison with Moore-Penrose and Drazin inverses.

$(I-\lambda K)^\dagger$ in Banach spaces and a canonical representation: Nashed and Votruba [934].

The interplay between $(I-\lambda K)^\dagger$ in the space $\mathcal{L}_2[0,1]$ and

$(I-\lambda K)^{\dagger}$ in the space $C[0,1]$: Moore and Nashed [894; Sec. 4], Kammerer and Nashed [640; pp. 559-560].

I4. For convergence rates of iterative methods (steepest descent, conjugate gradient and successive approximation) of integral equations of the first and second kinds, see Kammerer and Nashed [640], [641], [643]; for Cimmino's method for first kind equations, see [642]. For other papers on the successive approximation method and/or Picard's criterion for first kind equations, see Bialy [145], Fridman [416], [417], Diaz and Metcalf [318], Groetsch [507], Ivanov [1590], Mikhlin and Smolitsky [855], Bakušinskii and Strahov [79]; see also L2, O1 and O3.

For finite difference approximations and discretization, and use of generalized matrix inverses in numerical solutions of integral equations, see Pavel-Parvu [988], Korganoff and Pavel-Parvu [692], [693], Nashed [922]. For commutativity of moment discretization and least squares for integral equations of the first kind, see [922]. For convergence rates of approximate least-squares solutions by moment discretization, see Nashed and Wahba [937].

I5. For some approximations methods for operator equations that are useful for best approximate solutions of integral equations of the first and second kind see Nashed [915], [924]; see also O1 . For regularization-approximations of integral equations of the first kind and improperly-posed operator equations, see Nashed [920], [925]; see also L2 to L5 for further references. For quadrature and other approximations for Fredholm integral equations of the second kind with nonunique solution, see Moore and Nashed [894], K. E. Atkinson [68]; see also [920], [895]. For pointwise convergent approximations using quadrature, see [894].

For contrasts between various aspects of integral equations of the first and second kinds, see [925].

For generalized inverses of random linear operators, random integral operators and Green's functions, see [930], [923].

I6. For various topics in the theory and approximation methods for integral equations of the second kind, see Courant and Hilbert [282], Pogorzelski [1754], Anselone [52], Atkinson

[1722], Rall [1095], Bellman [1728], Mikhlin and Smolitsky [855]; [640] and [894]; we refer also to the extensive bibliography on integral equations by Ben Noble for relevant references.

I7. For some regularization methods and other numerical methods for numerical solution of integral equations of the first kind, in addition to those cited in I4, 5, we mention Tikhonov [1693], [1694], [1695], Tikhonov and Glasko [1696], Phillips [1016], Twomey [1334], [1700], Ivanov [1589], Hanson [534], Vainstein [1701], [1702], Baker et al. [1478], Bakusinskii [1479], Wahba [1706], Nashed and Wahba [935], [938], [939], Brynielsson [212], el-Tom [1324], Delves and Walsh [1533], Jennings [1593]; see also annotations under L, and the survey papers on regularization methods and ill-posed problems cited therein.

J. Generalized Inverses of Fredholm and Semi-Fredholm Operators and Other Topics in Operator Theory

J1. Generalized inverses and perturbation theory for Fredholm and semi-Fredholm operators. Substantial portion of the monographs by Kato [655] and Goldberg [454] are devoted to Fredholm operators and their generalizations. Introductory expositions are also given in Schetcher [1190], Bonic [1505]. Other expositions are given by Gramsch and Meise [1568], Palais [1644], Caradus, Pfaffenberger and Yood [1520].

Early fundamental contributions to the theory of Fredholm operators in Banach spaces are by Atkinson [66], [67], Yood [1399], Ruston [1172], [1673], Gohberg [1555], [1556], Gohberg and Krein [453], Kato [654], Ležański [1615], Sikorski [1679]-[1684] (for determinant theory of Ležański and Ruston and generalizations).

For more recent contributions to the theory of Fredholm operators, generalizations, and related topics, see Schechter [1189], Kaniel and Schechter [1598] (for spectral theory of Fredholm operators), Coburn and Lebow [1525] (for algebraic theory of Fredholm operators), Breuer [1511], [1512] and O'Neill [1447] (for Fredholm and semi-Fredholm operators, respectively, in von Neumann algebras), Gramsch [1567], [482], [481] (for Fredholm operators with a parameter and resolvents of elliptic operators, and other matters), Widom [1770] (for perturbing Fredholm operator to be an invertible operator), Williams [1771] (for closed Fredholm and semi-Fredholm operators and perturbations).

For Fredholm mapping and differential equations, see also Walker [1349], Mawhin [816]; see also H2-1 and H2-3.

For contributions to theory and perturbations of semi-Fredholm operators, see Schechter [1674], Yood [1399], Kato [654], Gohberg [1555], [1556], Gohberg and Krein [453] and others.

For pseudo Fredholm operators, see Tietz [1460].

For generalized inverses of Fredholm and semi-Fredholm operator and perturbation theory, see Nashed [924], Moore and Nashed [894]. For various aspects of perturbation theory of Fredholm or semi-Fredholm operators, see also Goldberg [454], [455], Kato [655], Saphar [1184], [1185], Caradus [226], [225], Widom [1770], Rakovščik [1660] and others.

Calkin algebras and connections with the set of Fredholm and semi-Fredholm operators: [1520], [1399], [925], [1484].

J2. Normally solvable operators (n. s. o.) and their perturbations: n. s. o. in normed linear spaces, ideals associated with them, perturbation theory, stability of certain properties of n. s. o., compact and other perturbations; topological properties of homomorphisms in Banach spaces, etc.: Atkinson [66], [67], Yood [1399], [1715], [1716], Dieudonné [319], Gohberg, Marcus and Fel'dman [1557], Gol'dman [1560], Gol'dman and Kračkovskii [1561], Rakovščik [1660], [1757], Whitley [1711], Goldberg [454], Jaunzems [611]; see also J1 and M.

Perturbations and generalized inverses of n. s. o. in Banach spaces, and approximation methods: Moore and Nashed [893], [894], Nashed [924]; see M.

Perturbations of closed operators and stability of index: Sz-Nagy [1692], [1691].

J3. The Riesz-Schauder theory for completely continuous operators and generalizations: Early important developments are in the independent work of Atkinson, Gohberg and Yood, all three papers appeared in 1951. See also Graves [1570], Ringrose [1138], Sheffield [1222], Altman [23], de Bruyn [210] for various generalizations.

J4. Riesz operators. An operator $T \in \mathcal{B}(X)$ is a Riesz operator if for any complex number $\lambda \neq 0$, $\lambda I - T$ is semi-Fredholm (cf. Lay [738] for this characterization). Other characterizations are given in Caradus [1519], Caradus et al. [1520], Bart [1484] and others. A Riesz operator is a generalization of a compact operator in that it abstracts as axioms the spectral properties of a compact operator. For Riesz operators and Fredholm perturbations, see Schechter [1674]. For other aspects of Riesz operators, see also the references cited in [738], [1519], and West [1710].

Riesz meromorphic functions, generalized inverses of values of Riesz meromorphic functions: Bart [1484], [1485]. Stability properties of finite meromorphic operator functions: Bart, Kaashoeck and Lay [90].

J5. Spectral theory: operator-valued analytic functions; meromorphic operator-valued functions; resolvents and operators of meromorphic type and other topics.

The monograph by Bart [1484] is devoted to meromorphic operator-valued functions; see also Derr and Taylor [314], Markus [1625].

Operator-valued analytic functions and spectral theory: Blum [1503], Mittenthal [1629], Bart [1484]. The recent thesis of Zettl [1408] deals with spectral analysis of homomorphism. For classical spectral theory of operators, see Taylor [1298], Lorch [1618], Dunford and Schwartz [350; vol. II], Yosida [1401].

See also Taylor [1297] for Mittag-Leffler expansions and spectral analysis, Schaeffer [1187] for singularities of analytic functions with values in a Banach space, and Schwartz [1198] for spectral operators. Spectral approximations for compact operators were developed recently by Osborn [1639].

Ascent, descent, nullity and defect, and related aspects of spectral analysis of operators; perturbation theory aspects: see Kato [654], [655], Taylor [1299], Lay [737], [738], Kaashoek [625], Kaashoek and Lay [626], Caradus [224], Taylor and Halberg [1300].

Holomorphic generalized inverses of operator-valued functions: Bart [1485], [1484]. Stability properties of finite meromorphic operator functions: Bart, Kaashoek and Lay [90].

J6. Other topics in operator theory relevant to generalized inverses; some monographs on algebras of operators: Diximier

[1532] deals with algebra of operators in Hilbert spaces. Douglas [1534], [1535] treats Banach algebra techniques in operator theory and Toeplitz operators, respectively. Caradus, Pfaffenberger and Yood devote their monograph [1520] to the study of Calkin algebras and algebras of operators on Banach spaces. For Banach algebras and related matters in generalized inverses, see G3. Kaashoeck and West [1596] treat locally compact semialgebras with applications to spectral theory of positive operators, Prössdorf [1658] considers several classes of singular operator equations.

Classes and properties of bounded linear operators with closed range; various characterizations: See Kato [655], Goldberg [454], Nashed [915], Ohwaki [970] and others.

For strictly singular operators, see for example Herman [1580], Whitley [1711], Porta [1652]. For several classes of singular operator equations, see Prössdorf [1658].

Operator ranges theorems: Douglas [328], Fillmore and Williams [394], Anderson and Trapp [49], Grabiner [1565]; see also the references in [394].

Ternary algebras: see E12(vi).

Nuclear spaces: Pietsch [1753], Gelfand and Vilenkin [444], Dunford and Schwartz [350; II].

Generalized inverses in reproducing kernel spaces: see F11.

K. Generalized Inverses in Nonlinear Analysis and Optimization (see [925]).

K1. Generalized inverses of nonlinear operators. Algebraic theory of generalized inverses and metric generalized inverses of nonlinear operators can be directly developed (see Nashed and Votruba [934; Sec. 1, pp. 43-46, and Sec. 8.2). These constructions are only useful when the sets are enriched with sufficient structure to permit the selection of meaningful "projectors". Other possibilities for defining generalized inverses of nonlinear maps were indicated in Nashed [915]. The direct theory of generalized inverses (with useful structures) for classes of nonlinear operators is still at its infancy, and much work need to be done before identifying appropriate nonlinear generalized inverses. Applications of generalized inverses in nonlinear analysis and optimization are for the most part confined to the use of generalized inverses of matrices and linear operators occurring in these problems in one of two ways. The nonlinear problem may often be written in the form of a sum of a nonlinear operator and a (noninvertible) linear operator as, for example, in bifurication or nonlinear alternative problems. Another way in which generalized inverses of linear operators is used in nonlinear problems is through the intermediacy of the (Fréchet) derivative; for various roles of derivatives in nonlinear analysis, see [916]. Various types of generalized inverses of the Fréchet (or other) derivatives may be brought to bear on the analysis and approximation of nonlinear and optimization problems.

K2. Generalized inverses in nonlinear alternative problems, $Nx = Lx$, where N is a nonlinear operator and L is a noninvertible linear operator. See [925; Sec. 6] for uses of generalized inverses in these problems. For various contributions to nonlinear alternative problems and/or related approximation methods, see Hale [518], [517], Bancroft, Hale and Sweet [82], Cesari [233], W. S. Hall [522], Locker [760], [761], Moore and Nashed [894], Sova [1255], Landesman and Lazer [721], Bartle [91], Graves [483], L. M. Hall [1571], Osborn and Sather [975], [1640], Fučik [1551] and others.

Aspects of bifurication theory which are related to generalized inverses: Vainberg and Trenogin [1338], Stakgold [1259], D. C. Lewis [746], Keller [665], Keller and Langford [666], Westreich [1374], Fučik et al. [423], [424].

Normal solvability for nonlinear operator equations, Fredholm alternative theory for nonlinear operators, surjectivity of operators involving noninvertible linear part, and nonlinear Fredholm operators, Browder [203], [204], [205], [206], Pohožaev [1029], [1030], [1031], [1032], Fučik, Nečas, Souček and Souček [423], Fučik [420], Fučik, Kučera and Nečas [421], de Figueriredo [393], Ize [1440] and others.

For use of generalized inverses in numerical solution of bifurication problems, see Atkinson [69], Demoulin and Chen [310] Moore and Nashed [894].

K3. Inverse mapping theorem, implicit function theorem and open mapping theorem for differentiable operators. For recasting and extensions of some of these classical theorems in functional analysis using generalized inverses, see Nashed [925; Sec. 7]. For various earlier contributions, see Graves

[1569], Hildebrandt and Graves [1581], Bartle [1488], Leach [741], [1613], Nevalinna and Nevalinna [1637], and Cesari [1734]. See also Newton's method (K5).

K4. Least-squares solutions of nonlinear operator equations; Gauss-Newton method, Marquardt and related algorithms: Fletcher [398], [399], Nashed [925; Sec. 8], Beltrami [104], Daniel [294], Pereyra [1004], Ben-Israel [109], [114], Golub and Saunders [475]; see also [125], [974], [982], [735].

K5. Newton's method and gradient methods for nonlinear systems or nonlinear operator equations, with singular derivatives:

For various results on Newton's method see Rall [1095], Ortega and Rheinboldt [974], Bartle [1726], Block [1733], Ostrowski [982], Kantorovich and Akilov [647].

For applications of 2-inverses, generalized inverses in Newton's method or related iterative methods for nonlinear equations, see [109], [114], [125; 1.10], [214], Leach [741], Altman [24], [25], Rheinboldt [1132; Theorem 3. 5], Fletcher [399]. For implicit use of generalized inverses in Newton and gradient methods for systems of equations and inequalities, see S. M. Robinson [1146] and Poljak [1755], respectively. Some recent results on a modified Newton's for the solution of ill-conditioned systems of nonlinear equations with applications to multiple shooting, see Deuflhard [1738].

K6. Constrained nonlinear optimization problems in Hilbert spaces; nonlinear programming, and related topics: See F4 and W.

K7. Other topics include nonlinear minimum-variance estimation, nonlinear discrete least-squares control theory, ridge analysis, etc. See various applications in V.

L. Generalized Inverses in Ill-Posed Problems and Regularization Methods: (see Nashed [920], [925] for details and other references).

L1. Survey papers on ill-posed problems and regularization methods with extensive bibliographies: Payne [991], [990], Tikhonov [1695], Nashed [918], [920], Turchin, Kozlov and

Malkevich [1699], Nedelkov [1636], Medgyessy [1627]. For classification of various approaches to regularization methods, see Payne [991] and Nashed [920] for Cauchy problems and operator equations, respectively. See also Lavrentiev [733].

L2. Some papers on regularization methods, error analysis and algorithms related to generalized inverses: Tikhonov [1693] [1694], Bakusinskiǐ [77], Bakušinskii and Aparcin [78], Ribière [1665], Nashed and Wahba [936], [938], Hilgers [583], [584], Gordonova and Morozov [1564], Franklin [413], Miller [856], Voevodin [1705], Turchin [1698], [1699], Korkina [695], Petrov and Hovanskiǐ [1010], Antohin [1477]. For finite-difference approximations to ill-posed problems, see, e.g., Gončarskiǐ, Leonov and Jagola [478]. For recent results on approximate regularized solutions to ill-posed linear integral and operator equations, see Nashed [920]. See also I4, 5, 7.

L3. Use of singular value decomposition in regularization: See [925] and the references cited therein; see also I1.

L4. Constrained minimization problems and related generalized inverses. Regularization methods based on stable minimization. See Nashed [920], [925], Ribière [1665], Nashed and Wahba [938] and others. For control theory, see X.

L5. Regularization method of generalized inverses in reproducing kernel Hilbert spaces: This method was initiated by Nashed and Wahba [938], [936]. See also [935], [937], [939], [1348] for various aspects of generalized inverses and reproducing kernel Hilbert spaces in connection with integral equations, regularization, constrained minimization problems and control theory.

L6. For implicit or explicit aspects of generalized inverses related to abstract spline functions via constrained minimization problems, see Anselone and Laurent [1476], Holmes [594], de Boor and Lynch [1508], Nashed [919], [920], Daniel and Schumaker [1530], Prenter [1656], Laurent [732] and others.

IV. Numerical Analysis and Approximation Theory for Generalized Inverse Matrices and Operators, Least Squares Solutions. Computational Methods and Packages

M. Perturbation Theory and Bounds for Generalized Inverses of Matrices and Linear Operators in Banach Spaces

M1. Continuity properties of the map $A \to A^\dagger$. The Moore-Penrose inverse of a matrix is not necessarily a continuous function of the matrix. The fact that $A^\dagger$ is a continuous function of A if the rank of A is kept fixed was observed by Penrose [1002; p. 408]. See also Nashed [924; Theorem 3. 5], Rosenberg [1163], Stewart [1271]. The investigation of continuity properties of the map $A \to A^\dagger_{P,Q}$ in Banach space leads to more subtle problems. This investigation was first undertaken by Moore and Nashed [894] as a by-product of a general setting for approximations of generalized inverse operators, which subsumes as special case necessary and sufficient conditions for continuity of the Moore-Penrose, oblique and other inverses (e.g. the Drazin inverse [335], [223]).

M2. Perturbation theory and error bounds for generalized inverses: $(A+T)^\dagger$ and $(A+T)^\dagger - A^\dagger$. For various contributions in the case of the Moore-Penrose inverse see Ben-Israel [111], Stewart [1271], Hanson and Lawson [535], Pereyra [1005]. A definitive treatment of perturbation theory for (the Moore-Penrose) generalized inverses of matrices was given by Wedin [1364], [1367]. Some of Wedin's results have been discovered independently by W. Kahan, and announced at the IFIPS Congress 1971 and at the Gatlinburg conference of 1972. The approximation theory setting of [894] also includes some new results for matrices.

Perturbation theory and error bounds for generalized inverses relative to general projectors are developed by Nashed [924]. The alternate approach by Moore and Nashed [894] to generalized inverse of perturbed operators includes also new results in this direction.

For expositions and other recent results on perturbation theory and error bounds see also Lawson and Hanson [735], Nashed [924] and Noble [962].

For related topics see also M3 and M4 .

M3. Generalized inverses of modified matrices. Formulas for the Moore-Penrose of the matrix obtained by modifying one or more elements of a given matrix, and related topics: Meyer [846], Cline [266], [267], Greville [493]. For earlier results on inverses of modified matrices, Woodbury [1713], Bartlett [92], Faddeev and Faddeeva [382], Sherman and Morrison [1224], Bennett [1501], Wilf [1379]. See also Ell and M2.

M4. Perturbation theory and error bounds for generalized inverses of linear operators in Banach space. A general setting is given by Moore and Nashed [894], where both uniform (in the operator norm) and pointwise (strong) convergence are considered, and a collectively compact approximation theory is developed in the context of generalized inverses. Applications are given to collectively compact pointwise convergent approximations to $(I-K)^{\dagger}$, where K is a compact operator, and quadrature approximations to least squares solutions of Fredholm integral equations of the second kind. These results extend various aspects of inverse operator approximations (cf. Anselone [52], Kantorovich and Akilov [647; Ch. XIV], Ostrowski [981], and others).

Bound for $\|B^{\dagger}_{P',Q'} - A^{\dagger}_{P,Q}\|$ and related decompositions; error bounds (in terms of the gap between two subspaces) for generalized inverses under change of projectors: Nashed [924].

M5. Perturbation theory for Fredholm, semi-Fredholm, normally solvable and related operators: See J.

M6. Error analysis and stability of linear least squares problems; round off error in normal equations; error analysis of ill-posed problems; experiments with error growth: See also M1, 2, N and Q .

van der Sluis [1246], [1686], [1687], Morozov [901], Gordonova [1563], Pereyra [1005], Lawson and Hanson [735], Korkina [695], Jennings and Osborne [615], Wilkinson [1712], Atta [70], Jordan [621], Wedin [1708], Björck [159], [160], [161], Abdelmalek [1] and others.

N. Direct Methods for Computing Matrix Generalized Inverses and Least Squares Solutions of Linear Equations

N1. Explicit formulas for the Moore-Penrose inverse and other generalized inverses (1, 1-2, 1-2, 3 inverses, etc.) see [1113], [125], [190], [1054, [17], Rohde [1151] for various formulas and references to the original contributions. See also N6 and N8.

N2. Computational methods of generalized inverses based on (rank) factorizations of matrices. Let A be an $m \times n$ matrix of rank r. Then $A = BC$, where B and C are of rank r, and $A^\dagger = C^\dagger B^\dagger$.
(The matrices B and C in this factorization are not unique; however $C^\dagger B^\dagger$ is invariant under the various factorizations.) Some useful factorizations include:

(i) $A = LU$, where L is an $m \times r$ lower trapezoidal matrix with 1's on the principal diagonal and 0's above the diagonal, and U is an $r \times n$ upper trapezoidal matrix. Then $A^\dagger = U^*(UU^*)^{-1}(L^*L)^{-1}L^*$.

(ii) $A = QR$, where Q has orthonormal columns and R is upper triangular. Then $A^\dagger = S^*(SS^*)^{-1}Q^*$.

(iii) Decomposition of A using singular values: $A = VDU^*$, where D is a diagonal matrix whose elements are singular values of A, the columns of V are orthogonal and so are the columns of U. Then $A^\dagger = UD^{-1}V^*$.

Direct methods which involve an explicit determination of the rank are developed on the basis of these and other decompositions. The main difficulty in generalized inverse computation is due to the discontinuous dependence of $A^\dagger$ on perturbations which alter the rank of A.

N3. Some practical methods for effecting the decompositions (in N2) and computations of the Moore-Penrose inverse. The matrix Q is found either by the Householder transformation (see Parlett [985], Golub [470], Hanson and Lawson [535]), or by a modified Gram-Schmidt orthonormalization (MGSO) procedure (Björck [159], Lawson and Hanson [735]). A detailed error analysis of the influence of rounding error for the MGSO procedure is given by Björck [159] and Abdelmalek [1], [2]. Experimental results are given by Rice [1133] and Jordan [621].

Practical ways for finding singular value decompositions

are based on the Householder transformation [596], [597] and the QR algorithm [410], [411]. These methods are compared in Golub [467], [470]. See also Lawson and Hanson [535], Lawson and Hanson [735], Noble [962]. For calculating singu lar values and their use in generalized inverse computation, see Golub and Kahan [471], and Lawson and Hanson [735].

Other matrix factorizations which are useful for generalized inverse computations include Cholesky's factorization of A^*A (see Wilkinson [1384], Pringle and Rayner [1054; 4.6], Golub and Saunders [475], Lawson and Hanson [735; 19.1], Bjerhammar [156; 4.6] and Noble's factorization [960]; see also [125].

For an expository account of orthogonal triangularization, QR algorithm, and the linear least squares problem, see also Stewart [1272].

Other practical methods for $A^\dagger$ and least squares solutions are included in the remaining sections of N .

Full-rank factorization can also be adapted to computation of (1, 3)-inverses, (1, 2, 3) and other inverses.

N4. Direct methods for the numerical solution of linear least squares problems. Solving the least squares problem for $Ax = b$ is equivalent to solving the normal equations $A^*Ax = A^*b$. However, since the conditioning of A^*A is worse than the conditioning of A (cf. Taussky [1295], Noble [961], [962], Stewart [1272]) the direct use of the normal equation is limited in practice only to the case when the normal equations are well conditioned. Using the full rank factorization in N2 the normal equations can be easily reduced to the equivalent form $B^*Ax = B^*b$. This form is useful if the factorization B^*A is not worse-conditioned than A . Methods for computing least squares solution based on factorizations which take conditioning into account have been studied by several authors: Golub [470], Björck [159], [160], [161], Björck and Golub [163], Businger and Golub [216], Golub and Wilkinson [477].

The Golub-Kahan algorithm [471] for reducing a general matrix to bidiagonal form is utilized by Paige [1642] to develop a bidiazonalization algorithm which is suitable for solving the linear least squares problem for large sparse matrices, and by Paige and Saunders [1643] for sparse indefinite system of linear equations.

Among the methods which are useful for solving the

normal equations when these equations are well conditioned, we mention the Gauss-Doolittle pivotal condensation method or square methods (cf. Rao and Mitra [1113; 11. 5], Rohde [1149]), and the Cholesky, Cholesky-Rubin method (cf. [735; 19.1], [156; 4. 6], Morkhov [900]). Morrison [905] and Riley [1136] develop also methods which use the normal equations to get the best approximate solutions. Some modified normal equations were studied by Rutishauser [1173]; see also Cvetkov [288] and Gale [433].

Some early work on direct numerical resolution (or iterative solution) of normal equations and calculation of errors include Benoit [131], Jensen [616], Birge [149], Andersen [35], [36], Stearn [1264]; see also Bjerhammar [156] for various references to early contributions.

N5. Iterative refinement (correction) of solutions, due to Wilkinson [1384] (see also Moler [888], Forsythe and Moler [1548], Stewart [1272]), is often recommended in conjunction with direct subroutines for solving a consistent system $Ax = b$. The kth correction, $\delta^{(k)}$, is obtained by solving $A\delta^{(k)} = r^{(k)}$, where $r^{(k)}$ is the kth residual $C = b - Ax^{(k)}$, where $x^{(k)}$ is the kth approximation. The $(k+1)^{st}$ approximation is $x^{(k+1)} = x^{(k)} + \delta^{(k)}$. The method provides a check on the conditioning of the problem and is easy to implement (double precision is only used in computing the residuals; the iteration is stopped if $\|\delta^{(k)}\| / \|x^{(k)}\|$ falls below a prescribed number).

Similar iterative refinement techniques for least squares problems have been developed by Golub and Wilkinson [477], Björck [160], [161] and Björck and Golub [163]; see also Stewart [1272; 5.4]. (See also Gill et al. [1554] for updating methods in matrix computations.) For iterative refinement of the solution of positive definite system of equations, see Martin, Peters and Wilkinson [809].

N6. Various other direct methods for computing $A^\dagger$ and projectors, numerical methods for linear least squares problems, rectangular, ill-conditioned or indefinite systems, algorithms and explicit representations for $A^\dagger$, etc: See Cho et al. [257], [17], [125], [190], [1113], Faddeev and Kublanovskaja and Faddeeva [383], [384], [385], Blum [167], Matveev [815], Mayne [819], [820], Kublanovskaja [704], [705], de Meersman [827], Tewarson [1301], [1302], [1303], [1304], Tewarson and Joshi

[1311], Tewarson and Ramnath [1314], Boot [174], Dragomir [331], Zlobec [1411], [1412], Zimmule [1410], Zielke [1409], Varah [1339], Klinger [676], Gentleman [445], LaBudde and Verma [213], Dalphin [291], Delaney and Speed [1309], Huang [1744], Žukovskiĭ and Lipcer [1774], Žukovskiĭ and Zaikin [1775], Bunch and Parlett [1514], Dahlquist, Sjöberg and Svensson [1529], Wedin [1708], Stallings and Boullion [1261], Narain [1446], Chen [242], [243].

(For related essays and some perspectives in numerical linear algebra, see Forsythe [1546], [1547] and Kahan [1597].)

N7. Methods based on Gauss elimination for $A^\dagger$ and least squares solutions; Gauss elimination with complete or partial pivoting and scaling. For a detailed exposition see Noble [962]. See also [960], [129], [19], [1009], [97], [900], [1391], [1389], [149], [156], [35].

For optimal matrix scaling, see Bauer [96] and Businger [1465]. For pivoting method for minimization of quadratic functions, see Beale [1493]. For modifying pivot elements in Gaussian elimination, see Stewart [1274].

N8. Miscellaneous methods and remarks

(i) Computation of generalized inverses when independent row or columns are identifiable: See, e.g., [1002], [1113], [962], [17], [1009], [125].

(ii) Generalized inverse ε-algorithm for matrices and projectors: Pyle [1087], [1088], Wynn [1396], [1397].

(iii) The Cayley-Hamilton theorem (cf., e.g., [101]) has been used by Decell [298] (see also [119]) to develop an expression for the generalized inverse of a matrix. Let $p(\lambda) = (-1)^n \sum_{j=0}^{n} a_j \lambda^{n-j}$, $a_0 = 1$, be the characteristic polynomial of AA^*. If $k = \max\{j: a_j \neq 0\}$, then $A^\dagger = -a_k^{-1} \sum_{j=0}^{k-1} a_j (AA^*)^{k-j-1}$; if $k = 0$, then $A^\dagger = 0$. An efficient method for generating the coefficients of the characteristic polynomial is developed in [382]. See also [17; Sec. 5.4] for a detailed account of the theory of this procedure for computing $A^\dagger$, also [190].

(iv) Recursive calculation of $A^{\dagger}$, Greville's method and extensions: [493], [125], [17], [675], [880]; see also S2 and X4 .

(v) Gauss-bordering method for 1-inverses and generalized inverses of specific rank: Zlobec and Chan (cf. [1415]).

(vi) Computation of 2-inverses with prescribed range and null spaces: [125; 7.4] and references cited therein.

(vii) The modified Gram-Schmidt orthonormalization method for computing the Moore-Penrose inverse: Rust, Burrus and Schneenburger [1171]; see also [17; 5.1], [475], [962].

(viii) Sparse matrices and large sparse linear least squares problems (see also some iterative methods in O): Reid [1662], [1663], Rose and Willoughby [1670], Paige [1642], Paige and Saunders [1643], Gentleman [1553], and the recent book by Tewarson on sparse matrices.

O. Iterative Methods for Generalized Inverses and Least Squares Solutions of Matrix and Operator Equations (see Nashed [915], [924]. Patterson [987] gives an excellent exposition of various iterative methods for the solution of a linear operator equation in Hilbert space. He does not treat, however, least squares solutions or generalized inverses.)

O1. Iterative methods for best approximate solutions of linear operator equations in Hilbert spaces (see also O2 for matrices):

1-1. Steepest descent and related gradient methods with convergence rates for closed and nonclosed ranges: Nashed [914], Kammerer and Nashed [641], McCormick and Rodrigue [823]. See also Fridman [417], Groetsch [502].

1-2. Conjugate gradient method: Kammerer and Nashed [643].

1-3. Successive approximations and related hyperpower iterative methods for linear operator equations: [915], [924], Petryshyn [1013], Showalter [1230], [1231] (see also [125]),

Bialy [145], Patterson [987], Lardy [727], [728], Krjanev [701], Nashed [914], Kammerer and Nashed [640], Altman [28]. For linear operator equations, see [924]. For successive approximations and related methods for operator equations of the first kind ($Kx = y$, where K is compact), see Landweber [722], Fridman [416], Diaz and Metcalf [318], Groetsch [507], Kammerer and Nashed [640], [642], Strand [1283], [1284]; see also I4, 5 and 7.

1-4. Summability methods, ergodic theory and averaging iterations for linear operator equations in Banach and Hilbert spaces (mostly for invertible operators): Bellman [1727], Niethammar and Schempp [959], Schempp [1191], Dotson [325], [326], Kwon and Redheffer [714], Oblomskaja [964], Groetsch [503], [504], Koliha [677], [678], [679], and others. For applications of the classical theory of summability to representations of generalized inverses, see [505], [503] and the remarks in [924; 6.3].

1-5. Methods of additive decompositions; splitting methods. For matrices: Keller [664], Korganoff [688], Korganoff and Pavel-Parvu [694], [989], Plemmons [1650], Berman and Plemmons [135], [137], Joshi [622], [1746], [1443], Marčuk and Kuznecov [1624], Linda Lawson [736], [1612].

For linear operators: Keller [664], Zlobec [1414], Kammerer and Plemmons [644]. (For example, let $A = M - N$ with $\mathcal{R}(A) = \mathcal{R}(M)$ and $\mathcal{N}(A) = \mathcal{N}(M)$. Such a splitting is called a proper splitting of the rectangular matrix A. Then $x_{n+1} = M^{\dagger}Nx_n + M^{\dagger}b$. Then $\{x_n\} \to A^{\dagger}b$ iff the spectral radius of $M^{\dagger}N$ is less than one.)

O2. Iterative methods for generalized inverses of matrices, projectors, and least squares solutions of algebraic linear equations

2-1. Gradient methods (steepest descent, conjugate gradient, gradient projection, etc.): Nashed [914], [915], [924], Kammerer and Nashed [643], Pyle [1086] (also [17; Sec. 5.3], [125]); Whitney and Meany [1377], Stewart [1273], Reid [1663], Tewarson [1303], and Hestenes [579], Tanabe [1292], Blum and Rodrigue [168], and others.

2-2. An iterative method of the form (*) $X_{n+1} = X_n(2I - AX_n)$ for computing the generalized inverse of a matrix was studied by Ben-Israel [108], [110] (see Joshi [623] for an argument which remedies a defect in the proof in [108]), Ben-Israel and Cohen [124], Petryshyn [1013] and Zlobec [1411]. This second-order iterative method which converges to $A^{\dagger}$ is a generalization of the well-known method of Schulz [1197] (see also Householder [597; p. 95] and Petryshyn [1012]) for the inverse of a nonsingular matrix or linear operator. A detailed error analysis and termination aspects, using singular value analysis, which reflect implicit difficulties with iterative methods, are given in Söderstrom and Stewart [1252]. The so-called hyperpower iterations are p^{th} order methods which generalize the iteration (*); see [124], [1013], [1411]; also Lonseth [765] and Altman [28]. Hyperpower iterative methods and Neumann-type expansions for (1, 2)-inverses of a matrix have also been given recently by Tanabe [1293].

2-3. Miscellaneous iterative methods for singular matrices and for computation of $A^{\dagger}$, projectors and matrix products, involving generalized inverses: Cho et al. [257], Tewarson [1309], [1303], Garnett et al. [436], Tannabe [1291], Marčuk and Kuznecov [798], [1624], Gupta [511], Abdennur [4], Vanderschel [1461], Decell and Kahng [303].

Cimmino's method: [640], [933], [597], [1346]; Kramarz's method: [1291], [438]. Projection methods of systems of linear equations: Harms [1439], Shen I-ming [1454]; [438], [798], [1306] and others.

For alternating direction methods in the presence of singular matrices, see Douglas and Pearcy [327], Kellogg and Spanier [667], Lovass-Nagy, Powers and Ullman [779]; also [798].

2-4. Miscellaneous iterative methods for incompatible systems, least squares solutions and minimization of quadratic functionals: See the book [798] by Marčuk and Kuznecov. Also Tewarson [1304], Tanabe [1292], Molčanev and Jakovlev [1752], Rodrigue [1453], Kuznecov [712], [713], Langhaar [726], Joshi [1443] and others. For least squares acceleration of iterative methods for solving equations, see Kaniel and Stein [645]. For eigenvalue problems and least squares problems, see Blum and Rodrigue [168]. See also splitting methods (O1-5), and summability methods (O1-4).

2-5. Four classical iterative methods for solutions of algebraic linear equations, which have been proposed and studied in the period 1937-40, converge in fact even in the case of incompatible systems. These are the methods of Cesari [232], Kaczmarz [627], Cimmino [256] and Jossa [624].

2-6. Iterative methods are of particular interest for very large sparse systems where the solution of normal equations or direct methods can be extremely tedious. In statistical problems where variances of the unknown quantities must be computed, iterative methods are of little importance. Iterative methods are useful for singular large sparse systems that arise often from discretization of Neumann problems and in several other contexts.

O3. Series, integral and limit representations of generalized inverses

3-1. The representation $A^{\dagger} = \lim_{\sigma \to 0} A^{*}(AA^{*} + \sigma I)^{-1} = \lim_{\sigma \to 0} (A^{*}A + \sigma I)^{-1} A^{*}$ for matrices and bounded linear operators with closed range: see, e.g., [17], [125], [915] for matrices, and [505], [915] for operators. For a bounded linear operator with arbitrary range $A^{\dagger}x = \lim_{\sigma \to 0} (A^{*}A + \sigma I)^{-1}A^{*}x$ for each x; convergence is uniform iff $\mathcal{R}(A)$ is closed; [915], [141]. For a closed densely-defined linear operator A in Hilbert space, let $A'_{\sigma} := A^{*}(AA^{*} + \sigma I)^{-1}$ and $A''_{\sigma} := (A^{*}A + \sigma I)^{-1}A^{*}$. Then $A'_{\sigma}x \to A^{\dagger}x$ for each $x \in \mathcal{D}(A^{\dagger}) = \mathcal{R}(A) + \mathcal{R}(A)^{\perp}$ and the closed minimal extension of A''_{σ} to an operator defined everywhere converges to $A^{\dagger}$ on $\mathcal{D}(A^{\dagger})$; in both cases convergence is uniform iff $\mathcal{R}(A)$ is closed (see Beutler and Root [141]).

For a square matrix A, $\mathcal{R}(A) = \mathcal{R}(A^2)$ iff $\lim_{\sigma \to 0} (A + \sigma I)^{-1}A$ exists, in which case the limit is $AA^{\#}$, where $A^{\#}$ is the group inverse. General results can be given for $\sigma(A + \sigma B)^{-1}$, where B is positive definite, A is nonnegative definite. Necessary and sufficient conditions for the existence of $\lim_{\sigma \to 0} \sigma^{m}(A + \sigma I)^{-1}A^{p}$ and other limits are also known in terms of the index of A. The Drazin inverse is obtained in the form

$A^d = \lim_{\sigma \to 0} (A^{k+1} + \sigma I)^{-1} A^k$, where $k \leqq$ index A. These results and their extensions to operators, as well as the various forms of series and integral representations, can be unified in the framework of spectral theory as mentioned in [924]; for various results see Langenhop [725], Meyer [848], Groetsch [506]; also [141], [140], [727], [728] and others.

3-2. Series representations: $A^\dagger = \sum_{n=0}^{\infty} \beta(I - \beta A^* A)^n A^*$ for $0 < \beta < 2\|A\|^{-2}$, $A^\dagger = \sum_{n=1}^{\infty} A^*(I + AA^*)^{-n}$ and others, for matrices, bounded or (for the second series) closed densely-defined linear operators in Hilbert space. Convergence is uniform iff $\mathfrak{R}(A)$ is closed. See Petryshyn [1013], Nashed [915], [924], [914], Lardy [728], Groetsch [505], Showalter [1230], Showalter and Ben-Israel [1231]. Neumann-type series representations of generalized inverses in Banach spaces are given in Nashed [915], [917], Votruba [1346] and Koliha [680], [681]; see also Tanabe [1293] for finite dimensional spaces; also [141], [125].

3-3. Integral representations, e.g., $A^\dagger = \int_0^\infty e^{-A^* A s} A^* ds$ and other variants: Showalter [1230], Showalter and Ben-Israel [1231], Nashed [915], [924], Groetsch [505]. For use of summability methods in obtaining series and integrals representations, see [505], [924]. (More unification is possible by using spectral theory as remarked in [924] and will be discussed elsewhere.)

P. Approximation Theory and Projection Methods for General Linear Operator Equations and Generalized Inverses (cf. Nashed [924]).

P1. Generalized inverse operator approximation theory and collectively compact approximation theory: Moore and Nashed [894], [895]; see also [920], [924].

P2. General approximation theory settings for operator equations: Anselone [52], Nashed and Moore [894]; Kantorovich and Akilov [647; Ch. 14], Rakovshchick [1094]; Stummel [1690], [1763], [1764].

P3. Projection methods for best approximate solutions of linear operator equations of the first and kinds, projection methods for ill-posed problems: Nashed [919], [924], [927], Prenter [1655], Savelova [1186], Vasin and Tanna [1340].

P4. General settings for projection methods and related theory: Petryshyn [1011], [1014], [1015], Polsky (cf. [924]), Hildebrandt and Wienholtz [582], also [924].

P5. Convergence rates of approximate least squares solutions of linear operator equations obtained by moment discretization: Nashed and Wahba [937].

P6. Projection and discretization-collocation methods for $x - \lambda Kx = y$ and operator equations of the second kind: Anselone [52], Atkinson [1722], Thomas [1316], Prenter [1654], Kantorovich and Akilov [647], Moore and Nashed [894] and others (see [924] and P2).

P7. Regularization methods: See L.

P8. Approximation theory for generalized inverses in reproducing kernel Hilbert spaces: Nashed and Wahba [937], [938], [939].

P9. Closest unitary, orthogonal and Hermitian operators to a given operator; best matrix approximation of a given rank. Let A be a given $m \times n$ matrix. The problem of finding the matrix B_r closest to A in the class of $m \times n$ matrices of rank r was solved for the Euclidean norm by Eckart and Young [357], using singular values of A. The best approximation is unique iff the kth and $(k+1)^{st}$ singular values of A are distinct. The error is $\|A - B_r\| = (\sum_{i=k+1}^{r} \alpha_i)^{\frac{1}{2}}$, where α_i is the ith singular value. Various other proofs in Golub and Kahan [471], Householder and Young [598], Gaches, Rigal and Rousset de Pina [431], Franck [412] and J. B. Keller [1602]. Mirsky [865] showed that the result is independent of the norm, provided the norm is unitarily invariant. A modification of this problem also arises in orthogonal approximation to an oblique structure in factor analysis. Then the best approximation is sought in the class of matrices CD where C is an arbitrary $m \times r$ matrix and D is given. These problems were solved by Green [488], J. B. Keller [1602] and Schönemann [1196] using Euclidean norm.

Fan and Hoffman [388] found the unitary and Hermitian matrices closest to a given matrix. Their result is independent of the norm, provided the norm is unitarily invariant.

These problems and others which deal with finding an operator B, in some specified class of operators $\mathcal{U}$, which is closest to a given operator, were recently analyzed and solved (in a unifying framework for several choices of $\mathcal{U}$) by J. B. Keller [1603].

Cvetkova [1737] constructed a positive definite matrix that is closest to a given matrix. Björck and Bowie [162] gave an iterative algorithm for computing the best estimate of an orthogonal matrix. Björck and Golub [164] gave numerical methods for computing angles between subspaces. For related papers, see also Seidel [1211] and Davis and Kahan [1531], and Varah [1703].

Q. Other Uses of Generalized Inverses and Related Topics in Numerical Analysis and Approximation Theory

Q1. Linear least squares problems with equality or inequality constraints. Numerical methods: Lawson and Hanson [735], Leringe and Wedin [745], Stoer [1277], Bartels, Golub and Saunders [1487], Björck [161], Golub and Saunders [475], characterizations of solutions and expressions using generalized inverses: Nashed [925], Albert [17], Leringe and Wedin [745] and others; see also F4.

Q2. Nonlinear least squares problems, least squares solutions of nonlinear operator equations, nonlinear optimization and mathematical programming: use of generalized inverses in theory and numerical methods, see K4, K5 and W.

Q3. Eigenvalue problems: Numerical solution of general (and ill-conditioned) eigenvalue problems for square or rectangular matrices: For recent results, see Burris [1431], Crawford [1434], Kaufman [663], Fix and Heiberger [1540], Ruhe [1167], [1166], Van Loan [1462], Varah [1703] and others. For eigenvalue problems in Hilbert space using generalized inverses, see Blum and Rodrigue [168], Rodrigue [1453], McCormick [822]; for recent results on the eigenvalue problem $\lambda Tx + Sx$ and differential equation using generalized inverses, see Wong [1773]. Inverse eigenvalue problems: Kublanovskaja [705], [707], [708], Hadler [514], [515].

Q4. Chebyshev and ℓ_1-solutions of (consistent or inconsistent) overdetermined systems; algorithms and theory: Abdelmalek [3], Barrodale and Roberts [89], Cazdow [221], Cheney and Goldstein [245], Duris [352], Goldstein and Cheney [461], Cheney [244], Goldstein, Levine and Hereshoff [462], Duris and Sreedharan [355], Meicler [830], Stiefel [1275], [1276], Rosen [1160], Temple [1459], Zuhovickiǐ [1418], [1419], [1420], [1421]. Solutions and least squares algorithms of overdetermined linear equations which minimize error in an abstract norm; theory and algorithms for best approximate solutions to $Ax = b$ in normed linear spaces: Sreedharan [1257], [1258], Anton and Duris [54], Borowsky [1429]; see also Nashed [924], [915], Nashed and Votruba [934] for theory and iterative methods.

Q5. Optimal quadrature formulae and generalized interpolation using generalized inverses: Duris [353], [354], Herring [570]; see also [693], [922].

Q6. Mathematical programming and generalized inverse techniques for solving singular and poorly conditioned systems of equations and ill-posed problems: Replogle, Holcomb and Burrus [1131], Rust and Burrus [1170], Torsti and Aurela [1697], Nashed [920] ([918] for additional references), Chen [1432] and others. Dynamic programming techniques for ill-conditioned or ill-posed linear systems: Bellman, Kalaba and Lockett [1495], [1496], [103], Bellman [101], Bellman, Glicksberg and Gross [102].

Q7. Other applications and implementation of orthogonal transformations to nonlinear algebraic equations, extremal problems and ill-posed problems: Kublanovskaja [706], [708], [709], Jennings [1467], Jennings and Osborne [614], Tsao [1326] and others.

R. Program Packages and Computer Codes for Computing Generalized Inverses and Least Squares Solutions

R1. Various codes for solving least squares problems are described by Lawson and Hanson [735], including ANS FORTRAN programs for:

Solution of the linear least squares problem by Householder orthogonal transformations, and by singular value decomposition.

Sequential solution of a least squares problem whose coefficient matrix has band structure.

Linear least squares problems with linear inequality constraints.

The set of programs is distinguished by the consistent use of orthogonal transformation methods to provide stable numerical procedures. The programs have been tested on IBM, CDC, UNIVAC, and Xerox computers. IMSL (International Mathematical and Statistical Libraries, Inc.), Houston, Texas, is acting as the distributing agent for these codes.

R2. Similar programs are also available at NAG (Numerical Algorithms Group, Oxford). EISPACK and a forthcoming package LIMPACK are developed at the Applied Mathematics Division at Argonne National Labs. Two ALGOL programs related to those in Lawson and Hanson are given in Wilkinson and Reinsch [1386]. An ALGOL procedure for computing the singular value decomposition is developed by Businger and Golub [1517] and is available in Golub and Reinsch [474]. A computer program in FORTRAN-II for computation of $A^\dagger$ based on orthonormalization method, written by Bhimasankaram, is also available at the Indian Statistical Institute, Calcutta.

Two ALGOL procedures were developed by Björck [161] for the iterative refinement of least squares solutions, based on a matrix decomposition obtained either by orthogonal Householder transformations or by a modified Gram-Schmidt orthogonalization; see also [160], [159] for related aspects.

A program for the solution of linear systems of equations and linear least squares problems in APL, written by Jenkins [613], is available from IBM-New York Scientific Center.

R3. An appraisal of least squares programs, as of 1967, from the point of view of the user was given by Longley [1617]. An evaluation of linear least squares computer programs and their accuracy, as of 1969, was made by Wampler [1354], [1355]. Various appraisals of direct and/or iterative methods for computing $A^\dagger$ were given by several authors [1009], [1228], [1229], [308], [159], [160], [161], [962], [1252].

R4. A subroutine package for calculating with B-splines was prepared by C. de Boor [1506]; see also [1507] for theoretical aspects. This is useful for sequential processing for banded least squares problems and data fitting and for nonlinear least squares spline approximations (see also Jupp [1747]). For a FORTRAN program, see Appendix C of [735].

V. Applications of Generalized Inverses

S. Applications of Generalized Inverses in Statistics. Statistical applications of generalized inverses have received considerable attention in the literature. This is perhaps the most developed area of applications of generalized inverses. Contributions to this area are amply covered as of 1971 in the books by Rao and Mitra [1113], Albert [17], and Pringle and Rayner [1054]. The earlier survey articles by Rhode [1152] and Chipman [250] provide also excellent accounts of many of the important statistical applications of generalized inverses. Albert's paper in this volume briefly surveys selected applications of the Moore-Penrose inverse in statistics, and picks up where Rhode left off in his 1968 article. Chipman's monograph (in this volume) provides an excellent self-contained exposition of the theory of estimation and theory of aggregation, and contains some new results as well as some refinements of earlier results, and includes an extensive bibliography.

It should be noted that statistical applications utilize a variety of subclasses of 1-inverses and not just the Moore-Penrose inverse.

S1. The general linear models (variably called also estimation in linear models, regression theory, and analysis of variance). This area represents the most developed application of generalized inverses to statistical problems.

Linear hypothesis, least square estimator (LSE), best linear unbiased estimation (BLUE), the Gauss-Markov estimator, the Gauss-Markov theorem, applications of the projection theorem to statistical estimation, estimation of parameters in various linear models, adjustment of LSE for removal or addition of an observation.

1-1. A treatment of the Gauss-Markov model and the associated problem of estimation and hypothesis testing are treated in their wide generality by Rao [1108], [1110], Rao and Mitra [1113], when the observations have a possibly singular covariance matrix. An extensive treatment of the sum of squares decomposition for fixed effects model is given in Albert [17; Ch. VI]. Some new results for the mixed model are given by Albert [19]. The linear model with singular variance matrix is discussed also in Pringle and Rayner [1054; Ch. 6, 7]. Chipman [251] gives an extensive treatment of the theory of estimation and the Gauss-Markov theorem. See also Luenberger [780; Ch. 3].

The paper by Goldman and Zelen [458] appears to be the first paper to have considered the Gauss-Markov model with a singular dispersion matrix. For basic and recent contributions to the Gauss-Markov theorem (singular covariance), see Zyskind and Martin [1424], Rao [1108], Mitra [875], Albert [18], Zyskind [1423], Drygas [336], [337], [338], [339], [340], [342], Lewis and Odell [752], Rao and Mitra [1112], [1114].

Gauss-Markov theorem (in Hilbert spaces): Beutler and Root [141].

When is a naive LSE a BLUE?, Rao [1103], Kruskal [699], Mitra and Rao [885], [1113; Ch. 8]; see also the recent papers by Haberman [1572] and Norlén [1638].

1-2. Other contributions to the Gauss-Markov theorem and the linear model: David and Neyman [295], Decell and Odell [304], Kruskal [698], Lewis and Odell [751], Martin [807], Martin and Zyskind [808], Mitra [871], Mitra and Rao [884], [885], Rao [1117], Rayner and Pringle [1120], Searle [1209], Harter [540], [541], Stein [1267], Tan [1290] and others. Sibuya [1237] notes that the Gauss-Markov theorem reduces to a duality of two types of g-inverses.

1-3. "Early" contributions to the generalized linear model

and regression (in which generalized linear model and regression (in which generalized inverses occur) include: Bose [178], Rao [1100], [1102], [1103], John [618], Chipman [249], Zelen [1406], Goldman and Zelen [458], Rohde [1149], Rohde and Harvey [1153], Zyskind and Martin [1424], Mitra and Rao [884], Chipman and Rao [252], Harris and Helvig [538], Khatri [669], Lucas [1622], Price [1051].

1-4. Various recent results on linear estimation in the general linear model (with or without constraints and weights): Adenstedt [1472], Adenstedt and Eisenberg [1473], Ahlers and Lewis [9], Bjerhammar [155], [156], Hall and Meyer [521], Hallum, Lewis and Boullion [523], [524], Mendel [833], Nelson, Lewis and Boullion [945], Rao [1109], [1111], Searle [1206], [1760], Anderson, P. O. [1425], Hallum [1438], Pringle and Rayner [1053], [1054], Healy [1576] and Kubácek [1609].

1-5. Conditions for optimality and validity of least-squares theory; specification of errors in the dispersion matrix and/or design matrix: Watson [1361], [1362], Rao [1103], Rao and Mitra [1112], Mitra and Moore [881]. Prediction and the efficiency of least squares: Watson [1707].

1-6. Classical and historical aspects, and early contributions to regression and multivariate analysis: Aitkin (1934), (1945), Anderson, T. W. (1958), Bleick (1940), David and Neyman (1938), Gauss (1809), (1821), (1889), see also [1961], Jeffreys (1932), Kolmogorov (1946), Legendre (1806), (1807), Linnik (1961), Plackett (1949), [1022], [1649], Yates and Hale [1714].

S2. Sequential (recursive) computation of least squares estimators, sequential algorithms, weighted estimation, stepwise regression: For expositions, see Albert [17; Ch. 8], Boullion and Odell [190; Ch. 6], Rohde [1152].

Applications of least squares to problems in estimations, filtering and prediction, missile tracking, etc. have led to modifications that keep up with the data in a sequential fashion. A general model was developed by Kalman [632]. Sequential least squares parameter scheme were obtained by Rainbolt [1093], Albert and Sittler [20], Rohde and Harvey [1153], Rohde [1152], Decell [299]; see also [1474]. Recent results on recursive estimation theory and applications are given by Bodin [1464]. See also Albert [15], Drygas [339], Goldstein [464], Golub [470], Golub and Styan [476], Chambers [1521], Hanson and Dyer [1573], Nayak and Foudriat [1634], Nayak [941], Sinha and Pille [1685] for computational algorithms for sequential (parameter) estimation, regression updating and on-line estimation using matrix generalized inverses.

For a use of the Abbreviated Doolittle method in linear estimation problems, see Rohde and Harvey [1153], who generalized a method of computing certain matrix products originally due to Aitkin.

Recursive calculation of $A^{\dagger}$: Greville [493] gave an interesting formula connecting $A^{\dagger}$ with $(A:a)^{\dagger}$, where A is $m \times n$ and a is $m \times 1$, and used the result in least squares polynomial curve fitting. Mitra and Bhimasankaram [880] (also [1113]) gave formulae for computing several types of 1-inverses of $(A:a)$ from corresponding 1-inverses of A, and reversely. Cline [265] gave formulae to obtain $A^{\dagger}$ from $(A:B)^{\dagger}$ which are mathematically elegant but not easily computable. The formulae in [880] are useful for revising least squares estimation in the following situations: (i) deletion or addition of an observation; (ii) deletion or addition of a parameter in the linear model.

S3. Distribution of quadratic forms, conditional distributions, singular multivariate normal distributions and associated predictive problems, independence of quadratic forms in normal variates, quadratic forms of correlated normal variates.

Early work is by Cochran (1934), Craig (1943), Hogg and Craig (1958), Sakamoto (1944), Graybill and Marsgalia (1957) and others [968], [969]. An early application of generalized inverses to conditional expectations was given by Marsaglia (1964); also Harris and Helvig [538], [539]. Fundamental contributions were made by Rao, and the results are given in [1113; Ch. 9]; see also [17], [19], [1110], [1152], [1209; Ch. 2], [41], [190; Ch. 6], [1054; Ch. 5], Good [1739], [486], Khatri [668], Khatri and Rao [672], Mitra [871], Mitra and Bhimasankaram [879], Rayner and Livingstone [1119], Shanbag [1215], [1761], Styan [1286].

S4. Variance component problems: Matrix reformulation of Henderson's methods, Rohde [1152], Rao and Mitra [1113; 10. 3],

Rohde and Tallis [1154], Searle [1208]; Golub [1562]. Estimation of variance and covariance components for general situation, Rao's MINQUE (Minimum norm quadratic unbiased estimator) theory, Rao [1107], [1109], Rao and Mitra [1113; 10. 3], Mitra [872], Searle [1760], P. W. M. John [618], J. A. John [617]. Statistical inference via the technique of analysis of variance using generalized inverses: Rao [1110; Ch. 4].

S5. Discriminant function in multivariate analysis for singular multivariate normal distribution: Rao [1110], Mitra [875], Rao and Mitra [1113; 10. 5], Rohde [1152]. Maximum likelihood estimation when the information matrix is singular: [1113; 10. 4], Albert [18].

S6. Orthogonal designs; optimal experimental designs for estimating parameters: Chernoff [246], Rohde [1152], Albert [17; 6. 6, 8. 8].

Fractional, singular and Hotelling's weighting designs: Banerjee [83], [84], [85], Banerjee and Federer [86], Hazra and Banerjee [557], Raghavarao [1092], Zacks [1403].

S7. Ridge analysis and generalizations: Bibby [147], Marquardt [801], Jones [1442], Hemmerle [1579], Hoerl [1582], [1583], [1584], Hoerl and Kennard [1585], [1586].

S8. For various other statistical applications of generalized inverses, see Anderson [39], Bhimasankaram [143], Bibby [147], Drygas [341], Duran and Odell [351], Gatley [439], Graybill [485], J. A. John [617], R. W. M. John [618], Khatri [669], Lewis [747], Lewis and Ahlers [748], Lewis and Odell [752], Magness and McGuire [785], Marchant and Jonnes [797], Marquardt [801], [802], Mathai [812], [813], [814], Mitra [868], [870], [871], [872], [877], Mitra and Rao [885], Moritz [896], [897], [898], [899], Oktaba [971], Osborne [979], Rao [1099], [1100], [1101], [1102], [1103], [1104], [1105], [1106], [1107], Rao and Mitra [1112], [1114], [1116], Rohde and Tallis [1154], Searle [1209], Stein [1267], Wani and Kabi [1356], Longley [1617], Quenouille [1659], Glassman [1437], Golub [470], Golub and Styan [476].

T. Applications of Generalized Inverses in Econometrics

1. Specification problems in econometrics and regression analysis

2. Theory of estimations

3. Theory of aggregation

We refer the interested reader to the paper by Chipman in this volume for extensive bibliography. In view of this we have not included in this Bibliography many of the papers cited by Chipman [251], [250]. See also J. S. Chipman, The aggregation problem in econometrics, Advances in Applied Probability, 7(1975), 72-83.

U. Uses of Generalized Inverses of Matrices and Operators in Probability, Stochastic Processes and Random Operator Equations

1. Markov chains. Finite Markov chains: Decell and Odell [304], [965] (see also [190]), Rohde [1152], Hartfiel [543], [1574]. For a comprehensive treatment of the role of the group inverse in the theory of finite Markov chains, see Meyer [849]. Markov renewal processes and chains: Hunter [602], Orey [973].

2. Stochastic matrices, convex sets and semigroup structure of stochastic matrices, generalized inverses and spectral inverses of stochastic matrices: [190], Wall [1351], [1352], Wall and Plemmons [1353], Montague and Plemmons [890], Hartfiel [544], [545], Schwarz [1200].

3. Doubly-stochastic matrices, semigroup structure: Farahat [389], Schwarz [1201]; doubly-stochastic generalized inverse of a doubly-stochastic matrix: Prasada Rao [1118].

4. The fixed point probability vector of a regular or erodic transition matrix: Boullion and Odell [190; 1.4], Decell and Odell [305], [965], sensitivity analysis, Worm [1393]; general background, Gantmacher [435], vol. 2.

5. Generalized inverses of random linear operators: For background on random linear operator equations, see Bharucha-Reid [1502]. Results on measurability of the generalized inverses of random linear operators are given by Nashed and Salehi [930] and Nashed [923]. These results make possible the extension of various approximation and iterative methods

to least squares solutions of random linear operator equations.

6. Stochastic processes, operator-valued measures and Markov processes. Interpolation and subordination of infinite dimensional stationary stochastic processes: Salehi [1177], [1178], [1181], Mandrekar and Salehi [789], Rosenberg [1162], [1164], Salehi and Schmidt [1182], Masani [810].

Hellinger integrals and applications in dection theory: Salehi [1176], [1177], [1178], [1179], [1180], Root [1155], [1156]. Interpolation of homogeneous random fields: [1180].

Multivariate and operator-valued wide-sense Markov processes: Mandrekar [788], Mandrekar and Salehi [790].

Square-integrability of operator-valued functions with respect to non-negative operator-valued measures: Rosenberg [1161], [1164], Robertson and Rosenberg [1142], Mandrekar and Salehi [791].

Singularity and Lebesgue-type decompositions: [792], [1142].

Generalized inverse of the spectral integral. When is a closed densely-defined operator from $\mathscr{H}^p$ to $\mathscr{H}^q$ a spectral integral?: See Rosenberg [1164; II] and the references cited therein.

Multivariate wide-sense stochastic processes and prediction theory: Beutler [138], Dolph and Woodbury [321].

Filtering: Falb [386], Kailath [629], Kalman [632], [634], Curtain [1736].

Multivariate prediction theory: Masani [810] and references cited therein.

V. Applications of Generalized Inverses to Linear Programming and Game Theory (see also [117]).

V1. For basic contributions to the structured problem in linear programming and the intersection projection method, see Pyle [1085], [1089], Cline [263], Cline and Pyle [273].

V2. Charnes, Cooper and Thompson [237] used generalized inverses and the associated solvability criteria in linear programming under certainty. Charnes and Cooper [235] used generalized inverses for structural sensitivity analysis in linear programming. Charnes and Kirby [240] used generalized inverses in modular design and convex programming. See also [234], [236], [237], [1113; 10.2.2].

V3. Interval linear programming (ILP) refers to the problem:

$$\max\langle c, x\rangle \quad \text{such that } \alpha \leq Ax \leq \beta \ .$$

Ben-Israel and Charnes [121] used 1-inverses to obtain explicit solution to the ILP problem in the case when A is of full row rank. Generalized inverses were also used as the basis for iterative method for solving the general ILP problem without this rank restriction by Robers [1758], Robers and Ben-Israel [1140], Zlobec [1413], Zlobec and Ben-Israel [1416], [1417]; see also [123] and Albert [17; p. 33], [125], [117; Sec. 3]. For some recent results, see Sposito [1762]. For applications of generalized inverse to parameter programming, see Salinetti [1183].

V4. For a characterization of equilibrium points of bi-matrix games in terms of the submatrices (of the payoff matrix) and the Moore-Penrose inverse, see [128]. For use of generalized inverse in optimal solutions of 2-persons, 0-sum games, see [1497].

V5. For use of generalized inverses for solvability and/or iterative methods for linear inequalities, see [1428], [1498], [588], [589].

W. Applications of Generalized Inverses in Nonlinear Programming and Nonlinear Least Squares and Optimization Problems

W1. Quadratic programming problems and numerical methods: Boot [1509], Abadie [1471], Nelson, Lewis and Boullion [944], Golub and Saunders [475], Nelson [942], Jennings [1592], Torsti and Aurela [1697], Worm [1392], Bartels, Golub and Saunders [1487], Rosen, Mangasarian and Ritter, eds. [1672]; see also Q1.

W2. Unconstrained and constrained optimization problems: numerical aspects (see also [925] and K5): Fletcher [398], [401], [1541], [1543], Fletcher and Lill [1544], Lill [1616], [1468], Tewarson [1307] and others.

W3. Multistage constrained quadratic optimization problems: Porter [1042], [1043], Minamide and Nakamura [862], [863], Holmes [593], [594], Nashed [920], [925], Laurent [732], Luenberger [780], Wahba and Nashed [1348], Worm [1392], F. J. Hall [519] and others.

W4. Least squares solutions of nonlinear systems of equations and operator equations: See [925] and K4.

W5. Nonlinear least squares. Algorithms for separable nonlinear least squares problems; Gauss-Newton and related algorithms; applications to parameter estimation and orthogonal regression.

5-1. The method of variable projections for separable nonlinear least squares problems and related algorithms and techniques: Golub and Pereyra [472], [473], Guttman et al. [1740], Kaufman [662], Barham [1427], Barham and Drane [1724], Nelson and Lewis [943], Ruhe and Wedin [1169], Lawton and Sylvestre [1748], Lawton et al. [1749], Scolnick [1203], [1204], Pérez and Scolnik [1007], Krogh [1608].

5-2. Other aspects of nonlinear least squares computation: Osborne [978], [979], [1641], Pereyra [1004], Ruhe and Wedin [1169], Stoer [1277], Schwartz [1199], Jupp [1747], Scolnick [1203], Shanno [1677], Meeter [828], Røeggen [1148].

5-3. Gauss-Newton iterative methods; Marquardt's modification and other related algorithms for nonlinear least problems (see also K4 and K5):

Levenberg [1614], Morrison [1632], Marquardt [802], Kowalik and Osborne [1605], Osborne [1641], Meeter [828], Fletcher [1542], Hartley [1575].

5-4. Theoretical and computational aspects of least squares estimation of nonlinear parameters: Kathovnik [1601], where generalized inverses are explicitly used, Bard [1481], [1482], Boggs [1504], Bus et al. [1515], Gelb, ed. [1552], Shanno [1677], [1678], Barham and Drane [1724], Guttman et al. [1740], Røeggen [1148], R. R. Meyer [1628].

W6. Generalized inverses in nonlinear programming. Generalized inverses are implicit also in various optimality criteria. The Moore-Penrose is used implicitly in Rosen [1671] where the relation between the Kuhn-Tucker and the gradient projection conditions for constrained maximization problems is developed, and in Mangasarian [1623], where it is shown that a similar relationship exists between Kuhn-Tucker type conditions and the gradient-projection type conditions for certain saddle point problems. Other implicit or explicit uses of generalized inverses in Kuhn-Tucker conditions and asymptotic Kuhn-Tucker conditions appear in Bazaraa et al. [1491], [1492], Beltrami [104] and Zlobec [1718], [1719]. See also Ryan [1470] for transformation methods in nonlinear programming.

W7. Gradient projection methods: Projectional properties of $AA^{\dagger}$ were used by Pyle [1085], [1086], [1089] in a gradient projection method for solving linear programming problems and computing generalized inverses, and by Rosen [1157], [1158] for nonlinear programming.

X. Generalized Inverses in Control Theory, System Theory, Filtering and Infinite Dimensional Estimation

X1. Optimal control problems; quadratic regular problem (QRP); minimum effort control problem (MECP). Generalized inverses were first applied to a QRP for a linear dynamical system by Kalman, Ho and Narenda [639]. Kalman [633] and Florentin [402] applied the generalized inverse in control problems by using its least squares properties in the MSE analysis. Beutler and Root [141] exhibit the pseudoinverses of a linear operator as unifying apparently disparate aspects of the QRP, providing generalizations beyond its customary form.

For various applications of generalized inverses (or, implicitly, representations thereof) of matrices and operators in MECP and in other aspects of control, system and filtering theory, see Anderson et al. [47], Barnett [88], Ho and Kalman [586], Kalman [632], [634], [635], [636], [638], Kishi [675], Kuo and Kazda [710], Lovass-Nagy and Powers [775], [1619], Matsuo and Akatsuka [1750], Minamide and Nakamura [861], [862], [863], [864], Porter [1042], [1043], Porter and Williams [1044], [1045], Nayak [941], Nayak and Foudriat [1634], Sinha and Pille [1685], Wahba and Nashed [1348]; Balakrishnan [80], [81], Bellman [101], [1494], Bellman, Glicksberg and Gross [102],

Bjerhammar [157], McGlothin [825], Meditch [826], Swerling [1289], Wells [1709], Sakawa [1175], Zadeh and Desoer [1404], IEEE issue [605], Bickart [148], and Schaffer [1188].

X2. Infinite dimensional linear estimation. Gauss-Markov theorem for nonsingular covariance operators and theory of linear unbiased minimum-variance estimators in Hilbert space: Beutler and Root [141], Root [1156].

X3. System identification and modeling: Beutler and Root [141], Root [1155], [1156]; time series modeling: Parzen [986]; other aspects: Sokovnin [1253], IEEE issue [605].

X4. Kalman filtering theory; optimal prediction and filtering in linear space; Wiener filtering computational techniques; sequential estimation in a state-variable model using generalized inverses, and other matters: Kalman [634], [632], Bjerhammar [156], Swerling [1289], Nayak [941], Nayak and Foudriat [1634]; Sinha and Pille [1685], Pratt [1048], Yoshikawa [1400], Moritz [899], Parzen [986]. Various theories of infinite dimensional filtering: Curtain [1736].

X5. Optimal approximation and estimation in reproducing kernel Hilbert spaces: Larkin [729], [730], [1611], Nashed and Wahba [936], [937], Parzen [986] and references cited in these papers.

Y. Applications of Generalized Inverses in Network and Electrical Circuit Theory (see also [117]).

Y1. Bott-Duffin inverse and its applications to electrical network analysis: The first occurrence of generalized inverse in electrical network theory was in 1953 in a fundamental paper by Bott and Duffin [180]. They introduced a kind of constrained generalized inverse which is now known as the Bott-Duffin inverse, and gave applications of its extremal properties to electrical network analysis. For details and related contributions, see [180], Duffin [346], [347], Ben-Israel and Charnes [120], Rao [1102], Rao and Mitra [1114], [1116]; also [125; 2.9, 2.12, 3.5].

Y2. For other applications of generalized inverses to n-port lumped linear electrical circuit theory (analysis, synthesis and duality), impedance matrix, quadratic error minimization, and compensation of multivariable control systems, see Manherz [794], Manherz and Hakimi [795], Manherz, Jordan and Hakimi [796], Sharpe and Spain [1216], Sharpe and Styan [1217], [1218], [1219], Bickart [148], Schaffer [1188], B. M. Anderson [38], Eisemann [363], Mitra [866], Nickolson [958], Olivares [972], Stern [1268], [1269], Zadeh and Desoer [1404], Zemanian [1407], Lovass-Nagy and Powers [1620].

Y3. A number of new operations in matrix and operator algebras were motivated by (and introduced in analogy with) series, parallel and hybrid connections in circuit theory. For example, the combination $A(A+B)^{\dagger}B$, introduced in Anderson and Duffin [44] in the case of matrices, is called the parallel sum of A and B, in analogy with the sum of two resistors r_1, r_2 connected in parallel: $r = (r_1^{-1} + r_2^{-1})^{-1} = r_1(r_1+r_2)^{-1}r_2$. Hybrid addition is also defined in analogy with hybrid connections in network theory. For these various new operations, their connections with generalized inverses, and applications and relations to shorted operators, linear programming, network synthesis, see Anderson [42], [43], Anderson and Duffin [44], Anderson, Duffin and Trapp [45], [46], Anderson and Schreiber [48], Anderson and Trapp [49], Duffin [347], [346], Duffin and Trapp [348], Trapp [1325], Mitra and Trapp [886]. Anderson et al. [47] discussed an application of the concept of parallel sum of a pair of matrices in some statistical problems connected with the estimation of parameter in a dynamic linear (control) system. The concept of parallel sum was extended and its elegance further demonstrated by Rao and Mitra [1113] who showed that most of the properties proved by Anderson and Duffin [44] are indeed true for much wider class of pairs of matrices (than nonnegative definite matrices (n. n. d.)) designated by the authors as "parallel summable". Mitra and Puri [882] explored additional properties of the parallel sum to obtain the condition for consistency together with the complete class of solutions (X) of the parallel sum equation $A \pm X = C$. A special case of this problem (when A, C and X are n. n. d.) was solved earlier in Mitra [873] and Bose and Mitra [179].

Y4. Applications of theory of simultaneous diagonalization of several Hermitian forms to (Foster-type) network synthesis and n-port problems, and other areas: Bose and Mitra [179], Bhimasankaram [143], Rao and Mitra [1113], Mitra [874], Mitra and Rao [884], Newcomb [954]; see also Guillemin [510] for classical aspects and quadratic forms in linear physical theory and network analysis and synthesis.

Z. Applications of Generalized Inverses in Pattern Recognition and Communication Theory

Z1. The thesis by Al-Alaoui [13] is devoted to applications of generalized inverse concepts and techniques to mean-square-error (MSE) problems in pattern recognition in the nonparametric case. The aim is to keep the attractive features of the MSE approach and try to combat its deficiencies. It deals with pattern recognition where the underlying probability densities of the different classes are unknown, and the results are suitable for the case of nonseparable classes. A new weighted MSE procedure for pattern classification is introduced and some results on redundancy and the generalized inverse solution are established. An algorithm for pattern classification and an adaptive constrained MSE procedure for pattern classification are developed. Constrained and weighted generalized inverses are used to reduce the error on the design set and to get better classification with the same number or fewer features than obtained by the MSE solutions. See also [1720], [1721].

Z2. Formulation of pattern recognition problems in the general inverse setting: Wee [1368], [1369], [1370], Smith and Yau [1689], Al-Alaoui [13].

Z3. MSE solution and minimum-squared-error approximation to the Bayes discriminant: Patterson and Womack [1646], Wee [1369], Fukunaga [425]. For surveys of MSE and other developments in pattern recognition, see Yau and Garnett [1398], Duda and Hart [345], Ho and Agrawala [587], Nagy [912]. For applications of generalized inverses to multiclass pattern classification, feature extraction and clustering, see [1368], [1371], [1398], [587], [13].

Z4. Algorithms. The Ho-Kashyap algorithm [588], [589]

provides a use of generalized inverse in an iterative algorithm for linear inequalities. The algorithm yields a separable solution in the two-class classification problem, if the features are linearly separable. Al-Alaoui's algorithm [13], [1721] applies to the separable and nonseparable cases. Descent algorithms for obtaining a two-class discriminant function are discussed in [345]; see also [914], [924], [1377] for steepest descent. Kishi's algorithm [675] is an adaptation of Greville's recursive formula for computing the Moore-Penrose inverse (cf. [125]). For related and other algorithms see also Wee and Fu [1371], Smith [1688] and [13], Kashyap [653], Sebestyen and Edie [1210].

Z5. Other references on related mathematical topics in pattern recognition include: Andrews [51], Chien and Fu [247], Duda and Hart [345], Fu [419], Fukunaga [425], Watanabe [1360], Whitney and Meany [1377], Young and Calvert [1402], Smith [1688].

Σ. Miscellaneous Applications of Generalized Inverses

1. Geodesic calculations, geodesy and photogrammetry, geometry and surveying: Bjerhammar [150], [151], [158], Krarup [696], Moritz [898]. Earth movement analysis: Pope and Stearn: [1040].

2. Hydrology: Bonnier and Korganoff [172], Korganoff [691].

3. Diophantine linear equations: Charnes and Granot [238]; see also [126; Sec. 4]. Mixed-integer linear equations: Bowman and Burdet [196]. Integer solutions to linear equations: Hurt and Waid [603]. Computation of generalized inverse matrices using residue arithmetic: Stallings and Boullion [1261].

4. Population dynamics; backward population projection: Greville and Keyfitz [501].

5. Continuous geometry: von Neumann [953].

6. Dynamics of elastic bodies: Milne [857].

7. Orbit determination: Solloway [1254]. Satellite altitude determination from celestial sightings: Grusas [508]. Aerospace vechile control and sequential estimation: Kishi [675]. Applications of Kalman filtering techniques to astronautical guidance: Battin [1489], Battin and Levine [1490].

8. Some applications in social and biometrical sciences and economics models: Horst [595], Charnes et al. [1523], Chipman [251], Searle [1209], [1760]; also T .

9. Data fitting: Greville [490], [492], Buchanan and Thomas [1513]; see also smoothing, and various related references.

10. Estimation in dynamical systems: Prochaska [1055]; see also X4.

11. Resolution of a remote sensing instrument and of linear equations, picture reconstruction by least squares, inference from inadequate data, reconstruction of functions from discrete mean values, etc.: For some perspectives, see for example, the following papers (a few of which involve generalized inverses): Abu-Shumays and Marinelli [5], Backus [75], Helstrom [1578], Korganoff [690], Krishnamurthy and Prabhu [697], Moritz [897], Nashed and Wahba [937], [935], Shaw [1220], [1221], Tewarson [1310], Tewarson and Narain [1312], [1313], Twomey [1700]; see also Bjerhammar [156] and the annotations on ill-posed problems (L).

12. Numerical inversion of the Laplace transform: Bellman, Kalaba and Lockett [103], Nashed and Wahba [940], also Schoenberg (1973), J. Math. Anal. Appl. 43, 823-828.

13. Ridge analysis and generalizations: See S7. Modular design problem and convex programming; chance constrained programming and other special topics in mathematical programming: See V. Graphs and generalized inverses; incidence matrices: See E16-11.

BIBLIOGRAPHY ON GENERALIZED INVERSES AND APPLICATIONS

[1] P Abdelmalek, N. N. (1971): Round-off error analysis for Gram-Schmidt method and solution of linear least squares problems, BIT, 11, 345-368.

[2] P Abdelmalek, N. N. (1974): On the solution of the linear least squares problems and pseudo-inverses, Computing, 13, 215-228.

[3] S Abdelmalek, N. N. (1974): On the discrete linear L_1 approximation and L_1 solutions of overdetermined linear equations, J. Approximation Theory, 11, 38-53.

[4] P Abdennur, Samir (1969): Generalized Inverses and Iterative Methods for Least Squares Solutions of Linear Systems, M. S. Thesis, Dept. of Math., American University of Beirut, Lebanon.

[5] T Abu-Shumays, I. K. and L. D. Marinelli (1971): A smoothing solution (unfolding) of a two dimensional density function from its measured spectrum, J. Computational Physics, 7, 219-238.

[6] S Afriat, S. N. (1956): On latent vectors and characteristic values of products of pairs of symmetric idempotents, Quart. J. Math., Oxford (2), 7, 76-78.

[7] S Afriat, S. N. (1957): Orthogonal and oblique projectors and the characteristics of pairs of vector spaces, Proc. Cambridge Philos. Soc., 53, 800-816.

[8] P Afriat, S. N. (1974): On the general matrix inverse, to appear.

Agrawala, A. K.: See Ho, Y.-C. (1968).

[9] S Ahlers, C. W. and T. O. Lewis (1972): Linear estimation with a positive semidefinite covariance matrix, Indust. Math., 21, 23-27.

Ahlers, C. W.: See Lewis, T. O. (1971).

[10] Aitkin, A. C. (1934): On least squares and linear com-
S binations of observations, Proc. Roy. Soc. Edinburgh, Ser. A, 55, 42-48.

[11] Aitkin, A. C. (1945): Studies in practical mathematics.
S IV: On linear approximations by least squares, Proc. Roy. Soc. Edinburgh, Ser. A, 62, 138-146.

[12] Aitkin, A. C. (1956): Determinants and Matrices, 9th
T edn., Oliver and Boyd, Edinburgh.

Akilov, G. P.: See Kantorovich, L. V. (1964).

[13] Al-Alaoui, M. A. (1974): Some Applications of General-
P ized Inverse to Pattern Recognition, Ph.D. Thesis, Dept. of Electrical Engineering, Georgia Institute of Technology, Atlanta, Ga.

[14] Albert, A. (1964): An introduction and beginner's guide
P to matrix pseudo inverses, ARCON-Advanced Research Consultants, Lexington, Mass.

[15] Albert, A. (1965): Real time computation of constrained
S least squares estimators. ARCON-Advanced Research Consultants, Lexington, Mass.

[16] Albert, A. (1969): Conditions for positive and nonnega-
P tive definiteness in terms of pseudoinverses, SIAM J. Appl. Math., 17, 434-440.

[17] Albert, A. (1972): Regression and the Moore-Penrose
P Pseudoinverse, Academic Press, New York, 1972, xiii + 180 pp.

[18] Albert, A. (1973): The Gauss-Markov theorem for regres-
P sion models with possibly singular covariances, SIAM J. Appl. Math., 24, 182-187.

[19] Albert, A. (1975): Statistical applications of pseudo-
P inverses, this Volume.

[20] Albert, A. and R. W. Sittler (1965): A method for comput-
S ing least squares estimators that keep up with the data,
SIAM J. Control, 3, 394-417.

[21] Alford, M. W. (1967): Some aspects of pseudoinverses
P and their applications, Short Course on Applications of
Modern Mathematics, 10 March 1967, University of
California at Los Angeles, 37 pp.

[22] Allgower, E. L. and P. M. Prenter (1974): On the branch-
T ing of solutions of quadratic differential equations,
Aequationes Math., 10, 81-96.

[23] Altman, M. (1953): On linear functional equations in
S locally convex linear topological spaces, Studia Math.,
13, 194-207.

[24] Altman, M. (1955): A generalization of Newton's method,
T Bull. Acad. Polon. Sci. Ser. Sci. Math. Astronom. Phys.,
cl. 3, 3, 189-193.

[25] Altman, M. (1957): On a generalization of Newton's
T method, Bull. Acad. Polon. Sci. Ser. Sci. Math.
Astronom. Phys., cl. 3, 5, 789-795.

[26] Altman, M. (1957): On the approximate solution of lin-
S ear algebraic equations, Bull. Acad. Polon. Sci. Ser.
Sci. Math. Astronom. Phys., cl. 3, 5, 365-370.

[27] Altman, M. (1957): An approximate method for solving
T linear equations in Hilbert space, Bull. Acad. Polon.
Sci. Ser. Sci. Math. Astronom. Phys., cl. 3, 5, 601-604.

[28] Altman, M. (1959): Approximation Methods in Functional
T Analysis, Lecture Notes, Math. 107c, California
Institute of Technology, Pasadena, California.

[29] Amburgey, J. K. (1968): A Theory for Rectangular
P Matrices, Ph.D. Dissertation, Texas Tech University, Lubbock, Texas.

[30] Amburgey, J. K., T. O. Lewis and T. L. Boullion (1968):
P On computing generalized characteristic vectors and values for a rectangular matrix, pp. 267-276 in Boullion and Odell (1968).

[31] Amburgey, J. K., T. O. Lewis and T. L. Boullion (1970):
P On squaring matrices and generalized inverses, Indust. Math., 20, pp. 33-59.

[32] Amburgey, J. K., T. O. Lewis, and T. L. Boullion (1971):
P A note on rectangular matrices, Texas J. Science, 23, 503-509.

Amburgey, J. K.: See Milnes, H. W. (1968), (1969).

[33] Amir-Moéz, A. R. (1971): Extreme Properties of Linear
S Transformations and Geometry in Unitary Spaces, Mathematics Series, Nos. 2 and 3, revised edition, Department of Mathematics, Texas Tech University, Lubbock, Texas, ix + 326 pp.

[34] Amir-Moéz, A. R. and T. G. Newman (1970): Geometry
P of generalized inverses, Math. Mag., 43, 33-36.

[35] Andersen, E. (1947): Solution of great systems of normal
S equations together with an investigation of Andrae's dot-figure. An arithmetical-technical investigation. Geod. Inst. Skr., 3rd ser., 11. Copenhagen.

[36] Andersen, E. (1950): Solution of great systems of normal
S equations. Bull. Géodésique, 15, 19-29.

[37] Andersen, E. (1955): Adjustment of observations by the
S method of least squares. Copenhagen.

[38] Anderson, B. M. (1973): Polynomial Matrices – Applica-
S tions to Synthesis and Analysis of Linear Multivariable Systems, Ph.D. Dissertation, Electrical Engineering, Northwestern University, Evanston, Illinois.

[39] Anderson, C. L. (1967): A Geometric Theory of Pseudo-
P inverses and Some Applications in Statistics, Master's Thesis, Southern Methodist University, Dallas, Texas.

[40] Anderson, Ned (1975): A generalization of the method of
S averages for overdetermined linear systems, Math. Comp., 29, 607-614.

[41] Anderson, T. W. (1958): An Introduction to Multivariate
T Statistical Analysis, Wiley, New York.

[42] Anderson, W. N., Jr. (1968): Series and Parallel Addition
P of Operators, Doctoral Dissertation, Carnegie-Mellon University, Pittsburgh, Pennsylvania.

[43] Anderson, W. N., Jr. (1971): Shorted operators, SIAM
P J. Appl. Math., 20, 520-525.

[44] Anderson, W. N., Jr. and R. J. Duffin (1969): Series
P and parallel addition of matrices, J. Math. Anal. Appl., 26, 576-594.

[45] Anderson, W. N. Jr., R. J. Duffin and G. E. Trapp (1972):
P Parallel subtraction of matrices, Proc. Nat. Acad. Sci. USA, 69, 2530-2531.

[46] Anderson, W. N., Jr., R. J. Duffin and G. E. Trapp (1975):
P Matrix operation induced by network connections, SIAM J. Control, 13, 446-461.

[47] Anderson, W. N. Jr., G. D. Kleindorfer, P. R. Kleindorfer
S and M. B. Woodroofe (1969): Consistent estimates of the parameters of a linear systems, Ann. Math. Stat., 40, 2064-2075.

[48] Anderson, W. N., Jr. and M. Schreiber (1972): On the
P infimum of two projections, Acta Sci. Math. Szeged, 33, 165-168.

[49] Anderson, W. N., Jr. and G. E. Trapp (1975): Shorted
P operators. II, SIAM J. Appl. Math., 28, 60-71.

[50] Anderssen, R. S. and P. Bloomfield (1973): Numerical
T differentiation procedures for non-exact data, Numer. Math., 22, 157-182.

[51] Andrews, H. C. (1972): Introduction to Mathematical
T Techniques in Pattern Recognition, Wiley, New York.

[52] Anselone, P. M. (1971): Collectively Compact Operator
S Approximation Theory, Prentice-Hall, Englewood Cliffs, New Jersey.

[53] Anselone, P. M. and J. Davis (1973): Perturbation of
T best approximation problems, Numer. Math., 21, 63-69.

[54] Anton, H. and C. S. Duris (1974): On an algorithm for
P best approximate solution to $Av = b$ in normed linear spaces, J. Approximation Theory, 8, 133-141.

[55] Aoki, M. (1967): Optimization of Stochastic Systems,
S Academic Press, New York.

[56] Arens, R. (1961): Operational calculus of linear relations,
S Pacific J. Math., 11, 9-23.

Aparcin, A. S.: See Bakušinski, A. B. (1975).

Arbib, M. A.: See Kalman, R. E. (1969), Padulo, L. (1974).

[57] Arghiriade, E. (1963): Sur les matrices qui sont permut-
P ables avec leur inverse géneralisée, Atti Accad. Naz. Lincei Rend. Cl. Sci. Fis. Mat. Natur., Ser. VIII, 35, 244-251.

[58] Arghiriade, E. (1967): Remarques sur l'inverse géneral-
P isée d'un produit de matrices, Atti Accad. Naz. Lincei Rend. Cl. Sci. Fis. Mat. Natur., Ser. VIII, 42, 621-625.

[59] Arghiriade, E. (1968): Sur l'inverse généralisee d'un
P opérateur linéaire dans les espaces de Hilbert, Atti Accad. Naz. Lincei Rend. Cl. Sci. Fis. Mat. Natur., Ser. VIII, 45, 471-477.

[60] Arghiriade, E. and E. Boros (1969): L'inverse généralisée
P d'un opérateur linéaire dans un espace à produit intérieur, Atti Acad. Naz. Lincei Rend. Cl. Sci. Fis. Mat. Natur. Ser. VIII, 46(1969), 646-649.

[61] Arghiriade, E. and A. Dragomir (1963): Une nouvelle
P definition de l'inverse géneralisée d'une matrice, Atti Accad. Naz. Lincei Rend. Cl. Sci. Fis. Mat. Natur., Ser. VIII, 35, 158-163.

[62] Arghiriade, E. and A. Dragomir (1969): Remarques sur
P quelques theoremes relatives a l'inverse generalisee d'un opérateur linéaire dans les espaces de Hilbert, Atti Accad. Naz. Lincei Rend. Cl. Sci. Fis. Mat. Natur., Ser. VIII, 46, 333-338.

[63] Arlt, D. (1966): Zusammenziebarkeit der allgemeinen
S linearen Gruppe des Raumes c_0 der Nullfolgen, Invent. Math., 1, 36-44.

[64] Arnowitz, S. and B. E. Eichinger (1975): Petrie matrices
P and generalized inverses, J. Math. Phys., 16, 1278-1283.

[65] Ashkenazi, V. (1965): Strength of a triangulation layout
T (Ill-conditioning analysis), Bull. Géodésique, 76, 125-134.

Atchison, T. A.: See Pye, W. C. (1973), (1973), (1973).

[66] Atkinson, F. V. (1951): The normal solubility of linear
P equations in normed spaces (Russian), Mat. Sbornik N. S., 28(70), 3-14 (Math. Rev., 13 (1952), p. 46).

[67] P Atkinson, F. V. (1953): On relatively regular operators, Acta Sci. Math. Szeged, 15, 38-56.

[68] S Atkinson, K. E. (1967): The solution of non-unique linear integral equations, Numer. Math., 10, 117-124.

[69] S Atkinson, K. E. (1975): The numerical solution of some bifurication problems, SIAM J. Numer. Anal., to appear.

[70] T Atta, S. E. (1957): Effect of propagated error on inverse of Hilbert matrix, J. Assoc. Comput. Mach., 4, 36-40.

[71] S Audley, D. R. and D. A. Lee (1974): Ill-posed and well-posed problems in system identification. System identification and time-series analysis, IEEE Trans. Automatic Control, AC-19, 738-747.

[72] T Autonne, L. (1902): Sur les groupes linéaires, réels, et orthogonaux, Bull. Soc. Math. France, 30, 121-133.

[73] T Autonne, L. (1917): Sur les matrices hypohermittennes et sur les matrices unitaires, Ann. Univ. Lyon Sect. A, 38, 1-77.

[74] S Azumaya, G. (1954): Strongly π-regular rings, J. Fac. Sci. Hokkaido Univ., Ser. 1, 13, 34-39.

[75] T Backus, G. E. (1970): Inference from inadequate and inaccurate data I, II, and III, Proc. Nat. Acad. Sci. USA, 65, 1-7, 281-289.

[76] T Baer, R. (1952): Linear Algebra and Projective Geometry, Academic Press, New York.

[77] S Bakušinskii, A. B. (1968): Regularization algorithms for linear equations with unbounded operators, Dokl. Akad. Nauk SSSR, 183, pp. 12-14; Soviet Math. Dokl., 9, pp. 1298-1300.

[78] T Bakušinskii, A. B. and A. S. Aparcin (1975): Methods of stochastic approximation type for the solution of linear ill-posed problems, Sibirsk Mat. Ž., 16, 12-18, 195.

[79] Bakušinskii, A. B. and V. N. Strahov (1968): The solu-
S tion of certain integral equations of the first kind by the method of successive approximations, Ž. Vyčisl. Mat. i Mat. Fiz., 8, 181-185.

[80] Balakrishnan, A. V. (1963): An operator theoretic formu-
S lation of a class of control problems and a steepest descent method of solution, J. Soc. Indust. Appl. Math. Ser. A: Control, 1, 109-127.

[81] Balakrishnan, A. V. (1970): Introduction to Optimization
S Theory in a Hilbert Space, Lecture Notes in Operations Research and Mathematical Systems, vol. 42, Springer-Verlag, New York.

[82] Bancroft, S., J. K. Hale and D. Sweet (1968): Alterna-
S tive problems for nonlinear functional equations, J. Differential Equations, 4, 40-56.

[83] Banerjee, K. S. (1966): On nonrandomized fractional
S weighing designs, Ann. Math. Statist., 37, 1836-1841.

[84] Banerjee, K. S. (1966): Singularity in Hotelling's
P weighing designs and a generalized inverse, Ann. Math. Statist., 37, 1021-1032. Correction, ibid., 40(1969), 719.

[85] Banerjee, K. S. (1972): Singular weighing designs and
P a reflexive generalized inverse, J. Amer. Statist. Assoc., 67, 211-212.

[86] Banerjee, K. S. and W. T. Federer (1968): On the
S structure and analysis of singular fractional replicates, Ann. Math. Statist., 39, 657-663.

Banerjee, K. S.: See Hazra, P. K. (1973).

[87] Baras, J. S., R. W. Brockett and P. A. Fuhrmann(1974):
T State-space models for infinite dimensional systems. System identification and time-series analysis, IEEE Trans. Automatic Control, AC-19, 693-700.

Barker, G. P.: See Schneider H. (1968).

[88] Barnett, S. (1971): Matrices in Control Theory, van
S Nostrand Reinhold, London, xii + 221 pp.

[89] Barrodale, I. and F. D. K. Roberts (1973): An improved
T algorithm for discrete ℓ_1 linear approximation, SIAM J. Numer. Anal., 10, 839-848.

[90] Bart, H., M. A. Kaashoek and D. C. Lay (1974): Stab-
T ility properties of finite memorphic operator functions. I. Nederl. Akad. Wet., Proc., Ser. A, 77, 217-231; II, ibid, 231-243; III, ibid, 244-259.

[91] Bartle, R. G. (1953): Singular points of functional
S equations, Trans. Amer. Math. Soc., 75, 366-384.

[92] Bartlett, M. S. (1951): An inverse matrix adjustment
S arising in discriminant analysis, Ann. Math. Statist., 22, 107-111.

[93] Barton, C. P. (1966): Pseudoinverses of Rectangular
P Matrices, Master's Thesis, Southern Methodist University, Dallas, Texas.

[94] Barwick, D. T. and J. D. Gilbert (1974): On general-
P ization of the reverse order law, SIAM J. Appl. Math., 27, 326-330.

[95] Baskett, T. S. and I. J. Katz (1969): Theorems on
P products of EPr matrices, Linear Algebra and Appl., 2, 87-103.

[96] Bauer, F. L. (1963): Optimally scaled matrices,
T Numer. Math., 45, 73-87.

[97] Bauer, F. L. (1965): Elimination with weighted row
S combinations for solving linear equations and least squares problems, Numer. Math., 7, 338-352. Republished, pp. 119-133 in Wilkinson and Reinsh (1971).

[98] T Bauer, F. L. (1967): Theory of Norms, Tech. Report No. CS 75, Computer Science Dept., Stanford University, Stanford, California.

[99] S Bauer, F. L. and Householder, A. S. (1960): Some inequalities involving the Euclidean condition of a matrix, Numer. Math., 2, 308-311.

[100] T Beckenbach, E. F. and R. Bellman (1971): Inequalities (3rd revised printing), Springer, New York.

[101] S Bellman, R. (1960): Introduction to Matrix Analysis, 2nd edition, McGraw-Hill, New York, 1970, xxiii + 403pp.

[102] S Bellman, R., I. Glicksberg and O. Gross (1954): On some variational problems occurring in the theory of dynamic programming, Rend. Circ. Mat. Palermo (ser. II), 3, 363-397.

[103] S Bellman, R., R. Kalaba and L. Lockett (1966): Numerical Inversion of the Laplace Transform, American Elsevier, New York.

Bellman, R.: See Beckenbach, E. F. (1971).

[104] T Beltrami, E. J. (1969): A constructive proof of the Kuhn-Tucker multiplier rule, J. Math. Anal. Appl., 26, 297-306.

[105] S Beltrami, E. J. (1970): An Algorithmic Approach to Nonlinear Analysis and Optimization, Academic Press, New York, xiv + 235pp.

[106] S Ben-Israel, A. (1964): On direct sum decompositions of Hestenes algebras, Israel J. Math., 2, 50-54.

[107] P Ben-Israel, A. (1965): A modified Newton-Raphson method for the solution of systems of equations, Israel J. Math., 3, 94-98.

[108] Ben-Israel, A. (1965): An iterative method for computing
P the generalized inverse of an arbitrary matrix, Math. Comp., 19, 452-455.

[109] Ben-Israel, A. (1966): A Newton-Raphson method for
P the solution of systems of equations, J. Math. Anal. Appl., 15, 243-252.

[110] Ben-Israel, A. (1966): A note on an iterative method for
P generalized inversion of matrices, Math. Comp., 20, 439-440.

[111] Ben-Israel, A. (1966): On error bounds for generalized
P inverses, SIAM J. Numer. Anal., 3, 585-592.

[112] Ben-Israel, A. (1967): On the geometry of subspaces
S in Euclidean N-spaces, SIAM J. Appl. Math., 15, 1184-1198.

[113] Ben-Israel, A. (1967): On iterative methods for solving
P nonlinear least squares problems over convex sets, Israel J. Math., 5, 211-224.

[114] Ben-Israel, A. (1968): Application of generalized in-
P verses in nonlinear analysis, pp. 183-202 in Boullion and Odell (1968).

[115] Ben-Israel, A. (1968): On decompositions of matrix
S spaces with applications to matrix equations, Atti Accad. Naz. Lincei Rend. Cl. Sci. Fis. Mat. Natur. Ser. VIII, 45, 54-60.

[116] Ben-Israel, A. (1968): On matrices of index zero or one,
P SIAM J. Appl. Math., 17, 1118-1121.

[117] Ben-Israel, A. (1975): Applications of generalized in-
P verses to programming, games and networks, this Volume.

[118] Ben-Israel, A. and A. Charnes (1961): Projection prop-
P erties and the Neumann-Euler expansions for the Moore-Penrose inverse of an arbitrary matrix, ONR Research Memo, 40, The Technological Institute, Northwestern University, Evanston, Ill.

[119] Ben-Israel, A. and A. Charnes (1963): Contributions
P to the theory of generalized inverses, SIAM J. Appl. Math., 11, 667-699.

[120] Ben-Israel, A. and A. Charnes (1963): Generalized in-
P verses and the Bott-Duffin network analysis, J. Math. Anal. Appl., 7, 428-435. Erratum, ibid. 18, 393.

[121] Ben-Israel, A. and A. Charnes (1968): An explicit solu-
P tion of a special class of linear programming problems, Operations Research, 16, 1167-1175.

[122] Ben-Israel, A. and A. Charnes (1968): On the intersec-
S tion of cones and subspaces, Bull. Amer. Math. Soc., 74, 541-544.

[123] Ben-Israel, A., A. Charnes and P. D. Robers (1968):
P On generalized inverses and interval linear programming, pp. 53-70 in Boullion and Odell (1968).

[124] Ben-Israel, A. and D. Cohen (1966): On iterative com-
P putation of generalized inverses and associated projections, SIAM J. Numer. Anal., 3, 410-419.

[125] Ben-Israel, Adi and T. N. E. Greville (1974): Generalized
P Inverses: Theory and Applications, Wiley-Interscience, New York, xi + 395 pp.

[126] Ben-Israel, A. and T. N. E. Greville (1975): Some topics
P in generalized inverses of matrices, this Volume.

[127] Ben-Israel, A. and Y. Ijiri (1963): A report on the
P machine calculation of the generalized inverse of an arbitrary matrix, ONR Research Memo., 110, Carnegie Institute of Technology, Pittsburgh, Pa.

[128] Ben-Israel, A. and M. J. Kirby (1969): A characteriza-
S tion of equilibrium points of bimatrix games, Atti Accad. Naz. Lincei Rend Cl. Sci. Fis. Mat. Natur., Ser. VIII, 46, 196-201.

[129] Ben-Israel, A. and S. J. Wersan (1963): An elimination
P method for computing the generalized inverse of an arbitrary complex matrix, J. Assoc. Comput. Mach., 10, 532-537.

Ben-Israel, A.: See Erdélyi, I. (1972), Garnett III, J. M. (1971), Hawkins, J. B. (1973), Robers, P. D. (1969), (1970), Showalter, D. W. (1970), Zlobec, S. (1970), (1973).

[130] Bennet, J. M. (1966): The pseudo-inverse and associ-
P ated topics, Technical Rept. No. 43, Basser Computing Dept., School of Physics, University of Sydney, 34 pp.

[131] Benoit, E. (1924): Note sur une methode de résolution
S des équations normales provenent de l'application de la méthod des moindres carrés a un système d'équations linéares en nombre inférieur à celui des inconnues. Bull. Géodésique, 1-4, 67-77.

[132] Berberian, S. K. (1972): The regular ring of a finite
S Baer *-ring, J. Algebra, 23, 35-65.

[133] Berge, Claude (1962): Théorie des graphes et ses ap-
T plications, Dunod, Paris, 1958. Translation: The Theory of Graphs and its Applications, Wiley, New York, 1962.

[134] Berman, A. (1974): Nonnegative matrices which are
P equal to their generalized inverse, Linear Algebra and Appl., 9, 261-265.

[135] Berman, A. and R. J. Plemmons (1972): Monotonicity
P and the generalized inverse, SIAM J. Appl. Math., 22, 155-161.

[136] Berman, A. and R. J. Plemmons (1974): Inverses of
S nonnegative matrices, Linear and Multilinear Algebra, 2, 161-172.

[137] Berman, A. and R. J. Plemmons (1974): Cones and iterative methods for best least squares solutions of linear systems, SIAM J. Numer. Anal., 11, 145-154.
P

[138] Beutler, F. J. (1964): Multivariate wide-sense processes and prediction theory, Ann. Math. Stat., 34, 424-438.
S

[139] Beutler, F. J. (1965): The operator theory of the pseudo-inverse, I. Bounded operators, J. Math. Anal. Appl., 10, 451-470.
P

[140] Beutler, F. J. (1965): The operator theory of the pseudo-inverse, II. Unbounded operators with arbitrary range, J. Math. Anal. Appl., 10, 471-493.
P

[141] Beutler, F. J. and W. L. Root (1973): The Operator Pseudoinverse in Control and Systems Identification, Computer, Information and Control Engineering Program, University of Michigan, Sept. 1973; also this Volume.
P

[142] Bhimasankaram, P. (1971): On generalized inverses of partitioned matrices, Sankhyā, Ser. A., 33, 311-314.
P

[143] Bhimasankaram, P. (1971): Some Contributions to the Theory, Application and Computation of Generalized Inverses of Matrices, Doctoral Dissertation, Indian Statistical Institute, Calcutta.
P

[144] Bhimasankaram, P. and S. K. Mitra (1969): On a theorem of Rao on g-inverses of matrices, Sankhyā, Ser. A., 31, 365-368.
P

Bhimasankaram, P.: See Mitra, S. K. (1970), (1971), (1971),

Bhimansharam, P.: See Rao, C. R. (1973).

[145] Bialy, H. (1959): Iterative Behandlung linearen Funktionalgleichungen, Arch. Rational Mech. Anal., 4, 166-176.
S

[146] Bibby, J. (1972): Review of Boullion and Odell, J. Roy. Statist. Soc., Ser. A., 135, 608-9.
P

[147] Bibby, J. (1974): Minimum mean square error estimation,
S ridge regression, and some unanswered questions, Progress in Statistics (European Meeting of Statisticians, Budapest, 1972), pp. 107-121, Colloq. Math. Soc. János Bolyani, vol. 9, North-Holland, Amsterdam.

[148] Bickart, T. A. (1968): On the compensation of multivari-
S ate control systems, Proc. IEEE, 58, 1258-1259.

[149] Birge, R. T. (1932): The calculation of errors by the
S method of least squares, Phys. Rev., 40, 207-227.

[150] Bjerhammar, A. (1951): Rectangular reciprocal matrices
P with special reference to geodetic calculations, Bull. Géodésique, 52, 188-220.

[151] Bjerhammar, A. (1951): Application of calculus of
P matrices to method of least squares; with special reference to geodetic calculations, Kungl. Tekn. Hogsk. Handl. (Trans. Royal Inst. of Technology) Stockholm, No. 49, 36 pp.

[152] Bjerhammar, A. (1957): A generalized matrix algebra.
P N. R. C. Can. Div. Appl. Phys., Ottawa.

[153] Bjerhammar, A. (1958): A generalized matrix algebra,
P Kungl. Tekn. Hogsk. Handl. (Trans. Royal Inst. of Technology) Stockholm, No. 124, 32 pp.

[154] Bjerhammar, A. (1967): Studies with generalized matrix
P algebra, Bull. Géodésique, 85, 193-210.

[155] Bjerhammar, A. (1971): Estimation with singular in-
P verses, Tellus, 23, 6.

[156] Bjerhammar, A. (1973): Theory of Errors and Generalized
P Inverse Matrices, Elsevier Scientific Publishing Company, Amsterdam, xii + 420 pp.

[157] Bjerhammar, A. (1975): A general model for optimal prediction and filtering in the linear space. Mathematical Geodesy, Part III (Internat. Summer School Math. Methods Phys. Geodesy, Ramasau, 1973), pp. 29-46. Methoden und Verfahren Math. Phys., Band 14, Bibliographisches Inst. Mannheim.
S

[158] Bjerhammar, A. (1975): Generalized matrix inverses. Mathematical Geodesy, Part III (Internat. Summer School Math. Methods Phys. Geodesy, Ramsau, 1973), pp. 47-81. Methoden und Verfahren Math. Phys., Band 14, Bibliographisches Inst. Mannheim.
P

[159] Björck, Å. (1967): Solving linear least squares problems by Gram-Schmidt orthogonalization, BIT, 7, 1-21.

[160] Björck, Å. (1967): Iterative refinement of linear least squares solutions I, BIT, 7, 257-278.
P

[161] Björck, Å. (1968): Iterative refinement of linear least squares solutions II, BIT, 8, 8-30.
P

[162] Björck, Å. and C. Bowie (1971): An iterative algorithm for computing the best estimate of an orthogonal matrix, SIAM J. Numer. Anal., 8, 358-364.
S

[163] Björck, Å. and G. H. Golub (1967): Iterative refinement of linear least squares solutions by Householder transformation, BIT, 7, 322-337.
P

[164] Björck, Å. and G. H. Golub (1973): Numerical methods for computing angles between linear subspaces, Math. Comp., 27, 579-594.
S

[165] Blattner, J. W. (1962): Bordered matrices, SIAM J. Appl. Math., 10, 528-536.
S

[166] Bleick, W. E. (1940): A least square accumulation
T theorem, Ann. Math. Statist., 11, 225-226.

Bloomfield, P.: See Anderssen, R. S. (1973).

[167] Blum, E. K. (1972): Numerical Analysis and Computa-
S tion, Addison-Wesley, Reading, Mass., 604 pp.

[168] Blum, E. K. and G. H. Rodrigue (1974): Solution of
P eigenvalue problems and least squares problems in Hilbert spaces by a gradient method, J. Comput. System Sci., 8, 220-237.

[169] Bodewig, E. (1956): Matrix Calculus, North Holland,
S Amsterdam.

[170] Bohnenblust, F. (1942): A characterization of complex
T Hilbert spaces, Portugal. Math., 3, 103-109.

[171] Bonnesen, T. and W. Fenchel (1934): Theorie der kon-
T vexen Körper, Springer, Berlin, viii + 164 pp.

[172] Bonnier, A. and A. Korganoff (1972): The determination
S of dispersion coefficients in non-homogeneous media in problems of salt water contamination of fresh ground water, Journal of Hydrology, 16, 39-47.

[173] de Boor, C. (1966): The Method of Projections as Ap-
S plied to the Numerical solution of Two Point Boundary Value Problems Using Cubic Splines, Doctoral Dissertation in Mathematics, University of Michigan, Ann Arbor.

[174] Boot, J. C. G. (1963): The computation of the general-
P ized inverse of singular or rectangular matrices, Amer. Math Monthly, 70, 302-303.

[175] Boot, J. C. G. (1965): Projection Matrices and the Gen-
P eralized Inverse, State University of New York, Buffalo, New York.

[176] P Boros, E. (1964): On the generalized inverse of an EPr matrix, An. Univ. Timisoara Ser. Sti. Mat. -Fiz., 2, 33-38.

[177] P Boros, E. and I. Sturz (1963): On quasi-inverse matrices, An. Univ. Timisoara Ser. Sti. Mat. -Fiz., 1, 59-66.

Boros, E.: See Arghiriade, E. (1969).

[178] S Bose, R. C. (1959): Analysis of Variance, Unpublished Lecture Notes, University of North Carolina, Chapel Hill, North Carolina.

[179] P Bose, N. K. and S. K. Mitra (1973): Applications of theory of simultaneous diagonalization of several Hermitian forms, Internat. J. Electronics, 35, 721-735.

[180] P Bott, R. and R. J. Duffin (1953): On the algebra of networks, Trans. Amer. Math. Soc., 74, 99-109.

[181] S Bouldin, R. (1970): The numerical range of a product, J. Math. Anal. Appl., 32, 459-467.

[182] S Bouldin, R. (1972): Numerical range for certain classes of operators, Proc. Amer. Math. Soc., 34, 203-206.

[183] P Bouldin, R. (1973): The product of operators with closed range, Tôhoku Math. J., 25, 359-363.

[184] P Bouldin, R. (1973): The pseudoinverse of a product, SIAM J. Appl. Math., 24, 489-495.

[185] P Bouldin, R. (1975): The product of operators with closed range, II, to appear.

[186] P Boullion, T. L. (1966): Contributions to the Theory of Pseudoinverses, Doctoral Dissertation, University of Texas, Austin, Texas.

[187] Boullion, T. L. and P. L. Odell, Editors (1968): Pro-
P ceedings of the Symposium on Theory and Applications of Generalized Inverses of Matrices, held at the Department of Mathematics, Texas Technological College, Lubbock, Texas, March, 1968, Texas Tech. Press, Lubbock, Texas, iii + 315 pp.

[188] Boullion, T. L. and P. L. Odell (1969): A generalization
S of the Wielandt inequality, Texas J. Sci., 21, 255-259.

[189] Boullion, T. L. and P. L. Odell (1969): A note on the
P Scroggs-Odell pseudoinverse, SIAM J. Appl. Math., 17, 7-10.

[190] Boullion, T. L. and P. L. Odell (1971): Generalized
P Inverse Matrices, Wiley-Interscience, New York, x + 108 pp.

[191] Boullion, T. L. and G. D. Poole (1970): A characteriza-
S tion of the general solution of the matrix equation AX + XB = C, Indust. Math., 20, 91-95.

Boullion, T. L.: See Amburgey, J. K. (1968), (1970), (1971), Hallum, C. R. (1973), (1973), Lewis, T. O. (1968), Milnes, H. W. (1968), (1969), Nelson, D. L. (1971), (1972), Poole, G. D. (1972), (1972), (1974), (1975), Pye, W. C. (1973), (1973), (1973), Shurbet, G. L. (1972), (1973), Singh, I. (1975), Stallings, W. T. (1970), (1972), (1972), (1974), Ward, J. F. (1971), (1971), (1972).

[192] Bounitzky, E. (1909): Sur la fonction de Green des
S equations differentielles linéaires ordinaires, J. Math. Pures Appl. (6), 5, 65-125.

[193] Bourbaki, N. (1953): Eléments de Mathématique. Livre
T V. Espaces Vectoriels Topologiques, Hermann & Cie, Paris.

[194] Bourbaki, N. (1958): Eléments de Mathématique. Livre
T II. Algèbre, Hermann & Cie, Paris.

[195] Bowdler, H. J., R. S. Martin, G. Peters and J. H.
S Wilkinson (1966): Solution of real and complex systems of linear equations, Numer. Math., 8, 217-239. Republished, pp. 93-110 in Wilkinson and Reinsch (1971).

Bowie, C.: See Björck, Å. (1971).

[196] Bowman, V. J. and C.-A. Burdet (1974): On the general
S solution to systems of mixed-integer linear equations, SIAM J. Appl. Math., 26, 120-125.

[197] Bradley, J. S. (1966): Adjoint quasi-differential opera-
P tors of Euler type, Pacific J. Math., 16, 213-237, errata, ibid., 587-588.

[198] Bradley, J. S. (1966): Generalized Green's matrices for
P compatible differential systems, Michigan Math. J., 13, 97-108.

[199] Brand, L. (1962): The solution of linear algebraic equa-
S tions, Math. Gaz., 46, 203-207.

[200] Brockett, R. W. (1970): Finite Dimensional Linear
S Systems, Wiley, New York.

Brockett, R. W.: See Baras, J. S. (1974).

[201] den Broeder, C. G., Jr. and A. Charnes (1957): Contri-
P butions to the theory of generalized inverses for matrices, Purdue University, Lafayette, Ind., (Reprinted as ONR Res. Memo No. 39, Northwestern University, Evanston, Illinois 1962).

[202] Browder, F. E. (1959): Functional analysis and partial
T differential equations, I. Math. Ann., 138, 55-79.

[203] Browder, F. E. (1970): On Fredholm alternative for non-
S linear operators, Bull. Amer. Math. Soc., 76, 933-998.

[204] Browder, F. E. (1971): Normal solvability for nonlinear
S mappings into Banach spaces, Bull. Amer. Math. Soc., 76, 993-998.

[205] Browder, F. E. (1971): Normal solvability and the
S Fredholm alternative for mappings into infinite dimensional manifolds, J. Funct. Anal., 8, 250-274

[206] Browder, F. E. (1972): Normal solvability and Φ-accre-
T tive mappings of Banach spaces, Bull. Amer. Math. Soc., 78, 186-192.

[207] Brown, B. and N. H. McCoy (1950): The maximal reg-
S ular ideal of a ring, Proc. Amer. Math. Soc., 1, 165-171.

[208] Brown, R. C. (1974): Generalized Green's functions and
P generalized inverses for linear differential systems with Stieltjes boundary conditions, J. Differential Equations, 16, 335-351.

[209] Brown, R. C. (1974): Duality theory for n^{th} order dif-
S ferential operators using Stieltjes boundary conditions: II. Nonsmooth coefficients and nonsingular measures, Technical Summary Report #1389, Mathematics Research Center, University of Wisconsin-Madison; to appear in Annali di Mat. (1975).

[210] de Bruyn, G. F. C. (1969): The existence of continuous
S inverse operators under certain conditions, J. London Math. Soc., 44, 68-70.

[211] Bryan, R. N. (1969): A linear differential system with
S general linear boundary conditions, J. Differential Equations, 5, 38-48.

[212] Brynielsson, L. (1974): On Fredholm integral equation of
T the first kind with convex constraint, SIAM J. Math. Anal., 5, 955-962.

[213] La Budde, C. D. and G. R. Verma (1969): On the compu-
P tation of a generalized inverse of a matrix, Quart. Appl. Math., 27, 391-395.

Burdet, C-A.: See Bowman, V. J. (1974).

[214] Burmeister, W. (1972): Inversionsfreie Verfahren zur
S Lösung nichtlinearer Operatorgleichungen, Zeit. Angew. Math. Mech., 52, 101-110.

[215] Burns, F., D. Carlson, E. Haynsworth and T. Markham
P (1974): Generalized inverse formulas using the Schur complement, SIAM J. Appl. Math., 26, 254-259.

Burrus, W. R.: See Rust, B. (1966), (1972), Replogle, J. (1967).

[216] Businger, P. A. and G. H. Golub (1965): Linear least
P squares solutions by Householder transformations, Numer. Math., 7, 269-276. Republished, pp. 111-118 in Wilkinson and Reinsch, (1971).

[217] Businger, P. A. and G. H. Golub (1969): Algorithm 358:
S Singular value decomposition of a complex matrix, Comm. Assoc. Comput. Mach., 12, 564-565.

[218] Butler, Kim Ki Hang (1974): A Moore-Penrose inverse for
P Boolean relation matrices. Combinatorial Mathematics (Proc. Second Australian Conf., Univ. Melbourne, Melbourne, 1973), pp. 18-28, Lecture Notes in Math., vol. 403, Springer Verlag, Berlin.

[219] Butler, T. and A. V. Martin (1962): On a method of
S Courant for minimizing functionals, J. Math. Phys., 41, 291-299.

[220] Buzbee, B. L., G. H. Golub and C. W. Nielson [1970):
S On direct methods for solving Poisson's equation, SIAM J. Numer. Anal., 7, 627-656.

Cabayan, H. S.: See Deschamps, G. A. (1972).

[221] Cadzow, J. A. (1974): A finite algorithm for the minimum
T ℓ_∞ solution to a system of consistent linear equations, SIAM J. Numer. Anal., 11, 1151-1165.

Calvert, T. W.: See Young, T. Y. (1974).

[222] Campbell, S. L. and C. D. Meyer, Jr. (1975): EP oper-
P ators and generalized inverses, Canad. Math. Bull, to appear.

[223] Campbell, S. L. and C. D. Meyer, Jr. (1975): Continu-
P ity properties of the Drazin pseudoinverse, Linear Algebra and Appl., 10, 77-83.

[224] Caradus, S. R. (1966): Operators with finite ascent and
T descent, Pacific J. Math., 18, 437-449.

[225] Caradus, S. R. (1974): Operator theory of the pseudo-
P inverse, Queen's papers in Pure and Applied Mathematics, No. 38. Queen's Univ., Kingston, Ont., ii + 67 pp.

[226] Caradus, S. R. (1974): Perturbation theory for general-
P ized Fredholm operators, Pacific J. Math., 52, 11-15.

[227] Caradus, S. R. (1975): An equational approach to prod-
P ucts of relatively regular operators, Aequationes Math., to appear.

[228] Caradus, S. R. (1975): The Drazin inverse for operators
P on Banach spaces, Compositio Math., 47, 409-412.

[229] Caradus, S. R. (1975): Mapping properties of relatively
P regular operators, Proc. Amer. Math. Soc., 47, 409-412.

[230] Carlson, D. (1975): Matrix decompositions involving the
S Schur complement, SIAM J. Appl. Math., 28, 577-587.

[231] Carlson, D., E. Haynsworth and T. Markham (1974): A
S generalization of the Schur complement by means of the Moore-Penrose inverse, SIAM J. Appl. Math., 26, 169-175.

Carlson, D.: See Burns, F. (1974).

Carmen, Maria: See Fernández, A. (1974).

[232] Cesari, L. (1937): Sulla risoluzione dei sistemi di
S equazioni lineari per approssimazioni successive, Rend. R. Accad. Naz. Lincei Cl. Sci. Fis. Math. Nat., Ser. 6A, 25, Rome.

[233] Cesari, L. (1964): Functional analysis and Galerkin's
T method, Michigan Math. J., 11, 385-414.

[234] Charnes, A. and W. W. Cooper (1961): Management
T Models and Industrial Applications of Linear Programming, Wiley, New York.

[235] Charnes, A. and W. W. Cooper (1968): Structural sensi-
S tivity analysis in linear programming and an exact product form left inverse, Naval Res. Logist. Quart., 15, 517-522.

[236] Charnes, A., W. W. Cooper, J. K. Devoe and D. B.
T Learner (1963): Demon: Decision Mapping via Optimum Go-No-Go Networks: A Model for New Products Marketing, Batten, Barton, Durstine and Osborne, New York.

[237] Charnes, A., W. W. Cooper and G. L. Thompson (1965):
S Constrained generalized medians and hypermedians as deterministic equivalents for two-stage linear programs under uncertainty, Management Sci., 12, 83-112.

[238] Charnes, A. and F. Granot (1973): Existence and repre-
S sentation of Diophantine and mixed Diophantine solutions to linear equations and inequalities, Center for Cybernetic Studies, The University of Texas, Austin, Texas.

[239] Charnes, A. and M. Kirby (1963): A linear programming
S application of a left inverse of a basis matrix, ONR Research Memo, 91, The Technological Institute, Northwestern University, Evanston, Illinois.

[240] Charnes, A. and M. Kirby (1965): Modular design, gen-
P eralized inverses and convex programming, Operations Res., 13, 836-847.

Charnes, A.: See Ben-Israel, A. (1961), (1963), (1963), (1968), (1968), (1968), den Broeder, C. G., Jr. (1957).

[241] Chen, C. T. (1975): Inertia theorem for general matrix
S equations, J. Math. Anal. Appl., 49, 207-210.

Chen, F. P.: See Vermuri, V. (1974).

[242] Chen, R. M. -M. (1971): New algorithms for computing
P pseudoinverses and the minimal least squares solutions, in Proc. 5th Annual Princeton Conf. Inform. Sci. and Syst., p. 356.

[243] Chen, R. M. -M. (1973): New matrix inversion algorithms
P based on exchange method, IEEE Transactions on Computers, C-22, 885-890.

Chen, Y. M. -M.: See Demoulin, Y. -M. J. (1974).

[244] Cheney, E. W. (1966): Introduction to Approximation
S Theory, McGraw-Hill Book Company, New York, xii + 259 pp.

[245] Cheney, W. and A. A. Goldstein (1958): Note on a paper
T by Zuhovickiĭ concerning the Tchebycheff problem for linear equations, SIAM J. Appl. Math., 6, 233-239.

Cheney, W.: See Goldstein, A. A. (1958).

[246] Chernoff, H. (1953): Locally optimal designs for esti-
S mating parameters, Ann. Math. Statist., 24, 586-602. (See especially Appendix A, pp. 598-601).

[247] Chien, Y. T. and K. S. Fu (1968): Selection and ordering of feature observations in a pattern recognition system, Information and Control, 2, 394-414.
T

[248] Ching, C. H. and C. K. Chui (1974): Uniqueness of solutions of an infinite system of equations, Rocky Mountain J. Math., 4, 699-706.
T

[249] Chipman, J. S. (1964): On least-squares with insufficient observations, J. Amer. Statist. Assoc., 54, 1078-1111.
P

[250] Chipman, J. S. (1969): Specification problems in regression analysis, pp. 114-176 in Boullion and Odell (1968).
P

[251] Chipman, J. S. (1975): Estimation and aggregation in econometrics - An application of the theory of generalized inverses, this Volume.
P

[252] Chipman, J. S. and M. M. Rao (1964): the treatment of linear restrictions in regression analysis, Econometrica, 32, 198-209.
P

[253] Chipman, J. S. and M. M. Rao (1964): Projections, generalized inverse and quadratic forms, J. Math. Anal. Appl., 9, 1-11.
P

[254] Chitwood, H. (1971): Generalized Green's Matrices for Linear Differential Systems, Ph. D. Dissertation, University of Tennessee, Knoxville, Tennessee, 64 pp.
P

[255] Chitwood, H. (1973): Generalized Green's matrices for linear differential systems, SIAM J. Math. Anal., 4, 104-110.
P

Chui, C. K.: See Ching, C. H. (1974).

[256] Cimmino, G. (1938): Calcolo approssimato per le soluzioni dei sistemi di equazioni lineari, La Ricerca Scientifica XVI, Serie II, Anno IX, Vol. 1, 326-333, Roma.
S

[257] Cho, C. Y., R. E. Cline, T. N. E. Greville, B. Noble,
P L. D. Pyle, J. B. Rosen and D. V. Steward (1966): Talks on generalized inverses and solutions of large, approximately singular linear systems, MRC Tech. Summary Rept. #644, Mathematics Research Center, University of Wisconsin, Madison.

[258] Clarkson, J. A. (1936): Uniformly convex spaces, Trans.
T Amer. Math. Soc., 40, 396-414.

[259] Cleveland, W. S. (1971): Projection with wrong inner
S product and its application to regression with correlated errors and linear filtering of time series, Ann. Math. Stat., 42, 616-624.

[260] Clifford, A. H. (1941): Semigroups admitting relative
S inverses, Ann. of Math., 42, 1037-1049.

[261] Clifford, A. H. and G. B. Preston (1961): The Algebraic
S Theory of Semigroups, I, Mathematical Surveys, 7, American Math. Soc., Providence, R. I.

[262] Cline, R. E. (1958): On the computation of the general-
P ized inverse $A^\dagger$, of an arbitrary matrix A, and the use of certain associated eigenvectors in solving the allocation problem, Preliminary Report, Statistical and Computing Laboratory, Purdue University, Lafayette, Indiana.

[263] Cline, R. E. (1963): Representations for the Generalized
P Inverse of Matrices with Applications in Linear Programming, Doctoral Dissertation, Purdue University, Lafayette, Indiana.

[264] Cline, R. E. (1964): Note on the generalized inverse of
P the product of matrices, SIAM Rev., 6, 57-58.

[265] Cline, R. E. (1964): Representations for the generalized
P inverse of a partitioned matrix, SIAM J. Appl. Math., 12, 588-600.

[266] Cline, R. E. (1965): Representations for the generalized
P inverse of sums of matrices, SIAM J. Numer. Anal., 2, 99-114.

[267] Cline, R. E. (1965): Representations for the generalized
P inverses of sums of matrices. II. Technical Summary Rept. #559, Mathematics Research Center, University of Wisconsin, Madison.

[268] Cline, R. E. (1965): An application of representations
P for the generalized inverse of a matrix, Technical Summary Rept. #592, Mathematics Research Center, University of Wisconsin, Madison.

[269] Cline, R. E. (1968): Inverses of rank invariant powers
P of a matrix, pp. 47-52 in Boullion and Odell (1968).

[270] Cline, R. E. (1968): Inverses of rank invariant powers
P of a matrix, SIAM J. Numer. Anal., 5, 182-197.

[271] Cline, R. E. and T. N. E. Greville (1970): An extension
P of the generalized inverse of a matrix, SIAM J. Appl. Math., 19, 682-688.

[272] Cline, R. E., R. J. Plemmons and G. H. Worm (1974):
P Generalized inverses of certain Toeplitz matrices, Linear Algebra and Appl., 8, 25-33.

[273] Cline, R. E. and L. D. Pyle (1973): The generalized in-
P verse in linear programming. An intersection projection method and the solution of a class of structured linear programming problems, SIAM J. Appl. Math., 24, 338-351.

Cline, R. E.: See Cho, C. Y. (1966), Plemmons, R. J. (1972).

[274] Cochran, W. G. (1934): The distribution of quadratic
T forms in a normal system with applications to analysis of covariance, Proc. Cambridge Philos. Soc., 30, 178-191.

[275] Coddington, E. A. and N. Levinson (1955): Theory of
T Ordinary Differential Equations, McGraw-Hill Book Co., New York, xii + 429 pp.

Cohen, D.: See Ben-Israel, A. (1966).

[276] Cole, R. H. (1968): Theory of Ordinary Differential
S Equations, Appleton-Century-Crofts, New York.

[277] Collar, A. R. (1951): On the reciprocal of a segment of
T a generalized Hilbert matrix, Proc. Cambridge Philos.
Soc., 47, 11-17.

[278] Conti, R. (1967): Recent trends in the theory of boundary
S value problems for ordinary differential equations, Boll.
U. M. I. (3), 22, 135-178.

[279] Cook, P. A. (1975): Estimates for the inverse of a
T matrix, Linear Algebra and Appl., 10, 41-53.

Cooper, W. W.: See Charnes, A. (1961), (1963), (1965), (1968).

[280] Coppel, W. A. (1974): Matrix quadratic equations, Bull.
T Austral. Math. Soc., 10, 377-401.

[281] Cottle, R. W. (1974): Manifestation of the Schur com-
S plement, Linear Algebra and Appl., 8, 189-211.

[282] Courant, R. and D. Hilbert (1931): Methoden der Math-
S ematischen Physik, Vol. I. Berlin: Springer. Methods
of Mathematical Physics, Wiley-Interscience, 1953.

[283] Craig, A. T. (1943): Note on the independence of cer-
T tain quadratic forms, Ann. Math. Statist., 14, 195-197.

Craig, A. T.: See Hogg, R. V. (1958).

[284] Croisot, R. (1953): Demi-groupes inversifs et demi-
T groupes réunions de semi-groupes simples, Ann. Sci.
École Norm. Sup., 70, 361-379.

[285] Crone, L. (1972): The singular value decomposition of
S matrices and cheap numerical filtering of systems of
linear equations, J. Franklin Institute, 294, 133-136.

[286] Crownover, R. M. (1972): Commutants of shifts on
S Banach spaces, Michigan Math. J., 19, 233-247.

[287] Cudia, D. F. (1963): Rotundity, pp. 73-97 in Convexity,
T Proc. Sympos. Pure Math., Vol. VII (V. Klee, Editor), Amer. Math. Soc., Providence, R.I., xv + 516 pp.

[288] Cvetkov, B. (1955): A new method of computation in the
S theory of least squares, Aust. J. Appl. Sci., 6, 274-280.

[289] Cvetkov, B. (1956): On the indefinite solution of a linear
S system by the principle of least squares, Emp. Sur. Rev., 13, 272-281.

[290] Dade, E. C., O. Taussky and H. Zassenhaus (1962): On
S the theory of orders, in particular on the semigroup of ideal classes and genera of an order of an algebraic number field, Math. Ann., 148, 31-64.

[291] Dalphin, J. F. (1973): Direct Computation of Generalized
P Inverses and Minimum Norm Least Squares Solutions for Some Linear Systems Expressed in Tensor Product Form, Ph.D. thesis, Dept. of Math., Clarkson College of Technology, Potsdam, New York.

[292] Dalphin, J. F. and V. Lovass-Nagy (1973): Best least
P squares solutions to finite difference equations using the generalized inverse and tensor product methods, J. Assoc. Comput. Mach., 20, 279-289.

[293] Dalphin, J. F., V. Lovass-Nagy and D. L. Powers (1972):
S Best least-squares solutions to discrete Neumann problems on some irregular regions, Technical Rept., Clarkson College of Technology, Potsdam, New York.

[294] Daniel, J. W. (1971): The Approximate Minimization of
S Functionals, Prentice-Hall, New Jersey.

[295] David, F. N. and J. Neyman (1938): Extension of the
S Markoff theorem on least squares, Statist. Res. Mem., 2, 105-116.

[296] Davis, D. L. and D. W. Robinson (1972): Generalized
P inverses of morphisms, Linear Algebra and Appl., 5, 319-328.

Davis, J.: See Anselone, P. M. (1973).

[297] Decell, H. P., Jr. (1965): An alternative form of the
P generalized inverse of an arbitrary complex matrix, SIAM Rev., 7, 356-358.

[298] Decell, H. P., Jr. (1965): An application of the Cayley-
P Hamilton theorem to generalized matrix inversion, SIAM Rev., 7, 526-528.

[299] Decell, H. P., Jr. (1965): A special form of the general-
P ized inverse of an arbitrary complex matrix, NASA TN D-2784, Washington, D. C.

[300] Decell, H. P., Jr. (1965): An application of generalized
P matrix inversion to sequential least squares parameter estimation, NASA TN D-2830, Washington, D. C.

[301] Decell, H. P., Jr. (1968): A characterization of the
S maximal subgroups of the semigroup of $m \times n$ complex matrices, pp. 177-182 in Boullion and Odell (1968).

[302] Decell, H. P., Jr. (1973): On the derivative of the gen-
P eralized inverse of a matrix, Linear and Multilinear Algebra, 1, 357-359.

[303] Decell, H. P., Jr. and S. W. Kahng (1966): An iterative
P method for computing the generalized inverse of a matrix, NASA TN D-3464, Washington, D. C.

[304] Decell, H. P., Jr. and P. L. Odell (1966): A note con-
S cerning a generalization of the Gauss-Markov theorem. Texas J. Sci., 18, 21-24.

[305] Decell, H. P., Jr. and P. L. Odell (1967): On the fixed
S point probability vector of regular or ergodic transition matrices, J. Amer. Statist. Assoc., 62, 600-602.

[306] Decell, H. P., Jr. and C. L. Wiginton (1968): A characterization of the maximal subgroups ot the semigroup ot
S n × n complex matrices, Czechoslovak Math. J., 18, 675-677.

Decell, H. P.: See Odell, P. L. (1967).

[307] Dekerlegand, R. J. (1967): Analysis of Generalized In-
P verse Computation Schemes, Master's Thesis, University of Southwestern Louisiana, Lafayette, Louisiana.

[308] Delaney, J. C. and G. G. Gaffney (1966): Efficiency of
P generalized matrix inversion methods, NASA-MSC Internal Note.

[309] Delaney, J. C. and F. M. Speed (1966): A new algorithm
P for calculating the generalized inverse of an arbitrary m × n matrix, NASA-MSC Internal Note.

[310] Demoulin, Yves-Marie J. and Y. M. Chen (1974): An
S iteration method for studying the bifurcation of solutions of nonlinear equations, $L(\lambda)u + \varepsilon R(\lambda, u) = 0$, Numer. Math., 23, 47-61.

[311] Dennis, J. B. (1959): Mathematical Programming and
T Electrical Networks, M. I. T. Technology Press, Cambridge, Mass., 186 pp.

[312] Dennis, J. E., Jr. (1972): Some computational tech-
S niques for the nonlinear least squares problems, in Numerical Solutions of Systems of Nonlinear Algebraic Equations, G. Byrne and C. Hall, editors, Academic Press, New York, pp. 157-183.

[313] Dent, B. A. and J. Newhouse (1959): Polynomials orthog-
T onal over a discrete domain, SIAM Rev., 1, 55-59.

[314] Derr, J. and A. E. Taylor (1962): Operators of meromor-
T phic type with multiple poles of the resolvent, Pacific J. Math., 12, 85-111.

[315] Deschamps, G. A. and H. S. Cabayan (1972): Antenna
S synthesis and solution of inverse problems by regularization methods, IEEE Transactions on Antenna and Propagation, AP-20, 268-274.

[316] Desoer, C. A. and B. H. Whalen (1963): A note on
P pseudoinverses, SIAM J. Appl. Math., 11, 442-447.

Desoer, C. A.: See Zadeh, L. A. (1962).

[317] Deutsch, E. (1971): Semi-inverses, reflexive semi-
P inverses, and pseudoinverses of an arbitrary linear transformation, Linear Algebra and Appl., 4, 313-322.

Devoe, J. K.: See Charnes, A. (1963).

[318] Diaz, J. B. and F. T. Metcalf (1970): On iterative pro-
S cedures for equations of the first kind, $Ax = y$, and Picard's criterion for the existence of a solution, Math. Comp., 24 (1970), 923-935.

[319] Dieudonné, J. (1943): Sur les homomorphismes d'espaces
S normés, Bull. Sci. Math. France, (2), 67, 72-84.

[320] Dokovic, D. (1965): On the generalized inverse for
P matrices, Glasnik Mat. -Fiz. Astron. Ser. II, Drustvo Mat. Fiz. Hrvatske, 20, 51-55.

[321] Dolph, C. L. and M. A. Woodbury (1952): On the rela-
S tion between Green's functions and covariances of certain stochastic processes and its application to unbiased linear prediction, Trans. Amer. Math. Soc., 72, 519-550.

[322] Dommanget, J. (1961): Peut-on generaliser la définition
P de l'inverse d'une cracovien rectangulaire proposee par A. Bjerhammar? Jury Central constitute par la collation des Grades Académiques, Lère Session, Bruxelles.

[323] Dommanget, J. (1963): L'inverse d'un cracovien rec-
P tangulaire: Son emploi dans la résolution des systèmes d'équations linéares, Publ. Sci. Tech. Ministère de l'Air (Paris) Notes Tech. 128, 11-41. (Math. Rev., 31, p. 614).

[324] Doolittle, M. H. (1878): On least squares solutions,
T U.S.G.S., Ann. Rep. App. 8, paper 3, 115-120.

[325] Dotson, W. G., Jr. (1970): On the Mann iterative pro-
T cess, Trans. Amer. Math. Soc., 149, 65-73.

[326] Dotson, W. G., Jr. (1970): On the solution of linear
T functional equations by averaging iteration, Proc. Amer. Math. Soc., 25, 504-506.

[327] Douglas, J., Jr. and C. M. Pearcy (1963): On con-
S vergence of alternating direction procedures in the presence of singular operators, Numer. Math., 5, 175-184.

[328] Douglas, R. G. (1966): On majorization, factorization,
T and range inclusion of operators on Hilbert space, Proc. Amer. Math. Soc., 17, 413-416.

[329] Doust, A. and V. E. Price (1964): The latent roots and
S vectors of a singular matrix, Comput. J., 7, 222-227.

[330] Downs, T. (1975): Some properties of the Souriau-
S Frame algorithm with application to the inversion of rational matrices, SIAM J. Appl. Math., 28, 237-251.

[331] Dragomir, P. (1963): On the Greville-Moore formula for
P calculating the generalized inverse matrix, An. Univ. Timisoara Ser. Sti. Math-Fiz., 1, 115-119.

[332] Dragomir, P. (1964): The generalized inverse of a bi-
P linear form, An. Univ. Timisoara Ser. Sti. Mat-Fiz., 2, 71-76.

[333] Dragomir, A. and M. Fildan (1969): L'inverse généralisé
P d'un opérateur lineaire, An. Univ. Timisoara, Ser. Sti. Mat.-Fiz., 7, 55-65.

Dragomir, A.: See Arghiriade, E. (1963), (1969).

[334] Drazin, M. P. (1956): Algebraic and diagonable rings,
S Canad. J. Math., 8, 341-354.

[335] Drazin, M. P. (1958): Pseudo-inverses in associative
P rings and semigroups, Amer. Math. Monthly, 65, 506-514.

[336] Drygas, H. (1969): Gauss-Markov estimation and mini-
S mum-bias estimation, Report No. 98 of Studiengruppe für Systemforschung, May, 1969.

[337] Drygas, H. (1969): On the theory of Gauss-Markov esti-
S mators, CORE discussion paper, Center for Operations Research and Econometrics, The Catholic University of Louvain, Heverlee (Belgium).

[338] Drygas, H. (1970): The Coordinate-Free Approach to
S Gauss-Markov Estimation, Springer-Verlag, Berlin, viii + 113 pp.

[339] Drygas, H. (1970): Stepwise estimation models and
S Bayesian analysis in regression models, CORE discussion paper, Center for Operations Research and Econometrics, Catholic University of Louvain, Belgium, July, 1970, 97 pp.

[340] Drygas, H. (1971): Consistency of the least squares
S and Gauss-Markov estimators in regression models, Z. Wahrscheinlichkeitstheorie verw. Gebeite, 17, 309-326.

[341] Drygas, H. (1972): The estimation of residual variance
S in regression analysis, Math. Operationsforsch. u. Statist., 3, 373-388.

[342] Drygas, H. (1974): A note on the Gauss-Markov esti-
S mation in multivariate linear models, Progress in Statistics (European Meeting of Statisticians, Budapest, 1972), pp. 181-190, Colloq. Math. Soc. János Bolyani, vol. 9, North-Holland, Amsterdam.

[343] Dück, Werner (1964): Einzelschrittverfahren zur Matri-
S zeninversion, Z. Angew Math. Mech., 44, 401-403.

[344] Dück, Werner (1966): Iterative Verfahren und Abänder-
S ungsmethoden zur Inversion von Matrizen, Wissenschaftl. Zeitschrift der Th. Chemnitz, Fakultät für Naturwissenschaften, Berlin.

[345] Duda, R. O. and P. E. Hart (1973): Pattern Classifica-
S tion and Scene Analysis, Wiley, New York.

[346] Duffin, R. J. (1971): Network models, pp. 65-91 in
S Mathematical Aspects of Electrical Network Analysis, SIAM-AMS Proc. Vol. III, (Wilf, H. S. and F. Harary, editors) Amer. Math. Soc., Providence, R.I.

[347] Duffin, R. J. (1974): Some problems of mathematics and
S science, Bull. Amer. Math. Soc., 80, 1053-1070.

[348] Duffin, R. J. and G. E. Trapp (1972): Hybrid addition of
P matrices - network theory concept, Applicable Analysis, 2, 241-254.

Duffin, R. J.: See Anderson, W. N., Jr. (1969), (1972), (1975), Bott, R. (1953).

[349] Dufour, H. M. (1971): Note au sujet de la résolution
S d'une système de moindres carrés indéterminée, Inst. Geogr. Nat., Paris.

[350] Dunford, N. and J. T. Schwartz (1958), (1963): Linear
T Operators, Parts I, II, Interscience, New York.

[351] Duran, B. and P. L. Odell (1974): Cluster Analysis.
T A Survey. Lecture Notes in Economics and Mathematical Systems, vol. 100, Springer-Verlag, Berlin-New York.

[352] Duris, C. S. (1968): An exchange algorithm for solving
S Haar or non-Haar overdetermined linear equations in the sense of Chebyshev, Proceedings of the Summer ACM Computer Conference, pp. 61-65.

[353] Duris, C. S. (1971): Optimal quadrature formulas using
P generalized inverses. Part I. General theory and minimum variance formulas, Math. Comp., 25, 495-504.

[354] Duris, C. S. (1971): Optimal quadrature formulas using
P generalized inverses. Part II, Sard 'best' formulas, Math Rept. 70-17, Drexel University, Philadelphia, Pa.

[355] Duris, C. S. and V. P. Sreedharan (1968): Chebyshev
P and ℓ^1-solutions of linear equations using least squares solutions, SIAM J. Numer. Anal., 5, 491-505.

Duris, C. S.: See Anton, H. (1974).

[356] Dwyer, P. S. and M. S. MacPhail (1948): Symbolic
T matrix derivatives, Ann. Math. Statist., 19, 517-534.

[357] Eckart, C. and G. Young (1936): The approximation of
S one matrix by another of lower rank, Psychometrika, 1, 211-218.

[358] Eckart, C. and G. Young (1939): A principal axis trans-
S formation for non-Hermitian matrices, Bull. Amer. Math. Soc., 45, 118-121.

Edie, J.: See Sebestyen, G. D. (1966).

[359] Egerváry, E. (1953): On a property of the projector
S matrices and its application to the canonical representation of matrix functions, Acta Sci. Math. Szeged, 15, 1-6.

[360] Egerváry, E. (1956): Generalized inverse of a matrix (in
P Hungarian), Publications of the Mathematics Institute of the Hungarian Academy of Sciences, vol. 1, 315-324.

[361] Egerváry, E. (1959): Über eine konstruktive Methode zur
T Reduktion einer Matrix auf die jordansche Normalform, Acta Math. Acad. Sci. Hungar, 10, 31-54.

[362] Egerváry, E. (1960): On rank diminishing operations and
S their applications to the solution of linear equations, Z. Angew Math. Phys., 11, 376-386.

Eichinger, B. E.: See Arnowitz, S. (1975).

[363] Eisemann, K. (1974): An application of generalized
P matrix inversion, IEEE Trans. Circuits and Systems, CAS-21, 701-702.

[364] Elliott, W. W. (1928): Generalized Green's functions
S for compatible differential systems, Amer. J. Math., 50, 243-258.

[365] Elliott, W. W. (1929): Green's functions for differential
S systems containing a parameter, Amer. J. Math., 51, 397-416.

[366] Embry, Mary R. (1973): Factorization of operators on
T Banach space, Proc. Amer. Math. Soc., 38, 587-590.

[367] Englefield, M. J. (1966): The commuting inverses of a
P square matrix, Proc. Cambridge Philos. Soc., 62, 667-671.

[368] Erdelsky, P. J. (1969): Projections in a Normed Linear
P Space and a Generalization of the Pseudo-Inverse, Doctoral Dissertation, California Institute of Technology, Pasadena, California.

[369] Erdélyi, I. (1966): On partial isometries in finite dimen-
S sional Euclidean spaces, SIAM J. Appl. Math., 14, 453-467.

[370] Erdélyi, I. (1966): On the "reverse order law" related to
P the generalized inverse of matrix products, J. Assoc. Comput. Mach., 13, 439-443.

[371] Erdélyi, I. (1967): The quasi-commuting inverses for a
S square matrix, Atti. Accad. Naz. Lencei. Rend. Cl. Sci. Fis. Math. Natur., Ser. VIII, 42, 626-633.

[372] Erdélyi, I. (1967): On the matrix equation $Ax = \lambda Bx$,
S J. Math. Anal. Appl., 17, 119-132.

[373] Erdélyi, I. (1968): Normal partial isometries closed
S under multiplication on unitary spaces, Atti. Accad. Naz. Lincei Rend. Cl. Sci. Fis. Mat. Natur., Ser. VIII, 43, 186-190.

[374] Erdélyi, I. (1968): Partial isometries defined by a spec-
S tral property on unitary spaces, Atti. Accad. Naz. Lincei Rend. Cl. Fis. Mat. Natur., Ser. VIII, 44, 741-747.

[375] Erdélyi, I. (1968): Partial isometries closed under mul-
S tiplication on Hilbert spaces, J. Math. Anal. Appl., 22, 546-551.

[376] Erdélyi, I. (1968): Partial isometrics and generalized
P inverses, pp. 203-217 in Boullion and Odell (1968).

[377] Erdélyi, I. (1972): A generalized inverse for arbitrary
P operators between Hilbert spaces, Proc. Cambridge Philos. Soc., 71, 43-50.

[378] Erdélyi, I. and A. Ben-Israel (1972): Extremal solutions
P of linear equations and generalized inversion between Hilbert spaces, J. Math. Anal. Appl., 39, 298-313.

[379] Erdélyi, I. and F. R. Miller (1970): Decomposition
S theorems for partial isometries, J. Math. Anal. Appl., 30, 665-679.

[380] Ernest, J. (1972): Left invertibility of closed operators
S modulo an α-compact operator, Tôhoku Math. J., 2nd series, 24, 529-537.

[381] Ernest, J. (1972): Operators with α-closed range,
S Tôhoku Math. J., 2nd series, 24, 45-49.

Evans, J. W.: See Hearon, J. Z. (1968), (1968).

[382] Faddeev, D. K. and V. N. Faddeeva (1963): Computa-
S tional Methods of Linear Algebra, Freeman, San Francisco.

[383] Faddeev, D. K., V. N. Kublanovskaja and V. N.
P Faddeeva (1968): Solution of linear algebraic systems with rectangular matrices, Proc. Steklov Inst. Math., 96, 93-111.

[384] Faddeev, D. K., V. N. Kublanovskaja and V. N.
P Faddeeva (1968): Linear algebraic systems with rectangular and ill-conditioned matrices (Russian), Modern numerical methods, No. 1, computational methods of linear algebra (Proc. Internat. Summer School on Numerical Methods, Kiev, (1966) (Russian) Vyčisl. Centr. Akad. Nauk SSSR Moscow, pp. 16-75 (Math. Rev. 39, #6887).

[385] Faddeev, D. K., V. N. Kublanovskaja and V. N.
P Faddeeva (1968): Sur les systèmes linéaires algébriques de matrices rectangulaires et mal-conditionnées, Colloques Internationaux du Centre National de la Recherche Scientifique, No. 165, Programmation en Mathématiques Numériques, Besançon, 7-14 Septembre, 1966, pp. 161-170.

Faddeeva, V. N.: See Faddeev, D. K. (1963), (1968), (1968), (1968).

[386] Falb, P. L. (1967): Infinite-dimensional filtering: the
T Kalman-Bucy filter in Hilbert space, Information and Control, 11, 102-137.

Falb, P. L.: See Kalman, R. E. (1969).

[387] Fan, Ky (1949): On a theorem of Weyl concerning
T eigenvalues of linear transformations, I, Proc. Nat. Acad. Sci. USA, 35, 652-655, II, ibid, 36(1950), 31-35.

[388] Fan, Ky and A. J. Hoffman (1955): Some metric inequal-
S ities in the space of matrices, Proc. Amer. Math. Soc., 6, 111-116.

[389] Farahat, H. K. (1966): The semigroup of doubly-stochas-
S tic matrices, Proc. Glasgow Math. Assoc., 7, 178-183.

Federer, W. T.: See Banerjee, K. S. (1968).

Fenchel, W.: See Bonnesen, T. (1934).

[390] Fernández, A. and Maria Carmen (1974): Conditions for
P the existence of an inverse of a non-square matrix (Spanish) Gac. Mat. (Madrid) (1), 26, 125-126.

[391] S Ficken, F. A. (1967): Linear Transformations and Matrices, Prentice-Hall, Englewood Cliffs, N. J., xiii + 398 pp.

[392] P Ficken, F. A. (1970): More on generalized inverses, Notices Amer. Math. Soc., 17, 929.

Ficken, F. A.: See Friedrichs, K. O. (1953).

[393] S de Figueiredo, D. G. (1974): On the range of nonlinear operators with linear asymptotes which are not invertible, Comment. Math. Univ. Carolinae, 15, 415-428.

[394] S Fillmore, P. A. and J. P. Williams (1971): On operator ranges, Advances in Math., 7, 254-281.

[395] S Fishel, B. (1975): Partial isometries which are sums of shifts, Proc. Cambridge Philos. Soc., 78, 107-110.

[396] P Fisher, A. G. (1967): On construction and properties of the generalized inverse, SIAM J. Appl. Math., 15, 269-272.

Fisher, M. E.: See Hartwig, R. E. (1969).

[397] S Fitzgerald, D. G. (1972): On inverses of products of idempotents in regular semigroups, Austral. J. Math., 13, 335-337.

[398] P Fletcher, R. (1970): Generalized inverses for nonlinear equations and optimization, in Numerical Methods for Nonlinear Algebraic Equations, P. R. Rabinowitz, Ed., pp. 75-85, Gordon and Breach, London.

[399] P Fletcher, R. (1968): Generalized inverse methods for the best least-squares solution of systems of nonlinear equations, Computer J., 10, 392-399.

[400] S Fletcher, R. (1969): A technique for orthogonalization, J. Inst. Math. Appl., 5, 162-166.

[401] S Fletcher, R., Ed. (1969): Optimization, Academic Press, New York-London.

[402] Florentin, J. J. (1961): Optimal control of continuous
S time, stochastic systems, J. Electronics Control, 10, 473-488.

[403] Förster, K. H. and M. A. Kaashoek (1975): The asymp-
S totic behaviour of the reduced minimum modulus of a Fredholm operator, Proc. Amer. Math. Soc., 49, 123-131.

[404] Forsythe, G. E. (1953): Solving linear algebraic equa-
S tions can be interesting, Bull. Amer. Math. Soc., 59, 299-329.

[405] Forsythe, G. E. (1970): The maximum and minimum of a
T positive definite quadratic polynomial on a sphere are convex functions of the radius, SIAM J. Appl. Math., 19, 551-554.

[406] Forsythe, G. E. and G. H. Golub (1965): On the
T stationary values of a second-degree polynomial on the unit sphere, SIAM J. Appl. Math., 13, 1050-1068.

[407] Foster, Manus (1961): An application of the Wiener-
S Kolmogorov smoothing theory to matrix inversion, SIAM J. Appl. Math., 9, 387-392.

[408] Foulis, D. J. (1963): Relative inverses in Baer *-semi-
P groups, Michigan Math. J., 10, 65-84.

[409] Frame, J. S. (1964): Matrix functions and applications,
P I, matrix operations and generalized inverses, IEEE Spectrum, 1, 209-220.

[410] Francis, J. (1961): The Q. R. transformation: A unitary
T analogue to the L. R. transformation - Part I, Computer J., 4, 265-271.

[411] Francis, J. (1962): The Q. R. transformation - Part II,
T Computer J., 4, 332-345.

[412] Franck, P. (1962): Sur la distance minimale d'une
P matrice régulière donnée au lieu des matrices singulières, Deux. Cong. Assoc. Francaise Calcul. et Traitement Information, Paris, 1961, Gauthier-Villars, Paris, 1962, 55-60. (Math Rev. 29 (1965), #2953).

[413] Franklin, J. N. (1974): On Tikhonov's method for ill-
S posed problems, Math. Comp., 28, 889-907.

[414] Fredholm, I. (1903): Sur une classe d'équations fonc-
P tionnelles, Acta Math., 27, 365-390.

[415] Friedman, B. (1956): Principles and Techniques of Ap-
T plied Mathematics, John Wiley and Sons, New York.

[416] Fridman, V. M. (1956): A method of successive approx-
S imation for Fredholm integral equations of the first kind, Uspehi Mat. Nauk, 11, 233-234.

[417] Fridman, V. M. (1962): On the convergence of the
P method of steepest descent type, Uspehi Mat. Nauk, 17, 201-204.

[418] Friedrichs, K. O. (1953): Functional Analysis and Ap-
S plications, (Notes of lectures given in 1949-50, Inst. Math. Sciences, New York University, New York, by F. Ficken).

[419] Fu, K. S. (1968): Sequential Methods in Pattern Recog-
T nition and Machine Learning, Academic Press, New York.

Fu, K. S.: See Chien, Y. T. (1968), Wee, W. G. (1968).

[420] Fučik, S. (1974): Surjectivity of operators involving
S linear noninvertible part and nonlinear compact perturbation, Funkcial. Ekvac., 17, 73-83.

[421] Fučik, S., M. Kučera and J. Nečas (1975): Ranges of
S nonlinear asymptotically linear operators, J. Differential Equations, 17, 375-394.

[422] Fučik, S. and J. Milota (1971): On the convergence of
T linear operators and adjoint operators, Comment Math. Univ. Carolinae, 12, 753-763.

[423] Fučik, S., J. Nečas, J. Souček and V. Souček (1973):
T Spectral Analysis of Nonlinear Operators, Lecture Notes in Mathematics, vol. 346, Springer-Verlag, Berlin-New York.

[424] Fučik, S., J. Nečas, J. Souček and V. Souček (1974):
T Krasnoselskii's main bifurication theorem, Arch. Rational Mech. Anal., 54, 328-339.

Fuhrmann, P. A.: See Baras, J. S. (1974).

[425] Fukunaga, K. (1972): Introduction to Statistical Pattern
T Recognition, Academic Press, New York.

[426] Fulkerson, D. R. and O. A. Gross (1965): Incidence
T matrices and interval graphs, Pacific J. Math., 15, 835-855.

[427] Gabriel, K. R. and M. Haber (1973): The Moore-Penrose
S inverse of a data matrix - a statistical tool with some meteorological applications, a preprint.

[428] Gabriel, R. (1965): Extension of generalized algebraic
S complement to arbitrary matrices (Romanian), Stud. Cerc. Mat., 17, 1567-1581, (Math. Rev., 35 (1968), #6703).

[429] Gabriel, R. (1969): Das verallgemeinerte Inverse einer
S Matrix deren Elemente einem beliebigen Körper angehoren, J. Reine Angew. Math., 234, 107-122, (Math. Rev., 41 (1971), #1753).

[430] Gabriel, R. (1970): Das verallgemeinerte Inverse einer
S Matrix über einem beliebigen Korper analytisch betrachtet, J. Reine Angew. Math., 244, 83-93.

Gabriel, R.: See Giurescu, C. (1964).

[431] Gaches, J., J.-L. Rigal, and X. Rousset de Pina (1965):
S Distance euclidienne d'une application linéaire σ au lieu des applications de rang r donne: détermination d'une meilleure approximation de rang r, C. R. Acad. Sci., Paris, 260, 5672-5674.

Gaffney, G. G.: See Delaney, J. C. (1966).

[432] Gainer, P. A. (1966): A method for computing the effect
S of an additional observation on a previous least squares estimate, NASA TN D-1599.

[433] Gale, L. A. (1955): A modified-equations method for
S the least squares solution of condition equations. Trans. Amer. Geophys. Union, 36, 779-791.

[434] Gallaher, L. J. and I. E. Perlin (1974): Use of Green's
T functions in the numerical solution of two-point boundary value problems, Proceedings of the Conference on the Numerical Solution of Ordinary Differential Equations (Univ. Texas, Austin, Texas, 1972), pp. 374-407, Lecture Notes in Math., Vol. 362, Springer-Verlag, Berlin-New York.

[435] Gantmacher, F. R. (1959): The Theory of Matrices,
T vols. I and II, Chelsea, New York.

[436] Garnett III, J. M., A. Ben-Israel and S. S. Yau (1971):
P A hyperpower iterative method for computing matrix products involving the generalized inverse, SIAM J. Numer. Anal., 8, 104-109.

Garnet, J. M.: See Yau, S. S. (1972).

[437] Gasquet, C. (1973): Perturbations de fonctions-spline,
S C. R. Acad. Sci. Paris, 276, 1465-1468.

[438] Gastinel, N. (1970): Linear Numerical Analysis,
S Hermann, Paris, and Academic Press, New York, xi + 341 pp.

[439] P Gatley, W. Y. (1962): Application of the Generalized Inverse Concept to the Theory of Linear Statistical Models, Doctoral Dissertation, Oklahoma State University, Stillwater, Oklahoma.

[440] T Gauss, C. F. (1809): Theoria Motus Corporum Coelestium in Sectionibus Conicis Solem Ambientium, Göttingen. In Carl Friedrich Gauss Werke, vol. 7, Königlichen Gesellschaft der Wissenschaften, Göttingen, 1906. English translation by C. H. Davis, Theory of the Motion of the Heavenly Bodies Moving about the Sun in Conic Sections, Little, Brown, Boston, 1857; reprinted by Dover, New York, 1963.

[441] T Gauss, C. F. (1821): Theoria Combinationis Observationum Erroribus Minimus Obnoxiae, Göttingen, in Carl Friedrich Gauss Werke, vol. 4, Königlichen Gesellschaft der Wissenschaften, Göttingen, 1880. Authorized French translation by J. Bertrand, Méthode de Moindres Carrés, Mallet-Bachelier, Paris, 1855; English translation from the French by H. F. Trotter, Gauss's Work (1803-1826) on the Theory of Least Squares, Statistical Technical Research Group, Dept. of Math., Princeton University, Tech. Rept. No. 5, Princeton, N.J., 1957.

[442] T Gauss, C. F. (1889): Abhandlungen zur Methode der kleinsten Quadrate, Borsch und Simon, Berlin.

[443] S Gavurin, N. K. (1963): Ill-conditioned systems of linear algebraic equations. USSR Comp. Math. and Mathematical Physics, 2, 407-418; Intern. J. of Computer Math., 1, 1964, 36-50. Translated from Ž. Vyčisl. Mat. i Mat. Fiz., 2(1962), 389-397.

[444] T Gelfand, I. M. and N. Ja. Vilenkin (1964): Generalized Functions, vol. 4: Applications of Harmonic Analysis (tr. by A. Feinstein), Academic Press, New York.

[445] S Gentleman, W. M. (1973): Least squares computations by Givens transformations without square roots, J. Inst. Math. Appl., 12, 329-336.

[446] Germain-Bonne, B. (1969): Calcul de pseudo-inverses,
P Rev. Française Informat. Recherche Opérationelle, 3, 3-14.

Gilbert, J. D.: See Barwick, D. T. (1974).

[447] Giles, R. and H. Kummer (1971): A matrix representation
T of a pair of projectors in a Hilbert space, Canad. Math. Bull., 14, 35-44.

[448] Gilfeather, F. (1969): Asymptotic convergence of op-
T erators in Hilbert space, Proc. Amer. Math. Soc., 22, 69-76.

[449] Gillman, L. and M. Jerison (1960): Rings of Continuous
S Functions, van Nostrand, New York.

[450] Giurescu, C. and R. Gabriel (1964): Some properties of
P the generalized matrix inverse and semi-inverse, An. Univ. Timisoara Ser. Sti. Mat.-Fiz., 2, 103-111.

[451] Glassey, C. R. (1966): An orthogonalization method of
P computing the generalized inverse of a matrix. Report ORC, 66-10, University of California College of Engineering, Operations Research Center, Berkley, Calif.

[452] Glazman, I. M. and Ju. I. Ljubich (1969): Finite Di-
T mensional Linear Analysis (Russian), Nauka, Moscow. English translation published by M. I. T. Press, 1974.

Glicksberg, I.: See Bellman, R. (1954).

[453] Gohberg, I. C and M. G. Krein (1957): Fundamental
S aspects of defect numbers, root numbers and indices of linear operators, Uspehi Mat. Nauk 12, 43-118. English translation in Amer. Math. Soc. Translations, Series 2, 13 (1960), 185-264.

[454] Goldberg, S. (1966): Unbounded Linear Operators,
S McGraw-Hill Book Co., New York, viii + 199 pp.

[455] Goldberg, S. (1974): Perturbations of semi-Fredholm
S operators by operators converging to zero compactly, Proc. Amer. Math. Soc., 45, 93-98.

[456] Goldberg, S. and A. Meir (1971): Minimum moduli of
S ordinary differential operators, Proc. London Math. Soc., 23, 1-15.

[457] Goldberger, A. S. (1961): Stepwise least squares resid-
S ual analysis and specification error, J. Amer. Statist. Assoc., 56, 998-1000.

[458] Goldman, A. J. and M. Zelen (1964): Weak generalized
P inverses and minimum variance linear unbiased estimation, J. Res. Nat. Bur. Standards, Sect. B, 68B, 151-172.

[459] Goldstein, A. A. (1956): On a method of descent in con-
S vex domains and its application to the minimal approximations of overdetermined systems of linear equations. Math. Preprint Series, Convair Astronautics, San Diego, California, No. 1.

[460] Goldstein, A. A. (1967): Constructive Real Analysis,
S Harper and Row, New York.

[461] Goldstein, A. A. and W. Cheney (1958): A finite algor-
S ithm for the solution of consistent linear equations and inequalities and for the Tchebycheff approximation of inconsistent linear equations, Pacific J. Math., 8, 415-428.

[462] Goldstein, A. A., N. Levine, and J. B. Hereshoff (1957):
S On "best" and "least qth" approximation of an overdetermined system of linear equations, J. Assoc. Comput. Mach., 4, 341-347.

Goldstein, A. A.: See Cheney, W. (1958).

[463] Goldstein, M. J. (1968): Solving systems of linear
P equations by using the generalized inverse, USL Report No. 876, US Navy Underwater Sound Laboratory, (AD667727).

[464] Goldstein, M. J. (1970): Linear least squares estima-
P tion using the generalized inverse, USE Proceedings, October, 1970.

[465] Goldstein, M. J. (1974): Reduction of the pseudoinverse
P of a Hermitian persymmetric matrix, Math. Comp., 28, 715-717.

[466] Goldstine, H. H. and J. von Neumann (1947): Numerical
T inverting of matrices of high order, Bull. Amer. Math. Soc., 53, 1021-1099.

[467] Golub, G. H. (1965): Numerical methods for solving
P linear least squares problems, Numer. Math., 7, 206-216.

[468] Golub, G. H. (1965): Numerical methods for solving
P linear least squares problems, Aplikace Matematiky, 10, 213-216.

[469] Golub, G. H. (1968): Least squares, singular values
P and matrix approximations, Aplikace Mathematiky, 13, 44-51.

[470] Golub, G. H. (1969): Matrix decompositions and sta-
S tistical calculations, pp. 365-397 in Statistical Calculations, ed. by R. C. Milton and J. A. Nelder, Academic Press, New York.

[471] Golub, G. H. and W. Kahan (1965): Calculating the
P singular values and pseudoinverse of a matrix, SIAM J. Numer. Anal., 2, 205-224.

[472] Golub, G. H. and V. Pereyra (1972): The differentiation
P of pseudoinverses and nonlinear least squares problems whose variables separate, Report STAN-CS-72-261, Stanford University, Computer Science Department. SIAM J. Numer. Anal. 10(1973), 413-432.

[473] Golub, G. H. and V. Pereyra (1975): Differentiation
P of pseudoinverses, separable nonlinear least squares problems, and other tales, this Volume.

[474] Golub, G. H. and C. Reinsch (1970): Singular value
P decomposition and least squares solutions, Numer. Math., 14, 403-420. Republished, pp. 134-151 in Wilkinson and Reinsch (1971).

[475] Golub, G. H. and M. A. Saunders (1969): Linear least squares and quadratic programming, pp. 229-256 in Integer and Nonlinear Programming, II, J. Abadie, ed., North Holland Publ. Co., Amsterdam.

[476] Golub, G. H. and G. P. H. Styan (1971): Numerical
S computations for univariate linear models, STAN-CS-236-71, Computer Science Dept., Stanford University. J. Statist. Comput. Stimulation, 2(1973), 253-274.

[477] Golub, G. H. and J. H. Wilkinson (1966): Note on
P iterative refinements of least squares solution, Numer. Math., 9, 139-148.

Golub, G. H.: See Björck, Å.(1967), (1973), Businger, P. A. (1965), (1969), Buzbee, B. L. (1970), Forsythe, G. E. (1965).

[478] Gončarskiĭ, A. V., A. S. Leonov, and A. G. Jagola
S (1974): Finite difference approximation of linear ill-posed problems. (Russian) Ž. Vyčisl. Mat. i Mat.-Fiz., 14, 15-24, 266.

[479] Good, I. J. (1969): Some applications of the singular
S decomposition of a matrix, Technometrics, 11, 823-831.

[480] Gotthardt, E. (1967): Vermittelnde Ausgleichung mit
S zusatzlichen Minimumbedingungen für Funktionen der Unbekanten, Z. Vermessungswes., 92(1), 11-17.

[481] Gramsch, B. (1973): Ein Zerlegungssatz für Resolventen
S elliptischer Operatoren, Math. Z., 133, 219-242.

[482] Gramsch, B. (1975): Inversion von Fredholmfunktionen
S bei stetiger und holomorpher Abhängigkeit von Parametern, Math. Ann., 214, 95-147.

Granot, F.: See Charnes, A. (1973).

[483] Graves, L. M. (1955): Remarks on singular points of
S functional equations, Trans. Amer. Math. Soc., 79, 150-157.

[484] Graybill, F. A. (1961): An Introduction to Linear Statistical Models, I, McGraw-Hill, New York.
S

[485] Graybill, F. A. (1969): Introduction to Matrices With Applications in Statistics, Wadsworth, Belmont, Calif.
S

[486] Graybill, F. A. and G. Marsaglia (1957): Idempotent
S matrices and quadratic forms in the general linear hypothesis, Ann. Math. Statist., 28, 678-686.

[487] Graybill, F. A., C. D. Meyer and R. J. Painter (1966):
P Note on the computation of the generalized inverse of a matrix, SIAM Rev., 8, 522-524.

[488] Green, B. (1952): The orthogonal approximation of an
T oblique structure in factor analysis, Psychometrika, 17, 429-440.

[489] Greub, W. and W. C. Rheinboldt (1960): Non-self-
S adjoint boundary value problems in ordinary differential equations, J. Res. Nat. Bur. Standards Sect. B, 64B, 83-90.

[490] Greville, T. N. E. (1957): On smoothing a finite table:
S A matrix approach, SIAM J. Appl. Math., 5, 137-154.

[491] Greville, T. N. E. (1957): The pseudoinverse of a rec-
P tangular or singular matrix and its application to the solution of systems of linear equations, SIAM News Letter, 5, 3-6.

[492] Greville, T. N. E. (1959): The pseudoinverse of a rec-
P tangular or singular matrix and its application to the solution of systems of linear equations, SIAM Rev., 1, 38-43.

[493] Greville, T. N. E. (1960): Some applications of the
P pseudoinverse of a matrix, SIAM Rev., 2, 15-22.

[494] Greville, T. N. E. (1961): Note on fitting of functions
S of several independent variables, SIAM J. Appl. Math., 9, 109-115. Erratum, ibid, (1961) 9, 317.

[495] Greville, T. N. E. (1966): Note on the generalized in-
P verse of a matrix product, SIAM Rev., 8, 518-521. Erratum, ibid, 9(1967), 249.

[496] Greville, T. N. E. (1967): Spectral generalized inverses
P of square matrices, Mathematics Research Center Technical Summary Report #823, University of Wisconsin, Madison, Wisconsin.

[497] Greville, T. N. E. (1968): Spectral generalized inverses
P of singular square matrices, Abstract in Notices Amer. Math. Soc., 15, 11.

[498] Greville, T. N. E. (1968): Some new generalzed in-
P verses with spectral properties, pp. 26-46 in Boullion and Odell (1968).

[499] Greville, T. N. E. (1973): The Souriau-Frame algorithm
P and the Drazin pseudoinverse, Linear Algebra and Appl., 6, 205-208.

[500] Greville, T. N. E. (1974): Solutions of the matrix equa-
P tion XAX = X, and relations between oblique and orthogonal projectors, SIAM J. Appl. Math., 26, 828-832.

[501] Greville, T. N. E. and N. Keyfitz (1974): Backward
P population projection by a generalized inverse, Theoretical Population Biology, 6, 135-142.

Greville, T. N. E.: See Ben-Israel, A. (1974), (1975), Cho, C. Y. (1966), Cline, R. E. (1970).

Grimshaw, M. E.: See Hamburger, H. L. (1951).

[502] Groetsch, C. W. (1974): Steepest descent and least
P squares solvability, Canadian Math. Bull, 17, 275-276.

[503] Groetsch, C. W. (1974): Some aspects of Mann's iter-
S ative method for approximating fixed points, Conference on Computing Fixed Points and Applications, Clemson University, June 1974.

[504] Groetsch, C. W. (1975): Ergodic theory and iterative
S solution of linear operator equations, Applicable Anal., to appear.

[505] Groetsch, C. W. (1975): Representations of the general-
P ized inverse, J. Math. Anal. Appl., 49, 154-157.

[506] Groetsch, C. W. (1975): A product integral representa-
P tion of the generalized inverse, Comment. Math. Univ. Carolinae, 16, 13-20.

[507] Groetsch, C. W. (1975): On existence criteria and ap-
S proximation procedure for integral equations of the first kind, Math. Comp., 29, 1105-1108.

Gross, O. A.: See Bellman, R. (1954), Fulkerson, D. R. (1965).

[508] Grusas, P. A. (1969): Satellite altitude determination
S from celestial sightings, Atrophys. J., 6, 1007-1012.

[509] Guedj, R. (1965): L'utilisation d'inverse généralises
P dans la résolution de systèmes linéaires de rang quelconque, Troisième Congr. de Calcul et de Traitement de l'Information Afcalti, 137-143.

[510] Guillemin, E. A. (1963): Theory of Linear Physical
T Systems, Wiley, New York, xvii + 586 pp.

[511] Gupta, N. N. (1972): On the convergence of an iterative
P method for the computation of generalized inverse and associated projections, Internat. J. Systems Sci., 2, 67-75.

[512] Gura, I. A. (1967): Notes on the pseudoinverse of a
P matrix, unpublished report.

[513] Ha, C. W. (1974): Approximation numbers of linear
S operators and nuclear spaces, J. Math. Anal. Appl., 46, 292-311.

Haber, M.: See Gabriel, K. R. (1973).

[514] Hadler, K. P. (1968): Eigenwertproblem, Linear Algebra
S and Appl., 1, 83-110.

[515] Hadler, K. P. (1969): Multiplicative inverses Eigen-
S wertproblem, Linear Algebra and Appl., 2, 65-86.

Hakimi, S. L.: See Manherz, R. K. (1968), (1969).

[516] Halany, A. and A. Moro (1968): A boundary value prob-
S lem and its adjoint, Ann. Mat. Pura Appl., 79, 399-411.

Halberg, C. J. A.: See Taylor, A. E. (1957).

[517] Hale, J. K. (1969): Ordinary Differential Equations,
S Wiley-Interscience, New York.

[518] Hale, J. K. (1971): Applications of Alternative Problems,
S Lecture Notes 71-1, Center for Dynamical Systems, Brown University, Providence, R. I.

Hale, J. K.: See Bancroft, S. (1968).

[519] Hall, F. J. (1973): The Fundamental Matrix of Constrained
P Minimization and Applications, Ph. D. Thesis, North Carolina State University, Raleigh, N. C.

[520] Hall, F. J. (1975): Generalized inverses of a bordered
P matrix of operators, SIAM J. Appl. Math., 29, 152-163.

[521] Hall, F. J. and C. D. Meyer, Jr. (1975): Generalized
P inverses of the fundamental bordered matrix used in linear estimation, Sankhyā, Series A, to appear.

[522] Hall, W. S. (1971): The bifurication of solutions in
S Banach spaces, Trans. Amer. Math. Soc., 161, 207-218.

[523] Hallum, C. R., T. O. Lewis and T. L. Boullion (1973):
P Estimation in the regression general linear model with a positive semidefinite covariance matrix, Communications in Statistics, 1, 157-166.

[524] Hallum, C. R., T. L. Boullion and P. L. Odell (1973):
S Parameter estimation and hypothesis testing in the restricted linear model, Indust. Math., 23, 1-25.

[525] Halmos, P. R. (1958): Finite-Dimensional Vector Spaces
T (2nd ed.), van Nostrand, Princeton, N. J., vii + 195pp.

[526] Halmos, P. R. (1968): A Hilbert Space Problem Book,
S van Nostrand, Princeton, N. J., xvii + 365 pp.

[527] Halmos, P. R. and J. E. McLaughlin (1963): Partial
S isometrics, Pacific J. Math., 13, 585-596.

[528] Halmos, P. R. and L. J. Wallen (1970): Powers of par-
S tial isometries, J. Math. Mech., 19, 657-663.

[529] Halperin, I. (1937): Closures and adjoints of linear
T differential operators, Ann. of Math., 38, 880-919.

[530] Hamburger, H. L. (1951): Non-symmetric operators in
S Hilbert space, Proceedings of the Symposium on Spectral Theory and Differential Problems, pp. 67-112. Oklahoma Agricultural and Mechanical College, Stillwater, Oklahoma.

[531] Hamburger, H. L. and M. E. Grimshaw (1951): Linear
S Transformations in N-Dimensional Vector Space, Cambridge Univ. Press, Cambridge.

[532] Hanna, M. P. (1972): Generalized overrelaxation and
T Gauss-Seidel convergence on Hilbert space, Proc. Amer. Soc., 35, 524-530.

[533] Hansen, G. W. and D. W. Robinson (1974): On the ex-
P istence of generalized inverses, Linear Algebra and Appl., 8, 95-104.

[534] Hanson, R. J. (1971): A numerical method for solving
S Fredholm integral equations of the first kind using singular values, SIAM J. Numer. Anal., 8, 616-622.

[535] Hanson, R. J. and C. L. Lawson (1969): Extensions and
P applications of the Householder algorithm for solving linear least squares problems, Math. Comp., 23, 787-812.

Hanson, R. J.: See Lawson, C. L. (1974).

[536] Harada, S. (1953): An existence proof of the generalized
S Green's function, Osaka Math. J., 5, 59-63.

[537] Harary, F. (1962): A graph theoretic approach to matrix
T inversion by partitioning, Numer. Math., 4, 128-135.

[538] Harris, W. A. and T. V. Helvig (1966): Applications of
P the pseudoinverses to modelling, Technometrics, 8, 351-357.

[539] Harris, W. A. and T. V. Helvig (1966): Marginal and
S conditional distributions of singular distributions, Publications of the Res. Inst. for Math. Sci., Kyoto University, Ser. A, 1, 199-204.

Hart, P. E.: See Duda, R. O. (1973).

[540] Harter, H. L. (1974): The method of least squares and
S some alternatives, I, Internat. Statist. Rev., 42, 147-174.

[541] Harter, H. L. (1974): The method of least squares and
S some alternatives, II, Internat. Statist. Rev., 42, 235-264.

[542] Hartfiel, D. J. (1971): The matrix equation AXB = X,
S Pacific J. Math., 36, 659-669.

[543] Hartfiel, D. J. (1974): A result concerning strongly er-
S godic nonhomogeneous Markov chains, Linear Algebra and Appl., 9, 169-174.

[544] Hartfiel, D. J. (1974): A study of convex sets of
S stochastic matrices induced by probability vectors, Pacific J. Math., 52, 405-418.

[545] Hartfiel, D. J. (1974): Concerning spectral inverses of
P stochastic matrices, SIAM J. Appl. Math., 27, 281-292.

[546] Hartfiel, D. J. (1975): Results on measures of irreduci-
S bility and full indecomposability, Trans. Amer. Math. Soc., 202, 357-368.

[547] Hartwig, R. E. (1972): The resultant and the matrix
S equation AX = XB, SIAM J. Appl. Math., 22, 538-544.

[548] Hartwig, R. E. (1974): Singular values and g-inverses
P of bordered matrices, to appear.

[549] Hartwig, R. E. (1974): AX - XB = C, resultants and gen-
P eralized inverses, SIAM J. Appl. Math., 28, 154-183.

[550] Hartwig, R. E. (1975): Block generalized inverses,
P Arch. Rational Mech. Anal., to appear.

[551] Hartwig, R. E. (1975): Rank factorization and g-inver-
P sion, to appear.

[552] Hartwig. R. E. (1975): 1 - 2 inverses and the invari-
P ance of $BA^{\dagger}C$, Linear Algebra and Appl., 11, 271-275.

[553] Hartwig, R. E. and M. E. Fisher (1969): Asymptotic
T behavior of Toeplitz matrices and determinants, Arch. Rational Mech. Anal., 32, 190-225.

Harvey, J. R.: See Rohde, C. A. (1965).

[554] Harwood, W. R., V. Lovass-Nagy and D. L. Powers
P (1970): A note on the generalized inverses of some partitioned matrices, SIAM J. Appl. Math., 19, 555-559.

[555] Hausdorff, F. (1932): Zur Theorie der linearen
S metrischen Räume, J. Reine Angew Math., 167, 294-311.

[556] Hawkins, J. B. and A. Ben-Israel (1973): On generalized matrix functions, Linear and Multilinear Algebra, 1, 163-171.
P

Haynsworth, E.: See Burns, F. (1974), Carlson, D. (1974).

[557] Hazra, P. K. and K. S. Banerjee (1973): On the augmentation procedure in singular weighing designs, J. Amer. Statist. Assoc., 68, 392-393.
P

[558] Hearon, J. Z. (1967): Construction of EPr generalized inverses by inversion of nonsingular matrices, J. Res. Nat. Bur. Standards, Sect. B, 71B, 57-60.
P

[559] Hearon, J. Z. (1967): A generalized matrix version of Rennie's inequality, J. Res. Nat. Bur. Standards, Sect. B, 71B, 61-64.
P

[560] Hearon, J. Z. (1967): Polar factorization of a matrix, J. Res. Nat. Bur. Standards, Sect. B, 71B, 65-67.
S

[561] Hearon, J. Z. (1967): Partially isometric matrices, J. Res. Nat. Bur. Standards, Sect. B, 71B, 225-228.
S

[562] Hearon, J. Z. (1967): Symmetrizable generalized inverses of symmetrizable matrices, J. Res. Nat. Bur. Standards, Sect. B, 71B, 229-231.
P

[563] Hearon, J. Z. (1967): On the singularity of a certain bordered matrix, SIAM J. Appl. Math., 15, 1413-1421.
P

[564] Hearon, J. Z. (1968): Generalized inverses and solutions of linear systems, J. Res. Nat. Bur. Standards, Sect. B, 72B, 303-308.
P

[565] Hearon, J. Z. (1968): Idempotent matrices with nilpotent difference, Mathematics Magazine, 41, 80-84.
S

[566] Hearon, J. Z. and J. W. Evans (1968): On spaces and
P maps of generalized inverses, J. Res. Nat. Bur. Standards, Sect. B, 72B, 103-107.

[567] Hearon, J. Z. and J. W. Evans (1968): Differentiable
P generalized inverses, J. Res. Nat. Bur. Standards, Sect. B, 72B, 109-113.

[568] Heath, M. T. (1974): The numerical solution of ill-
P conditioned systems of linear equations. Report based on the author's University of Tennessee M. S. Thesis. Report No. ORNL-4957. Oak Ridge National Laboratory, Oak Ridge, Tenn., 54 pp.

Helvig, T. V.: See Harris, W. A. (1966), (1966).

[569] Henderson, C. R. (1953): Estimation of variance and
T covariance components, Biometrics, 9, 226-252.

Hereshoff, J. B.: See Goldstein, A. A. (1957).

[570] Herring, G. P. (1967): A note on generalized interpola-
P tion and the pseudoinverse, SIAM J. Numer. Anal., 4, 548-556.

[571] Hestenes, M. R. (1951): Quadratic forms in Hilbert
S space, Pacific J. Math., 1, 525-581.

[572] Hestenes, M. R. (1958): Inversion of matrices by bior-
P thogonalization and related results, SIAM J. Appl. Math., 6, 51-90.

[573] Hestenes, M. R. (1961): Relative Hermitian matrices,
S Pacific J. Math., 11, 225-245.

[574] Hestenes, M. R. (1961): Relative self-adjoint operators
P in Hilbert space, Pacific J. Math., 11, 1315-1357.

[575] Hestenes, M. R. (1962): A ternary algebra with appli-
S cations to matrices and linear transformations, Arch. Rational Mech. Anal., 11, 138-194.

[576] Hestenes, M. R. (1969): Quadratic variational theory,
T Control Theory and the Calculus of Variations, A. V. Balakrishnan, ed., Academic Press, New York.

[577] Hestenes, M. R. (1973): On a ternary algebra, Scripta
S Mathematica, 29, 253-272.

[578] Hestenes, M. R. (1975): A role of the pseudoinverse in
P analysis, this Volume.

[579] Hestenes, M. R. (1975): Pseudoinverses and conjugate
P gradients. Collection of articles honoring Alston S. Householder, Comm. ACM, 18, 40-43.

[580] Hestenes, M. and R. Redheffer (1974): On the minimi-
S zation of certain quadratic functionals. I. Arch. Rational Mech. Anal., 56, 1-14; II., ibid 56, 807-817.

[581] Hilbert, D. (1912): Grundzüge einer allgemeinen Theorie
S der linearen Integralgleichungen, B. G. Teubner, Leipzig and Berlin. xxvi + 282 pp. Reprint of six articles which appeared originally in the Göttingen Nachrichten (1904, pp. 49-51; 1904, pp. 213-259; 1905, pp. 307-338; 1906, pp. 157-227; 1906, pp. 439-480; 1910, pp. 355-417).

Hilbert, D.: See Courant, R. (1931).

[582] Hildebrandt, S. and E. Wienholtz (1964): Constructive
S proofs of representation theorems in separable Hilbert space, Comm. Pure Appl. Math., 17, 369-373.

[583] Hilgers, J. W. (1974): Non-Iterative Methods for Solv-
P ing Operator Equations of the First Kind, Ph.D. Thesis, University of Wisconsin-Madison (1973); also MRC Tech. Summary Rept. #1413, Mathematics Research Center, University of Wisconsin-Madison.

[584] Hilgers, J. W. (1974): Approximating the optimal regularization parameter, MRC Technical Summary Report #1472, Mathematics Research Center, University of Wisconsin.
T

[585] Hilgers, J. W. (1975): On the equivalence of regularization and certain reproducing kernel Hilbert space approaches for solving first kind problems, to appear.
P

[586] Ho, B. L. and R. E. Kalman (1966): Effective construction of linear state-variables models from input/output functions, Regelungstechnik, 14, 545-548.
S

[587] Ho, Y.-C. and A. K. Agrawala (1968): On pattern classification algorithms: Introduction and survey, Proceedings of the IEEE, 56, 2101-2114.
S

[588] Ho, Y.-C. and R. L. Kashyap (1965): An algorithm for linear inequalities and its applications, IEEE Transactions on Electronic Computers, EC-14, 683-688.
S

[589] Ho, Y.-C. and R. L. Kashyap (1966): A class of iterative procedures for linear inequalities, SIAM J. Control, 4, 112-115.
S

Ho, Y.-C.: See Kalman, R. E. (1963).

Hoffman, A. J.: See Fan, Ky (1955).

[590] Hogan, D. A. and C. E. Langenhop (1975): Invertibility in a Banach algebra, Indiana Univ. Math. J., 24, 965-977.
S

[591] Hogg, R. V. and A. T. Craig (1958): On the decomposition of certain χ-square variables, Ann. Math. Statist., 29, 608-610.
T

Holcomb, B. D.: See Replogle, S. (1967).

[592] Hölder, E. (1935): Die lichtensteinsche Methode für die Entwicklung des zweiten Variation, angewandt auf das Problem von Lagrange, Prace Mathematycano-Fizyczne, 43, 307-346.
T

[593] Holmes, R. B. (1972): A Course on Optimization and
S Best Approximation, Springer-Verlag, Berlin, viii + 233 pp.

[594] Holmes, R. B. (1972): R-splines in Banach spaces: I.
S Interpolation of linear manifolds, J. Math. Anal. Appl., 40, 574-593.

[595] Horst, P. (1963): Matrix Algebra for Social Scientists,
S Chapters 17-20, Holt, Rinehart, and Winston, Inc., New York.

[596] Householder, A. S. (1958): Unitary triangularization of
S a nonsymmetric matrix, J. Assoc. Comput. Mach., 5, 339-342.

[597] Householder, A. S. (1964): The Theory of Matrices in
S Numerical Analysis, Blaisdell, New York, xi + 257 pp.

[598] Householder, A. S. and G. Young (1938): Matrix approx-
T imation and latent roots, Amer. Math. Monthly, 45, 165-171.

Householder, A. S.: See Bauer, F. L. (1960).

Hovanskiĭ, A. V.: See Petrov, A. P. (1974).

[599] Hsiao, G. C. and W. L. Wendland (1975): A finite ele-
S ment method for some integral equations of the first kind, to appear.

[600] Hsu, P. U. (1946): On a factorization of pseudoorthog-
T onal matrices, Quart. J. Math. Oxford Ser., 17, 162-165.

[601] Hung, C. H. and T. L. Markham (1975): The Moore-
P Penrose inverse of a partitioned matrix $M = \begin{pmatrix} A & B \\ C & D \end{pmatrix}$, Linear Algebra and Appl., 11, 73-86.

[602] Hunter, J. J. (1969): On moments of Markov renewal
S processes, Adv. Appl. Probability, 1, 188-210.

[603] Hurt, M. F. and C. Waid (1970): A generalized inverse
P which gives all the integral solutions to a system of linear equations, SIAM J. Appl. Math., 19, 547-550.

[604] Hurwitz, W. A. (1912): On the pseudo-resolvent to the
P kernel of an integral equation, Trans. Amer. Math. Soc., 13, 405-418.

[605] IEEE Transactions on Automatic Control, Special Issue
S on the Linear-Quadratic-Gaussian Problem, AC-16, 1971.

[606] Ijiri, Y. (1965): On the generalized inverse of an inci-
P dence matrix, SIAM J. Appl. Math., 13, 827-836.

Ijiri, Y.: See Ben-Israel, A. (1963).

[607] Ikramov, H. D. (1974): The eigenvalues of a pseudo-
P inverse matrix. (Russian, English Summary), Vestnik Moskov. Univ. Ser. I Mat. Meh., 29, 5-8.

[608] Ilioni (1974): The conjugate directions and conjugate
S gradients methods for linear equations with self-adjoint operators, Boll. Un. Mat. Ital., (4) 9, 16-22.

[609] Ivanov, V. K. (1974): The value of the regularization
S parameter in ill-posed control problems (Russian). Differencial'nye Uravenija, 10, 2279-2285.

Jagola, A. G.: See Gončarskiĭ, A. V. (1974).

[610] Jameson A. (1968): Solution of equation $AX + XB = C$
S by inversion of an $m \times m$ or $n \times n$ matrix, SIAM J. Appl. Math., 16, 1020-1023.

[611] Jaunzems, A. Ja. (1974): The stability of openness and
P normal solvability of linear operators. (Russian, Latvian and English summaries) Latvian Mathematical Yearbook, 14, 233-236, Izdat. "Zinatre" Riga, 1974.

[612] Jeffreys, H. (1932): On the theory of errors and least
S squares, Proc. Roy. London Soc., 138, 48-55.

[613] Jenkins, M. A. (1970): The solution of linear systems
S of equations and linear least squares problems in APL, IBM-New York Scientific Center, Tech. Rept. No. 320-2989, June, 1970, 14 pp.

[614] Jennings, L. S. and M. R. Osborne (1970): Applications
S of orthogonal matrix transformations to the solution of systems of linear and nonlinear equations, The Australian National University, Tech. Rept. No. 37, 45 pp.

[615] Jennings, L. S. and M. R. Osborne (1974): A direct
P error analysis for least squares, Numer. Math., 22, 325-332.

[616] Jensen, H. (1944): An attempt at a systematic classifi-
S cation of some methods for solution of normal equations, Geod. Inst. Medd., 18, Copenhagen.

Jerrison, M.: See Gillman, L. (1960).

[617] John, J. A. (1970): Use of generalized inverse matrices
P in MANOVA, J. Roy. Statist. Soc., Ser. B, 32, 137-143.

[618] John, P. W. M. (1964): Pseudoinverses in the analysis
P of variance, Ann. Math. Statist., 35, 895-896.

[619] Jones, John, Jr. (1972): Solution of certain matrix equa-
S tions, Proc. Amer. Math. Soc., 31, 333-339.

[620] Jones, John, Jr. (1965): On the Lyapunov stability
S criteria, SIAM J. Appl. Math., 13, 941-945.

Jonnes, L.: See Marchant, R. (1969).

Jordan, B. W.: See Manherz, R. K. (1968).

[621] Jordan, T. L. (1966): Experiments in error growth associ-
S ated with some linear least-squares procedures, Math. Comp., 21, 579-588.

[622] Joshi, V. N. (1970): A note on the solution of rectangular
S systems by iteration, SIAM Review, 12, 463-466.

[623] Joshi, V. N. (1973): Remarks on iterative methods for
P computing the generalized inverse, Studia Sci. Math. Hung., 8, 457-461.

Joshi, V. N.: See Tewarson, R. P. (1972).

[624] Jossa, F. (1940): Risoluzione progressiva di un sistema
T di equazioni lineari. Analogia con un problema meccanico. Rend. Accad. Sci. Fis. Mat. Napoli, (4) 10, 346-352.

[625] Kaashoek, M. A. (1967): Ascent, descent, nullity and
T defect: a note on a paper of A. E. Taylor, Math. Ann., 172, 105-115.

[626] Kaashoek, M. A. and D. C. Lay (1972): Ascent, de-
T scent, and community perturbations, Trans. Amer. Math. Soc., 169, 35-47.

Kaashoek, M. A.: See Bart, H. (1974), Förster, K. H. (1975).

Kabe, D. G.: See Wani, J. K. (1970).

[627] Kaczmarz, S. (1937): Angenäherte Auflösung von Syste-
T men linearer Gleichungen, Bull. Int. Acad. Pol. Sci., A, 355-357.

[628] Kadets, M. I. and B. S. Mityagin (1973): Complemented
S subspaces in Banach spaces, Russian Math. Surveys, 28, 77-95.

Kahan, W.: See Golub, G. (1965).

Kahng, S. W.: See Decell, H. P., Jr. (1966).

[629] Kailath, T. (1974): A view of three decades of linear
S filtering theory, IEEE Trans. Information Theory, IT-20, 146-181.

[630] Kakutani, S. (1939): Some characterizations of Euclidean
T spaces, Japan J. Math., 16, 93-97.

Kalaba, R: See Bellman, R. (1966).

[631] Kallina, C. (1969): A Green's function approach to per-
S turbations of periodic solutions, Pacific J. Math., 29, 325-334.

[632] Kalman, R. E. (1960): A new approach to linear filtering
S and prediction problems, Trans. ASME Ser. D, Jour. Basic Engineering, 82, 35-44.

[633] Kalman, R. E. (1960): Contributions to the theory of
S optimal control, Bol. Soc. Mat. Mexicana, 5, 102-119.

[634] Kalman, R. E. (1961): New results in linear filtering and
S prediction theory, Trans. ASME Ser. D, Jour. of Basic Engineering, 83, 95-107.

[635] Kalman, R. E. (1963): Mathematical description of linear
T dynamical systems, SIAM J. Control, 1, 152-192.

[636] Kalman, R. E. (1963): New methods in Wiener filtering
S theory, in Proc. 1st Symp. on Engineering Applications of Random Function Theory and Probability, pp. 270-388, Wiley, New York.

[637] Kalman, R. E. (1975): Algebraic aspects of the general-
P ized inverse, this Volume.

[638] Kalman, R. E., P. L. Falb and M. A. Arbib (1969):
S Topics in Mathematical System Theory, McGraw-Hill, New York, ix + 358 pp.

[639] Kalman, R. E., Y.-C. Ho and K. S. Narenda (1963):
S Controllability of linear dynamic systems, Contributions to Differential Equations, vol. I, pp. 189-213, Wiley-Interscience, New York.

Kalman, R. E.: See Ho, B. L. (1966).

[640] Kammerer, W. J. and M. Z. Nashed (1971): Iterative
P methods for best approximate solutions of linear integral equations of the first and second kinds, MRC Report No. 1117, Mathematics Research Center, University of Wisconsin, Madison, J. Math. Anal. Appl., 40 (1972), 547-573.

[641] Kammerer, W. J. and M. Z. Nashed (1971): Steepest
P descent for singular linear operators with nonclosed range, Applicable Anal., 1, 143-159.

[642] Kammerer, W. J. and M. Z. Nashed (1971): A general-
P ization of a matrix iterative method of G. Cimmino to best approximate solution of linear integral equations of the first kind, Atti Accad. Naz. Lincei Rend. Cl. Sci. Fis. Mat. Natur. Ser. VIII, 51, 20-25.

[643] Kammerer, W. J. and M. Z. Nashed (1972): On the con-
P vergence of the conjugate gradient method for singular linear operator equations, SIAM J. Numer. Anal., 9, 165-171.

[644] Kammerer, W. J. and R. J. Plemmons (1975): Direct
P iterative methods for least-squares solutions to singular operator equations, J. Math. Anal. Appl., 49, 512-526.

[645] Kaniel, S. and J. Stein (1974): Least-square accelera-
S tion of iterative methods for linear equations, J. Optimization Theory Appl., 14, 431-437.

[646] Kantorovitz, S. (1965): Classification of operators by
T means of their operational calculus, Trans. Amer. Math. Soc., 115, 194-224.

[647] Kantorovich, L. V. and G. P. Akilov (1964): Functional
T Analysis in Normed Spaces (Translated from Russian by D. E. Brown), Pergamon Press, Oxford, England, xiii + 773 pp.

[648] Kantorovich, L. V. and V. I. Krylov (1958): Approximate
T Methods of Higher Analysis, Interscience, New York, xii + 681 pp.

[649] Kaplansky, I. (1948): Regular Banach algebras, J.
S Indian Math. Soc., N. S. 12, 57-62.

[650] Kaplansky, I. (1955): Any ortho-complemented complete
S modular lattice is a continuous geometry, Ann. of Math., 61, 524-541.

[651] Kaplansky, I. (1968): Rings of Operators, W. A. Benjamin,
T Inc., New York.

[652] Kara, H. (1971): Wide-sense Martingale Approach to
S Discrete-Time Optimal Estimation, Ph.D. Dissertation, Michigan State University, East Lansing. Reprinted as Simulation, Estimation, and Control in Power Systems, Prog. Rep. No. 3, Division of Engineering Research, Michigan State University, 1971.

[653] Kashyap, R. L. (1970): Algorithms for pattern classifi-
S cation, in Adaptive Learning and Pattern Recognition Systems, J. M. Mendel and K. S. Fu, eds., Academic Press, New York.

Kashyap, R. L.: See Ho, Y.-C. (1965), (1966).

[654] Kato, T. (1958): Perturbation theory for nullity, defici-
P ency and other qualities of linear operators, J. Analyse Math., 11, 261-322.

[655] Kato, T. (1966): Perturbation Theory for Linear Operators,
S die Grundlehren der mathematischen Wissenschaften, vol. 132, Springer-Verlag, Berlin-Heidelberg-New York, 592 pp.

[656] Katz, I. J. (1965): Weigmann type theorems for EPr
S matrices, Duke Math. J., 32, 423-427.

[657] Katz, I. J. (1970): Remarks on a paper of Ben-Israel,
P SIAM J. Appl. Math., 18, 511-513.

[658] Katz, I. J. (1972): Remarks on two recent results in
S matrix theory, Linear Algebra and Appl., 5, 109-112.

[659] Katz, I. J. (1975): Theorems on products of EPr
P matrices. II. Linear Algebra and Appl., 10, 37-40.

[660] Katz, I. J. and M. H. Pearl (1966): On EPr and normal
P EPr matrices, J. Res. Nat. Bur. Standards, Sect. B, 70B, 47-77.

[661] Katz, I. J. and M. H. Pearl (1966): Solutions of the
S matrix equations A = XA = AX, J. London Math. Soc., 41, 443-452.

Katz, I. J.: See Baskett, T. S. (1969).

[662] Kaufmann, Linda (1974): Variable Projection Method for
S Solving Separable Nonlinear Least Squares Problems, Ph.D. Thesis, Stanford University, Palo Alto, California.

[663] Kaufman, Linda (1974): The LZ-algorithm to solve the
S generalized eigenvalue problem, SIAM J. Numer. Anal., 11, 997-1024.

Kazda, L. F.: See Kuo, M. C. Y. (1967).

[664] Keller, H. B. (1965): On the solution of singular and
S semi-definite linear systems by iteration, SIAM J. Numer. Anal., 2, 281-290.

[665] Keller, H. B. (1970): Nonlinear bifurication, J. Differ-
T ential Equations, 3, 417-435.

[666] Keller, H. B. and W. F. Langford (1974): Iterations,
T perturbations and multiplicities for nonlinear bifurcation problems, Arch. Rational Mech. Anal., 48, 83-108.

[667] Kellogg, R. B. and J. Spanier (1965): On optimal alter-
S nating direction parameters for singular matrices, Math. Comp., 19, 448-452.

Keyfitz, N.: See Greville, T. N. E. (1974).

[668] Khatri, C. G. (1963): Further contributions to
S Wishartness and independence of second degree polynomials in normal vectors, J. Indian Statist. Assoc., 1, 61-70.

[669] Khatri, C. G. (1968): Some results for the singular
P multivariate regression models, Sankhyā, Ser. A, 30, 267-280.

[670] Khatri, C. G. (1969): A note on some results on generalized inverse of a matrix, J. Indian Statist. Assoc., 7, 38-45.
P

[671] Khatri, C. G. (1970): A note on a commutative g-inverse of a matrix, Sankhyā, Ser. A, 32, 299-310.
P

[672] Khatri, C. G. and C. R. Rao (1968): Solution to some functional equations and their applications to characterization of probability distribution, Sankhyā, Ser. A, 30, 167-180.
S

[673] Kim, J. B. (1966): On singular matrices, J. Korean Math. Soc., 3, 1-2.
P

[674] Kirby, M. J. L. (1965): Generalized Inverses and Chance-Constrained Programming, Doctoral Dissertation in Applied Mathematics, Northwestern University, Evanston, Ill., June, 1965.
P

Kirby, M. J. L.: See Ben-Israel, A. (1969), Charnes, A. (1963), (1965).

[675] Kishi, F. H. (1964): On Line Computer Control Techniques and their Application to Re-entry Aerospace Vehicle Control, pp. 245-257 in Advances in Control Systems Theory and Applications. (C. T. Leondes, Editor), Academic Press, New York.
S

Kleindorfer, G. D.: See Anderson, W. N., Jr. (1969).

Kleindorfer, P. R.: See Anderson, W. N., Jr. (1969).

[676] Klinger, A. (1968): Approximate pseudoinverse solutions to ill-conditioned linear systems, J. Optimization Theory Appl., 2, 117-128.
P

[677] Koliha, J. J. (1972): On the iterative solution of linear
S operator equations with selfadjoint operators, J. Austral. Math. Soc., 13, 241-255.

[678] Koliha, J. J. (1973): Convergent and stable operators
S and their generalizations, J. Math. Anal. Appl., 43, 778-794.

[679] Koliha, J. J. (1973): Ergodic theory and averaging iter-
S ations, Canadian J. Math., 25, 14-23.

[680] Koliha, J. J. (1974): Series representation of pseudo-
P inverses and partial inverses of operators, Bull. Amer. Math. Soc., 80, 325-328.

[681] Koliha, J. J. (1974): Power convergence and pseudo-
P inverses of operators between banach spaces, J. Math. Anal. Appl., 48, 446-469.

[682] Koliha, J. J. (1975): The product of relatively regular operators, Comment. Math. Univ. Carolinae, 16, to appear.

[683] Koliha, J. J. (1975): Convergence of an operator series,
P to appear.

[684] Kolmogorov, A. N. (1946): On the motivation of the
S method of least squares (Russian), Uspehi Mat. Nauk., 1, 57-70.

[685] Kolomý, J. (1974): Normal solvability and solvability
S of nonlinear equations, Theory of Nonlinear Operators, Proceedings of a Summer School, pp. 155-167, Akademie-Verlag, Berlin.

[686] Kolomý, J. (1974): Normal solvability, solvability and
S fixed point theorems, Colloq. Math., 29, 761-764.

[687] Koop, J. C. (1963): Generalized inverse of a singular
P matrix, Nature, 198, 1019-1020 and 200, 716.

[688] Korganoff, A. (1961): Functions of a normed vector
P space applied to the iterative solution of rectangular and square matrix non-linear equations of any given form, Nordisk Symposium, Oslo, August 18-22, 1961.

[689] Korganoff, A. (1962): Les polynômes d'interpolation de
P matrices carrées a coefficients matriciels et les méthodes iteratives de résolution numérique des équations de matrices carrées de forme quelconque, Proc. IFIP Congress 1962, 102-106.

[690] Korganoff, A. (1964): The inversion of rectangular
P matrices in the resolution of ill-conditioned linear systems, Proc. Nordsam Congress, Helsinki, 2, 179-190.

[691] Korganoff, A. (1970): Sur la résolution de problèmes
S inverses en hydrogéologie, Bull. Int. Assoc. Sci. Hydrol., 15, 67-78.

[692] Korganoff, A. and M. Pavel-Parvu (1964): Interprétation
P a l'aide des pseudo-inverses de la solution d'équations matricielles linéaires provenant de la discretisation d operateurs differentiels et intergraux, 83e Congrès de L'Association Française Pour L'Avancement des Sciences, Lille (July, 1964).

[693] Korganoff, A. and M. Pavel-Parvu (1967): Méthodes de
P calcul numérique-2. Eléments de théorie des carrées et rectangles en analyse numérique, Dunod, Paris, xx + 441 pp. English translation to be published by Gordon and Breach, under the title: Norms and Pseudoinverses in Numerical Analysis.

[694] Korganoff, A. and M. Pavel-Parvu (1973): Iterative
P methods for pseudo-inverse computation: Methods of additive decomposition, a preprint.

Korganoff, A.: See Bonnier, A. (1972), Pavel-Parvu, M. (1969).

[695] Korkina, L. F. (1974): Estimation of error in the solu-
S tion of ill-posed problems (Russian), Ž. Vyčisl. Mat. i Mat. Fiz., 14, 584-597, 811.

[696] Krarup, T. (1969): A contribution to the mathematical
S foundation of physical geodesy, Danish Geod. Inst. Publ., 44. Copenhagen.

Krein, M. G.: See Gohberg, I. C. (1957).

[697] Krishnamurthy, E. V. and S. S. Prabhu (1974): Iterative
S solution of a class of linear equations with application to reconstruction of three dimensional object arrays, TR-325, Computer Science Center, University of Maryland, College Park, Maryland.

[698] Kruskal, W. (1960): The coordinate-free approach to
S Gauss-Markov estimation and its application to missing and extra observations, Proc. of the Fourth Berkeley Symp. Math. Statist. and Prob., 1, 435-451.

[699] Kruskal, W. (1968): When are Gauss-Markoff and least
P squares estimators identical? A coordinate-free approach, Ann. Math. Statist., 39, 70-75.

[700] Kruskal, W. (1975): The geometry of generalized in-
P verse, J. Royal Stat. Soc., Ser. B, 37, 272-283.

[701] Krjanev, A. V. (1974): An iteration method for the solution of ill-posed problems (Russian) Ž. Vyčisl. Mat. i Mat. Fiz., 14, 25-35, 266.

Krylov, V. I.: See Kantorovich, L. V. (1958).

[702] Kshirsagar, A. M. (1972): Multivariate Analysis, Marcel
S Dekker, New York, N.Y., 552 pp.

[703] Kubik, K. (1970): The estimation of the weights of
S measured quantities within the method of least squares. Bull. Géodésique, 95, 21-32.

[704] Kublanovskaja, V. N. (1966): On the computation of
P the generalized matrix inverse and projections, USSR Comp. Math. Math. Phys., 6, 179-188, Ž. Vyčisl. Mat. i Mat. Fiz., 6, 326-332.

[705] Kublanovskaja, V. N. (1970): On an approach to the
P solution of inverse eigenvalues problems, (Russian) Automatic Programming and Numerical Methods of Analysis, 18, 138-149.

[706] Kublanovskaja, V. N. (1971): Application of orthogonal
S transformations to the solutions of nonlinear systems, Numerical Methods and Functional Analysis, 23, 53-71.

[707] Kublanovskaja, V. N. (1971): Application of a normali-
S zation process to the solution of the inverse eigenvalue problem for matrices, (Russian) Numerical Methods and Functional Analysis, 23, 72-83.

[708] Kublanovskaja, V. N. (1972): Application of Orthor-
S thogonal Transformation to the Solution of Algebraic Problems, Doctoral Dissertation, Leningrad State University, 15 pp.

[709] Kublanovskaja, V. N. (1972): Applications of the or-
S thogonal transformations to the solution of one extremal problem, IFIP Information Processing, North-Holland Publishing Company, Amsterdam, pp. 1311-1316.

Kublanovskaja, V. N.: See Faddeev, D. K. (1968), (1968), (1968).

Kučera, M.: See Fučik, S. (1975).

Kummer, H.: See Giles, R. (1971).

[710] Kuo, M. C.Y. and L. F. Kazda (1967): Minimum energy
S problems in Hilbert function space, J. Franklin Inst., 283, 38-54.

[711] Kurepa, S. (1968): Generalized inverse of an operator
P with a closed range, Glasnik Mat., 3(23), 207-214.

[712] Kuznecov, Ju. A. (1969): Iterative methods and mini-
S mization of a functional, Computing Methods in Transport Theory (Proc. First All-Union Sympos., Novosibirsk, 1967) (Russian), pp. 96-109, Atomizdat, Moscow.

Kuznecov, Ju. A.: See Marchuk, G. I.

[713] Kuznecov, V̌. C. (1967): Solution of a system of linear
S equations, Ž. Vyčisl. Mat. i. Mat. Fiz., 7, 157-160.

[714] Kwon, Y. K. and R. M. Redheffer (1969): Remarks on
S linear equations in Banach space, Arch. Rational Mech. Anal., 32, 247-254.

[715] Lancaster, P. (1969): Theory of Matrices, Academic
S Press, New York, xii + 316 pp.

[716] Lancaster, P. (1970): Explicit solutions of linear matrix
S equations, SIAM Review, 12, 544-566.

[717] Lanczos, C. (1958): Linear systems in self-adjoint
S form, Amer. Math. Monthly, 65, 665-679.

[718] Lanczos, C. (1961): Linear Differential Operators,
S D. van Nostrand Co., Princeton, N. J., 124-129.

[719] Landesman, E. M. (1967): Hilbert space methods in
P elliptic partial differential equations, Pacific J. Math., 21, 113-131.

[720] Landesman, E. M. (1968): A generalized Lax-Milgram
S theorem, Proc. Amer. Math. Soc., 19, 339-344.

[721] Landesman, E. M. and A. C. Lazer (1970): Nonlinear
S perturbations of linear elliptic boundary value problems at resonance, J. Math. Mech., 19, 609-623.

[722] Landweber, L. (1951): An iteration formula for Fredholm
S integral equations of the first kind, Amer. J. Math., 73, 615-624.

[723] Langenhop, C. E. (1967): On generalized inverse of
P matrices, SIAM J. Appl. Math., 15, 1239-1246.

[724] Langenhop, C. E. (1971): On the index of a square
P matrix, SIAM J. Appl. Math., 21, 191-194.

[725] Langenhop, C. E. (1971): The Laurent expansion for a
S nearly singular matrix, Linear Algebra and Appl., 4, 329-340.

Langenhop, C. E.: See Hogan (1975).

Langford, W. F.: See Keller, H. B. (1974).

[726] Langhaar, H. L. (1969): Two numerical methods that
S converge to the methods of least squares, J. Franklin Institute, 228, 165-173.

[727] Lardy, L. J. (1973): Some iterative methods for linear
P operator equations with applications to generalized inverses, Tech. Rept. 73-65, Dept. of Math., University of Maryland, College Park, November 1973.

[728] Lardy, L. J. (1974): A series representation for the
P generalized inverse of a closed linear operator, Tech. Rept. 74-18, Dept. of Math., University of Maryland, College Park, April 1974.

[729] Larkin, F. M. (1972): Gaussian measure in Hilbert space
S and applications in numerical analysis, Rocky Mountain J. Math., 2, 379-421.

[730] Larkin, F. M. (1959): Optimal approximation in Hilbert
S spaces with reproducing kernel functions, Math. Comp., 24, 911-921.

[731] Lass, J. and C. B. Solloway (1961): A note on the secu-
T lar equation on the product of two matrices, Amer. Math. Monthly, 68, 906-907.

[732] Laurent, P.-J. (1972): Approximation et Optimisation,
S Enseignement des Sciences, No. 13, Hermann, Paris.

[733] Lavrentiev, M. M. (1967): Some Improperly Posed
T Problems of Mathematical Physics, Springer Tracts in Natural Philosphy, Vol. II, Springer-Verlag, Berlin-New York.

[734] Lawson, C. L. (1971): Applications of singular value
S analysis, in Mathematical Software (Proceedings of Symposium, Purdue Univ., April, 1970), J. R. Rice, ed., pp. 347-356, Academic Press, New York.

[735] Lawson, C. L. and R. J. Hanson (1972): Solving Least
P Squares Problems, Prentice-Hall, Englewood Cliffs, New Jersey, 340 pp.

Lawson, C. L.: See Hanson, R. J. (1969).

[736] Lawson, Linda M. (1973): Computational Methods for
P Generalized Inverse Matrices, Ph. D. Dissertation, University of Tennessee, Knoxville, Tennessee, June 1973.

[737] Lay, D. C. (1966): Studies in Spectral Theory Using
S Ascent, Descent, Nullity and Defect. Doctoral Dissertation, University of California, Los Angeles, January 1966.

[738] Lay, D. C. (1970): Spectral analysis using ascent,
S descent, nullity and defect, Math. Ann., 184, 197-214.

[739] Lay, D. C. (1975): Spectral inverses of generalized in-
P verses of linear operators, SIAM J. Appl. Math., 29, 103-109.

Lay, D. C.: See Bart, H. (1974), Kaashoek, M. A. (1972).

[740] Lazer, A. C. (1972): Application of a lemma on bilinear
T forms to a problem in nonlinear oscillations, Proc. Amer. Math. Soc., 33, 89-94.

Lazer, A. C.: See Landesman, E. M. (1970).

[741] Leach, E. B. (1961): A note on inverse function theorems,
P Proc. Amer. Math. Soc., 12, 694-697.

Learner, D. B.: See Charnes, A. (1963).

Lee, D. A.: See Audley, D. R. (1974).

[742] Legendre, A. M. (1806): Nouvelles Méthodes pour la
T Détermination des Orbites des Comètes, Paris.

[743] Legendre, A. M. (1810): Méthode des moindres carrés,
S pour trouver le milieu le plus probable entre les resultats de différentes observations. Mem. Inst. Fr., 149-154.

Leonov, A. S.: See Gončarskiĭ, A. V. (1974).

[744] Lent, A. H. (1971): Wiener-Hopf Operator and Factor-
P izations, Doctoral Dissertation in Applied Mathematics, Northwestern University, Evanston, Ill., xii + 122 pp.

[745] Leringe, Ö. and P.-Å. Wedin (1970): A comparison
P between different methods to compute a vector x which minimizes $\|Ax - b\|_2$ when $Gx = h$, Dept. of Computer Sciences, Lund University, Lund.

Levine, N.: See Goldstein, A. A. (1957).

Levinson, N.: See Coddington, E. A. (1955).

[746] Lewis, D. C. (1956): On the role of first integrals in
S the perturbation of periodic solutions, Ann. of Math., 63, 535-548.

[747] Lewis, T. O. (1966): Application of the Theory of Gen-
P eralized Matrix Inversion to Statistics, Doctoral Dissertation, University of Texas, Austin, Texas.

[748] Lewis, T. O. and C. W. Ahlers (1971): Comparison of
S two linear estimators with no assumptions on rank, Indust. Math., 21, 23-27.

[749] Lewis, T. O., T. L. Boullion and P. L. Odell (1968):
P A bibliography on generalized matrix inverses, pp. 283-315 in Boullion and Odell (1968).

[750] Lewis, T. O. and T. G. Newman (1968): Pseudo-inverses
P of positive semi-definite matrices, SIAM J. Appl. Math., 16, 703-708.

[751] Lewis, T. O. and P. L. Odell (1967): A generalization
P of the Gauss-Markov theorem, J. Amer. Statist. Assoc., 61, 1063-1066.

[752] Lewis, T. O. and P. L. Odell (1971): Estimation in
S Linear Models, Prentice-Hall, Englewood Cliffs, N. J.

Lewis, T. O.: See Ahlers, C. W. (1972), Amburgey, J. K. (1968), (1970), (1971), Hallum, C. R. (1973), Milnes, H. W. (1968), (1969), Nelson, D. L. (1970), (1971), (1972), Shurbet, G. L. (1969), (1972), (1973), Ward, J. F. (1971), (1971), (1972).

[753] Liber, A. E. (1954): On the theory of generalized groups,
T Dokl. Akad. Nauk SSSR (N. S.), 97, 25-28.

[754] Linkwitz, K. (1969): Einige Bemerkungen zum Fehler-
S fortpflanzungsgesetz und über die Einführung von Ersatzbeobachtungen, Z. Vermessungswes., 94(2), 57-71.

[755] Lin, C. S. (1974): Regularization of closed operators,
S Canad. Math. Bull., 17, 67-71.

[756] Linnik, Y. (1961): Method of Least Squares and Princi-
S ples of the Theory of Observations, translated from Russian by R. C. Elandt, Pergamon Press, New York.

[757] Liskovec, O. A. (1974): The method of ε-quasisolutions
S for equations with a closed operator, (Russian) Dokl. Akad. Nauk SSSR, 18, 201-203.

Livingstone, D.: See Rayner, A. A. (1965).

[758] S Ljapin, E. S. (1958): Inversion of elements in semigroups, Leningrad. Gos. Ped. Inst. Učen. Zap., 166, 65-74.

[759] S Ljapin, E. S. (1963): Semigroups, Translations of Math. Monographs, Amer. Math. Soc., 3, 447 pp.

Ljubich, Ju. I.: See Glazman, I. M. (1969).

[760] S Locker, J. (1967): An existence analysis for nonlinear equations in Hilbert space, Trans. Amer. Math. Soc., 128, 403-413.

[761] S Locker, J. (1970): An existence analysis for nonlinear boundary value problems, SIAM J. Appl. Math., 19, 199-207.

[762] S Locker, J. (1971): The method of least squares for boundary value problems, Trans. Amer. Math. Soc., 154, 57-68.

[763] T Locker, J. (1973): Self-adjointness for multi-point differential operators, Pacific J. Math., 45, 561-570.

[764] P Locker, J. (1975): On constructing least squares solutions to two-point boundary value problems, Trans. Amer. Math. Soc., 203, 175-183.

Lockett, J.: See Bellman, R. (1966).

[765] S Lonseth, A. T. (1954): Approximate solutions of Fredholm-type integral equations, Bull. Amer. Math. Soc., 60, 415-430.

[766] P Losey, G. (1964): On the structure of π-regular semigroups, Proc. Amer. Math. Soc., 15, 955-959.

[767] P Loud, W. S. (1966): Generalized inverses and generalized Green's functions, SIAM J. Appl. Math., 14, 342-369.

[768] Loud, W. S. (1970): Some examples of generalized
P Green's functions and generalized Green's matrices, SIAM Review, 12, 194-210.

[769] Lovass-Nagy, V. (1973): On the calculation of the
P Moore-Penrose generalized inverse of a non-normal square matrix, Math. Dept. Tech. Rept., Clarkson College of Technology, Potsdam, N.Y.

[770] Lovass-Nagy, V. and D. L. Powers (1970): On functions
T of partitioned matrices with nondiagonalizable sub-matrices, Linear Algebra and Appl., 3, 257-262.

[771] Lovass-Nagy, V. and D. L. Powers (1971): On the com-
P muting reciprocal inverse of some partitioned matrices, Linear Algebra and Appl., 4, 183-190.

[772] Lovass-Nagy, V. and D. L. Powers (1972): A note on
T block diagonalization of some partitioned matrices, Linear Algebra and Appl., 5, 339-346.

[773] Lovass-Nagy, V. and D. L. Powers (1972): A nonitera-
P tive method for computing the Moore-Penrose of an arbitrary matrix, Math. Dept. Tech. Rept., Clarkson College of Technology, Potsdam, N. Y.

[774] Lovass-Nagy, V. and D. L. Powers (1973): A relation
P between the Moore-Penrose and commuting recriprocal inverses, SIAM J. Appl. Math., 24, 44-49.

[775] Lovass-Nagy, V. and D. L. Powers (1973): A note on
P the "Y-inverse" of a matrix, Int. J. Control, 18, 1113-1115.

[776] Lovass-Nagy, V. and D. L. Powers (1973): On exact and
P approximate solutions of rectangular systems of differential equations, Math. Dept. Tech. Rept., Clarkson College of Technology, Potsdam, N.Y.

[777] Lovass-Nagy, V. and D. L. Powers (1974): On under-
P and over-determined initial value problems, Internat. J. Control, 19, 653-656.

[778] Lovass-Nagy, V. and D. L. Powers (1974): On a relation
P among generalized inverses, with application to the Moore-Penrose inverse of certain Toeplitz matrices, Indust. Math., 24, 67-76.

[779] Lovass-Nagy, V., D. L. Powers and F. D. Ullman
P (1973): A modified ADI method for computing the "best least-squares solution" of an incompatible system $(A \times I + I \times B)x = g$, Linear Algebra and Appl., 7, 179-185.

Lovass-Nagy, V.: See Dalphin (1972), (1973), Harwood, W. R. (1970).

[780] Luenberger, D. G. (1969): Optimization by Vector Space
S Methods, Wiley, New York.

[781] Lynch, R. E., J. R. Rice and D. H. Thomas (1964):
T Direct solution of partial difference equations by tensor product methods, Numer. Math., 6, 185-199.

[782] Lynn, M. S. and W. P. Timlake (1968): The use of
T multiple deflations in the numerical solution of singular systems of equations with applications to potential theory, SIAM J. Numer. Anal., 5, 303-322.

MacDonald, J. R.: See Powell, D. R. (1972).

[783] MacDuffee, C. C. (1956): The Theory of Matrices,
S Chelsea, New York, v + 110 pp.

MacPhail, M. S.: See Dwyer, P. S. (1948).

[784] Magill, K. D. and S. Subbiah (1974): Green's relations
S for regular elements of semigroups of endomorphisms, Canad. J. Math., 26, 1484-1497.

[785] Magness, T. A. and T. B. McGuire (1962): Comparison
S of least squares and minimum variance estimates on regression parameters, Ann. Math. Statist., 33, 462-470.

[786] Malanjuk, L. B. (1974): Estimation of the error in the
S solution of equations of the first kind (Russian), Ukrain. Mat. Ž., 26, 253-255, 286.

[787] Malik, H. J. (1968): A note on generalized inverses,
P Naval Res. Logist. Quart., 15, 605-612.

[788] Mandrekar, V. (1968): On multivariate wide-sense
T Markov processes, Nagoya Math. J., 33, 7-19.

[789] Mandrekar, V. and H. Salehi (1970): Subordination of
S infinite-dimensional stationary stochastic processes,
Ann. Inst. Henri Poincaré, 6, 115-130.

[790] Mandrekar, V. and H. Salehi (1970): Operator-valued
S wide-sense Markov processes and solutions of infinite-dimensional linear differential systems by white noise,
Math. Systems Theory, 4, 340-356.

[791] Mandrekar, V. and H. Salehi (1970): The square-
S integrability of operator-valued functions with respect to a non-negative operator-valued measure and the Kolmogorov isomorphism theorem, Indiana Univ. Math. J., 20, 545-563.

[792] Mandrekar, V. and H. Salehi (1971): On singularity and
S Lebesgue type decomposition for operator-valued measures, J. Multivariate Analysis, 1, 167-185.

[793] Mandrekar, V. and H. Salehi (1972): A factorization
S theorem of D. Lowdenslager, Proc. Amer. Math. Soc., 31, 185-188.

[794] Manherz, R. K. (1968): Applications of the Generalized
P Matrix Inverse in Network Theory, Doctoral Disseration, Northwestern University, Evanston, Illinois.

[795] Manherz, R. K. and S. L. Hakimi (1969): The generalized
P inverse in network analysis and quadratic-error minimization problems, IEEE Trans. Circuit Theory, CT-16, 559-562.

[796] Manherz, R. K., B. W. Jordan and S. L. Hakimi (1968):
P Analog methods for computation of the generalized inverse, IEEE Trans. Aut. Control, AC-13, 582-585.

[797] Marchant, R. and L. Jonnes (1969): Résultats de re-
S cherches dans le domaine de l'étude statistique des erreurs de nivellement, Bull. Géodésique, 94, 365-378.

[798] Marchuk, G. I. and Ju. A. Kuznecov (1971): Iterative
P Methods and Quadratic Functionals (Russian). Moscow.

[799] Marchuk, G. I. and V. G. Vasilev (1970): On an approxi-
S mate solution for operator equations of the first kind, Soviet Math. Dokl., 11, 1562-1566.

[800] Marcus, M. and H. Minc (1964): A Survey of Matrix
S Theory and Matrix Inequalities, Allyn and Bacon, Boston, Mass, xvi + 180 pp.

Marinelli, L. D.: See Abu-Shumays, I. K. (1971).

Markham, T.: See Burns, F. (1974), Carlson, D. (1974), Hung, C. H. (1975).

[801] Marquardt, D. W. (1963): An algorithm for least-squares
S estimation of nonlinear parameters, SIAM J. Appl. Math., 11, 431-444.

[802] Marquardt, D. W. (1970): Generalized inverses, ridge
P regression, biased linear estimation, and nonlinear estimation, Technometrics, 12, 591-611.

[803] de Marr, R. (1974): Nonnegative idempotent matrices,
S Proc. Amer. Math. Soc., 45, 185-188.

[804] Marsaglia, G. (1964): Conditional means and variances
P of normal variables with singular covariance matrix, J. Amer. Statist. Assoc., 59, 1203-1204.

[805] Marsaglia, G. (1967): Bounds on the rank of the sum of
T matrices, Proceedings of the 4th Prague Conference (1965) on Information Theory, Statistical Decision Functions, and Random Processes, pp. 455-462, Academic Press, New York.

[806] Marsaglia, G. and G. P. H. Styan (1972): When does
S rank (A+B) = rank(A) + rank(B)? Canadian Math. Bull., 15, 451-452.

Marsaglia, G.: See Graybill, F. A. (1957).

Martin, A. V.: See Butler, T. (1962).

[807] Martin, F. B. (1968): Contributions to the Theory of
P Estimation in the General Linear Model, Ph. D. Thesis, Iowa State University, Ames, Iowa.

[808] Martin, F. B. and G. Zyskind (1966): On combinability
T of information from uncorrelated linear models by simple weighting, Ann. Math. Statist., 37, 1338-1347.

Martin, F. B.: See Zyskind, G. (1969).

[809] Martin, R. S., G. Peters and J. H. Wilkinson (1966):
S Iterative refinement of the solution of a positive definite system of equations, Numer. Math., 8, 203-216. Republished, pp. 31-44 in Wilkinson and Reinsch (1971).

Martin, R. S.: See Bowdler, H. J. (1966).

[810] Masani, P. (1966): Recent trends in multivariate pre-
S diction theory, Technical Summary Rept. #637, Mathematics Research Center, The University of Wisconsin-Madison; also appeared in Multivariate Analysis, pp. 351-382, Academic Press, New York, 1966.

[811] Massera, J. L. and J. J. Schäffer (1966): Linear Differ-
T ential Equations and Functional Analysis, Academic Press, New York.

[812] Mathai, A. M. (1965): An approximate method of analysis
S for a two-way layout, Biometrics, 21, 376-385.

[813] Mathai, A. M. (1966): Pseudoinverses in normal equa-
P tions, Estadistica, 24, 620-628.

[814] Mathai, A. M. (1973): Analysis of multiple classification, Estadistica.
S

[815] Matveev, A. A. (1974): A certain algorithm for the pseudoinversion of matrices. (Russian), Ž. Vyčisl. Mat. i Mat. Fiz., 14, 483-487, 534.
P

[816] Mawhin, J. (1974): Fredholm mappings and solutions of linear differential equations at singular points, Séminaires de Mathématique Appliquee et Mécanique, Rapport No. 72. Institut de Mathématique Pure et Appliquée, Université Catholique de Louvain, Louvain-la-Neuve, 12 pp.
S

[817] Maxwell, J. C. (1892): Treatise of Electricity and Magnetism, 3rd Ed., vol. I, Oxford University Press, Oxford, England.
T

[818] May, K. O. (1972): Carl Friedrich Gauss, in Dictionary of Scientific Bibliography, edited by C. C. Gillispie, vol. 5, pp. 298-315.
T

[819] Mayne, D. Q. (1966): An algorithm for calculation of the pseudoinverse of a singular matrix, Comput. J., 9, 312-317.
P

[820] Mayne, D. Q. (1969): On the calculation of pseudoinverses, IEEE Trans. Automatic Control, 14, 204-205.
P

[821] McCormick, S. F. (1972): An iterative procedure for the solution of constrained nonlinear equations with application to optimization problem, Rept. No. IEC-002-72, The Claremont Colleges, Claremont, Calif.
T

[822] McCormick, S. F. (1972): A general approach to one-step iterative methods with application to eigenvalue problems, J. Comput. Syst. Sci., 6, 354-372.
S

[823] McCormick, S. F. and G. H. Rodrigue (1975): A uniform approach to gradient methods for linear operator equations, J. Math. Anal. Appl., 49, 275-285.
P

[824] McCoy, N. H. (1939): Generalized regular rings, Bull.
S Amer. Math. Soc., 45, 175-178.

McCoy, N. H.: See Brown, B. (1950).

[825] McGlothin, G. E. (1974): Orthogonal decomposition of
S Green's identity with application to eigenvalue assignment in distributed systems, Tech. Rept., Dept. of Electrical Eng., American University of Beirut, Lebanon.

McGuire, T. B.: See Magness, T. A. (1962).

McLaughlin, J. E.: See Halmos, P. R. (1963).

Meany, R. K.: See Whitney, T. M. (1967).

[826] Meditch, J. S. (1967): Orthogonal projection and discrete
S optimal linear smoothing, SIAM J. Control, 5, 74-89.

[827] de Meersman, R. (1971): Algorithms for pseudo-inverses
P of linear mappings: A unified approach, Bull. Soc. Math. Belg., 23, 283-294.

[828] Meeter, D. A. (1966): On a theorem used in nonlinear
S least squares, SIAM J. Appl. Math., 14, 1176-1179.

[829] Meicler, M. (1966): Weighted Generalized Inverses with
P Minimal p and q Norms, Doctoral Dissertation, University of Texas, Austin, Texas.

[830] Meicler, M. (1968): Chebyshev solution of an inconsist-
P ent system of $N + 1$ linear equations in N unknowns in terms of its least squares solution, SIAM Review, 10, 373-375.

[831] Meicler, M. (1969): A characterization of the general
S solution of the matrix equation $AX - XB = C$ in terms of the eigenvectors of either A or B, Texas J. Science, 21, 195-198.

[832] Meicler, M. and P. L. Odell (1967): Weighted generalized
P inverses, NASA NAS 9-5384, 1-13.

Meicler, M.: See Newman, T. G. (1968).

Meir, A.: See Goldberg, S. (1971).

[833] S Mendel, J. M. (1974): Discrete Techniques of Parameter Estimation: The Equation Error Formulation, Marcel Dekker, Inc., New York, N. Y., 408 pp.

[834] S Merchant, R. (1951): La généralisation de la méthode des moindres carrés par les procédés de l'algèbre matricelle, Inst. Geogr. Mil., Bruxelles.

[835] T Merris, R. (1974): An identity involving generalized matrix functions, Linear and Multilinear Algebra, 2, 123-125.

[836] S Merris, R. (1974): Two problems involving Schur functions, Linear Algebra and Appl., 10, 155-162.

Metcalf, F. T.: See Diaz, J. B. (1970).

[837] P Meyer, C. D. (1968): Pseudoinverses and the Hermite Form, Doctoral Thesis, Colorado State University, Fort Collins, Colorado.

[838] P Meyer, C. D., Jr. (1969): On ranks of pseudoinverses, SIAM Review, 11, 382-385.

[839] S Meyer, C. D., Jr. (1969): On the rank of the sum of two rectangular matrices, Canadian Math. Bull., 12, 508.

[840] P Meyer, C. D., Jr. (1969): On the construction of solutions to the matrix equations $AX = A$ and $YA = A$, SIAM Review, 11, 612-615.

[841] P Meyer, C. D., Jr. (1970): Generalized inverses of triangular matrices, SIAM J. Appl. Math., 18, 401-406.

[842] P Meyer, C. D., Jr. (1970): Generalized inverses of block triangular matrices, SIAM J. Appl. Math., 19, 741-750.

[843] P Meyer, C. D., Jr. (1970): Some remarks on EPr matrices and generalized inverses, Linear Algebra and Appl., 3, 275-278.

[844] Meyer, C. D., Jr. (1971): Representations for (1)- and
P (1, 2)-inverses for partitioned matrices, Linear Algebra and Appl., 4, 221-232.

[845] Meyer, C. D., Jr. (1972): The Moore-Penrose inverse
P of a bordered matrix, Linear Algebra and Appl., 5, 375-382.

[846] Meyer, C. D., Jr. (1973): Generalized inversion of mod-
P ified matrices, SIAM J. Appl. Math., 24, 315-323.

[847] Meyer, C. D., Jr. (1973): Generalized inverses and
P ranks of block matrices, SIAM J. Appl. Math., 25, 597-602.

[848] Meyer, C. D., Jr. (1974): Limits and the index of a
P square matrix, SIAM J. Appl. Math., 26, 469-478.

[849] Meyer, C. D., Jr. (1975): The role of the group gener-
P alized inverse in the theory of finite Markov chains, SIAM Review, 17, 443-464.

[850] Meyer, C. D., Jr. and R. J. Painter (1970): Note on a
P least squares inverse for a matrix, J. Assoc. Comput. Mach., 17, 110-112.

Meyer, C. D., Jr.: See Campbell, S. L. (1975), (1975), Graybill, F. A. (1966), Hall, F. J. (1975).

[851] Miamee, A. G. and H. Salehi (1974): Factorization of
S positive operator valued functions on a Banach space, Indiana Univ. Math. J., 24, 103-113.

[852] Mihalyffy, L. (1971): An alternative representation of
P the generalized inverse of partitioned matrices, Linear Algebra and Appl., 4, 95-100.

[853] Mikhlin, S. G. (1965): The Problem of the Minimum of
S a Quadratic Functional, Translated from the Russian by A. Feinstein, Holden-Day, San Francisco, California.

[854] Mikhlin, S. G. (1971): The Numerical Performance of
S Variational Methods. Translated from the Russian by R. S. Anderssen. Wolters-Noordhoff Publishing, Groningen, xxiii + 373 pp.

[855] Mikhlin, S. G. and K. L. Smolitsky (1967): Approximate
S Methods for Solution of Differential and Integral Equations, American Elsevier, New York, N.Y.

[856] Miller, K. (1970): Least square methods for ill-posed
S problems with a prescribed bound, SIAM J. Math. Anal., 1, 52-74.

Miller, F. R.: See Erdélyi, I. (1970).

[857] Milne, R. D. (1968): An oblique matrix pseudoinverse,
P SIAM J. Appl. Math., 16, 931-944.

[858] Milnes, H. W. (1968): Characteristic vectors for rec-
S tangular matrices, Indust. Math., 19, 63-80.

[859] Milnes, H. W., J. Amburgey, T. O. Lewis and T. L.
P Boullion (1968): Special eigenvalue property of $A^\dagger$ for rectangular matrices, pp. 98-113 in Boullion and Odell (1968).

[860] Milnes, H. W., J. Amburgey, T. O. Lewis and T. L.
P Boullion (1969): Spectral eigenvalue property of $A^\dagger$ for rectangular matrices, Indust. Math., 19, 81-88.

Milnes, H. W.: See Shurbet, G. L. (1969), Tipton, A. R. (1972).

Milota, J.: See Fučik, S. (1971).

[861] Minamide, N. and K. Nakamura (1969): Minimum error
S control problem in Banach space, Research Reports of Automatic Control Lab., Faculty of Engineering, Nagoya University, vol. 16, 51-58 (April 1969).

[862] Minamide, N. and K. Nakamura (1970): A restricted
P pseudoinverse and its applications to constrained minima, SIAM J. Appl. Math., 19, 167-177.

[863] Minamide, N. and K. Nakamura (1971): A note on the
S Ho-problem in Hilbert space, J. Franklin Institute, 291, 181-194.

[864] Minamide, N. and K. Nakamura (1972): Linear bounded
S phase coordinate control problems under certain regularity and normality conditions, SIAM J. Control, 10, 82-92.

Minc, H.: See Marcus, M. (1964).

[865] Mirsky, L. (1960): Symmetric gauge functions and uni-
S tarily invariant norms, Quart. J. Math. Oxford (2), 11, 50-59.

[866] Mitra, Sanjit K. (1969): On the Analysis and Synthesis
T of Linear Active Networks, Wiley, New York.

[867] Mitra, S. K. (1968): The subalgebra generated by a
P square matrix and generalized inverse, Tech. Report #11/68, Research and Training School, Indian Statist. Inst., Calcutta.

[868] Mitra, S. K. (1968): On a generalized inverse of a matrix
P and applications, Sankhyā, Ser. A, 30, 107-114.

[869] Mitra, S. K. (1968): A new class of g-inverse of square
P matrices, Sankhyā, Ser. A, 30, 323-330.

[870] Mitra, S. K. (1970): A density free approach to matrix
S variate beta distribution, Sankhyā, Ser. A, 32, 81-88.

[871] Mitra, S. K. (1971): Restricted generalized inverse and
P the theory of minimum bias estimation in a Gauss-Markoff model, Tech. Report No. Math-Stat/2/71, Research and Training School, Indian Statistical Institute, Calcutta.

[872] Mitra, S. K. (1971): Another look at Rao's minque of
S variance components, Bull. Inst. Internat. Statist., 44, Book 2, 279-283.

[873] Mitra, S. K. (1972): Simultaneous diagonalization of
S two or more Hermitian forms, Proc. Fifteenth Midwest Symposium on Circuit Theory, Rolla, Missouri, 2, XII. 3.1-XII. 3.10.

[874] Mitra, S. K. (1972): Fixed rank solutions of linear
P matrix equations, Sankhyā, Ser. A., 34, 387-392.

[875] Mitra, S. K. (1973): Unified least squares approach to
P linear estimation in a general Gauss-Markov model, SIAM J. Appl. Math., 25, 671-680.

[876] Mitra, S. K. (1973): Common solutions to a pair of
P linear matrix equations $A_1XB_1 = C_1$ and $A_2XB_2 = C_2$, Proc. Cambridge Phil. Soc., 74, 213-216.

[877] Mitra, S. K. (1973): Statistical proofs of some proposi-
S tions on non-negative definite matrices, 39th Session of International Statistical Institute, Vienna.

[878] Mitra, S. K. and P. Bhimasankaram (1970): Some re-
S sults on idempotent matrices and a matrix equation connected with the distribution of quadratic forms, Sankhyā, Ser. A, 32, 353-356.

[879] Mitra, S. K. and P. Bhimasankaram (1971): A character-
P isation of the Moore-Penrose inverse and related results, Sankhyā, Ser. A, 33, 411-416.

[880] Mitra, S. K. and P. Bhimasankaram (1971): Generalized
P inverse of partitioned matrices and recalculation of least squares estimates for data or model changes, Sankhyā, Ser. A, 33, 395-410.

[881] Mitra, S. K. and Betty Moore (1973): Gauss-Markov
P estimation with an incorrect dispersion matrix, Sankhyā, Ser. A, 35, 139-152.

[882] Mitra, S. K. and M. L. Puri (1973): On parallel sum
P and difference of matrices, J. Math. Anal. Appl., 44, 222-228.

[883] Mitra, S. K. and C. R. Rao (1968): Simultaneous re-
P duction of a pair of quadratic forms, Sankhyā, Ser. A, 30, 313-322.

[884] Mitra, S. K. and C. R. Rao (1968): Some results in
S estimation and tests of linear hypothesis under the Gauss-Markoff model, Sankhyā, Ser. A, 30, 281-290.

[885] Mitra, S. K. and C. R. Rao (1969): Conditions for op-
S timality and validity of simple least squares theory, Ann. Math. Statist., 40, 1617-1624.

[886] Mitra, S. K. and G. E. Trapp (1975): On hybrid addition
P of matrices, Linear Algebra and Appl., 10, 19-35.

Mitra, S. K.: See Bhimasankaram, P. (1969), Bose, N. K. (1973), Rao, C. R. (1971), (1971), (1972), (1972), (1973).

Mityagin, B. S.: See Kadets, M. I. (1973).

[887] Mizel, V. J. and M. M. Rao (1962): Nonsymmetric pro-
S jections in Hilbert space, Pacific J. Math., 12, 343-357.

[888] Moler, C. B. (1967): Iterative refinement in floating
T point, J. Assoc. Comput. Mach., 14, 316-321.

[889] Mond, B. (1969): Generalized inverse extension of
P matrix inequalities, Linear Algebra and Appl., 2, 393-399.

[890] Montague, J. S. and R. J. Plemmons (1972): Convex
P matrix equations, Bull. Amer. Math. Soc., 78, 965-968.

Moore, Betty: See Mitra, S. K. (1973).

[891] Moore, E. H. (1920): On the reciprocal of the general
P algebraic matrix (abstract), Bull. Amer. Math. Soc., 26, 394-395.

[892] Moore, E. H. (1935): General Analysis, Memoirs
P American Philosophical Society, I: esp. pp. 147-209, Philadelphia.

[893] Moore, R. H. and M. Z. Nashed (1973): Approximation
P of generalized inverses of linear operators in Banach spaces, pp. 425-429 in Approximation Theory, ed. by G. G. Lorentz, Academic Press, New York.

[894] Moore, R. H. and M. Z. Nashed (1974): Approximations
P to generalized inverses of linear operators, SIAM J. Appl. Math., 27, 1-16.

[895] Moore, R. H. and M. Z. Nashed (1975): Approximation-
P projection methods for generalized inverses in Banach spaces, in preparation.

[896] Moritz, H. (1966): An extension of error theory with ap-
S plication to spherical harmonics, Bull. Géodésque, 81, 225-234.

[897] Moritz, H. (1967): Reconstruction of functions from
T discrete mean values, Bull. Géodésique, 84, 89-108.

[898] Moritz, H. (1970): Least-squares estimation in physical
S geodesy, Deutsche Geodät Kommiss. Bayer. Akad. Wiss. Reihe A, Heft 69, 34 pp.

[899] Moritz, H. (1972): Advanced least squares methods,
S Rept. AFCRL-72-0363, Ohio State Univ., Columbus, Ohio.

[900] Morkhov, Y. V. (1967): Loss of accuracy in working out
P systems of normal equations by Gauss and Cholesky method, Bull. Géodésique, 84, 123-129.

Moro, A.: See Halany, A. (1968).

[901] Morozov, V. A. (1969): Pseudosolutions, Ž. Vyčisl. Mat.
P i Mat. Fiz., 9, 1387-1391.

[902] Morris, G. L. (1967): Characterizations of Generalized
P Inverses for Matrices, Doctoral Dissertation, Texas Technological College, Lubbock, Texas.

[903] Morris, G. L. and P. L. Odell (1968): A characterization
P for generalized inverses of matrices, SIAM Rev., 10, 208-211.

[904] Morris, G. L. and P. L. Odell (1968): Common solutions
S for n matrix equations with applications, J. Assoc. Comput. Mach., 15, 272-274.

[905] Morrison, D. D. (1960): Remarks on the unitary triangu-
T larization of a nonsymmetric matrix, J. Assoc. Comput. Mach., 7, 185-195.

Morrison, W. L.: See Sherman, I. V. (1958).

[906] Munn, W. D. (1961): Pseudoinverses in semigroups,
P Proc. Cambridge Philos. Soc., 57, 247-250.

[907] Munn, W. D. and R. Penrose (1955): A note on inverse
S semigroups, Proc. Cambridge Philos. Soc., 51, 396-399.

[908] Murray, F. J. (1937): On complementary manifolds and
S projections in spaces L_p and ℓ_p, Trans. Amer. Math. Soc., 41, 138-152.

[909] Murray, F. J. and J. von Neumann (1936): On rings of
S operators, I, Ann. of Math., 37 (vol. III, No. 2), 116-229.

[910] Myller, A. (1906): Gewöhnliche Differentialgleichungen
T höherer Ordnung in ihre Beziehung zu den Integralgleichungen, Inaugural-Dissertation (Gottingen).

[911] Nagasaka, H. (1965): Error propagation in the solution
T of tridiagonal linear equations, Information Processing in Japan, 5, 38-44.

[912] Nagy, G. (1968): State of the art in pattern recognition,
S Proceedings IEEE, 56, 836-861.

Nakamura, K.: See Minamide, N. (1969), (1970), (1971), (1972).

[913] Nanda, V. C. (1967): A generalization of Cayley's
S theorem, Math. Zeitschr., 101, 331-334.

Narain, P.: See Tewarson, R. P. (1974), (1975).

Narenda, K. S.: See Kalman, R. E. (1963).

[914] Nashed, M. Z. (1970): Steepest descent for singular
P operator equations, SIAM J. Numer. Anal., 7, 358-362.

[915] Nashed, M. Z. (1971): Generalized inverses, normal
P solvability, and iteration for singular operator equations, pp. 311-359 in Nonlinear Functional Analysis and Applications, ed. by L. B. Rall, Academic Press, New York.

[916] Nashed, M. Z. (1971): Differentiability and related
T properties of nonlinear operators: Some aspects of the role of differentials in nonlinear functional analysis, pp. 109-309 in Nonlinear Functional Analysis and Applications, ed. by L. B. Rall, Academic Press, New York.

[917] Nashed, M. Z. (1971): A retrospective and prospective
P survey of generalized inverse of operators, Notes. (A revised version is under preparation).

[918] Nashed, M. Z. (1972): Some aspects of regularization
S and approximation of solutions of ill-posed operator equations, Proceedings of the 1972 Army Numerical Analysis Conference, pp. 163-181, ARO-D Rept. 72-3, Army Research Office, Durham, N.C.

[919] Nashed, M. Z. (1973): On applications of generalized
P splines and generalized inverses in regularization and projection methods, pp. 415-419, Proceedings of 1973 Annual National Conference of the Association for Computing Machinery.

[920] Nashed, M. Z. (1974): Approximate regularized solu-
P tions to improperly posed linear integral and operator equations, pp. 289-322, Constructive and Computational Methods for Differential and Integral Equations (Symposium Proceedings, Research Center for Applied Science, Indiana University, Bloomington, Indiana, February 17-20, 1974), ed. by D. Colton and R. G. Gilbert, Lecture Notes in Mathematics, vol. 430, Springer-Verlag, Berlin-New York.

[921] Nashed, M. Z. (1975): On a functional equation which
S characterizes polynomial operators with applications to uniqueness, Applicable Anal., 5, to appear.

[922] Nashed, M. Z. (1974): On moment-discretization and
P least squares solution of integral equations of the first kind, Tech. Summary Rept. #1371, Mathematics Research Center, The University of Wisconsin-Madison. To appear in J. Math. Anal. Appl., 1975.

[923] Nashed, M. Z. (1975): Measurability of generalized
P inverses of random linear operators in Banach spaces and random generalized Green's matrices, Notices Amer. Math. Soc., 22, 31.

[924] Nashed. M. Z. (1975): Perturbations and approximations
P for generalized inverses and linear operator equations, this Volume.

[925] Nashed, M. Z. (1975): Aspects of generalized
P inverses in analysis and regularization, this Volume.

[926] Nashed, M. Z. (1976): On nonsingular inner inverses,
P to appear.

[927] Nashed, M. Z. (1976): Regularization and numerical
P analysis of ill-posed operator equations (an expanded version of an invited address delivered to the 710th meeting of the American Mathematical Society, November 16, 1973), Bull. Amer. Math. Soc., to appear.

[928] Nashed, M. Z.: Regularization methods — A survey,
S in preparation.

[929] Nashed, M. Z.: Sources of ill-posed problems in ap-
S plications, in preparation.

[930] Nashed, M. Z. and H. Salehi (1973): Measurability of
P generalized inverses of random linear operators, SIAM J. Appl. Math., 25, 681-692.

[931] Nashed, M. Z. and G. F. Votruba (1974): A unified
P approach to generalized inverses of linear operators: I. Algebraic, topological and projectional properties, Bull. Amer. Math. Soc., 80, 825-830.

[932] Nashed, M. Z. and G. F. Votruba (1974): A unified
P approach to generalized inverses of linear operator: II. Extremal and proximinal properties, Bull. Amer. Math. Soc., 80, 831-835.

[933] Nashed, M. Z. and G. F. Votruba (1974): Convergence
P of a class of iterative methods of Cimmino-type to weighted least square solutions, Notices Amer. Math. Soc., 21, A-245.

[934] Nashed, M. Z. and G. F. Votruba (1975): A unified
P operator theory of generalized inverses, this Volume.

[935] Nashed, M. Z. and G. Wahba (1973): Approximate reg-
P ularized pseudosolutions of linear operator equations in reprducing kernel spaces, TSR #1265, Mathematics Research Center, The University of Wisconsin-Madison.

[936] Nashed, M. Z. and G. Wahba (1973): Generalized inverses in reproducing kernel spaces: An approach to
P regularization of linear operator equations, MRC. Tech. Summary Rept. 1200, Mathematics Research Center, University of Wisconsin-Madison.

[937] Nashed, M. Z. and G. Wahba (1974): Rates of converg-
P ence of approximate least squares solutions of linear integral and operator equations, Math. Comp., 28, 69-80.

[938] Nashed, M. Z. and G. Wahba (1974): Regularization and
P approximation of linear operator equations in reproducing kernel spaces, Bull. Amer. Math. Soc., 80, 1213-1218.

[939] Nashed, M. Z. and G. Wahba (1974): Generalized in-
P verses in reproducing kernel spaces: An approach to regularization of linear operator equations, SIAM J. Math. Anal., 5, 974-987.

[940] Nashed, M. Z. and G. Wahba (1975): Some exponentially
S decreasing error bounds for the numerical inversion of the Laplace transform, J. Math. Anal. Appl., to appear.

Nashed, M Z.: See Kammerer, W. J. (1971), (1971), (1971), (1972), Moore, R. H. (1973), (1974), (1975), Wahba, G. (1973).

[941] Nayak, R. P. (1972): Development of Sequential Filters
P Using Pseudoinverses, Ph.D. Dissertation, Electrical Engineering, Marquette University, Milwaukee, Wis.

Nečas, J.: See Fučik, S. (1973), (1974), (1975).

[942] Nelson, D. L. (1970): Quadratic Programming Techniques
P Using Matrix Pseudoinverses, Ph. D. Dissertation, Texas Tech University, Lubbock, Texas.

[943] Nelson, D. L. and T. O. Lewis (1970): A method for the
S solution of nonlinear least squares problems when some of the parameters are linear, Texas J. Science, 21, No. 4, 480-481.

[944] Nelson, D. L., T. O. Lewis and T. L. Boullion (1971):
P A quadratic programming technique using matrix pseudo-inverses, Indust. Math., 21, 1-21.

[945] Nelson, D. L., T. O. Lewis and T. L. Boullion (1972):
P A generalized inverse method for regression analysis with linear restrictions, Indust. Math., 22, 1-10.

[946] Neuberger, J. W. (1965): The lack of self-adjointness
T in three-point boundary-value problems, Pacific J. Math., 18, 165-168.

[947] Neudecker, H. (1969): A note on Kronecker matrix prod-
S ucts and matrix equations systems, SIAM J. Appl. Math., 17, 603-606.

[948] Neudecker, H. (1969): Some theorems on matrix differ-
S entiation with special reference to Kronecker matrix products, J. Amer. Statist. Assoc., 64, 953-963.

[949] von Neumann, J. (1932): Über adjungierte Funktional-
S operation, Ann. of Math., 33, 294-310.

[950] von Neumann, J. (1936): On regular rings, Proc. Nat.
P Acad. Sci. USA, 22, 707-713.

[951] von Neumann, J. (1937): Some matrix-inequalities and
T metrization of metric-space, Tomsk Univ. Rev., 1, 286-300. Republished in John von Neumann Collected Works, vol. IV, pp. 205-219, MacMillan, New York, 1962.

[952] von Neumann, J. (1950): Functional Operators, Vol. II.
T The Geometry of Orthogonal Spaces, Ann. of Math. Studies No. 29, Princeton Univ. Press, Princeton, N.J.

[953] von Neumann, J. (1960): Continuous Geometry,
T Princeton Univ. Press, Princeton, N.J., 299 pp.

von Neumann, J.: See Goldstine, H. H. (1947),
Murray, F. J. (1936).

[954] Newcomb, Robert W. (1960): On the simultaneous
S diagonalization of two semidefinite matrices, Quart. J. Appl. Math., 19, 144-146.

Newhouse, J. : See Dent, B. A. (1959).

[955] Newhouse, S. E. (1966): Introduction to matrix gener-
P alized inverses and their applications, NASA Tech. Memo. X-55415, Goddard Space Flight Center, Maryland.

[956] Newman, T. G. and P. L. Odell (1969): On the concept
P of a p-q generalized inverse of a matrix, SIAM J. Appl. Math., 17, 520-525.

[957] Newman, T. G., P. L. Odell and M. Meicler (1968):
P The concept of a p-q generalized inverse, pp. 276-282 in Boullion and Odell (1968).

Newman, T. G.: See Amir-Moéz, A. R. (1970), Lewis, T. O. (1968).

Neyman, J.: See David, F. N. (1938).

[958] Nickolson, H. (1972): Properties of the generalized in-
P verse matrix in the electrical-network problem, Electronics Letter, 8, 267-268.

Nielson, C. W.: See Buzbee, B. L. (1970).

[959] Niethammer, W. and W. Schempp (1970): On the con-
T struction of iteration methods for linear equations in Banach spaces by summation methods, Aequationes Math., 5, 273-284.

[960] Noble, B. (1966): A method for computing the general-
P ized inverse of a matrix, SIAM J. Numer. Anal., 3, 582-584.

[961] Noble, B. (1969): Applied Linear Algebra, Prentice-Hall
S Inc., Englewood Cliffs, N. J., xvi + 523 pp.

[962] Noble, B. (1975): Methods for computing the Moore-
P Penrose generalized inverse, and related matters, this Volume.

Noble, B.: See Cho, C. Y. (1966).

[963] Noll, W. (1969): Quasi-reversibility in a staircase
S diagram, Proc. Amer. Math. Soc., 23, 1-4.

[964] Oblomskaja, L. (1968): Methods of successive approxi-
S mation for linear equations in Banach spaces, USSR Computational Math. and Math. Phys., 8, 239-253.

[965] Odell, P. L. and Decell, H. P. (1967): On computing the
S fixed point probability vector of regular or ergodic transition matrices, J. Assoc. Comput. Mach., 14, 765-768.

Odell, P. L.: See Boullion, T. L. (1968), (1969), (1969), (1971), Decell, H. P., Jr. (1966), (1967), Duran, B. (1974), Hallum, C. R. (1973), Lewis, T. O. (1967), (1968), (1971), Meicler, M. (1967), Morris, G. L. (1968), (1968), Newman, T. G. (1968), (1969), Scroggs, J. E. (1966).

[966] Oettli, W. and W. Prager (1964): Computability of ap-
S proximate solution of linear equations with given error bounds for coefficients and right-hand sides, Numer. Math., 6, 405-409.

[967] Oettli, W., W. Prager and J. H. Wilkinson (1965): Admissible
S solutions of linear systems with not sharply defined coefficients, SIAM J. Numer. Anal., 2, 291-299.

[968] Ogasawara, T. and M. Takahashi (1951): Independence
T of quadratic forms in normal system, J. Sci. Hiroshima University, 15, 1-9.

[969] Ogawa, J. (1949): On the independence of linear and
T quadratic forms of a random sample from a normal population, Ann. Inst. Statist. Math., 1, 83-108.

[970] Ohwaki, S. (1974): On linear operators with closed
S range, Proc. Japan Acad., 50, 97-99.

[971] Oktaba, W. (1969): Generalized inverses of matrices
P in a fixed model, Biometrische Z., 11, 228-251.

[972] Olivares, J. E. (1968): Generalized inverse of the tie-
P set and cut-set matrix for networks with complete graphs, Proc. Hawaii International Conference on System Sciences, 583-586.

[973] Orey, S. (1964): Potential kernels for recurrent Markov
T chains, J. Math. Anal. Appl., 8, 104-132.

[974] Ortega, J. M. and W. C. Rheinboldt (1970): Iterative
S Solution of Nonlinear Equations in Several Variables, Academic Press, New York, xx + 572 pp.

[975] Osborn, J. E. and D. Sather (1975): Alternative prob-
S lems for nonlinear equations, J. Differential Equations, 17, 12-31.

[976] Osborne, E. E. (1961): On least squares solutions of
P linear equations, J. Assoc. Comput. Mach., 8, 628-636.

[977] Osborne, E. E. (1965): Smallest least squares solutions
P of linear equations, SIAM J. Numer. Anal., 2, 300-307.

[978] Osborne, M. R. (1972): Some aspects of nonlinear least
S squares calculations, Proceedings of the Dundee Conference on Optimization, Academic Press, New York.

[979] Osborne, M. R. (1970): A class of nonlinear regression
S problems, in Data Representation (R. S. Anderssen and M. R. Osborne, editors), pp. 94-101, Australia National University.

Osborne, M. R.: See Jennings, L. S. (1970), (1974).

[980] Ostrowski, A. M. (1962): A regularity condition for a
S class of partitioned matrices, Math. Comp., 15, 23-27.

[981] Ostrowski, A. M. (1967): General criteria for the in-
S verse of an operator, Amer. Math. Monthly, 74, 824-
826.

[982] Ostrowski, A. M. (1973): Solution of Equations in
S Euclidean and Banach Spaces, 3rd edition, Academic
Press, New York, xx + 412 pp.

[983] Padulo, L. and M. A. Arbib (1974): A Unified State-
T Space Approach to Continuous and Discrete Systems,
W. B. Saunders Co., Philadelphia, Pa., 799 pp.

Painter, R. J.: See Graybill, F. A. (1966), Meyer, C. D., Jr. (1970).

[984] Papini, P. L. (1974): Some questions related to the con-
S cept of orthogonality in Banach spaces. Orthogonal
projections, Boll. Un. Mat. Ital., (4) 9, 386-401.

[985] Parlett, B. N. (1967): The LU and QR algorithms, in
T Mathematical Methods for Digital Computers, pp. 116-
130, vol. 2, ed. by A. Ralston and H. S. Wilf, Wiley,
New York.

[986] Parzen, E. (1974): Some recent advances in time series
T modelling. System identification and time-series analysis, IEEE Trans. Automatic Control, AC-19, 723-730.

[987] Patterson, W. M. 3rd (1974): Iterative Methods for the
S Solution of a Linear Operator Equation in a Hilbert
Space – A Survey, Lecture Notes in Mathematics, vol.
394, Springer-Verlag, Berlin-New York, 183 pp.

[988] Pavel-Parvu, M. (1965): Traitement numérique des
S équations intégrals de Fredholm, Colloq. sur les Équa-
tions intégrals, anime par A. Korganoff, C. R. du
Quatrieme Congres de l'Afcalti, Dunod, 357-370.

[989] Pavel-Parvu, M and A. Korganoff (1969): Iteration func-
S tions for solving polynomial matrix equations, in Constructive Aspects of the Fundamental Theory of Algebra, pp. 225-280, ed. by B. Dejon and P. Henrici. Proceedings of a Symposium conducted at the IBM Research Laboratory, Zurich, June 1967. Wiley-Interscience, London-New York.

Pavel-Parvu, M.: See Korganoff, A. (1964), (1967), (1973).

[990] Payne, L. E. (1973): Some general remarks on improp-
T erly posed problems for partial differential equations, in Symposium on Non-Well-Posed Problems and Logarithmic Convexity, pp. 1-30, Lecture Notes in Mathematics, Vol. 316, Springer-Verlag, Berlin-New York.

[991] Payne, L. E. (1975): Improperly Posed Problems in
S Partial Differential Equations (An expanded version of a series of lectures given at an NSF Regional Conference, The University of New Mexico, May 20-24, 1974), to be published by SIAM in the CBMS Regional Conference Series in Applied Mathematics.

Pearcy, C. M.: See Douglas, J., Jr. (1963).

[992] Pearl, M. H. (1957): On Cayley's parameterization,
T Canad. J. Math., 9, 553-562.

[993] Pearl, M. H. (1959): A further extension of Cayley's
T parameterization, Canad. J. Math., 11, 48-50.

[994] Pearl, M. H. (1959): On normal and EPr matrices,
S Michigan Math. J., 6, 1-5.

[995] Pearl, M. H. (1959): On normal and EPr matrices,
S Michigan Math. J., 6, 89-94.

[996] Pearl, M. H. (1961): On normal EPr matrices,
S Michigan Math. J., 8, 33-37.

[997] Pearl, M. H. (1966): On generalized inverse of matrices,
P Proc. Cambridge Philos. Soc., 62, 673-677.

[998] Pearl, M. H. (1967): A decomposition theorem for
S matrices, Canad. J. Math., 19, 344-349.

[999] Pearl, M. H. (1968): Generalized inverses of matrices
P with entries taken from an arbitrary field, Linear Algebra and Appl., 1, 571-587.

Pearl, M. H.: See Katz, I. J. (1966), (1966).

[1000] Pease, M. C. (1965): Methods of Matrix Algebra,
S Academic Press, New York-London.

[1001] Pennington, R. H. (1970): Introductory Computer
T Methods and Numerical Analysis (2nd edition), MacMillan Co., New York, xi + 452 pp.

[1002] Penrose, R. (1955): A generalized inverse for matrices,
P Proc. Cambridge Philos. Soc., 51, 406-413.

[1003] Penrose, R. (1956): On best approximate solutions of
P linear matrix equations, Proc. Cambridge Philos. Soc., 52, 17-19.

Penrose, R.: See Munn, W. D. (1955).

[1004] Pereyra, V. (1967): Iterative methods for solving non-
S linear least squares problems, SIAM J. Numer. Anal., 4, 27-36.

[1005] Pereyra, V. (1969): Stability of general systems of
P linear equations, Aequationes Math., 2, 194-206.

[1006] Pereyra, V. and J. B. Rosen (1964): Computation of the
P pseudoinverse of a matrix of unknown rank, Tech. Rep. CS 13, Computer Sciences Dept., Stanford University. (Comp. Rev., 6 (1965), 259, #7948).

Pereyra, V.: See Golub, G. H. (1972), (1975).

[1007] Pérez, A. and M. D. Scolnik (1972): Derivatives of
P pseudoinverses and constrained nonlinear regression problems, manuscript.

Perlin, I. E.: See Gallaher, L. J. (1974).

[1008] Perlis, S. (1952): Theory of Matrices, Addison-Wesley,
T Cambridge, Mass.

[1009] Peters, G. and J. H. Wilkinson (1970): The least
P squares problem and pseudo-inverses, Computer J., 13, 309-316.

Peters, G.: See Bowdler, H. J. (1966), Martin, R. S. (1966).

[1010] Petrov, A. P. and A. V. Hovanskiǐ (1974): Estimation of
S the error of solution of linear problems when there are accuracies in the operators and in the right-hand sides of the equations (Russian), Ž. Vyčisl. Mat. i Mat. Fiz., 14, 479-483.

[1011] Petryshyn, W. V. (1962): Direct and iterative methods
S for the solution of linear operator equations in Hilbert spaces, Trans. Amer. Math. Soc., 105, 136-175.

[1012] Petryshyn, W. V. (1965): On the inversion of matrices
S and linear operators, Proc. Amer. Math. Soc., 16, 893-901.

[1013] Petryshyn, W. V. (1967): On generalized inverses and
P on the uniform convergence of $(I - \beta K)^n$ with application to iterative methods, J. Math. Anal. Appl., 18, 417-439.

[1014] Petryshyn, W. V. (1968): On projectional-solvability
S and the Fredholm alternative for equations involving linear A-proper operators, Arch. Rational Mech. Anal., 30, 270-284.

[1015] Petryshyn, W. V. (1975): On the approximation-solva-
S bility of equations involving A-proper and pseudo A-proper mappings, Bull. Amer. Math. Soc., 81, 223-312.

[1016] Phillips, D. L. (1962): A technique for the numerical
S solution of certain integral equations of the first kind, J. Assoc. Comput. Mach., 9, 84-97.

[1017] Phillips, D. L. (1964): Three applications of subroutine
P ANF 105 - Generalized solution of matrix equations, Argonne National Laboratory, Appl. Math. Division, Tech. Memorandum No. 76, July 30, 1964, 6 pp.

[1018] Phillips, R. S. (1940): On linear transformations, Trans.
S Amer. Math. Soc., 48, 516-541.

[1019] Pietsch, A. (1960): Zur Theorie der σ-Transformation in
P lokalkonvexen Vektorräumen, Math. Nach., 21, 347-369.

[1020] de Pillis, J. (1973): Gauss-Seidel convergence for op-
T erators on Hilbert space, SIAM J. Numer. Anal., 10, 112-122.

[1021] Plackett, R. L. (1949): An historical note on the method
S of least squares, Biometrika, 36, 458-460.

[1022] Plackett, R. L. (1950): Some theorems in least squares,
S Biometrika, 37, 149-157.

[1023] Plemmons, R. J. (1971): Generalized inverses of Boolean
P relation matrices, SIAM J. Appl. Math., 20, 426-433.

[1024] Plemmons, R. J. (1972): Graphs and nonnegative
S matrices, Linear Algebra and Appl., 5, 283-292.

[1025] Plemmons, R. J. (1972): Monotonicity and iterative
P approximations involving rectangular matrices, Math. Comp., 26, 853-858.

[1026] Plemmons, R. J. (1973): Regular nonnegative matrices,
S Proc. Amer. Math. Soc., 39, 26-32.

[1027] Plemmons, R. J. (1974): Linear least squares by elim-
S ination and MGS, J. Assoc. Comput. Mach., 21, 581-585.

[1028] Plemmons, R. J. and R. E. Cline (1972): The generalized
P inverse of a nonnegative matrix, Proc. Amer. Math. Soc., 31, 46-50.

Plemmons, R. J.: See Berman, A. (1972), (1974), (1974), Cline, R. E. (1974), Kammerer, W. J. (1975), Montague, J. S. (1972), Wall, J. R. (1972).

[1029] Pohožaev, S. I. (1969): Normal solvability of nonlinear
S operators, Dokl. Akad. Nauk SSSR, 184, 40-43. Translated in Soviet Math. Dokl., 19 (1969), 35-38.

[1030] Pohožaev, S. I. (1969): On nonlinear operators which
S have a weakly closed range of values and quaslinear elliptic equations, Mat. Sb., 78 (120), 237-259. Translated in Math. USSR Sb., 7 (1969), 227-250.

[1031] Pohožaev, S. I. (1969): Normal solvability of nonlinear
S mappings in uniformly convex Banach spaces, Funkcional. Anal. i Prilozen, 3, 80-84. Translated in Functional Anal. Appl., 3, 147-151.

[1032] Pohožaev, S. I. (1971): On nonlinear operators whose
S ranges are subspaces, Dokl. Akad. Nauk SSSR, 12, Soviet Math. Dokl., 12 (1971), 168-172.

[1033] Poole, G. D. (1973): Geometry of minimum norm and
P least squares solutions to matrix equations, Delta, 3, 12-17.

[1034] Poole, G. D. (1973): A graph theoretic approach to
P pseudoinverses, Texas J. Science, 24, 439-444.

[1035] Poole, G.D. and T. L. Boullion (1972): The Drazin inverse
P for certain power matrices, Indust. Math., 22, 35-37.

[1036] Poole, G. D. and T. L. Boullion (1972): Weak spectral
P inverses which are partial isometries, SIAM J. Appl. Math., 23, 171-172.

[1037] Poole, G. and T. L. Boullion (1974): A survey on M-
S matrices, SIAM Review, 16, 419-427.

[1038] Poole, G. and T. L. Boullion (1975): The Drazin inverse
P and a spectral inequality of Marcus, Minc and Moyls,
J. Optimization Theory and Appl., 15, 503-508.

Poole, G. D.: See Boullion, T. L. (1970), Singh, I. (1975).

[1039] Popa, C. (1962): Note on the inversion of singular
P matrices, Lucrar Sti. Inst. Ped. Timisoara Mat. -Fiz.,
149-153.

[1040] Pope, A. J. and J. L. Stearn (1964): Matrix algebra
S applied to earth movement analysis, Washington, D. C.

[1041] La Porte, M. and J. Vignes (1974): Méthode numérique
S de détection de la singularite d'une matrice, Numer.
Math., 23, 73-81.

[1042] Porter, W. A. (1966): Modern Foundation of System
S Engineering, MacMillan, New York, 493 pp.

[1043] Porter, W. A. (1971): A basic optimization problem in
S linear systems, Math. Systems Theory, 5, 20-44.

[1044] Porter, W. A. and J. P. Williams (1966): A note on the
S minimum effort control problem, J. Math. Anal. Appl.,
13, 251-264.

[1045] Porter, W. A. and J. P. Williams (1966): Extensions of
S the minimum effort control problem, J. Math. Anal. Appl.,
13, 536-549.

[1046] Powell, D. R. and J. R. MacDonald (1972): A rapidly
S convergent iterative method for the solution of the generalized nonlinear least squares problem, Comput. J.,
15, 148-155.

[1047] Powell, M. J. D. and J. K. Reid (1968): On applying
P Householder transformations to linear least squares
problems, IFIP, Edinburgh.

Powers, D. L.: See Dalphin, J. F. (1972), Harwood, W. R. (1970), Lovass-Nagy, V. (1970), (1971), (1972), (1972), (1973), (1973), (1973), (1974), (1974).

Prabhu, S. S.: See Krishnamurthy, E. V. (1974).

Prager, W.: See Oettli, W. (1964), (1965).

[1048] Pratt, W. K. (1972): Generalized Wiener filtering com-
S putation techniques, IEEE Trans. on Computers, C-21, 636-641.

Prenter, P. M.: See Allgower, E. L. (1974).

[1049] Presić, S. B. (1963): Certaines équations matricielles,
S Publications de la Faculté d'Électrotechnique de l'Université à Belgrade Série. Math. et Phys., 121, 31-32.

[1050] Preston, G. B. (1954): Inverse semigroups, J. London
S Math. Soc., 29, 396-403.

Preston, G. B.: See Clifford, A. H. (1961).

[1051] Price, C. M. (1964): The matrix pseudoinverse and
P minimal variance estimates, SIAM Review, 6, 115-120.

Price, V. E.: See Doust, A. (1964).

[1052] Prijatelj, N. and I. Vidav (1971): On special *-regular
S rings, Michigan Math. J., 18, 213-221.

[1053] Pringle, R. M. and A. A. Rayner (1970): Expressions
P for generalized inverses of a bordered matrix with application to the theory of constrained linear models, SIAM Review, 12, 107-115.

[1054] Pringle, R. M. and A. A. Rayner (1971): Generalized
P Inverse Matrices with Applications to Statistics, Charles Griffin and Co., London.

Pringle, R. M.: See Rayner, A. A. (1967).

[1055] Prochaska, B. J. (1971): Applications of generalized inverse to estimation in dynamical systems, Trans. ASME, Ser. C.J., Dynamics Systems, Measurement and Control, 93, 252-256.
P

[1056] Przeworska-Rolewicz, D. (1960): Sur les équations involutives d'ordre n, Bull. de l'Acad. Pol. d. Sci., 8, 741-746.
T

[1057] Przeworska-Rolewicz, D. (1961): Sur les équations involutives et leur applications, Studia Math., 20, 95-117.
T

[1058] Przeworska-Rolewicz, D. (1962): Sur les opérations satisfaisantes a l'identité polynomiale, Studia Math., 22, 43-58.
T

[1059] Przeworska-Rolewicz, D. (1963): Équations avec opérations algébriques, Studia Math., 22, 337-367.
S

[1060] Przeworska-Rolewicz, D. (1965): Sur les équations avec opérations presque algébriques, Studia Math., 25, 121-131.
S

[1061] Przeworska-Rolewicz, D. (1969): A characterization of algebraic derivative, Bull. de l'Acad. Pol. d. Sci. 17, 11-13.
T

[1062] Przeworska-Rolewicz, D. (1969): A mixed boundary value problem with an algebraic derivative, Bull. de l'Acad. Pol. d. Sci., 20, 645-648.
T

[1063] Przeworska-Rolewicz, D. (1970): Algebraic derivative and abstract differential equations, Anais da Academia Brasileira de Ciencias, 42, 403-409.
T

[1064] Przeworska-Rolewicz, D. (1970): Algebraic derivative and definite integrals, Bull. de l'Acad. Pol. d. Sci., 20, 641-644.
S

[1065] Przeworska-Rolewicz, D. (1972): Algebraic derivatives
T and initial value problems, Bull. de l'Acad. Pol. d. Sci., 20, 629-633.

[1066] Przeworska-Rolewicz, D. (1972): Generalized linear
S equations of Carleman type, Bull. de l'Acad. Pol. d. Sci., 20, 635-639.

[1067] Przeworska-Rolewicz, D. (1972): Concerning left in-
S vertible operators, Bull. de l'Acad. Pol. d. Sci., 20, 837-839.

[1068] Przeworska-Rolewicz, D. (1975): Right invertible op-
S erators and their applications. 5 TMP, Wissenschaftliche Schriftenreiche, pp. 288-296.

[1069] Przeworska-Rolewicz, D. (1973): Algebraic Analysis
T and Differential Equations (Polish), WAT, Warsaw.

[1070] Przeworska-Rolewicz, D. (1973): Algebraic theory of
T partial differential equations with variable coefficients, Funktionenteoretische Eigenschaften der Lösungen partieller Differentialgleichungen, Math. Institut der Universität Bonn, 83.X, Bonn.

[1071] Przeworska-Rolewicz, D. (1973): Algebraic theory of
S right invertible operators, Studia Math., 48, 129-144.

[1072] Przeworska-Rolewicz, D. (1973): Equations with Trans-
T formed Argument. An Algebraic Approach. (Chapter IV), Elsevier Scientific Publ. Comp. and PWN-Polish Scientific Publishers, Amsterdam-Warsaw.

[1073] Przeworska-Rolewicz, D. (1973): Pseudocategories,
S paraalgebras and perturbations of linear operators. Preprint. Institute of Mathematics, Polish Academy of Sciences, Warsaw.

[1074] Przeworska-Rolewicz, D. (1973): Right invertible op-
S erators and functional-differential equations with involutions, Demonsratio Math., 5, 165-177.

[1075] Przeworska-Rolewicz, D. (1973): Concerning boundary
S value problems for equations with right invertible operators, Demonstratio Math., 7, 365-380.

[1076] Przeworska-Rolewicz, D. (1973): Extension of operation-
T al calculus, Control and Cybernetics, 2, 5-14.

[1077] Przeworska-Rolewicz, D. (1974): On linear differential
S equations with transformed argument solvable by means of right invertible operators, Annales Polonici Math., 29, 141-148.

[1078] Przeworska-Rolewicz, D. and S. Rolewicz (1964): On
S operators with finite d-characteristic, Studia Math., 24, 257-270.

[1079] Przeworska-Rolewicz, D. and S. Rolewicz (1965): On
S quasi-Fredholm ideals, Studia Math., 26, 67-71.

[1080] Przeworska-Rolewicz, D. and S. Rolewicz (1967): On
S d and d_{Ξ}-characteristic of linear operators, Annales Polonici Math., 19, 117-121.

[1081] Przeworska-Rolewicz, D. and S. Rolewicz (1968): Equa-
S tions in Linear Spaces, Monog. Mat. 47 PWN Polish Scientific Publishers, Warsaw, 380pp.

Puri, M. L.: See Mitra, S. K. (1973).

[1082] Pye, W. C., T. L. Boullion and T. A. Atchison (1973):
S A note on the upper triangular Toeplitz matrix, Indust. Math., 23, 53-60.

[1083] Pye, W. C., T. L. Boullion and T. A. Atchison (1973):
P The pseudoinverse of a centrosymmetric matrix, Linear Algebra and Appl., 6, 201-204.

[1084] Pye, W. C., T. L. Boullion and T. A. Atchison (1973):
P The pseudoinverse of a composite matrix of cirulants, SIAM J. Appl. Math., 24, 552-555.

[1085] Pyle, L. D. (1960): The Generalized Inverse in Linear
P Programming, Doctoral Dissertation, Purdue University, Lafayette, Indiana.

[1086] Pyle, L. D. (1964): Generalized inverse computations
P using the gradient projection method, J. Assoc. Comput. Mach., 11, 422-428.

[1087] Pyle, L. D. (1967): A generalized inverse ϵ-algorithm
P for construction projection matrices, with applications, Numer. Math., 10, 86-102.

[1088] Pyle, L. D. (1968): Remarks on a generalized inverse
P ϵ-algorithm for matrices, pp. 218-238 in Boullion and Odell (1968).

[1089] Pyle, L. D. (1972): The generalized inverse in linear
P programming - Basic structure, SIAM J. Appl. Math., 22, 335-355.

Pyle, L. D.: See Cho, C. Y. (1966), Cline, R. E. (1973).

[1090] Rabson, G. (1969): The generalized inverse in set theory
P and matrix theory, Dept. of Mathematics, Clarkson College of Technology, Potsdam, N. Y.

[1091] Rado, R. (1956): Note on generalized inverses of
P matrices, Proc. Cambridge Philos. Soc., 52, 600-601.

[1092] Raghavarao, D. (1964): Singular weighing designs, Ann.
S Math. Statist., 35, 673-680.

[1093] Rainbolt, M. B., Sequential least-squares parameter
S estimation, NASA-MSC Unpublished Report.

[1094] Rakovshchick, L. (1966): Approximate solutions of equa-
S tions with normally resolvable operators, U. S. S. R. Computational Math. and Math. Phy., 6, 3-11.

[1095] Rall, L. B. (1969): Computational Solution of Nonlinear
S Operator Equations, Wiley, New York, viii + 225 pp.

[1096] Rall, L. B. (1975): The Fredholm pseudoinverse – An
P analytic episode in the history of generalized inverses, this Volume.

[1097] Ramanujan, P. B. (1970): On operators of class (N, k),
T Proc. Cambridge Philos. Soc., 68, 141-142.

Ramnath, B.: See Tewarson, R. P. (1969).

[1098] Rao, C. R. (1955): Analysis of dispersion for multiply
S classified data with unequal numbers in cells, Sankhyā, Ser. A, 15, 253-280.

[1099] Rao, C. R. (1961): A study of large sample test criteria
S through properties of efficient estimates, Sankhyā, Ser. A, 23, 25-40.

[1100] Rao, C. R. (1962): A note on a generalised inverse of a
P matrix with applications to problems in mathematical statistics, J. Roy. Statist. Soc., Ser. B, 24, 152-158.

[1101] Rao, C. R. (1965): On the theory of least squares when
S parameters are stochastic and its application to analysis of growth curves, Biometrika, 52, 447-458.

[1102] Rao, C. R. (1966): Generalized inverse for matrices and
P its applications in mathematical statistics, Research Papers in Statistics, Festschrift for J. Neyman, pp. 263-279, Wiley, New York.

[1103] Rao, C. R. (1967): Least squares theory using an esti-
S mated dispersion matrix and its application to measurement of signals, in Proc. Fifth Berkeley Symposium on Math. Stat. and Prob., 1, 355-372, University of California Press.

[1104] Rao, C. R. (1967): Calculus of generalized inverse of
P matrices, Part I: General theory, Sankhyā, Ser. A, 29, 317-342.

[1105] Rao, C. R. (1968): A note on a previous lemma in the
S theory of least squares and some further results, Sankhyā, Ser. A, 30, 245-252.

[1106] Rao, C. R. (1970): Estimation of heteroscedastic vari-
S ables in linear models, J. Amer. Statist. Assoc., 65, 161-172.

[1107] Rao, C. R. (1970): Estimation of variance and covariance
S components, Tech. Report No. Math-Stat/44/70, Research and Training School, Indian Statistical Institute, Calcutta.

[1108] Rao, C. R. (1971): A unified theory of linear estimation
P Sankhyā, Ser. A, 33, Part 4, 371-394.

[1109] Rao, C. R. (1971): Estimation of variance and covariance
S components-minque theory, Discussion Paper No. 58, Indian Statistical Institute, New Delhi.

[1110] Rao, C. R. (1971): Linear Statistical Inference and Its
S Applications, 2nd edition, Wiley, New York.

[1111] Rao, C. R. (1972): A note on the IPM method in the uni-
S fied theory of linear estimation, Sankhyā, Ser. A, 34, Part 3, 285-288.

[1112] Rao, C. R. and S. K. Mitra (1971): Further contributions
P to the theory of generalized inverse of matrices and its applications, Sankhyā, Ser. A, 33, 289-300.

[1113] Rao, C. R. and S. K. Mitra (1971): Generalized Inverse
P of Matrices and its Applications, John Wiley and Sons, Inc., New York, xiv + 240 pp.

[1114] Rao, C. R. and S. K. Mitra (1972): Generalized Inverse
P of a matrix and its applications, Proc. Sixth Berkeley Symposium on Math. Stat. and Prob., 1, 601-620.

[1115] Rao, C. R., S. K. Mitra and P. Bhimasankaram (1972):
P Determination of a matrix by its subclasses of g-inverses, Sankhyā, Ser. A, 34, 5-8.

[1116] Rao, C. R. and S. K. Mitra (1973): Theory and applica-
P tion of constrained inverse of matrices, SIAM J. Appl. Math., 24, 473-488.

Rao, C. R.: See Khatri, C. G. (1968), Mitra, S. K. (1968), (1968), (1969), (1972).

[1117] Rao, K. K. (1966): A simpler proof of Gauss-Markov
P theorem when the regression matrix is of less than full rank, Amer. Math. Monthly, 73, 394-395.

Rao, M. M.: See Chipman, J. S. (1964), (1964), Mizel, V. J. (1962).

[1118] Rao, P. S. N. V. Prasada (1973): On generalized in-
P verses of doubly stochastic matrices, Sankhyā, Ser. A, 35, 103-105.

[1119] Rayner, A. A. and D. Livingstone (1965): On the distri-
S bution of quadratic forms in singular normal variates, South African J. Agricultural Sci., 8, 357-370.

[1120] Rayner, A. A. and R. M. Pringle (1967): A note on gen-
P eralized inverses in the linear hypothesis not of full rank, Ann. Math. Statist., 38, 271-273.

Rayner, A. A.: See Pringle, R. M. (1970), (1971).

Redheffer, R.: See Hestenes, M. R. (1974), Kwon, Y. K. (1969).

Reed, M. B.: See Seshu, S. (1961).

[1121] Rege, M. B. (1973): A note on the Azumaya algebras,
S J. Indian Math. Soc. (N.S.), 37, 217-225.

Reid, J. K.: See Powell, M. J. D. (1968).

[1122] Reid, W. T. (1931): Generalized Green's matrices for
S compatible systems of differential equations, Amer. J. Math., 53, 443-459.

[1123] Reid, W. T. (1932): A boundary value problem associ-
S ated with the calculus of variations, Amer. J. Math., 54, 769-790.

[1124] Reid, W. T. (1963): Riccati matrix differential equations
S and non-oscillation criteria for associated linear differential systems, Pacific J. Math., 12, 665-685.

[1125] Reid, W. T. (1964): Principal solutions of non-oscillatory linear differential systems, J. Math. Anal. Appl.,
S 9, 397-423.

[1126] Reid, W. T. (1965): A matrix equation related to a non-
S oscillation criterion and Liapunov stability, Quart. Appl. Math., 23, 83-87.

[1127] Reid, W. T. (1967): Generalized Green's matrices for
P two-point boundary problems, J. Soc. Indust. Appl. Math., 15, 856-870.

[1128] Reid, W. T. (1968): Generalized inverses of differential
P and integral operators, pp. 1-25 in Boullion and Odell (1968).

[1129] Reid, W. T. (1970): Ordinary Differential Equations,
S Wiley-Interscience, New York, xv + 551 pp.

[1130] Reid, W. T. (1972): Riccati Differential Equations,
S Academic Press, New York.

Reinsch, C.: See Golub, G. H. (1970), Wilkinson, J. H. (1971).

[1131] Replogle, J., B. D. Holcomb and W. R. Burrus (1967):
S The use of mathematical programming for solving singular and poorly conditioned systems of equations, J. Math. Anal. Appl., 20, 310-324.

[1132] Rheinboldt, W. C. (1968): A unified convergence theory
S for a class of iterative processes, SIAM J. Numer. Anal., 5, 42-63.

Rheinboldt, W. C.: See Greub, W. (1960), Ortega, J. M. (1970).

[1133] Rice, J. R. (1966): Experiments on Gram-Schmidt
T orthogonalization, Math. Comp., 20, 325-328.

[1134] Rice, J. R. and K. H. Usow (1968): The Lawson al-
T gorithm and extensions, Math. Comp., 22, 118-127.

Rice, J. R.: See Lynch, R. E. (1964).

Richardson, H.: See Stearn, J. L. (1962).

[1135] Rickart, C. E. (1946): Banach algebras with an ad-
S joint operation, Ann. of Math., 47, 528-550.

Rigal, J.-L.: See Gaches, J. (1965).

[1136] Riley, J. D. (1956): Solving systems of linear equa-
S tions with a positive definite, symmetric, but possibly ill-conditioned matrix, Math. Tables Aids Comput., 9, 96-101.

[1137] Rinehart, R. F. (1955): The equivalence of definitions
T of a matrix function, Amer. Math. Monthly, 62, 395-414.

[1138] Ringrose, J. R. (1957): Precompact linear operators in
T locally convex spaces, Proc. Cambridge Philos. Soc., 53, 581-591.

[1139] Rivlin, T. J. (1963): Overdetermined systems of lin-
S ear equations, SIAM Review, 5, 52-66.

[1140] Robers, P. D. and A. Ben-Israel (1969): An interval
S programming algorithm for discrete linear L_1 approximation problems, J. Approximation Theory, 2, 323-336.

[1141] Robers, P. D. and A. Ben-Israel (1970): A suboptimi-
S zation method for interval linear programming: A new method for linear programming, Linear Algebra and Appl., 3, 383-405.

Robers, P. D.: See Ben-Israel, A. (1968).

[1142] Robertson, J. B. and M. Rosenberg (1968): The de-
S composition of matrix-valued measures, Michigan Math. J., 15, 353-368.

[1143] Robert, P. (1968): On the group-inverse of a linear
P transformation, J. Math. Anal. Appl., 22, 658-669.

Roberts, F. D. K.: See Barrodale, I. (1973).

[1144] Robinson, D. W. (1962): On the generalized inverse
P of an arbitrary linear transformation, Amer. Math. Monthly, 69, 412-416.

[1145] Robinson, D. A. (1973): A historical perspective:
P Gauss and generalized inverses, a preprint.

Robinson, D. W.: See Davis, D. L. (1972), Hansen, G. W. (1974).

[1146] Robinson, S. M. (1972): Extension of Newton's method
S to nonlinear functions with values in a cone, Numer. Math., 19, 341-347.

[1147] Rockafellar, R. T. (1970): Convex Analysis, Princeton
T University Press, Princeton, N. J., xviii + 451 pp.

Rodrigue, G. H.: See Blum, E. K. (1974), McCormick S. F. (1975).

[1148] Røeggen, I. (1973): On least squares estimation of
S nonlinear parameters with particular emphasis on matrix evaluation, Phys. Norveg., 7, 33-38.

[1149] Rohde, C. A. (1964): Contributions to the Theory,
P Computation and Applications of Generalized Inverses, Doctoral Dissertation, North Carolina State University, Raleigh, N.C., May, 1964.

[1150] Rohde, C. A. (1965): Generalized inverses of par-
P titioned matrices, SIAM J. Appl. Math., 13, 1033-1035.

[1151] P Rohde, C. A. (1966): Some results on generalized inverses, SIAM Review, 8, 201-205.

[1152] P Rohde, C. A. (1968): Special applications of the theory of generalized matrix inversion to statistics, pp. 239-266 in Boullion and Odell (1968).

[1153] S Rohde, C. A. and J. R. Harvey (1965): Unified least squares analysis, J. Amer. Statist. Assoc., 60, 523-527.

[1154] S Rohde, C. A. and G. M. Tallis (1969): Exact first and second order moments of estimates of components of covariance, Biometrika, 56, 517-525.

Rolewicz, S.: See Przeworska-Rolewicz, D. (1964), (1965), (1967), (1968).

[1155] S Root, W. L. (1971): On the structure of a class of system identification problems, Automatica, 7, 219-231.

[1156] S Root, W. L. (1973): On the modelling and estimation of communication channels, in Multivariate Analysis-III, R. Krishnaiah, ed., Academic Press, New York.

Root, W. L.: See Beutler, F. J. (1973).

[1157] T Rosen, J. B. (1960): The gradient projection method for nonlinear programming, Part I: Linear constraints, SIAM J. Appl. Math., 8, 181-217.

[1158] T Rosen, J. B. (1961): The gradient projection method for nonlinear programming, Part II: Nonlinear constraints, SIAM J. Appl. Math., 9, 514-532.

[1159] S Rosen, J. B. (1964): Minimum and basic solutions to singular linear systems, SIAM J. Appl. Math., 12, 156-162.

[1160] T Rosen, J. B. (1967): Chebyshev solution of large linear systems, J. Comput. Syst. Sci., 1, 29-43.

Rosen, J. B.: See Cho, C. Y. (1966), Pereyra, V. (1964).

[1161] S Rosenberg, M. (1964): The square-integrability of matrix-valued functions with respect to a non-negative Hermitian measure, Duke Math. J., 291-298.

[1162] S Rosenberg, M. (1969): Mutual subordination of multivariate stationary processes over any locally compact abelian group, Z. Wahrscheinlichkeitstheorie und verw. Gebiete, 12, 333-343.

[1163] P Rosenberg, M. (1969): Range decomposition and generalized inverse of nonnegative Hermitian matrices, SIAM Review, 11, 568-571.

[1164] S Rosenberg, M. (1974): Operators as spectral integrals of operator-valued functions from the study of multivariate stationary stochastic processes, J. Multivariate Anal., 4, 166-209. Part II, to appear.

Rosenberg, M.: See Robertson, J. B. (1968).

[1165] T Rosenbloom, P. C. (1956): The method of steepest descent, in Numerical Analysis, pp. 127-176. Proc. 65h Symposium in Applied Mathematics, McGraw-Hill Book Co., New York.

Rousset de Pina, X.: See Gaches, J. (1965).

[1166] S Ruhe, A. (1970): An algorithm for the numerical determination of the structure of a general matrix, BIT, 10, 196-216.

[1167] S Ruhe, A. (1970): Properties of a matrix with a very ill-conditioned eigenproblem, Numer. Math., 15, 57-60.

[1168] Ruhe, A. (1970): Perturbation bounds for means of
S eigenvalues and invariant subspaces, BIT, 10, 343-354.

[1169] Ruhe, A. and P. Å. Wedin (1974): Algorithms for
S separable nonlinear least squares problems, Univ. of Umeå, Dept. Information Processing, Rept. UMINF-47.74.

[1170] Rust, B. W. and W. R. Burrus (1972): Mathematical
S Programming and the Numerical Solution of Linear Equations, American Elsevier, New York.

[1171] Rust, B. W., W. R. Burrus and C. Schneeberger (1966):
P A simple algorithm for computing the generalized inverse of a matrix, Comm. ACM, 9, 381-386.

[1172] Ruston, A. F. (1954): Operators with a Fredholm
T theory, J. London Math. Soc., 29, 318-326.

[1173] Rutishauser, H. (1968): Once again: The least squares
S problem, Linear Algebra and Appl., 1, 479-488.

[1174] Sakamoto, H. (1944): On independence of statistics
T (in Japanese), Res. Memoirs Instit. Stat. Math., 1, 1-25.

[1175] Sakawa, Y. (1972): Optimal filtering in linear distrib-
T uted parameter systems, Internat. J. Control, 16, 115-127.

[1176] Salehi, H. (1967): The Hellinger square-integrability
S of matrix-valued measures with respect to a non-negative hermitian measure, Ark. Mat., 7, 299-303.

[1177] Salehi, H. (1967): Applications of the Hellinger inte-
S grals to q-variate stationary stochastic processes, Ark. Mat., 7, 305-311.

[1178] Salehi, H. (1968): On the Hellinger integrals and in-
S terpolation of q-variate stationary stochastic processes, Ark. Mat., 8, 1-6.

[1179] Salehi, H. (1968): Applications of Hellinger's inte-
S grals in detection theory, J. Math. Anal. Appl., 21, 264-276.

[1180] Salehi, H. (1969): Interpolation of a q-variate homog-
T eneous random fields, J. Math. Anal. Appl., 25, 653-662.

[1181] Salehi, H. (1969): On interpolation of q-variate sta-
T tionary stochastic processes, Pacific J. Math., 28, 183-191.

[1182] Salehi, H. and J. K. Schmidt (1972): Interpolation of
T q-variate weakly stationary stochastic processes over a locally compact abelian group, J. Multivariate Analysis, 2, 307-331.

Salehi, H.: See Mandrekar, V. (1970), (1970), (1970), (1971), (1972), Miamee, A. G. (1974), Nashed, M. Z. (1973).

[1183] Salinetti, G. (1974): The generalized inverse in param-
P eter programming, Calcolo, 11, 351-363.

[1184] Saphar, P. (1964): Contribution à l'étude des appli-
S cations linéaires dans un espace de Banach, Bull. Soc. Math. France, 92, 364-384.

[1185] Saphar, P. (1965): Sur les applications linéaires dans
S un espace de Banach II, Ann. Sci. École Norm. Sup., (3) 82, 205-240.

Sather, D.: See Osborn, J. E. (1975).

Saunders, M. A.: See Golub, G. H. (1969).

[1186] Savelova, T. I. (1974): Projection methods for the solu-
P tion of linear ill-posed equations, Ž. Vyčisl. Mat. i Mat. Fiz., 14, 1027-1031, 1078.

[1187] Schaeffer, H. H. (1960): On the singularities of analytic
S functions with values in Banach space, Arch. Math., 11, 40-43.

Schäffer, J. J.: See Massera, J. L. (1966).

[1188] Schaffer, W. S. (1967): Multivariable control systems
P synthesis utilizing the generalized inverse of a matrix, Proc. IEEE, 55, 1202-1203.

[1189] Schechter, M. (1967): Basic theory of Fredholm operators,
S Ann. Scoula Norm. Sup. Pisa (3) 21, 361-380.

[1190] Schechter, M. (1971): Principles of Functional Analysis,
T Academic Press, New York.

[1191] Schempp, W. (1970): Iterative solution of linear operator
S equations in Hilbert space and optimal Euler methods, Arch. Math. (Basel) 21, 390-395.

Schempp, W.: See Niethammer, W. (1970).

[1192] Schmidt, E. (1907): Zur Theorie der linearen und nicht-
T linearen Integralgleichungen, I. Entwicklung willkürlicher Funktionen nach Systemen vorgeschriebener, Math. Ann., 63, 433-476.

[1193] Schmidt, E. (1907): Zur Theorie der linearen und nicht-
T linearen Integralgleichungen, II. Auflösung der allgemeinen linearen Integralgleichung, Math. Ann., 64, 161-174.

Schmidt, J. K.: See Salehi, H. (1972).

Schneeberger, C: See Rust, B. (1966).

[1194] Schneider, H. (1956): A matrix problem concerning pro-
S jections, Proc. Edinburgh Math. Soc., 10, 129-130.

[1195] Schneider, H. and G. P. Barker (1968): Matrices and
T Linear Algebra, Holt, Rinehart, and Winston, Inc.

[1196] Schönemann, P. H. (1966): A generalized solution of
S the orthogonal procrustes problem, Psychometrika, 31, 1-10.

Schreiber, M.: See Anderson, W. N., Jr. (1972).

[1197] Schulz, G. (1933): Iterative Berechnung der reziproken
T Matrix, Z. Angew Math. Mech., 13, 57-59.

[1198] Schwartz, J. T. (1954): Perturbations of spectral op-
S erators, and applications, Pacific J. Math., 4, 415-458.

[1199] Schwartz, L. E. (1974): A globally convergent algor-
S ithm for nonlinear least squares, Progress in Statistics (European Meeting of Statisticians, Budapest, 1972), pp. 667-676, Colloq. Math. Soc. János Bolyai, vol. 9, North-Holland, Amsterdam.

Schwartz, J. T.: See Dunford, N. (1958), (1963).

[1200] Schwarz, Š. (1964): On the structure of the semigroup
S of stochastic matrices, Magyar Tud. Akad. Mat. Kutató Inst. Közl., 9, 297-311.

[1201] Schwarz, Š. (1967): A note on the structure of the
S semigroup of doubly-stochastic matrices, Mat. Časopis Sloven. Akad. Vied., 17, 308-316.

[1202] Schwerdtfeger, H. (1950): Introduction to Linear
S Algebra and the Theory of Matrices, P. Noordhoff, Groningen, 1950, 288 pp.

[1203] Scolnik, H. D. (1970): On the Solution of Nonlinear
P Least Squares Problems, Doctoral Thesis, University of Zürich.

[1204] Scolnik, H. D. (1972): On the solution of nonlinear
P least squares problems, Proc. IFIP-71, Numerical Math., pp. 1258-1265, North-Holland Publishing Co., Amsterdam.

Scolnik, H. D.: See Pérez, A. (1972).

[1205] P Scroggs, J. E. and P. L. Odell (1966): An alternative definition of the pseudoinverse of a matrix, SIAM J. Appl. Math., 14, 796-810.

[1206] P Searle, S. R. (1965): Additional results concerning estimator function and generalized inverse matrices, J. Roy. Statist. Soc., Ser. B, 27, 486-490.

[1207] S Searle, S. R. (1966): Matrix Algebra for the Biological Sciences (Including Applications in Statistics), Chapters 6, 9, and 10, Wiley, New York.

[1208] S Searle, S. R. (1968): Another look at Henderson's methods of estimating variance components, Biometrics, 24, 749-778.

[1209] S Searle, S. R. (1971): Linear Models, Wiley-Interscience, New York.

[1210] T Sebestyen, G. D. and J. Edie (1966): An algorithm for nonparametric pattern recognition, IEEE Trans. Electronic Computers, vol. EC-15, 908-915.

[1211] S Seidel, J. J. (1955): Angles and distances in N-dimensional Euclidean and non-Euclidean geometry, I, II, III, Nederl. Akad. Wetensch. Proc., Ser. A, 17, 329-335, 336-340, 535-541.

[1212] T Seneta, E. (1973): Nonnegative Matrices. An Introduction to Theory and Applications, Halsted Press, New York.

[1213] S Seshu, S. and M. B. Reed (1961): Linear Graphs and Electrical Networks, Addison-Wesley, Reading, Mass.

[1214] S Shanbag, D. N. (1968): Some remarks on Khatri's result in quadratic forms, Biometrika, 55, 593-595.

[1215] S Shanbag, D. N. (1970): On the distribution of a quadratic form, Biometrika, 57, 222-223.

[1216] S Sharpe, G. E. and B. Spain (1960): On the solution of networks by equicofactor matrix, IRE Transactions on Circuit Theory, CT-7, 230-239.

[1217] P Sharpe, G. E. and G. P. H. Styan (1965): Circuit duality and the general network inverse, IEEE Transactions on Circuit Theory, 12, 22-27.

[1218] P Sharpe, G. E. and G. P. H. Styan (1965): A note on the general network inverse, IEEE Transactions on Circuit Theory, 12, 632-633.

[1219] S Sharpe, G. E. and G. P. H. Styan (1967): A note on equicofactor matrices, Proc. IEEE, 55, 1226-1227.

[1220] S Shaw, C. B., Jr. (1972): Improvement of the resolution of an instrument by numerical solution of an integral equation, J. Math. Anal. Appl., 37, 83-112.

[1221] S Shaw, C. B., Jr. (1973): Best accessible estimation: Convergence properties and limiting form of direct and reduced versions, J. Math. Anal. Appl., 44, 531-552.

[1222] P Sheffield, R. D. (1956): On Pseudo-Inverses of Linear Transformations in Banach Spaces, Oak Ridge Nat. Lab. Report, #2133. Ph.D. Thesis, University of Tennessee, Knoxville, Tennessee, 1956.

[1223] P Sheffield, R. D. (1958): A general theory of linear systems, Amer. Math. Monthly, 65, 109-111.

[1224] S Sherman, I. V. and W. J. Morrison (1950): Adjustment of an inverse matrix corresponding to a change in one element of a given matrix, Ann. Math. Statist., 21, 124-127.

[1225] S Shinozaki, N. and M. Sibuya (1974): Consistency of a pair of matrix equations with an application, Keio Engineering Reports, vol. 27, No. 10, Faculty of Engineering, Keio Univ., Yokohama, Japan.

[1226] Shinozaki, N. and M. Sibuya (1974): The reverse
P order law $(AB)^{-} = B^{-}A^{-}$, Linear Algebra and Appl., 9, 29-40.

[1227] Shinozaki, N. and M. Sibuya (1974): Product of pro-
S jectors, Scientific Center Report, IBM, Tokyo, Japan.

[1228] Shinozaki, N., M. Sibuya and K. Tanabe (1972):
P Numerical algorithms for the Moore-Penrose inverse of a matrix: Direct Methods, Ann. Inst. Statist. Math., 24, 193-203.

[1229] Shinozaki, N., M. Sibuya and K. Tanabe (1972):
P Numerical algorithms for the Moore-Penrose inverse of a matrix: Iterative Methods, Ann. Inst. Statist. Math., 24, 621-629.

[1230] Showalter, D. W. (1967): Representation and computation
P of the pseudoinverse, Proc. Amer. Math. Soc., 18, 584-587.

[1231] Showalter, D. W. and A. Ben-Israel (1970): Repre-
P sentation and computation of the generalized inverse of a bounded linear operator between Hilbert spaces, Atti Accad. Naz. Lincei Rend. Cl. Sci. Fis. Mat. Natur., Ser. VIII, 48, 120-130.

[1232] Shurbet, G. L., T. O. Lewis and T. L. Boullion (1972):
S Consistency of systems of linear matrix equations, Indust. Math., 22, 55-64.

[1233] Shurbet, G. L, T. O. Lewis and T. L. Boullion (1973):
S P-potent and R-circulant matrices, Texas J. Science, 25, No. 4.

[1234] Shurbet, G. L., T. O. Lewis and H. W. Milnes (1969):
P Recovery of linear transformations using collinear invariant points and pseudoinverses, Texas J. Science, 20, 361-366.

[1235] Sibuya, C. L. (1937): Über die analytische Theorie
S der quadratischen Formen III, Annals of Mathematics, 38, 212-291. (Particularly pp. 217-229).

[1236] Sibuya , M. (1969): Generalized inverses of
P matrices. Part I (Japanese), Ann. Inst. Statist. Math., Tokyo, 17, 109-131.

[1237] Sibuya, M. (1970): Subclasses of generalized inverses
P of matrices, Annals of the Institute of Statistical Mathematics, Tokyo, 22, 543-556.

[1238] Sibuya, M. (1971): Generalized inverses of mappings,
P Sankhyā, Ser. A, 33, 301-310. (Correction, ibid. 35, 1973).

[1239] Sibuya, M. (1972): Commutative generalized inverses
P of a square matrix, IBM Japan Scientific Center Report, G318-1904, Sept. 1972, 21 pp.

[1240] Sibuya, M. (1973): The Azumaya-Drazin pseudoinverse
P and the spectral inverses of a matrix, Sankhyā, Ser. A, 35, 95-102.

Sibuya, M.: See Shinozaki, N. (1972), (1972), (1974), (1974), (1974).

[1241] Siegel, C. L. (1941): Equivalence of quadratic forms,
S Amer. J. Math., 63, 658-680.

[1242] Singer, I. (1970): Best Approximation in Normed Lin-
S ear Spaces by Elements of Linear Subspaces, Springer-Verlag, Berlin.

[1243] Singer, I. (1974): The Theory of Best Approximation
S and Functional Analysis, CBMS Regional Conference Series in Applied Mathematics, vol. 13, SIAM Publication, Philadelphia, Pa.

[1244] Singh, I., G. Poole and T. L. Boullion (1975): A
P class of Hessenberg matrices with known pseudoinverse and Drazin inverse, Math. Comp., 29, 615-619.

Sittler, R. W.: See Albert, A. (1965).

[1245] S Skornyakov, L. A. (1964): Complemented Modular Lattices and Regular Rings, Oliver and Boyd, London.

[1246] S van der Sluis, A. (1975): Stability of the solutions of linear least squares problems, Numer. Math., 23, 241-254.

[1247] S Smith, D. K. (1969): A Dynamic Component Suppression Algorithm for the Acceleration of Vector Sequences, Doctoral Dissertation, Purdue University, Lafayette, Indiana.

[1248] T Smithies, F. (1937): The eigenvalues and singular values of integral equations, Proc. London Math. Soc., 43, 255-279.

[1249] T Smithies, F. (1958): Integral Equations, Cambridge University Press, x + 172 pp.

[1250] P Smogorshewsky, A. (1940): Les fonctions de Green des systems differéntials linéaires dans un domaine à une seule dimension, Recueil Math., 7 (49), 179-196.

Smolitsky, K. L.: See Mikhlin, S. G. (1967).

[1251] S Sobczyk, A. (1941): Projections in Minkowski and Banach spaces, Duke Math. J., 8, 78-106.

[1252] P Söderstrom, T. and G. W. Stewart (1974): On the numerical properties of an iterative method for computing the Moore-Penrose generalized inverse, SIAM J. Numer. Anal., 11, 61-74.

[1253] P Sokovnin, V. M. (1973): Pseudoinverse operators and the identification of linear nonstationary objects. (Russian) Control methods and models, No. 6, pp. 63-67, 136. Redakcionno-Izdat. Otdel Rizsk. Politehn. Inst., Riga.

[1254] Solloway, C. B. (1964): Elements of the Theory of Orbit
S Determination, Jet Prop. Lab., California Inst. Tech., California.

Solloway, C. B.: See Lass, J. (1961).

Souček, J.: See Fučik, S. (1973), (1974).

Souček, V.: See Fučik, S. (1973), (1974).

[1255] Sova, M. (1973): Abstract semilinear equations with
S small parameters, Proceedings of Equadiff. III (Brno, 1972) Czechoslovak Academy of Sciences, Prague, pp. 71-79.

Spain, B.: See Sharpe, G. E. (1960).

Spanier, J.: See Kellog, R. B. (1965).

[1256] Speed, F. M. (1974): An application of the generalized
P inverse to the one-way classification, Amer. Statist., 28, 16-18.

Speed, F. M.: See Delaney, J. C. (1966).

[1257] Sreedharan, V. P. (1969): Solutions of overdetermined
S linear equations which minimize error in an abstract norm, Numer. Math., 13, 146-151.

[1258] Sreedharan, V. P. (1971): Least squares algorithms for
S finding solutions of overdetermined linear equations which minimize error in an abstract norm, Numer Math., 17, 387-401.

Sreedharan, V. P.: See Duris, C. S. (1968).

[1259] Stakgold, I. (1971): Branching of solutions of nonlinear
S equations, SIAM Review, 13, 289-332. Erratum, ibid, 14 (1972), 492.

[1260] Stallings, W. T. and T. L. Boullion (1970): A spectral
P inverse for rectangular matrices, Indust. Math., 20, 61-70.

[1261] Stallings, W. T. and T. L. Boullion (1972): Computation
P of pseudoinverse matrices using residue arithmetic, SIAM Review, 14, 152-163.

[1262] Stallings, W. T. and T. L. Boullion (1972): The pseudo-
P inverse of an r-circulant matrix, Proc. Amer. Math. Soc., 34, 385-388.

[1263] Stallings, W. T. and T. L. Boullion (1974): A strong
P spectral inverse for r-circulant matrix, SIAM J. Appl. Math., 27, 322-325.

[1264] Stearn, J. L. (1951): Iterative solutions of normal equa-
S tions, Bull. Géodésique, 331-339.

[1265] Stearn, J. L. and H. Richardson (1962): Adjustment of
S conditions with parameters and error analysis, Bull. Géodésique, 64.

Stearn, J. L.: See Pope, A. J. (1964).

Stein, J.: See Kaniel, S. (1974).

[1266] Stein, P. (1952): Some general theorems on iterants,
T J. Res. Nat. Bur. Standards, Sect. B, 48, 82-83.

[1267] Stein, R. A. (1972): Linear Model Estimation Projection
P Operators, and Conditional Inverses, Ph. D. Thesis, Iowa State University, Ames, Iowa.

[1268] Stern, T. E. (1962): Extremum relations in nonlinear
S networks and their applications to mathematical programming, Journées d'Études sur le Contrôle Optimum et les Systèmes Nonlinéaires, Saclay, France, Institut National des Sciences et Techniques Nucléaires, pp. 135-156.

[1269] Stern, T. E. (1965): Theory of Nonlinear Networks and
S Systems, Addison-Wesley, Reading, Mass., xiv + 594 pp.

Steward, D. V.: See Cho, C. Y. (1966).

[1270] S Stewart, G. W. (1966): Perturbation bounds for the linear least squares problem, Computing Technological Center, Oak Ridge, Tenn., preprint.

[1271] P Stewart, G. W. (1969): On the continuity of the generalized inverses, SIAM J. Appl. Math., 17, 33-45.

[1272] S Stewart, G. W. (1972): Introduction to Matrix Computations, Academic Press, New York.

[1273] T Stewart, G. W. (1973): Conjugate direction methods for solving systems of linear equations, Numer. Math., 21, 285-297.

[1274] T Stewart, G. W. (1974): Modifying pivot elements in Gaussian elimination, Math. Comp., 28, 537-542.

Stewart, G. W.: See Söderstrom, T. (1974).

[1275] T Stiefel, E. (1959): Über diskrete und lineare Tschebycheff-Approximationen, Numer. Math., 1, 1-28.

[1276] T Stiefel, E. (1960): Note on Jordan elimination, linear programming and Tchebycheff approximation, Numer. Math., 2, 1-17.

[1277] S Stoer, J. (1971): On the numerical solution of constrained least-squares problems, SIAM J. Numer. Anal., 8, 382-411.

[1278] S Stojakovic, M. (1953): Sur les matrices quasiinverses et les matrices quasi-unités, Comptes Rendus, 236, 877-879.

[1279] S Stojakovic, M. (1954): Sur une propriété des matrices quasiinverses, Bull. Soc. Math. Phys. (Belgrade), 6, 155-158.

[1280] Stojaković, M. (1954): Une théorie générale axiomatique
T des déterminants, Bull. Soc. Math. Phys. (Belgrade), 6, 41-55.

[1281] Stojakovic, M. (1957): Quelques remarques sur les
S hypermatrices, Publ. Inst. Math. Yugoslav. Akad. of Sc., 11, 33-42.

[1282] Stojakovic, M. (1960): Sur l'inversion des matrices dans
S la méthode des moindres carrés, Annuaire de la Faculté des Lettres et Sciences à Novi Sad, 5, 425-430.

[1283] Strand, O. N. (1972): Theory and Methods for Operator
P Equations of the First Kind, Ph. D. Thesis, Colorado State University, Boulder, Colorado.

[1284] Strand, O. N. (1974): Theory and methods related to the
S singular-function expansion and Landweber's iteration for integral equations of the first kind, SIAM J. Numer. Anal., 11, 798-925.

Strahov, V. N.: See Bakušinskii, A. B. (1968).

Sturz, I.: See Boros, E. (1963).

[1285] Styan, G. P. H. (1969): Hadamard products and multi-
S variate analysis (abstract), Ann. Math. Statist., 40, 1149-1150.

[1286] Styan, G. P. H. (1970): Notes on the distribution of
S quadratic forms in singular normal variables, unpublished manuscript.

Styan, G. P. H.: See Golub, G. H. (1971), Marsaglia, G. (1972), Sharpe, G. E. (1965), (1965), (1967).

Subbiah, S.: See Magill, K. D. (1974).

[1287] Suzuki, Y. (1964): On the use of some extraneous infor-
S mation in the estimation of the components of regression, Annals of the Institute of Statistical Mathematics, Tokyo, 16, 161-174.

[1288] Sweet, D. (1970): An alternative method based on ap-
S proximate solutions, Math. Systems Theory, 4, 306-315.

Sweet, D.: See Bancroft, S. (1968).

[1289] Swerling, P. (1971): Modern state estimation methods
S from the viewpoint of the method of least squares, IEEE Trans. Aut. Control, AC-16, 707-719.

Takahashi, M.: See Ogasawara, T. (1951).

Tallis, G. M.: See Rohde, C. A. (1969).

[1290] Tan, W. Y. (1971): Note on an extension of the Gauss-
S Markov theorems to multivariate linear regression models, SIAM J. Appl. Math., 20, 24-29.

[1291] Tanabe, K. (1971): Projection method for solving a singu-
P lar system of linear equations and its applications, Numer. Math., 17, 203-214.

[1292] Tanabe, K. (1974): Characterization of linear stationary
P iterative processes for solving a singular system of linear equations, Numer. Math., 22, 349-359.

[1293] Tanabe, K. (1974): Newmann-type expansion of reflexive
P generalized inverse of a matrix and the hyperpower iterative method, Linear Algebra and Appl., 10, 163-175.

Tanabe, K.: See Shinozaki, N. (1972), (1972).

[1294] Tanna, V. P. (1974): The stability of the method of the
S residual in the solution of ill-posed problems, Vysš. Učebn. Zaved. Matematika, 7, 75-80.

Tanna, V. P.: See Vasin, V. V. (1974).

[1295] Taussky, O. (1950): Note on the condition of matrices,
S Math. Tables Aids Comput., 4, 111-112.

[1296] Taussky, O. (1964): Matrices C with C^n approaching
S 0, J. Algebra, 1, 5-10.

[1297] Taylor, A. E. (1960): Mittag-Leffler expansions and
S spectral analysis, Pacific J. Math., 10, 1049-1066.

[1298] Taylor, A. E. (1964): Introduction to Functional
T Analysis, John Wiley and Sons, New York.

[1299] Taylor, A. E. (1966): Theorems on ascent, descent,
S nullity and defect of linear operators, Math. Ann., 163, 18-49.

[1300] Taylor, A. E. and C. J. A. Halberg, Jr. (1957):
T General theorems about a bounded linear operator and its conjugate, J. Reine Angew. Math., 198, 93-111.

Taylor, A. E.: See Derr, J. (1962).

[1301] Tewarson, R. P. (1967): A direct method for gener-
P alized matrix inversion, SIAM J. Numer. Anal., 4, 499-507.

[1302] Tewarson, R. P. (1968): A computational method for
P evaluating generalized inverses, Computer J., 10, 411-413.

[1303] Tewarson, R. P. (1969): On some representations of
P generalized inverses, SIAM Review, 11, 272-276.

[1304] Tewarson, R. P. (1969): A least squares iterative
P method for singular equations, Computer J., 12, 388-392.

[1305] Tewarson, R. P. (1969): On computing generalized
P inverses, Computing, 4, 139-152.

[1306] Tewarson, R. P. (1969): On projection methods for
P solving linear systems, Computer J., 12, 78-81.

[1307] Tewarson, R. P. (1970): On the use of generalized
P inverses in function minimization, Computing, 6, 241-248.

[1308] Tewarson, R. P. (1971): On two direct methods for computing generalized inverses, Computing, 7, 236-239.

[1309] Tewarson, R. P. (1971): An iterative method for
P computing generalized inverses, Int. J. Comp. Math., 3, 65-74.

[1310] Tewarson, R. P. (1973): Solution of equations in
P remote sensing and picture reconstruction, Computing, 10, 221-230.

[1311] Tewarson, R. P. and V. N. Joshi (1972): On solving
S ill-conditions systems of linear equations, Trans. of N. Y. Acad. of Sci. Series II, 34, 565-571.

[1312] Tewarson, R. P. and P. Narain (1974): Generalized
P inverses and resolution in the solution of linear equations, Computing 13, 81-88.

[1313] Tewarson, R. P. and P. Narain (1975): Solution of
P linear equations resulting from satellite remote soundings, J. Math. Anal. Appl., to appear.

[1314] Tewarson, R. P. and B. Ramnath (1969): Some com-
S ments on the solution of linear equations, Nord. Tids. Inf. Bhld. (BIT), 9, 167-173.

[1315] Theil, H. (1971): Principles of Econometrics, Wiley,
T New York.

Thomas, D. H.: See Lynch, R. E. (1964).

[1316] Thomas, K. S. (1974): On the approximate solution of
S operator equations, Numer. Math. 23, 231-239.

Thompson, G. L. See Charnes, A. (1965).

[1317] Thompson, R. C. (1972): Principal submatrices IX:
S Interlacing inequalities for singular values, Linear Algebra and Appl., 5, 1-12.

[1318] Thompson, R. C. (1975): Singular value inequalities
S for matrix sums and minors, Linear Algebra and Appl., 11, 251-269.

[1319] Tienstra, J. M. (1947): An extension of the technique
T of the method of least squares to correlated observations, Bull. Géodésique, 6, 301-335.

[1320] Tienstra, J. M. (1948): The foundation of the calculus
T of observations and the method of least squares, Bull. Géodésique, 10, 289-306.

Timlake, W. P.: See Lynn, M. S. (1968).

[1321] Tipton, A. R. and H. W. Milnes (1972): Least squares
P solution of linear equations, Indust. Math. 22, 11-16.

[1322] Todd, J. (1951): Computational problems concerning
T the Hilbert matrix, J. Research Nat. Bur. Standards, B. Mathematics and Math. Physics, 65, 19-22.

[1323] Todd, J. (1954): The condition of the finite segments
T of the Hilbert matrix, Nat. Bur. Standards Appl. Math. Series, 39, 109-116.

[1324] el-Tom, M. E. A. (1974): On spline function approxi-
S mations to the solution of Volterra integral equations of the first kind, Nordisk Tidskr. Informationsbehandling (BIT), 14, 288-297.

[1325] Trapp, G. E. (1970): Algebraic Operations Derived
P from Electrical Networks, Doctoral Dissertation, Carnegie-Mellon University, Pittsburgh, Pennsylvania.

Trapp, G. E.: See Anderson, W. N., Jr. (1972), (1975), (1975), Duffin, R. J. (1972), Mitra, S. K. (1975).

Trenogin, V. A.: See Vainberg, M. M. (1974).

[1326] Tsao, N. K. (1974): A note on implementing the
S Householder transformation, SIAM J. Numer. Anal., 12, 53-58.

[1327] Tseng, Y. Y. (1933): The Characteristic Value Problem
P of Hermitian Functional Operators in a Non-Hilbertian Space, Doctoral Dissertation (Published by the University of Chicago Libraries, 1936), Chicago, Illinois.

[1328] Tseng, Y. Y. (1949): Generalized inverses of unbounded operators between two unitary spaces, Dokl. Akad. Nauk, SSSR (N.S.), 67, 431-434 (Math. Reviews 11 (1950), p. 115).
P

[1329] Tseng, Y. Y. (1949): Properties and classifications of generalized inverses of closed operators, Dokl. Akad. Nauk, SSSR (N.S.), 67, 607-610 (Math. Reviews, 11 (1950), p. 115).
P

[1330] Tseng. Y. Y. (1949): Sur les solutions des équations opératrices fonctionelles entre les espaces unitaires, solutions extrêmales, solutions virtuelles, C. R. Acad. Sci. Paris, 228, 640-641.
P

[1331] Tseng, Y. Y. (1956): Virtual solutions and generalized inversions, Uspehi Mat. Nauk (N.S.), 11, 213-215 (Math. Reviews 18 (1957), p. 749).
P

[1332] Tucker, D. H. (1969): Boundary value problems for linear differential systems, SIAM J. Appl. Math., 17, 769-783.
S

[1333] Turbin, A. F. (1974): Formulae for the computation of the semi-inverse and pseudoinverse matrices (Russian), Ž. Vyčisl. Mat. i Mat. Fiz., 14, 772-776, 815.
P

[1334] Twomey, S. (1963): On the numerical solution of Fredholm integral equations of the first kind by the inversion of linear system produced by quadrature, J. Assoc. Comput. Mach., 10, 97-101.
S

Ullman, F. D.: See Lovass-Nagy, V. (1973).

[1335] P Urquhart, N. S. (1968): Computation of generalized inverse of matrices which satisfy specified conditions. SIAM Review, 10, 216-218.

[1336] P Urquhart, N. S. (1969): The nature of the lack of uniqueness of generalized inverse matrices, SIAM Review, 11, 268-271.

Usow, K. H.: See Rice, J. R. (1968).

[1337] T Vagner, V. V. (1952): Generalized groups, Dokl. Akad. Nauk SSSR, 124, 1119-1122.

[1338] T Vainberg, M. M. and V. A. Trenogin (1974): Theory of Branching of Solution of Nonlinear Equations, Noordhoff International Publishing, Leyden, 485 pp.

[1339] S Varah, J. M. (1973): On the numerical solution of ill-conditioned linear systems with applications to ill-posed problems, SIAM J. Numer. Anal., 10, 257-267.

Vasilev, V. G.: See Marchuk, G. I. (1970).

[1340] S Vasin, V. V. and V. P. Tanna (1974): Necessary and sufficient conditions for the convergence of projection methods for linear unstable problems. (Russian) Dokl. Akad. Nauk SSSR, 215, 1032-1034. Soviet Math. Dokl., 15(1974), 628-631.

Verma, G. R.: See La Budde, C. D. (1969).

[1341] S Vermuri, V. and F. P. Chen (1974): An initial value method for Fredholm integral equation of the first kind, J. Franklin Inst., 297, 187-200.

[1342] S Vetter, W. J. (1975): Vector structures and solutions of linear matrix equations, Linear Algebra and Appl., 10, 181-188.

Vidav, I.: See Prijatelj, N. (1971).

Vignes, J.: See La Porte, M. (1974).

Vilenkin, N. Ja.: See Gelfand, I. M. (1964).

[1343] Villumsen, P.V. (1965): On the solution of normal
S equations, BIT, 5, 203-210.

[1344] Vlasov, L. P. (1973): Approximative properties of sets
S in normed linear spaces, Russian Math. Surveys, 28, 1-66.

[1345] Vogt, A. (1972): On the linearity of form isometries,
S SIAM J. Appl. Math., 22, 553-560.

[1346] Votruba, G. F. (1963): Generalized Inverses and
P Singular Equations in Functional Analysis, Doctoral Dissertation, University of Michigan, Ann Arbor, Michigan.

Votruba, G. F.: See Nashed, M. Z. (1974), (1974), (1974), (1975).

[1347] Vuličevič, B. D. (1973): The inversion of rectangular
P matrices that are partitioned into blocks. (Russian) Collection of articles dedicated to the memory of the Balkan mathematicians Constantin Carathéodory, Iosip Plemelj, Demitrie Pompeiu and Georghe Titeica in connection with the first centenary of their birth, Math. Balkanica, 3, 600-604.

[1348] Wahba, G. and M. Z. Nashed (1973): The approximate
P solution of a class of constrained control problems, Proceedings of the Sixth Hawaii International Conference on System Sciences (January 9-11, 1973), pp. 112-115, ed. by Art Lew.

Wahba, Grace: See Nashed, M. Z. (1972), (1972), (1973), (1974), (1974), (1975).

Waid, C.: See Hurt, M. F. (1970).

[1349] Walker, H. F. (1971): On the null-space of first
T order elliptic partial differential operators in R^n, Proc. Amer. Math. Soc., 30, 278-286.

[1350] Walker, H. F. (1972): A Fredholm theory for a class
S of first order elliptic partial differential operators in R^n, Trans. Amer. Math. Soc., 165, 75-86.

[1351] Wall, J. R. (1971): Stochastic Matrices and Matrices
P of Monotone Kind, Doctoral Thesis, University of Tennessee, Knoxville, Tennessee.

[1352] Wall, J. R. (1975): Generalized inverses of stochastic
P matrices, Linear Algebra and Appl., 10, 147-154.

[1353] Wall, J. R. and R. J. Plemmons (1972): Spectral
P inverses of stochastic matrices, SIAM J. Appl. Math., 22, 22-26.

Wallen, L. J.: See Halmos, P. R. (1970).

[1354] Wampler, R. H. (1969): An evaluation of linear least
P squares computer programs, J. Research Nat. Bur. Standards, Sect. B, 73B, 59-90.

[1355] Wampler, R. H. (1970): A report on the accuracy of
P some widely used least squares computer programs, J. Amer. Stat. Assoc., 65, 549-565.

[1356] Wani, J. K. and D. G. Kabe (1970): On some least
P square problems, J. Indian Statist. Assoc., 8, 32-36.

[1357] Ward, J. F., T. L. Boullion and T. O. Lewis (1971):
P On weighted pseudoinverses with singular weights, SIAM J. Appl. Math., 2, 480-482.

[1358] Ward, J. F., T. L. Boullion and T. O. Lewis (1971):
P A note on the oblique matrix pseudoinverse, SIAM J. Appl. Math., 20, 173-175.

[1359] Ward, J. F., T. L. Boullion and T. O. Lewis (1972):
P Weak spectral inverses, <u>SIAM J. Appl. Math</u>., 22, 514-518.

[1360] Watanabe, S. (1965): Karhunen-Loéve expansion and
T factor analysis; Theoretical remarks and applications, Information Theory, Statistical Decision Functions, Random Processes, <u>Trans. 4th Prague Conf</u>., pp. 635-640.

[1361] Watson, G. S. (1967): Linear least square regression,
S <u>Ann. Math. Statist.</u>, 38, 1679-1699.

[1362] Watson, G. S. (1972): Prediction and the efficiency
S of least squares, <u>Biometrika</u>, 59, 91-98.

[1363] Wedderburn, J. H. M. (1934): <u>Lectures on Matrices</u>,
T American Math. Soc. Colloq. Publ., Vol. XVIII, Providence, R. I.

[1364] Wedin, P.-Å. (1969): On pseudoinverses of perturbed
P matrices, Dept. of Computer Sciences, Lund University, Lund, 56 pp.

[1365] Wedin, P.-Å. (1972): Perturbation bounds in connec-
P tion with singular value decomposition, <u>BIT</u>, 12, 99-111.

[1366] Wedin, P.-Å. (1972): The non-linear least squares
S problem from a numerical point of view. I. Geometrical properties, Dept. of Computer Sciences, Lund University, Lund, Part II, to appear.

[1367] Wedin, P.-Å. (1973): Perturbation theory for pseudo-
P inverses, <u>BIT</u>, 13, 217-232.

Wedin, P.-Å.: See Leringe, Ö. (1970), Ruhe, A. (1974).

[1368] P Wee, W. G. (1968): Generalized inverse approach to adaptive multiclass pattern classification, IEEE Trans. Electron. Comput., C-17, 1157-1164.

[1369] S Wee, W. G. (1970): On feature selection in a class of distribution-free pattern classifiers, IEEE Trans. on Information Theory, IT-16, 47-55.

[1370] P Wee, W. G. (1971): A generalized inverse approach to clustering pattern selection and classification, IEEE Transactions on Information Theory, IT-17, 262-269.

[1371] P Wee, W. G. and K. S. Fu (1968): An extension of the generalized inverse algorithm to multiclass pattern classification, IEEE Trans. on Systems Science and Cybernetics, SSS-6, 192-194.

Weinholtz, E.: See Hildebrandt, S. (1964).

Wendland, W. L.: See Hsaio, G. C. (1975).

Wersan, S. J.: See Ben-Israel, A. (1963).

[1372] T Westfall, W. D. A. (1905): Zur Theorie der Integralgleichungen, Inaugural-Dissertation, (Göttingen).

[1373] S Westfall, W. D. A. (1909): Existence of the generalized Green's function, Ann. of Math. (2), 10, 177-181.

[1374] S Westreich, D. (1972): Banach space bifurcation theory, Trans. Amer. Math. Soc., 171, 135-156.

[1375] T Weyl, H. (1923): Reparticion de couriente en una red conductora, Revista Matematica Hispano-Americana 5, 153-164.

[1376] T Weyl, H. (1950): Inequalities between the two kinds of eigenvalues of a linear transformation, Proc. Nat. Acad. Sci. USA, 36, 49-51.

Whalen, B. H.: See Desoer, C. A. (1963).

[1377] S Whitney, T. M. and R. K. Meany (1967): Two algorithms related to the method of steepest descent, SIAM J. Numer. Anal., 4, 109-120.

[1378] S Whittle, P. (1963): Prediction and Regulation by Linear Least-Square Methods, The English Universities Press Ltd., London. Reviewed by P. Masani, J. Amer. Statist. Assoc., 61 (1966), 268-273.

Wiginton, C. L.: See Decell, H. P., Jr. (1968).

[1379] T Wilf, H. S. (1959): Matrix inversion by annihilation of rank, J. Soc. Indust. Appl. Math., 7, 149-151.

[1380] S Wilkinson, G. N. (1958): Estimation of missing values for the analysis of incomplete data, Biometrics, 14, 257-286.

[1381] T Wilkinson, J. H. (1958): The calculation of the eigenvectors of co-diagonal matrices, Comput. J., 1, 148-152.

[1382] T Wilkinson, J. H. (1962): Calculation of the eigenvalues of a symmetric tridiagonal matrix by the method of disection, Numer. Math., 4, 362-367.

[1383] S Wilkinson, J. H. (1963): Rounding Errors in Algebraic Processes, Notes on Applied Science No. 32, HMS Office, London, 161 pp.; Prentice-Hall, Englewood Cliffs, N. J., 1963.

[1384] S Wilkinson, J. H. (1965): The Algebraic Eigenvalue Problem, Oxford University Press, London, xviii + 662 pp.

[1385] S Wilkinson, J. H. (1967): The Solution of ill-conditioned linear equations, pp. 65-93 in Mathematical Methods for Digital Computers, Vol. II, ed. by A. Ralston and H. Wilf, Wiley, New York

[1386] P Wilkinson, J. H. and C. Reinsch (editors) (1971): Handbook for Automatic Computation, Vol. II: Linear Algebra, Springer-Verlag, Berlin, viii + 439 pp.

Wilkinson, J. H.: See Bowdler, H. J. (1966), Golub, G. (1966), Martin, R. S. (1966), Oettli, W. (1965), Peters, G. (1970).

Williams, J. P.: See Fillmore, P. A. (1971), Porter W. A. (1966), (1966).

[1387] S Williamson, J. (1935): A polar representation of singular matrices, Bull. Amer. Math. Soc., 41, 118-123.

[1388] T Williamson, J. (1939): Note on a principal axis transformation for non-Hermitian matrices, Bull. Amer. Math. Soc., 45, 920-922.

[1389] P Willner, L. B. (1967): An elimination method for computing the generalized inverse, Math. Comp., 21, 227-229.

[1390] S Wimmer, H. and A. D. Ziebur (1972): Solving the matrix equation $\sum_{p=1}^{r} f_p(A)Xg_p(B) = C$, SIAM Review, 14, 318-323.

[1391] T Wolf, H. (1950): Der modernisierte Gaussche Algorithmus, Z. Vermessungswes., 75 (11), 329-337.

Woodbury, M. A.: See Dolph, C. L. (1952).

Woodroofe, M. B.: See Anderson, W. N., Jr. (1969).

[1392] P Worm, G. H. (1971): Explicit Solutions for Quadratic Minimization Problems, Doctoral Dissertation, University of Tennessee, Knoxville, Tennessee.

[1393] P Worm, G. H. (1975): Sensitivity analysis of fixed point vectors, J. Amer. Stat. Assoc., 69, 961-967.

Worm, G. H.: See Cline, R. E. (1974).

[1394] P Wyler, O. (1964): Green's operators, Ann. Mat. Pura Appl., 66, 251-264.

[1395] S Wyler, O. (1965): On two-point boundary problems, Ann. Mat. Pura. Appl., 67, 127-142.

[1396] S Wynn, P. (1966): Upon a conjecture concerning a method for solving linear equations and certain other matters, Tech. Summary Report #626, Mathematics Research Center, University of Wisconsin-Madison.

[1397] P Wynn, P. (1971): Upon the generalized inverse of a formal power series with vector valued coefficients, Compositio Math., 23, 453-460.

[1398] S Yau, S. S. and J. M. Garnett (1972): Least-mean-square approach to pattern classification, in Frontiers of Pattern Recognition, M. S. Watanabe, ed., Academic Press, New York, pp. 575-587.

Yau, S. S.: See Garnett III, J. M. (1971).

[1399] S Yood, B. (1951): Properties of linear transformation preserved under addition of a completely continuous transformation, Duke Math. J., 18, 599-612.

[1400] P Yoshikawa, T. (1971): A kind of pseudoinverse of a matrix and discrete-time Kalman filter, Japanese System and Control, 15, 691-701.

[1401] T Yosida, K. (1958): Functional Analysis (2nd edition), Springer-Verlag, Berlin-New York, xi + 458 pp.

Young, G.: See Eckart, C. (1936), (1939), Householder, A. S. (1958).

[1402] S Young, T. Y. and T. W. Calvert (1974): Classification, Estimation, and Pattern Recognition, American Elsevier, New York, xiv + 366 pp.

[1403] T Zacks, S. (1964): Generalized least squares estimators for random fractional replication designs, Ann. Math. Statist., 35, 696-704.

[1404] S Zadeh, L. A. and C. A. Desoer (1963): Linear System Theory, McGraw-Hill, New York, xxi + 628 pp.

[1405] S Zarantonello, E. H. (1968): Differentioids, Advances in Mathematics, 2, 187-306.

Zassenhauss, H.: See Dade, E. C. (1962).

[1406] P Zelen, M. (1962): The role of constraints in the theory of least squares, MRC Technical Summary Report #314, Mathematics Research Center, University of Wisconsin-Madison.

Zelen, M.: See Goldman, A. J. (1964).

[1407] T Zemanian, A. H. (1974): Passive operator networks, IEEE Trans. Circuits and Systems, CAS-21, 184-193.

[1408] P Zettl, H. (1974): Spekraldarstellung von Homomorphismen, Diplomarbeit angefertigt bei Professor D. Kölzow in Erlangen.

Ziebur, A. D.: See Wimmer, H. (1972).

[1409] T Zielke, G. (1970): Numerische Berechnung von benachbarten inversen Matrizen und linearen Gleichungssystemen, Vieweg, Berlin.

[1410] P Zimmule, D.: Techniques for computing the pseudo-inverse of a matrix, preprint.

[1411] P Zlobec, S. (1967): On computing the generalized inverse of a linear operator, Glasnik Mat., 2 (22), 265-271.

[1412] P Zlobec, S. (1970): An explicit form of the Moore-Penrose inverse of an arbitrary complex matrix, SIAM Review, 12, 132-134.

[1413] P Zlobec, S. (1970): Contributions to Mathematical Programming and Generalized Inversion, Doctoral Dissertation, Northwestern University, Evanston, Illinois.

[1414] P Zlobec, S. (1972): On computing the best least squares solutions in Hilbert spaces, preprint, Dept. of Math., McGill University, Montreal.

[1415] P Zlobec, S. (1974): The Gauss-bordering method for generalized inversion of matrices, preprint.

[1416] P Zlobec, S. and A. Ben-Israel (1970): On explicit solutions of interval linear programs, Israel J. Math., 8, 12-22.

[1417] P Zlobec, S. and A. Ben-Israel (1973): Explicit solutions of interval linear programs, Operations Res., 21, 390-393.

[1418] S Zuhovickiĭ, S. I. (1951): An algorithm for the solution of the Čebyšev approximation in the case of a finite system of incompatible linear equations (Russian), Dokl. Akad. Nauk SSSR, 79, 561-564.

[1419] S Zuhovickiĭ, S. I. (1953): On the best approximation in the sense of P. L. Čebyšev for finite systems of inconsistent linear equations (Russian), Mat. Sb., 33, 327-342.

[1420] S Zuhovickiĭ, S. I. (1961): A new numerical scheme of the algorithm for Čebyšev approximation of an incompatible system of linear equations and a system of linear inequalities (Russian), Dokl. Akad. Nauk SSSR, 139, 534-537.

[1421] S Zuhovickiĭ, S. I. (1962): The approximation of an incompatible system of linear equations by minimizing the sum of the moduli of all the deviations (Russian), Dokl. Akad. Nauk SSSR, 143, 1030-1033.

[1422] S Zyskind, G. (1967): On canonical forms, non-negative covariance matrices and best and simple, least square linear estimator in linear models, Ann. Math. Statist., 38, 1092-1110.

[1423] P Zyskind, G. (1973): Error structures, projections and conditional inverses in linear model theory, paper presented for the International Symposium on Statistical Design and Linear Models, Colorado State University, (March 19-23, 1973), Boulder, Colorado.

[1424] P Zyskind, G. and F. B. Martin (1969): On best linear estimation and a general Gauss-Markov theorem in linear models with arbitrary nonnegative covariance structure, SIAM J. Appl. Math., 17, 1190-1202.

Zyskind, G.: See Martin, F. B. (1966).

Supplement 1

Additional Doctoral Dissertations

(See Annotations for a complete listing)

[1425] S Anderson, P. O. (1969): Estimation of Simultaneous Systems of Linear Equation with Nonlinear Constraints Among the Coefficients, Ph.D. Thesis, Department of Statistics, Stanford University, Stanford, California.

[1426] T Bailes, G. L., Jr. (1973): Right Inverse Semigroups, Ph.D. Thesis, Clemson University, Clemson, South Carolina.

[1427] S Barham, R. H. (1970): Parameter Reclassification in Nonlinear Least Squares, Ph.D. Thesis, Department of Statistics, Southern Methodist University, Dallas, Texas.

[1428] S Berman, A. (1971): Linear Inequalities in Matrix Theory, Ph.D. Thesis in Applied Mathematics, Northwestern University, Evanston, Illinois.

[1429] P Borowsky, M. S. (1972): On Best Approximate Solutions of $Av = b$, Ph.D. Thesis, Drexel University, Philadelphia, Pennsylvania.

[1430] Brown, M. R. S. (1970): Generalizations of Inverse
S Semigroups, Ph. D. Thesis, Colorado State University, Fort Collins, Colorado.

[1431] Burris, C. H., Jr. (1974): The Numerical Solution of
P the Generalized Eigenvalue Problem for Rectangular Matrices, Ph. D. Thesis, University of New Mexico, Albuquerque, New Mexico.

[1432] Chen, Li-An L. (1969): Solving Improperly Posed
S Problems by Mathematical Programming Techniques, Ph. D. Thesis, New York University, New York.

[1433] Cohoon, D. K. (1969): Existence of a Continuous Right
T Inverse for Linear Partial Differential Operators in Spaces of Indefinitely Differentiable Functions, Ph. D. Thesis, Purdue University, Lafayette, Indiana.

[1434] Crawford, C. R. (1971): The Numerical Solution of the
S Generalized Eigenvalue Problem, Ph. D. Thesis, University of Michigan, Ann Arbor, Michigan.

[1435] Deutsch, E. (1969): Vectorial and Matricial Norms,
S Ph. D. Thesis, Polytechnic Institute of Brooklyn, New York.

[1436] Funderlic, R. E. (1970): Norms and Semi-Inverses,
P Ph. D. Thesis, University of Tennessee, Knoxville, Tennessee.

[1437] Glassman, B. A. (1971): A Least Squares Decomposition
S Theorem with Applications to Data Compaction, Ph. D. Thesis, University of California, Los Angeles, California.

[1438] Hallum, C. R. (1972): Estimation and Testing in Re-
P stricted General Linear Models, Ph. D. Thesis, Texas Tech University, Lubbock, Texas.

[1439] Harms, D. W. (1974): A Direct Method Based on Pro-
S jections for Solving Systems of Linear Equations, Ph. D. Thesis, Iowa State University, Ames, Iowa.

[1440] Ize, G. (1974): Bifurication Theory for Fredholm Oper-
S ators, Ph. D. Thesis, New York University, New York.

[1441] Johnson, D. A. (1972): On the p-q Generalized In-
P verse, Ph. D. Thesis, Texas Tech University, Lubbock, Texas.

[1442] Jones, P. K. (1973): Generalized Ridge Regression,
P Ph. D. Thesis, Department of Statistics, University of Iowa, Iowa City, Iowa.

[1443] Joshi, V. N. (1971): Contributions to the Theory of the
P System of Equations, Ph. D. Thesis, Department of Applied Mathematics, State University of New York at Stony Brook, New York.

[1444] Mackichan, B. B. (1969): A Generalization to Over-
T determined Systems of the Notion of Diagonal Operators, Ph. D. Thesis, Stanford University, Stanford, California.

[1445] Morrel, B. B. (1969): Inversion of Matrices of Oper-
S ators, Ph. D. Thesis, University of Virginia, Charlottesville, Virginia.

[1446] Narain, P. (1974): Solutions of Linear Equations and
P Generalized Inverses, Ph. D. Thesis, Department of Applied Mathematics and Statistics, State University of New York at Stony Brook, New York.

[1447] O'Neill, M. J. (1969): Semi-Fredholm Operators in
T von Neumann Algebras, Ph. D. Thesis, University of Kansas, Lawrence, Kansas.

[1448] Patterson, W. M., III (1973): A Survey of Iterative
S Methods for the Solution of a Linear Operator Equation in Hilbert Space, Ph. D. Thesis, Syracuse University, Syracuse, New York.

[1449] Perry, C. R. (1972): Linear Transformations Restricted
S to Subspaces and Inequalities Among Their Proper Values, Ph. D. Thesis, Texas Tech University, Lubbock, Texas.

[1450] Poole, G. D. (1972): Weak Spectral Inverses, Ph. D.
P Thesis, Texas Tech University, Lubbock, Texas.

[1451] Price, K. H. (1970): Projections and Approximation in
S a Normed Linear Space, Ph. D. Thesis, University of Texas at Austin, Texas.

[1452] Pye, W. C. (1973): Pseudoinverses and Higher Order
P G-transforms, Ph. D. Thesis, Texas Tech University, Lubbock, Texas.

[1453] Rodrigue, G. H. (1972): A Variational Method for the
S Numerical Solution of Algebraic Problems, Ph. D. Thesis, University of Southern California, Los Angeles, California.

[1454] Shen I-ming (1971): Acceleration of Projection Method
S for Solving Linear Systems of Equations, Ph. D. Thesis, Department of Computer Science, Iowa State University, Ames, Iowa.

[1455] Shurbet, G. L. (1973): Systems of Matrix Equations, Ph.D.
P Thesis, Texas Tech University, Lubbock, Texas.

[1456] Stallings, W. T. (1971): Pseudoinverses and r-Circu-
P lants, Ph. D. Thesis, Texas Tech University, Lubbock, Texas.

[1457] Stewart, G. W. III (1969): Some Topics in Numerical
S Analysis, Ph. D. Thesis, University of Tennessee, Knoxville, Tennessee.

[1458] Stein, J. (1971): Singular Quadratic Functionals, Ph. D.
S Thesis, University of California, Los Angeles, California.

[1459] Temple, M. G. (1972): Contributions to the Theory of
S Chebyshev Solutions of Overdetermined Inconsistent Systems of Linear Equations, Ph. D. Thesis, Drexel University, Philadelphia, Pennsylvania.

[1460] Teitz, E. M. (1970): Pseudo-Fredholm Operators, Ph.D.
T Thesis, Yeshiva University, New York, New York.

[1461] Vanderschel, D. J. (1970): A Theory of Approximate In-
S verses for the Solution of Matrix Equations by Iteration, Ph.D. Thesis, Rice University, Houston, Texas.

[1462] Van Loan, C. F. (1974): Generalized Singular Values
S with Algorithms and Applications, Ph.D. Thesis, University of Michigan, Ann Arbor, Michigan.

[1463] Weber, J. K., Jr. (1970): Complementation in Locally
T Convex Spaces L_{p^+} and L_{p^-}, Ph.D. Thesis, Duke University, Durham, North Carolina.

[1464] Bodin, L. (1974): Recursive Fix-Point Estimation Theory
S and Applications, Doctoral Dissertation, Department of Statistics, University of Uppsala, Uppsala, Sweden.

[1465] Businger, P. A. (1967): Matrix Scaling with Respect To
T the Maximum-Norm, the Sum-Norm, and the Euclidean-Norm, Thesis TNN71, The University of Texas, Austin, 119 pp.

[1466] Cohen, C. (1969): An Investigation of the Geometry of
S Subspaces for Some Multivariate Statistical Models, Ph.D. Thesis, Dept. of Indust. Eng., University of Illinois, Urbana, Illinois.

[1467] Jennings, L. S. (1973): Orthogonal Transformations and
S Improperly-Posed Problems, Ph.D. Thesis, Australian National University, Canberra, Australia. Abstract in Bull. Austr. Math. Soc., 9(1973), 303-304.

[1468] Lill, Shirley A. (1971): A Class of Methods for Nonlinear
S Programming, Ph.D. Thesis, University of Leeds.

[1469] Pringle, R. M. (1969): Generalized Inverse Matrices
P with Applications to Statistics, Ph. D. Thesis, University of Natal, South Africa.

[1470] Ryan, D. M. (1971): Transformation Methods in Non-
S linear Programming, Ph. D. Thesis, Australian National University, Canberra, Australia.

Supplement 2

[1471] Abadie, J., editor (1970): Integer and Nonlinear Pro-
S gramming, North-Holland Publishing Co., Amsterdam, 544 pp. + x.

[1472] Adenstedt, R. K. (1975): Asymptotically efficient esti-
S mators for a constant regression with vector-valued stationary residuals, Ann. Statist., 3, 1109-1121.

[1473] Adenstedt, R. K. and B. Eisenberg (1974): Linear esti-
S mation of regression coefficients, Quart. Appl. Math., 32, 317-327.

[1474] American Statistical Association (1965): Regression pro-
S cedures for missile trajectory estimation, Proc. of the 105th Regional Meeting, Florida State University.

[1475] Amir-Moéz, A. R. (1974): Quasi-singular values of
T linear transformations, Rend. Circ. Mat. Palermo (2), 22, 1974, 314-316.

[1476] Anselone, P. M. and P. J. Laurent (1968): A general
S method for the construction of interpolating or smoothing splines, Numer. Math., 12, 66-82.

[1477] Antohin, Ju. T. (1967): Incorrect problems in Hilbert
S space and stable method of solving them, Differencial'nye Uravnenija, 3, 1135-1156.

[1478] Baker, C. T. H., L. Fox, D. F. Mayers and K. Wright
T (1964): Numerical solution of Fredholm integral equations of the first kind, Comput. J., 7, 141-148.

[1479] S Bakušinskii, A. B. (1965): On a numerical method of solving Fredholm's integral equation of the first kind, Ž. Vyčisl. Mat. i Mat. Fiz., 5, 744-749.

[1480] P Bakušinskii, A. B. (1967): A general method for constructing algorithms for a linear ill-posed equation in Hilbert space, Ž. Vyčisl. Mat. i Mat. Fiz., 7, 672-677. USSR Comp. Math. Math. Phys., 7, 279-287.

[1481] S Bard, Y. (1970): Comparison of gradient methods for the solution of nonlinear parameter estimation problems, SIAM J. Numer. Anal., 7, 157-186.

[1482] T Bard, Y. (1974): Nonlinear Parameter Estimation, Academic Press, New York, 352 pp.

[1483] T Barnes, B. A. (1969): The Fredholm elements of a ring, Canad. J. Math., 21, 84-95.

[1484] S Bart, H. (1973): Meromorphic Operator Valued Functions, Mathematical Centre Tracts 44, Mathematisch Centrum, Amsterdam, 126 pp.

[1485] P Bart, H. (1974): Holomorphic relative inverses of operator valued functions, Math. Ann., 208, 179-194.

[1486] T Bartels, R. H. (1971): A stabilization of the simplex method, Numer. Math., 16, 414-434.

[1487] S Bartels, R. H., G. H. Golub and M. A. Saunders (1970): Numerical techniques in mathematical programming, pp. 123-176 in Rosen, Mangasarian and Ritter (1970).

[1488] S Bartle, R. G. (1958): On the openness and inversion of differentiable mappings, Ann. Acad. Scient. Fennicae A. I. 257, 8 pp.

[1489] T Battin, R. H. (1964): Astronautical Guidance, McGraw-Hill, New York.

[1490] S Battin, R. H. and G. Levine (1969): Application of Kalman filtering techniques to the Apollo program, MIT Inst. Lab. Tech. Rept. E 2401, April, 1969.

[1491] Bazaraa, M. S., J. J. Goode and M. Z. Nashed (1974):
T On cones of tangents with applications to mathematical programming, J. Optimization Theory and Appl., 13, 389-426.

[1492] Bazaraa, M. S., J. J. Goode, M. Z. Nashed and C. M.
T Shetty (1975): Nonlinear programming without differentiability in Banach spaces: Necessary and sufficient constraint qualifications, Applicable Anal., 5, to appear.

[1493] Beale, E. M. L. (1970): Computational methods for
S least squares, pp. 213-226 in Abadie (1970).

[1494] Bellman, R. (1967): Introduction to the Mathematical
S Theory of Control Processes, Volume 1, Academic Press, New York.

[1495] Bellman, R., R. Kalaba and J. Lockett (1965): Dynamic
S programming and ill-conditioned linear systems, J. Math. Anal. Appl., 10, 206-215.

[1496] Bellman, R., R. Kalaba and J. Lockett (1965): Dynamic
S programming and ill-conditioned linear systems - II, J. Math. Anal. Appl., 12, 393-400.

[1497] Ben-Israel, A. (1968): On optimal solutions of 2-person
S 0-sum games, Atti Accad. Naz. Lincei Rend. Cl. Sci. Fis. Mat. Natur. Ser. VIII, 44, 274-278.

[1498] Ben-Israel, A. (1969): Linear equations and inequalities
T on finite dimensional, real or complex, vector spaces: A unified theory, J. Math. Anal. Appl., 27, 367-389.

[1499] Ben-Israel, A. (1970): On Newton's method in nonlinear
S programming, pp. 339-352 in Proc. Princeton Sympos. Math. Programming, H. W. Kuhn, editor, Princeton University Press, Princeton, N.J., vi + 620 pp.

[1500] Ben-Israel, A., A. Charnes, A. P. Hunter, Jr. and
S P. D. Robers (1970): On the explicit solution of a special class of linear economic models, Operations Research, 18, 462-470.

[1501] Bennett, J. M. (1965): Triangular factors of modified
T matrices, Numer. Math., 7, 217-221.

[1502] Bharucha-Reid, A. T. (1972): Random Integral Equations,
T Academic Press, New York, xiii + 267 pp.

Bhaskara Rao, K. P. S.: See Prasada Rao, P. S. S. N. V. (1975).

Björck, Å.: See Dahlquist, G. (1974).

[1503] Blum, E. (1955): A theory of analytic functions in
T Banach algebras, Trans. Amer. Math. Soc., 78, 343-370.

[1504] Boggs, D. H. (1972): A partial-step algorithm for the
T nonlinear estimation problem, AIAA Journal, 10, 675-679.

[1505] Bonic, R. A. (1969): Linear Functional Analysis, Gordon
T and Breach, New York.

[1506] de Boor, C. (1971): Subroutine package for calculating
T with B-splines, Rept. No. LA-4728-MS, Los Alamos Scientific Lab., 12 pp.

[1507] de Boor, C. (1972): On calculating with B-splines, J.
T Approximation Theory, 6, 50-62.

[1508] de Boor, C. and R. E. Lynch (1966): On splines and
T their minimum properties, J. Math. Mech., 15, 953-970.

[1509] Boot, J. C. G. (1964): Quadratic Programming, North-
S Holland Publishing Co., Amsterdam.

[1510] Brent, R. P. (1972): Algorithms for Minimization without
T Derivatives, Prentice-Hall, Englewood Cliffs, New Jersey.

[1511] Breuer, M. (1968): Fredholm theories in von Neumann
T algebras I, Math. Ann., 178, 243-254.

[1512] T Breuer, M. (1969): Fredholm theories in von Neumann algebras II, Math. Ann., 180, 313-325.

[1513] S Buchanan, J. E. and D. H. Thomas (1968): On least-squares fitting of two-dimensional data with a special structure, SIAM J. Numer. Anal., 5, 252-257.

[1514] S Bunch, J. R. and B. N. Parlett (1972): Direct methods for solving symmetric indefinite systems of linear equations, SIAM J. Numer. Anal., 8, 639-655.

[1515] S Bus, J. C. P., B. van Domselaar and J. Kok (1975): Nonlinear least squares estimation, Numerieke Wiskunde, No. NW 17/75, Mathematisch Centrum, Amsterdam.

[1516] S Businger, P. A. (1970): Updating a singular value decomposition, BIT, 10, 376-385.

[1517] S Businger, P. A. and G. H. Golub (1967): An Algol procedure for computing the singular value decomposition, Stanford Univ. Rept. No. CS-73, Stanford, California.

[1518] T Calkin, J. (1941): Two-sided ideals and congruences in the ring of bounded operators in Hilbert space, Ann. Math. (2), 42, 839-873.

[1519] T Caradus, S. R. (1966): Operators of Riesz type, Pacific J. Math., 18, 61-71.

[1520] S Caradus, S. R., W. E. Pfaffenberger and B. Yood (1974): Calkin Algebras and Algebras of Operators on Banach Spaces, Marcel Dekker, Inc., New York.

[1521] S Chambers, J. M. (1971): Regression updating, J. Amer. Statist. Assoc., 66, 744-748.

[1522] T Chan, C. Y. and E. C. Young (1975): Singular matrix solutions for time-dependent fourth order quasilinear matrix differential inequalities, J. Differential Equations, 18, 386-392.

Charnes, A.: See Ben-Israel (1970).

[1523] S Charnes, A., W. Raike and J. Stutz (1975): V-positivity poverses and the economic global unicity theorems of Gale and Nikaido, Z. Operations Res. Ser. A-B, 19, A115-A121.

[1524] P Cline, R. E. and R. J. Plemmons (1975): ℓ_2-solutions to underdetermined linear systems, to appear.

[1525] S Coburn, L. A. and A. Lebow (1966): Algebraic theory of Fredholm operators, J. Math. Mech., 15, 577-584.

[1526] T Coburn, L. A., R. G. Douglas and I. M. Singer (1972): An index theorem for Wiener-Hopf operators on the discrete quarter plane, J. Differential Geom., 6, 587-595.

[1527] S Cochran, J. A. (1974): Square-integrable kernels and growth estimates for their singular values, Bull. Amer. Math. Soc., 80, 661-663.

[1528] S Dahlquist, G. and Å. Björck (1974): Numerical Methods, Prentice-Hall, Englewood Cliffs, N. J.

[1529] S Dahlquist, G., B. Sjöberg and P. Svensson (1968): Comparison of the method of averages with the method of least squares, Math. Comp., 22, 833-845.

[1530] S Daniel, J. W. and L. L. Schumaker (1974): On the closedness of the linear image of a set, with applications to generalized spline functions, Applicable Anal., 4, 191-205.

[1531] S Davis, Chandler and W. M. Kahan (1970): The rotation of eigenvectors by a perturbation, III, SIAM J. Numer. Anal., 7, 1-46.

[1532] T Dixmier, J. (1969): Les Algèbras d'operateurs dans l'espace Hilbertien, Ganthier-Villars, Paris.

[1533] Delves, L. M. and J. Walsh, editors (1974): Numerical Solution of Integral Equations, Clarendon Press, Oxford, 339 pp.
T

von Domselaar, B.: See Bus, J. C. P. (1975).

[1534] Douglas, R. G. (1972): Banach Algebra Techniques in Operator Theory, Academic Press, New York.
T

[1535] Douglas, R. G. (1974): Banach Algebra Techniques in the Theory of Toeplitz Operators, Regional Conference Series in Mathematics No. 15, Amer. Math. Soc., Providence, Rhode Island.
T

Douglas, R. G.: See Coburn, L. A. (1972).

[1536] Dragomir, A. and P. Dragomir (1974): Sur l'inverse généralisée d'un produit de matrices. Papers presented at the Fifth Balkan Mathematical Congress (Belgrade, 1974), Math. Balkanica, 4, 141-150.
P

Dragomir, P.: See Dragomir, A. (1974).

[1537] Dunn, J. C. (1973): Inversion of normal operators by polynomial interpolation, Proc. Amer. Math. Soc. 40, 225-228.
T

Eisenberg, B.: See Adenstedt, R. K. (1974).

[1538] Eldén, L. (1974): Numerical methods for the regularization of Fredholm integral equations of the first kind, Linköping University, Dept. Math. Rept. 7, December 1974.
S

[1539] Fan, Ky (1975): Two applications of a consistency theorem for systems of linear inequalities, Linear Algebra and Appl., 11, 171-180.
T

Feldman, I. A.: See Gohberg, I. C. (1960), (1971).

[1540] Fix, G. and R. Heiberger (1972): An algorithm for the ill-conditioned generalized eigenvalue problem, SIAM J. Numer. Anal., 9, 78-88.
S

[1541] Fletcher, R. (1970): A class of methods for nonlinear
P programming with termination and convergence properties, pp. 157-175 in Abadie (1970).

[1542] Fletcher, R. (1971): A modified Marquardt subroutine
S for nonlinear least squares, Rept. No. R-6799, Atomic Energy Research Estab., Harwell, Berkshire, England, 241 pp.

[1543] Fletcher, R. (1975): An ideal penalty function for con-
S strained optimization, J. Inst. Math. Appl., 15, 319-342.

[1544] Fletcher, R. and Shirley A. Lill (1970): A class of
P methods of nonlinear programming II. Computational experience, pp. 67-92 in Rosen, Mangasarian and Ritter (1970).

[1545] Forster, K. -H. (1967): Über lineare, abgeschlossene
S Operatoren, die analytisch von einem Parameter abhängen, Math. Z., 95, 251-258.

[1546] Forsythe, G. E. (1967): Today's methods of linear
T algebra, SIAM Rev., 9, 489-515.

[1547] Forsythe, G. E. (1970): Pitfalls in computation, or why
T a math book isn't enough, Amer. Math. Monthly, 77, 931-956.

[1548] Forsythe, G. E. and C. B. Moler (1967): Computer
T Solution of Linear Algebraic Systems, Prentice-Hall, Inc., Englewood Cliffs, N.J., 148 pp.

Foudiat, E. C.: See Nayak, R. P. (1974).

Fox, L.: See Baker, C. T. H. (1964).

[1549] Franklin, J. N. (1968): Matrix Theory, Prentice-Hall,
S Inc., Englewood Cliffs, N. J., 292 pp.

[1550] Franklin, J. N. (1970): Well-posed stochastic exten-
P sions of ill-posed linear problems, J. Math. Anal. Appl. 31, 682-716.

[1551] S Fučik, S. (1974): Nonlinear equations with noninvertible linear part, Czechoslovak Math. J., 24, 467-495.

[1552] S Gelb, A., editor (1974): Applied Optimal Estimation, written by the Technical Staff, the Analytic Sciences Corporation, M.I.T. Press, Cambridge, Mass.

[1553] P Gentleman, W. M. (1972): Basic procedures for large, sparse or weighted linear least squares problems, Univ. of Waterloo Rept. CSRR-2068, Waterloo, Ontario, Canada, 14pp.

[1554] T Gill, P. E., G. H. Golub, W. Murray and M. A. Saunders (1972): Methods for modifying matrix factorizations, Stanford Univ. Rept. No. CS-322, Stanford, Calif., 60 pp., Math. Comp., 28(1974), 505-535.

Glasko, V. B.: See Tikhonov, A. N. (1964).

[1555] S Gohberg, I. C. (1951): On linear equations in normed spaces (Russian), Dokl. Akad. Nauk SSSR, 76, 477-480.

[1556] S Gohberg, I. C. (1951): On linear equations depending on a parameter (Russian), Dokl. Akad. Nauk SSSR, 78, 629-632.

[1557] S Gohberg, I. C., A. S. Markus and I. A. Fel'dman (1960): Normally solvable operators and ideals associated with them, Byl. Akad. Stiince RSS Moldoven, 10(76), 51-69 (Russian). Amer. Math. Soc. Transl. (2), 61(1967), 63-84.

[1558] T Gohberg, I. C. and I. A. Fel'dman (1971): Convolution Equations and the Projection Method for Their Solution, "Nauka", Moscow; Amer. Math. Soc., Providence, R.I., 1974.

[1559] T Gohberg, I. C. and M. G. Kreĭn (1965): Introduction to the Theory of Linear Nonselfadjoint Operators, "Nauka", Moscow, 1965; English transl. Transl. Math. Monographs, vol. 18, Amer. Math. Soc., Providence, R.I., 1969.

[1560] S Gol'dman, M. A. (1955): On the stability of the property of normal solvability of linear equations (Russian), Dokl. Akad. Nauk SSSR, 100, 201-204.

[1561] S Gol'dman, M. A. and S. N. Kračkovskiĭ (1967): Perturbation of homomorphisms by operators of finite rank, Soviet Math. Dokl., 8, 670-673.

[1562] S Golub, G. H. (1963): Comparison of the variance of minimum variance and weighted least squares regression coefficients, Ann. Math. Statist., 34, 984-991.

Golub, G. H.: See Bartels, R. H. (1970), Businger, P. A. (1967), Gill, P. E. (1972).

Goode, J. J.: See Bazaraa, M. S. (1974), (1975).

[1563] S Gordonova, V. I. (1970): Estimates of the roundoff error in the solution of a system of conditional equations, Stanford Univ. Rept. No. CS-164, (translated from Russian by Linda Kaufmann), Stanford University, Stanford, California.

[1564] S Gordonova, V. I. and V. A. Morozov (1973): Numerical algorithms for parameter choice in the regularization method, Zh. Vychisl. Mat. i Mat. Fiz., 13, 539-545.

[1565] T Grabiner, Sandy (1974): Ranges of products of operators, Canad. J. Math., 26, 1430-1441.

[1566] S Gramsch, B. (1966): σ-Transformationen in lokalbeschränkten Vektorräumen, Math. Ann., 165, 135-151.

[1567] S Gramsch, B. (1967): Ein Schema zur Theorie Fredholmschen Endomorphismen und eine Andwendung auf die Idealkette der Hilbertraumen, Math. Ann., 171, 263-272.

[1568] T Gramsch, B. and R. Meise (1968): Vorlesung über Fredholm Operatoren, Lecture Notes, Mainz.

[1569] S Graves, L. M. (1950): Some mapping theorems, Duke Math. J., 17, 111-114.

[1570] T Graves, L. M. (1955): A generalization of the Riesz theory of completely continuous transformations, Trans. Amer. Math. Soc., 79, 141-149.

Graves, L. M.: See Hildebrandt, T. H. (1927).

[1571] S Grimm, L. J. and L. M. Hall (1975): An alternative theorem for singular differential systems, J. Differential Equations, 18, 411-422.

[1572] P Haberman, S. J. (1975): How much do Gauss-Markov and least square estimates differ? A coordinate-free approach, Ann. Statist., 3, 982-990.

Hale, R. W.: See Yates, F. (1939).

Hall, L. M.: See Grimm, L. J. (1975).

[1573] S Hanson, R. J. and P. Dyer (1971): A computational algorithm for sequential estimation, Comput. J., 14, 285-290.

[1574] T Hartfiel, D. J. (1975): Two theorems generalizing the mean transition probability results in the theory of Markov chains, Linear Algebra and Appl., 11, 181-187.

[1575] S Hartley, H. O. (1961): The modified Gauss-Newton method for fitting of nonlinear regression functions by least squares, Technometrics, 3, 269-280.

[1576] S Healy, M. J. R. (1968): Multiple regression with a singular matrix, Applied Statist., 17, 110-117.

[1577] S Healy, M. J. R. (1968): Inversion of a positive semi-definite symmetric matrix (Algorithm AS 7), Applied Statist., 17, 198-199.

Heiberger, R.: See Fix, G. (1972).

[1578] S Helstrom, C. W. (1967): Image restoration by the method of least squares, J. Opt. Soc. Amer., 57, 297-303.

[1579] S Hemmerle, W. J. (1975): An explicit solution for generalized ridge regression, Technometrics, 17, 309-314.

[1580] S Herman, R. H. (1968): On the uniqueness of the ideals of compact and strictly singular operators, Studia Math., 29, 161-165.

[1581] T Hildebrandt, T. H. and L. M. Graves (1927): Implicit functions and their differentials in general analysis, Trans. Amer. Math. Soc., 29, 127-153.

Hocking, R. R.: See LaMotte, L. R. (1970).

[1582] S Hoerl, A. E. (1959): Optimum solution of many variable equations, Chemical Engineering Progress, 55, 69-78.

[1583] S Hoerl, A. E. (1962): Application of ridge analysis to regression problems, Chemical Engineering Progress, 58, 54-59.

[1584] S Hoerl, A. E. (1964): Ridge analysis, Chemical Engineering Progress, 60, 67-78.

[1585] S Hoerl, A. E. and R. W. Kennard (1970): Ridge regression: Biased estimation for nonorthogonal problems, Technometrics, 12, 55-67.

[1586] S Hoerl, A. E. and R. W. Kennard (1970): Ridge regressions: Applications to nonorthogonal problems, Technometrics, 12, 69-82.

[1587] P Hung, C. H. and T. L. Markham (1975): The Moore-Penrose inverse of a partitioned matrix $M = \begin{pmatrix} A & 0 \\ B & C \end{pmatrix}$, Czechoslovak Math. J., 25, 354-361.

Hunter, A. P., Jr.: See Ben-Israel (1970).

[1588] S Ivanov, V. K. (1962): On linear problems which are not well-posed, Dokl. Akad. Nauk SSSR, 145, 270-272.

[1589] S Ivanov, V. K. (1967): Fredholm integral equations of the first kind, Differencial'nye Uravnenija, 3, 410-421.

[1590] Ivanov, V. K. (1968): The application of Picard's method
S to the solution of integral equations of the first kind, Bul. Inst. Politenn. Iasi. (N. S.), 14, 71-78.

[1591] James, R. C. (1972): Quasicomplements, J. Approximation Theory, 6, 147-160.
T

[1592] Jennings, L. S. (1972): An improperly posed quadratic
S programming problem, pp. 152-171 in Optimization, edited by R. S. Anderssen, L. S. Jennings and D. M. Ryan, University of Queensland Press, St. Lucia, Queensland.

[1593] Jennings, L. S. (1973): Regularization techniques for
T first kind integral equations, in Error, Approximation and Accuracy, University of Queensland Press.

[1594] Johnson, W. B. (1973): On quasi-complements, Pacific
T J. Math., 48, 113-118.

[1595] Kaashoek, M. A. (1965): Stability theorems for closed
S linear operators, Indag. Math., 27, 452-466.

[1596] Kaashoek, M. and T. T. West (1974): Locally Compact
T Semi-Algebras with Applications to Spectral Theory of Positive Operators, North-Holland Mathematics Studies 9, North-Holland Publishing Co., Amsterdam, 102pp + x.

[1597] Kahan, W. (1966): Numerical linear algebra, Canadian
T Math. Bull., 9, 757-801.

Kahan, W.: See Davis, Chandler (1970).

Kailath, T.: See Morf, M. (1975).

Kalaba, R.: See Bellman, R. (1965), (1965).

[1598] Kaniel, S. and M. Schechter (1963): Spectral theory for
S Fredholm operators, Comm. Pure Appl. Math., 16, 423-448.

[1599] Kaplansky, I. (1949): Normed algebras, Duke Math. J., 16, 399-418.
T

[1600] Kaplansky, I. (1951): Projections in Banach algebras, Ann. Math., 53, 235-249.
T

[1601] Kathovnik, V. J. (1969): The method of least squares in a problem of estimation of parameters of a nonlinear model, in Cybernetics and Computing Techniques, No. 5: Complex Control Systems (in Russian), pp. 79-91.
S

[1602] Keller, J. B. (1962): Factorization of matrices by least squares, Biometrika, 49, 239-242.
S

[1603] Keller, J. B. (1975): Closest unitary, orthogonal and Hermitian operators to a given operator, Math. Magazine, 48, 192-197.
P

Kennard, R. W.: See Hoerl, A. E. (1970), (1970).

[1604] Khalilov, Z. (1947): Linear singular equations in normed rings (Russian), Doklady Akad. Nauk SSSR, 58, 1613-1616.
S

Kok, J.: See Bus, J. C. P. (1975).

[1605] Kowalik, J. S. and M. R. Osborne (1968): Methods for Unconstrained Optimization Problems, Elsevier, New York.
T

Kozlov, V. P.: See Turchin, V. F. (1970).

Kračkovskiĭ, S. N.: See Gol'dman, M. A. (1967).

Krasnosel'skiĭ, M. A.: See Krein, M. G. (1948).

[1606] Krein, M. G., M. A. Krasnosel'skii and D. C. Mil'man (1948): On the defect of linear operators in Banach space and some geometric problems (Russian), Sbornik Trud. Inst. Mat. Akad. Naak Ukr. SSR, 11, 97-112.
S

Krein, M. G.: See Gohberg, I. C. (1965).

[1607] T Krishnamurthy, V. (1960): On the state of a linear operator and its adjoint, Math. Ann., 141, 153-160.

[1608] P Krogh, F. T. (1974): Efficient implementation of a variable projection algorithm for nonlinear least squares problems, Comm. ACM, 17, 167-169.

[1609] T Kubácek, L. (1975): On a generalization of orthogonal regression, Apl. Mat., 20, 87-95.

Kuznecov, Ju. A.: See Marčuk, G. I. (1974).

[1610] S LaMotte, L. R. and R. R. Hocking (1970): Computational efficiency in the selection of regression variables, Technometrics, 12, 83-93.

[1611] T Larkin, F. M. (1969): Estimation of a non-negative function, BIT, 9, 30-52.

Laurent, P. J.: See Anselone, P. M. (1968).

[1612] P Lawson, Linda Marie (1975): Computational methods for generalized inverse matrices arising from proper splittings, Linear Algebra and Appl., 12, 111-126.

[1613] T Leach, E. B. (1963): On a related function theorem, Proc. Amer. Math. Soc., 14, 687-689.

Lebow, A.: See Coburn, L. A. (1966).

[1614] T Levenberg, K. (1944): A method for the solution of certain non-linear problems in least squares, Quart. Appl. Math., 2, 164-168.

Levine, G.: See Battin, R. H. (1969).

[1615] S Leżański, T. (1953): The Fredholm theory of linear equations in Banach spaces, Studia Math., 13, 244-276.

[1616] Lill, Shirley A. (1972): Generalisation of an exact
S method for solving constrained problems to deal with inequality constraints, in Numerical Methods for Nonlinear Optimisation, F. A. Lootsma, editor, Academic Press, New York.

Lill, Shirley A.: See Fletcher, R. (1970).

Lockett, J.: See Bellman, R. (1965), (1965).

[1617] Longley, J. (1967): An appraisal of least squares pro-
S grams from the point of view of the user, J. Amer. Statist. Assoc., 62, 819-841.

[1618] Lorch, E. R. (1962): Spectral Theory, Oxford University
S Press, New York.

[1619] Lovass-Nagy, V. and D. L. Powers (1975): Matrix
P generalized inverses in handling of control problems containing input derivatives, Internat. J. Systems Sci., 6, 693-696.

[1620] Lovass-Nagy, V. and D. L. Powers (1975): On rectan-
P gular systems of differential equations and their applications to circuit theory, J. Franklin Inst., 299, 399-407.

[1621] Lovass-Nagy, V. and D. L. Powers (1975): On matrix
S solution of discretized Dirichlet and Neumann problems, Inter. J. Systems Sci., 6, 397-400.

[1622] Lucas, H. L. (1962): Unpublished lecture notes on the
S Linear Model and its Analysis, North Carolina State University, Raleigh, N. C.

Lynch, R.: See de Boor, C. (1966).

Malkevich, M. S.: See Turchin, V. F. (1970).

[1623] Mangasarian, O. L. (1963): Equivalence in nonlinear
S programming, Naval Res. Logist. Quart., 10, 299-306.

Mangasarian, O. L.: See Rosen, J. B. (1970).

[1624] Marčuk, G. I. and Ju. A. Kuznecov (1974): Iteration
P methods for the solution of systems of linear equations with singular matrices (Russian English summary) Proceedings of the Third Conference on Basic Problems of Numerical Mathematics (Prague, 1973), Acta Univ. Carolinae-Math. et Phys., 15, 87-95.

Markham, T. L.: See Hung, C. H. (1975).

[1625] Markus, A. S. (1968): On holomorphic operator-func-
T tions (Russian), Dokl. Akad. Nauk SSSR, 119, 1099-1102.

Markus, A. S.: See Gohberg, I. C. (1960).

[1626] Marsaglia, G. and G. P. H. Styan (1974/75): Equal-
P ities and inequalities for ranks of matrices, Linear and Multilinear Algebra, 2, 269-292.

Mayers, D. F.: See Baker, C. T. H. (1964).

[1627] Medgyessy, P. (1971): Inkorrekt Matematikai Problé-
S mák Vizsgálának Jelen Állásáról, Különös Tekinettel I. Fajú Operátoregyenletek Megoldására. MTA III. Osztály Közleményei, 20, 97-131.

Meise, R.: See Gramsch, B. (1968).

[1628] Meyer, R. R. (1970): Theoretical and computational
T aspects of nonlinear regression, pp. 465-486 in Rosen, Mangasarian and Ritter (1970).

Mil'man, D. C.: See Krein, M. G. (1948).

[1629] Mittenthal, L. (1968): Operator valued analytic func-
S tions and generalizations of spectral theory, Pacific J. Math., 24, 119-132.

Moler, C. B.: See Forsythe, G. E. (1967).

[1630] Monahan, J. E. (1967): Unfolding measured distribu-
T tions, in Scintillation Spectroscopy of Gamma Radiation, vol. 1, Gordon and Breach Science Publishers, London.

[1631] Morf, M. and T. Kailath (1975): Square-root algorithms
T for least squares estimation, IEEE Trans. Aut. Control, AC-20, 487-497.

Morozov, V. A.: See Gordonova, V. I. (1973).

[1632] Morrison, D. D. (1960): Methods for nonlinear least
S squares problems and convergence proofs, Proc. of Seminar on Tracking Programs and Orbit Determination, (J. Lorell and F. Yagi, Cochairmen), 1-9, Jet Propulsion Lab., Pasadena, California.

Murray, W.: See Gill, P. E. (1972).

[1633] Naimark, M. A. (1964): Normed Rings, P. Noordhoff,
T Groningen, The Netherlands.

Nashed, M. Z.: See Bazaraa, M. S. (1974), (1975).

[1634] Nayak, R. P. and E. C. Foudriat (1974): Sequential
P parameter estimation using pseudoinverse, IEEE Trans. Aut. Control, AC-19, 80-83.

[1635] Naylor, A. W. and G. R. Sell (1971): Linear Operator
T Theory in Engineering and Science, Holt, Rinehart and Winston, New York.

[1636] Nedelkov, I. P. (1972): Improper problems in compu-
T tational physics, Comput. Phys. Commun., 4, 157-164.

[1637] Nevanlinna, F. and R. Nevanlinna (1973): Absolute
T Analysis, Springer-Verlag, New York-Heidelberg, vi + 270 pp.

Nevanlinna, R.: See Nevanlinna, F. (1973).

[1638] S Norlén, U. (1975): The covariance matrix for which least squares is best linear unbiased, Scand. J. Statist., 2, 85-90.

[1639] T Osborn, J. E. (1975): Spectral approximation for compact operators, Math. Comp., 29, 712-725.

[1640] T Osborn, J. E. and D. Sather (1975): Alternative problems and monotonicity, J. Differential Equations, 18, 393-410.

[1641] S Osborne, M. R. (1975): Some special nonlinear least squares problem, SIAM J. Numer. Anal., 12, 571-592.

Osborne, M. R.: See Kowalik, J. S. (1968).

[1642] S Paige, C. C. (1974): Bidiagonalization of matrices and solution of linear equations, SIAM J. Numer. Anal., 11, 197-209.

[1643] S Paige, C. C. and M. A. Saunders (1975): Solution of sparse indefinite systems of linear equations, SIAM J. Numer. Anal., 12, 617-629.

[1644] T Palais, R. S. (1965): Seminar on the Atiyah-Singer Index Theorem, Princeton University Press, Princeton, N. J.

[1645] T Papini, P. L. (1975): Some questions related to the concept of orthogonality in Banach spaces. Proximity maps; bases, Boll. Un. Mat. Ital. (4), 11, 44-63.

Parlett, B. N.: See Bunch, J. R. (1972).

[1646] T Patterson, J. D. and B. F. Womack (1966): An adaptive pattern classification system, IEEE Trans. Syst. Sci. Cyb., SSC-2, 62-67.

[1647] T Pelczynski, A. (1960): Projection in certain Banach spaces, Studia Math., 19, 209-228.

Pfaffenberger, W.: See Caradus, S. R. (1974).

[1648] Picard, E. (1910): Sur un théorème générale relatif aux
S équations integrales de première espèce et sur quelques problèmes de physique mathématique, R. C. Mat. Palermo, 29, 615-619.

Pille, W.: See Sinha, N. K. (1971).

[1649] Plackett, R. L. (1960): Principles of Regression Anal-
T ysis, Clarendon Press, Oxford.

[1650] Plemmons, R. J. (1974): Direct iterative methods for
S linear systems using weak splittings, Proceedings of the Third Conference on Basic Problems of Numerical Mathematics (Prague 1973), Acta Univ. Carolinae-Math. et Phys., 15, 117-120.

Plemmons, R. J.: See Cline, R. E. (1975).

[1651] Poole, G. D. (1975): Generalized M-matrices and
S applications, Math. Comp., 29, 903-910.

[1652] Porta, H. (1969): Two-sided ideals of operators, Bull.
S Amer. Math. Soc., 75, 599-602.

Powers, D. L.: See Lovass-Nagy, V. (1975), (1975), (1975).

[1653] Prasada Rao, P. S.-S. N. V. and K. P. S. Bhaskara
P Rao (1975): On generalized inverses of Boolean matrices, Linear Algebra and Appl., 11, 135-153.

[1654] Prenter, P. M. (1973): A collocation method for the
T numerical solution of integral equations, SIAM J. Numer. Anal., 10, 570-581.

[1655] Prenter, P. M. (1973): A least squares solution to
S equations of the first kind, a preprint.

[1656] Prenter, P. M. (1975): Splines and Variational Methods,
T Wiley-Interscience, New York.

[1657] T Proceedings of a Symposium on Inverse Semigroups and their Generalizations (Northern Illinois Univ., Dekalb, Ill., 1973).

[1658] S Prössdorf, S. (1974): Einige Klassen singularen Gleichungen, Akademie-Verlag Berlin, Mathematische Reihe, Band 46, Birkhäuer, Verlag, Basel and Stuttgart, 353 pp.

[1659] T Quenouille, M. H. (1950): Computational devices in the application of least squares, J. Roy. Statist. Soc., Ser. B., 12, 256-272.

Raike, W.: See Charnes, A. (1975).

[1660] S Rakovščik, L. S. (1973): The stability of certain properties of normally solvable operators under compact approximations. (Russian), Deposition No. 6710-73, Vsesojuz. Inst. Naučn. i Tehn. Informacii (VINITI), Moscow, 1973: abstract in Sibirsk. Mat. Ž., 25(1974), 457.

[1661] S Raphael, R. (1975): Some remarks on regular and strongly regular rings, Canad. Math. Bull., 17, 709-712.

[1662] T Reid, J. K., ed. (1971): Large Sparse Sets of Linear Equations, Academic Press, New York, 284 pp.

[1663] T Reid, J. K. (1972): The use of conjugate gradients for systems of linear equations possessing property A, SIAM J. Numer. Anal., 9, 325-332.

[1664] T Reinsch, C. H. (1971): Smoothing by spline functions, Numer. Math., 16, 451-454.

[1665] S Ribière, G. (1967): Régularization d'opérateurs, Rev. Franç. Inform. Rech. Opér., 1, 57-79.

[1666] T Rickart, C. E. (1947): The singular elements of a Banach algebra, Duke Math. J., 14, 1066-1077.

[1667] Rickart, C. E. (1960): General Theory of Banach Algebras, van Nostrand, Princeton, New Jersey.
T

[1668] Riesz, F. (1918): Über lineare Functionalgleichungen, Acta Math., 41, 71-98.
T

[1669] Riesz, F. and B. Sz-Nagy (1955): Functional Analysis, Frederick Ungar, New York.
T

Ritter, K.: See Rosen, J. B. (1970).

Robers, P. D.: See Ben-Israel (1970).

[1670] Rose, D. J. and R. A. Willoughby, editors (1972): Sparse Matrices and Their Applications, Plenum Press, New York.
T

[1671] Rosen, J. B. (1965): Gradient projection as a least squares solution of Kuhn-Tucker conditions, Computer Sciences Department, The University of Wisconsin-Madison, July, 1965.
S

[1672] Rosen, J. B., O. L. Mangasarian and K. Ritter, editors (1970): Nonlinear Programming, Academic Press, New York.
S

[1673] Ruston, A. F. (1951): On the Fredholm theory of integral equations for operators belonging to the trace class of a general Banach space, Proc. London Math. Soc., 2nd series, 53, 109-124.
S

Sather, D.: See Osborn, J. E. (1975).

Saunders, M. A.: See Bartels, R. H. (1970), Gill, P. E. (1972), Paige, C. C. (1975).

[1674] Schechter, M. (1968): Riesz operators and Fredholm perturbations, Bull. Amer. Math. Soc., 74, 1139-1144.
S

Schechter, M.: See Kaniel, S. (1963).

[1675] Schneider, H. (1956): The elementary divisors, associ-
T ated with 0, of a nonsingular M-matrix, Proc. Edinburgh Math. Soc. (2), 10, 108-122.

[1676] Schock, E. (1975): Approximation numbers of bounded
T operators, J. Math. Anal. Appl., 51, 440-448.

Schumaker, L. L.: See Daniel, J. W. (1974).

Sell, G. R.: See Naylor, A. W. (1971).

[1677] Shanno, D. F. (1970): An accelerated gradient projec-
S tion method for linearly constrained nonlinear estimation, SIAM J. Appl. Math., 18, 322-334.

[1678] Shanno, D. C. (1970): Parameter selection for modified
T Newton methods for function minimization, SIAM J. Numer. Anal., 7, 366-372.

Shetty, C. M.: See Bazaraa, M. S. (1975).

[1679] Sikorski, R. (1953). On Leżański's determinants of
S linear equations in Banach spaces, Studia Math, 14, 24-48.

[1680] Sikorski, R. (1957): On determinants of Leżański and
S Rustin, Studia Math., 16, 99-112.

[1681] Sikorski, R. (1959): Determinant systems, Studia Math.,
S 18, 161-186.

[1682] Sikorski, R. (1959): On Leżański endomorphisms,
T Studia Math., 18, 187-188.

[1683] Sikorski, R. (1961): Remarks on Leżański's determinants,
T Studia Math., 20, 145-161.

[1684] Sikorski, R. (1961): The determinant theory in Banach
T spaces, Coll. Math., 7, 141-198.

Singer, I. M.: See Coburn, L. A. (1972).

[1685] S Sinha, N. K. and W. Pille (1971): On line parameter estimation using matrix pseudoinverse, Proc. Inst. Elec. Eng., 118, August 1971.

Sjöberg, B.: See Dahlquist, G. (1968).

[1686] S van der Sluis, A. (1969): Condition numbers and equilibration of matrices, Numer. Math., 14, 14-23.

[1687] T van der Sluis, A. (1970): Stability of solutions of linear algebraic systems, Numer. Math., 14, 246-251.

[1688] T Smith, F. W. (1972): Small-sample optimality of design techniques for linear classifiers of Gaussian patterns, IEEE Trans. Inform. Theory, IT-18, 118-126.

[1689] S Smith, S. E. and S. S. Yau (1972): Linear sequential pattern classification, IEEE Trans. Inform. Theory, IT-18, 673-678.

[1690] T Stummel, F. (1973): Discrete convergence of mappings, pp. 285-310 in Topics in Numerical Analysis (Proc. Roy. Irish Acad. Conf., University College, Dublin, 1972), Academic Press, London.

Stutz, J.: See Charnes, A. (1975).

Styan, G. P. H.: See Marsglia, G. (1974/75).

Svensson, P.: See Dahlquist, G. (1968).

[1691] T Sz.-Nagy, B. (1951): Perturbations des transformations linéaires fermées, Acta Sci. Math., 14, 125-137.

[1692] T Sz.-Nagy, B. (1952): On the stability of the index of unbounded linear transformations, Acta Math. Hung., 3, 49-51.

Sz.-Nagy, B.: See Riesz, F. (1955).

Thomas, D. H.: See Buchanan, J. E. (1968).

[1693] S Tikhonov, A. N. (1963): Solution of incorrectly formulated problems and the regularization method, Dokl. Akad. Nauk SSR, 151, 501-504. Soviet Math. Dokl., 4, 1035-1038.

[1694] S Tikhonov, A. N. (1963): Regularization of incorrectly posed problems, Dokl. Akad. Nauk SSR, 153, 49-52. Soviet Math. Dokl., 4, 1624-1627.

[1695] S Tikhonov, A. N. (1967): Methods for the solution of incorrect problems (Russian), Vych. Metody i Program., 8, 3-33.

[1696] T Tikhonov, A. N. and V. B. Glasko (1964): An approximate solution of Fredholm integral equations of the first kind, Zhurnal Vychislitel'noi Matematiki i Matematicheskoi Fiziki, 4, 564-571. USSR Comp. Math. Math. Phys., 4, 236-247.

[1697] S Torsti, J. J. and A. M. Aurela (1972): A fast quadratic programming method for solving ill-conditioned systems of equations, J. Math. Anal. Appl., 38, 193-204.

[1698] T Turchin, V. F. (1968): Selection of an ensemble of smooth functions for the solution of the inverse problem, Zh. Vychisl. Mat. i Mat. Fiz., 8, 230-238. USSR Comp. Math. Phys., 8, 328-339.

[1699] T Turchin, V. F., V. P. Kozlov and M. S. Malkevich (1970): The use of mathematical-statistics methods in the solution of incorrectly posed problems, Usp. Fiz. Nauk, 102, 345-386, Soviet Phys. Usp., 13 (1971), 681-703.

[1700] T Twomey, S. (1965): The application of numerical filtering to the solution of integral equations encountered in indirect sensing measurements, J. Franklin Inst., 279, 95-109.

[1701] T Vainstein, L. A. (1972): Filtering of noise in a numerical solution of integral equations of the first kind, Soviet Phys. Dokl., 17, 519-521.

[1702] S Vainstein, L. A. (1972): Numerical solution of integral equations of the first kind using a priori information on the function to be determined, Soviet Phys. Dokl., 17, 532-534.

[1703] T Varah, J. M. (1970): Computing invariant subspaces of a general matrix when the eigensystem is poorly conditioned, Math. Comp., 24, 137-149.

[1704] T Vladimirskiĭ, Ju. N. (1974): Remarks on compact approximation in Banach spaces. (Russian) Sibirsk. Mat. Ž., 15, 200-204, 238.

[1705] T Voevodin, V. V. (1969): The method of regularization, Zh. Vychisl. Mat. i Mat. Fiz., 9, 673-675. USSR Comput. Math. Math. Phys., 9, 228-232.

[1706] T Wahba, Grace (1973): Convergence rates of certain approximate solutions to Fredholm integral of the first kind, J. Approximation Theory, 7, 167-185.

Walsh, J.: See Delves, L. M. (1974).

[1707] S Watson, G. S. (1972): Prediction and the efficiency of least squares, Biometrika, 59, 91-98.

[1708] S Wedin, P.-Å. (1973): On the almost rank deficient case of the linear least squares problem, BIT, 13, 344-354.

[1709] S Wells, C. H. (1967): Minimum norm control of discrete systems, IEEE Int. Conv. Rec., 15, 55-64.

[1710] T West, T. T. (1966): The decomposition of Riesz operators, Proc. London Math. Soc., 16, 737-752.

West, T. T.: See Kaashoek, M. (1974).

[1711] S Whitley, R. J. (1964): Strictly singular operators and their conjugates, Trans. Amer. Math. Soc., 113, 252-261.

[1712] Wilkinson, J. (1965): Error analysis of transformations
T based on the use of matrices of the form $I - 2ww^H$, pp. 77-101 in Error in Digital Computation, vol. II, L. B. Rall, ed., Wiley, New York.

Willoughby, R. A.: See Rose, D. J. (1972).

Womack, B. F.: See Patterson, J. D. (1966).

[1713] Woodbury, Max (1950): Inverting modified matrices,
T Memorandum Report 42, Statistical Research Group, Princeton, N. J.

Wright, K.: See Baker, T. H. (1964).

[1714] Yates, F. and R. W. Hale (1939): The analysis of
S Latin squares when two or more rows, columns, or treatments are missing, J. Roy. Statist. Soc., Suppl., 6, 67-79.

Yau, S. S.: See Smith, S. E. (1972).

[1715] Yood, B. (1954): Topological properties of homomorphisms
S between Banach algebras, Amer. J. Math., 76, 155-167.

[1716] Yood, B. (1954): Difference algebras of linear trans-
T formations on Banach spaces, Pacific J. Math., 4, 615-636.

Yood, B.: See Caradus, S. R. (1974).

Young, E. C. (1975): See Chan, C. Y. (1975).

[1717] Żelazko, W. (1973): Banach Algebras, Elsevier, New
T York.

[1718] Zlobec, S. (1970): Asymptotic Kuhn-Tucker conditions
S for mathematical programming problems in a Banach space, SIAM J. Control, 8, 505-512.

[1719] T Zlobec, S. (1971): Extensions of asymptotic Kuhn-Tucker conditions in mathematical programming, SIAM J. Appl. Math., 21, 448-460.

Supplement 3

Akatsuka, Y.: See Matsuo, T. (1965).

[1720] P Al-Alaoui, M. A. (1975): Application of constrained generalized inverses to pattern classification, submitted to J. Pattern Recognition.

[1721] P Al-Alaoui, M. A. (1975): A new weighted generalized inverse algorithm for pattern classification, submitted to IEEE Trans. Information Theory.

[1722] T Atkinson, K. E. (1972): A Survey of Numerical Methods for the Solution of Fredholm Integral Equations of the Second Kind, SIAM, to appear.

[1723] S Ballentine, C. B. (1975): Products of EP matrices, Linear Algebra and Appl., to appear.

[1724] S Barham, R. H. and W. Drane (1972): An algorithm for least squares estimation of nonlinear parameters when some of the parameters are linear, Technometrics, 14, 757-766.

[1725] S Bartels, R. H. and G. W. Stewart (1972): Algorithm, 432, solution of the matrix equation $AX + XB = C$, Comm. ACM, 15, 820-826.

[1726] T Bartle, R. G. (1955): Newton's method in Banach space, Proc. Amer. Math. Soc., 6, 827-831.

[1727] T Bellman, R. (1950): A note on the summability of formal solutions of linear integral equations, Duke Math. J., 17, 53-55.

[1728] T Bellman, R. (1968): A new technique for the numerical solution of Fredholm integral equations, Computing, 3, 131-138.

[1729] Ben-Israel, A. (1969): A note on partitioned matrices
S and equations, SIAM Rev., 11, 247-250.

[1730] Bergmann, P. G., R. Penfield, R. Schiller and
S H. Zatzkis (1950): The Hamiltonian of the general theory of relativity with electromagnetic field, Phys. Rev., 80, 81-88.

[1731] Berman, A. and R. J. Plemmons (1974): Matrix group
P monotonicity, Proc. Amer. Math. Soc., 46, 355-359.

[1732] Brenner, J. L. and M. J. S. Lim (1974): The matrix
T equations $A = XYZ$ and $B = ZYX$ and related ones, Canad. Math. Bull., 17, 179-183. Correction, ibid., 17(1974), 426.

[1733] Block, H. D. (1953): Construction of solutions and
T propagation of errors in nonlinear problems, Proc. Amer. Math. Soc., 4, 715-722.

[1734] Cesari, L. (1966): The implicit function theorem in
T functional analysis, Duke Math. J., 33, 417-440.

[1735] Cochran, J. A. (1975): Growth estimates for the singu-
S lar values of square-integrable kernels, Pacific J. Math., 56, 51-58.

[1736] Curtain, Ruth (1975): A survey of infinite-dimensional
T filtering, SIAM Rev., 17, 395-411.

[1737] Cvetkova, T. A. (1974): The construction of positive
S definite matrices that are very close to a given matrix. (Russian) Ukrain. Mat. Ž., 26, 418-420, 432. Ukrainian Math. J., 26, 348-349 (1975).

[1738] Deuflhard, P. (1974): A modified Newton method for the
P solution of ill-conditioned systems of nonlinear equations with application to multiple shooting, Numer. Math., 22, 289-315.

Drane, W.: See Barham, R. H. (1972).

[1739] S Good, I. J. (1963): On the independence of quadratic expression, J. Roy. Statist. Soc., Ser. B, 25, 377-382.

[1740] P Guttman, I., V. Pereyra and H. D. Scolnik (1973): Least squares estimation for a class of nonlinear models, Technometrics, 15, 209-218.

[1741] P Harley, T. D. (1964): Pseudo-estimates vs. pseudo-inverses for singular sample covariance matrices, Philadelphia Meeting of the American Statistical Association, September 1964.

[1742] S Hartwig, R. E. and I. J. Katz (1975): Products of EP elements in reflexive semigroups, Linear Algebra and Appl., to appear.

[1743] S Horn, R. A. (1975): On inequalities between Hermitian and symmetric forms, Linear Algebra and Appl., 11, 277-289.

[1744] S Huang, H. Y. (1974): A direct method for the general solution of a system of linear equations, J. Optimization Theory Appl., 16, 429-445.

[1745] S Hurwicz, L. (1950): Prediction and least squares, in Statistical Inference in Dynamic Economic Models, edited by Tjalling C. Koopmans, pp. 266-300, John Wiley and Sons, Inc., New York.

[1746] S Joshi, V. N. (1973): On the class of matrices which are convergent for a given matrix, Studia Sci. Math. Hungar., 8, 447-456.

[1747] S Jupp, D. (1971): Nonlinear least square spline approximation, Tech. Rept., The Flinders Univ. of South Africa.

Katz, I. J.: See Hartwig, R. E. (1975).

[1748] S Lawton, W. H. and E. A. Sylvestre (1971): Elimination of linear parameters in nonlinear regression, Technometrics, 13, 461-478.

[1749] Lawton, W. H., E. A. Sylvestre and M. S. Maggio
S (1972): Self modeling nonlinear regression, Technometrics, 14, 513-532.

Lim, M. J. S.: See Brenner, J. L. (1974).

Lipcer, R. Š.: See Žukovskii, E. L. (1975).

Maggio, M. S.: See Lawton, W. H. (1974).

[1750] Matsuo, T. and Y. Akatsuka (1965): Optimal control
S of linear discrete systems using the generalized inverse of a matrix, Tech. Rept. vol. 13, Faculty of Engineering, Institute of Automatic Control, Nagoya University, Japan, pp. 97-107.

[1751] Mitra, S. K. and C. R. Rao (1974): Projections under
P seminorms and generalized Moore-Penrose inverses, Linear Algebra and Appl., 9, 155-167.

[1752] Molčanov, I. N. and M. F. Jakovlev (1975): Iteration
P processes for the solution of a certain class of incompatible systems of linear algebraic equations. (Russian) Ž. Vyčisl. Mat. i Mat. Fiz., 15, 547-558, 809.

Penfield, R.: See Bergmann, P. G. (1950).

Pereyra, V.: See Guttman, I. (1973).

[1753] Pietsch, A. (1972): Nuclear Locally Convex Spaces.
T Translated from the second German edition by William H. Ruckle. Ergebnisse der Mathematik und ihrer Grenzgebiete, Band 66. Springer-Verlag, New York-Heidelberg, 1972. ix + 193 pp.

Plemmons, R. J.: See Berman, A. (1974).

[1754] Pogorzelski, W. (1966): Integral Equations and Their
T Applications, vol. 1, MacMillan (Pergamon), New York, and Polish Scientific Publishers, Warsaw.

[1755] Poljak, B. T. (1964): Gradient methods for solving
S equations and inequalities, Ž. Vyčisl. Mat. i Mat. Fiz., 4, 995-1005.

[1756] Przeworska-Rolewicz, D. (1976): Algebraic Analysis
S and Differential Equations, Marcel Dekker Inc., New York, to appear.

[1757] Rakovščik, L. S. (1972): Stability of index and semi-
S stability of the defect numbers under compact perturbations (Russian), Sibirsk. Mat. Ž., 13, 630-637.

Rao, C. R.: See Mitra, S. K. (1974).

[1758] Robers, P. D. (1968): Interval Linear Programming,
S Ph.D. Thesis, Dept. of Industrial Engineering and Management Science, Northwestern University, Evanston, Illinois.

[1759] Rozenbljum, G. V. (1972): Estimates for the singular
S values of a certain class of integral operators. (Russian), Problems of Mathematical Analysis, No. 3: Integral and Differential Operators. Differential Equations (Russian), pp. 111-118, Izdat Leningrad Univ., Leningrad. (MR 50 #3013).

Schiller, R.: See Bergmann, P. G. (1950).

Scolnik, H. D.: See Guttman, I. (1973).

[1760] Searle, S. R. (1974): Prediction, mixed models, and
S variance components, pp. 229-266 in Reliability and Biometry: Statistical Analysis of Lifelength (Proc. Conf. Florida State Univ., Tallahassee, Fla., 1973), Soc. Indust. Appl. Math., Philadelphia, Pa.

[1761] Shanbhag, D. N. (1966): On the independence of qua-
S dratic forms, J. Roy. Statist. Soc., Ser. B, 18, 582-583.

[1762] Sposito, V. A. (1973): Solutions of a special class of
P linear programming problems, Operations Res., 21, 386-388.

Stewart, G. W.: See Bartels, R. H. (1972).

[1763] S Stummel, F. (1970): Diskrete Konvergenz linearer Operatoren, I. Math. Ann., 190, 45-92.

[1764] S Stummel, F. (1971): Discrete Konvergenz linearer Operatoren, II. Math. Z., 120-231-264.

Sylvestre, E. A.: See Lawton, W. H. (1971), (1972).

[1765] S Tasche, M. (1975): Algebraische Operatorenrechnung für einen rechtsinvertierbaren Operator, Wiss. Z. der Univ. Rostock, to appear.

[1766] S Tasche, M. (1975): Operatorenrechnung in einer Algebra. Beiträge zur Analysis, to appear.

Therianos, S.: See Thompson, R. C. (1973).

[1767] S Thompson, R. C. and S. Therianos (1973): The eigenvalues and singular values of matrix sums and products. VII. Canad. Math. Bull., 16, 561-569.

[1768] S van der Vaart, H. R. (1964): Generalization of Wilcoxen statistic for the case of k samples, by E. H. Yen, Statistica Neerlandica, 18, 303-305.

[1769] T Vaǐinkko, G. M. (1974): Discretely compact sequences. (Russian), Ž. Vyčisl. Mat. i Mat. Fiz., 14, 572-583, 811.

[1770] S Widom, H. (1975): Perturbing Fredholm operators to obtain invertible operators, J. Functional Analysis, 20, 26-31.

[1771] S Williams, V. (1975): Closed Fredholm and semi-Fredholm operators, essential spectra and perturbations, J. Functional Analysis, 20, 1-25.

[1772] S Wintner, A. and F. D. Murnaghan (1931): On a polar representation of nonsingular matrices, Proc. Nat. Acad. Sci. U.S.A., 17, 676-678.

[1773] Wong, K. T. (1974): The eigenvalue problem $\lambda Tx + Sx$,
S J. Differential Equations, 16, 270-280.

Zaikin, P. N.: See Žukovskiĭ, E. L. (1975).

Zatzkis, H.: See Bermann, P. G. (1950).

[1774] Žukovskiĭ, E. L. and R. Š. Lipcer (1975): The computation of pseudoinverse matrices. (Russian), Ž. Vyčisl. Mat. i Mat. Fiz., 15, 489-492, 542.
P

[1775] Žukovskiĭ, E. L. and P. N. Zaikin (1975): Numerical statistical algorithms for finding the quasisolutions of conditional systems of linear algebraic equations. (Russian), Ž. Vyčisl. Mat. i Mat. Fiz., 15, 559-572, 809.
S

[1776] Jefferson, Thomas et al. (1776): Declaration of Independence, In Congress, July 4.
USA

Index

D

L

M

R

S